GENE TRANSFER AND EXPRESSION IN MAMMALIAN CELLS

New Comprehensive Biochemistry

Volume 38

General Editor

G. BERNARDI
Paris

ELSEVIER
Amsterdam · Boston · Heidelberg · London · New York · Oxford
Paris · San Diego · San Francisco · Singapore · Sydney · Tokyo

Gene Transfer and Expression in Mammalian Cells

Editor

Savvas C. Makrides

EIC Laboratories, Inc.,
Norwood, Massachusetts,
USA

2003
ELSEVIER
Amsterdam · Boston · Heidelberg · London · New York · Oxford
Paris · San Diego · San Francisco · Singapore · Sydney · Tokyo

ELSEVIER SCIENCE B.V.
Sara Burgerhartstraat 25
P.O. Box 211, 1000 AE Amsterdam, The Netherlands

Library of Congress Cataloging in Publication Data
A catalog record from the Library of Congress has been applied for.

ISBN HARDBOUND: 0-444-51371-X
       PAPER BACK: 0-444-51370-1
ISBN: 0-444-80303-3(Series)
ISSN: 0167-7306

♾ The paper used in this publication meets the requirements of ANSI/NISO Z39.48-1992 (Permanence of Paper).
Printed in Hungary.

Lives of great men all remind us
We can make our lives sublime,
And, departing, leave behind us
Footprints in the sands of time.

Footprints, that perhaps another,
Sailing o'er life's solemn main,
A forlorn and shipwrecked brother,
Seeing, shall take heart again.

Henry Wadsworth Longfellow, "A Psalm of Life"

Preface

Expression of heterologous genes in mammalian cells is at the core of many scientific and commercial endeavors, including molecular cloning, biochemical and biophysical characterization of proteins, production of proteins for therapeutic applications, protein engineering, development of cell-based biosensors, production of diagnostic reagents, drug screening, vaccination and gene therapy. In recent years, advances in the development and refinement of mammalian expression vectors to meet diverse experimental objectives have been remarkable. Our goal here is to provide a comprehensive volume that integrates a wide, but not all-inclusive, spectrum of research topics in gene expression in mammalian cells. The book covers several broad, related areas: The development of expression vectors for production of proteins in cultured cells, in transgenic animals, vaccination, and gene therapy; progress in methods for the transfer of genes into mammalian cells and the optimization and monitoring of gene expression; advances in our understanding and manipulation of cellular biochemical pathways that have a quantitative and qualitative impact on mammalian gene expression; gene and protein targeting; and the large-scale production and purification of proteins from cultured cells. The knowledge collected in this volume defines the impressive progress made in many aspects of mammalian gene expression, and also reminds us of how much remains to be overcome in this important field.

The focus of each chapter is delineated clearly, and every effort has been made to avoid duplication of subject matter. Nevertheless, some overlap between sections on related topics is unavoidable. The advantage of this is that topics are examined from different perspectives, and each chapter can be read as an independent review. It is my hope that the assembly of this diverse material into a single volume will provide a useful and stimulating guide to workers in this field as well as to the broader community of biological scientists interested in gene expression in mammalian cells.

I am grateful to the authors for their willingness to share their knowledge and for their enthusiasm and suggestions about the content of the volume. I thank all the scientists who carefully reviewed the manuscripts and offered thoughtful suggestions for improvements. I am indebted to Elsevier Science for funding the volume, and to Brechtje M. de Leij, my Publishing Editor. My thanks to Joyce V. Makrides for her help with the labyrinth of computer software, and I am grateful to EIC Laboratories for providing generous support and a nurturing environment that made this project possible.

Special thanks to Alkis C. Makrides, *il miglior fabbro*.

Savvas C. Makrides, Ph.D.

List of contributors[*]

Mari Aker 381
Department of Medicine, Division of Medical Genetics, Box 357720, University of Washington, Seattle, WA 98195, USA

Jawed Alam 291
Ochsner Clinic Foundation, Division of Research, Molecular Genetics, 1516 Jefferson Hwy, New Orleans, LA 70121, USA

Fernando Almazán 151
Department of Molecular and Cell Biology, Centro Nacional de Biotecnología, CSIC, Campus Universidad Autónoma, Cantoblanco, 28049 Madrid, Spain

Raul Andino 169
Department of Microbiology and Immunology, University of California, 513 Parnassus Ave., Box 0414, San Francisco, CA 94143, USA

Stacey M. Arnold 411
Department of Biological Chemistry, University of Michigan Medical School, Ann Arbor, MI 48109-0650, USA

Alexandra Baer 551
Gesellschaft für Biotechnologische Forschung/German Research Centre for Biotechnology, Epigenetic Regulation, D-38124 Braunschweig, Mascheroder Weg 1, Germany

Qing Bai 27
Department of Molecular Genetics and Biochemistry, School of Medicine, University of Pittsburgh, Pittsburgh, PA, USA

Michel Bessodes 279
Université René Descartes Paris 5, Pharmacologie Moléculaire et Structurale, CNRS FRE2463, 4, avenue de l'Observatoire, 75270 Paris Cedex 06, France

Pascal Bigey 349

Faculté de Pharmacie Unité INSERM/CNRS de Pharmacologie Chimique et Génétique 4, avenue de l'Observatoire, 75270 Paris Cedex 06, France

---

[*] Authors' names are followed by the starting page number(s) of their contribution(s).

Jürgen Bode   551
Gesellschaft für Biotechnologische Forschung/German Research Centre for Biotechnology, Epigenetic Regulation, D-38124 Braunschweig, Mascheroder Weg 1, Germany

Denis Bourbeau   109
Institut de Recherche en Biotechnologie, Conseil National de Recherche Canada, 6100 Avenue Royalmount, Montréal, QC, Canada H4P 2R2

Edward A. Burton   27
Department of Molecular Genetics and Biochemistry, School of Medicine, University of Pittsburgh, Pittsburgh, PA, USA

Marie Carrière   349
Faculté de Pharmacie Unité INSERM/CNRS de Pharmacologie Chimique et Génétique 4, avenue de l'Observatoire, 75270 Paris Cedex 06, France

Miles W. Carroll   125
Oxford BioMedica (UK) Ltd., Oxford Science Park, Oxford OX4 4GA, UK

J. Patrick Condreay   137
Department of Gene Expression and Protein Biochemistry, GlaxoSmithKline Discovery Research, 5 Moore Drive, Research Triangle Park, NC 27709, USA

Julia L. Cook   291
Ochsner Clinic Foundation, Division of Research, Molecular Genetics, 1516 Jefferson Hwy, New Orleans, LA 70121, USA

Pierre Cordelier   71
Department of Pathology and Cell Biology, Jefferson Medical College, 1020 Locust Street, Room 251, Philadelphia, Pennsylvania 19107, USA

Shane Crotty   169
Department of Microbiology and Immunology, University of California, 513 Parnassus Ave., Box 0414, San Francisco, CA 94143, USA

Dale A. Cumming   433
Acton Biotech Consulting, 57 Hammond Street, Acton, MA 01720-3204, USA

Nico P. Dantuma   535
Microbiology and Tumor Biology Center, Karolinska Institutet, Box 280, S-171 77 Stockholm, Sweden

William A. Dunn, Jr.   513
Department of Anatomy and Cell Biology, University of Florida, College of Medicine, Gainesville, FL 32610, USA

Yann Echelard   625
GTC Biotherapeutics, Inc, Five Mountain Road, Framingham, MA 01701, USA

David W. Emery   381
Department of Medicine, Division of Medical Genetics, Box 357720, University of Washington, Seattle, WA 98195, USA

Luis Enjuanes   151
Department of Molecular and Cell Biology, Centro Nacional de Biotecnología, CSIC, Campus Universidad Autónoma, Cantoblanco, 28049 Madrid, Spain

Ellen Ernst   551
Gesellschaft für Biotechnologische Forschung/German Research Centre for Biotechnology, Epigenetic Regulation, D-38124 Braunschweig, Mascheroder Weg 1, Germany

Virginie Escriou   279
Université René Descartes Paris 5, Pharmacologie Moléculaire et Structurale, CNRS FRE2463, 4, avenue de l'Observatoire, 75270 Paris Cedex 06, France

Xiangdong Fang   397
Division of Medical Genetics, University of Washington, Seattle, WA 98195, USA

David Fink   27
Department of Molecular Genetics and Biochemistry, School of Medicine, University of Pittsburgh, Pittsburgh, PA, USA

Martin Fussenegger   457, 589
Institute of Biotechnology, Swiss Federal Institute of Technology, ETH Zurich, CH-8093 Zurich, Switzerland

Mehdi Gasmi   251
Ceregene, Inc., 9381 Judicial Drive #130, San Diego, CA 92121, USA

Steven L. Giardina   641
Purification Laboratory, Biopharmaceutical Development Program, SAIC Frederick, Inc., National Cancer Institute at Frederick, Building 320, PO Box B, Frederick, MD 21702-1201, USA

Pierre-Alain Girod   359
Laboratory of Molecular Biotechnology, Center for Biotechnology UNIL-EPFL and Institute of Biotechnology, University of Lausanne, 1015 Lausanne, Switzerland

Joseph C. Glorioso   27
Department of Molecular Genetics and Biochemistry, School of Medicine, University of Pittsburgh, Pittsburgh, PA, USA

William F. Goins   27
Department of Molecular Genetics and Biochemistry, School of Medicine, University of Pittsburgh, Pittsburgh, PA, USA

Sandra Götze   551
Gesellschaft für Biotechnologische Forschung/German Research Centre for Biotechnology, Epigenetic Regulation, D-38124 Braunschweig, Mascheroder Weg 1, Germany

Henry V. Huang   189
Department of Molecular Microbiology, Washington University School of Medicine, Campus Box 8230, 660 South Euclid Avenue, Saint Louis, Missouri 63110-1093, USA

Yves Hüsemann   551
Gesellschaft für Biotechnologische Forschung/German Research Centre for Biotechnology, Epigenetic Regulation, D-38124 Braunschweig, Mascheroder Weg 1, Germany

Angela Inácio   495
Instituto Nacional de Saúde Dr. Ricardo Jorge, Lisboa, Portugal

Martin Jordan   309, 337
Swiss Federal Institute of Technology Lausanne (EPFL), Faculty of Basic Sciences, Institute of Chemical and Biological Process Science, Center of Biotechnology, CH-1015 Lausanne, Switzerland

Randal J. Kaufman   411
Howard Hughes Medical Institute, University of Michigan Medical School, 4570 MSRB II, 1150 W. Medical Center Dr., Ann Arbor, MI 48109-0650, USA

Hitto Kaufmann   457
Walter and Eliza Institute of Medical Research, Post Office, Royal Melbourne Hospital Victoria, 3050, Australia

Gregory Kennedy   55
McArdle Laboratory for Cancer Research, University of Wisconsin-Madison Medical School, 1400 University Avenue, Madison, WI 53706, USA

Thomas A. Kost   137
Department of Gene Expression and Protein Biochemistry, GlaxoSmithKline Discovery Research, 5 Moore Drive, Research Triangle Park, NC 27709, USA

Gerald R. Kovacs   125
Office of Biodefense Research Affairs, National Institutes of Allergy and Infectious Disease, 6610 Rockledge Drive, Bethesda, MD 20892, USA

Marilyn Kozak   471
Department of Biochemistry, Robert Wood Johnson Medical School, University of Medicine and Dentistry of New Jersey, 675 Hoes Lane, Piscataway, NJ 08854, USA

Julien Landré   71
Department of Pathology and Cell Biology, Jefferson Medical College, 1020 Locust Street, Room 251, Philadelphia, Pennsylvania 19107, USA

Qiliang Li   397
Division of Medical Genetics, University of Washington, Seattle, WA 98195, USA

Stephen A. Liebhaber   495
Departments of Genetics and Medicine, University of Pennsylvania, Room 428 CRB, 415 Curie Boulevard, Philadelphia, PA 19104, USA

Kenneth Lundstrom   207
Regulon Inc., Biopole Epalinges, Les Croisettes, CH-1066 Epalinges, Lausanne, Switzerland

Savvas C. Makrides   1, 9
EIC Laboratories, Inc., 111 Downey Street, Norwood, MA 02062, USA

Wayne A. Marasco   573
Department of Cancer Immunology and AIDS, Dana-Farber Cancer Institute; Department of Medicine, Harvard Medical School, 44 Binney St., Boston, MA 02115, USA

Bernard Massie   109
Institut de Recherche en Biotechnologie, Conseil National de Recherche Canada, 6100 Avenue Royalmount, Montréal, QC, Canada H4P 2R2

Christophe Masson   279
Université René Descartes Paris 5, Pharmacologie Moléculaire et Structurale, CNRS FRE2463, 4, avenue de l'Observatoire, 75270 Paris Cedex 06, France

Maria G. Masucci   535
Microbiology and Tumor Biology Center, Karolinska Institutet, Box 280, S-171 77 Stockholm, Sweden

Alexei Matskevitch   71
Department of Pathology and Cell Biology, Jefferson Medical College, 1020 Locust Street, Room 251, Philadelphia, Pennsylvania 19107, USA

Hayley J. McKee 71
Department of Pathology and Cell Biology, Jefferson Medical College, 1020 Locust Street, Room 251, Philadelphia, Pennsylvania 19107, USA

Harry M. Meade 625
GTC Biotherapeutics, Inc, Five Mountain Road, Framingham, MA 01701, USA

Nicolas Mermod 359
Laboratory of Molecular Biotechnology, Center for Biotechnology UNIL-EPFL and Institute of Biotechnology, University of Lausanne, 1015 Lausanne, Switzerland

Christian Mielke 551
Medizinische Poliklinik Dept. Clinical Chemistry University of Wuerzburg, Klinikstraße 6-8, D-97070 Würzburg, Germany

Carmen N. Nichols 71
Department of Pathology and Cell Biology, Jefferson Medical College, 1020 Locust Street, Room 251, Philadelphia, Pennsylvania 19107, USA

Pamela A. Norton 265
Department of Biochemistry and Molecular Pharmacology, Jefferson Center for Biomedical Research, Thomas Jefferson University, Doylestown, PA 18901, USA

Javier Ortego 151
Department of Molecular and Cell Biology, Centro Nacional de Biotecnología, CSIC, Campus Universidad Autónoma, Cantoblanco, 28049 Madrid, Spain

Catherine J. Pachuk 265
Nucleonics Inc., Malvern, PA, 19355, USA

Giorgio Palù 231
Department of Histology, Microbiology and Medical Biotechnologies, University of Padova, Via A. Gabelli n.63, 35121 Padova, Italy

Cristina Parolin 231
Department of Histology, Microbiology and Medical Biotechnologies, University of Padova, Via A. Gabelli n.63, 35121 Padova, Italy

Kenneth R. Peterson 397
Department of Biochemistry & Molecular Biology, University of Kansas Medical Center, Kansas City, KS 66160, USA

Holly L. Prentice 1
Biogen, Inc., 14 Cambridge Center, Cambridge, MA 02142, USA

David Robinson   605
Biocatalysis and Fermentation Development, Bioprocess R & D, Merck Research Laboratories, Merck & Co. Inc., Mail Stop R80Y-110, P.O. Box 2000, Rahway, NJ 07065, USA

Daniel Scherman   279, 349
Université René Descartes Paris 5, Pharmacologie Moléculaire et Structurale, CNRS FRE2463, 4, avenue de l'Observatoire, 75270 Paris Cedex 06, France

Sondra Schlesinger   189
Department of Molecular Microbiology, Washington University School of Medicine, Campus Box 8230, 660 South Euclid Avenue, Saint Louis, Missouri 63110-1093, USA

Jost Seibler   551
Artemis Pharmaceuticals GmbH, Neurather Ring 1, D51063 Köln, Germany

George Stamatoyannopoulos   381, 397
Department of Medicine, Division of Medical Genetics, Box 357720, University of Washington, Seattle, WA 98195, USA

David S. Strayer   71
Department of Pathology and Cell Biology, Jefferson Medical College, 1020 Locust Street, Room 251, Philadelphia, Pennsylvania 19107, USA

Marlene S. Strayer   71
Department of Pathology and Cell Biology, Jefferson Medical College, 1020 Locust Street, Room 251, Philadelphia, Pennsylvania 19107, USA

Bill Sugden   55
McArdle Laboratory for Cancer Research, University of Wisconsin-Madison Medical School, 1400 University Avenue, Madison, WI 53706, USA

Wilfried Weber   589
Institute of Biotechnology, Swiss Federal Institute of Technology, ETH Hoenggerberg, CH-8093 Zurich, Switzerland

Martyn K. White   71
Department of Pathology and Cell Biology, Jefferson Medical College, 1020 Locust Street, Room 251, Philadelphia, Pennsylvania 19107, USA

Flossie Wong-Staal   251
Department of Medicine MS0665, University of California San Diego, La Jolla, CA 92093, USA

Florian M. Wurm   309, 337
Swiss Federal Institute of Technology Lausanne (EPFL), Faculty of Basic Sciences, Institute of Chemical and Biological Process Sciences, Center of Biotechnology, CH-1015 Lausanne, Switzerland

Xiao Xiao   93
Department of Molecular Genetics and Biochemistry, University of Pittsburgh School of Medicine, Room W1244 BST, Pittsburgh, PA 15261, USA

Liangzhi Xie   605
Fermentation and Cell Culture, Merck Research Laboratories, Merck & Co., Inc. West Point, PA 19486, USA

Yué Zeng   109
Institut de Recherche en Biotechnologie, Conseil National de Recherche Canada, 6100 Avenue Royalmount, Montréal, QC, Canada H4P 2R2

Weichang Zhou   605
Fermentation and Cell Culture, Merck Research Laboratories, Merck & Co., Inc. West Point, PA 19486, USA

Quan Zhu   573
Department of Cancer Immunology and AIDS, Dana-Farber Cancer Institute; Department of Medicine, Harvard Medical School, 44 Binney St., Boston, MA 02115, USA

Contents

Chapter 3.2. Virus-based vectors for gene expression in mammalian cells:
Epstein-Barr virus
Gregory Kennedy and Bill Sugden   55

Chapter 3.12. Virus-based vectors for gene expression in mammalian cells:
Retrovirus
Cristina Parolin and Giorgio Palù

Chapter 3.13. Virus-based vectors for gene expression in mammalian cells: Lentiviruses
Mehdi Gasmi and Flossie Wong-Staal 251

Chapter 4. Methods for DNA introduction into mammalian cells
Pamela A. Norton and Catherine J. Pachuk 265

Chapter 11.   Chromatin insulators and position effects
David W. Emery, Mari Aker and George Stamatoyannopoulos 381

Chapter 12.   Locus Control Regions
Xiangdong Fang, Kenneth R. Peterson, Qiliang Li and George
Stamatoyannopoulos 397

Chapter 15. Metabolic engineering of mammalian cells for higher protein yield
Hitto Kaufmann and Martin Fussenegger 457

Other volumes in the series

S.C. Makrides (Ed.) *Gene Transfer and Expression in Mammalian Cells*

CHAPTER 1

Why choose mammalian cells for protein production?

Savvas C. Makrides[1] and Holly L. Prentice[2]

[1]*EIC Laboratories, Inc., 111 Downey Street, Norwood, MA 02062, USA; Tel.: +781-769-9450;*
Fax: +781-551-0283; E-mail: savvas@eiclabs.com
[2]*Biogen, Inc., 14 Cambridge Center, Cambridge, MA 02142, USA; Tel.: +617-679-3320;*
Fax: +617-679-3200; E-mail: holly_prentice@biogen.com

There are many different types of hosts to use for production of natural or recombinant proteins: mammalian cells [1] (Table 1); bacteria, including Gram-negative [2], Gram-positive [3,4], and L-form [5,6]; filamentous fungi and yeast, including *Saccharomyces cerevisiae* and *Pichia pastoris* [7]; insect, including *Drosophila melanogaster*, *Aedes albopictus*, *Spodoptera frugiperda*, and *Bombyx mori* [8]; *Dictyostelium* [9]; *Xenopus* oocytes [10], and other types of cells, as well as plant tissue culture [11], transgenic animals ([12] and Chapter 24) and transgenic plants [12,13]. In addition, progress continues in the development of cell-free systems consisting of purified components [14,15].

The choice of a suitable host cell or expression system for protein production depends on many considerations, such as cell growth characteristics, ability to effect extracellular expression, post-translational modifications, folding and biological activity of the protein of interest, as well as regulatory and economic issues in the large-scale production of therapeutic proteins. The economics of the selection of a particular expression system requires a cost breakdown in terms of process, design, and other considerations [16]. The relative merits of bacterial, yeast, insect, and mammalian expression systems were examined in an earlier review [17]. Since then, expression technology has evolved to meet the range of research and commercial objectives [18]. One of the most exciting developments in the field of gene expression is the use of metabolic engineering to modify biochemical pathways, with the objective of endowing host cells with a spectrum of new properties for robust growth and enhanced protein production (Chapter 15).

Key advantages of mammalian cells over other hosts are the ability to carry out proper protein folding, and complex *N*-linked and authentic *O*-linked glycosylation of mammalian proteins. Also, mammalian cells posses an extensive post-translational modification machinery, including the ability to produce "mature" proteins through proteolytic processing.

Protein folding in mammalian cells is achieved during secretion from the endoplasmic reticulum (ER) to Golgi to the extracellular medium. In the ER, oxidized glutathione promotes thiol-disulfide bond exchange, and molecular chaperones also mediate correct disulfide bond formation. Proteins traversing the secretion pathway are protected from intracellular proteases [19], in contrast to bacteria where proteolytic degradation is a major problem [2]. In bacteria, the reducing environment

Table 1

Selected mammalian cell lines for gene expression

Cell line	Cell type/Origin	Characteristics	Primary use	Expression level* (mg/l)	Reference
COS	Fibroblast. African green monkey CV1 kidney cells	Transformed with an origin-defective SV40. Cells express the SV40 T antigen, allowing episomal amplification of plasmids containing the SV40 *ori*.	Transient expression	0.1–5	[38–40]
HEK-293T	Epithelial. Human embryonic kidney. Transformed with Ad5	Cells express the SV40 T antigen; episomal amplification of plasmids containing the SV40 *ori*.	Transient and stable expression	1–50	[41,42]
HEK-293E	Epithelial. Human embryonic kidney. Transformed with Ad5	Cells express EBNA-1; episomal amplification of plasmids containing the *OriP* from EBV	Transient and stable expression	1–50	[42,43]
BHK-21	Fibroblast. Baby hamster kidney		Vaccine production and stable expression	10	[44]
MDCK	Epithelial. Madin-Darby canine kidney	Cells become polarized to maintain two morphologically and functionally distinct plasma membrane domains. Useful for studying the mechanisms responsible for polarized localization of plasma membrane components	Vaccine production and stable expression	10–15	[45–47]
Vero	Epithelial. African green monkey kidney		Vaccine production		[45]
PER.C6	Retinoblast.Human fetal retinoblast immortalized with Ad5	Glycan structures are more "human-like." High-level gene expression without amplification	Stable expression. Production of vaccines and recombinant adenovirus vectors	300–500	[48–50]
CHO-K1 "Super CHO"	Epithelial. Chinese hamster ovary	Cells transfected with transferrin and IGF-1, capable of autocrine growth in protein-free medium without the addition of exogenous growth factors	Stable expression		[51]

CHO DUKX-B11	Epithelial. Chinese hamster ovary	Deficient in DHFR activity	Stable expression. Gene amplification using wild type dhfr as a selectable and amplifiable marker	10–1000	[52]
CHO DG44	Epithelial. Chinese hamster ovary	Mutant cells containing large deletions in the dhfr locus, eliminating the possibility of reversion	Stable expression. Gene amplification using dhfr.	10–1000	[53]
Sp2/0-Ag14	Myeloma, murine		Production of monoclonal antibodies		[54,55]
NS0	Myeloma, murine	Phenotypically deficient in GS. Can be used in combination with MSX, an inhibitor of GS activity, for selection and gene amplification	Production of monoclonal antibodies		[55–58]
Hybridoma	Fusion of a myeloma cell and an immune B lymphoblast (spleen) expressing a specific antibody gene		Preparation of immortalized cell lines for production of recombinant monoclonal antibodies	10–100	[59,60]

*Expression levels listed here are approximations. In general, production yields depend significantly on several factors, including the specific protein under study, gene amplification state, optimization of culture conditions, etc.

of the cytoplasm inhibits the formation of stable disulfide bonds and disfavors correct folding of complex proteins, often resulting in the formation of inclusion bodies. Although inclusion bodies have advantages, for example, they effectively concentrate the protein, they also require cumbersome solubilization and renaturation procedures that may be inefficient for all but the smallest proteins. A variety of strategies have been used to overcome this limitation [2], including the use of recently developed strains that favor production of proteins with complex cysteine connectivities [20]. These approaches, however, may be ineffective for high-yield bacterial expression of proteins with a large number of disulfide bonds. For example, a genetically engineered soluble form of the human complement receptor type 1 (sCR1) contains 60 disulfide bonds. Although biologically active small fragments of this protein have been produced in bacteria following arduous refolding protocols, to date the full-length sCR1 has been produced only in mammalian cells [21]. Biochemical pathways and their components involved in protein folding and post-translational modifications in mammalian cells are discussed in Chapter 13.

Glycosylation modulates several biochemical and biophysical properties of proteins, including protein folding, secretion, thermostability, antigenicity, catalytic efficiency, recognition and clearance ([22] and Chapter 14). These glycoprotein attributes are particularly important in human therapy where, for example, pharmacokinetic properties [23] and receptor targeting [24] may be dependent on the presence of specific sugars on the protein, or the presence of non-authentic glycosyl derivatives may confer immunogenicity. Although the glycosylation process in prokaryotes exhibits a diversity of glycan compositions and linkage units that rivals that in eukaryotes [25], prokaryotic glycoproteins in general lack the antennae of eukaryotic *N*-glycans. Some mammalian glycoproteins retain their activity when produced in bacteria. Most, however, must be produced in mammalian hosts to exhibit full, authentic biological activity. Yeast [17] and insect cells [8] do not functionalize proteins with complex oligosaccharides found in mammalian cells, although it is possible to metabolically engineer insect cells for *N*-glycoprotein sialylation by the insertion of mammalian glycosyltransferase genes [26]. A comparative study of the properties of human interferon-γ produced in Chinese hamster ovary (CHO) cells, the mammary gland of transgenic mice, and baculovirus-infected Sf9 insect cells, demonstrated the significant influence of host cell type on the type of incorporated *N*-glycans [27].

Other *post-translational modifications* that may be important for protein functionality require the use of mammalian host cells. A large number of different types of protein modifications has been documented [28,29], including phosphorylation, fatty acid acetylation (palmitoylation, myristoylation, isoprenylation), N-terminal acetylation, C-terminal α-amidation, methylation, and others. Some (most?) of these covalent modifications also occur in *Escherichia coli*, but they are of minor importance in the large-scale production of recombinant proteins. More important is the ability of mammalian cells to perform proteolytic processing that is necessary for maturation of specific proteins, e.g., insulin, insulin-like growth factor, relaxin, and other proteins. Finally, mammalian cells are preferred for the production of large proteins that require oligomerization of multiple chains, such as antibodies.

Until recently, key disadvantages of mammalian cell culture were its relatively high cost and complicated purification processes necessary for recovery of the secreted recombinant proteins. The high cost of production is mainly due to the use of fetal bovine serum, an expensive medium supplement that also increases the potential risk of virus, prion and mycoplasma contamination (see Chapter 23). The development of serum- and protein-free media has mitigated the cost of cell culture and has also simplified the purification process by eliminating contaminating serum proteins from growth media, in addition to reducing regulatory risks. An interesting approach to further minimizing the cost of culturing cytokine-dependent mammalian cells uses receptor engineering to produce cells that can grow in the presence of an exogenously added, inexpensive protein [30]. Mammalian cell lines that are used for the production of recombinant proteins and vaccines are listed in Table 1. In addition, a wealth of information on the characteristics of many other mammalian cell types, their culture and biochemical analysis can be found in a three-volume laboratory manual [31].

What are the "unmet needs" of mammalian gene expression systems? Media formulation continues to be an important area, with the goals of reducing costs, enhancing production yields, as well as increasing product sialylation (e.g., [32]). Unfortunately, commercial improvements in media formulations are often kept secret. New methods to improve the notoriously cumbersome process of preparing stable cell lines are also of major interest. Efforts in this area include the construction of vectors containing an internal ribosome entry site (IRES) followed by a quantitative selectable reporter marker or cell surface protein. The gene of interest is placed upstream of the IRES sequence. Cell populations or clonal cell lines expressing specific amounts of a desired protein are identified by fluorescent activated cell sorting (FACS) based on the level of expression of the gene downstream of the IRES [33]. Other approaches are directed at the construction of new expression vectors that effect gene targeting to transcriptionally active areas of the host chromosomal DNA (see Chapters 2 and 20). Lower yields and poorer quality of biopharmaceutical products that result from necrosis and apoptosis in bioreactors may be avoided using novel techniques that protect cells from apoptosis [34]. Improvements in bioprocess technologies for on-line monitoring of cultures will probably minimize variability in metabolism across different cell culture processes [35]. In this latter area, commercial activity is robust and impressive. For example, BioProcessors Corp. (Woburn, MA; http://www.bioprocessors.com) is developing a microfluidic platform for the growth of cells on micro fabricated devices that permit unprecedented level of environmental control. Such tools will enable the performance of massively parallel cell culturing experiments while simultaneously varying multiple environmental parameters, such as pH, temperature, and media conditions.

In summary, mammalian cells remain the host of choice for the production of proteins that require authentic glycosylation, proper folding, and other post-translational modifications. Nevertheless, expression technology for all host cell

6

types continues to evolve rapidly, with no end in sight. For example, bacterial strains have been developed that are deleted in all known cell envelope proteases [36], and other strains can facilitate the proper folding of proteins with a large number of cysteines [20]. The glycosylation apparatus of insect cells continues to catch up with the mammalian apparatus [26], and novel metabolic engineering approaches have produced mammalian cells with robust growth and anti-apoptotic properties ([37] and Chapter 15). We dare not predict the future.

Acknowledgements

We thank Edward G. Hayman for his critical reading of the manuscript.

Abbreviations

Ad5   adenovirus serotype 5
DHFR   dihydrofolate reductase
EBNA-1   Epstein-Barr virus Nuclear Antigen-1
EBV   Epstein-Barr virus
GS   glutamine synthetase
IGF-1   insulin-like growth factor I
MSX   methionine sulfoximine
MTX   methotrexate
ori   origin of replication

References

1. Chu, L. and Robinson, D.K. (2001) Curr. Opin. Biotechnol. 12, 180–187.
2. Makrides, S.C. (1996) Microbiol. Rev. 60, 512–538.
3. Bolhuis, A., Tjalsma, H., Smith, H.E., de Jong, A., Meima, R., Venema, G., Bron, S., and van Dijl, J.M. (1999) Appl. Environ. Microbiol. 65, 2934–2941.
4. Connell, N.D. (2001) Curr. Opin. Biotechnol. 12, 446–449.
5. Grichko, V.P. and Glick, B.R. (1999) Can. J. Chem. Eng. 77, 973–977.
6. Gumpert, J. and Hoischen, C. (1998) Curr. Opin. Biotechnol. 9, 506–509.
7. Punt, P.J., van Biezen, N., Conesa, A., Albers, A., Mangnus, J., and van den Hondel, C. (2002) Trends Biotechnol. 20, 200–206.
8. Altmann, F., Staudacher, E., Wilson, I.B.H., and Marz, L. (1999) Glycoconj. J. 16, 109–123.
9. Cubeddu, L., Moss, C.X., Swarbrick, J.D., Gooley, A.A., Williams, K.L., Curmi, P.M.G., Slade, M.B., and Mabbutt, B.C. (2000) Protein Expr. Purif. 19, 335–342.
10. Marino, M.H. (1996) In: Cleland, J.L. and Craik, C.S. (eds.) Protein engineering. Principles and Practice, Wiley-Liss, New York, pp. 219–247.
11. James, E. and Lee, J.M. (2001) Adv. Biochem. Eng. Biotechnol. 72, 127–156.
12. Larrick, J.W. and Thomas, D.W. (2001) Curr. Opin. Biotechnol. 12, 411–418.
13. Giddings, G. (2001) Curr. Opin. Biotechnol. 12, 450–454.
14. Shimizu, Y., Inoue, A., Tomari, Y., Suzuki, T., Yokogawa, T., Nishikawa, K., and Ueda, T. (2001) Nat. Biotechnol. 19, 751–755.

15. Kim, D.-M. and Swartz, J.R. (2001) Biotechnol. Bioeng. 74, 309–316.
16. Datar, R.V., Cartwright, T., and Rosen, C.-G. (1993) Biotechnology 11, 349–357.
17. Marino, M.H. (1989) BioPharm 2, 18–33.
18. Andersen, D.C. and Krummen, L. (2002) Curr. Opin. Biotechnol. 13, 117–123.
19. Gething, M.-J. and Sambrook, J. (1992) Nature 355, 33–45.
20. Levy, R., Weiss, R., Chen, G., Iverson, B.L., and Georgiou, G. (2001) Protein Expr. Purif. 23, 338–347.
21. Makrides, S.C., Nygren, P.-Å., Andrews, B., Ford, P.J., Evans, K.S., Hayman, E.G., Adari, H., Levin, J., Uhlén, M., and Toth, C.A. (1996) J. Pharmacol. Exp. Ther. 277, 534–542.
22. Helenius, A. and Aebi, M. (2001) Science 291, 2364–2369.
23. Fukuda, M.N., Sasaki, H., Lopez, L., and Fukuda, M. (1989) Blood 73, 84–89.
24. Makrides, S.C. (1998) Pharmacol. Rev. 50, 59–87.
25. Schäffer, C., Graninger, M., and Messner, P. (2001) Proteomics 1, 248–261.
26. Seo, N.-S., Hollister, J.R., and Jarvis, D.L. (2001) Protein Expr. Purif. 22, 234–241.
27. James, D.C., Goldman, M.H., Hoare, M., Jenkins, N., Oliver, R.W.A., Green, B.N., and Freedman, R.B. (1996) Protein Sci. 5, 331–340.
28. Uy, R. and Wold, F. (1977) Science 198, 890–896.
29. Wilkins, M.R., Gasteiger, E., Gooley, A.A., Herbert, B.R., Molloy, M.P., Binz, P.A., Ou, K.L., Sanchez, J.C., Bairoch, A., Williams, K.L., Hochstrasser, D.F. (1999) J. Mol. Biol. 289, 645–657.
30. Kawahara, M., Natsume, A., Terada, S., Kato, K., Tsumoto, K., Kumagai, I., Miki, M., Mahoney, W., Ueda, H., Nagamune, T. (2001) Biotechnol. Bioeng. 74, 416–423.
31. Spector, D.L., Goldman, R.D., and Leinwand, L.A. (Eds.), (1998) Cells-A Laboratory Manual. Cold Spring Harbor: Cold Spring Harbor Laboratory Press.
32. Gu, X. and Wang, D.I. (1998) Biotechnol. Bioeng. 58, 642–648.
33. Liu, X., Constantinescu, S.N., Sun, Y., Bogan, J.S., Hirsch, D., Weinberg, R.A., and Lodish, H.F. (2000) Anal. Biochem. 280, 20–28.
34. Sauerwald, T.M., Betenbaugh, M.J., and Oyler, G.A. (2002) Biotechnol. Bioeng. 77, 704–716.
35. Larson, T.M., Gawlitzek, M., Evans, H., Albers, U., and Cacia, J. (2002) Biotechnol. Bioeng. 77, 553–563.
36. Meerman, H.J. and Georgiou, G. (1994) Biotechnology 12, 1107–1110.
37. Fussenegger, M. and Betenbaugh, M.J. (2002) Biotechnol. Bioeng. 79, 509–531.
38. Gluzman, Y. (1981) Cell 23, 175–182.
39. Mellon, P., Parker, V., Gluzman, Y., and Maniatis, T. (1981) Cell 27, 279–288.
40. Edwards, C.P. and Aruffo, A. (1993) Curr. Opin. Biotechnol. 4, 558–563.
41. Kim, C.H., Oh, Y., and Lee, T.H. (1997) Gene 199, 293–301.
42. Durocher, Y., Perret, S., and Kamen, A. (2002) Nucleic Acids Res. 30, e9.
43. Cachianes, G., Ho, C., Weber, R.F., Williams, S.R., Goeddel, D.V., and Leung, D.W. (1993) BioTechniques 15, 255–259.
44. Burger, C., Carrondo, M.J.T., Cruz, H., Cuffe, M., Dias, E., Griffiths, J.B., Hayes, K., Hauser, H., Looby, D., Mielke, C., Moreira, J.L., Rieke, E., Savage, A.V., Stacey, G.N., Welge, T. (1999) Appl. Microbiol. Biotechnol. 52, 345–353.
45. Govorkova, E.A., Kodihalli, S., Alymova, I.V., Fanget, B., and Webster, R.G. (1999) Dev. Biol. Stand. 98, 39–51.
46. Robertson, J.S. (1999) Dev. Biol. Stand. 98, 7–11.
47. Pei, D. and Yi, J. (1998) Protein Expr. Purif. 13, 277–281.
48. Fallaux, F.J., Bout, A., Vandervelde, I., Vandenwollenberg, D.J.M., Hehir, K.M., Keegan, J., Auger, C., Cramer, S.J., Vanormondt, H., Vandereb, A.J., Valerio, D., Hoeben, R.C. (1998) Hum. Gene Ther. 9, 1909–1917.
49. Pau, M.G., Ophorst, C., Koldijk, M.H., Schouten, G., Mehtali, M., and Uytdehaag, F. (2001) Vaccine 19, 2716–2721.
50. Xie, L.Z., Pilbrough, W., Metallo, C., Zhong, T.Y., Pikus, L., Leung, J., Aunins, J.G., and Zhou, W.C. (2002) Biotechnol. Bioeng. 80, 569–579.
51. Pak, S.C.O., Hunt, S.M.N., Bridges, M.W., Sleigh, M.J., and Gray, P.P. (1996) Cytotechnology 22, 139–146.

8

52. Urlaub, G. and Chasin, L.A. (1980) Proc. Natl. Acad. Sci. USA 77, 4216–4220.
53. Urlaub, G., Käs, E., Carothers, A.M., and Chasin, L.A. (1983) Cell 33, 405–412.
54. Shulman, M., Wilde, C.D., and Kohler, G. (1978) Nature 276, 269–270.
55. Sauer, P.W., Burky, J.E., Wesson, M.C., Sternard, H.D., and Qu, L. (2000) Biotechnol. Bioeng. 67, 585–597.
56. Galfre, G. and Milstein, C. (1981) Methods Enzymol. 73, 3–46.
57. Bebbington, C.R., Renner, G., Thomson, S., King, D., Abrams, D., and Yarranton, G.T. (1992) Biotechnology 10, 169–175.
58. Barnes, L.M., Bentley, C.M., and Dickson, A.J. (2001) Biotechnol. Bioeng. 73, 261–270.
59. Köhler, G. and Milstein, C. (1975) Nature 256, 495–497.
60. Little, M., Kipriyanov, S.M., Le Gall, F., and Moldenhauer, G. (2000) Immunol. Today 21, 364–370.

S.C. Makrides (Ed.) *Gene Transfer and Expression in Mammalian Cells*

CHAPTER 2

Vectors for gene expression in mammalian cells

Savvas C. Makrides

*EIC Laboratories, Inc., 111 Downey Street, Norwood, MA 02062, USA; Tel.: +781-769-9450;
fax: +781-551-0283; E-mail: savvas@eiclabs.com*

1. Introduction

Achievement of robust and regulated protein production in mammalian cells is a complex process that requires careful consideration of many factors, including transcriptional and translational control elements, RNA processing, gene copy number, mRNA stability, the chromosomal site of gene integration, potential toxicity of recombinant proteins to the host cell, and the genetic properties of the host. Some of these topics are covered in detail elsewhere [1] and in other chapters in this book, therefore, only brief discussion will be provided here. Gene transfer into mammalian cells may be effected either by infection with virus that carries the recombinant gene of interest, or by direct transfer of plasmid DNA (Chapters 4 and 5). This chapter provides an overview of the molecular architecture of non-viral vectors for high-level protein production. Virus-based vectors for gene therapy, protein production, vaccine development and other applications are summarized in Table 1 and discussed in Chapters 3.1–3.13. In addition, inducible vector systems are examined in Chapter 22. Due to space limitations, many original publications regrettably could not be included, and the reader is referred to cited reviews and other chapters in this book.

2. Transient gene expression

Transient gene expression is typically used for rapid production of small quantities of protein for initial characterization, testing of vector functionality, and optimization of different combinations of promoters and other elements in expression vectors. Newly developed transient expression systems facilitate high-level production of recombinant proteins on a larger scale [2]. There are several cell types used for transient expression, including COS, baby hamster kidney (BHK), and human embryonic kidney (HEK)-293 cells, as well as genetically modified HEK-293 cells (see Table 1 in Chapter 1). COS cells were generated by transfection of African green monkey kidney CV1 cells with an origin-defective SV40 [3]. COS cells express the SV40 T antigen, which allows replication of plasmids containing the SV40 origin of replication. This host/vector system facilitates high-level plasmid amplification and protein production, followed by lysis of the cells several days from the time of transfection. Transient gene expression, therefore, permits rapid production of recombinant proteins, but does not enable preparation of "permanent" cell lines. Thus, transfection of the gene of interest must be repeated, as

Table 1

Virus-based vectors for gene delivery and expression in mammalian cells

Virus	Family	Vector features
DNA viruses Herpes simplex virus (HSV)	*Herpesviridae*. A heterogeneous family of viruses that contain linear dsDNA (130–230 kb) and infect man and many other vertebrates. Virions are enveloped, 180–200 nm in diameter. An icosadeltahedral capsid 100–110 nm in diameter contains 162 capsomers	The HSV-1 genome is 152 kb long and accomodates ~30 kb exogenous DNA. Broad mammalian host and cell type range. Potential for gene therapy. Difficulties with vector targeting and long-term transgene expression in certain tissues
Epstein-Barr virus (EBV)	*Herpesviridae*. See above	EBV has a large (~172 kb) dsDNA genome. Maintenance as a plasmid requires the viral origin of replication (*oriP*) and the viral gene encoding the *trans*-acting factor EBNA-1. *OriP*-based vectors can be maintained extrachromosomally in human, monkey, bovine, canine, and feline cells, but not in murine and rat cells in culture. Used as recombinant DNA shuttle vectors, screening of cDNA libraries, and production of recombinant proteins. EBV can accommodate up to 180 kb DNA. Potential for gene therapy
Simian virus 40 (SV40)	*Polyomaviridae*. This family was previously considered to be a subfamily of *Papovaviridae*. Small, antigenically distinct viruses that replicate in nuclei of infected cells; most have oncogenic properties. Virions are nonenveloped, 45–55 nm in diameter. The icosahedral capsids contain three virus-encoded proteins, VP1-3, with 72 pentameric capsomers, surrounding a molecule of circular dsDNA (5.2 kb)	Integrates in host genome, and provides stable transgene expression. In the presence of SV40 ori and large T antigen it replicates episomally at high copy number. Transduces both dividing and nondividing cells. Broad mammalian host range. Used in gene therapy. Nonimmunogenic, high yield and transduction efficiency. Principal limitation is the size of packageable insert (5 kb)
Adenovirus (Ad)	*Adenoviridae*. Viruses that replicate in the cell nuclei of mammals and birds. Virions are nonenveloped, 70–100nm in diameter; the icosahedral capsids are composed of 252 capsomers, of which 240 are hexons and 12 are pentons. Contain linear dsDNA (30–38 kb). No integration into host genome. The family includes two genera	Broad mammalian host range. Used in gene therapy. Infects both dividing and non-dividing cells. Immunogenic and toxic. Vector is maintained as a nuclear episome, which may lead to loss of DNA during cell division. New Ad vectors deleted in most viral genes are less immunogenic and accomodate ~35 kb insert

Adeno-associated virus (AAV)	*Parvoviridae*. Small viruses containing linear ssDNA (~5.0 kb), which converts to dsDNA after infection. Virions are nonenveloped, 18–26 nm in diameter, composed of three capsid proteins, VP1-3. The particle is icosahedral, and the capsid consists of 60 protein subunits. The inverted terminal repeats (ITRs) can pair to form hairpins, which are required for replication and packaging. Replication and assembly occur in the nucleus of infected cells. The family includes two subfamilies, each containing three genera. AAV (a member of the genus *Dependovirus*) normally requires a helper virus (Ad or herpes virus) to proceed through replication and lytic infection	AAV replication requires extra genes from a helper virus to mediate infection, but vectors can also be constructed that do not require the input of helper virus. Broad mammalian host range. Used in gene therapy. Vectors transduce cells through both episomal transgene expression and by random chromosomal integration. Infects both dividing and non-dividing cells with minimal cell-mediated immune response or toxicity. Prevalence of neutralizing antibodies against wild-type AAV may limit vector re-administration. Major limitation is the packaging capacity (~5 kb) that precludes the use of large genes, but which may be increased through viral DNA heterodimerization, concatemerization, or AAV/Ad hybrid vector constructs
Vaccinia virus (VV)	*Poxviridae*. Virions are enveloped, 200–400 nm long. Replication occurs in the cytoplasm of infected cells. Capsids are of complex symmetry and contain linear, dsDNA (130–300 kb) with a hairpin loop at each end. The family includes two subfamilies containing eight and three genera, respectively	Used for expression of heterologous genes and for vaccination. Broad mammalian host range. Vector can accomodate 25 kb exogenous DNA
Baculovirus	*Baculoviridae*. Insect, arachnid and crustacean viruses with a large circular dsDNA genome (90–160 kb), which is packaged in a rod-shaped capsid. Baculoviruses are divided into two genera: the nucleopolyhedroviruses (NPVs) and granuloviruses (GVs)	Mammalian promoters in baculovirus vectors enable heterologous gene expression in mammalian cells. Broad host range, no overt cytotoxicity, may be used for transient and stable gene expression. Its rapid inactivation by human complement is disadvantageous for *in vivo* gene delivery. Protein fusions to the amino terminus of the membrane glycoprotein gp64 may facilitate surface display applications, complement inactivation, and virus targeting to specific cell types. Vector can accommodate 40 kb exogenous DNA
RNA viruses Coronavirus	*Coronaviridae*. Viruses contain positive-sense, capped and polyadenylated ssRNA (27–32 kb). Virions are enveloped, 60–220 nm in diameter. The family includes two genera, *Coronavirus* and *Torovirus*	Virus replicates in cytoplasm without DNA intermediate, making its integration into host genome unlikely. Potential for vaccine development and gene therapy

(Continued)

Table 1

Continued

Virus	Family	Vector features
Poliovirus	*Picornaviridae*. Nonenveloped viruses, 27–30 nm in diameter, with one molecule of positive-sense polyadenylated ssRNA (7.2–8.5 kb) enclosed in a capsid of icosahedral symmetry with 60 protomers. Each protomer consists of four polypeptides, VP1-4. Replication occurs in the cytoplasm. The family includes six genera.	Primarily used for vaccination.
Sindbis virus (SIN)	*Togaviridae*. Virions are enveloped, spherical, 60–70 nm in diameter. The capsid is of icosahedral symmetry. The family consists of two genera, *Alphavirus* and *Rubivirus*. Alphaviruses (SIN and SFV) contain one molecule of linear, positive-sense, capped with 7-methylguanosine, polyadenylated ssRNA (11–12 kb)	Mosquito-borne, with broad host range including mammals, birds, reptiles and amphibia. Used for expression of heterologous genes, production of retrovirus vectors, detection and identification of other human viruses, construction of libraries of sequences inserted into SIN replicons to identify specific protease-cleavage sites, and development of high-throughput cloning systems. Potential applications include the control of mosquito-transmitted diseases, and vaccination for infectious diseases and cancer
Semliki Forest virus (SFV)	*Togaviridae*. Genus Alphavirus. See above	Used for expression of heterologous genes, production of retrovirus vectors, vaccination and potentially in gene therapy. Broad host range. Cloning capacity is ~8 kb. In DNA-based SFV vectors expression is RNA polymerase II-dependent
Retrovirus (RV)	*Retroviridae*. Virions are about 100 nm in diameter, enveloped, and contain two identical molecules of linear, positive-sense ssRNA (each monomer 7–13 kb), which have a 5′ cap and 3′ poly(A). RVs possess RNA-dependent DNA polymerases (reverse transcriptases). Upon entry into the host cell, the virion genomic RNA is reverse-transcribed into DNA, which is integrated into the host chromosomal DNA. The preintegration complex requires disruption of the nuclear membrane during mitosis to access the chromatin, thus they transduce only dividing cells. The family includes seven genera, according to recent taxonomic criteria [65]	Used in gene therapy. Accomodates ~9 kb insert. Host range: *ecotropic* virus replicates in cells derived from the host species; *amphotropic* virus replicates in a range of mammalian host cells. Minor immune response. Safety concerns. RV long terminal repeat (LTR) (used as the promoter) attenuates transgene expression in transduced cells. In general, RV-mediated high-level and tissue-specific transgene expression using non-LTR promoters is difficult to achieve

| Lentivirus (LV) | *Retroviridae*. LVs rely on active transport of the preintegration complex through the nuclear pores for translocation into the nucleus of the target cell. They transduce dividing and non-dividing cells | Replication-deficient vectors derived from human immunodeficiency virus (HIV) and from non-human lentiviruses that may not be infectious to humans. Cloning capacity is $\sim 9\,$kb. Potential for gene therapy. Minor immune response. Vector improvements include minimizing HIV sequences and eliminating viral accessory proteins for enhanced transduction efficiency and safety. In self-inactivating LVs, a deletion in the U3 region of the 3′ LTR results in transcriptional inactivation of the 5′ LTR after integration, enabling transgene expression to be regulated solely by an internal promoter, without reducing viral titers. This diminishes the risk of vector mobilization and recombination, and facilitates high-level targeted transgene expression |

Virus vector systems are reviewed in Chapters 3.1–3.13. Other RNA virus vectors are examined by Palese [66]. ds, double-stranded; ss, single-stranded.

necessary. In contrast, stable transformants may be prepared by a more labor-intensive procedure, as discussed below. Virus-based vectors that are useful for transient gene expression include adenovirus, adeno-associated virus, Epstein-Barr virus, Semliki Forest virus, baculovirus, Sindbis virus, lentivirus, Herpes simplex virus, and vaccinia virus (Table 1).

3. Stable gene expression

In contrast to transient gene expression, preparation of stable cell lines usually depends on integration of plasmid into the host chromosome. Transformants must be cloned in order to ensure that all cells in the culture are genetically identical. Typically, DNA-transfected cells are maintained in non-selective medium for about two days, followed by transfer to selective medium. Marker-containing cells that survive the selection are allowed to proliferate, and single transformants are then isolated and characterized using a variety of techniques, including cloning cylinders, soft agar, limiting dilution, or flow cytometry. It is also possible, however, to generate stable cell lines that harbor vectors extrachromosomally. For example, vectors that carry the Epstein-Barr virus nuclear antigen-1 (*EBNA-1*) and the origin of replication (*oriP*) can be maintained episomally in primate and canine cell lines but not in rodent cell lines [4]. An episomal replicating vector has been described that does not express any viral proteins, thus avoiding cell transformation [5]. The vector contains the SV40 origin of replication and the scaffold/matrix attachment region (S/MAR) (Chapters 10 and 20) from the human interferon-β gene. The vector was shown to replicate at very low copy numbers (below 20) in CHO cells and was stably maintained without selection for more than 100 generations [5].

The host cell (see Table 1 in Chapter 1) may have a significant impact on gene expression levels. For example, myeloma cells, such as NS0 and Sp2/0, have been used mainly for high-level production of monoclonal antibodies. An epithelial cell line, Madin-Darby canine kidney, was shown to be capable of producing large amounts of protein, comparable to those obtained from CHO amplification systems [6]. The human cell line PER.C6 [7] has recently generated considerable interest for commercial production of therapeutic proteins. Amplifiable gene expression using CHO cells has been widely used for protein production (Chapter 7). The two most widely used amplification systems rely on the dihydrofolate reductase and glutamine synthetase genes. Typically, the selectable marker and the cDNA are under the control of separate transcription units. By growing cells in increasing concentrations of selection drugs it is possible to amplify the copy number of the cotransfected (and cointegrated) gene of interest and concomitantly elevate the amount of protein produced. An alternative method for high-level production of recombinant proteins in CHO cells utilizes an expression vector that produces both selectable marker and cDNA from a single primary transcript via differential splicing [8].

Generation of stable cell lines, particularly the selection of amplified and high-expressing clonal cells, involves screening of large numbers of transfected cells, both during the initial transfection as well as at each subsequent amplification step. This arduous exercise is necessitated by the wide variation in the level of expression and amplification of the transfected gene in different cells, an outcome that reflects the chromosomal site of plasmid integration (reviewed in [1]). An alternative strategy for efficient preparation of stable cell lines is site-specific gene integration using recombination systems (Chapter 20) such as Cre/*loxP* and FLP/*FRT*. Cre (cyclization recombination) recombinase of bacteriophage P1 recombines DNA at 34-bp sites called *loxP* (locus of crossover of P1). The FLP recombinase from the 2-μm circle of *Saccharomyces cerevisiae* recognizes *FRT* (the FLP recombination target). It should be possible to engineer a cell line using a reporter gene to select a transcriptionally active chromosomal locus. Such a cell line could then be used for the routine excision and replacement of the reporter construct with the gene of interest. A commercially available vector–host system makes use of the FLP/*FRT* elements (Flp-In^{TM} expression vectors; Invitrogen, Carlsbad, CA). In this case, different mammalian cell lines were engineered to contain a single FRT site integrated at a transcriptionally active locus. These cells can be used with targeting vectors to prepare recombinant cell lines containing the gene of interest.

Other integrases that hold promise for the engineering of mammalian stable cell lines include those derived from phages R4 and ϕC31 of *Streptomyces spp.* [9]. These enzymes function in mammalian cells with no added cofactors. Unlike Cre and FLP, which catalyze reversible recombination between two identical sites, R4 and ϕC31 integrases mediate unidirectional site-specific recombination between two attachment sites with dissimilar sequences, at higher net integration frequencies than is possible with Cre and FLP [9]. Olivares *et al.* [10] used the integrase from ϕC31 to achieve site-specific integration of the gene encoding the human blood clotting Factor IX into the chromosomes of mice, resulting in the stable production of normal levels of the protein. Recent work using DNA shuffling and screening aims at the generation of phage integrases that exhibit improved integration frequency and sequence specificity in human cells [11].

An alternative vector system for gene expression involves receptor-mediated endocytosis of recombinant protein vehicles that target cell-surface receptors ([12] and references therein). The construct in this case comprises a modified β-galactosidase gene containing an insertion of a viral peptide that binds the integrin $\alpha_v\beta_3$, and an amino-terminal DNA-condensing poly-L-lysine domain. The construct is expressed in *Escherichia coli*, and when the purified protein is mixed with plasmid DNA, it facilitates transfection of cells expressing $\alpha_v\beta_3$ receptors [12]. This approach exploits the cell-targeting specificity of viruses without the disadvantages of virus-based vectors.

4. Genetic elements of mammalian expression vectors

Vectors for protein production in mammalian cells comprise a variety of genetic elements with distinct functionalities (Fig. 1): (1) a constitutive or inducible promoter that is capable of robust transcriptional activity; (2) a transcription terminator that stabilizes the transcript and prevents transcription interference; (3) optimized mRNA processing and translational signals that include the Kozak sequence,

16

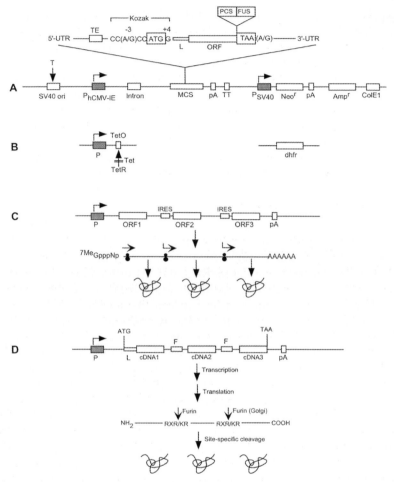

Fig. 1. Configuration of model genetic elements in mammalian expression vectors. The combination of different elements (not drawn to scale) may vary in order to meet specific objectives. SV40 ori facilitates transient gene expression in COS cells. Promoters (P) facilitate constitutive (A) or inducible (B) expression. The optimal translational initiation sequence (Kozak) and termination tetranucleotide are shown. The ColE1 origin and the Amp^r gene allow plasmid replication and selection, respectively, in bacteria. The Neo^r gene facilitates selection, and the *dhfr* gene allows both selection and gene amplification in cells. Multiple gene expression utilizes polycistronic constructs (C) where IRES elements enable ORFs to be translated from a single transcript (see Section 8). Alternatively, a monocistronic construct (D) contains in-frame cDNAs joined by linkers encoding recognition sites (Arg-X-Arg/Lys-Arg) for the endoprotease furin, thus facilitating the post-synthetic cleavage of different proteins (see Section 8). *Abbreviations*: Amp^r, ampicillin-resistance gene (*β*-lactamase); ColE1, prokaryotic origin of replication; dhfr, dihydrofolate reductase (methotrexate resistance); F, furin-recognition sequence; FUS, fusion moiety; hCMV-IE, human cytomegalovirus immediate early enhancer/promoter; IRES, internal ribosome entry site; L, leader (targeting sequence); MCS, multiple cloning site; Neo^r, neomycin-resistance gene (aminoglycoside phosphotransferase, *aph*); ORF, open reading frame; ori, origin of replication; P, promoter; pA, polyadenylation signal; PCS, protease cleavage site; T, SV40 large tumor (T) antigen; TE, translational enhancer; Tet, tetracycline; TetO, tetracycline operator; TetR, tetracycline repressor protein; TT, transcription terminator; UTR, untranslated region.

translation termination codon, mRNA cleavage and polyadenylation signals, as well as mRNA splicing signals for higher levels of expression; (4) prokaryotic origin of replication and selection marker for vector propagation in bacteria; and (5) selection markers for the preparation of stable cell lines and for gene amplification. The inclusion of the SV40 origin of replication facilitates transient gene expression in COS cells. Other genetic elements for specific applications include sequences for gene or protein targeting, signal peptides for protein secretion, fusion moieties and protease cleavage sites (see Section 6), and ribosome- or protease-recognition sites that facilitate the expression of multiple genes from polycistronic (Fig. 1C) or monocistronic (Fig. 1D) constructs, respectively (see Section 8). An extensive list of mammalian expression vectors has been published [1].

4.1. Transcriptional control elements

Regulation of transcription in eukaryotic genomes involves the coordinated interaction of multiple genetic elements, a remarkably complex process that is understood in great detail [13]. The promoter is defined as the region proximal to the transcription start site. Transcription initiation is mediated through interactions of transcription factors with their cognate promoter and enhancer elements. Enhancers are sequences which may be located thousands of bases upstream or downstream of the promoter that enhance transcriptional activity when bound by transcription factors. In addition, upstream activation sequences, located within a few hundred bases of the promoter, influence transcription activity. The variability in expression levels observed in different clones during the preparation of stable cell lines is caused by several factors, collectively referred to as position effects. These include the proximity of the target gene to heterochromatin, orientation/location relative to other endogenous genes, and proximity to chromosomal structural elements. Chromatin elements that may abrogate position effects include S/MARs (Chapter 10), chromatin insulators (Chapter 11), and Locus Control Regions (Chapter 12).

Promoters: Promoters used for gene expression in mammalian cells are listed in Table 2. Some promoters are transcriptionally active in a wide range of cell types and tissues. Most, however, exhibit tissue selectivity, a property that must be carefully considered prior to the construction of expression vectors for high-level production of proteins. Strong constitutive promoters, which drive expression in many cell types, include the adenovirus major late promoter, the human cytomegalovirus immediate early promoter (hCMV-IE), the SV40 and Rous Sarcoma virus promoters, the murine 3-phosphoglycerate kinase promoter, the translation elongation factor 1α (EF-1α) promoter, and the human ubiquitin C promoter. Tissue-selective promoters [14] may facilitate gene targeting and expression in specific organs and tissues.

Promoters can be divided into two classes, those that function constitutively, and those that are regulated by induction or derepression (Chapter 22). Promoters used for high-level production of proteins in mammalian cells should be strong and, preferably, active in a wide range of cell types to permit qualitative and quantitative evaluation of the recombinant protein. Inducible promoters should exhibit a minimal level of basal transcriptional activity, and be capable of

Table 2

Selected promoter elements for gene expression in mammalian cells

Promoter	Source	Properties
SV40	Simian virus 40	Constitutive expression; in some cell lines inducible with phorbol ester. Broad host and cell type range. In COS cell lines expressing the T antigen, high vector copy number is achieved
hCMV-IE mCMV-IE	Human and mouse cytomegalovirus immediate-early promoter genes	High-level constitutive expression. Broad host and cell type range
RSV-LTR	Rous sarcoma virus long terminal repeat	High-level constitutive expression in murine and avian cell lines
MMTV-LTR	Mouse mammary tumor virus	Inducible with glucocorticoids. Moderate level of transcriptional induction
MoMLV-LTR	Moloney murine leukemia virus	Moderate to strong transcriptional induction
Ad2MLP-TPL	Adenovirus major late promoter and tripartite leader	High-level constitutive expression. Broad host range
hUbC	Human ubiquitin C gene	High-level constitutive expression in a broad range of tissues and cell types
hEF-1α	Human translation elongation factor 1α subunit gene	High-level constitutive expression. Broad host and cell type range
mPGK	Mouse phosphoglycerate kinase gene	High-level constitutive expression. Broad host and cell type range
mMT-I	Mouse metallothionein I gene	Inducible with Cd^{++}, Zn^{++}, phorbol esters. "Leaky" promoter
hMT-II	Human metallothionein II gene	Inducible with Cd^{++}, Zn^{++}, phorbol esters. "Leaky" promoter
hMT-IIA (mutant)	Human metallothionein II gene	Inducible with Cd^{++}, Zn^{++}, phorbol esters. High inducibility, low basal activity
hIFN-α	Human interferon-α gene	Inducible with virus
β-actin	Chicken β-actin gene	High-level constitutive expression in a broad range of tissues and cell types
β-globin	β-globin gene	Tissue-specific for adult β erythroid cells. Potential gene therapy applications
ET-1	Endothelin-1 gene	Endothelium-specific
vWf	von Willebrand factor gene	Endothelium-specific
Endoglin	Gene encoding human endoglin (CD105), a component of the TGF-β complex	Endothelium-specific
GFAP	Gene encoding human glial fibrillary acidic protein	Brain-specific
Synapsin I	Rat synapsin I gene	Brain-specific

A list of promoters that are tissue-specific, tumor-selective, treatment-responsive or cell cycle-regulated has been published [14].

substantial induction with a non-toxic inducer in a simple and cost-effective manner. Weaker promoters may be desirable in specific applications. For example, in studies of intracellular targeting of antibodies, the powerful EF-1α promoter led to aggregation of expressed antibody, an effect that was avoided by using a weaker promoter [15]. Inducible promoters are desirable for the production of proteins that may be toxic to the host cell, for the study of gene regulation during development in transgenic animals, and for experimental and therapeutic applications of gene transfer.

Introns: Most genes from higher eukaryotes contain introns, which are removed during RNA processing. Genomic constructs have been shown to be expressed more efficiently in transgenic animals than identical constructs lacking introns [16], an effect thought to be due to an enhanced rate of RNA polyadenylation and nuclear transport coupled to RNA splicing [17]. Although many cDNA constructs lacking introns can be expressed efficiently in mammalian cells, the inclusion of introns can enhance expression 10- to 20-fold, and some sequences, such as the β-globin cDNA, show a virtual requirement for the presence of an intron [18]. The placement of introns at the 3′ end of the transcription unit may lead to aberrant splicing [19,20], therefore, it is preferable to place introns at the 5′ end of the open reading frame. A synthetic intron, SIS, generated by the fusion of an adenovirus splice donor site and an immunoglobulin G splice acceptor site was very active in a variety of cell types [21]. The insertion of the human EF-1α first intron downstream of the human or murine CMV-IE promoters strongly enhanced the level of reporter gene expression in several cell lines [22]. The use of introns, however, demands careful vector design, as it is possible that cryptic splicing signals may cause aberrant processing of the mRNA transcript, resulting in reduced expression levels and defective protein products (e.g., [23,24]).

Polyadenylation signals: Most eukaryotic nascent mRNAs possess a poly(A) tail ($n \approx 200$) at their 3′ ends, which is added during cleavage of the primary transcript and a coupled polyadenylation reaction [25]. The poly(A) tract is important for mRNA stability and translatability [26]. The signals for polyadenylation of mammalian mRNAs are well defined: One component consists of a highly conserved AAUAAA sequence, which is located about 20–30 nucleotides upstream of the 3′ end of the mRNA, and the other element consists of an unconserved GU-rich sequence immediately downstream of the polyadenylation site [27]. Among the more efficient poly(A) signal sequences to insert in mammalian expression vectors are those derived from bovine growth hormone, mouse β-globin, the SV40 early transcription unit, and the herpes simplex virus thymidine kinase gene.

Transcription terminators: Continued transcription from a promoter through a second transcription unit reduces expression of the second gene, a phenomenon known as transcriptional interference [28]. This has been documented in bacteria, yeast, mammalian and plant cells, but the mechanism is poorly understood. The placement of transcription termination signals between two transcription units, along with the designing of gene orientation, can minimize transcriptional interference. Prokaryotic transcription terminators are well characterized, and their incorporation in expression vectors has multiple beneficial effects on gene expression [29]. In eukaryotes, a consensus sequence consisting of ATCAAA(A/T) TAGGAAGA has been identified in the termination region of nine genes [30].

4.2. Translational control elements

Optimal expression of eukaryotic cDNAs requires careful consideration of several structural features, including the nucleotide context around the translation initiation codon, and the 5'- and 3'-untranslated regions (UTRs), which are involved in many posttranscriptional processes that control mRNA localization, stability and translation efficiency [31]. In addition, codon usage can have a significant impact on the translation efficiency of some heterologous genes in mammalian cells.

5'-Untranslated region: Based on comparison of eukaryotic mRNA sequences and systematic mutagenesis of specific genes, Kozak proposed the "scanning model" of translation initiation in higher eukaryotes (Chapter 16). The initiation complex, consisting of the 40S ribosomal subunit and cap-binding proteins, forms at the mRNA 5' terminal cap (m^7GpppN) followed by movement of the ribosome to the "correct" initiating AUG codon, which is surrounded by an optimal consensus sequence, GCC(A/G)CCaugG (the Kozak sequence, [32]). The purines A or G in position -3 (i.e. three nucleotides upstream from the AUG codon) and G immediately following the AUG codon are the most influential in facilitating optimal translation initiation. The presence of AUG codons in the 5'-UTR of the transcript can severely depress translation initiation at the "authentic" start codon, although the extent of inhibition depends on sequences surrounding the upstream AUG. The design of plasmids for the expression of heterologous genes should therefore consider the 5' sequence context of the gene of interest and, preferably, avoid the presence of upstream AUGs in the 5' UTR.

Another concern in the design of expression vectors involves the potential ability of the 5'-UTR to form secondary structure. GC-rich regions have the potential to form stable hairpin structures, which can inhibit translation initiation, a phenomenon that has been documented in eukaryotic [32] and prokaryotic [29] expression systems. One remedy to this potential problem is the removal of the 5'-UTR prior to the insertion of cDNAs into expression vectors, with the caveat that the 5'-UTR may contain translational enhancer elements, such as the SP163 element of the vascular endothelial growth factor mRNA [33]. The SP163 sequence has been shown to enhance translation of different mRNAs 25- to 40-fold in several mammalian cell types [33].

3'-Untranslated region: mRNA destabilization can be influenced by specific sequences present in the 3'-UTR (see Section 7 and Chapter 17). In addition, translational regulation of certain mRNAs is mediated by protein-binding AU-rich elements located in the 3'-UTR (Chapter 17).

Termination codon: Translational termination in mammalian genes may be modulated by nucleotides additional to those of the trinucleotide stop codons. Statistical analysis of the context of termination codons in 5208 mammalian genes showed a highly significant bias in the position immediately following the stop codon [34]. Experimental evidence determined that the base following the stop codon influences the efficiency of translation termination both *in vitro* and *in vivo*. Thus, tetranucleotides with a purine in the fourth position are more effective as termination signals than those with a pyrimidine [34].

Codon usage: Genes from both prokaryotic and eukaryotic organisms exhibit a nonrandom usage of synonymous codons. In general, highly expressed genes exhibit a greater degree of codon bias than do poorly expressed ones, and the frequency of use of synonymous codons is strongly correlated with the abundance of their cognate tRNAs within each pool of isoaccepting tRNAs. These observations led to the hypothesis that the main reason for codon bias is translational efficiency. An alternative view holds that the abundance of tRNAs is probably a consequence of and not a reason for codon bias and, furthermore, the primary reason for codon bias is selection for the accuracy of protein synthesis on the ribosome [35]. In *E. coli*, transcripts of heterologous genes enriched with codons that are rarely used by *E. coli* may not be translated efficiently, or may result in polypeptides containing misincorporated amino acid residues (reviewed in [29,36]). Similarly, mammalian codon usage can adversely affect translation efficiency of heterologous genes. In some cases, codon optimization has been demonstrated to enhance expression levels of the target genes by 10- to 50-fold (reviewed in [1]).

5. Selectable markers

In addition to the presence of a selectable marker for vector propagation in bacteria, mammalian expression vectors contain markers for the selection of transfected cells, preparation of stable cell lines and for gene amplification. There are several amplifiable genes, the most commonly used ones being dihydrofolate reductase and glutamine synthetase. A compilation of markers, including their mechanism of action, has been published [1].

6. Signal peptides and fusion moieties

Mammalian proteins that are targeted for secretion are translocated from the endoplasmic reticulum (ER) to Golgi to the extracellular medium [37,38] (Chapter 13). Secreted proteins are synthesized as precursor proteins possessing a signal or leader peptide composed of 15–30 amino acid residues, usually located at the amino-terminus, which is subsequently cleaved by a signal peptidase in the ER lumen. Signal peptides typically consist of three regions: a positively charged amino-terminal region (N-region), a central hydrophobic region (H-region), and a polar carboxy-terminal region (C-region) followed by the signal peptidase cleavage site [39,40]. Signal peptides are usually interchangeable and have been widely used to effect protein secretion in mammalian cells. Moreover, signal sequences from bacteria (e.g., [41]) and yeast (e.g., [42]) are recognized by mammalian cells. However, signal peptides can vary significantly in their ability to promote protein secretion. In fact, the presence of a signal sequence *per se* does not necessarily ensure protein secretion, as has been documented in both prokaryotic (reviewed in [29]) and eukaryotic cells (e.g., [43]). There are many examples of efficient signal peptides including those derived from erythropoietin [44], tissue plasminogen activator [45], interleukin-2 [46], albumin [47], and immunoglobulin sequences [48].

Fusion partners, have a wide range of applications in both prokaryotic [29,49] and eukaryotic [1,49] expression systems. Fusion moieties are used as affinity handles for the facile isolation and purification of proteins (Chapter 25); as reporters (Chapter 6) in studies of promoter activity or localization of proteins in cellular compartments; as protein dimerization domains; as immunogens for the production of antibodies; to target antibodies (Chapter 21); to increase expression, folding, solubility, and secretion of proteins; or to display polypeptides on the surface of cells for vaccine development, protein–protein interactions, drug screening, and other potential applications. Fusion moieties have also been used to increase the half-life of target proteins for potential therapeutic applications ([50] and references therein). Several factors must be carefully considered in the design of fusion proteins. For example, a linker may be inserted between two protein partners in order to optimize protein folding and stability. The length and sequence composition of the linker can impact protein folding. An affinity tag is often fused to the N-terminus of a protein to facilitate purification. This is less desirable, in theory, than a C-terminal tag, which has the advantage that only fully translated proteins can be purified. The latter strategy, however, requires that the C-terminus be structurally accessible. Interestingly, the widely used $(His)_6$ tag has been shown recently to modify the properties of certain proteins expressed in *E. coli* [51]. Separation of the fusion moiety from the target protein is facilitated by a protease cleavage site engineered between the two components. Selection of an appropriate protease enables regeneration of the native terminus of the target protein upon proteolytic digestion. Fusion proteins are examined in a recent comprehensive review [49].

7. mRNA and protein stability

Turnover of mRNA is an important posttranscriptional mechanism for the physiological control of gene expression (Chapter 17). The potential ability to extend significantly the half-life of transcripts offers an attractive means to enhance protein production in mammalian cells. One determinant of eukaryotic mRNA lability is an AU-rich sequence in the 3'-UTR of many unstable mammalian mRNAs [52]. Insertion of an AU-rich element into the 3'-UTR of a stable mRNA destabilizes the chimeric transcript [53]. The optimal sequence for this destabilizing determinant is UUAUUUAUU [53] or UUAUUUA(U/A)(U/A) [54]. Removal of these sequences from the 3'-UTR of unstable mRNAs can prolong the half-life of transcripts and enhance protein production.

Synthetic 5' secondary structures have been shown to increase mRNA half-lives in *E. coli*. In seeking to maximize transcript stability and protein production in mammalian cells, investigators have substituted the UTRs of stable mRNAs, such as β-globin, for the UTRs of target transcripts. This strategy, effective in specific cases, may not have universal application, as mRNA degradation is effected by multiple pathways in mammalian cells. Thus, in addition to exonucleolytic activity at both the 5′ and 3′ termini, determinants of mRNA half-life have been mapped to the coding regions of several mRNA species. Furthermore, mRNA stability is modulated by a variety of cell-specific proteins that act in *trans* to destabilize or stabilize transcripts (Chapter 17). The use of a specific UTR for the purpose of stabilizing a heterologous transcript in mammalian cells assumes the presence of the cognate UTR-binding proteins in the same cells. At present our knowledge of the distribution of such proteins in different mammalian cell lines used for protein production is incomplete.

Levels of heterologous proteins are also affected by protein degradation pathways (Chapter 18). Recent work has shown that the Gly-Ala repeat of the Epstein-Barr virus nuclear antigen-1 is a *cis*-acting transferable element that inhibits ubiquitin/proteasome-dependent proteolysis. It has been suggested that the viral Gly-Ala repeat might be used for the prolongation of protein half-life in gene therapy (Chapter 19).

8. Coordinated expression of multiple genes

Coordinated expression of two or more heterologous genes is an important requirement in many applications, including establishment of stable mammalian cell lines that require coexpression of the gene of interest and a selectable marker; characterization of antibody responses in DNA immunization protocols; coexpression of genes for positive-negative (suicide) selections in gene therapy; gene trapping for the identification of developmentally regulated genes; gene targeting; *in vitro* and *in vivo* imaging using reporter genes; and coordinated constitutive or inducible high-level expression of several genes in mammalian cells (for references see [1]). As briefly outlined below, a variety of methods exist for the coordinated expression of two or more genes. The suitability of each strategy will depend on the experimental context:

(a) Different expression vectors may be used, each carrying a different gene of interest. This approach is widely used for the production of equimolar amounts of target proteins or protein chains, e.g., antibodies, and it also allows for the ability to evaluate different protein ratios for optimal results.

(b) A single vector can be constructed containing multiple genes each with its own promoter. This type of construct may be subject to promoter interference, a problem that is usually avoided using transcription terminators.

(c) In a translational fusion, two proteins are genetically joined in-frame (see Section 6). The success of this strategy depends on the accessibility of the termini of the two fusion partners. Potential problems related to steric hindrance, misfolding, instability and loss of activity of one or both of the protein partners are addressed through the insertion of an appropriate peptide linker between the joined proteins. A key advantage of this approach is the production of stoichiometric amounts of both proteins.

(d) Two or more genes may be connected via virus-derived elements, known as internal ribosome entry sites (IRES), which facilitate ribosome binding to the second and subsequent transcription units (e.g., [55]) (Fig. 1C). The use of IRES elements, however, presents its own problems, as often the first gene is favored, and the efficiency of translation initiation from different IRES elements varies substantially [56]. Moreover, tissue tropism determinants of IRES activity are poorly understood, and evidence exists for internal ribosome entry dependence on cellular factors that are differentially expressed in different cell types (reviewed in [57]). Consequently, for specific applications, e.g., *in vivo* imaging in transgenic animals [58], translational processivity may vary among different tissues. Incidentally, although many studies have reported the presence of IRES elements in cellular mRNAs, the experimental evidence in this body of work has been questioned [59] and continues to be vigorously debated [60] (Chapter 16).

(e) Another approach utilizes monocistronic transcripts [61] (Fig. 1D). In this case, the construct contains several in-frame cDNAs joined by linkers encoding cleavage sites for furin, a Golgi-localized ubiquitous endoprotease. The encoded polypeptides are post-synthetically cleaved and processed into biologically active proteins. Processing of the fusion protein, however, may be suboptimal in cells with low levels of furin. This type of construct could potentially utilize tissue-specific delivery and transcription elements, as well as cleavage sites for tissue-specific endoproteases, rather than furin, to achieve a high level of targeting [61].

(f) The use of the 2A sequence from the foot and mouth disease virus, which functions as a ribosome slippage site [62,63]. This sequence, previously thought to be an autocatalytic proteolytic cleavage site [64], facilitates the stoichiometric production of two joined open reading frames.

Acknowledgements

I thank Holly Prentice for her critical reading of the manuscript. James Barsoum and William Sisk made things happen, and I am much obliged to them.

Abbreviations

BHK	baby hamster kidney
CHO	Chinese hamster ovary
EBNA	Epstein-Barr virus nuclear antigen
ER	endoplasmic reticulum
HEK	human embryonic kidney
IRES	internal ribosome entry site
LTR	long terminal repeat
UTR	untranslated region

References

1. Makrides, S.C. (1999) Protein Expr. Purif. 17, 183–202.
2. Meissner, P., Pick, H., Kulangara, A., Chatellard, P., Friedrich, K., and Wurm, F.M. (2001) Biotechnol. Bioeng. 75, 197–203.
3. Mellon, P., Parker, V., Gluzman, Y., and Maniatis, T. (1981) Cell 27, 279–288.
4. Yates, J.L., Warren, N., and Sugden, B. (1985) Nature 313, 812–815.
5. Piechaczek, C., Fetzer, C., Baiker, A., Bode, J., and Lipps, H.J. (1999) Nucleic Acids Res. 27, 426–428.
6. Pei, D. and Yi, J. (1998) Protein Expr. Purif. 13, 277–281.
7. Pau, M.G., Ophorst, C., Koldijk, M.H., Schouten, G., Mehtali, M., and Uytdehaag, F. (2001) Vaccine 19, 2716–2721.
8. Lucas, B.K., Giere, L.M., DeMarco, R.A., Shen, A., Chisholm, V., and Crowley, C.W. (1996) Nucleic Acids Res. 24, 1774–1779.
9. Olivares, E.C., Hollis, R.P., and Calos, M.P. (2001) Gene 278, 167–176.
10. Olivares, E.C., Hollis, R.P., Chalberg, T.W., Meuse, L., Kay, M.A., and Calos, M.P. (2002) Nat. Biotechnol. 20, 1124–1128.
11. Sclimenti, C.R., Thyagarajan, B., and Calos, M.P. (2001) Nucleic Acids Res. 29, 5044–5051.
12. Alcalá, P., Feliu, J.X., Arís, A., and Villaverde, A. (2001) Biochem. Biophys. Res. Commun. 285, 201–206.
13. Lemon, B. and Tjian, R. (2000) Genes Dev. 14, 2551–2569.
14. Nettelbeck, D.M., Jérôme, V., and Müller, R. (2000) Trends Genet. 16, 174–181.
15. Persic, L., Righi, M., Roberts, A., Hoogenboom, H.R., Cattaneo, A., and Bradbury, A. (1997) Gene 187, 1–8.
16. Choi, T., Huang, M., Gorman, C., and Jaenisch, R. (1991) Mol. Cell. Biol. 11, 3070–3074.
17. Huang, M.T.F. and Gorman, C.M. (1990) Nucleic Acids Res. 18, 937–947.
18. Buchman, A.R. and Berg, P. (1988) Mol. Cell. Biol. 8, 4395–4405.
19. Wise, R.J., Orkin, S.H., and Collins, T. (1989) Nucleic Acids Res. 17, 6591–6601.
20. Huang, M.T.F. and Gorman, C.M. (1990) Mol. Cell. Biol. 10, 1805–1810.
21. Petitclerc, D., Attal, J., Théron, M.C., Bearzotti, M., Bolifraud, P., Kann, G., Stinnakre, M.-G., Pointu, H., Puissant, C., Houdebine, L.-M. (1995) J. Biotechnol. 40, 169–178.
22. Kim, S.-Y., Lee, J.-H., Shin, H.-S., Kang, H.-J., and Kim, Y.-S. (2002) J. Biotechnol. 93, 183–187.
23. Hall, J., Hirst, B.H., Hazlewood, G.P., and Gilbert, H.J. (1992) Biochim. Biophys. Acta 1130, 259–266.
24. Zaboikin, M.M. and Schuening, F.G. (1998) Hum. Gene Ther. 9, 2263–2275.
25. Proudfoot, N. (1996) Cell 87, 779–781.
26. Gray, N.K. and Wickens, M. (1998) Annu. Rev. Cell Develop. Biol. 14, 399–458.
27. Proudfoot, N. (1991) Cell 64, 671–674.
28. Proudfoot, N.J. (1986) Nature 322, 562–565.
29. Makrides, S.C. (1996) Microbiol. Rev. 60, 512–538.
30. Maa, M.-C., Chinsky, J.M., Ramamurthy, V., Martin, B.D., and Kellems, R.E. (1990) J. Biol. Chem. 265, 12513–12519.
31. Pesole, G., Mignone, F., Gissi, C., Grillo, G., Licciulli, F., and Liuni, S. (2001) Gene 276, 73–81.
32. Kozak, M. (1999) Gene 234, 187–208.
33. Stein, I., Itin, A., Einat, P., Skaliter, R., Grossman, Z., and Keshet, E. (1998) Mol. Cell. Biol. 18, 3112–3119.
34. McCaughan, K.K., Brown, C.M., Dalphin, M.E., Berry, M.J., and Tate, W.P. (1995) Proc. Natl. Acad. Sci. USA 92, 5431–5435.
35. Fedorov, A., Saxonov, S., and Gilbert, W. (2002) Nucleic Acids Res. 30, 1192–1197.
36. Kane, J.F. (1995) Curr. Opin. Biotechnol. 6, 494–500.
37. Ellgaard, L., Molinari, M., and Helenius, A. (1999) Science 286, 1882–1888.
38. Sakaguchi, M. (1997) Curr. Opin. Biotechnol. 8, 595–601.
39. Perlman, D. and Halvorson, H.O. (1983) J. Mol. Biol. 167, 391–409.

40. von Heijne, G. (1990) J. Membr. Biol. 115, 195–201.
41. Clément, J.-M. and Jehanno, M. (1995) J. Biotechnol. 43, 169–181.
42. Kamiya, T., Sugio, S., Yamanouchi, K., and Kagitani, Y. (1996) Tohoku J. Exp. Med. 180, 297–308.
43. Farrell, P.J., Behie, L.A., and Iatrou, K. (2000) Proteins: Structure, Function and Genetics 41, 144–153.
44. Herrera, A.M., Musacchio, A., Fernandez, J.R., and Duarte, C.A. (2000) Biochem. Biophys. Res. Commun. 273, 557–559.
45. Chapman, B.S., Thayer, R.M., Vincent, K.A., and Haigwood, N.L. (1991) Nucleic Acids Res. 19, 3979–3986.
46. Liu, J., O'Kane, D.J., and Escher, A. (1997) Gene 203, 141–148.
47. Maeda, Y., Soda, M., Ito, K., and Sato, K. (1997) Biochem. Mol. Biol. Int. 42, 825–832.
48. Lo, K.-M., Sudo, Y., Chen, J., Li, Y., Lan, Y., Kong, S.-M., Chen, L.L., An, Q., and Gillies, S.D. (1998) Protein Eng. 11, 495–500.
49. Hearn, M.T.W. and Acosta, D. (2001) J. Mol. Recognit. 14, 323–369.
50. Makrides, S.C., Nygren, P.-Å., Andrews, B., Ford, P.J., Evans, K.S., Hayman, E.G., Adari, H., Levin, J., Uhlén, M., Toth, C.A. (1996) J. Pharmacol. Exp. Ther. 277, 534–542.
51. Rumlová, M., Benedíková, J., Cubínková, R., Pichová, I., and Ruml, T. (2001) Protein Expr. Purif. 23, 75–83.
52. Chen, C.-Y.A. and Shyu, A.-B. (1995) Trends Biochem. Sci. 20, 465–470.
53. Zubiaga, A.M., Belasco, J.G., and Greenberg, M.E. (1995) Mol. Cell. Biol. 15, 2219–2230.
54. Lagnado, C.A., Brown, C.Y., and Goodall, G.J. (1994) Mol. Cell. Biol. 14, 7984–7995.
55. Fussenegger, M., Mazur, X., and Bailey, J.E. (1998) Biotechnol. Bioeng. 57, 1–10.
56. Hennecke, M., Kwissa, M., Metzger, K., Oumard, A., Kröger, A., Schirmbeck, R., Reimann, J., and Hauser, H. (2001) Nucleic Acids Res. 29, 3327–3334.
57. Martinez-Salas, E. (1999) Curr. Opin. Biotechnol. 10, 458–464.
58. Contag, C.H. and Bachmann, M.H. (2002) Annu. Rev. Biomed. Eng. 4, 235–260.
59. Kozak, M. (2001) Mol. Cell. Biol. 21, 1899–1907.
60. Schneider, R. and Kozak, M. (2001) Mol. Cell. Biol. 21, 8238–8246.
61. Gäken, J., Jiang, J., Daniel, K., van Berkel, E., Hughes, C., Kuiper, M., Darling, D., Tavassoli, M., Galea-Lauri, J., Ford, K., Kemeny, M., Russell, S., Farzaneh, F. (2000) Gene Ther. 7, 1979–1985.
62. Donnelly, M.L.L., Hughes, L.E., Luke, G., Mendoza, H., ten Dam, E., Gani, D., and Ryan, M.D. (2001) J. Gen. Virol. 82, 1027–1041.
63. Donnelly, M.L.L., Luke, G., Mehrotra, A., Li, X.J., Hughes, L.E., Gani, D., and Ryan, M.D. (2001) J. Gen. Virol. 82, 1013–1025.
64. Donnelly, M.L.L., Gani, D., Flint, M., Monaghan, S., and Ryan, M.D. (1997) J. Gen. Virol. 78, 13–21.
65. Goff, S.P. (2001) In: Knipe, D.M., Howley, P.M., Griffin, D.E., Lamb, R.A., Martin, M.A., Roizman, B. and Straus, S.E. (eds.) Fields Virology Fourth Edition, Lippincott Williams & Wilkins, Philadelphia, PA, pp. 1871–1939.
66. Palese, P. (1998) Proc. Natl. Acad. Sci. USA 95, 12750–12752.

S.C. Makrides (Ed.) *Gene Transfer and Expression in Mammalian Cells*

Virus-based vectors for gene expression in mammalian cells: Herpes simplex virus

Edward A. Burton[1,*], Qing Bai[1,**], William F. Goins[1,],
David J. Fink[1,2,4], and Joseph C. Glorioso[1,3]

[1]*Department of Molecular Genetics and Biochemistry, University of Pittsburgh School of Medicine,*
Pittsburgh, PA, USA
[2] *Neurology, School of Medicine, University of Pittsburgh, Pittsburgh, PA, USA*
[3]*E1240 Biomedical Sciences Tower, 200 Lothrop Street, Pittsburgh, PA 15261, USA,*
E-mail: glorioso@pitt.edu
[4]*Geriatric Research, Education and Clinical Center (GRECC) Pittsburgh VA Healthcare System,*
Pittsburgh, PA, USA

1. Introduction

Herpes simplex virus (HSV) based vectors are flexible and efficient vehicles for introducing experimental and therapeutic transgenes into a variety of tissues. The natural neurotropism and latency of the wild-type vector can be exploited to generate vectors with particular utility for neuroscience and neurological applications. In this chapter, we provide a basic overview of HSV biology, and then discuss ways of generating non-toxic vectors from the wild-type virus. The chapter concludes with a brief summary of published applications of these vectors, both within and outside the nervous system. Sadly, the limited available space precludes exhaustive citation of original publications, particularly with reference to basic HSV biology, where a complex story has been pieced together from the painstaking work of many investigators. Several reviews are cited, which contain many of the original references; the excellent chapters on HSV in *Field's Virology* cover the subject in an exhaustive and rigorous fashion [1].

2. Herpes Simplex Virus—relevant basic biology

HSV is a neurotropic enveloped double-stranded DNA virus. The virus naturally infects human hosts, causing recurrent cold sores (peripheral nervous system and skin) and severe encephalitis (central nervous system).

---

*Current address: Department of Clinical Neurology, University of Oxford, Radcliffe Infirmary, Woodstock Road, Oxford, OX2 6HE, UK.
** Current address: Department of Human Anatomy and Genetics, University of Oxford, South Parks Road, Oxford, OX1 3QX, UK.

28

2.1. Structure of HSV-1

The mature HSV virion consists of the following components (Fig. 1A—reviewed in [1]):

- A trilaminar lipid envelope in which are embedded 10 viral glycoproteins—these are responsible for several functions including receptor-mediated cellular entry.
- A matrix of proteins, the tegument, which forms a layer between the envelope and the underlying capsid. Functions of the tegument proteins include: induction of viral gene expression; shutoff of host protein synthesis immediately following infection; virion assembly functions.
- An icosadeltahedral capsid, typical of the Herpesvirus family.
- A core of toroidal dsDNA.

Viral genes encode the majority of the proteins and glycoproteins of the mature virion. The HSV genome consists of 152 kb of dsDNA arranged as long and short unique segments (U_L and U_S) flanked by repeated sequences (ab, b'a', a'c', ca) [2,3]

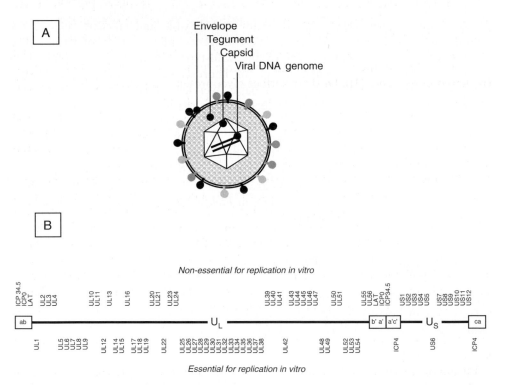

Fig. 1. The structure of HSV. (A) A schematic depiction of a mature HSV-1 virion is shown to illustrate key structural components, discussed in Section 2.1. (B) The HSV-1 genome is illustrated diagrammatically (not to scale) to illustrate genomic locations of the various viral genes discussed in the text. Genes that are essential and non-essential for viral replication *in vitro* are indicated.

(Fig. 1B). Eighty-four viral genes are encoded, approximately half of which are essential for viral replication in a permissive tissue culture environment. Non-essential ('accessory') genes often encode functions that are important for specific virus–host interactions *in vivo*, for example immune evasion, replication in non-dividing cells or shutdown of host protein synthesis. The importance of this observation is that non-essential genes may be deleted in the generation of gene therapy vectors, allowing the insertion of exogenous genetic material [4,5]. In addition, deletion of specific accessory genes may limit viral replication to certain cellular subsets.

2.2. Mechanisms of HSV-1 cell entry

HSV-1 has evolved a complex series of virion components that allow the mature particle to deliver its DNA and protein payload directly into the cytoplasm of infected cells [6,7] (see Fig. 6A). Electron microscopic studies show that initial cell attachment is followed by fusion of the viral envelope with the cell membrane, and entry of the tegument proteins and nucleocapsid into the cytoplasm [6]. The fusion process results in the appearance of viral surface components in the plasmalemma [8]. Approximately, one-fifth of the 84 viral genes [2,3] encode virion surface and envelope components. Ten viral glycoproteins are embedded in the trilaminar envelope of the mature virion, and are available for interaction with cellular targets (reviewed in reference [1]). Of these surface glycoproteins, gB [9], gD [10,11], gH [12] and gL [13] are essential to enable viral replication in cell culture.

The first phase in the entry cascade consists of viral attachment ('adsorption') to cells [6]. Attachment of HSV-1 to cells occurs through binding to glycosaminoglycan (GAG) moieties of cell surface proteoglycans [14]. The virion surface glycoproteins implicated in the attachment process are gC [15,16], and gB [9,15].

Following the initial adsorption event, a secondary binding event occurs between viral gD and a cellular receptor [17,18]. Two structurally unrelated HSV-1 secondary receptors have been identified. These bind gD following gC/gB/HS-dependent adsorption, resulting in commitment to virion–cell fusion. Herpesvirus Entry Mediator A (HVEM/HveA) is a member of the TNFα/NGF receptor superfamily, and has a limited expression pattern [19]. Herpesvirus Entry Mediator C (HveC— also called poliovirus receptor related protein-1 (PRRP-1) and nectin-1) is a type 1 transmembrane glycoprotein, and a member of the immunoglobulin (Ig) superfamily [20]. HveC is a widely expressed protein isoform resulting from alternative splicing of the primary transcript from the nectin-1 gene [20]. The important domain for gD binding lies within the Ig family V-homology portion of the ectodomain of nectin-1/HveC [21].

Fusion of the viral envelope with the cell membrane, allowing release of the virion contents into the cell, occurs after the binding of gD to its cognate cell surface receptor [6]. Three HSV glycoproteins, gB [22], gH [13] and gL [23] have been implicated in this final step of the entry cascade; the details of viral envelope–cell fusion are not clear at present.

2.3. Regulation of viral genes during lytic infection

Following cell entry, viral genes are expressed in a tightly regulated, interdependent temporal sequence [24,25] (Fig. 2—reviewed in [1]). Transcription of the five immediate-early (IE) genes, ICP0, ICP4, ICP22, ICP27 and ICP47 commences on viral DNA entry to the nucleus. Expression of these genes is regulated by promoters that are responsive to VP16, a viral structural protein, present in the tegument layer of the mature virion. Following envelope–plasmalemma fusion, VP16 enters the cytoplasm of the cell with the nucleocapsid; it is transported to the host cell nucleus with the viral DNA genome. VP16 is a potent trans-activator that associates with cellular transcription factors and binds to cognate DNA motifs within the IE

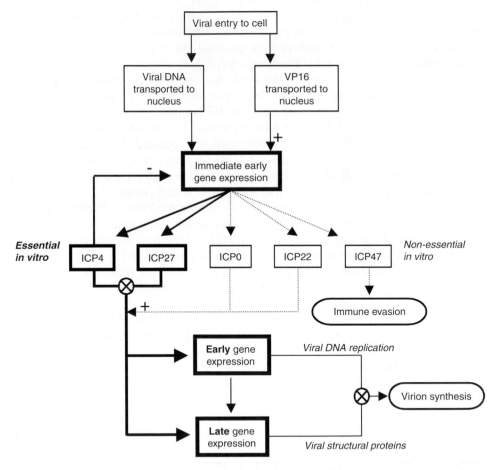

Fig. 2. Regulation of HSV-1 gene expression during lytic infection. This flow chart illustrates the important regulatory events in gene expression occurring during lytic HSV-1 infection. In order to proceed to the later stages of infection, during which the viral genome is replicated and new virions assembled, both of the essential immediate early genes ICP4 and ICP27 must be expressed. Deletion of one or other essential IE gene results in a replication-defective virus that may be used as a vector. See Sections 2.3 and 3.2.2.

promoter sequences. Expression of IE genes initiates a cascade of viral gene expression. Transcription of early (E) genes, which primarily encode enzymes involved in DNA replication, is followed by expression of late (L) genes mainly encoding structural components of the virion [24,25] (reviewed in reference [1]). Of the IE gene products, only ICP4 and ICP27 are essential for expression of E and L genes, and hence viral replication [26].

2.4. The viral life cycle in vivo

The life cycle of HSV-1 *in vivo* is complex (Fig. 3). Following primary cutaneous or mucosal inoculation, the virus undergoes lytic replication in the infected epithelia. Viral particles are released at the site of the primary lesion; they may enter sensory neurons whose axon terminals innervate the affected area. The nucleocapsid and

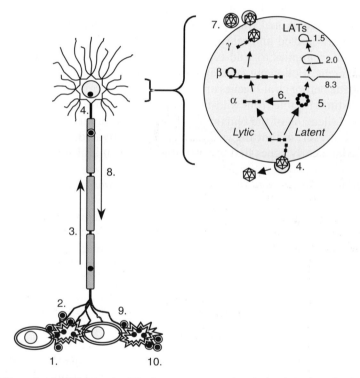

Fig. 3. The life cycle of HSV-1 *in vivo*. The key events occurring during infection of a human host are depicted schematically: 1. Lytic cycle of replication at epithelial port of entry; 2. Virions released from epithelia enter sensory nerve terminals; 3. Nucleocapsid and tegument undergoes retrograde axonal transport to soma; 4. Viral DNA enters neuronal nucleus and either initiates lytic cascade of gene expression or becomes latent; 5. During latency, viral genome remains episomal and nuclear. Only the LAT genes are expressed; 6. Immunosuppression, intercurrent illness or other stimulus 'reactivates' lytic infection; 7. Virions formed by budding from nuclear membrane; 8. Nucleocapsid and glycoproteins transported separately by anterograde axonal transport; 9. Virion assembly and egress from nerve terminal; 10. Recurrent epithelial infection at or near site of primary lesion. See Section 2.4.

tegument are carried by retrograde axonal transport from the site of entry to the neuronal soma in the dorsal root ganglia or trigeminal ganglia, where the viral genome and VP16 enter the nucleus [27,28]. At this point, one of two chains of events may ensue. First, the lytic replicative cycle described above may take place. This pathway results in neuronal cell death and egress of infectious particles. Alternatively, the viral DNA can enter the latent state. During latency, the viral genome persists as a stable episomal element, sometimes for the lifetime of the host. The DNA adopts a chromatin-like structure, and it is not extensively methylated. IE, E and lytic L gene expression is not readily demonstrable during latency. However, a set of non-translated RNA species, the latency-associated transcripts (LATs), is produced and detectable in the nuclei of latently infected neurons [29] (and see below). At a time point that may be remote from the establishment of latency, alterations in the host–virus interaction may cause 'reactivation' of the viral infection. IE genes are expressed and the lytic cascade of gene expression follows, resulting in the pro-duction of mature virions. The nucleocapsid and glycoproteins are transported by separate anterograde axonal transport pathways to the peripheral nerve terminals, where they are assembled and released.

This description applies to wild-type virus. The life cycle of replication-defective vectors is different. Following cellular entry, IE gene mutations tend to abort the infective cascade before viral E genes are expressed. The viral DNA adopts a state that is almost indistinguishable from wild-type latency, with production of the LATs and transcriptional silencing of the remainder of the viral genome. This state differs from true latency, however, in that the virus cannot be reactivated, as it is genetically null for the immediate early genes that are essential for the initial stages of reactivation.

2.5. Regulation of latency and the latency-associated transcripts

The processes regulating the establishment and maintenance of latency are poorly understood. During latent infection, only the LAT genes are expressed abundantly from the viral genome [29]. A primary 8.3 kb LAT transcript of low steady-state abundance [30] undergoes post-transcriptional processing to form other LAT species. The 2.0 kb LAT is a stable lariat intron that is spliced from the primary 8.3 kb transcript [31]. The 2.0 kb LAT intron is located within the nuclei of infected cells, and can be readily detected during either lytic or latent infection [29]. The 1.5 kb LAT is also a stable lariat intron [32]; it is formed by the removal of a second intron from within the 2.0 kb LAT sequence [31,33,34]. The 1.5 kb LAT intron has only been detected during viral latency [34,35], implying that the relevant splicing event is specific to both neurons and latent infection. The unusual stability of the LATs has been ascribed to peculiarities in the lariat structure adjacent to the branch point and splice acceptor region [36–38]. The presumed 6.3 kb spliced exon product of LAT has not been detected, and may be unstable.

Despite the extensive characterisation of the origin and nature of the LATs, relatively little is known about their function, in particular their role in the establishment of, or reactivation from, latent HSV infection. The most striking conclusion from studies on LAT-null HSV strains is that LAT deletion does not completely prevent either process [39]. Various more subtle deficits have been reported, however. First, there is some evidence that LAT-null mutants are less efficient at establishing latent infection than wild type or rescuant virus [40]. Second, latent LAT-null mutants show different reactivation kinetics compared with controls [41], although this may be attributable in some instances to the reduced efficiency with which latent infection is established in the first place. Finally, LAT-null mutants appear more toxic to neurons than LAT-expressing viruses [42,43], resulting in more cell death following infection, which may be apoptotic in nature [42]. Enhanced cytotoxicity may account for the reduced efficiency with which LAT-null viruses establish latency. The molecular events underlying these LAT-null phenotypes are poorly characterised. The regulation of viral genes is disturbed in LAT-null mutants; the steady-state levels of the IE gene products ICP0, ICP4 and ICP27 [44] and the early gene product TK are increased following infection with LAT-null mutants. It is unclear whether these changes in gene expression are primarily responsible for the phenotypic effects apparent after LAT deletion. Furthermore, it is unknown whether LAT functions by encoding a protein, or through an RNA-mediated process.

Expression of the LATs is driven by two promoters, latency active promoters 1 and 2—referred to as 'LAP1' and 'LAP2' [45]. LAP1 is a typical RNA polymerase II promoter with a TATA element, which directs transcription starting at position −736 with respect to the 5′ end of the 2.0 kb LAT intron. LAP1 is primarily responsible for LAT expression during latent infection, as determined by deletion analysis [46]. LAP2 is a GC-rich promoter, typical of eukaryotic 'housekeeping' promoters [45,46]. It is situated 3′ to LAP1 and deletion analysis suggests that it is primarily responsible for LAT expression during lytic infection [46]. The situation arising in the intact virus is, however, complex. Several studies suggest that sequences contained within LAP2 facilitate the sustained transcription from LAP1 that occurs during latency [47,48]. Furthermore, it has been possible during latency to drive transcription of a gene placed at an ectopic locus within HSV from LAP2, but not LAP1 [45,47]. Finally, sequences contained within LAP2 seem able to direct high-level and sustained latent-phase transcription from some heterologous promoter elements [48,49]. Clearly, however, the LAT promoters are able to drive transcription of the LAT genes from the virus for prolonged periods *in vivo*, and as such might prove highly valuable for long-term transgene expression in the nervous system using HSV vectors (see below).

3. Using HSV-1 to make gene therapy vectors

3.1. Advantages of HSV as a vector

Various aspects of the basic biology of HSV-1 are attractive when considering the design of neurological gene therapy vectors:

- HSV has a broad host cell range.
- HSV vectors are highly infectious.

- Non-dividing cells of the CNS and PNS may be efficiently transduced and made to express transgenes [4].
- Of the 84 known viral genes, approximately half are non-essential for growth in tissue culture. This means that multiple or very large therapeutic transgenes can be accommodated, by replacing dispensable viral genes [5,50].
- Recombinant replication-defective HSV-1 may readily be prepared to high titre and purity without contamination from wild-type recombinants [51].
- The latent behaviour of the virus may be exploited for the stable long-term expression of therapeutic transgenes in neurons [47,52].
- The abortive gene expression cascade produced when a replication-defective vector enters a cell results in a state that is similar to latency; the main difference is that the virus cannot reactivate. This enables chronic transgene expression in both neuronal and non-neuronal cells, without the danger of viral reactivation [53].

Harnessing the advantageous features of HSV biology for gene delivery, whilst eliminating the pathogenicity inherent in the wild-type virus has been a major goal of vector construction.

There are two important steps in eliminating pathogenicity. First, viral replication must be prevented in cells to which transgenes are to be delivered, because the result of viral replication is inevitably a lytic infection cycle resulting in cell death. Second, even in the absence of replication, some viral gene products are cytotoxic, and so their expression must be prevented to minimise cytotoxicity.

3.2. Eliminating viral replication

Broadly speaking, there are three ways that lytic infection can be either eliminated or targeted to certain cell types (Fig. 4):

3.2.1. Conditionally replicating vectors
Deletion of some non-essential genes results in viruses that retain the ability to replicate *in vitro*, but are compromised *in vivo*, in a context-dependent manner. For example, deletion of the gene encoding ICP34.5 results in a virus that may replicate *in vitro*, but not in neurons *in vivo* [54]. The virus, however, retains the ability to undergo lytic replication in rapidly dividing cancer cells. ICP34.5 mutants have been used to treat patients with brain tumours in phase I clinical trials, in the hope that the virus will destroy the tumour cells and spare normal brain tissue. Although these mutants appear non-toxic at present, it is not yet clear whether this therapeutic strategy is efficacious [55].

3.2.2. Replication-defective vectors
Deletion of one or other of the essential IE genes (ICP4, ICP27) results in a virus that cannot replicate [26,56], except in cells that complement the null mutations by providing ICP4 or ICP27 *in trans* [26,56]. In appropriate complementing cell lines, however, the virus is able to replicate. By using this method, it is possible to prepare high titre viral stocks that are free from contaminating replication-competent

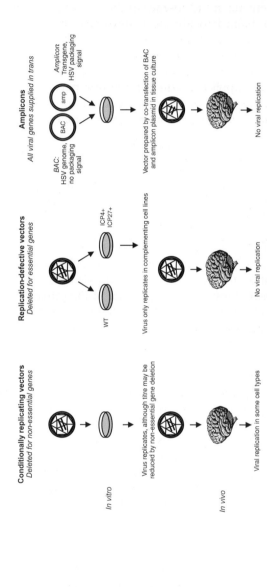

Fig. 4. Eliminating viral replication in the generation of **HSV** gene therapy vectors. Three broad strategies for eliminating replication of **HSV** are illustrated schematically. (i) Conditionally replicating vectors replicate only in certain cell types, such as malignant cells, and have been developed for application in malignant glioma. (ii) Replication-defective vectors use complementing cell lines to enable replication of HSV bearing null mutations in one or more essential IE genes. In the absence of complementation *in vivo*, these viruses do not replicate. (iii) Amplicons use HSV packaging signals within plasmids to enable their concatemeric packaging within HSV-like particles that are generated by co-transfection of BACs encoding the viral genome devoid of packaging signals. See Section 3.2.

viruses [51]. In addition, the genetic manipulation of these viruses is relatively straightforward, exploiting the recombinogenic properties of HSV-1 to introduce exogenous sequences by homologous recombination [4,5]. *In vivo*, these viruses undergo abortive cascades of lytic gene transcription, resulting in a state that is very similar to latency. The genomes may persist for long periods in neuronal and non-neuronal cells, but cannot reactivate in the absence of the essential IE genes [53,57]. These vectors may be further refined to prevent cytotoxicity resulting from non-essential IE gene expression.

3.2.3. Amplicons

The entire viral genome may be supplied *in trans*, generating particles that contain very few viral gene sequences. In this instance, the desired transgene cassette is placed in a plasmid containing the viral genomic packaging/cleavage signals, in addition to both viral and bacterial origins of replication—an 'amplicon' plasmid [58]. Defective HSV-like particles are generated by double transfection of eukaryotic cells with: (i) the amplicon plasmid, and (ii) a bacterial artificial chromosome (BAC) containing the viral genome, but devoid of packaging and eukaryotic replication signals [59]. Concatemerised plasmid DNA is packaged into disabled particles that contain HSV structural proteins and surface glycoproteins. The HSV BAC is a recent advance on the prior practice of using a series of cosmids or a helper virus to supply viral functions. Although a perceived advantage of the amplicon system is that no viral coding sequence is delivered, it has proven difficult in practice to produce a pure preparation of vector with clinically useful yields.

Our laboratories have amassed considerable experience in the generation, use and propagation of replication-defective vectors. The remainder of this discussion therefore concentrates on the replication-defective system.

3.3. Minimising toxicity from replication-defective vectors

Deletion of essential IE genes prevents viral replication and thus eliminates the component of viral cytopathogenicity that is attributable to viral replication. However, IE gene products, with the exception of ICP47, are toxic to host cells [60]. Infection with an ICP4 null mutant results in extensive cell death in the absence of viral replication [4,26]. This is caused by over-expression of other IE gene products, some of which are negatively regulated by ICP4 [26]. To prevent cytotoxicity, a series of vectors has been generated that are deleted for multiple IE genes. Quintuple mutants, null for ICP0, ICP4, ICP22, ICP27 and ICP47, have been produced, are entirely non-toxic to cells and the genomes are able to persist for long periods of time [60]. However, vectors grow poorly in culture and express transgenes at very low levels in the absence of ICP0 [60,61]. Retention of the gene encoding the trans-activator ICP0 allows efficient expression of viral genes and transgenes, and allows the virus to be prepared to high titre. Recent work has shown that the post-translational processing of ICP0 in neurons is different to that in glia [62]. It appears that, although ICP0 mRNA is efficiently expressed in both cell types, ICP0 undergoes proteolytic degradation in neurons. It might be predicted that a vector carrying an intact ICP0 gene would not be toxic to neurons, but may be advantageous for oncological applications, where ICP0 toxicity may be desirable. Deletion of ICP47 restores the expression and priming of MHC class I molecules to the surface of the cells [63]. This may potentially confer advantages in the gene therapy of malignancy, although the utility of this modification is unclear at present. For most other applications, where immune evasion is desirable, triple mutants (ICP4⁻:ICP22⁻:ICP27⁻) have been used. These vectors show minimal cytotoxicity *in vitro* and *in vivo*, are efficient vehicles for transgene delivery and can be grown efficiently in cells that complement the absence of ICP4 and ICP27 *in trans* [4,52,64].

3.4. Inserting transgenes into replication-defective vectors

As the 152-kb HSV-1 genome is too large to manipulate by conventional cloning techniques, insertion of transgenes into the replication-defective HSV vectors is achieved by homologous recombination in eukaryotic cells in cell culture. The transgene cassette is inserted into a shuttle plasmid that contains sequence from the targeted viral locus. In the resulting shuttle vector, the transgene is flanked on either side by 1–2 kb of viral sequence. The plasmid DNA is linearised and transfected into cells that complement the deleted IE genes from the defective virus. The cells are co-transfected with viral genomic DNA. Plaques form as viral genes are expressed and virions are generated. The recombination rate between linearised plasmid and purified viral DNA ranges from 0.1 to 1% of the plaques, when the calcium phosphate method is used for the transfection. Virus is prepared from the plaques, and the viral DNA is screened for recombinants.

Two additional features have been built into this system to simplify the isolation of recombinant plaques (Fig. 5):

- The replication-defective vectors discussed above have been designed to express reporter genes in certain important loci. Recombination of the transgenic cassette into these loci results in loss of reporter gene activity, which is readily assayed. This allows rapid screening of plaques for putative reporter-negative recombinants, which are then subjected to secondary screening by Southern blot analysis [5,65].
- The viral DNA may be cleaved at the site of the desired recombination event by using restriction endonucleases. Unique *Pac*I and *Pme*I sites have been engineered into appropriate viral genes for this purpose. Following digestion, only recombination or re-ligation events can yield DNA capable of being incorporated into infectious virus particles in the complementing cell line. This technique substantially reduces the non-recombinant background. In most cases, the proportion of viral plaques that represent recombinants rises to 10–50% using this technique [65]. By eliminating native *lacZ*⁺ viral DNA from the transfection, virtually all plaques are formed by *lacZ*⁻ viruses. These may be either recombinants or simply re-ligations of the cut DNA. Both of these grow as clear plaques in culture, facilitating further isolation and screening, as it can be difficult to isolate a clear plaque from a blue-staining background.

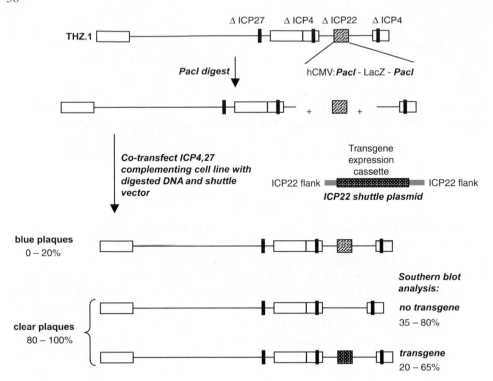

Fig. 5. Technique for introduction of a transgene expression cassette into replication-defective HSV vector. The *Pac*I digest technique of Krisky *et al.* [65] is illustrated. The vector backbone is engineered to contain unique *Pac*I restriction enzyme sites flanking the targeted genetic locus, which contains a reporter gene. Following digestion and co-transfection with the shuttle plasmid, recombination and re-ligation events generate three possible products. The desired product is screened from background by the absence of reporter gene activity in the viral plaque, and formally identified by Southern blot analysis of the viral DNA. The proportion of reporter-negative plaques is greatly enhanced by the digest, as is the proportion of reporter-negative clones that contain the transgene expression cassette.

3.5. Vector targeting

The broad host cell range of the unmodified vector is advantageous when developing a system with sufficient flexibility to allow use in multiple circumstances. However, in some situations, it may be desirable or necessary to restrict the expression of therapeutic transgenes to pre-defined sub-populations of cells. One way of achieving this goal might be through re-directing the tropism of HSV vectors. Various strategies have been considered for modifying the tropism of HSV.

3.5.1. Targeted adsorption
As binding of viral gB/gC to GAGs is the first step in the entry cascade, it is possible that targeted adsorption of virions to pre-determined cell subtypes might be achieved through the modification of gC and gB, or by their replacement with a ligand for another receptor (Fig. 6B). This was tested by generating engineered virions carrying

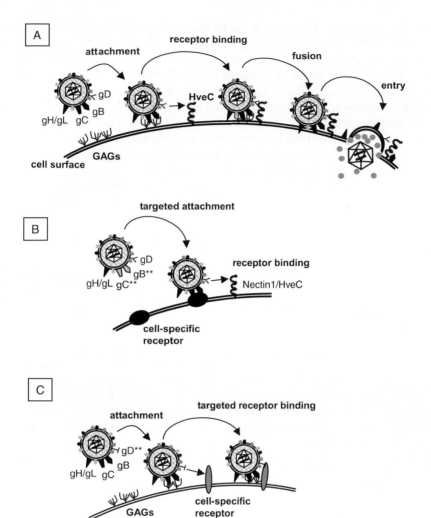

Fig. 6. Engineering viral entry into the cell. (A) The wild-type viral entry cascade is depicted. Initial binding to cell surface heparan sulphate moieties by viral gC and gB is followed by a secondary binding event between viral gD and a cell surface receptor, HveC. This commits the virus to cell entry and results in fusion between the viral envelope and the plasmalemma. The latter step is dependent on viral gB and gH/L. (B) One means of engineering cell entry would be through targeted attachment. In this scheme, Heparan sulphate binding is eliminated from gC and gB by protein engineering, and a new cell receptor-specific binding activity induced by addition of heterologous sequence into gC. These engineered gB and gC molecules are labelled gC** and gB**. The result is cell-specific binding in a distribution dictated by that of the new cellular receptor. The secondary binding and fusion events then proceed as per wild-type virus, but this only occurs in cells to which viral binding is targeted. (C) An alternative to targeted attachment would be targeting of entry by restriction of the secondary binding event. Engineering of gD might eliminate its attachment to the ubiquitously expressed cellular receptors HveA and HveC, and induce a new binding activity to a cellular receptor. The engineered gD is labelled gD**. Binding would proceed normally to all cells expressing heparan sulphate, but entry would be restricted to those cells expressing the new gD ligand.

the gB:pK⁻ mutation; gC was replaced by gC-erythropoietin (EPO) fusion proteins [66]. Initial studies confirmed that replacement of different parts of the N-terminal ectodomain of gC with EPO was possible, and that the fusion proteins were expressed at the cell surface in transient assays. Viruses were generated by recombining the chimeric genes encoding gC-EPO into the gC locus of a vector containing the gBpK⁻ mutation. The viruses expressed the fusion proteins at their surface, and showed greatly reduced GAG binding, and acquisition of specific binding to the new ligand, the EPO receptor [66]. It has not been possible so far to demonstrate targeted viral entry through the trans-plasmalemmal pathway with these vectors, as it was necessary in the initial experiments to use cells that are not permissive for HSV entry to distinguish virus–EPO receptor binding from virus–HveC binding. Experiments designed to assess whether re-directed surface adsorption can be followed by viral entry into the targeted cells are ongoing.

3.5.2. Targeted entry

Modification of the viral entry/attachment glycoprotein D might present an alternative to engineering the cell attachment receptors gC and gB. If gD could be modified in such a way that prevented its binding to HveA and HveC, retained its cell entry signalling function and generated a new binding specificity, it might be possible to limit HSV entry to cells in which the appropriate cell surface receptor was expressed (Fig. 6C). Thus, the virion could bind to all GAG-expressing cells, but the fusogenic signal to the virion would only be generated in the context of the target cell expressing the new gD-binding receptor. This possible targeting strategy demands an intimate knowledge of crucial gD amino acid residues involved in HveA and HveC binding, and represents a formidable protein engineering challenge. We have started to address these issues through mutagenesis studies; progress towards engineered gD-mediated cell entry targeting is still in early stages. A different approach to targeted entry may be simpler to effect. A soluble adapter molecule could provide both the cell-specific binding and the gD viral fusion trigger signal. A bi-specific adapter molecule might contain the HveC V-domain at its N-terminus and a ligand for a cellular receptor at its C terminus. It follows that use of such an adapter would be expected to cause both cell-specific attachment and triggering of virus–cell fusion, whilst preventing non-targeted virion binding to cellular HveA/C by occupying ligand binding sites on viral gD. We are currently evaluating several such adapter molecules.

3.5.3. Pseudotyping HSV using surface determinants from another virus

The host range of a virus may be modified using the surface receptors of another virus. This approach has been successfully used to broaden the host range of HIV vectors, using the G-protein of vesicular stomatitis virus (VSV) as a surface receptor [67]. Pseudotyped HSV vectors have been made using VSV proteins. A mutant virus was utilised, in which the short unique segment of the HSV genome U_S3-8 (encoding gD in addition to non-essential glycoproteins gE, gG, gJ and gI) was deleted. This vector was transiently rescued using an expression plasmid for VSV-G or HSV/VSV fusion proteins. It was demonstrated that VSV-G chimeras containing the transmembrane domain of gD, or a truncated gB transmembrane domain, were incorporated into the viral envelope efficiently, and that the wild-type VSV-G protein was incorporated rather less efficiently [68]. The latter, however, was able to partially rescue the gD-deficient phenotype, whereas the chimeric proteins were non-functional [68]. The poor efficiency of phenotypic rescue of the U_S3-8 null virus by VSV-G may be attributable to either: (i) inefficient incorporation of the foreign viral glycoprotein, or (ii) acid degradation of HSV in the endosome compartment (recall that HSV usually enters cells at physiological pH by fusing with the cell membrane, whereas VSV enters cells at acid pH from within the endosome compartment). This issue was further studied by generating a recombinant virus, in which the VSV-G expression cassette was incorporated into the genome of the U_S3-8 deleted virus. The recombinant particle entered cells possessing the VSV receptor, although an abortive infection ensued, culminating in endosomal degradation of the virion at low pH (Goins, Anderson, Laquerre and Glorioso, manuscript in preparation). The use of lysomotropic agents that raise endosomal pH, such as chloroquine [69], enabled release of viral contents from the endosome into the cytoplasm, resulting in plaque formation. The mechanism underlying this observation is currently being examined in more detail.

3.6. Use of latency promoters to drive transgene expression

Various strategies have been contemplated, whereby the latency promoters might be exploited to drive expression of a transgene long-term in neurons *in vivo*. First, isolated elements of the LAP system have been utilised at ectopic positions within the vector genome; this approach has enabled transcription of a gene placed at an ectopic locus within HSV from LAP2, but not LAP1 [45,47]. Second, use of hybrid promoters has been examined. In this scheme, the latency elements of the LAP2 promoter have been linked to strong viral IE promoters to achieve high-level sustained latent-phase expression [49]. Finally, as the LATs are not an absolute requirement for the establishment of latency, it has been possible to insert transgenes into the LAT loci, replacing the LAT sequences, to utilise all of the relevant *cis*-acting regulatory elements contained within the LAT locus to drive transgene expression [70]. Using these different approaches, long-term expression of transgenes has proved feasible in the peripheral nervous system in both sensory [47,49] and motor [70] neurons. Indeed, the LAP2 promoter was initially identified by its ability to drive virus-delivered reporter gene activity in latent infection of trigeminal ganglia *in vivo* for up to 300 days [45].

3.7. Vector production

One advantage of HSV is the ease with which large quantities of pure vector stocks are manufactured, enabling cost-effective and efficient production of clinically useful quantities of vector. Recent work has enabled characterisation of the optimal culture conditions for different types of disabled vector strain, to facilitate the large-scale preparation of replication-defective vector [51]. There is an inverse relationship between deletion of immediate early genes and vector yield from tissue culture. Thus, the least cytotoxic (and therefore most likely to be clinically useful) vectors produce the lowest yield from standard viral preparations. By a series of painstaking optimisation steps for cell confluence, pH and culture time, and multiplicity, it was possible to enhance the yield of disabled vectors to produce clinically useful quantities of even the most attenuated vectors, by use of a large-scale culture procedure [51]. Interestingly, mixtures of protease inhibitors that are used to prevent protein degradation during *in vitro* protein synthesis contain components that reduce the yield of HSV vectors in culture [71]. These data will have important implications as the development of clinical applications for HSV vectors progresses, and the demand for large-scale production increases.

4. Applications of HSV vectors in the nervous system

Evidence showing utility of replication-defective HSV vectors in gene delivery to the nervous system has been demonstrated in a series of recent reports. These are briefly considered below.

4.1. The peripheral nervous system

The latent life cycle of the wild-type virus occurs naturally in the sensory ganglia of peripheral nerves. Viral functions are optimised for the delivery and long-term expression of genetic material in this tissue; deletion of functions allowing reactivation and pathogenicity would be expected to yield a safe and efficient system for chronic peripheral nerve gene expression.

 In vitro, a triple IE mutant (ICP4, ICP22, ICP27 null) virus was able to efficiently infect dorsal root ganglion cultures, and to drive sustained transgene expression [4]. There was little or no evidence of vector toxicity in these neurons, although supporting cells present in the cultures showed a cytopathic response to infection, with evidence of cell death and metabolic disturbance [4]. This may be at least partially attributable to the difference in post-translational processing of ICP0 between neurons and glia [62]. The electrophysiological responses of primary sensory neurons in culture have been examined following infection with HSV vectors [72]. In contrast to wild-type virus, which abolishes transmembrane Na^+ currents following infection, transduction with a replication-defective vector has no effect on the electrophysiological profile of these cells. *In vivo*, peripheral inoculation allows replication-defective vector to exploit the same retrograde axonal transport system used by wild-type virus to reach the sensory ganglia [45,47,49,70,73]. In contrast to the *in vitro* setting, glial toxicity is not seen, as the virus does not replicate within the ganglia and is unable to enter Schwann cells. Peripheral inoculation with replication-defective vectors apparently gives rise to less efficient nerve transduction than with replication-competent virus [49]. This is probably a manifestation of the initial round of lytic replication in epithelia, which amplifies the virus dose delivered to nerves. Indeed, with an elevated dose of replication-deficient vector it is possible to demonstrate efficient nerve transduction via peripheral cutaneous infection [45,49]. In addition, motor neurons may be targeted by viral entry through their peripheral termini in muscle [74].

Several applications for HSV vectors have been investigated in the peripheral nervous system. First, trophic factors have been delivered to support neurons in models of neuropathic disease (Fig. 7A). In an initial series of experiments, vectors producing nerve growth factor (NGF) were generated [47]. Biologically active NGF, capable of inducing differentiation of PC12 cells, was produced in cultured neurons, for several weeks. The NGF was sufficient to protect primary DRG neuron cultures from an oxidative insult with H_2O_2, probably by induction of anti-oxidative genes. *In vivo*, a replication-defective HSV vector expressing neurotrophin-3 (NT-3) was able to protect sensory neurons from a toxic insult, pyridoxine poisoning, which normally results in a sensory neuropathy. HSV-delivered NT-3 mediated significant protection of proprioceptive nerve function; effects were noted at the morphologic, physiological and behavioural levels [75]. In addition, protection of the peripheral nervous system from the chronic metabolic stress of diabetes mellitus has been demonstrated. In one study, the effects of chronic hyperglycaemia on urinary bladder function were ameliorated by HSV-mediated NGF expression in a model of diabetic cystopathy [76]. In a second study, the sensory neuropathy resulting from diabetes was prevented in a streptozo-tocin mouse model, by peripheral administration of a HSV vector encoding NGF [77].

Chronic pain is a second application investigated using transgene delivery to PNS sensory neurons by HSV vectors [78] (Fig. 7B). Pre-proenkephalin is normally expressed in the spinal cord, where it is post-translationally processed to yield leu- and met-enkephalin. The latter are neurotransmitters that are stored in synaptic vesicles in spinal interneurons, released upon neuronal activity controlled by local and distant pain-modulating networks and act at opioid receptors located on primary afferent fibres and second order neurons, which mediate pain transmission in the dorsal horn of the spinal cord. Activation of these receptors leads to inhibition of neurotransmitter release from the primary afferents, and inhibits activation of second order neurons within the CNS both directly and indirectly. Conditionally replicating and replication-defective HSV vectors, which express pre-proenkephalin, have been made. These have been shown to express enkephalin *in vitro* and *in vivo*. Inoculation of the footpads of mice with the conditionally replicating vector demonstrated that HSV-delivered pre-proenkephalin was processed and stored at CNS terminals of sensory axons, released in response to nerve activity, and acted at opioid receptors. Sensitisation of unmyelinated or small myelinated pain-transmitting nerve fibres with pharmacological agents was abolished [78]. More recently, a replication-defective HSV-enkephalin vector has been shown to reduce the chronic phase of pain in a formalin-footpad model of pain in rats [79]. The results raise the exciting possibility of novel approaches to the treatment of chronic pain, by intrasynaptic delivery of analgesic molecules that are associated with significant side effects when delivered systemically.

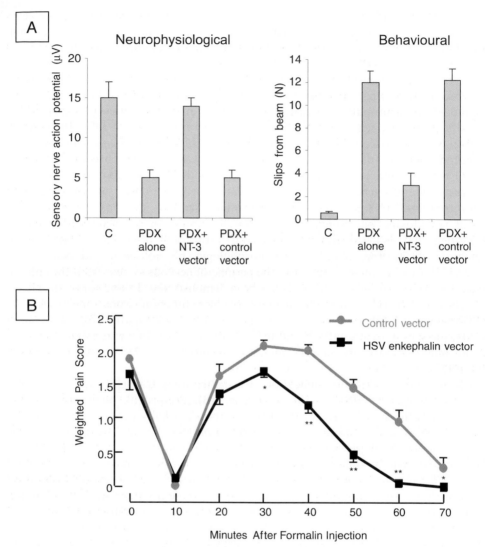

Fig. 7. Application of replication-defective HSV vectors in models of peripheral nervous system disease. (A) Rat sensory neurons, transduced by peripheral inoculation *in vivo* with a NT-3 expressing HSV vector, became resistant to intoxication with pyridoxine (PDX). Two measures of sensory neuronal viability and function are shown. The amplitudes of the evoked sensory nerve action potentials provide a physiological marker of large sensory nerve fibre function. The proprioception-dependent performance of the rats when walking on a beam provides a sensitive behavioural measure of large sensory fibre function. Following PDX poisoning, both parameters were much better preserved in animals pre-treated with the NT-3 vector, than in animals treated with a control vector. Abbreviations: C, negative control; HSV, control HSV vector expressing *LacZ*, but not NT-3. (B) Injection of the rat footpad with formalin results in pain–response behaviour that may be scored. Following an initial transient nociceptive response, pain behaviour reappears and lasts for approximately 1 h, reflecting a more chronic pain mechanism. Pre-treatment of rats with a pre-proenkephalin-expressing HSV vector reduces the chronic pain behaviour in this model without affecting the initial nociceptive response.

4.2. The central nervous system

In contrast to viral latency observed in the PNS, infection of the CNS with wild-type HSV-1 results in a rapidly fatal haemorrhagic encephalitis, which is dependent upon viral replication; replication-defective HSV vectors do not cause this dramatic effect. However, eliminating expression of multiple IE genes appears crucial in minimising CNS neuron toxicity. Thus, a single IE mutant (ICP4−) virus was toxic to cultured cortical neurons [80], which showed minimal evidence of toxicity or metabolic disturbance when infected with a triple IE mutant vector (ICP4−:ICP22−:ICP27−) [4]. The same appears true in the brain *in vivo*. Thus, use of a single IE mutant (ICP27−) gave rise to cell death and an inflammatory response following intra-parenchymal injection [72], whereas a triple IE mutant [4] caused a small degree of tissue damage that was similar to that seen with saline injection, and was presumably partially mechanical in origin [81]. Importantly, there is no evidence that direct introduction of disabled HSV into the cerebral parenchyma can effect reactivation of latent wild-type virus [82].

Viral DNA persists long-term following intracerebral inoculation with a replication-defective HSV vector [83]. Use of viral promoters other than the latency promoters gives rise to short-term transgene expression in CNS neurons, as might be expected [81,84]. It is known that, following acute infection, replication-competent and neuro-attenuated vectors persist in CNS neurons where they transcribe the LAT genes [85]. Stable CNS gene expression using the latency promoters has not yet been reported from a replication-defective vector.

Delivery of biologically active neuroprotective transgene products has been demonstrated in models of CNS disease, using replication-defective HSV vectors [84,86,87] (Fig. 8). HSV vector-mediated expression of the anti-apoptotic protein bcl-2, either alone or in combination with glial derived neurotrophic factor, was able to protect cells of the substantia nigra from the neurotoxin, 6-hydroxydopamine (6-OHDA) [84,86]. In addition, prior injection of the bcl-2 vector into the spine protected motor neurons from undergoing cell death following proximal root avulsion [87]. These experiments establish proof of principle that gene delivery to the CNS using HSV vectors may be effective and result in a beneficial effect on a pathological phenotype.

Recently, the use of the meninges as a depot site for synthesis and secretion of anti-inflammatory cytokines into the cerebrospinal fluid (CSF) has been reported [88]. Replication-defective HSV vectors expressing interferon-γ established stable infection of the leptomeninges and secreted the cytokine in detectable quantities into the CSF. Subsequent experiments using an interleukin-4 expressing HSV vector showed that the presence of the vector-derived anti-inflammatory cytokine was sufficient to ameliorate the pathological phenotype arising from experimental allergic encephalitis, an animal model of autoimmune CNS inflammation. Furthermore, secretion of fibroblast growth factor II (FGF-II) into the CSF from the meninges following transduction with an FGF-II-expressing HSV vector was associated with reduced inflammation and enhanced myelin repair processes in the same animal model [89].

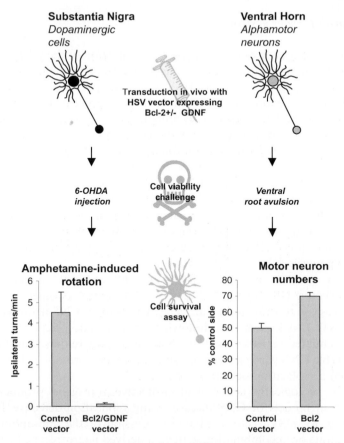

Fig. 8. CNS neuroprotection using replication-defective HSV vectors. Replication-defective HSV vectors expressing the anti-apoptotic molecule Bcl-2, and the neurotrophic agent BDNF were used to demonstrate delivery of biologically active neuroprotective molecules to the CNS *in vivo*. Two models of nerve degeneration within the CNS were studied. First, treatment of the substantia nigra of the midbrain with 6-OHDA results in loss of dopaminergic neurons, with resulting ipsilateral rotational behaviour in response to amphetamine treatment. Second, avulsion of the ventral root of the spinal cord results in loss of motor neurons within the ventral horn. Both of these pathological responses were dramatically attenuated by prior treatment of appropriate areas of the CNS with HSV vectors expressing Bcl-2, either alone or in combination with GDNF.

Finally, HSV vectors have been used experimentally to modify the neurochemical phenotype of groups of CNS neurons. Amygdala transduction in rats was demonstrated by histochemistry, after inoculation with a *LacZ*-expressing replication-defective HSV vector. Intra-amygdala injection of a pre-proenkephalin-expressing HSV vector altered the behavioural response to pain compared to transduction with the isogenic control vector [90]. The effect was specific, as it was reversed by naloxone. The behavioural changes, therefore, resulted from expression of enkephalin within the amygdala. These studies demonstrate the utility of HSV vectors in the experimental investigation of neural circuitry *in vivo*.

4.3. Malignant glioma

Malignant glioma is a common, fatal malignancy of the CNS. A series of replication-defective HSV vectors has been described, which deliver anti-cancer transgenes to malignant glioma cells (reviewed in [91]). The genes were selected for both: (i) the individual toxicities of their products to rapidly dividing cells, and (ii) predicted synergy in combination with one another. In a series of experiments using animal models of malignant glioma, it was shown that the combination of transgenes significantly augmented the response to gene therapy (Fig. 9). The simultaneous

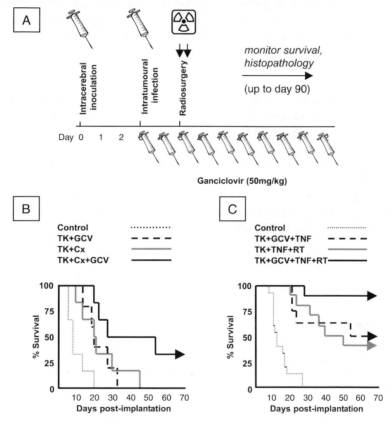

Fig. 9. Replication-defective HSV vectors expressing multiple transgenes, in an animal model of malignant glioma. (A) An orthotopic xenotransplant model was used in these studies. U-87 human glioma cells were implanted into the corpus striatum of athymic mice on day 1. A gene therapy vector or control was then inoculated at the same stereotactic coordinates on day 3, at which time GCV or control treatment was started. Radiotherapy was given on day 5 depending on the treatment protocol, and survival of the animals monitored until day 70. (B) Kaplan–Meier survival curves for a cohort of animals treated with a vector that co-expressed TK and Connexin-43. The enhanced survival of the animals, attributable to co-expression of the anti-tumour genes, is apparent. (C) Kaplan–Meier survival curves for a cohort of animals treated with a vector that co-expressed TK and TNFα. The synergistic effect of TNFα and radiotherapy on animal survival is apparent. Abbreviations: TK, HSV thymidine kinase; Cx, connexin-43; RT, radiotherapy.

delivery of multiple genes at high multiplicity and infectivity is an imaginative use of the properties of HSV vectors.

The HSV thymidine kinase-ganciclovir (TK/GCV) suicide gene therapy system is well described and has been applied in various cancer gene therapy paradigms, with varying degrees of success. In many cases, non-transduced cells surrounding the transduced cells are killed following GCV administration. This phenomenon is referred to as 'bystander lysis' [92]. *In vitro*, bystander lysis is largely attributable to uptake of activated GCV by HSV-TK-negative cells. GCV monophosphate may pass from cell to cell through gap junctions. These are intercellular channels formed by a number of proteins including connexin-43. Gliomas are often defective in connexin expression and intercellular gap junctions. On this basis, connexin-43 was incorporated into an anti-tumour HSV vector. Connexin alone had an *in vivo* effect on animal survival that was comparable with that of HSV-TK/GCV, whereas the combination of HSV-TK/GCV and connexin proved to be synergistic, by enhancing bystander lysis of non-transduced tumour cells [92] (Fig. 9B).

In parallel experiments, viruses were generated that deliver HSV-TK with a secreted factor (tumour necrosis factor alpha—TNFα) designed to effect destruction of neighbouring tumour tissue. The molecule is too toxic to deliver systemically, but the ability of HSV vectors to accommodate multiple transgenes readily enabled its incorporation into a locally administered suicide gene therapy paradigm. Analysis of these vectors showed that TNFα expression had a direct anti-tumour effect *in vivo* and *in vitro* [93]. *In vivo*, the effect of gamma irradiation was augmented greatly by the simultaneous expression of TNFα, and combination therapy with TNFα, gamma irradiation and HSV-TK/GCV led to the ablation of many of the tumours [94] (Fig. 9C).

5. Applications outside the nervous system

5.1. Skeletal muscle

Gene therapy for treatment of hereditary muscular diseases presents a number of challenges. First, the tissue compartment is vast; a large tissue mass is distributed over an immense area of the body, necessitating the use of large quantities of vector in conjunction with some means of systemic delivery. Second, the tissue is post-mitotic. The commonest hereditary muscular disease, Duchenne muscular dystrophy (DMD), presents additional problems for gene therapy. DMD is an X-linked recessive disorder affecting 1 in 3500 live male births. The phenotype is a manifestation of mutations that prevent expression of a muscle cytoskeletal protein, dystrophin. The cDNA encoding dystrophin is 14 kb in length, which is too large for the limited cloning capacity of some vector systems.

In vitro, transduction of myoblasts and myotubes has been demonstrated with a single IE mutant (ICP4⁻) HSV vector expressing a reporter gene, but the single mutant virus caused cytotoxicity in infected cells [95]. *In vivo*, transduction of many muscle fibres in newborn muscle was demonstrated, but the transduction efficiency was much lower in adult fibres [95]. Two problems have been identified. First, the maturation-dependent reduction in transduction efficiency is partly attributable to the muscle basal lamina acting as a physical barrier to viral particle access. Second, vector-related cytotoxicity from single IE mutant virus, and an immune response directed against viral antigens resulted in short-lived transgene expression. Improved transduction and reporter gene expression was demonstrated following construction of triple IE mutant vectors (ICP4:ICP22:ICP27) both *in vitro* and *in vivo*. Vectors expressing full-length or truncated dystrophin were generated [50]. *In vitro*, these directed dystrophin synthesis in dystrophin-null myotubes. As expected, the triple mutant was less toxic than the single mutant. Importantly, the cellular localization of dystrophin at the cytoplasmic surface of the sarcolemma was restored when the full-length construct was delivered; diffuse cytoplasmic staining was evident with expression of the truncated form [50]. *In vivo*, sub-sarcolemmal dystrophin staining was evident in dystrophin-null muscle, but only in a small area adjacent to the site of vector injection [50]. Thus, the large cloning capacity of HSV vectors solves one of the problems of DMD gene therapy, namely the size of the dystrophin expression cassette. The problems posed by systemic HSV delivery have yet to be addressed, although cell-mediated gene delivery might allow partial solution. Several cell populations have been described that are capable of systemic engraftment and fusion with myofibres. It may be possible in the future to use these populations to achieve widespread genetic correction in muscle. *Ex-vivo* transduction of these cells with HSV vectors encoding multiple transgenes could plausibly be utilised to introduce immunosuppressive, myogenic-inducing and other therapeutic transgenes, simultaneously.

5.2. Stem cells

Various stem cell populations have shown potential as tissue-repopulating and gene delivery vehicles. The ability to transfer genes into human stem cell populations may permit cell-mediated delivery of therapeutic transgenes, or even allow the fate of the cells to be specified, which is an exciting prospect worthy of further study. Infection of stem cell populations has been demonstrated using replication-defective HSV vectors. $CD34^+$ human mobilised peripheral blood and monkey bone marrow cells were infected with a replication-defective HSV-1 vector [4,96]. Twelve hours following infection, almost all of the cells expressed a reporter gene encoded by the virus. The infected culture showed similar cell viability to an uninfected sample at days 2, 4 and 7 post-infection. Transduced monkey $CD34^+$ cells were transplanted into monkeys with skin autografts, which were subjected to biopsy 5 days later. Vector-transduced endothelial cells were detected in the walls of nascent vasculature within the skin graft, and mononuclear cells from the peripheral blood and bone marrow showed reporter gene expression for over 3 weeks following transplantation. GCV treatment induced progressive necrosis of the vasculature supplying the autograft, which detached after 5 days treatment [96]. This indicates that a functional and potentially therapeutic gene product may be introduced into stem cells using an HSV vector; the possibility of allowing HSV-TK-expressing stem cells to contribute to tumour neovascularisation, followed by GCV ablation is exciting. The utility of HSV-1 vectors in transducing $CD34^+$ cells purified from human umbilical cord blood is currently being evaluated. It is worth noting that the non-integrating nature of HSV-1 vectors will preclude their use in unmodified form for stably expressing genes in cells whose final destinations are many cell divisions away from the transduction event. In some respects, this may be advantageous for cell fate specification, where transient expression of differentiation signals may be optimal.

5.3. Arthritis and secreted proteins

The development of gene therapy vectors allowing long-term local modulation of cytokine activity in joints would represent a major advance in the treatment of inflammatory arthritis, such as rheumatoid arthritis (RA). Initial experiments using replication-defective HSV vectors engineered to express IL-1 receptor antagonist or TNFα soluble receptor (which antagonise the activity of the appropriate pro-inflammatory cytokines) confirmed that these products were detectable within the synovial fluid after intra-articular inoculation of the vector. Following vector inoculation, significant improvement in inflammatory markers was noted in an experimental model of arthritis, generated by intra-articular over-expression of interleukin-1β using retrovirus-transduced synovial cells [97].

Establishment of latent infection in the neurons innervating joints was attempted; it was hoped that transduced sensory nerve terminals would give rise to chronic, sustained production and secretion of transgene product into the joint space. Interestingly, neurons innervating the joint space were inconsistently infected and the vector did not persist in synovial cells. However, intra-articular inoculation of an NGF-expressing HSV vector resulted in a significant increase in joint lavage NGF levels. In addition, NGF was detected in the blood plasma and was demonstrable at increased levels up to one year after the initial inoculation [53]. Subsequent analysis showed that the primary site of transduction was joint ligament tissue. It appears, therefore, that replication-defective HSV vectors are able to persist in non-neuronal 'depot' tissues and express transgenes long-term. Methods of exploiting this observation are currently being investigated, for the development of novel therapeutic strategies to target several diseases. First, the sustained intra-articular delivery of soluble inhibitors of inflammation is likely to represent an important advance in chronic arthritis. Second, the expression of NGF in the circulation may ameliorate the chronic peripheral neuropathies associated with many systemic diseases. Finally, this technology may enable the delivery of circulating proteins to replace crucial factors missing in genetic diseases like the haemophilias.

6. Conclusions

Gene delivery vectors can be generated from HSV by a series of genetic manipulations; these modifications remove pathogenic functions, while retaining features inherent to HSV biology that may enhance vector interactions with host tissues. Current vector development issues are centred on vector manufacture/ production, cellular and tissue targeting, further eliminating pathogenicity and viral gene expression and regulating transgene expression. These vectors have proven useful in experimental gene delivery in a variety of settings, and the many successful applications to animal models of disease imply that these reagents might find clinical applications in the near future. At the time of writing, HSV vectors targeting peripheral neuropathy, chronic pain and malignant glioma are poised to enter clinical trials. We are optimistic that these hitherto intractable neurological problems may be amenable to gene therapy using replication-defective HSV vectors.

Abbreviations

BAC	bacterial artificial chromosome
CNS	central nervous system
CSF	cerebrospinal fluid
DMD	Duchenne muscular dystrophy
EPO	erythropoietin
FGF	fibroblast growth factor
GAG	glycosaminoglycan
GCV	ganciclovir
GDNF	glial cell line-derived neurotrophic factor
HSV	Herpes simplex virus
HveA/C	Herpesvirus entry mediator A/C
IE	immediate-early
LAP	latency-active promoter
LAT	latency-associated transcript
NGF	nerve growth factor
NT-3	neurotrophin-3
PNS	peripheral nervous system
TK	thymidine kinase
TNF	tumour necrosis factor
VSV	vesicular stomatitis virus

References

1. Roizman, B. and Sears, A.E. (1996) In: Fields, B.N., Knipe, D.M. and Howley, P.M. (eds.) Fields Virology, Lippincott-Raven, Philadelphia, pp. 2231–2295.
2. McGeoch, D.J., Dalrymple, M.A., Davison, A.J., Dolan, A., Frame, M.C., McNab, D., Perry, L.J., Scott, J.E., and Taylor, P. (1988) J. Gen. Virol. 69, 1531–1574.
3. McGeoch, D.J., Dolan, A., Donald, S., and Brauer, D.H. (1986) Nucleic Acids Res. 14, 1727–1745.
4. Krisky, D.M., Wolfe, D., Goins, W.F., Marconi, P.C., Ramakrishnan, R., Mata, M., Rouse, R.J., Fink, D.J., and Glorioso, J.C. (1998) Gene Ther. 5, 1593–1603.
5. Krisky, D.M., Marconi, P.C., Oligino, T.J., Rouse, R.J., Fink, D.J., Cohen, J.B., Watkins, S.C., and Glorioso, J.C. (1998) Gene Ther. 5, 1517–1530.
6. Morgan, C., Rose, H.M., and Mednis, B. (1968) J. Virol. 2, 507–516.

7. Spear, P.G. (1993) Sem. Virol. 4, 167–180.
8. Para, M.F., Baucke, R.B., and Spear, P.G. (1980) J. Virol. 34, 512–520.
9. Cai, W.H., Gu, B., and Person, S. (1988) J. Virol. 62, 2596–2604.
10. Fuller, A.O. and Spear, P.G. (1987) Proc. Natl. Acad. Sci. USA 84, 5454–5458.
11. Highlander, S.L., Sutherland, S.L., Gage, P.J., Johnson, D.C., Levine, M., and Glorioso, J.C. (1987) J. Virol. 61, 3356–3364.
12. Desai, P.J., Schaffer, P.A., and Minson, A.C. (1988) J. Gen. Virol. 69, 1147–1156.
13. Hutchinson, L., Browne, H., Wargent, V., Davis Poynter, N., Primorac, S., Goldsmith, K., Minson, A.C., and Johnson, D.C. (1992) J. Virol. 66, 2240–2250.
14. Spear, P.G., Shieh, M.T., Herold, B.C., WuDunn, D., and Koshy, T.I. (1992) Adv. Exp. Med. Biol. 313, 341–353.
15. Herold, B.C., WuDunn, D., Soltys, N., and Spear, P.G. (1991) J. Virol. 65, 1090–1098.
16. Laquerre, S., Argnani, R., Anderson, D.B., Zucchini, S., Manservigi, R., and Glorioso, J.C. (1998) J. Virol. 72, 6119–6130.
17. Johnson, D.C. and Ligas, M.W. (1988) J. Virol. 62, 4605–4612.
18. Campadelli Fiume, G., Arsenakis, M., Farabegoli, F., and Roizman, B. (1988) J. Virol. 62, 159–167.
19. Montgomery, R.I., Warner, M.S., Lum, B.J., and Spear, P.G. (1996) Cell 87, 427–436.
20. Geraghty, R.J., Krummenacher, C., Cohen, G.H., Eisenberg, R.J., and Spear, P.G. (1998) Science 280, 1618–1620.
21. Krummenacher, C., Rux, A.H., Whitbeck, J.C., Ponce de Leon, M., Lou, H., Baribaud, I., Hou, W., Zou, C., Geraghty, R.J., Spear, P.G., Eisenberg, R.J., Cohen, G.H. (1999) J. Virol. 73, 8127–8137.
22. Manservigi, R., Spear, P.G., and Buchan, A. (1977) Proc. Natl. Acad. Sci. USA 74, 3913–3917.
23. Roop, C., Hutchinson, L., and Johnson, D.C. (1993) J. Virol. 67, 2285–2297.
24. Honess, R.W. and Roizman, B. (1975) Proc. Natl. Acad. Sci. USA 72, 1276–1280.
25. Honess, R.W. and Roizman, B. (1974) J. Virol. 14, 8–19.
26. DeLuca, N.A., McCarthy, A.M., and Schaffer, P.A. (1985) J. Virol. 56, 558–570.
27. Cook, M.L. and Stevens, J.G. (1973) Infect. Immun. 7, 272–288.
28. Bak, I.J., Markham, C.H., Cook, M.L., and Stevens, J.G. (1977) Brain Res. 136, 415–429.
29. Stevens, J.G., Wagner, E.K., Devi Rao, G.B., Cook, M.L., and Feldman, L.T. (1987) Science 235, 1056–1059.
30. Zwaagstra, J.C., Ghiasi, H., Slanina, S.M., Nesburn, A.B., Wheatley, S.C., Lillycrop, K., Wood, J., Latchman, D.S., Patel, K., Wechsler, S.L. (1990) J. Virol. 64, 5019–5028.
31. Alvira, M.R., Goins, W.F., Cohen, J.B., and Glorioso, J.C. (1999) J. Virol. 73, 3866–3876.
32. Wu, T.T., Su, Y.H., Block, T.M., and Taylor, J.M. (1996) J. Virol. 70, 5962–5967.
33. Wechsler, S.L., Nesburn, A.B., Watson, R., Slanina, S.M., and Ghiasi, H. (1988) J. Virol. 62, 4051–4058.
34. Spivack, J.G., Woods, G.M., and Fraser, N.W. (1991) J. Virol. 65, 6800–6810.
35. Wagner, E.K., Flánagan, W.M., Devi Rao, G., Zhang, Y.F., Hill, J.M., Anderson, K.P., and Stevens, J.G. (1988) J. Virol. 62, 4577–4585.
36. Wu, T.T., Su, Y.H., Block, T.M., and Taylor, J.M. (1998) Virology 243, 140–149.
37. Zabolotny, J.M., Krummenacher, C., and Fraser, N.W. (1997) J. Virol. 71, 4199–4208.
38. Krummenacher, C., Zabolotny, J.M., and Fraser, N.W. (1997) J. Virol. 71, 5849–5860.
39. Ho, D.Y. and Mocarski, E.S. (1989) Proc. Natl. Acad. Sci. USA 86, 7596–7600.
40. Thompson, R.L. and Sawtell, N.M. (1997) J. Virol. 71, 5432–5440.
41. Steiner, I., Spivack, J.G., Lirette, R.P., Brown, S.M., MacLean, A.R., Subak Sharpe, J.H., and Fraser, N.W. (1989) EMBO J. 8, 505–511.
42. Perng, G.C., Jones, C., Ciacci Zanella, J., Stone, M., Henderson, G., Yukht, A., Slanina, S.M., Hofman, F.M., Ghiasi, H., Nesburn, A.B., Wechsler, S.L. (2000) Science 287, 1500–1503.
43. Thompson, R.L. and Sawtell, N.M. (2001) J. Virol. 75, 6660–6675.
44. Mador, N., Goldenberg, D., Cohen, O., Panet, A., and Steiner, I. (1998) J. Virol. 72, 5067–5075.
45. Goins, W.F., Sternberg, L.R., Croen, K.D., Krause, P.R., Hendricks, R.L., Fink, D.J., Straus, S.E., Levine, M., and Glorioso, J.C. (1994) J. Virol. 68, 2239–2252.
46. Chen, X., Schmidt, M.C., Goins, W.F., and Glorioso, J.C. (1995) J. Virol. 69, 7899–7908.

47. Goins, W.F., Lee, K.A., Cavalcoli, J.D., O'Malley, M.E., DeKosky, S.T., Fink, D.J., and Glorioso, J.C. (1999) J. Virol. 73, 519–532.

48. Berthomme, H., Lokensgard, J., Yang, L., Margolis, T., and Feldman, L.T. (2000) J. Virol. 74, 3613–3622.

49. Palmer, J.A., Branston, R.H., Lilley, C.E., Robinson, M.J., Groutsi, F., Smith, J., Latchman, D.S., and Coffin, R.S. (2000) J. Virol. 74, 5604–5618.

50. Akkaraju, G.R., Huard, J., Hoffman, E.P., Goins, W.F., Pruchnic, R., Watkins, S.C., Cohen, J.B., and Glorioso, J.C. (1999) J. Gene Med. 1, 280–289.

51. Ozuer, A., Wechuck, J.B., Goins, W.F., Wolfe, D., Glorioso, J.C., and Ataai, M.M. (2002) Biotechnol. Bioeng. 77, 685–692.

52. Wolfe, D., Goins, W.F., Yamada, M., Moriuchi, S., Krisky, D.M., Oligino, T.J., Marconi, P.C., Fink, D.J., and Glorioso, J.C. (1999) Exp. Neurol. 159, 34–46.

53. Wolfe, D., Goins, W.F., Kaplan, T.J., Capuano, S.V., Fradette, J., Murphey-Corb, M., Robbins, P.D., Cohen, J.B., and Glorioso, J.C. (2001) Mol. Ther. 3, 61–69.

54. Markert, J.M., Malick, A., Coen, D.M., and Martuza, R.L. (1993) Neurosurgery 32, 597–603.

55. Rampling, R., Cruickshank, G., Papanastassiou, V., Nicoll, J., Hadley, D., Brennan, D., Petty, R., MacLean, A., Harland, J., McKie, E., Mabbs, R., Brown, M. (2000) Gene Ther. 7, 859–866.

56. Samaniego, L.A., Webb, A.L., and DeLuca, N.A. (1995) J. Virol. 69, 5705–5715.

57. Ramakrishnan, R., Poliani, P.L., Levine, M., Glorioso, J.C., and Fink, D.J. (1996) J. Virol. 70, 6519–6523.

58. Spaete, R.R. and Frenkel, N. (1982) Cell 30, 295–304.

59. Stavropoulos, T.A. and Strathdee, C.A. (1998) J. Virol. 72, 7137–7143.

60. Samaniego, L.A., Neiderhiser, L., and DeLuca, N.A. (1998) J. Virol. 72, 3307–3320.

61. Samaniego, L.A., Wu, N., and DeLuca, N.A. (1997) J. Virol. 71, 4614–4625.

62. Chen, X., Li, J., Mata, M., Goss, J., Wolfe, D., Glorioso, J.C., and Fink, D.J. (2000) J. Virol. 74, 10132–10141.

63. York, I.A., Roop, C., Andrews, D.W., Riddell, S.R., Graham, F.L., and Johnson, D.C. (1994) Cell 77, 525–535.

64. Wu, N., Watkins, S.C., Schaffer, P.A., and DeLuca, N.A. (1996) J. Virol. 70, 6358–6369.

65. Krisky, D.M., Marconi, P.C., Oligino, T., Rouse, R.J., Fink, D.J., and Glorioso, J.C. (1997) Gene Ther. 4, 1120–1125.

66. Laquerre, S., Anderson, D.B., Stolz, D.B., and Glorioso, J.C. (1998) J. Virol. 72, 9683–9697.

67. Emi, N., Friedmann, T., and Yee, J.K. (1991) J. Virol. 65, 1202–1207.

68. Anderson, D.B., Laquerre, S., Goins, W.F., Cohen, J.B., and Glorioso, J.C. (2000) J. Virol. 74, 2481–2487.

69. Nash, T.C. and Buchmeier, M.J. (1997) Virology 233, 1–8.

70. Lachmann, R.H. and Efstathiou, S. (1997) J. Virol. 71, 3197–3207.

71. Wechuck, J.B., Goins, W.F., Glorioso, J.C., and Ataai, M.M. (2000) Biotechnol. Prog. 16, 493–496.

72. Howard, M.K., Kershaw, T., Gibb, B., Storey, N., MacLean, A.R., Zeng, B.Y., Tel, B.C., Jenner, P., Brown, S.M., Woolf, C.J., Anderson, P.N., Coffin, R.S., Latchman, D.S. (1998) Gene Ther. 5, 1137–1147.

73. Marshall, K.R., Lachmann, R.H., Efstathiou, S., Rinaldi, A., and Preston, C.M. (2000) J. Virol. 74, 956–964.

74. Huard, J., Goins, W.F., Akkaraju, G.R., Krisky, D., Oligino, T., Marconi, P., and Glorioso, J.C. (1998) In: Quesenberry, P.J., Stein, G.S., Forget, B. and Weissman, S. (eds.) Stem Cell Biology and Gene Therapy, John Wiley and Sons, New York, pp. 179–200.

75. Chattopadhyay, M., Wolfe, D., Huang, S., Goss, J., Glorioso, J.C., Mata, M., and Fink, D.J. (2002) Ann. Neurol. 51, 19–27.

76. Goins, W.F., Yoshimura, N., Phelan, M.W., Yokoyama, T., Fraser, M.O., Ozawa, H., Bennett, N.J., de Groat, W.C., Glorioso, J.C., Chancellor, M.B. (2001) J. Urol. 165, 1748–1754.

77. Goss, J.R., Goins, W.F., Lacomis, D., Mata, M., Glorioso, J.C., and Fink, D.J. (2002) Diabetes 51, 2227–2232.

54

78. Wilson, S.P., Yeomans, D.C., Bender, M.A., Lu, Y., Goins, W.F., and Glorioso, J.C. (1999) Proc. Natl. Acad. Sci. USA 96, 3211–3216.
79. Goss, J.R., Mata, M., Goins, W.F., Wu, H.H., Glorioso, J.C., and Fink, D.J. (2001) Gene Ther. 8, 551–556.
80. Johnson, P.A., Yoshida, K., Gage, F.H., and Friedmann, T. (1992) Brain Res. Mol. Brain Res. 12, 95–102.
81. Marconi, P., Simonato, M., Zucchini, S., Bregola, G., Argnani, R., Krisky, D., Glorioso, J.C., and Manservigi, R. (1999) Gene Ther. 6, 904–912.
82. Wang, Q., Guo, J., and Jia, W. (1997) Gene Ther. 4, 1300–1304.
83. Bloom, D.C., Maidment, N.T., Tan, A., Dissette, V.B., Feldman, L.T., and Stevens, J.G. (1995) Brain Res. Mol. Brain Res. 31, 48–60.
84. Yamada, M., Oligino, T., Mata, M., Goss, J.R., Glorioso, J.C., and Fink, D.J. (1999) Proc. Natl. Acad. Sci. USA 96, 4078–4083.
85. Drummond, C.W., Eglin, R.P., and Esiri, M.M. (1994) J. Neurol. Sci. 127, 159–163.
86. Natsume, A., Mata, M., Goss, J., Huang, S., Wolfe, D., Oligino, T., Glorioso, J.C., and Fink, D. (2001) Exp. Neurol. 169, 231–238.
87. Yamada, M., Natsume, A., Mata, M., Oligino, T., Goss, J., Glorioso, J., and Fink, D.J. (2001) Exp. Neurol. 168, 225–230.
88. Martino, G., Poliani, P.L., Marconi, P.C., Comi, G., and Furlan, R. (2000) Gene Ther. 7, 1087–1093.
89. Ruffini, F., Furlan, R., Poliani, P.L., Brambilla, E., Marconi, P.C., Bergami, A., Desina, G., Glorioso, J.C., Comi, G., Martino, G. (2001) Gene Ther. 8, 1207–1213.
90. Kang, W., Wilson, M.A., Bender, M.A., Glorioso, J.C., and Wilson, S.P. (1998) Brain Res. 792, 133–135.
91. Burton, E.A. and Glorioso, J.C. (2001) Drug Discov. Today 6, 347–356.
92. Marconi, P., Tamura, M., Moriuchi, S., Krisky, D.M., Niranjan, A., Goins, W.F., Cohen, J.B., and Glorioso, J.C. (2000) Mol. Ther. 1, 71–81.
93. Moriuchi, S., Oligino, T., Krisky, D., Marconi, P., Fink, D., Cohen, J., and Glorioso, J.C. (1998) Cancer Res. 58, 5731–5737.
94. Niranjan, A., Moriuchi, S., Lunsford, L.D., Kondziolka, D., Flickinger, J.C., Fellows, W., Rajendiran, S., Tamura, M., Cohen, J.B., Glorioso, J.C. (2000) Mol. Ther. 2, 114–120.
95. Huard, J., Goins, W.F., and Glorioso, J.C. (1995) Gene Ther. 2, 385–392.
96. Gomez Navarro, J., Contreras, J.L., Arafat, W., Jiang, X.L., Krisky, D., Oligino, T., Marconi, P., Hubbard, B., Glorioso, J.C., Curiel, D.T., Thomas, J.M. (2000) Gene Ther. 7, 43–52.
97. Oligino, T., Ghivizzani, S., Wolfe, D., Lechman, E., Krisky, D., Mi, Z., Evans, C., Robbins, P., and Glorioso, J. (1999) Gene Ther. 6, 1713–1720.

S.C. Makrides (Ed.) *Gene Transfer and Expression in Mammalian Cells*

Virus-based vectors for gene expression in mammalian cells: Epstein-Barr virus

Gregory Kennedy and Bill Sugden

McArdle Laboratory for Cancer Research, University of Wisconsin-Madison Medical School, 1400 University Avenue, Madison, WI 53706, USA; Tel.: +1 (608) 262-6697; Fax: +1 (608) 262-2824; E-mail: sugden@oncology.wisc.edu

1. Introduction

Epstein-Barr virus (EBV) is a large DNA tumor virus identified as a causative agent for infectious mononucleosis and a risk factor in the development of Burkitt's lymphoma and nasopharyngeal carcinoma more than 20 years ago. In the following years EBV has been causally associated with B-cell lymphomas in immunocompromized patients, and with portions of Hodgkin's disease and gastric carcinoma. In the vast majority of humanity, however, lifelong infection with EBV is benign and free of any symptoms. EBV maintains its genome in cells extrachromosomally and this feature, even though EBV is a pathogen, has fostered interest in using vectors derived from it for gene therapy in a variety of diseases. Here we shall review what is known of the mechanism and elements by which EBV replicates and possible applications of its derived vectors.

2. Replication of the EBV genome

2.1. Replication during the latent phase of EBV's life cycle

The life cycle of EBV is bipartite within its human host. Upon primary transmission to a susceptible person, EBV can cause an acute infection in which B-lymphocytes are usually latently infected and a few may evolve to become lytic hosts for the virus. Infected cells can support a latent infection in which few viral genes are expressed and the cell survives; infected cells also can rarely evolve to support a lytic infection in which most viral genes are expressed, viral DNA is amplified, progeny virions are assembled, and the cell dies to release virus. Lifelong infection is apparently maintained in small, non-proliferating B-cells, perhaps, memory B-cells [1]. Epithelial cells can be infected and also may serve as lytic hosts. In latently infected cells the EBV genome is maintained in the nucleus as an extrachromosomal element [2,3]. A region of the plasmid genome located 8 kbp from one end of this 165 kbp genome, termed the origin of plasmid replication (*oriP*), supports extrachromosomal replication of plasmid DNA in B-lymphoblasts transformed by EBV (Fig. 1) [3,4].

OriP consists of two *cis*-acting domains: an element of dyad symmetry (DS) and a family of repeats (FR) [5,6]. The DS contains four low affinity, binding sites for

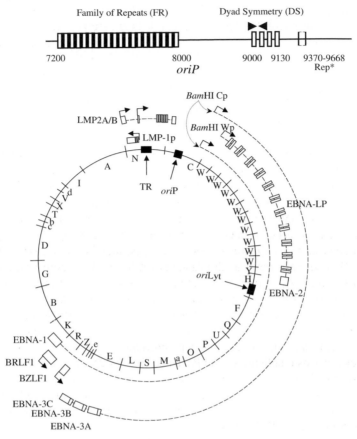

Fig. 1. The genome of the B95-8 strain of EBV is depicted [9]. TR represents the terminal repeats found on linear virion DNA, which are joined upon infection of cells. *OriP* and *oriLyt* are the identified origins of replication used, respectively, during the latent and the lytic phase of EBV's life cycle. An expanded representation of *oriP* is depicted with the possible minor origin Rep* shown relative to this major latent *oriP* [20]. The approximate map locations of FR, DS, and Rep* are given as nucleotide positions from the B95-8 strain below the figure. The open reading frame of BZLF-1 and BRLF-1, two genes required for lytic cycle replication are represented, respectively. EBNA-1 is required for replication of the viral genome during the latent stage of the viral life cycle and is shown in relation to the other nuclear antigens expressed during latent infection.

Epstein-Barr nuclear antigen-1 (EBNA-1), a viral protein critical to the life-cycle of the virus, two of which are located within a 65-base pair extended palindrome for which the element was named [7–9]. Early studies demonstrated that DNA synthesis initiates at or near the DS within the B95-8 strain of EBV and within *oriP* in an engineered, subgenomic plasmid, whereas replication forks pause at the FR [10–12]. The FR also functions as a transcriptional enhancer element and, when bound by EBNA-1, mediates up to 100-fold increase in transcription of both native and heterologous promoters [13–15].

OriP alone in *cis* can mediate efficient replication of plasmids derived from one half of EBV's genome [16]. However, it has also been shown that a mutant of EBV lacking the DS element replicates extrachromosomally in lymphoblastoid cell lines [17]. The initiation of DNA synthesis in this variant deleted for DS apparently occurs over broad regions of the EBV genome. These results are reminiscent of known mammalian origins such as those found within the hamster dihydrofolate reductase gene or the human β-globin locus from which replication has been shown to initiate over zones consisting of thousands of base pairs of DNA [18,19]. Potential initiation zones 20 kbp away from DS have been mapped by two-dimensional gel electrophoresis in the viral genome deleted for DS [17]. In addition a site nearby DS, termed Rep*, has been found to substitute inefficiently for DS in the presence of FR when EBNA-1 is present in *trans* [20]. Rep* is located between nucleotides 9370 and 9668 of the EBV genome and is just downstream of *oriP* (Fig. 1). In these experiments, when EBNA-1 was provided in *trans*, plasmids with the FR and three copies of Rep* supported replication more efficiently than those with one copy of Rep* [20]. Given that DS-independent replication of the EBV genome has been demonstrated, it would seem that at least some of the functions of *oriP* are redundant; for example, Rep* and the broad origin regions identified by Norio *et al.*, may contribute to the replication of EBV when DS and perhaps, EBNA-1 fail to function [17].

EBNA-1 binds site-specifically to *oriP* and is the only viral factor required in *trans* for the efficient replication of the EBV genome [21–23]. EBNA-1 when bound to *oriP* also regulates transcription of genes with promoters in the vicinity of *oriP* [11,13,14]. EBNA-1 lacks intrinsic helicase or ATPase activity [24,25]. The host cell contributes all replication machinery other than EBNA-1 to the synthesis of EBV DNA during latent infection which must include, not only enzymes required for elongation, but also for the initiation of DNA synthesis. Both EBV and its derived plasmids replicate once per cell cycle [26–28]. Such semiconservative replication is typical of licensed replicons, the synthesis of which is dependent upon assembling large complexes of proteins containing the minichromosome maintenance (MCM) proteins and the origin replication complexes (ORCs) as well as other enzymatic proteins required for DNA synthesis [29]. Consistent with other licensed replicons, recent experiments using chromatin immunoprecipitation have shown that both ORC and MCM proteins are assembled on the DS probably in an EBNA-1-dependent fashion [16,30]. It is not yet clear how *oriP* plasmids are efficiently maintained in cells. It is clear they lack formed centromeric elements and the FR contributes to the retention of plasmid DNA in EBNA-1-positive cells [31,32]. In addition, EBNA-1, the EBV genome, and yeast artificial chromosomes containing *oriP* and encoding EBNA-1, all have the ability to associate with metaphase chromosomes in human cells [33–37]. This association may contribute to segregating the viral genome to daughter cells during mitosis. A possible clue to a means by which EBV and *oriP* segregate comes from recent experiments which have identified proteins that associate with telomeres as also binding DS in the presence of EBNA-1 [38]. These proteins include telomere-associated proteins, telomeric binding factor-2 (TRF-2), a TRF-2 interacting protein (hRap1), and a poly ADP-polymerase (Tankyrase). It is not clear how these proteins might contribute to synthesis initiating at the DS. Were they to bind FR in the presence of EBNA-1, they might contribute to maintenance of these replicons by sequestering the replicons to sites favoring their non-random partitioning.

2.2. Replication during the lytic phase of EBV's life cycle

Infection with EBV, as well as other members of the γ-herpes virus family, rarely leads to the production of viral particles. Instead, a state of latency is usually established in which the viral DNA is transmitted to the daughter cells with only a subset of viral genes being expressed. Nevertheless, reactivation from this latent state to the productive, lytic phase of the viral life-cycle can occur. In productive infection most of EBV's approximately 80 genes are expressed. Viral DNA is synthesized largely with viral enzymes and the complex virus particle is assembled from its virus-encoded constituents. As a consequence, the host cell is destroyed and viral particles are released. In the case of EBV, the physiological conditions that lead to a reactivation of the latent virus are not clearly defined, but the expression of two viral genes BZLF1 and BRLF-1 is required for productive infection to ensue [39,40].

The *cis*-acting region of the EBV genome required for lytic DNA synthesis, termed *oriLyt*, has been identified and characterized (Fig. 1) [41]. *OriLyt* is complex and contains multiple components required for DNA synthesis and additional sequences to modulate this synthesis (Fig. 1) [42,43]. The binding of the BZLF-1 protein to *oriLyt* is essential for its synthetic function while one component of *oriLyt* can be functionally substituted by a transcriptional enhancer element. The product of DNA synthesis mediated in *cis* by *oriLyt* and in *trans* by at least seven viral proteins [44] is a linear polymer of DNA generated presumably by a rolling circle mechanism. These polymers are cleaved at the terminal repeats, which make up the ends of virion DNA in the process of being packaged into progeny viral particles. Plasmids derived from EBV containing both *oriP* and *oriLyt* replicate once per cell cycle in normally proliferating EBV-positive cell lines and are amplified 100–1000-fold when the host cells are induced to support the viral productive cycle [41]. This amplification could be used in engineered vectors to produce large amounts of encoded proteins in EBV-positive cells.

3. EBV-based vectors for gene expression in cell culture

EBV-derived vectors have been used successfully in cell culture as recombinant DNA shuttle vectors, as vectors for gene expression, as expression vectors for screens of cDNA libraries, and as vectors for efficient production of recombinant proteins. Recombinant DNA shuttle vectors have made the enumeration and analysis of mutational lesions in mammalian cells feasible. Mutations can be fixed within a gene of interest in an *oriP* shuttle vector by its replication in mutagenized mammalian cells. The shuttle vector can then be enriched and amplified in *Escherichia coli* making characterization of the mutations facile. However, a difficulty with the use of available shuttle vectors for the study of mutagenesis has been the observation that the process of transfection of DNA into mammalian cells can itself be highly mutagenic [45–48]. Plasmid DNA isolated from a variety of mammalian cells shortly after transfection of papovavirus-based shuttle vectors was found to contain a high frequency (approximately 1%) of mutations [47]. Additionally, time course experiments performed by Lebkowski and colleagues indicated that the mutations likely arose shortly after arrival of the plasmid in the nucleus [47]. In contrast, vectors derived from EBV have been shown to have a much lower rate of spontaneous mutations in an encoded gene; for example, the herpes simplex virus thymidine kinase gene undergoes two mutations in 10^5 alleles to two mutations in 10^6 alleles per 10 or 60 population doublings, respectively [49]. These apparent spontaneous mutation rates are reminiscent of those observed for the cell's hypoxanthine (guanine) phophoribosyltransferase (HPRT) gene [50]. These findings support the interpretation that synthesis and repair of *oriP* plasmids is directed by cellular machinery. Consistent with this interpretation is the recent finding that chromatin immunoprecipitation with an antibody to a member of the MRE11 complex involved in DNA repair enriches *oriP* sequences in EBV-positive cells in the S-phase of the cell cycle [51]. It seems likely that this association of *oriP* with DNA repair proteins during the S-phase of the cell cycle contributes to the high-fidelity for replication of the plasmid.

A second general use of EBV-derived vectors has been to express their encoded genes in recipient cells. *OriP* vectors can be maintained extrachromosomally in cells derived from different species including dog, monkey, human, and perhaps hamster cells [4,52]. These data indicate that *oriP*-based vectors are useful for expression of genes in a wide variety of animal species. These vectors are also stable. In fact, *oriP* vectors encoding EBNA-1 and a selectable marker can be stably maintained extrachromosomally for at least 60 population doublings under selection [12,49]. In addition EBNA-1 in conjunction with the transcriptional enhancer, FR can increase expression of genes up to 10 kbp away on these vectors [14,53,54]. All of these functions taken together make the use of EBV-derived vectors for the expression and analysis of genes in cells quite attractive. A potential limitation of the use of EBV-derived plasmids for gene expression is the fact that expression of EBNA-1 in cells may result in unwanted phenotypes that complicate the analysis of the genes under study. For example, EBNA-1 can inhibit apoptosis mediated by p53 in the absence of other viral genes (Kennedy, Komano, and Sugden, manuscript in preparation). Additionally, the ability of EBNA-1 to activate transcription of some promoters within 10-kbp of *oriP* could affect experiments designed to characterize these promoters [53]. On the other hand *oriP* vectors have been powerful tools to dissect the control of promoters, such as within the β-globin locus [55]. The activities of EBNA-1 must be considered when designing experiments that use EBV-derived vectors because its properties may affect the outcome of the experiment.

A third general use of *oriP*-derived vectors is to isolate cDNAs that encode proteins which complement defects in recipient cells. Various *oriP* plasmids have been designed as vectors for cloning and expressing libraries of cDNAs and been found to be useful [56–58]. Upon transfection of mammalian cells with a mixture of these *oriP* plasmids, cells can be shown to harbor a diverse group of plasmids within them. Importantly, the ratio between plasmids in one cell can be shifted in favor of a vector with a particular insert, when selection for this insert is applied. Up to 100-fold enrichment of a low-abundance plasmid from a mixture of vectors has been demonstrated when using *oriP*-derived vectors during one round of selection [58–60]. Sequential rounds of such enrichment have allowed the functional cloning, for example, of formerly unidentified human DNA repair genes [61,62].

OriP-derived vectors can also be used for the production of recombinant proteins in mammalian cell lines [63]. One virtue of producing recombinant proteins in mammalian cells is that the proteins possess all relevant post-translational modifications required for function. Using an *oriP*-based vector and the 293 human embryonic kidney cell line stably expressing EBNA-1 (293E), Durocher *et al.* were able to achieve a high level of gene expression from transient transfections (up to 20% of total cellular proteins were encoded by the transgene) [63]. They were also able to purify the reporter gene secreted alkaline phosphatase (SEAP) from the culture medium and obtained levels of protein exceeding 20 mg/l.

4. EBV-based vectors for gene therapy

4.1. Replication-competent vectors

EBV-based plasmids replicate extrachromosomally, dependent upon the presence of only one viral protein, EBNA-1. Such plasmids can replicate in all examined human cells and, in established cell lines, are lost at a rate of 2–5% per cell generation in the absence of selection independent of the number of plasmids per cell [12,13]. These features of EBV-based plasmids make them attractive as vectors in cell culture and for gene therapy. EBV-derived plasmids are used widely in cell culture now and are being developed and tested in animal models for their possible use in human patients in the future.

One obvious advantage of extrachromosomal vectors is that they obviate the difficulties intrinsic to those that integrate. In particular retroviral vectors are insertional mutagens [64]. Regulation of gene expression by the provirus is often dominated by cellular controls in the vicinity of their sites of integration. Retroviral vectors also have size constraints imposed by their needing to be packaged into virions. On the other hand, vectors that integrate into the host's cell DNA have the obvious advantage of being faithfully maintained in a proliferating host cell. EBV-derived vectors are not mutagens, can express encoded genes tissue-specifically, but will be lost slowly in the absence of selection.

Given the strengths and weaknesses of EBV-derived vectors, researchers are developing them for gene therapy. One major hurdle to the development of vectors derived from EBV to be used in patients is the host range of EBNA-1/*oriP* replicons. These replicons are synthesized and maintained extrachromosomally in human, monkey, bovine, canine, and feline cells in culture but not in murine and rat cell lines [3,4]. Experiments on plasmid replication in hamster cell lines are equivocal [52]. Much work on testing vectors is performed in small animal models, in particular in mice and rats, such that the animal models are likely to undervalue or even misrepresent the potential usefulness of EBV-vectors. Work done by Michelle Calos, Carl Schildkraut, Ann Kirchmaier, Richard Longnecker, John Yates, and their colleagues indicates that sequences other than DS can support the initiation of DNA synthesis while the FR in *cis* and EBNA-1 in *trans* are crucial for maintaining these plasmids extrachromosomally in proliferating cells [17,20,65,66]. Some of these sequences, from the cell or even from within EBV's genome unlike DS can support DNA synthesis in murine cells.

Even though *oriP* vectors cannot replicate in murine cells, vectors that are competent for replication in human cell lines can provide efficient gene expression in murine models. Cui *et al.* injected 10 µg of plasmid DNA into the tail vein of mice under high-pressure [67]. The plasmid DNA used in this study encoded EBNA-1 under the control of a constitutive promoter with *oriP* in *cis*. A reporter gene was delivered by these plasmids and expression from the EBV-based plasmid was compared to a control plasmid lacking the EBV-elements. EBV-based plasmids gave a 100-fold increase in reporter expression up to 35 days post-injection. The efficient expression of the reporter gene reflects either the maintenance function, transcriptional activation, or both functions mediated by EBNA-1 bound to *oriP* [6,14]. The contribution by *oriP* to replication in human cells would be lacking in these experiments in mice.

EBV-based vectors can support replication of large, human chromosomal DNA that may contribute signals required for DNA synthesis in murine cells. Stoll *et al.* have developed large plasmids that can replicate extrachromosomally in cell culture and can be used to express human genes in mice [68]. They have found that a 19-kb plasmid containing FR and encoding EBNA-1 and the full-length human alpha-1-antitrypsin protein replicates extrachromosomally in human cells. Additionally, following injection of large quantities of DNA into the tail vein of mice, the plasmid sustained expression of the alpha-1-antitrypsin gene at physiologically relevant levels (> 300 µg/ml) for as long as 9-month duration of the experiment. Given that expression probably occurred in hepatocytes, which only proliferate in the setting of normal development and injury, it is unclear whether this vector supported replication in the animal model.

Because most human tissues are not replicating, the use of vectors competent for replication may not be necessary for the treatment of benign diseases. On the other hand, malignant diseases are characterized by a high proliferative index and may be the application for which replication-competent vectors are most useful. A recent finding that only 1–10% of cells that take up an EBV-based plasmid will support its stable replication may pose a potential limitation to the utility of EBV-derived plasmids [69]. Therefore, to be effective for inhibiting proliferation and inducing regression of established tumors, a significant toxic effect on adjacent non-gene modified tumor cells and sometimes also on more distant tumor cells may be required [70]. A therapy that has such an effect is said to give rise to a "bystander effect". Suicide genes, the expression of which renders target cells susceptible to killing by a prodrug such as ganciclovir or 5-fluorocytosine (5-FC), are notable for giving such bystander effects [70,71]. In fact, vectors expressing suicide genes such as the herpes simplex virus-1 thymidine kinase gene (HSV-1 tk) have rendered cells

100–300 times more susceptible to killing by ganciclovir than were those cells transfected with a similar vector lacking EBNA-1 and *oriP* [72]. Results *in vitro* demonstrated that while only 30–40% of cells took up the EBV-derived vector, nearly 100% of cells were killed, including both the transfected and untransfected cells. On the other hand, despite similar transfection efficiency, up to 40% of the cells transfected with the control plasmid survived. To extend these observations, subcutaneous tumors were established in a SCID mouse and transfected *in vivo* using a polyamidoamine dendrimer (PAMAM) with either control or EBV-derived plasmids. Those mice treated with ganciclovir whose tumors were transfected with the EBV-derived plasmid had a 50% increased survival relative to those mice whose tumors were transfected with control plasmid. Additionally, after ganciclovir treatment, tumors from mice transfected with the EBV-derived plasmid were reduced in size by 2–3-fold relative to those mice transfected with control plasmid [72].

In summary, vectors competent for replication are attractive but perhaps unnecessary for the treatment of benign disease. However, the maintenance function of *oriP* may contribute to prolonged plasmid retention in non-proliferating cells. Likewise, the ability of EBNA-1 to enhance transcription of genes expressed from a variety of promoters may contribute to efficient expression of these genes. A potentially useful application of replication-competent vectors may exist in the treatment of malignant diseases and those benign diseases involving tissues that actively replicate, such as mucosal surfaces and hematologic tissues. However, a major limitation to our understanding of the role for replication-competent vectors in the treatment of disease arises from the lack of an appropriate animal model in which to study the biology of the vectors and diseases together.

4.2. Replication-deficient vectors

Many applications for gene therapy will require targeting of genes for long-term expression in non-dividing cells. Such an example is the expression of alpha-1-antitrypsin in liver cells of patients with alpha-1-antitrypsin disease. A successful application of gene therapy to benign conditions would require genes to be expressed in tissues for long periods of time resulting in the expression of physiological levels of protein if the treatments are to be useful. Because the cells are not replicating, a plasmid that is capable of replication is not required but one that is maintained extrachromosomally as an active transcriptional template would be beneficial. EBV-based vectors satisfy such requirements and can provide higher levels of gene expression from these non-dividing cells than can vectors lacking the EBV-elements [68]. Findings by Stoll *et al.* [68] and Cui *et al.* [67], are further supported by experiments from Tu *et al.* demonstrating that plasmid DNA, which encodes the luciferase reporter gene from the CMV promoter enhanced by the FR are maintained extrachromosomally [73]. Additionally, plasmids with the FR gave increased levels of luciferase over plasmids lacking the FR. In these experiments, EBNA-1 was provided on a plasmid DNA separate from the reporter plasmid and a single injection of the plasmid DNAs expressing EBNA-1 and the reporter gene led to high levels of expression 31 days post-injection. The authors demonstrated that the plasmids were maintained extrachromosomally and did not replicate. Interestingly, both the EBV-based vectors and the vectors lacking the EBV-elements were present at similar levels 14 days post-injection. However, the EBV-derived vectors gave ninefold more gene expression even 62 days post-injection [73]. This finding demonstrates that the EBV-derived vectors that do not replicate in murine tissues do provide advantages for gene expression. These advantages could either be in homing or maintaining the plasmid DNAs in the nucleus of the cells, in increasing transcription from the localized plasmid template, or both.

Using EBV-derived vectors for expression specifically in tumors that are caused by EBV may be another application for replication-deficient vectors in the form of a replication-deficient, *oriP*-based adenoviral vector. The β-galactosidase reporter gene has been delivered to EBV-positive nasopharyngeal carcinoma cells or EBV-negative cells on an *oriP*-based adenoviral vector [74]. EBV-positive cells infected with this virus expressed significantly higher levels of reporter gene demonstrating the cell type-specific effect. When the tumor suppressor gene p53 was substituted for β-galactosidase, the infected cells underwent apoptosis more frequently than did those cells infected with a control virus. EBV-positive and -negative tumors were established in SCID mice by subcutaneous injection of cells in subsequent experiments. Once established these tumors were treated with either an *oriP*-positive or -negative virus from which β-galactosidase was expressed. After intratumoral injection of the *oriP*-negative virus, X-gal staining was positive in both EBV-negative and -positive tumors. However, when the tumors were injected with the *oriP*-containing virus, X-gal did stain the EBV-positive tumors but not the EBV-negative tumors. Systemic expression of β-galactosidase was detected by staining the liver with X-gal. No positive cells were detected in the liver of normal animals injected with the *oriP*-containing virus confirming that tumor-specific expression of genes can be obtained using *oriP* vectors from which target genes are expressed [74]. Although these "all or nothing" responses seem surprising, they likely reflect large differences in the expression of β-galactosidase; *oriP* vectors in the presence of virally provided EBNA-1 can yield 100 times more expression than in its absence [53]. Additionally, the authors confirm that functions other than those associated with replication contribute to gene expression by EBV-derived vectors.

4.3. Vectors that accommodate large inserts for gene therapy

As knowledge about the regulation of genes in specific tissues has grown over the years, interest in exploiting some of the regulatory elements that provide this tissue specificity has also increased. For example, promoter elements of the human cytokeratin K18 gene direct expression of its gene in the epithelia of lung, liver, kidney, and intestine [75]. These regulatory elements have been exploited for the tissue-specific expression of the cystic fibrosis (CF) transmemebrane regulator gene for the therapy of CF [76]. A potential limitation of using such genomic elements may be presented by the size of these elements. Most cDNAs have not presented a particular problem for the delivery of genes on plasmid or retroviral vectors due to size restraints. However, typical mammalian genes range from tens to hundreds of kilobase pairs in size and even the genomic promoter elements can be several kilobase pairs in length [76]. Therefore, if delivery of genomic DNA or other genomic elements is required for the effective gene therapy of benign or malignant disease states, vectors that can accommodate such inserts will need to be developed. Any vector using genomic DNAs for gene therapy must accommodate such large inserts. EBV can accommodate up to 180 kb of DNA [77,78]. If virus particles are not needed, EBV's plasmid replicon can accommodate even larger DNAs. In fact, a library carrying human genomic inserts averaging 150–200 kbp in size has been constructed in a vector containing *oriP* and established in a human B-cell line expressing EBNA-1 [79]. Additionally, yeast artificial chromosomes up to 660 kbp in size have been maintained on *oriP* vectors in human cells expressing EBNA-1 [33].

The entire EBV-genome has also been cloned onto the prokaryotic F-plasmid making its genetic manipulation feasible [80]. Using a temperature-dependent, recombinase-positive strain of *E. coli*, the EBV-genome can be manipulated through homologous recombination in bacteria. Introduction of the plasmid DNA into the human embryonic kidney-293 (HEK-293) results in packaging and release of infectious virus upon induction of the lytic cycle. The virus obtained can then be used to infect B-lymphocytes, *in vitro*. These types of vectors may be particularly useful to introduce large genes into B-lymphocytes on an extrachromosomally replicating plasmid [81].

5. Cell-specific gene expression

A major hurdle to the development of gene therapies is posed by the inability to target the genes to appropriate cells. This problem is best illustrated in the therapy of malignant disease when the gene to be delivered is toxic to the cells in which it is expressed. For example, delivery of a constitutively expressed p53 to cells is likely to result in apoptosis for both benign and malignant cells. Therefore, effort has been focused on delivering genes that will result in cell-type specific expression.

Because of their broad host range, robust growth in culture, and capacity to infect mitotically quiescent cells, adenoviral vectors have garnered a great deal of interest in the field of gene therapy. Novel hybrid adenoviruses have used expression of the bacteriophage P1 site-specific recombinase Cre to recombine the linear adenoviral genome into a replication competent circular genome. Tan *et al.* have used a two-virus delivery system in which one virus encoded the therapeutic gene while the other virus encoded Cre recombinase [82]. Expression of Cre recombinase from the first vector catalyzed recombination between the *loxP* sites in the second vector, causing excision and circularization of the intervening DNA. The excised, circularized DNA was engineered to contain elements required for stable maintenance of an EBV plasmid, *oriP* and an expression cassette for EBNA-1, as well as a gene to be stably expressed in infected cells. Wang *et al.* used a similar approach in which the Cre recombinase and the EBV-elements along with the therapeutic gene were expressed on the same viral construct [83]. Cre recombinase was expressed from a tissue-specific promoter in this vector and recombination resulted in the excision and inactivation of Cre. Circularization of the vector resulted in expression of EBNA-1 and replication of the circular plasmid in a tissue-specific fashion.

Such hybrid vectors overcome some of the limitations of adenovirus for delivery. Specifically, it has been noticed that gene expression by adenovirus constructs diminishes over a few days. This diminution of protein expression likely results from immunological clearance of infected cells based on the immune recognition of late gene products of the virus [84]. In experiments by Tan *et al.*, the late gene expression was reduced by the recombination event [82]. It is unclear if this reduction in late gene expression could result in a corresponding diminution of the immune response. One virtue of using EBV-based vectors is that EBNA-1 has a domain consisting of glycine and alanine repeats that enables this protein to escape proteasome degradation and antigen presentation via the MHC class I pathway [85,86] (see Chapter 19). Those proteins of EBV that are recognized by cytotoxic T-lymphocytes are not required for the synthesis, maintenance, or transcriptional activation from *oriP* (for a review on the immunology of EBV infection see [87]). Cell-specific gene expression based upon an EBV-derived vector may elicit a less robust immune response leading to prolonged gene expression. Furthermore, gene expression could be enhanced based upon the presence on the plasmid of the transcriptional enhancer FR.

An approach that may be particularly useful in treatments of blood cell disease may be the *ex vivo* transfer of genes on EBV-derived vectors. Hacein-Bey-Abina *et al.*, demonstrated that X-linked severe combined immunodeficiency could be treated successfully in four of five patients by infecting CD-34+ bone marrow cells *in vitro* with a retrovirus expressing γc chain, which is an essential component of five cytokine receptors necessary for the development of T cells and natural killer cells, followed by introduction of the infected cells into the patient [88]. Four of five patients treated by such an approach showed clinical improvement by a decrease in the number and severity of infections. Improvement could also be demonstrated by an increase in the number of circulating T- and NK-cells. EBV-based vectors could certainly be used in such experimental procedures wherein bone marrow cells of patients could be established *in vitro* expressing EBNA-1 and the γc chain from an *oriP*-based plasmid. Once established *in vitro* the cells could be reintroduced into the patient for correction of the defect. Using a retroviral approach to stable transduction of bone marrow cells, such as that used by Hacein-Bey-Abina *et al.* [88], will give rise to a small but real risk of malignant transformation in some fraction of infected cells due to insertional mutagenesis. Substitution of an extrachromosomally replicating EBV-derived plasmid would eliminate this risk. Furthermore, expression from a retrovirus will be affected by epigenetic events such as DNA methylation leading to loss of gene expression. Gene expression from an EBV-based plasmid would also diminish overtime due to a slow loss in the absence of selection. A second potential limitation to the approach of using an *oriP*-derived vector for the *in vitro* transfection of primary bone marrow cells may arise from the fact that primary cells are quite difficult to transfect. Furthermore, Elizabeth Leight has recently demonstrated that only 1–10% of cells transfected with an *oriP*-plasmid expressing the hygromycin-resistant gene give rise to drug-resistant colonies [69]. In these experiments, Leight demonstrated that the establishment of drug-resistant colonies is dependent upon a stochastic, infrequent, epigenetic event. The low frequency of such an event would make the isolation of stably transfected bone marrow cells *in vitro* a difficult task if a very low percentage of cells actually take up the plasmid after transfection.

6. A case study for the future

Cystic fibrosis (CF) is the most common hereditary, lethal, autosomal, recessive disease in the United States. The disorder is caused by mutations of the CF transmembrane conductance regulator (CFTR) gene resulting in disordered ion transport in the airways and the ducts of exocrine glands such as the pancreas. This defective ion transport leads to underhydrated, hyperviscous exocrine secretions and impaired mucociliary clearance. The most devastating clinical manifestations of CF are within the pulmonary and gastrointestinal systems [89]. Pulmonary manifestations include chronic pneumonia in the early stages. The sputum microbiology is characteristic for CF with *Staphylococcus aureus* and *Haemophilus influenzae* predominating in the young patients while *Pseudomonas aeruginosa* dominates after years of chronic antibiotic therapy. Lung function is gradually lost leading to respiratory failure with transplantation as the only possible treatment. Neonates with CF classically present with a bowel obstruction, caused by the inability to pass stool secondary to the thick secretions that obstruct the bowel lumen. Exocrine pancreatic insufficiency complicates CF in more than 90% of patients and leads to protein and fat malabsorption, as well as vitamins E and K deficiency. The average lifespan of a CF patient is 28 years [89].

CF is a good candidate for gene therapy as the disease is recessive and the affected tissues are fairly accessible for gene delivery. Major limitations to the advancement of gene therapy for CF have included lack of an ideal vector from which the gene can be expressed long term. Adenoviruses have been used in the early human experiments but have failed to maintain expression of CFTR at 30 days post-infection [90]. Readministration of the virus in one study did not generally lead to an increase in gene expression. A third administration gave no increase in CFTR expression. One potential limitation to most of the studies performed may arise from the fact that the CFTR gene is delivered in a cDNA-form and not in the context of its genomic DNA [90,91]. The genomic DNA may provide tissue-specificity by virtue of the presence of its own promoter.

Huertas *et al.*, attempted to overcome this potential limitation by cloning the genomic DNA encoding CFTR onto a yeast artificial chromosome in *cis* with EBNA-1 and *oriP* [92]. The authors were able to demonstrate that cells arising after fusion between mouse cells and yeast spheroplasts retained the YAC in circular form with a rate of loss between 2 and 5% per cell generation, consistent with the previously published literature [12]. The human CFTR gene was expressed in only some CFTR-negative cell lines and at levels less than 25% of the mouse CFTR in cell lines with endogenous CFTR. It is not clear from these experiments why the human transgene is expressed at lower levels than the endogenous murine CFTR. It is also unclear if tissue-specificity for the promoter element of the CFTR gene exists because the human transgene was expressed even in fused cells in which the endogenous CFTR expression could not be detected. However, these cells were derived from a murine colon carcinoma and the CFTR gene is likely expressed in the normal colon so the transcriptional machinery required for production of the CFTR gene may be intact.

EBV-derived vectors may be ideal for delivery of the CFTR gene for several reasons. First, EBNA-1 and *oriP* can mediate synthesis of a vector in dividing human cells. Second, EBNA-1 is not usually a target for cell-mediated cytotoxicity making diminution in gene expression as a result of immune clearance less likely. Third, EBV-based vectors can be efficiently delivered in cationic-based liposomal–DNA complexes resulting in long-term expression *in vivo* [73]. Such a preparation may be ideal for delivery via nebulized inhalation or via direct topical application [91]. Finally, the rate of loss of an *oriP*-based plasmid in established cell lines is < 5% per cell generation independent of plasmid number [4,12]. This low rate would result in long-term expression of the transgene and the transactivation function of EBNA-1 bound to the FR should increase expression of the transgene.

7. Conclusions

Our understanding of how *oriP* and EBNA-1 contribute to replication of plasmid DNA has increased significantly over the past 10 years and promises to increase further in the next 5 years. As our knowledge of the mechanism by which EBV-derived plasmids replicate increases, the utility of such plasmids is likely to increase as well. EBV-derived plasmids have already proven to be useful tools *in vitro* with such applications as gene expression, cDNA library expression, recombinant shuttle vectors, and recombinant protein production being particularly successful. Vectors derived from EBV may hold promise for the treatment of both benign and malignant diseases by gene therapy. Their ability to replicate extrachromosomally using cellular replicative machinery with only two required viral elements makes them particularly attractive for future applications. Limitations of the current studies arise from the lack of appropriate experimental models in which to test the EBV-derived vectors. However, data collected thus far using the available resources appear promising. A better understanding of the mechanism by which the EBV-elements, *oriP* and EBNA-1, support sustained plasmid replication should lead to the development of more efficient and safer vectors in the future.

Acknowledgements

The authors would like to thank Elizabeth R. Leight for her critical review of this manuscript. This work was supported by Public Health Service grants CA-22443,

CA-14520, and T32-CA-009614-13. Bill Sugden is an American Cancer Society Research Professor.

Abbreviations

CF	cystic fibrosis
CFTR	the CF transmembrane conductance regulator
DS	dyad symmetry element
EBNA-1	Epstein-Barr nuclear antigen-1
EBV	Epstein-Barr virus
FR	family of repeats
MCM	minichromosome maintenance
ORC	origin recognition complex
OriLyt	origin of lytic replication
OriP	origin of plasmid replication

References

1. Babcock, G.J., Decker, L.L., Volk, M., and Thorley-Lawson, D.A. (1998) Immunity 9, 395–404.
2. Lindahl, T., Adams, A., Bjursell, G., Bornkamm, G.W., Kaschka-Dierich, C., and Jehn, U. (1976) J. Mol. Biol. 102, 511–530.
3. Yates, J., Warren, N., Reisman, D., and Sugden, B. (1984) Proc. Natl. Acad. Sci. USA 81, 3806–3810.
4. Yates, J.L., Warren, N., and Sugden, B. (1985) Nature 313, 812–815.
5. Lupton, S. and Levine, A.J. (1985) Mol. Cell. Biol. 5, 2533–2542.
6. Reisman, D., Yates, J., and Sugden, B. (1985) Mol. Cell. Biol. 5, 1822–1832.
7. Ambinder, R.F., Shah, W.A., Rawlins, D.R., Hayward, G.S., and Hayward, S.D. (1990) J. Virol. 64, 2369–2379.
8. Rawlins, D.R., Milman, G., Hayward, S.D., and Hayward, G.S. (1985) Cell 42, 859–868.
9. Baer, R., Bankier, A.T., Biggin, M.D., Deininger, P.L., Farrell, P.J., Gibson, T.J., Hatfull, G., Hudson, G.S., Satchwell, S.C., Seguin, C. (1984) Nature 310, 207–211.
10. Wysokenski, D.A. and Yates, J.L. (1989) J. Virol. 63, 2657–2666.
11. Gahn, T.A. and Schildkraut, C.L. (1989) Cell 58, 527–535.
12. Kirchmaier, A.L. and Sugden, B. (1995) J. Virol. 69, 1280–1283.
13. Sugden, B. and Warren, N. (1989) J. Virol. 63, 2644–2649.
14. Reisman, D. and Sugden, B. (1986) Mol. Cell. Biol. 6, 3838–3846.
15. Langle-Rouault, F., Patzel, V., Benavente, A., Taillez, M., Silvestre, N., Bompard, A., Sczakiel, G., Jacobs, E., and Rittner, K. (1998) J. Virol. 72, 6181–6185.
16. Schepers, A., Ritzi, M., Bousset, K., Kremmer, E., Yates, J.L., Harwood, J., Diffley, J.F., and Hammerschmidt, W. (2001) EMBO J. 20, 4588–4602.
17. Norio, P., Schildkraut, C.L., and Yates, J.L. (2000) J. Virol. 74, 8563–8574.
18. Dijkwel, P.A. and Hamlin, J.L. (1995) Mol. Cell. Biol. 15, 3023–3031.
19. Aladjem, M.I., Groudine, M., Brody, L.L., Dieken, E.S., Fournier, R.E., Wahl, G.M., and Epner, E.M. (1995) Science 270, 815–819.
20. Kirchmaier, A.L. and Sugden, B. (1998) J. Virol. 72, 4657–4666.
21. Yates, J.L., Camiolo, S.M., and Bashaw, J.M. (2000) J. Virol. 74, 4512–4522.
22. Lee, M.A., Diamond, M.E., and Yates, J.L. (1999) J. Virol. 73, 2974–2982.

23. Leight, E.R. and Sugden, B. (2000) Rev. Med. Virol. 10, 83–100.
24. Frappier, L. and O'Donnell, M. (1991) J. Biol. Chem. 266, 7819–7826.
25. Middleton, T. and Sugden, B. (1992) J. Virol. 66, 489–495.
26. Adams, A. (1987) J. Virol. 61, 1743–1746.
27. Yates, J.L. and Guan, N. (1991) J. Virol. 65, 483–488.
28. Shirakata, M., Imadome, K.I., and Hirai, K. (1999) Virology 263, 42–54.
29. Donaldson, A.D. and Blow, J.J. (1999) Curr. Opin. Genet. Dev. 9, 62–68.
30. Chaudhuri, B., Xu, H., Todorov, I., Dutta, A., and Yates, J.L. (2001) Proc. Natl. Acad. Sci. USA 98, 10085–10089.
31. Krysan, P.J., Haase, S.B., and Calos, M.P. (1989) Mol. Cell. Biol. 9, 1026–1033.
32. Middleton, T. and Sugden, B. (1994) J. Virol. 68, 4067–4071.
33. Simpson, K., McGuigan, A., and Huxley, C. (1996) Mol. Cell. Biol. 16, 5117–5126.
34. Marechal, V., Dehee, A., Chikhi, B.R., Piolot, T., Coppey, M.M., and Nicolas, J.C. (1999) J. Virol. 73, 4385–4392.
35. Harris, A., Young, B.D., and Griffin, B.E. (1985) J. Virol. 56, 328–332.
36. Petti, L., Sample, C., and Kieff, E. (1990) Virology 176, 563–574.
37. Hung, S.C., Kang, M.S., and Kieff, E. (2001) Proc. Natl. Acad. Sci. USA 98, 1865–1870.
38. Deng, Z., Lezina, L., Chen, C.J., Shtivelband, S., So, W., and Lieberman, P.M. (2002) Mol. Cell 9, 493–503.
39. Countryman, J. and Miller, G. (1985) Proc. Natl. Acad. Sci. USA 82, 4085–4089.
40. Feederle, R., Kost, M., Baumann, M., Janz, A., Drouet, E., Hammerschmidt, W., and Delecluse, H.J. (2000) EMBO J. 19, 3080–3089.
41. Hammerschmidt, W. and Sugden, B. (1988) Cell 55, 427–433.
42. Gruffat, H., Renner, O., Pich, D., and Hammerschmidt, W. (1995) J. Virol. 69, 1878–1886.
43. Baumann, M., Feederle, R., Kremmer, E., and Hammerschmidt, W. (1999) EMBO J. 18, 6095–6105.
44. Fixman, E.D., Hayward, G.S., and Hayward, S.D. (1992) J. Virol. 66, 5030–5039.
45. Razzaque, A., Mizusawa, H., and Seidman, M.M. (1983) Proc. Natl. Acad. Sci. USA 80, 3010–3014.
46. Razzaque, A., Chakrabarti, S., Joffee, S., and Seidman, M. (1984) Mol. Cell. Biol. 4, 435–441.
47. Lebkowski, J.S., DuBridge, R.B., Antell, E.A., Greisen, K.S., and Calos, M.P. (1984) Mol. Cell. Biol. 4, 1951–1960.
48. Calos, M.P., Lebkowski, J.S., and Botchan, M.R. (1983) Proc. Natl. Acad. Sci. USA 80, 3015–3019.
49. Drinkwater, N.R. and Klinedinst, D.K. (1986) Proc. Natl. Acad. Sci. USA 83, 3402–3406.
50. DeMars, R., Jackson, J.L., and Biehrke-Nelson, D. (1981) In: Hook, E.B. and Porter, I.H. (eds.) Population and Biological Aspects of Human Mutation, Academic, New York, pp. 209–234.
51. Maser, R.S., Mirzoeva, O.K., Wells, J., Olivares, H., Williams, B.R., Zinkel, R.A., Farnham, P.J., and Petrini, J.H. (2001) Mol. Cell. Biol. 21, 6006–6016.
52. Krysan, P.J. and Calos, M.P. (1993) Gene 136, 137–143.
53. Gahn, T.A. and Sugden, B. (1995) J. Virol. 69, 2633–2636.
54. Mackey, D. and Sugden, B. (1999) Mol. Cell. Biol. 19, 3349–3359.
55. Gui, C.Y. and Dean, A. (2001) Mol. Cell. Biol. 21, 1155–1163.
56. Peterson, C. and Legerski, R. (1991) Gene 107, 279–284.
57. Margolskee, R.F., Kavathas, P., and Berg, P. (1988) Mol. Cell. Biol. 8, 2837–2847.
58. Belt, P.B., Groeneveld, H., Teubel, W.J., van de Putte, P., and Backendorf, C. (1989) Gene 84, 407–417.
59. Belt, P.B., Jongmans, W., de Wit, J., Hoeijmakers, J.H., van de Putte, P., and Backendorf, C. (1991) Nucleic Acids Res. 19, 4861–4866.
60. Pan, L.C., Margolskee, R.F., and Blau, H.M. (1992) Somat. Cell Mol. Genet. 18, 163–177.
61. Henning, K.A., Li, L., Iyer, N., McDaniel, L.D., Reagan, M.S., Legerski, R., Schultz, R.A., Stefanini, M., Lehmann, A.R., Mayne, L.V. (1995) Cell 82, 555–564.
62. Legerski, R. and Peterson, C. (1992) Nature 360, 610.
63. Durocher, Y., Perret, S., and Kamen, A. (2002) Nucleic Acids Res. 30, E9.
64. Wong-Staal, F. and Buchschacher, G.L., Jr. (2000) Blood 95, 2499–2504.
65. Krysan, P.J. and Calos, M.P. (1991) Mol. Cell. Biol. 11, 1464–1472.

66. Haan, K.M., Aiyar, A., and Longnecker, R. (2001) J. Virol. 75, 3016–3020.
67. Cui, F.D., Kishida, T., Ohashi, S., Asada, H., Yasutomi, K., Satoh, E., Kubo, T., Fushiki, S., Imanishi, J., Mazda, O. (2001) Gene Ther. 8, 1508–1513.
68. Stoll, S.M., Sclimenti, C.R., Baba, E.J., Meuse, L., Kay, M.A., and Calos, M.P. (2001) Mol. Ther. 4, 122–129.
69. Leight, E.R. and Sugden, B. (2001) Mol. Cell. Biol. 21, 4149–4161.
70. Djordjevic, B. (2000) Bioessays 22, 286–290.
71. Dilber, M.S. and Gahrton, G. (2001) J. Intern. Med. 249, 359–367.
72. Maruyama-Tabata, H., Harada, Y., Matsumura, T., Satoh, E., Cui, F., Iwai, M., Kita, M., Hibi, S., Imanishi, J., Sawada, T., Mazda, O. (2000) Gene Ther. 7, 53–60.
73. Tu, G., Kirchmaier, A.L., Liggitt, D., Liu, Y., Liu, S., Yu, W.H., Heath, T.D., Thor, A., and Debs, R.J. (2000) J. Biol. Chem. 275, 30408–30416.
74. Li, J.H., Chia, M., Shi, W., Ngo, D., Strathdee, C.A., Huang, D., Klamut, H., and Liu, F.F. (2002) Cancer Res. 62, 171–178.
75. Moll, R., Franke, W.W., Schiller, D.L., Geiger, B., and Krepler, R. (1982) Cell 31, 11–24.
76. Chow, Y.H., O'Brodovich, H., Plumb, J., Wen, Y., Sohn, K.J., Lu, Z., Zhang, F., Lukacs, G.L., Tanswell, A.K., Hui, C.C., Buchwald, M., Hu, J. (1997) Proc. Natl. Acad. Sci. USA 94, 14695–14700.
77. Sclimenti, C.R. and Calos, M.P. (1998) Curr. Opin. Biotechnol. 9, 476–479.
78. Bloss, T.A. and Sugden, B. (1994) J. Virol. 68, 8217–8222.
79. Sun, T.Q., Fernstermacher, D.A., and Vos, J.M. (1994) Nat. Genet. 8, 33–41.
80. Delecluse, H.J., Hilsendegen, T., Pich, D., Zeidler, R., and Hammerschmidt, W. (1998) Proc. Natl. Acad. Sci. USA 95, 8245–8250.
81. Delecluse, H.J. and Hammerschmidt, W. (2000) Mol. Pathol. 53, 270–279.
82. Tan, B.T., Wu, L., and Berk, A.J. (1999) J. Virol. 73, 7582–7589.
83. Wang, X., Zeng, W., Murakawa, M., Freeman, M.W., and Seed, B. (2000) J. Virol. 74, 11296–11303.
84. Kafri, T., Morgan, D., Krahl, T., Sarvetnick, N., Sherman, L., and Verma, I. (1998) Proc. Natl. Acad. Sci. USA 95, 11377–11382.
85. Levitskaya, J., Coram, M., Levitsky, V., Imreh, S., Steigerwald-Mullen, P.M., Klein, G., Kurilla, M.G., and Masucci, M.G. (1995) Nature 375, 685–688.
86. Levitskaya, J., Sharipo, A., Leonchiks, A., Ciechanover, A., and Masucci, M.G. (1997) Proc. Natl. Acad. Sci. USA 94, 12616–12621.
87. Moss, D.J., Burrows, S.R., Silins, S.L., Misko, I., and Khanna, R. (2001) Philos. Trans. R. Soc. Lond. B Biol. Sci. 356, 475–488.
88. Hacein-Bey-Abina, S., Le Deist, F., Carlier, F., Bouneaud, C., Hue, C., De Villartay, J.P., Thrasher, A.J., Wulffraat, N., Sorensen, R., Dupuis-Girod, S., Fischer, A., Davies, E.G., Kuis, W., Leiva, L., Cavazzana-Calvo, M. (2002) N. Engl. J. Med. 346, 1185–1193.
89. Boucher, R.C. (1998) In: Fauci, A.S., Braunwald, E., Isselbacher, K.J., Wilson, J.D., Martin, J.B., Kasper, D.L., Hauser, S.L. and Longo, D.L. (eds.) Harrison's Principles of Internal Medicine. McGraw-Hill, New York, pp. 1448–1451.
90. Harvey, B.G., Leopold, P.L., Hackett, N.R., Grasso, T.M., Williams, P.M., Tucker, A.L., Kaner, R.J., Ferris, B., Gonda, I., Sweeney, T.D., Ramalingam, R., Kovesdi, I., Shak, S., Crystal, R.G. (1999) J. Clin. Invest. 104, 1245–1255.
91. Flotte, T.R. and Laube, B.L. (2001) Chest 120, 124S–131S.
92. Huertas, D., Howe, S., McGuigan, A., and Huxley, C. (2000) Hum. Mol. Genet. 9, 617–629.

S.C. Makrides (Ed.) *Gene Transfer and Expression in Mammalian Cells*

Virus-based vectors for gene expression in mammalian cells: SV40

David S. Strayer, Pierre Cordelier, Julien Landré, Alexei Matskevitch, Hayley J. McKee, Carmen N. Nichols, Martyn K. White, and Marlene S. Strayer

Department of Pathology and Cell Biology, Jefferson Medical College, 1020 Locust Street, Room 251, Philadelphia, Pennsylvania 19107, USA. Tel.: +1 (215) 923-7689; Fax: +1 (215) 503-1156
E-mail: david.strayer@mail.tju.edu

Abstract

Recombinant gene delivery vectors derived from *Tag*-deleted SV40 viruses (rSV40s) are potentially useful tools for transduction of many cell types, both in culture and in animals. Characteristically, these vectors: (i) can be made to very high titers ($> 10^{12}$ infectious units/ml); (ii) infect almost all nucleated mammalian cell types, whether resting or dividing, with high efficiency such that selection is not necessary; (iii) integrate rapidly into the cellular genome and provide for permanent transgene expression without detectable diminution over time; (iv) lack immunogenicity, neither imparting antigenicity to transduced cells nor eliciting neutralizing immune responses against themselves and thus; (v) can be administered multiple times *in vivo* to normal, immunocompetent animals; (vi) can carry up to $\approx 5\,$kb of foreign DNA; and (vii) are safe. Recombinant SV40-derived vectors have been used with pol II promoters to deliver transgenes that encode intracellular, cell membrane, and secreted proteins. They have also been used with pol III promoters to deliver untranslated RNAs, including ribozymes, antisense, RNA decoys, and small inhibitory RNAs. Experimental therapeutic applications of these vectors have included inhibiting HIV in culture and *in vivo*; delivering transgenes encoding important missing enzymes, in animal models of human diseases; protecting cells from free radical-induced damage; immunizing against lentiviral antigens; and manipulating the balance of neurotransmitters in the brain. These vectors are limited by their cloning capacity of 5 kb and by the fact that they do not easily express commonly used marker genes such as fluorescent proteins and β-galactosidase. Currently, human trials of rSV40 vectors are being planned.

1. Introduction

Gene therapy proposes to deliver genetic material, be it DNA or RNA, to influence the course of a disease. The diversity of genetic and acquired human diseases, and the multiplicity of potential approaches to treating them, mean that one approach will

Table 1
Vector properties that limit effectiveness for gene delivery

Problem	Basis for the problem	Consequence
Antigenicity	Infected cells express viral antigens (cytotoxic T cells)	Transduced cells are eliminated by host immune system
	Virion capsid or envelope is antigenic (neutralizing antibody)	Only one administration of a given recombinant virus is possible
Low virus titers	Limitations of packaging cells	Insufficient virus to transduce targets Virus production may be cumbersome and expensive
Transgene expression wanes with time	Promoter inactivation	Need for readministration
	Elimination of vector DNA	Need for readministration
	Need for activating stimulus	Need for restimulation
Low transduction efficiency	Many reasons	Little therapeutic effect
		Selection may be needed Transduction may have to be done *ex vivo*
Limited transgene size	Virus capsid/genome size	Cannot carry large transgenes
Restricted infectivity	Some cells lack receptors	Some cells cannot be transduced
	Requirement for cell division	Resting cells cannot be transduced
Vector sequestration	Extensive phagocytosis	Limited exposure of cells beyond major phagocytic organs
Vector inactivation	Preformed antibodies (etc.)	Vector removed before it reaches cells, so delivery is ineffective

not suit all purposes. The only limits to the potential applications of gene therapy are those of the human imagination.

Despite its original promise for the treatment of disease, however, gene therapy has stumbled thus far, and can point to precious few clear clinical successes [1,2]. Although there are a number of problems that have limited certain applications of gene therapy, the major stumbling block has been that delivery systems have been unable to do their job: vectors have not delivered therapeutic genes to the necessary locations in sufficient quantities, to express their transgenes long enough, or with sufficient effectiveness, to treat disease. The reasons for these failures are listed in Table 1. Vector immunogenicity limits the ability of transduced cells to express the therapeutic transgene for more than a few weeks and also prevents repeat administration of vectors *in vivo*. Poor transduction efficiency, or transduction limited to dividing cells, has made use of some vectors *in vivo* virtually impossible. Low titers of some vectors, compared to the numbers of target cells that need to be transduced, have restrained progress in many areas.

Against this background, gene delivery vehicles derived from simian virus 40 (SV40) were developed. They are not subject to most of the problems that have limited other vectors, and therefore show considerable promise for therapeutic gene delivery. They, like all vectors, have weaknesses that have limited the numbers of investigators that have worked with them. Nonetheless, rSV40s have been extremely effective in many settings and show great promise for clinical gene therapy.

2. The biology of wild type SV40

The effectiveness of rSV40s in delivering genes is a direct function of the biology of wild type (wt) SV40. SV40 has probably been more extensively studied than any individual virus except HIV. It is a nonenveloped virus of the polyomavirus family, approximately 40 nm in diameter.

2.1. Wt SV40 genome organization

The circular genome of wtSV40 is 5.25 kb of double stranded DNA. It lacks the terminal repeats that as are characteristic of many viruses with linear genomes. The organization of the wtSV40 genome is shown in Fig. 1. Briefly, the regulatory region of the genome, shown here between 06:00 and 07:00 h, includes the origin of replication (ori), packaging sequences (ses), early (EP) and late (LP) promoters and enhancer (enh) [3,4]. The two early genes, *Tag* and *tag*, are transcribed counterclockwise (as shown). The late genes are transcribed from the opposite strand, clockwise (as shown). Both sets of transcripts end near the unique *Bam*HI site [5].

2.2. Host range and cell entry

The natural host for SV40 is monkeys [6], but it infects virtually all mammals, including humans. SV40 binds its cellular receptor, MHC-I, and enters cells via caveolae. Since antibody against MHC class I prevents SV40 entry into cells [7,8], the virus probably recognizes a common epitope(s) of MHC-I, or a closely associated protein. This method of cell entry differs greatly from pinocytosis or phagocytosis, such as are characteristic of many other viruses. After virus entry, MHC-I are shed, to be regenerated later [9]. The caveolae deliver the virus *intact* to microtubular conduits that transport the viral particle directly to the nucleus [10]. SV40 virions uncoat in the nucleus. No cytoplasmic phase in early SV40 infection has been clearly identified [11–13]. In this manner, SV40 virions completely bypass cellular antigen processing and presentation machinery, even in dendritic cells and macrophages. Under normal circumstances, wtSV40 antigens—both Tag and tag, and the capsid proteins—are presented to the immune system when they are produced by the cell.

2.3. Tag

Large T antigen is key to understanding the biology of wtSV40 and, at the same time, the effectiveness of rSV40 vectors. Its ability to immortalize cells by binding

74

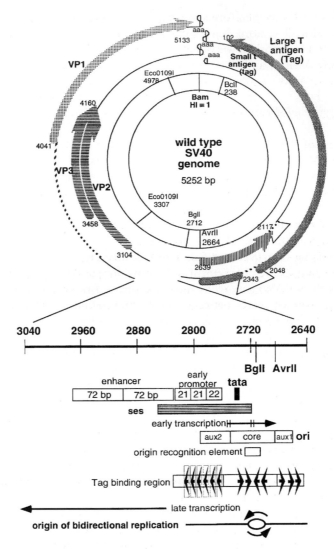

Fig. 1. Map of the wtSV40 genome. Map of the wtSV40 genome. The wtSV40 genome, Baylor strain, is a circular, double-stranded DNA species of 5252 bp in size. It is shown here with the *Bam*HI recognition site arbitrarily at position #1. The early genes, *tag* and *Tag*, are transcribed counterclockwise, driven by the early promoter (shown below). The late genes, VP1, VP2, VP3, are transcribed clockwise, driven by the late promoter. Important restriction sites are indicated, as well as the translational start sites, splice sites, etc. Below, the regulatory region of the SV40 genome contains the origin of replication (ori), the SV40 encapsidation sequences (ses), and the enhancer. The several transcriptional start sites for the early transcripts are indicated, as well as the complex structure of the ori. Binding sites for Tag protein are indicated, along with the direction in which transcription (boxed arrows, which combine Tag and Sp1 binding sites) and bidirectional virus genome replication are initiated. Modified from references [5,49].

and inactivating p53 and pRb is well documented [14,15]. Tag protein binds the SV40 ori and is required for virus genome replication [16]. It also activates the SV40 late promoter after virus genome replication, and is necessary for transcription of the capsid genes [17]. Therefore, an SV40 genome lacking *Tag* can only replicate in cells that supply the requisite Tag protein *in trans*. Further, the capsid proteins will not be produced except in the presence of Tag. As a consequence, *Tag*-deleted SV40 genomes will neither replicate nor make capsid proteins in cells that do not produce Tag.

2.4. Persistence of wtSV40

Wild type SV40 will preferentially replicate its DNA, produce daughter virions, and lyse infected cells. The wtSV40 genome may persist in infected cells either by integration into the genome or as a self-replicating episome. The latter course requires continued production of Tag. Many studies have characterized the integration of the wtSV40 genome into cellular DNA. These studies can be summarized as follows: (a) no genomic site is favored for integration; (b) the circular SV40 genome opens at different sites each time it integrates; and (c) there is no clear or strong homology or pattern of similarity between SV40 and cellular DNA sequences at integration sites [18–20].

2.5. Immunogenicity

Tag is the main wtSV40 antigen seen by the immune system on wtSV40-infected cells. Although principally a nuclear protein, Tag is made in very large amounts, and some inserts into cell membranes where it is seen by humoral and cellular immune systems [21,22]. Virus capsid proteins are antigenic also, but are only made if Tag is expressed: transactivation of SV40 late genes by cell transcription factors, as occurs with adenoviral genes [23], has not been described. Therefore, deleting Tag should render resulting viruses replication incompetent and at the same time protect infected cells from immune elimination.

By virtue of its unusual mode of entry, phagolysosomes containing SV40 particles do not form. Virus capsid antigens are thus apparently not processed at the time of virus entry. As a consequence, these proteins are not presented to the immune system until they are produced during virus assembly, late in infection. These steps do not occur if Tag is not present. If an SV40 lacks Tag, it should thus elicit very little immune response, either against the capsid proteins that are not produced, or against the absent Tag.

2.6. Virus replication

Wild type SV40 replicates to very high titers. Infected cells make several logs of daughter virus for each complete infecting virion. The virus particle is also very stable: it can be band-purified, lyophilized, etc., without losing infectivity appreciably.

3. Making recombinant SV40 viruses and using them for gene delivery

With this background, the characteristics of rSV40s as gene delivery vehicles can be understood.

3.1. Producing rSV40 vectors

There have been several approaches to making SV40-derived vectors. They can be produced *in tube*, exploiting the fact that SV40 encapsidation signals (ses) are the only DNA sequences required to package these vectors in the SV40 capsid. At the same time, the major capsid protein, VP1, can self-assemble into capsids. Such vectors can accommodate large amounts of foreign DNA [24–26], and have been used successfully to deliver transgenes [27,28]. These approaches exploit the important influence of the viral capsid on the early phase of infection: delivering the virus intact to the nucleus. The yields achieved with these vectors assembled *in tube* are not yet comparable to those achieved if packaging cells are used to mimic wtSV40 packaging (below).

A different approach to modifying SV40 for gene delivery is shown in Fig. 2 [29,30]. Starting with wtSV40 genome cloned into a carrier plasmid, we replaced the coding region of the Tag gene (2.3 kb) with a polylinker, downstream from the SV40-EP. The SV40-EP overlaps the ori, and so must be retained.

Recombinant virus genomes carrying a transgene (+ additional promoters, etc.) are excised from carrier plasmids. They must be recircularized in order to be able to produce progeny virions. Resulting circular rSV40 genomes are then transfected into packaging cells. COS-7 cells have an integrated wtSV40 genome that is defective at the ori (and so are unable to produce virus by themselves [31]). They do, however, supply Tag *in trans*, and, as indicated below, also supply the capsid proteins [16]. An rSV40 made this way may be isolated by lysing the COS-7 cells. Vectors are amplified by *infecting* COS-7 cells with the rSV40 virus prepared from the COS-7 cell lysate. Helper virus is not used, and no further transfection is needed.

Resulting rSV40 viruses may be titered in many ways. We use *in situ* PCR, thus measuring the amount of virus that can deliver its DNA into cells, i.e., infectious units (IU) rather than virus particles [32]. Purification of rSV40 can be done by ultracentrifugation, using published protocols [32]. Lysates of rSV40-infected COS-7 cells are treated with mild detergents, which disaggregate clumps of virus particles and dissociate complexes formed between virions and cellular debris. Yields of purified rSV40 vectors obtained by this procedure usually exceed 10^{12} IU/ml, and are often greater than 10^{13} IU/ml.

3.2. Transgene expression

The SV40-EP, which cannot be deleted, can be used to drive transgene expression. Alternatively, it can be supplemented by more powerful promoters (e.g., cytomegalovirus immediate early promoter (CMV-IEP), Rous sarcoma virus long

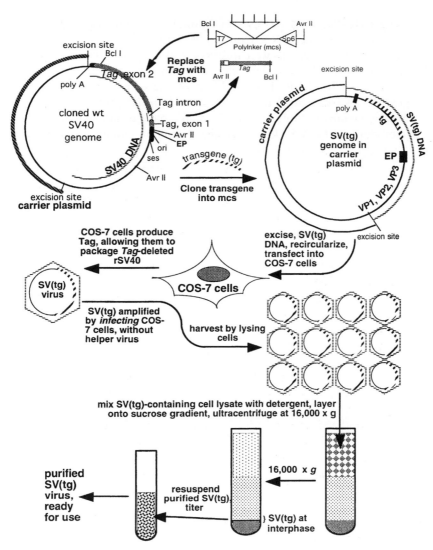

Fig. 2. Producing rSV40 vectors. Producing recombinant, replication-crippled SV40. To make rSV40 constructs the Tag (and tag) gene was excised from the cloned SV40 genome in pBR322 (a plasmid called pBSV-1), to be replaced by a polylinker. A transgene shown here as HBsAg, is then cloned into unique restriction sites in the polylinker. Infectious, replication-defective virus is made by excising the resulting rSV40 genome from its carrier plasmid (here, pBR322), gel-purifying and recircularizing it by religating it to itself. The resulting, circularized rSV40 genome is packaged as virus by transfecting it into COS-7 cells. Replication-incompetent rSV40 stocks may be expanded by infecting COS-7 cells with the resulting rSV40 virus. No helper virus is needed. No subsequent transfection is involved. rSV40 may then be harvested from cell *lysates*, to be band--purified as described [50]. In this illustration, "ses" represents the packaging sequences, as defined by others [4].

terminal repeat (RSV-LTR)). To use conditional promoters (e.g., HIV-1 LTR) or pol III promoters (e.g., met t-RNA promoter, U6 small nuclear RNA promoter), multiple polyadenylation signals were inserted to block transcription from the SV40-EP; the additional promoters were cloned downstream of these. Most of our constructs use the CMV-IEP (Fig. 3). Most of the constructs we use to express untranslated RNAs (see below) use one of several pol III promoters.

Deleting *Tag* creates a "hole" of ≈ 2.3 kb. We have found that the virus capsid can accommodate up to ≈ 5.7 kb. Thus, the maximum cloning capacity of vectors

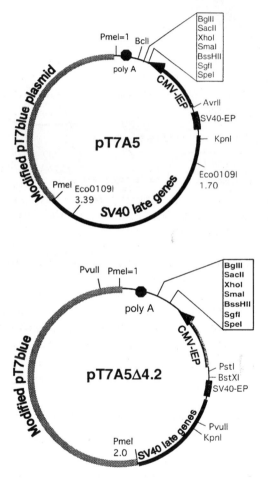

Fig. 3. Two plasmid cloning targets used to make rSV40 vectors. *Tag*-deleted rSV40 genomes were cloned as *Pme*I fragments into a modified pT7Blue plasmid. The upper plasmid, pT7A5, carries the CMV-IEP as its functional promoter, with a polylinker immediately downstream of the CMV-IEP. Relevant restriction sites and the polyadenylation signal (PolyA, shown as a STOP sign) are indicated. To accommodate larger inserts (e.g., CFTR), a larger deletion was made by removing the EcoO109I fragment at the indicated sites. This generated pT7A5(Δ4.2). It, too, is carried in a modified pT7Blue plasmid as a *Pme*I fragment. Some residual SV40 late (capsid) gene sequences remain in pT7A5(Δ4.2), but all coding sequences for the capsid genes are gone. For both plasmids, the location of the SV40-EP is shown.

made from SV40 by simply deleting *Tag* is ≈ 2.8 kb. If it is desired to deliver larger transgenes, the late genes may also be deleted. Vectors produced from constructs lacking the SV40 capsid genes are made at equal titer, compared to those containing the capsid genes. The capsid proteins needed to package these recombinant viruses are entirely encoded by the COS-7 cells, rather than the vector genome itself (unpublished). They are not expressed constitutively by the COS-7 cells but rather are activated following reimplantation of the rSV40 genome (unpublished).

3.3. Manipulating rSV40 vectors

Normally, rSV40 vectors are stored in suspension at $-80\,°C$. However, unlike most other gene delivery vehicles, rSV40s can be lyophilized with minimal loss of infectivity. In lyophilized form these vectors are relatively stable (Table 2), losing infectivity gradually with prolonged storage at room temperature or $4\,°C$.

3.4. Transduction efficiency and persistence of transgene expression

rSV40 is highly efficient as a gene delivery vector in both resting and dividing cells. Selection is not needed to generate populations of transgene-expressing cells: we have never selected for transduced cells in culture. A single exposure to most rSV40 vectors, at virus:cell ratios (multiplicity of infection, MOI) of 10–100 generally will transduce $>98\%$ of cells in a culture. The effectiveness of transduction does not depend upon the stage of the cell cycle, and is independent of the cell type: immortal cell lines, primary lymphocytes, hepatocytes, neurons, and other types of cells are transduced at comparable efficiencies.

   In vivo, transduction efficiencies have also been very high. For example, seven daily injections of rats via the hepatic portal vein at approximate MOI of 1.5 transduced $\approx 85\%$ of hepatocytes, as judged by immunostaining [33]. A single injection of human thymus implanted into SCID mice, at MOI ≈ 10, gave comparable percentages of cells expressing the transgene [34].

Table 2
Effect of lyophilization on stability of rSV40 vectors

Storage temperature (°C)	Recovered infectivity following storage	
	Aqueous suspension (%)	Lyophilized (%)
−80	80	70
−20	40	70
4	< 10	60
37	< 10	50

SV(HBS) was lyophilized or kept as an aqueous suspension, at the temperatures indicated, for 1 week. Recovered infectivity was measured by *in situ* PCR (see reference [32]). Percentages are expressed relative to the infectivity of the virus suspension before the manipulations.

Both *in vivo* and *in vitro*, transgene expression has generally persisted as long as experiments have been carried out. Thus, continuously dividing cells in culture generally express rSV40-delivered transgenes >4 months after transduction. Rats transduced *in vivo* have continued to express their transgenes >16 months after transduction. Waning transgene expression, which is a problem with both retroviral and lentiviral vectors, has not been seen with these vectors.

The persistence of transgene expression was studied in several settings. After transduction *in vitro*, cell lines and stimulated primary cells that were continuously dividing continued to express their transgenes for many months. Mice whose livers were transduced, partially resected and then allowed to regenerate showed similar levels of transgene expression in both the regenerated and pre-operative livers. Replication-crippled rSV40s persist indefinitely *in vitro* and *in vivo* despite cell division [35].

The most likely explanation for such persistence is integration of the vector DNA into the cellular genome. Southern analyses of DNA from rat livers transduced by intraportal injection *in vivo* showed that vector genomes incorporated into cell DNA was noted within days of transduction, and persisted throughout the observation period of >16 months. Episomal vector DNA could no longer be detected by 4 days following transduction. Vector integration sites were cloned and sequenced. rSV40 DNA integrated randomly into the cellular genome. The circular rSV40 genome was found to have opened at different points at each integration site. Quantitative Southern analysis showed 3.05 copies of the transgene per cell. Thus, rSV40s integrated into cellular DNA, whether the cells are resting or dividing. The vector genomes integrated randomly and within days of gene delivery.

3.5. Immunogenicity

There are two different aspects of immunogenicity that pertain to gene delivery vectors: (a) the immunogenicity that they impart to transduced cells, which allows the immune system to target transduced cells for destruction and so limits the durability of transduction; and (b) the immunogenicity of the vector particle itself, which elicits neutralizing antibodies that limit the ability of a vector to be administered more than once. Most discussions of the immunogenicity of gene delivery vectors, focus almost exclusively on the former issue, since all gene delivery vectors (except for SV40, see below) elicit neutralizing antibodies.

Immunogenicity imparted by vectors to transduced cells is largely associated with adenoviral vectors. Although the transcriptional activators in the adenovirus E1 region are deleted, cellular proteins may transactivate adenoviral capsid and other genes. Transduced cells thus synthesize virus antigens and are rapidly eliminated by the immune system [36]. Removing additional adenovirus genome may mitigate, but does not eliminate, this difficulty [37,38].

The ability of SV40 to deliver enduring transgene expression to immunocompetent animals has been tested extensively. Long-term transgene expression has been observed. Further, extensive histologic studies failed to demonstrate any evidence of inflammatory or immune reactions against rSV40-transduced cells in any animal, at any time, in any tissue [29,30,39,40].

The question of whether rSV40 vectors elicit neutralizing antibodies was tested directly. We made SV(HBS), which carries the cDNA for hepatitis B surface antigen (HBsAg). This vector was administered to normal BALB/c mice × 8, monthly. High titers of antibody vs. HBsAg were found by ELISA. Using a plaque-reduction assay to test for neutralizing antibody, no neutralizing antibody vs. SV40 was found in any animal at any time, even after eight intraperitoneal injections [39]. This characteristic is striking among gene delivery vectors; it uniquely allows one or more rSV40 vectors to be administered multiple times without loss of transduction efficiency.

It also allows rSV40 vectors to be used for immunization. That is, as in the study described above with HBsAg, a rSV40 vector carrying an immunogenic transgene can be administered multiple times, to prime and/or boost immune responses against a transgene product. We have found that this approach, particularly when combined with rSV40s carrying immunostimulatory cytokines, provides very high levels of immune responsiveness—particularly cytotoxic lymphocyte responses—against targeted antigens [41,42].

3.6. Uses of rSV40 gene delivery vectors

rSV40 vectors to date have been used experimentally *in vitro* and *in vivo*. These applications are summarized in Table 3. They fall into several broad categories, studies done: (i) *in vitro* delivering transgenes whose protein products are largely intracellular; (ii) *in vitro* using transgenes whose products are untranslated transcripts; (iii) *in vitro* to deliver proteins that are secreted; (iv) *in vivo* to immunize against foreign antigens; and (v) *in vivo* to rectify an inherited or acquired defect.

3.6.1. Delivering in vitro transgenes encoding intracellular proteins
Most studies done with rSV40s in cultured cells have focused on inhibiting HIV. This work necessarily, therefore, delivered transgenes designed to prevent completion of the HIV replicative cycle at one or more points in that cycle. Targets in this process have been both HIV gene products [43] and cellular processes that, when inhibited, prevent HIV replication [44]. Along these lines, we exploited the high efficiency of rSV40-delivered gene transfer to use combinations of rSV40 vectors to transduce the same cells multiply in order to deliver combination genetic therapy. This approach was very successful in inhibiting HIV: when two or more transgenes were combined they usually provided synergistic protection and inhibited even very high challenge doses of HIV [45].

3.6.2. Studies in vitro with transgenes encoding untranslated RNAs
In these experiments, we delivered ribozymes that targeted an endogenous transcript in hepatocytes (mutant α1-antitrypsin), and showed that levels of mutant α1AT mRNA could be decreased by >85% using a rSV40-delivered ribozyme [46]. Other studies demonstrated that a polymeric TAR RNA decoy delivered using a conditional promoter (the HIV-1 LTR) effectively inhibited HIV replication [47].

Table 3
Published studies describing rSV40 gene transfer

Test setting	Administration	rSV40 vector(s)	Durability (assay period)	Tropism; cells tested	Vector antigenicity	Transduction efficiency	Ref.
In vivo (mice)	IV	SVluc	≥376 days (1–376 days)	Many organs	No inflammation	NR	[27,28]
Bone marrow *ex vivo* (mice)	Reinfusion *in vivo*, 450R	SVluc	≥376 days (21–376 days)	PBMC	NR	≈50%	[28]
In vitro	No stimulation	SVluc, SV(HBS)	2 days (1–2 days)	PBMC	N/A	≈50%	[41]
In vivo	Monthly, IV or SQ injection	SV(HBS)	NR	NR	Ab vs. HBsAg, but not SV40	NR	[42]
In vitro	×1	SV(HBS)	≥6 weeks (0–6 weeks)	Human, simian CD34+ cells	N/A	≈50%	[34]
In vitro	×3	SV(Aw)§	≥6 weeks (0–6 weeks)	SupT1 cells	N/A	>90%	[33]
In vitro	×3	SV(Aw)§	≥6 weeks (0–6 weeks)	Human CD34+ cells	N/A	>90%	[34]
In vitro	×3	SV(α_1ATRz)§	≥3 weeks (0–3 weeks)	Human hepatocytes	No inflammation	≈90%	[32]
Bone marrow *ex vivo* (mice)	SV(Aw)	≥3 months	Human CD34+ cells	NR	>90%	[34]
In vitro	Reinfusion *in vivo*, 350R	MDR§	NR	NIH-3T3 cells	N/A	95%	[25]
In vitro	×1	β-globin§	NR	Human rbc progenitors	N/A	NR	[26]
In vivo (rats)	×1 to ×7	SV(BUGT)	16 months (16 months)	Liver	None	>80%	BVS
In vitro	×3	SV(Aw), SV(HE), SV(RevM10), SV(PolyTAR), SV(IE8), SV(RT3)	6 weeks (6 weeks)	SupT1 cells	N/A	>98%	DSS MT
In vitro	×3	SV(HE)	4 weeks (4 weeks)	SupT1 cells	N/A	>98%	Bou2
In vitro	×3	SV(PolyTAR)	4 weeks (4 weeks)	SupT1 cells	N/A	>98%	GCJ
In vitro	×3	SV(AT)	4 weeks (4 weeks)	SupT1 cells, PBL	N/A	>98%	PC1
In vitro	×3	SV[HIVLTR](IFN)	4 weeks (4 weeks)	SupT1 cells, PBL	N/A	>98%	PC2
In vivo (mice)	IP, ×1 to ×6	SVgp130	N/A	N/A	None; Ab and CTL vs. SIV Env	N/A	HJ1
In vivo (mice)	×1	SV(Aw)	2 weeks (2 weeks)	Human thymocytes	N/A	80–90%	HG

*Unpublished data. SV(HBS) carries hepatitis B virus surface antigen; SVluc carries luciferase. These are marker genes.
§Specific rSV40 viruses carrying transgenes with expected therapeutic activities are mentioned for illustration only (e.g., SV(Aw) carries a single chain antibody against HIV-1 integrase). NR, not reported; N/A, not applicable; IV, intravenous injection; SQ, subcutaneous inoculation.

The former study used the methionine tRNA promoter, a pol III promoter, to drive ribozyme expression. Additional recent, unpublished, studies have used other pol III promoters and have been very effective: rSV40 vectors delivered other RNA decoys using U6 small nuclear RNA promoter and U16 nucleolar RNA promoter; antisense RNAs were delivered to neurons of the rat brain using the VA adenoviral cytoplasmic RNA promoter.

3.6.3. Work done in vitro to deliver proteins that are secreted

These studies are entirely unpublished. rSV40s have been used to deliver cDNAs for secreted cytokines. Both cytokines designed to stimulate the immune response and cytokines designed to inhibit HIV and other viruses have been delivered effectively in this fashion.

3.6.4. Immunizing against foreign antigens in vivo

These studies have been done in both mice and nonhuman primates. Work done with HBsAg was described above [39]. More recently, rSV40 vectors have been used to deliver lentiviral antigens. A manuscript in press describes the effectiveness of these vectors in eliciting very strong cytotoxic lymphocyte responses [41] against SIV envelope antigens. SVgp130-binding antibody responses also reached very high levels after 3–4 immunizations. Unpublished studies documented the ability of these vectors to elicit strong cytotoxic lymphocyte responses against the same antigen in rhesus macaques (R.P. Johnson and D.S. Strayer, unpublished). More recently, we combined such lentivirus envelope-bearing rSV40s with rSV40s carrying immuno-stimulatory cytokines IL-12 and IL-15 to elicit immune responses against HIV antigens [42].

3.6.5. Studies performed in vivo to rectify an inherited or acquired defect

Several such studies have been completed. Gunn rats are a model of the human Crigler–Najjar disease. These animals lack bilirubin UDP-glucuronysyl transferase (BUGT) and therefore have very high levels of serum bilirubin. After treatment with SV(BUGT) intraportally, we observed almost complete rectification of their serum bilirubin levels [33].

The first demonstration that gene delivery in vivo could inhibit HIV in vivo was reported using rSV40 vectors. SCID mice received human fetal thymus tissue, implanted under their renal capsules. After implantation, SV(Aw), carrying a transgene that inhibits HIV-1 integrase (IN), was injected into the human thymus. The implants were challenged in vivo by direct inoculation of HIV-1$_{59}$, a clinical isolate of HIV. HIV replication within the implanted tissues was inhibited 80–90% by in vivo treatment with SV(Aw). When the implanted tissues were analyzed for transgene expression, 80–90% of the cells within the implant expressed the anti-IN transgene [34].

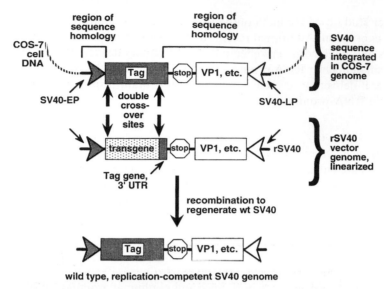

Fig. 4. Potential mechanism by which recombination would generate wtSV40 during packaging of rSV40 vectors. Homology between the rSV40 genome to be packaged, and the wtSV40 genome in the COS-7 cells could regenerate Tag+, wtSV40 by homologous recombination. Areas of DNA sequence homology that would help align the wtSV40 and rSV40 genomes for double cross-over are shown: 5′ to *Tag* gene in the COS cells (5′ to the transgene in rSV40), and 3′ to *Tag* gene in COS cells, and 3′ to the transgene in rSV40.

3.7. Safety

Concerns regarding the safety of rSV40 vectors involve questions of possible contamination with wtSV40, and intrinsic concerns relating to rSV40 gene delivery.

3.7.1. Recombination in packaging

In order to regenerate *Tag+*, wtSV40 during rSV40 packaging in COS-7 cells, a double cross-over would be required. Such a double cross-over event would require homology between wtSV40 genome in COS-7 cells and SV40 sequences that remain in the rSV40 vector (as in Fig. 4).

We found, however, that the point at which the wtSV40 genome opened to integrate into COS-7 cells is in the *Tag* gene, just *upstream* of the *Bcl*I restriction site used to excise *Tag* (Fig. 5). Downstream wtSV40 sequences are thus at the other end of the integrated SV40 genome; so there is no homology between rSV40 and COS-7 genomes 3′ to the transgene, making double crossover to reacquire *Tag* very unlikely. Further, wtSV40 was not found in any of our rSV40 preparations, as tested by virus plaque assay, PCR, and Western blot analyses for Tag.

Thus, much analysis has been done in an attempt to determine and understand the likelihood of recombination during packaging to regenerate Tag+, wtSV40. Within very low tolerance limits, no such recombination has been seen. Considering how wtSV40 inserted into the COS-7 cell DNA, such recombination is unlikely.

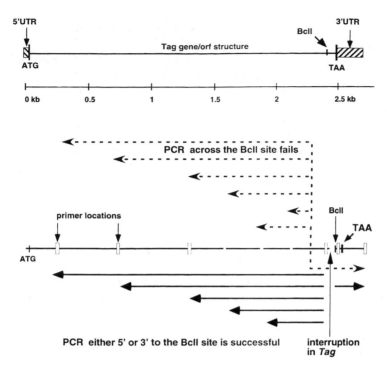

Fig. 5. The wtSV40 integration point in COS-7 cell genome maps to the *Tag* gene. Using PCR primers at locations indicated (magenta lines), we found that the *Tag* gene in COS-7 cells was interrupted, i.e., the circular wtSV40 genome opened at that point to integrate into the cellular DNA. This break point is upstream to the *Bcl*I site that was used to insert a polylinker in place of *Tag*. Thus, rSV40 genomes share no homology to COS-7 wtSV40 sequences 3′ to the transgene.

3.7.2. Insertional mutation

An additional concern applies to any integrating vector: insertion of the vector into the cellular genome may activate or inactivate a critical gene. Like oncoretroviruses, rSV40s integrate randomly into the cellular genome. Any vector that integrates may disrupt a critical gene, causing the cell to become dysfunctional or die. Practically, however, this has not proven to be a substantial problem: neither appreciable cell death (from insertional inactivation of a critical gene) nor tumors (from insertional inactivation of a tumor suppressor gene) have been reported in clinical trials of gene delivery using retroviral vectors.

Insertional activation is potentially more problematic. Depending on the nature of the activated gene, cellular proliferation and tumor development could potentially result. Oncoretroviruses, like MuLV, may cause tumors by such a mechanism. For retroviruses long terminal repeat (LTR) sequences are always adjacent to integration sites. When retroviruses integrate, the structure of the integrated viral genome is very similar at each integration site: the linearity of the retroviral genome and the need for the LTR in order to integrate assure that the 5′ and 3′ LTRs are the viral sequences that abut the cellular DNA at insertion points. Retroviral LTR possess outward-directed promoter activity. The potential for activation of expression of an undesired gene is thus quite real. Nonetheless, functionally significant gene activation (or inactivation) has not been reported in clinical trials with MuLV vectors.

Unlike retroviral vectors, SV40 lacks LTR and the SV40 sequences at each integration site are different. Cellular genes would thus be activated only if virus sequences at an integration site were just downstream of the early promoter. Thus, the possibility of insertional activation of a cellular gene, while real, is less for SV40 than for MuLV vectors.

Empirically, studies of rSV40 gene delivery in rodents and nonhuman primates have involved extensive tissue examinations and (for the primates) examination of blood chemistries and peripheral blood cell populations. No abnormalities in any of these indices have ever been seen. Furthermore, gene delivery is not to be undertaken lightly, and, at this point, only individuals with serious diseases are reasonable candidates for gene therapy. The risk of experimental gene delivery with any vector must be considered in relation to the gravity of the disease being treated.

The safety of rSV40 vectors—or any other gene delivery vehicle—cannot be assumed. All data to date suggest that recombinant SV40-derived vectors do not pose safety concerns.

3.8. Limitations

All vectors have strengths and weaknesses. Recombinant SV40 vectors are no exception to this rule. Although their advantages are formidable, they have distinct limitations. These fall into three categories: (i) physical constraints; (ii) restrictions in effective expression of certain transgenes; and (iii) appropriateness for certain applications.

3.8.1. Physical constraints

As indicated above, the native genome size of wtSV40 is 5.25 kb. We have found that rSV40 genomes of 5.7 kb can be packaged effectively. We have also found that rSV40 genomes ≥ 6 kb cannot be packaged using the system described here (unpublished data). Other investigators, using different approaches, have been able to deliver much larger transgenes with rSV40 vectors [27].

Beyond this limitation, the randomness of integration by rSV40 vector genomes means that the likelihood that a transgene will be fully functional (i.e., that the promoter + transgene will be intact in an integrand) is inversely proportional to the percentage of the vector genome that is occupied by the promoter and transgene together. Thus, we have effectively packaged the Fanconi A cDNA (5 kb) driven by the native SV40 early promoter (0.25 kb) in an rSV40 vector (total size, 5.6 kb). Although the virus genome expresses FancA during packaging in COS-7 cells, effective correction of the Fanconi phenotype in target, FancA $-/-$ cells has not been seen at significant levels as yet (unpublished data). By contrast, the full length cDNA for human cystic fibrosis transmembrane conductance regulator (CFTR, 4.2 kb) has been readily expressed in rSV40s and rectifies the chloride channel defect in CFTR $-/-$ cells (unpublished data).

3.8.2. Restrictions in effective expression of certain transgenes

Simply put, some transgenes do not express well—or in some cases, at all—using rSV40 vectors that were made as we have described. With the exception of luciferase, which we have found is expressed inconsistently in this system, invertebrate, bacterial and fungal DNAs have proven very difficult to express detectably. These include most of the commonly used marker genes: the several fluorescent proteins, β-galactosidase, chloramphenicol acetyl transferase, etc.

The reason(s) for the lack of expression are not clear but preliminary studies suggest that at least partial expression of red fluorescent protein (rfp) may be possible if vectors are grown in medium that contains 5-azacytidine (unpublished data). This observation, if confirmed, would indicate that DNA methylation may be involved in the inability to express rfp. The extent to which the same mechanism is responsible for poor expression of the other transgenes remains to be tested.

Some mammalian transgenes have not expressed well in rSV40 vectors. For example, human factor IX, deficiency of which is responsible for hemophilia B, has proven very difficult to express using rSV40 vectors. Some transgenes whose expression has been problematic with one promoter (usually CMV-IEP) can be expressed readily using other promoters. Thus, galactocerebrosidase, which is deficient in Krabbe's disease, could not be expressed using an rSV40 with CMV-IEP as a promoter, but was expressed well when SV40-EP was used to drive expression (unpublished data).

The vast majority of mammalian genes, all viral genes, and all untranslated DNAs tested can be delivered and expressed effectively with these vectors. Some transgenes, as indicated, cannot—at least, not yet. These observations are summarized in Table 4.

3.8.3. Appropriateness for certain applications

As has been indicated above, rSV40 vectors deliver permanent transgene expression to a vast array of different cell types. Not all gene delivery applications are suitable for such vectors. Some applications require only one delivery and only short-lived transgene expression. Although our studies with inducible expression systems (using tetracycline-inducible elements) have been encouraging, rSV40 vectors that can turn transgene expression on and off using tetracycline or another inducing agent are not yet available. Thus, short-term transgene expression would be best delivered by other vectors.

As well, because germ line cells lack MHC-I they would not be expected to be transduced by rSV40s. Preliminary work done using mouse embryos confirms that rSV40s cannot deliver genes to these cells. Consequently, they cannot be used as tools for producing alterations in the germ line of humans or experimental animals.

4. Conclusions and future perspectives

Small molecule medications are generally effective because they can achieve effective blood concentrations inaccessible to most gene therapy vectors. Further, small molecule drugs can be—and frequently are—administered multiple times.

Table 4

Limitations of rSV40 vectors for delivering transgenes

Type of product	Example (+promoter)	Effectiveness of delivery	Limitation
Intracellular mammalian proteins	Single chain antibodies (+CMV-IEP)	Excellent expression	None
	galC (+2 promoters)	CMV-IEP: poor	Changing promoters improved expression
		SV40-EP: good	
	CFTR (+SV40-EP)	Good	Large cDNA (4.2 kb)
	FancA (+SV40-EP)	Poor	At 5 kb, transgene is probably too large
Secreted mammalian proteins	α1-Antitrypsin (+CMV-IEP)	Blood levels $\approx 3\,\mu g/ml$ in mice	None
	IL-12 + SV40-EP	Stimulates T cell production of γ-IFN	None
	Factor IX + CMV-IEP or SV40-EP	No detectable factor IX production	Reason for lack of expression unknown
Viral proteins	HIV-1 gp120 (+CMV-IEP)	Good expression at the cell membrane	None
	HBsAg (+SV40-EP)	Good expression at the cell membrane	None
Untranslated transcripts + pol III promoters	HIV-1 TAR (+U6)	Strongly inhibits HIV	None
	siRNAs (+VA)	Good targeting of multiple mRNAs	None
	Rz vs. α1-antitrypsin (+met tRNA promoter)	>90% decline in liver cell α1-AT mRNA	None
Bacterial, invertebrate proteins	gfp (+CMV-IEP, SV40-EP, or RSV-LTR)	Poor	Possible methylation during packaging

Effectiveness of rSV40 vectors in delivering different types of transgenes. CFTR, cystic fibrosis transmembrane conductance regulator; GalC, galactocerebrosidase; gfp, green fluorescent protein; HBsAg, hepatitis B surface antigen; Rz, ribozyme; U6, small nuclear RNA promoter; VA, adenovirus cytoplasmic small RNA promoter.

Of the available gene delivery systems, only rSV40 vectors approach both of these characteristics: their extremely high titers ($>10^{13}$/ml in many cases) are sufficient to allow transduction of a large percentage of cells in targeted organs or systemically; and their lack of immunogenicity allows repeat, and repeated, delivery. rSV40 are extremely efficient and provide permanent transgene expression safely and without detectable diminution over time. Many important applications, from most genetic diseases to protection from infectious diseases (e.g., AIDS) require such versatility.

Chemotherapy of many diseases from cancer to infections requires the use of multiple drugs to which the offending agent has been demonstrated to be sensitive. Because of their high transduction efficiency, rSV40 vectors uniquely can deliver variable combination gene therapy regimens without necessitating devising new vectors for each different combination.

Current collaborations have been aimed at advancing these vectors to clinical applications, and are proceeding apace. Much work remains to be done before we can understand the level of safety and efficacy of recombinant SV40-derived vehicles for gene delivery in humans, and recent events must give pause [48]. We are optimistic that rSV40 vectors, like the field of gene therapy, have a bright future.

Acknowledgements

The authors are deeply indebted to many investigators and colleagues, too numerous to mention. Hundreds, if not thousands, of researchers have studied, and continue to study, the biology of wild type SV40. Without their efforts, we would have been unable to move rSV40 vectors forward as rapidly as we have. Further, thousands of people worldwide have devoted untold energies to trying to make clinical gene therapy a reality. The fact that this has not happened in a practical sense is a reflection of the difficulty of the task to which they have applied their efforts. Their dedication has both guided and instructed us: our efforts could not have been made without them. Collaborators in these studies, and colleagues whose input and help were invaluable include Drs. Scott Cairns, J. Roy Chowdhury, Ling-Xun Duan, Mark Feitelson, Puri Fortes, Harris Goldstein, Roger Pomerantz, the late Nava Sarver, Bernard Sauter, Maria Vera, John A. Zaia and Mark A. Zern. People who have worked in the Strayer laboratory and whose efforts have moved this work forward deserve particular thanks: Luca D'Agostino, Charles Ko, Maria Lamothe, Joe Milano, and Lev Yurgenev, and Drs. Francisco Branco, Sandra Calarota, Geetha Jayan, Rumi Kondo, Barry Morse, Linda Starr-Spires, Danlan Wei and Min Yu. Work described here was supported by NIH grants AI41399, AI48244, AI46253, RR13156 and CA44800.

Abbreviations

α1AT	alpha-1-anti-trypsin
BUGT	bilirubin UDP-glucuronysyl transferase
CFTR	cystic fibrosis transmembrane conductance regulator
CMV-IEP	cytomegalovirus immediate early promoter
enh	enhancer
EP	early promoter
FancA	Fanconi A gene
galC	galactocerebrosidase
gfp	green fluorescent protein
HBsAg	hepatitis B surface antigen
IU	infectious units
LP	late promoter

LTR	long terminal repeat
MHC-I	major histocompatibility antigen, type I
MOI	multiplicity of infection (virus:cell ratio)
ori	origin of replication
PCR	polymerase chain reaction
pRb	protein product of the retinoblasoma gene
rfp	red fluorescent protein
RSV	Rous sarcoma virus
Rz	ribozyme
SCID	severe combined immune deficient
ses	SV40 encapsidation sequences
siRNA	small inhibitory RNA (also called RNAi)
Tag	large T antigen
tag	small t antigen
TAR	Recognition element for HIV protein Tat
VA	Adenovirus VA promoter (pol III)
wt	wild type

References

1. Cavazzana-Calvo, M., Hacein-Bey, S., de Saint Basile, G., Gross, F., Yvon, E., Nusbaum, P., Selz, F., Hue, C., Certain, S., Casanova, J.-L., Bousso, Pl, Le Deist, F., Fischer, A. (2000) Science 288, 669–672.
2. Isner, J.M. (2002) Nature 415, 234–239.
3. Fried, M. and Prives, C. (1986) Cancer Cells 4, 1–16.
4. Oppenheim, A., Sandalon, Z., Peleg, A., Shaul, O., Nicolis, S., and Ottolenghi, S. (1992) J. Virol. 66, 5320–5328.
5. Cole, C.N. (1996) In: Fields, B.N. and Howley, P.M., et al., Fields Virology, Lippincott-Raven, Philadelphia, pp. 1997–2025.
6. Sweet, B.H. and Hilleman, M.R. (1960) Proc. Soc. Exp. Biol. Med. 105, 420–427.
7. Breau, W.C., Atwood, W.J., and Norkin, L.C. (1992) J. Virol. 66, 2037–2045.
8. Stang, E., Kartenbeck, J., and Parton, R.G. (1997) Mol. Biol. Cell 8, 47–57.
9. Norkin, L.C. (1999) Immunol. Rev. 168, 13–22.
10. Pelkmans, L., Kartenbeck, J., and Helenius, A. (2001) Nat. Cell Biol. 3, 473–483.
11. Hummeler, K., Tomassini, N., and Sokol, F. (1970) J. Virol. 6, 87–93.
12. Frost, E. and Bourgaux, P. (1975) Virology 68, 245–255.
13. Mackay, R.L. and Consigli, R.A. (1976) J. Virol. 19, 620–636.
14. Lane, D.P. and Crawford, L.V. (1979) Nature 278, 261–263.
15. DeCaprio, J.A., Ludlow, J.W., Figge, J., Shew, J.Y., Huang, C.M., Lee, W.H., Marsilio, E., Paucha, E., and Livingston, D.M. (1988) Cell 54, 275–283.
16. Myers, R.M. and Tjian, R. (1980) Proc. Natl. Acad. Sci. USA 77, 6491–6495.
17. Tornow, J., Polvino-Bodnar, M., Santangelo, G., and Cole, C.N. (1985) J. Virol. 53, 415–424.
18. Bullock, P., Forrester, W., and Botchan, M. (1984) J. Mol. Biol. 174, 55–84.
19. Yano, O., Hirano, H., Karasaki, Y., Higashi, K., Nakamura, H., Akiya, S., and Gotoh, S. (1991) Cell Struct. Funct. 16, 63–71.
20. Gilbert, D.M. and Cohen, S.N. (1990) Mol. Cell. Biol. 10, 4345–4355.
21. Butel, J.S. and Jarvis, D.L. (1986) Biochim. Biophys. Acta 865, 171–195.
22. Gooding, L.R. (1977) J. Immunol. 118, 920–927.

23. Parks, C.L. and Shenk, T. (1997) J. Virol. 71, 9600–9607.
24. Sandalon, Z., Dalyot-Herman, N., Oppenheim, A.B., and Oppenheim, A. (1997) Hum. Gene Ther. 8, 843–849.
25. Sandalon, Z. and Oppenheim, A. (1997) Virology 237, 414–421.
26. Fang, B., Koch, P., Bouvet, M., Ji, L., and Roth, J.A. (1997) Anal. Biochem. 254, 139–143.
27. Rund, D., Dagan, M., Dalyot-Herman, N., Kimchi-Sarfaty, C., Schoenlein, P.V., Gottesman, M.M., and Oppenheim, A. (1998) Hum. Gene Ther. 9, 649–657.
28. Dalyot-Herman, N., Rund, D., and Oppenheim, A. (1999) J. Hematother. Stem Cell Res. 8, 593–599.
29. Strayer, D.S. (1996) J. Biol. Chem. 271, 24741–24746.
30. Strayer, D.S. and Milano, J. (1996) SV40 mediates stable gene transfer *in vivo*. Gene Ther. 3, 581–587.
31. Gluzman, Y. (1981) Cell 23, 175–182.
32. Strayer, D.S., Duan, L.-X., Ozaki, I., Milano, J., Bobraski, L.E., and Bagasra, O. (1997) BioTechniques 22, 447–450.
33. Sauter, B.V., Parashar, B., Chowdhury, N.R., Kadakol, A., Ilan, Y., Singh, H., Milano, J., Strayer, D.S., and Chowdhury, J.R. (2000) Gastroenterology 119, 1348–1357.
34. Goldstein, H., Pettoello-Mantovani, M., Anderson, C.M., Cordelier, P., Pomerantz, R.J., and Strayer, D.S. (2002) J. Infect. Dis., in press.
35. Strayer, D.S., Branco, F., Zern, M.A., Yam, P., Calarota, S., Nichols, C.N., Zaia, J.A., Rossi, J., Li, H., Parashar, B., Ghosh, S., and Chowdhury, J.R. (2002) Mol. Ther., in revision.
36. Yang, Y., Li, Q., Ertl, H.C., and Wilson, J.M. (1995) J. Virol. 69, 2004–2015.
37. Engelhardt, J.F., Litzky, L., and Wilson, J.M. (1994) Hum. Gene Ther. 5, 1217–1229.
38. Dedieu, J.F., Vigne, E., Torrent, C., Jullien, C., Mahfouz, I., Caillaud, J.M., Aubailly, N., Orsini, C., Guillaume, J.M., Opolon, P., Delaere, P., Perricaudet, M., Yeh, P. (1997) J. Virol. 71, 4626–4637.
39. Kondo, R., Feitelson, M.A., and Strayer, D.S. (1998) Gene Ther. 5, 575–582.
40. Strayer, D.S. and Zern, M.A. (1999) Sem. Liver Dis. 19, 71–81.
41. McKee, H.J. and Strayer, D.S. (2002) Vaccine, in press.
42. McKee, H.J., Vera, M., Fortes, P., and Strayer, D.S. (2002) Proc. 5th. Annu. Conf. Vaccine Res. 5, 51.
43. BouHamdan, M., Duan, L.-X., Pomerantz, R.J., and Strayer, D.S. (1999) Gene Ther. 6, 660–666.
44. BouHamdan, M., Strayer, D.S., Wei, D., Mukthar, M., Duan, L.-X., Hoxie, J., and Pomerantz, R.J. (2001) Gene Ther. 8, 408–418.
45. Strayer, D.S., Branco, F., Landré, J., BouHamdan, M., Shaheen, F., and Pomerantz, R.J. (2002) Mol. Ther. 5, 33–41.
46. Zern, M., Ozaki, I., Duan, L.-X., Pomerantz, R., Liu, S.-L., and Strayer, D.S. (1999) Gene Ther. 6, 114–120.
47. Jayan, G.C., Cordelier, P., Patel, C., BouHamdan, M., Johnson, R.P., Lisziewicz, J., Pomerantz, R.J., and Strayer, D.S. (2001) Gene Ther. 8, 1033–1042.
48. Marshall, E. (2000) Science 287, 973–974.
49. Fried, M. and Prives, C. (1986) Cancer Cells 4, 1–16.
50. Rosenberg, B.H., Deutsch, J.F., and Ungers, G.E. (1981) Growth and purification of SV40 virus for biochemical studies. J. Virol. Methods 3, 167–176.

S.C. Makrides (Ed.) *Gene Transfer and Expression in Mammalian Cells*

Virus-based vectors for gene expression in mammalian cells: Adeno-associated virus

Xiao Xiao

Department of Molecular Genetics and Biochemistry, University of Pittsburgh School of Medicine,
Room W1244 BST, Pittsburgh, PA 15261, USA; Tel.: +1 (412) 648-9487; Fax: +1 (412) 648-9610.
E-mail: xiaox@pitt.edu

1. Biology of adeno-associated virus

Adeno-associated virus (AAV) is a non-pathogenic and replication-defective member of the parvoviridae family. It is a small, non-enveloped DNA virus (Fig. 1) consisting of a single-stranded genome of approximately 4.7 kilobases (kb) and an icosahedron protein coat [1]. Currently, up to eight different serotypes of AAVs have been isolated from primates [2–11]. AAV has no etiological association with any known diseases. Its propagation requires the co-infection of an unrelated virus, such as adenovirus or herpesvirus, to provide essential helper functions. In the absence of a helper virus, however, AAV establishes a latent infection by integrating its DNA into the host cell chromosomes or maintaining its genome as free episomal DNA. Its latent infection life cycle provides an important basis for the use of AAV as a stable gene transfer vector, while its non-pathogenicity and replication defectiveness offer additional safety features for *in vivo* gene therapy applications.

AAV virion has a diameter about 20 to 22 nm and a buoyant density of 1.39–1.41 g/cm^3 in CsCl gradient. The viral particle is highly stable and resistant to certain proteases, detergents and organic solvents. Exposure to repeated freeze-and-thaw, lyophilizing, heating up to 56 °C and pH changes from 3 to 9 does not affect the viral infectivity. These characteristics are important for the prolonged shelf life of AAV vector as a "gene drug". Each AAV contains only one copy of a single-stranded linear DNA genome, which is packaged either as a positive strand or a negative strand with equal infectivity [12]. Since single-stranded DNA is transcriptionally inactive, the conversion of single-stranded viral genome to double-stranded DNA template is considered a rate-limiting step for AAV gene expression as well as for AAV-mediated transgene expression [13]. The viral DNA is encapsidated by sixty molecules of AAV capsid proteins, i.e. VP1 (87 kD), VP2 (73 kD) and VP3 (61 kD). These three proteins overlap extensively in their amino acid sequences and differ only at their N-termini. Such overlap makes it difficult to genetically modify one capsid protein without affecting the others. VP3 is the most abundant coat protein and accounts for about 90% of the protein mass. AAV particles can be readily purified to high titers and high purity either by density gradient centrifugation or by chromatography techniques.

The AAV genome is made of three essential components: *Rep* (replication) gene, *Cap* (capsid) gene and inverted terminal repeats (ITR) (Fig. 2). The non-structural

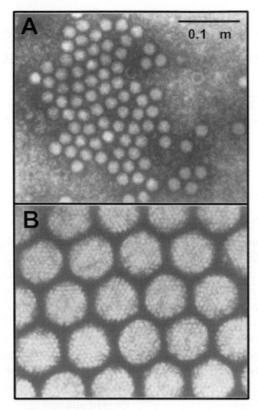

Fig. 1. Electron microscopy of AAV2 (panel A) and its helper virus adenovirus 2 (panel B). Note the few empty AAV particles that lack viral DNA genome and show a ring-like image. The black bar shows the scale of 0.1 µm.

protein gene *Rep* is primarily responsible for the functions of viral replication and gene regulation. The *Rep* gene encodes four overlapping non-structural proteins, Rep78, Rep68, Rep52, and Rep40, according to their respective molecular weights in kilo-daltons. Rep78 and Rep68 proteins are synthesized respectively from the unspliced and the spliced transcripts of promoter p5. Similarly, the two smaller proteins Rep52 and Rep40 are the products of the unspliced and spliced transcripts of promoter p19. Like the capsid proteins, all Rep proteins share the same C-terminus as well as a major portion of their internal protein sequences. AAV Rep proteins exhibit multiple functions. Besides regulating the gene expression of AAV and its host cells and helper viruses, the primary functions of the Rep proteins are to mediate the replication of AAV DNA [14] as well as to facilitate the packaging of viral genomes [15]. Rep78 and Rep68 exhibit the following activities that are directly involved in AAV DNA replication as well as the process of DNA packaging: (1) *sequence-specific binding activity* to ITR, the origin of AAV DNA replication; (2) *site-specific and strand-specific endonuclease activity* on the terminal resolution site (trs) of the ITR; (3) *Helicase activity* of three Rep proteins (Rep78, Rep68 and

A

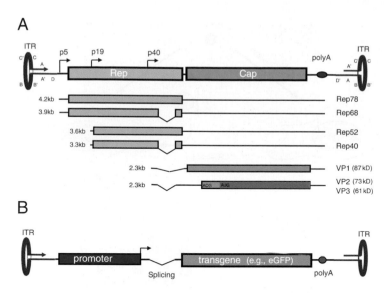

B

Fig. 2. (A) The genome structure and gene products of AAV. A wild-type AAV genome contains two ITRs that flank two genes, the replication gene (*Rep*) and the viral coat protein gene (*Cap*). These two genes are driven by three promoters respectively located at map positions 5, 19 and 40. The ITR can form a T-shaped palindromic structure containing three stems (A & A′; B & B′ and C & C′). The non-palindromic D and D′ sequences are respectively localized on either side of the genome. The 3′ ITR can serve as a primer for the synthesis of the second-strand DNA. (B) The genome structure of a recombinant AAV vector. The only DNA sequence from the AAV is the 145-bp ITR that flanks the transgene cassette including a promoter, an RNA splicing site, the transgene-coding sequence (for example, the enhanced green fluorescent protein, eGFP) and a polyadenylation site. The total size of an AAV vector DNA from ITR to ITR cannot surpass the 5-kb limit for efficient packaging.

Rep52); and (4) *ATPase activity*. On the other hand, the three viral structural proteins VP1, VP2 and VP3 are the products of the *Cap* gene controlled by promoter p40, which is situated within the coding sequence of the *Rep* gene. The capsid proteins VP2 and VP3 are translated from the same spliced mRNA using different start codons. VP2 uses a less efficient upstream ACG codon, while VP3 uses a robust downstream AUG start codon for more efficient protein synthesis (see Chapter 16). The VP1 protein, however, is synthesized from an alternative spliced mRNA (Fig. 2).

The ITR is approximately 145 base pairs (bp) in length, and is located on both ends of the single-stranded viral genome (Fig. 2). The ITR is the only *cis*-acting sequence required for AAV DNA replication (origin), packaging, chromosomal integration and provirus rescue (a reverse process of integration) [16]. However, the ITR is not required for high-level expression of viral *Rep* and *Cap* genes in a productive life cycle, although the ITR displays specific enhancer functions for AAV promoters. Each ITR can form a T-shaped hairpin structure through its A, A′, B, B′, C, C′ palindromic sequences. The ITR at the 3′ end of the viral genome serves as a primer for AAV DNA replication, while the non-palindromic D sequence of the ITR contains viral DNA packaging signals. ITRs from different serotypes of AAV all

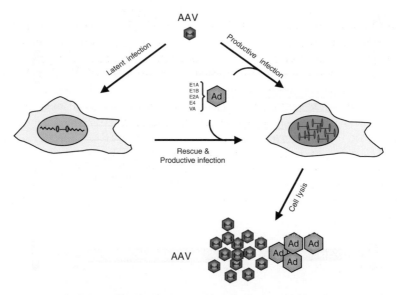

Fig. 3. The two life cycles (latent and productive) of AAV. In the presence of a helper virus, such as an adenovirus, AAV infects the host cells and assumes a productive life cycle. The adenovirus provides five genes for essential helper functions. In the absence of a helper virus, however, AAV establishes a latent infection by integrating its viral DNA into the chromosomes or maintaining it as an episome. The AAV can re-enter a productive life cycle from the latently infected cells upon superinfection of a helper virus.

display similar T-shaped secondary structures despite significant differences in DNA sequences.

AAV has both latent and productive life cycles (Fig. 3). In the absence of a helper virus, AAV proceeds to establish a latent infection in the host cells. The viral particle binds to and enters the cells through a primary receptor, identified as the heparan sulfate proteoglycan for AAV2 [17], and as the sialic acid moiety for AAV5 [18]. A few co-receptors have also been identified such as integrin $\alpha V\beta 5$ and fibroblast growth factor receptor [19,20]. After entry into the host cell, the viral particle migrates into the nucleus, uncoats and releases its viral DNA. The latter is then converted into double-stranded genome either by second-strand synthesis [13] or by annealing of the plus and the minus strands of viral DNA molecules [21]. The viral genomes persist in the host cells either as free episomes or as integrated proviruses in the chromosomes. The integration by wild-type AAV2 DNA has site-specificity on the q13.3-qter region, termed AAVS1, on human chromosome 19 [22,23]. The site-specific integration is apparently mediated by the AAV Rep proteins, which bind to AAV ITR as well as to a consensus sequence on the chromosomal target site [24]. During co-infection of a helper virus such as adenovirus, however, AAV enters the productive life cycle. Its viral genes, both *Rep* and *Cap*, are highly activated, leading to efficient viral DNA replication and packaging into the preformed empty particles. Similarly, the latently infected AAV provirus can also be activated and proceed to the productive life cycle upon super-infection of a helper virus. This process is termed "rescue." It is an important mechanism not only for the wild-type AAV to propagate from the latently infected cells, but also for the vector DNA to propagate in the packaging cell lines under appropriate conditions. Numerous unrelated viruses, e.g., adenovirus, herpes virus, vaccinia virus and papilloma virus, can all serve as a helper virus for AAV propagation with variable efficiencies. Genotoxic stress to the host cells also triggers AAV replication in a helper-virus independent manner. AAV DNA replication uses the host cellular DNA replication machinery. Helper viruses provide helper functions by making a more suitable cellular environment rather than being directly involved in AAV propagation. The most commonly used helper viruses are adenovirus (Ad) and herpes simplex virus (HSV). Five Ad genes, *E1A, E1B, E4, E2A* and VA RNA, are sufficient to provide full helper functions for the AAV productive life cycle [25].

2. Production of recombinant AAV vectors

Recombinant AAV vectors can be produced at very high titers (more than 10^{13} viral particles/ml) by several different methods. All these methods share the same three essential components: (1) AAV vector component that contains the foreign transgene(s) flanked by the 145-bp AAV ITRs. (2) AAV *Rep* and *Cap* genes that provide non-structural and structural proteins for vector DNA replication and packaging. However, these genes themselves cannot be packaged due to the lack of the *cis*-acting ITRs. (3) Helper functions from adenovirus or herpesvirus that facilitate efficient AAV propagation. When the three components above are introduced either transiently or stably into a suitable host cell, such as human 293 or HeLa cells, the AAV vector DNA is replicated and packaged into viral particles.

Currently, a widely used AAV vector production method is the helper virus-free triple-plasmid transfection method [26] (Fig. 4). It uses an AAV vector plasmid containing the transgene flanked by ITRs, a packaging plasmid containing *Rep/Cap* genes, and a helper plasmid containing a few essential helper genes from adenovirus. These three plasmids are co-transfected into human 293 cells to generate high-titer AAV vectors completely free of helper adenovirus contamination. Since the packaging plasmid and the Ad helper plasmid are common components and suitable for any specific vector plasmid constructs, such a production method is versatile, productive, and cost-effective for laboratory applications, particularly when a large number of different vectors are needed. However, the transfection step is a rate-limiting factor that makes it inconvenient for large-scale vector production for clinical gene therapy applications. An alternative AAV vector production method is the use of packaging cell lines that harbor both the AAV vector DNA and the AAV *Rep/Cap* genes. After infection with wild-type adenovirus, the cell lines produce high-titer AAV vectors [27]. Similarly, other packaging cell lines harboring only the AAV vector component, when infected with a helper herpesvirus that carries the AAV *Rep/Cap* genes, can also efficiently produce AAV vectors [28]. Although, cost effective for large-scale vector production, the packaging cell line methods are time

98

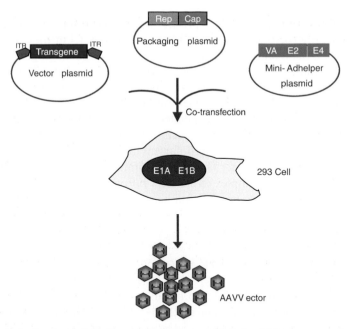

Fig. 4. The helper-virus-free, triple-plasmid transfection method for AAV vector production. After co-transfection, only the recombinant AAV vectors will be produced. Three of the five essential adenovirus helper genes are provided by the helper plasmid (VA RNA, E2 and E4 genes), while the other two adenovirus genes (E1A and E1B) are located in the host 293 cells. There is no adenovirus DNA replication or packaging involved.

consuming, especially when multiple vectors are involved. In addition, these methods inevitably generate pathogenic helper viruses that contaminate the AAV vector stocks, although the helper adenovirus or herpesvirus could be removed during purification. AAV vector purification and concentration can be achieved according to the physical and biochemical properties of AAV viral particles. A commonly used method is the CsCl or the non-ionic iodixanol [29] density gradient centrifugation, which is based on the buoyant density of the viral particles. This method can separate the full particles from the empty particles. On the other hand, chroma-tographic methods are used to purify AAV by their biochemical properties, such as surface charge, the specific binding activities and virion size. Chromatography is less time-consuming and yields higher purity. For example, heparan sulfate-based affinity chromatography is an effective purification method for AAV serotype-2 [29], while sialic acid ligand is used for the affinity purification of AAV serotype 5. Other chromatographic methods have also been developed. A combination of density gradient centrifugation and chromatography can remove empty particles and obtain high purity. High-titer and high-quality AAV vectors offer more defined gene transfer reagents for both preclinical and clinical gene therapy studies.

The titers of recombinant AAV vectors can be determined by different methods with different titer definitions. Unlike the lytic adenovirus and herpesvirus, AAV infection does not lead to plaque formation. Therefore, no such titer as plaque forming unit (pfu) exists for either wild-type AAV or an AAV vector. The commonly used definitions for recombinant AAV vector titers are as follows:

(a) *Viral particle titer* usually indicates the number of virions containing the vector DNA genome, regardless of their infectivity and functionality. The viral particle titer is determined either by DNA hybridization or by quantitative PCR analysis. This titer is also referred to as *Dnase-resistant particles (Drp) or viral genome (vg) numbers*. A less frequently used *viral particle titer* reflects the number of total viral physical particles, both full (DNA-containing) and empty (without DNA). Such titer is determined by an ELISA assay using a monoclonal antibody that recognizes all assembled AAV-2 viral particles.

(b) *Infectious unit (i.u.) titer* refers to the number of virions capable of infecting the host cells and ready to replicate if AAV *Rep* genes and Ad helper genes are simultaneously provided. This titer, however, does not reflect the activity of transgene expression. It is usually determined by the *infectious center assay* (also called *replication center assay*), where each viral DNA molecule is amplified after replication in individual host cells. The cells are then transferred to a nylon membrane. The replicated vector DNA in each single cell is detected and quantitated by hybridization with vector-specific probes.

(c) *Transducing unit (t.u.) titer* refers to the number of virions capable of delivering the transgene functions into the host cells. For example, one transducing unit of an AAV vector that contains the green fluorescent protein (GFP) gene imparts fluorescence to the host cell. However, different host cells give rise to different apparent titers. A commonly used indicator host cell is 293, which is co-infected with adenovirus at low multiplicity of infection (m.o.i.). The latter expedites and intensifies the transgene expression of AAV vectors for easy detection. HeLa cell is another commonly used indicator cell line. Generally, for a given vector preparation, the highest numerical titer is the viral particle titer, followed in quantity by infectious unit titer and then transducing unit titer.

3. In vitro and in vivo gene transfer

AAV vectors have been increasingly used for gene transfer and gene therapy, particularly *in vivo*, because of the safety, high efficiency and long-term gene transfer by the vector system. The AAV vector is the only viral vector system derived from a non-pathogenic and defective virus. Furthermore, the vector does not contain any AAV viral genes. It also lacks cytotoxicity to the host cells even with very high m.o.i. The sole AAV DNA sequence in the vector is the 145-bp ITR. The deletion of all the AAV genes renders additional safety features preventing immune and other complications caused by undesirable viral gene expression. Another advantage of AAV is its broad host range, capable of infecting a wide variety of cells from different tissues, including lung, brain, spinal cord, liver, cardiac and skeletal muscles, eye, guts, joints, bone marrow, islets, hair follicles, etc. The vectors can be readily used to infect cells *in vitro* or directly injected *in vivo* for gene transfer in both dividing and non-dividing cells. The *in vivo* delivery methods can be either local injection [30] or systemic administration through the blood circulation [31]. Unlike the retrovirus long terminal repeats, the AAV ITR has minimal transcriptional activities. AAV vectors have been successfully equipped with tissue-specific promoters [32] or inducible promoters [33] for regulated transgene expression. Furthermore, AAV vectors render persistent gene transfer by integration into the host cell chromosomes, or by maintenance as episomal molecules [34,35]. Finally, the apparent lack of cellular immune response, i.e., the cytotoxic T lymphocyte (CTL) reaction against AAV vector-transduced cells, also contributes to the long-term *in vivo* gene transfer [30].

However, AAV vectors have several intrinsic limitations. First, because AAV is derived from one of the smallest DNA viruses, its size limitation prevents the packaging of genes larger than 4.7 kb. Second, the AAV vector DNA is single-stranded requiring conversion to double-stranded template in order for the transgenes to be expressed. This process is particularly slow *in vivo* and often needs weeks to reach a plateau of transgene expression. Third, AAV integration is an inefficient process, particularly *in vivo*. Therefore, rapid cell division can lead to the dilution and eventually the loss of the vector DNA before it integrates into the chromosomes. However, in non-dividing cells, AAV vector DNA can persist for a prolonged period (months and even years) without integration. Furthermore, the recombinant AAV vectors have lost the characteristic of wild-type AAV for site-specific integration on human chromosome 19. Random integration may pose a potential risk of mutagenesis, although the probability for this is very small.

AAV vectors are primarily used to infect various tissues *in vivo*. Although, all major organs/tissues and cell types have been investigated with AAV vectors carrying different genes, extensive studies have been done on such tissues as lung [36], central nervous system (CNS) [37], muscle [30] and liver [38]. One of the pioneer studies is the use of an AAV vector that carried the cystic fibrosis transductance regulator (*CFTR*) gene for the treatment of cystic fibrosis in the respiratory tract [36]. An early example of AAV vectors stably infecting non-dividing cells *in vivo* involved the brain, where AAV vectors carrying the *Escherichia coli* β-galactosidase reporter gene (*LacZ*) or a tyrosine hydroxylase (TH) therapeutic gene effectively infected neurons. The AAV-TH vector successfully achieved therapeutic efficacy in the Parkinson's rat and monkey models [37,39,40]. Numerous studies involving growth factor genes and other therapeutic genes also showed prolonged gene transfer in various areas of the brain and spinal cord and demonstrated neuro-protective effects. AAV-mediated gene delivery into CNS has corrected enzymatic deficiency in the brain. It is worth mentioning that AAV vectors show much higher affinity to neurons than to other cell types in the brain. Stereotactic injection of AAV into brain usually leads to the transduction of neurons. This phenomenon could be partially explained by the abundance of AAV receptors on neurons [41] and also by the promoters of choice.

Muscle, a largely non-dividing and easily accessible tissue, is another favorite target of AAV-mediated gene therapy, not only for the treatment of genetic muscle diseases [42,43], but also for the treatment of metabolic deficiencies. In early studies, direct intramuscular injection of an AAV-*LacZ* vector rendered highly efficient and long-term (about 2 years) transgene expression. Thus far, AAV is the most efficient vector system in muscle tissues. One of the major reasons is the small viral particle size that can readily bypass the extracellular matrix (ECM) barrier on mature myofibers. This ECM barrier is a rate-limiting step in muscle for viruses with large particle sizes. In addition, the expression of a foreign protein, e.g., *LacZ*, in the muscle of immunocompetent animals, did not trigger cell-mediated immune response against the vector-transduced cells [30]. This phenomenon occurred in a vast majority of the *in vivo* gene transfer studies using AAV vectors containing various transgenes and involving different target tissues. Based on the promising results from reporter genes in muscle, therapeutic genes for muscular dystrophies have also been packaged into AAV vectors to treat these devastating diseases in animal models [31,42,43]. AAV vectors either carrying a human sarcoglycan gene or miniature versions of human dystrophin gene have achieved long-term therapeutic efficacy in a limb girdle muscular dystrophin hamster (*Bio14.6*) model and a Duchenne muscular dystrophy mouse (*mdx*) model, respectively [31,42,43]. A human clinical trial for limb girdle muscular dystrophy caused by alpha-sarcoglycan deficiency was performed using AAV as the vector system. Cardiac muscle can also be efficiently transduced with AAV vectors either through coronary circulation or through direct intracardiac injection [44]. In addition, muscle tissue is also used as a platform to produce and secrete therapeutic proteins for metabolic diseases. For example, AAV-mediated muscle gene transfer of clotting factor IX has been successfully investigated not only in small and large animal models, but also in a human clinical trial for hemophilia B [45]. AAV-mediated muscle gene transfer of alpha-1 antitrypin (AAT) has also succeeded in small animals to achieve extremely high-levels of AAT in blood circulation [46]. Human clinical trials will be performed in the near future (T. Flotte, personal communication).

The liver is another major target organ of gene therapy with AAV vectors, primarily for metabolic diseases such as enzyme deficiencies. The liver is also utilized as a platform for the production of other therapeutic gene products into the blood circulation. Like muscle, the liver is largely made of non-dividing cells. AAV vector genomes are maintained as stable episomal DNA. AAV vectors administered systemically into the blood circulation are primarily taken up by the liver, particularly in hepatocytes. Vector enrichment in the liver may be due to the nature of the liver as a metabolic sink and its large fenestration. Vector delivery through the portal vein or liver arterial circulation increases local vector concentration and enhances transduction efficiency. A major form of AAV vector genomes in the liver are maintained as stable episomal DNA without chromosomal integration [47]. Numerous success stories of AAV-mediated liver gene therapy have been reported. For example, an AAV vector carrying beta-glucuronidase gene, when delivered into blood circulation and infected liver cells, essentially cured the MPS VII mouse model of Sly disease, a childhood genetic lysosomal storage disease caused by beta-glucuronidase deficiency [48]. AAV-mediated liver gene transfer of clotting factor IX gene has also achieved profound therapeutic efficacy in both small and large animal models of hemophilia B [49]. A liver-directed hemophilia B human clinical trial using AAV factor IX vector is ongoing. All human clinical trials with AAV vectors have shown no detectable toxicity and apparent side effects, demonstrating that AAV is a safe *in vivo* gene therapy vector system.

4. Immunological aspects of AAV vectors

The immunology of AAV vector-mediated *in vivo* gene transfer is intriguing. As expected, direct *in vivo* administration of AAV vectors causes humoral immune response to the AAV coat proteins simply because they are foreign antigens. However, AAV vector-mediated *in vivo* gene transfer usually does not trigger cellular immune responses, i.e., the CTL reaction, against the vector-transduced cells that express the foreign transgene products. It is understandable that AAV gene delivery into the brain does not elicit an immune response because the CNS is considered to be an immune-privileged site. However, vector injection into the peripheral tissues also fails to trigger a CTL reaction against the vector-transduced cells that express the foreign protein. One of the best examples is the intramuscular injection of AAV vectors carrying the *LacZ* gene into immunocompetent mice. High levels of *LacZ* expression in muscle did not cause CTL responses against *LacZ*-expressing myocytes [30]. Additionally, intravenous or intraportal vein administration of AAV vectors in the liver rendered long-term transduction of liver cells without cellular immune responses. These phenomena occurred in a vast majority of the *in vivo* gene transfer studies using AAV vectors containing various transgenes and involving different target tissues. A mechanistic study has suggested that AAV vectors could not transduce the professional antigen presenting cells (APC). As a result, the host immune system failed to detect the transgene products, thus leading to immune ignorance [50]. Adoptive T-cell transfer or injection into the host animal of an adenovirus that contains the same transgene as the AAV vector would trigger a CTL immune response and clear the AAV-transduced muscle cells. The slow onset and the lack of sudden robust transgene expression from AAV vectors may also contribute to the immune ignorance. Another interesting phenomenon is the segregation of humoral and cellular immune response against the transgene products. Intramuscular injection of an AAV vector containing human clotting factor IX into hemophiliac dogs triggered very strong antibody reactions against the secreted human protein, but failed to trigger a CTL response against the muscle cells producing human factor IX from the vector transgene [51]. Intraportal injection of an AAV vector containing human factor VIII gene into the mouse liver also triggered strong antibody reaction for up to 10 months. Interestingly, the antibody titer diminished to undetectable levels after 10 months, while the human factor VIII protein became detectable again in the blood, suggesting the absence of a CTL reaction against the liver cells producing the human protein, and the eventual immune tolerance by the host [52].

However, whether a CTL immune response will be elicited by an AAV vector and to what extent, is also dependent on the nature of the transgene products, the route of administration, the immune alertness of the target tissue and the tissue-specificity of the promoters. For instance, an AAV vector carrying a gene of human papillomavirus (HPV) 16 E7 peptide fused with the tuberculosis bacterial hsp70, a strong antigen as a molecular adjuvant, induced strong CTL immune responses intramuscularly and systemically [53]. The route of vector administration is also important. An AAV vector containing a strong secretable antigen (chicken ovalbumin), when injected into immunocompetent mice intraperitoneally, intravenously, or subcutaneously, developed potent ovalbumin-specific CTLs as well as anti-ovalbumin antibodies. However, intramuscular injection of the same vector only developed strong humoral but minimal CTL response to the chicken protein [54]. A target tissue that has pre-existing inflammation or an elevated immune alertness is also capable of triggering a CTL response to an AAV vector, which does not trigger such immune response in normal tissues. For example, an AAV-*LacZ* vector injection into healthy young or adult muscle would not cause a CTL response. However, the same vector when injected into the mdx mouse muscle, which suffers muscular dystrophy and chronic inflammation, triggered a mild CTL response against the *LacZ*-positive myofibers. Similarly, injection of AAV-*LacZ* vector into the acutely damaged muscle after cardiotoxin treatment also caused a mild CTL response.

By contrast, AAV vectors carrying human mini-dystrophin gene, a therapeutic gene, did not cause any CTL immune response against the human protein in the mdx mouse muscle [55]. Furthermore, tissue-specific promoters usually minimize the CTL responses because they minimize the non-specific transgene expression in antigen-presenting cells [56]. Finally, previous immunology studies, particularly the cellular immune responses, were almost exclusively based on AAV vectors that are derived from AAV serotype 2. Recently, up to eight different serotypes of AAV have been identified and their derivative vector systems have been developed. It remains to be elucidated if AAV vectors of different serotypes exhibit different capacities to infect antigen-presenting cells and display different cellular immunogenicity. However, it is well known that *in vivo* delivery of any AAV vectors would elicit strong antibody reactions against the AAV coat proteins. These neutralizing antibodies make it difficult to repeatedly deliver the transgene with the vector of the same serotype. However, repeated gene delivery can be achieved with a different serotype of AAV vectors. Transient immune suppression during vector administration could also minimize the antibody reactions and make repeated delivery possible.

5. Development of new AAV vectors

Compared to other viral vector systems, such as those based on retrovirus and adenovirus, the AAV vector is relatively new with potential for further development. As discussed earlier, an intrinsic limitation of AAV is the packaging size that excludes the foreign DNA insertion larger than 4.7 kb. Genes with large coding sequences and promoters with large regulatory elements could not be easily packaged into AAV vectors. Numerous strategies have been developed to overcome this shortcoming. One strategy is to streamline and truncate the target gene to create a "minigene" that can fit AAV packaging capacity. One such example is the truncation of the dystrophin gene, the largest gene so far identified. The dystrophin gene has a genomic sequence of 2.5 million bp, a mRNA of 14 kb and a protein-coding sequence of 11 kb. Novel minigenes that are smaller than 4 kb have been successfully created and demonstrated to be highly functional for treating Duchenne muscular dystrophy in an mdx mouse model [43]. The use of B-domain-deleted clotting factor VIII is another such example. However, for most of the large genes that are difficult to truncate, alternative strategies are required. A recent new technology termed the split AAV vectors (SAVE), based on the mechanism of AAV vector heterodimer formation, has doubled the vector packaging size by joining two separate AAV vectors that each packages half of the gene expression cassette [57,58]. Another new development is the use of two AAV vectors that package DNA inserts with overlapping sequences. Homologous recombination through the overlapping sequences join the two vectors into an intact gene expression cassette [59].

To overcome the problem of repeated vector delivery and to use an AAV other than AAV2 with different cell- and tissue-tropism, numerous serotypes (type 1–8) of AAV have been identified. Characterizations of vectors derived from new serotypes have shown that they display different tissue-tropisms. For example, AAV1 is much more efficient than AAV2 in skeletal muscle, liver and CNS, while AAV5 and AAV4 are superior in rat retina [60]. AAV8 is the most efficient serotype in liver [11]. While AAV2 renders localized gene delivery in the CNS, AAV5 exhibits a more widespread infection in a much larger area in the brain [61]. AAV1, 5 and 6 also showed much more efficient gene delivery into the airway epithelial cells than the AAV2 vectors [59,62]. Initially, AAV vectors of different serotypes contained their own specific ITRs and Rep and Cap genes, making it inconvenient to compare the same transgene in different vectors. To overcome this problem, pseudotyping or cross-packaging techniques were developed [60], where the vector plasmid is based on AAV2 ITR and the packaging plasmid contains the Rep genes from AAV2, but is coupled with specific Cap genes from individual serotypes. Thus, a single vector plasmid can be cross-packaged by different packaging plasmids, making it more effective to compare different serotypes with the same vector DNA.

Another drawback of AAV vector-mediated gene transfer is the slow onset of transgene expression that often takes weeks to plateau. This hurdle limits the use of AAV vectors for certain applications that require prompt transgene expression. The slow process is mainly due to the single-stranded nature of the AAV vector genome. The conversion of single-stranded to double-stranded DNA is one of the major rate-limiting steps in AAV-mediated gene transfer [13]. Another rate-limiting step involves the intracellular trafficking of the vector DNA [63]. A number of chemical and physical factors that may help to overcome these rate-limiting steps have been explored. The most frequently used intervention is the genotoxic and stress treatment of the target cells to expedite the second-strand DNA synthesis [13,64]. Methods that facilitate the intracellular trafficking and inhibit virus particle degradation also greatly enhance AAV vector transduction efficiency [63].

Finally, AAV vectors that package a hairpin genome—a double-stranded or self-complimentary DNA—exhibit not only stronger but also much faster transgene expression *in vivo* and *in vitro* [65]. The double-stranded AAV vectors also enable efficient transduction of certain cell types (e.g., 3T3 fibroblast and B16 melanoma), which are otherwise refractory to transduction by the typical single-stranded AAV vectors (Xiao *et al.*, unpublished observations).

Targeted gene transfer is a major goal in AAV vector development. Targeting includes: (1) targeted infection into desirable cell and tissue types while avoiding non-specific uptake of vector; and (2) targeted integration of AAV vectors into the specific site (AAVS1) on human chromosome 19 while avoiding non-specific insertions. To achieve targeted infection, the AAV viral particles are either indirectly modified by biochemical methods, such as the cross-linking of specific binding ligands, e.g., antibodies [66], or directly modified by genetic engineering of the capsid protein gene [67]. Recently, new technologies have been explored that implement molecular evolution and gene shuffling of the AAV capsid protein to create novel serotypes with new tissue tropisms. Efforts have also been made to restore the property of site-specific integration, which is mediated by the AAV Rep proteins Rep78 and Rep68, to the recombinant AAV vectors that lack any AAV viral coding sequences. A variety of tactics have been investigated to re-introduce the *Rep* gene or gene products along with the AAV vectors into the host cells [68]. Innovative technical advancements will undoubtedly greatly enhance the basic as well as translational research on AAV vectors, and expedite the application of AAV vectors in clinical studies.

Acknowledgements

The author would like to thank Mr. Michael Y. Xiao and Dr. Juan Li for their critical reading and suggestions.

Abbreviations

AAV	adeno-associated virus
Ad	adenovirus
Cap	AAV capsid gene
CTL	cytotoxic T lymphocyte
ITR	inverted terminal repeat
i.u.	infectious unit
m.o.i.	multiplicity of infection
Rep	AAV replication gene
t.u.	transducing unit
v.g.	viral genome copy number

106

References

1. Xie, Q., Bu, W., Bhatia, S., Hare, J., Somasundaram, T., Azzi, A., and Chapman, M.S. (2002) Proc. Natl. Acad. Sci. U.S.A. 99, 10405–10410.
2. Hoggan, M.D., Blacklow, N.R., and Rowe, W.P. (1966) Proc. Natl. Acad. Sci. U.S.A. 55, 1467–1474.
3. Clarke, J.K., McFerran, J.B., McKillop, E.R., and Curran, W.L. (1979) Arch. Virol. 60, 171–176.
4. Samulski, R.J., Berns, K.I., Tan, M., and Muzyczka, N. (1982) Proc. Natl. Acad. Sci. U.S.A. 79, 2077–2081.
5. Chiorini, J.A., Yang, L., Liu, Y., Safer, B., and Kotin, R.M. (1997) J. Virol. 71, 6823–6833.
6. Xiao, W., Chirmule, N., Berta, S.C., McCullough, B., Gao, G., and Wilson, J.M. (1999) J. Virol. 73, 3994–4003.
7. Rutledge, E.A., Halbert, C.L., and Russell, D.W. (1998) J. Virol. 72, 309–319.
8. Bantel-Schaal, U., Delius, H., Schmidt, R., and zur Hausen, H. (1999) J. Virol. 73, 939–947.
9. Chiorini, J.A., Kim, F., Yang, L., and Kotin, R.M. (1999) J. Virol. 73, 1309–1319.
10. Hildinger, M., Auricchio, A., Gao, G., Wang, L., Chirmule, N., and Wilson, J.M. (2001) J. Virol. 75, 6199–6203.
11. Gao, G.P., Alvira, M.R., Wang, L., Calcedo, R., Johnston, J., and Wilson, J.M. (2002) Proc. Natl. Acad. Sci. U.S.A. 99, 11854–11859.
12. Samulski, R.J., Chang, L.S., and Shenk, T. (1987) J. Virol. 61, 3096–3101.
13. Ferrari, F.K., Samulski, T., Shenk, T., and Samulski, R.J. (1996) J. Virol. 70, 3227–3234.
14. Im, D.S. and Muzyczka, N. (1990) Cell 61, 447–457.
15. Smith, R.H. and Kotin, R.M. (1998) J. Virol. 72, 4874–4881.
16. Xiao, X., Xiao, W., Li, J., and Samulski, R.J. (1997) J. Virol. 71, 941–948.
17. Summerford, C. and Samulski, R.J. (1998) J. Virol. 72, 1438–1445.
18. Walters, R.W., Yi, S.M., Keshavjee, S., Brown, K.E., Welsh, M.J., Chiorini, J.A., and Zabner, J. (2001) J. Biol. Chem. 276, 20610–20616.
19. Qing, K., Mah, C., Hansen, J., Zhou, S., Dwarki, V., and Srivastava, A. (1999) Nat. Med. 5, 71–77.
20. Summerford, C., Bartlett, J.S., and Samulski, R.J. (1999) Nat. Med. 5, 78–82.
21. Nakai, H., Storm, T.A., and Kay, M.A. (2000) J. Virol. 74, 9451–9463.
22. Kotin, R.M., Siniscalco, M., Samulski, R.J., Zhu, X.D., Hunter, L., Laughlin, C.A., McLaughlin, S., Muzyczka, N., Rocchi, M., Berns, K.I. (1990) Proc. Natl. Acad. Sci. U.S.A. 87, 2211–2215.
23. Samulski, R.J., Zhu, X., Xiao, X., Brook, J.D., Housman, D.E., Epstein, N., and Hunter, L.A. (1991) EMBO J 10, 3941–3950.
24. Weitzman, M.D., Kyostio, S.R., Kotin, R.M., and Owens, R.A. (1994) Proc. Natl. Acad. Sci. U.S.A. 91, 5808–5812.
25. Berns, K.I. (1990) Microbiol. Rev. 54, 316–329.
26. Xiao, X., Li, J., and Samulski, R.J. (1998) J. Virol. 72, 2224–2232.
27. Clark, K.R., Voulgaropoulou, F., Fraley, D.M., and Johnson, P.R. (1995) Hum. Gene Ther. 6, 1329–1341.
28. Conway, J.E., Rhys, C.M., Zolotukhin, I., Zolotukhin, S., Muzyczka, N., Hayward, G.S., and Byrne, B.J. (1999) Gene Ther. 6, 986–993.
29. Zolotukhin, S., Byrne, B.J., Mason, E., Zolotukhin, I., Potter, M., Chesnut, K., Summerford, C., Samulski, R.J., and Muzyczka, N. (1999) Gene Ther. 6, 973–985.
30. Xiao, X., Li, J., and Samulski, R.J. (1996) J. Virol. 70, 8098–8108.
31. Greelish, J.P., Su, L.T., Lankford, E.B., Burkman, J.M., Chen, H., Konig, S.K., Mercier, I.M., Desjardins, P.R., Mitchell, M.A., Zheng, X.G., Leferovich, J., Gao, G.P., Balice-Gordon, R.J., Wilson, J.M., Stedman, H.H. (1999) Nat. Med. 5, 439–443.
32. Walsh, C.E., Liu, J.M., Xiao, X., Young, N.S., Nienhuis, A.W., and Samulski, R.J. (1992) Proc. Natl. Acad. Sci. U.S.A. 89, 7257–7261.
33. Haberman, R.P., McCown, T.J., and Samulski, R.J. (1998) Gene Ther. 5, 1604–1611.
34. Flotte, T.R., Afione, S.A., and Zeitlin, P.L. (1994) Am. J. Respir. Cell. Mol. Biol. 11, 517–521.

35. Duan, D., Sharma, P., Yang, J., Yue, Y., Dudus, L., Zhang, Y., Fisher, K.J., and Engelhardt, J.F. (1998) J. Virol. 72, 8568–8577.
36. Flotte, T.R., Solow, R., Owens, R.A., Afione, S., Zeitlin, P.L., and Carter, B.J. (1992) Am. J. Respir. Cell. Mol. Biol. 7, 349–356.
37. Kaplitt, M.G., Leone, P., Samulski, R.J., Xiao, X., Pfaff, D.W., O'Malley, K.L., and During, M.J. (1994) Nat. Genet. 8, 148–154.
38. Xiao, W., Berta, S.C., Lu, M.M., Moscioni, A.D., Tazelaar, J., and Wilson, J.M. (1998) J. Virol. 72, 10222–10226.
39. During, M.J., Samulski, R.J., Elsworth, J.D., Kaplitt, M.G., Leone, P., Xiao, X., Li, J., Freese, A., Taylor, J.R., Roth, R.H., Sladek, J.R., Jr., O'Malley, K.L., Redmond, D.E., Jr. (1998) Gene Ther. 5, 820–827.
40. Bankiewicz, K.S., Eberling, J.L., Kohutnicka, M., Jagust, W., Pivirotto, P., Bringas, J., Cunningham, J., Budinger, T.F., and Harvey-White, J. (2000) Exp. Neurol. 164, 2–14.
41. Bartlett, J.S., Samulski, R.J., and McCown, T.J. (1998) Hum. Gene Ther. 9, 1181–1186.
42. Xiao, X., Li, J., Tsao, Y.P., Dressman, D., Hoffman, E.P., and Watchko, J.F. (2000) J. Virol. 74, 1436–1442.
43. Wang, B., Li, J., and Xiao, X. (2000) Proc. Natl. Acad. Sci. U.S.A. 97, 13714–13719.
44. Kaplitt, M.G., Xiao, X., Samulski, R.J., Li, J., Ojamaa, K., Klein, I.L., Makimura, H., Kaplitt, M.J., Strumpf, R.K., Diethrich, E.B. (1996) Ann. Thorac. Surg. 62, 1669–1676.
45. Kay, M.A., Manno, C.S., Ragni, M.V., Larson, P.J., Couto, L.B., McClelland, A., Glader, B., Chew, A.J., Tai, S.J., Herzog, R.W., Arruda, V., Johnson, F., Scallan, C., Skarsgard, E., Flake, A.W., High, K.A. (2000) Nat. Genet. 24, 257–261.
46. Song, S., Morgan, M., Ellis, T., Poirier, A., Chesnut, K., Wang, J., Brantly, M., Muzyczka, N., Byrne, B.J., Atkinson, M., Flotte, T.R. (1998) Proc. Natl. Acad. Sci. U.S.A. 95, 14384–14388.
47. Nakai, H., Yant, S.R., Storm, T.A., Fuess, S., Meuse, L., and Kay, M.A. (2001) J. Virol. 75, 6969–6976.
48. Daly, T.M., Ohlemiller, K.K., Roberts, M.S., Vogler, C.A., and Sands, M.S. (2001) Gene Ther. 8, 1291–1298.
49. High, K.A. (2001) Ann. NY. Acad. Sci. 953, 64–74.
50. Jooss, K., Yang, Y., Fisher, K.J., and Wilson, J.M. (1998) J. Virol. 72, 4212–4223.
51. Monahan, P.E., Samulski, R.J., Tazelaar, J., Xiao, X., Nichols, T.C., Bellinger, D.A., Read, M.S., and Walsh, C.E. (1998) Gene Ther. 5, 40–49.
52. Chao, H. and Walsh, C.E. (2001) Blood 97, 3311–3312.
53. Liu, D.W., Tsao, Y.P., Kung, J.T., Ding, Y.A., Sytwu, H.K., Xiao, X., and Chen, S.L. (2000) J. Virol. 74, 2888–2894.
54. Brockstedt, D.G., Podsakoff, G.M., Fong, L., Kurtzman, G., Mueller-Ruchholtz, W., and Engleman, E.G. (1999) Clin. Immunol. 92, 67–75.
55. Wang, B., Li, J., and Xiao, X. (2000) Proc. Natl. Acad. Sci. U.S.A. 97, 13714–13719.
56. Cordier, L., Gao, G.P., Hack, A.A., McNally, E.M., Wilson, J.M., Chirmule, N., and Sweeney, H.L. (2001) Hum. Gene Ther. 12, 205–215.
57. Sun, L., Li, J., and Xiao, X. (2000) Nat. Med. 6, 599–602.
58. Yan, Z., Zhang, Y., Duan, D., and Engelhardt, J.F. (2000) Proc. Natl. Acad. Sci. U.S.A. 97, 6716–6721.
59. Halbert, C.L., Allen, J.M., and Miller, A.D. (2002) Nat. Biotechnol. 20, 697–701.
60. Rabinowitz, J.E., Rolling, F., Li, C., Conrath, H., Xiao, W., Xiao, X., and Samulski, R.J. (2002) J. Virol. 76, 791–801.
61. Davidson, B.L., Stein, C.S., Heth, J.A., Martins, I., Kotin, R.M., Derksen, T.A., Zabner, J., Ghodsi, A., and Chiorini, J.A. (2000) Proc. Natl. Acad. Sci. U.S.A. 97, 3428–3432.
62. Zabner, J., Seiler, M., Walters, R., Kotin, R.M., Fulgeras, W., Davidson, B.L., and Chiorini, J.A. (2000) J. Virol. 74, 3852–3858.
63. Duan, D., Yue, Y., Yan, Z., Yang, J., and Engelhardt, J.F. (2000) J. Clin. Invest. 105, 1573–1587.
64. Qing, K., Wang, X.S., Kube, D.M., Ponnazhagan, S., Bajpai, A., and Srivastava, A. (1997) Proc. Natl. Acad. Sci. U.S.A. 94, 10879–10884.

65. McCarty, D.M., Monahan, P.E., and Samulski, R.J. (2001) Gene Ther. 8, 1248–1254.
66. Bartlett, J.S., Kleinschmidt, J., Boucher, R.C., and Samulski, R.J. (1999) Nat. Biotechnol. 17, 181–186.
67. Wu, P., Xiao, W., Conlon, T., Hughes, J., Agbandje-McKenna, M., Ferkol, T., Flotte, T., and Muzyczka, N. (2000) J. Virol. 74, 8635–8647.
68. Owens, R.A. (2002) Curr. Gene Ther. 2, 145–159.

S.C. Makrides (Ed.) *Gene Transfer and Expression in Mammalian Cells*

Virus-based vectors for gene expression in mammalian cells: Adenovirus

Denis Bourbeau[1], Yué Zeng[1], and Bernard Massie[1,2,3]

[1]*Institut de Recherche en Biotechnologie, Conseil National de Recherche Canada, 6100 Avenue Royalmount, Montréal, QC, Canada H4P 2R2; Tel.: +1 (514) 496-6131; Fax: +1 (514) 496-5143*
E-mail: bernard.massie@nrc.ca
[2]*INRS-IAF, Université du Québec, Laval, QC, Canada H7N 4Z3*
[3]*Département de Microbiologie et Immunologie, Faculté de Médecine, Université de Montréal, Montréal, QC, Canada H3C 3J7*

1. Introduction

Adenoviruses (Ad) were first described in the 1950s as causal agents of upper respiratory tract infections and were subsequently associated with only minor pathologies. There are more than 100 Ad serotypes currently identified both in mammals and birds. The Ad virion is a 70–100 nm icosahedral particle composed of 252 capsomers and containing a double-stranded DNA linear genome of 25–45 kb depending on the serotype. Their biology and molecular structure were more extensively characterized using human Ads from serotypes 2 and 5 [1]. As a result, the majority of adenoviral vectors (AdV) were derived from these serotypes.

Following receptor interactions, the virus is internalized by means of clathrin-dependent endocytosis and is rapidly transported to the nucleus. The Ad2 genome is shown schematically in Fig. 1A. It contains *cis*-acting elements at each end, the inverted terminal repeats (ITRs), which are involved in DNA replication and packaging (at the left ITR). Viral genes are expressed in two phases delineated by the onset of DNA replication; the early phase, which lasts about 6–8 h, and the late phase, which culminates around 30 h post-infection with the production of more than 10^4 infectious virions per cell. Genes from the early phase are subdivided into six transcription units: E1A, E1B, E2A, E2B, E3, and E4. Each region gives rise to a set of alternatively spliced transcripts from which several proteins are synthesized, whereas two delayed early units produce only one protein, respectively, pIX and IV2a. The roles of the early gene products can be generalized as follows. The E1 genes are involved essentially in activating expression and diverting the cellular machinery toward viral DNA replication and preventing apoptosis. The E2 region bears genes directly involved in viral DNA replication. The genes from E3 region are involved in hampering the host defense system. Finally, the six ORFs forming the E4 region are involved in viral mRNA processing and transport as well as regulation of transcription and apoptosis. Although the function of some of these gene products is unknown, most of them are multifunctional while many share complex interactions with each other. Late genes are expressed from five families of transcripts (L1–5) and encode proteins involved in encapsidation and maturation of viral particles (reviewed in [2]).

A Adeno virus genome

Genome

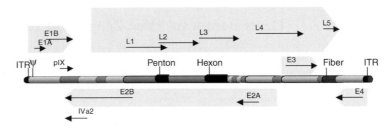

B Adeno virus vectors

First Generation

Second Generation

Third Generation

Replicative

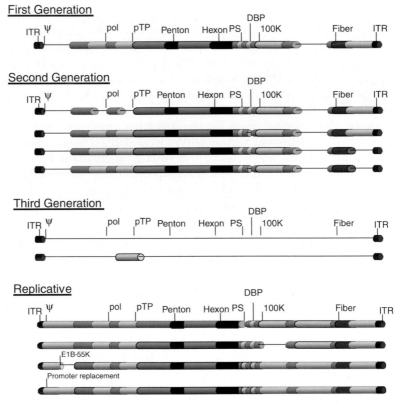

Fig. 1. (A) Simplified schema of the transcriptional map of Ad2/5 (36 kb). Arrows represent viral transcripts, shaded arrows represent transcriptional units. E, early expression; L, late expression. The ITRs and the packaging signal (ψ) are indicated. Light-shaded portions of the genome represent DNA that was deleted in several recombinant AdV. (B) Schema of the adenovirus vectors currently reported in the literature. Deletions characterizing the first, second, third generations, and replicative AdV are presented as follows, in order of appearance. 1st generation: ΔE1ΔE3. 2nd generation: Δ(E1, E2B, pTP, E3), Δ(E1, E2A, E3), Δ(E1, E3, E4), Δ(E1, E2A, E3, E4). 3rd generation: all genes deleted, all genes deleted but pTP.

In the last two decades Ads have been extensively used for gene transfer into mammalian cells (reviewed in [3]). Several features make AdV attractive vehicles for gene transfer and protein production. First, AdV can easily be produced at high titers; second, they can transduce a large variety of cell types; third, they do not require cell division for transgene expression; and fourth, their genome remains episomal, therefore they do not permanently alter the target cell. AdVs have been extensively engineered in order to be adapted to several applications of gene transfer in mammalian cells and gene therapy *in vivo*. In this chapter, we aim to provide the reader with an overview of the AdV platforms currently developed, with means to engineer, produce, and characterize them.

2. Adenoviral vectors

To effect the transfer of exogenous genes into mammalian cells, the Ad has been modified in order to: (1) eliminate its replication capacity; (2) increase its cloning size capacity; (3) reduce the host immune response; and (4) retarget it to specific cell receptors. The most widely used AdV is the so-called first generation that is deleted in the E1 and the E3 regions. Later, other modifications were made in order to meet specific objectives. The various AdV engineered to date are depicted in Fig. 1B.

2.1. First generation AdV

A deletion that removed the E1A and E1B genes distinguishes the first generation AdV from the wild type Ad. The E1 deletion renders the AdV replication dependent on cell lines such as the 293, A549-E1, or PER.C6, which complement the E1 functions [4]. The E3 region is not required for viral production *in vitro* and thus its deletion increases the cloning capacity without having to complement its function in producing cells. Those vectors can accommodate exogenous DNA of up to 8 kb in size.

Construction of first generation AdV is routinely performed by recombination between a digested AdV genome and a transfer plasmid that contains the left arm portion of the AdV, the transgene cassette, and a portion of homologous sequence for recombination to occur. Recombination is performed in mammalian cells, generally in 293 cells, which are readily transfected by the calcium phosphate procedure. The recombination requires the co-transfection of both the transfer plasmid and the viral genome. This procedure, although favored for its simplicity, is inefficient. Recombination is a rare event and incompletely digested AdV may give rise to non-recombinant AdV. Our laboratory is relying on co-expression

Replicative AdV: ΔPS, Δ100K, ΔE1B-55K, E1A promoter substitution. Abbreviations: ITR, inverted terminal repeat; pol, Ad polymerase; pTP, preterminal protein; PS, Ad protease; DBP, DNA-binding protein; E1, E2, E3, E4, early phase transcriptional units; L1, L2, L3, L4, L5, late phase transcriptional units.

of the green fluorescent protein (GFP) to select recombinant plaques under the microscope [5]. Several of our transfer vectors carry a cassette, which joins the GFP with the transgenes through an internal ribosome entry site (IRES). In this way, GFP expression is linked to transgene expression, and recombinant plaques can be distinguished from non-recombinant ones. Other strategies have been developed in order to minimize contamination with non-recombinant AdV and to reduce the time required for AdV production. These strategies rely on a viral genome that is carried as a plasmid or cosmid, and recombinant AdV can be produced by several means: (i) subcloning directly the transgene into the plasmid containing the viral genome; (ii) using a recombinase such as Cre or Flp that facilitates specific and efficient introduction of transgenes into the viral genome; or (iii) using bacterial recombination for introduction of the transgene into the viral genome.

Strategies of direct cloning into AdV genome serve the purpose of eliminating non-recombinant plaques, therefore plaque purification. Cloning in large plasmids is difficult; thus cosmids were designed to facilitate the subcloning steps [6]. The strategy making use of recombinases replaces the subcloning step through efficient homologous recombination either in bacteria or in mammalian cells. Such strategies require AdV that are engineered to contain recombination sequences, such as loxP or frt, which are substrates for Cre and FLp, respectively, and cell lines or bacteria expressing the specific recombinase [6,7].

Recently, a novel strategy based on positive selection was developed for producing recombinant AdV. The strategy consists of inserting the Ad protease gene in the shuttle plasmid, to complement a protease (PS) deletion in the virus backbone used for recombination. Since the virus backbone cannot form viral particles in the absence of the PS, the only infectious particles produced are those that have recombined with the complementing shuttle plasmid. Thus, following transfection of permissive cells with the shuttle plasmid and infection with an AdV deleted for its PS gene, pure recombinant plaques are produced within 2 weeks with an efficiency of 100%. For the construction of recombinants one at a time, this positive selection method is at least as efficient as the best methods developed so far. Furthermore, this represents the first method allowing for easy construction of AdV libraries with high diversities [8].

2.2. Second generation AdV

First generation AdV can sustain transgene expression for years in post-mitotic cells of immunodeficient animals. However, their expression will persist only a few weeks in actively growing cells, due to the dilution of the viral genome by cell division, or in immunocompetent animals, due to the host immune system that destroys the transduced cells. Despite the absence of E1 genes required to initiate expression of viral genes, de novo viral gene expression occurs to some extent. Viral proteins are then exposed to the cell surface and elicit an immune response. In order to minimize this unwanted expression, AdV have been further crippled by deletion of genes in the E2 or E4 regions. The production of these second generation AdV has required the development of complementation cell lines expressing the corresponding deleted genes.

Deletions were engineered in E2A, E2B, DNA polymerase and/or preterminal protein (pTP), as well as E4, and in most cases *in vivo* studies showed significant reduction of the immune response. In the case of E2B, the deletion of either the polymerase or pTP alone decreased hepatic toxicity and prolonged vector stability, while combining both deletions resulted in further improvement [9]. In the case of E4 deletions, the generation of Ad E1$^-$ E2A$^-$ E4$^-$ demonstrated that further crippling of AdV by removing E4 genes reduced vector toxicity but also reduced vector stability [10]. In fact, it appears that ORF3 is required for efficient expression of transgene under the control of several promoters used in AdV, particularly the CMV promoter [11,12]. The mechanism responsible for the reduced expression and stability of second generation AdV deleted in E4 remains to be elucidated. Nonetheless, most studies using E4-deleted AdV have reported decreased transgene stability, unless the ORF3 or ORF6 have been reintroduced. Moreover, reintroducing ORF3 in E4-deleted AdV restores its stability while retaining a reduced toxicity. On the other hand, ORF6 reintroduction restores both stability and hepatotoxicity [13]. It is noteworthy that either ORF3 or ORF6 can complement the complete E4 deletion.

2.3. Third generation AdV

Because second generation AdV were still sub-optimal in escaping the immune response, and there was a need for larger exogenous DNA insert capacity, the more drastic approach of eliminating all viral genes was undertaken. Fully deleted vectors have been given several names: third generation, helper-dependent, high capacity, or gutless. These AdV offer three main advantages: (1) no expression of viral genes is possible, thus reducing the immune response; (2) a cloning capacity of up to 36 kb; and (3) they retain all of the advantages of first generation AdV with respect to transduction efficiency. However, they are more difficult to produce since it is almost impossible to generate a stable cell line that would complement all of the Ad genes, due to the toxicity associated with the over-expression of most of the Ad proteins. The helper functions required for the complementation of third generation AdV are most often provided by co-infection with a helper virus, but they can also be provided by co-transfection of the adenoviral genome as plasmid [14], or through a baculovirus/Ad hybrid [14–16]. The dependency on a helper Ad results in decreased viral yields and contamination with helper viruses. A few strategies have been designed to overcome these drawbacks. The strategy that is currently favored is to remove the helper virus-packaging signal through Cre/loxP recombinase [17], and more recently the Flp/frt recombinase system [18]. These systems require that third generation AdV be amplified with a helper Ad bearing loxP or frt sequences flanking its packaging signal. In 293 cells expressing Cre or Flp recombinases, the packaging signal of the helper virus is deleted resulting in viral DNA that still can produce viral proteins, but cannot be packaged into viral particles. Although the cleavage by Cre or FLP is not 100% and some helper viruses are still packaged, this procedure reduces contamination to less than 0.01% after CsCl purification in the best stocks. While helper AdV are a very small proportion of the preparation, high-dose injections such as 10^{12} particles would still deliver more than 10^8 contaminating particles. This might be sufficient to elicit an immune response. Therefore, further improvement in third generation AdV purity is required.

Another issue about the production of third generation AdV is the requirement of an optimal genome size of 26–38 kb for efficient particle formation. In order to generate optimized third generation AdV, the large amount of deleted DNA must be replaced with stuffer DNA that is carefully chosen. Thus, the stuffer DNA should have the following desirable characteristics: (1) it should be preferentially of mammalian origin; (2) it should be free of repeats, retrovirus elements, and genes; (3) contiguous human DNA should be fragmented to prevent integration; and (4) a matrix attachment region might confer stability [19]. Other *cis*-acting elements, such as a 400-bp fragment from the right end of Ad, which appears to confer growth advantage, can contribute to improved yields of gutless AdV [19].

While efforts are still ongoing to further optimize their production, several studies performed with third generation AdV have provided very promising results in reducing immune response and prolonging stability. The hurdles remain essentially in the production of clinical grade third generation viruses due to the complexity of the process.

2.4. Replicative AdV

While a major concern in the use of AdV for gene replacement therapy is reduction of toxicity and host immune response, there are applications, such as vaccination or cancer gene therapy, where toxicity and immunogenicity are rather an asset. For such applications the increased transgene expression through an increase in copy number following viral DNA replication can also be advantageous. This can be done by deleting genes that do not affect DNA replication, such as the PS or the 100K (late genes), while interfering with the production of infectious particles [20,21]. Although the full potential of such AdV has yet to be demonstrated, they could have applications for recombinant protein production, vaccination, cancer gene therapy, and production of adeno-associated viruses (AAV).

Other types of replicative AdV were developed more specifically for cancer gene therapy. Because transduction efficiency of 100% is not achievable, yet complete eradication of tumor cells is required for successful treatment; the ability of transduced tumor cells to kill bystander tumor cells is of paramount importance. One approach to achieving this goal is to engineer AdV that can replicate and form mature particles only in tumor cells. This was achieved using AdV deleted either in the E1B-55 kDa and/or the E1B-19 kDa gene. Thus, after infecting a few tumor cells, the progeny of conditional replicative oncolytic AdV can spread and destroy neighboring tumor cells while sparing normal cells. Other strategies rely on a tumor-specific promoter to drive E1A gene expression [22]. The above strategies have also been combined with cytokine or suicide genes to further improve the efficacy of conditional replicative AdV.

3. Gene delivery

3.1. Cell entry pathway

Effective vectors are those which deliver the transgene specifically to the appropriate tissues. Strategies for targeting AdV to specific cell types became possible following the improved understanding of the viral entry mechanisms, including the identification of the viral proteins and cellular receptors that are involved in virus–cell interactions.

3.1.1 Ad capsid proteins involved in virus entry

Two capsid proteins, the fiber and the penton base, play essential roles in virus–cell interaction and virus internalization. Five penton base subunits are associated with each fiber, itself composed of a homotrimer of fiber proteins. The fiber is responsible for the assembly and stabilization of the virion. It can be divided into three domains with distinct functions: (1) the short amino-terminal tail is involved in its association with the penton base protein through an FNPVYP motif. This interaction anchors the fiber to the Ad capsid. Also located in this domain is a nuclear localization signal (KRXR), which directs the intracellular trafficking of newly synthesized fibers to the cell nucleus where Ad particles are assembled. (2) The central shaft domain, whose length varies among different serotypes, is characterized by sequence repeats of approximately 15 residues. The shaft domain, which makes the carboxy terminal end of the fiber extending away from the virion, therefore provides optimal condition for fiber–receptor recognition. (3) The carboxy terminal knob domain forms an $\alpha\beta$ sandwich structure, which contains two anti-parallel β-sheets named R and V. These sheets are each composed of four strands linked by loops in which several residues are involved in the binding of the virus to the coxsackie virus and Ad receptor (CAR). Another distinct function carried out by the knob domain is the initiation of fiber trimerization.

The penton base subunit contains a flexible loop structure consisting of two stretches of α helices. The loop of the five penton base units forms a protrusion on the protein surface, exposing an arginine–glycine–aspartic acid (RGD) motif (aa 340–342), which interacts with the cellular integrin α_v, and a leucine–aspartic acid–valine (LDV) sequence (aa 287–289) also identified as an integrin-binding motif. In addition, the penton base plays an important role in the internalization of viral particles and release of capsids from the endosome.

3.1.2 Cell receptors

CAR, a 46-kDa transmembrane glycoprotein widely expressed in different tissues, is the primary receptor for Ad transduction. It belongs to the immunoglobulin (Ig) superfamily and contains two extracellular immunoglobulin-like domains, a hydrophobic transmembrane region and an intracellular cytoplasmic domain. Functional analyses of recombinant forms of CAR have demonstrated that the first immunoglobulin-like domain is sufficient for Ad binding. It should be noted that certain subgroup B or D Ads do not use CAR as receptor for their entry. Also, an

α sialic acid residue of some glycoproteins is used by Ad37 as receptor for their cell entry [23].

Interaction of the fiber with CAR leads to virus attachment onto the cell surface, but, in general, this is not sufficient for rapid virus uptake. It is the association of the cellular integrins $\alpha_v\beta_3$ or $\alpha_v\beta_5$ with the viral penton base that facilitates virus internalization. Moreover, fiber-less particles have been demonstrated to keep their infectious capacity for monocytic cells via an integrin-dependent pathway [24]. Recent studies have suggested that heparan sulfate glycosaminoglycans are sufficient to mediate Ad2/5 binding on the cell surface and infection [25].

The $\alpha 2$ domain of the major histocompatibility complex class I protein (MHC-I) is also reported to function as an alternative receptor [26]. However, when CAR and MHC-I proteins were co-expressed on the same cell, only CAR had high affinity for the Ad5 fiber. MHC-I-dependent Ad5 attachment may only occur when CAR expression is low.

3.1.3 Mechanisms of early steps of cell infection by Ad

Cellular uptake of the Ad particle is a two-step process: (1) an initial interaction of the fiber with cellular receptors like CAR or MHC-I [27]; and (2) the internalization of the virus mediated by the interaction of the RGD motif of the penton base with other cellular receptors, such as the integrins $\alpha_v\beta_3$ and $\alpha_v\beta_5$. This latter interaction triggers the phosphorylation of several signaling molecules, and induces a signaling cascade that promotes actin polymerization, which provides the mechanical force necessary for the internalization of clathrin-coated pit containing AdV.

Viral proteins begin to be dismantled during the endocytotic uptake from the plasma membrane and during subsequent changes in the endosomes. After the endosomal membrane is lysed, the uncoated particle escapes to the cytoplasm with the help of the penton base protein, and the core is translocated to the nucleus through nuclear pores. The Ad genome then enters the nucleoplasm where viral replication takes place.

3.2. Targeting of AdV

Active research is ongoing to develop targeting vectors to selectively transduce the tissue of interest for the following reasons: the Ad wide tropism leads to indiscriminate transduction of both target and bystander cells. Transduction of non-target cells not only dilutes the amount of therapeutic viruses delivered to target cells but also results in undesirable transgene expression. Moreover, because some cells have no or low levels of CAR, targeting AdV to a specific cellular receptor would overcome the problem of poor transduction efficiency. Finally, preventing AdV interaction with antigen-presenting cells should reduce the humoral and cytotoxic T lymphocyte response caused by AdV *in vivo*.

Since the fibers of different Ad serotypes display various amino acid sequences, it was suggested that they might recognize different receptors and consequently have different tropism. Therefore, several studies were undertaken using either chimeric fibers or exchanging fibers from different serotypes in order to alter AdV tropism. For example, Shayakhmetov *et al.* [28] demonstrated that the chimeric vector Ad5/ F35 can successfully infect human $CD34^+$ cells by CAR- and α_v integrin-independent pathways. However, effective targeting requires the abolition of vector interaction with its natural receptors, and the redirection of the vector to another type of receptor, which is specific to the target cells. Two strategies have been developed to realize this goal: (1) a conjugate system that allows AdV interaction with bridging molecules for the binding to a new receptor; and (2) a genetic approach in which the virus capsid proteins are modified to recognize only the target receptor (Fig. 2A).

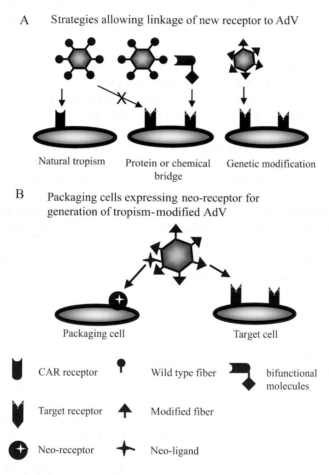

Fig. 2. (A) Schema depicting two strategies to retarget AdV. Left, the natural tropism of AdV and its inability to interact with certain target cells. Middle, the use of a bifunctional molecule bridging the virus to the target cell. Right, the genetic modification of AdV as the alternate strategy. (B) Special packaging cells need to be engineered for production of genetically modified AdV that may fail to infect "normal" packaging cells.

3.2.1 Conjugate targeting system

This system makes use of a bispecific molecule, which binds to the AdV and to the targeted receptor. The AdV specificity is usually provided by a soluble form of CAR, the binding domain of MHC, or neutralizing antibodies against one of the capsid protein. Other conjugation strategies include biotinylation, and coating with polyethyleneglycol (PEGylation) or a multivalent hydrophilic polymer. The cellular specificity can be generated with an antibody or a high-affinity ligand specific for a receptor or an adhesion molecule expressed on the targeted cell surface, such as CD40, α_v integrin, the receptors for fibroblast growth factor (FGF), tumor necrosis factor α (TNFα), insulin-like growth factor (IGF), epidermal growth factor (EGF), and several tumor markers. The bispecific molecules were made by chemically linking two molecules, or by engineering recombinant fusion proteins.

Any receptor can be targeted with the above strategies because any high-affinity antibody can be used as ligand. An additional advantage of the PEGylation and multivalent polymeric modification of AdV is that vectors are protected against neutralizing antibodies, thus avoiding some of the immune responses. However, some limitations remain to be overcome: there are several steps involved in the large-scale production and purification of AdV and bispecific molecules in the development of stable forms of the complexes for storage and for intravenous administration. Thus, the development of other approaches, such as genetic retargeting, may be more promising.

3.2.2 Genetic targeting system

This approach relies on genetic modifications to incorporate targeting ligands to the coat proteins of AdV. One challenge is to determine which locations are appropriate for incorporating high-affinity ligands. Wickham *et al.* [29] have constructed AdV whose fibers contain an RGD motif or a polylysine in its C-terminus. These viruses increased transduction in multiple cell types lacking high levels of CAR. The HI loop of fiber, the RGD loop of penton base, and an exposed loop of hexon were also chosen for the insertion of high-affinity peptides. Modifications in these locations were demonstrated to lead to virus attachment and cell entry in a CAR-independent manner. Another issue concerns the size of the peptide to be inserted. The incorporation of a ligand into Ad capsid proteins must not alter the conformation of these proteins, and should not adversely affect their other normal functions. Ideal ligands are small peptides specific for the targeted cellular receptor. Published findings suggest that less than 25–30 residues can be inserted into the C-terminus of the fiber, because longer sequences destabilize the fiber trimer [30]. It is also reported that ligands added into the HI loop of the fiber should not exceed 63 amino acids [27]. Full re-targeting requires the elimination of AdV interaction with its natural receptors. Several amino acid residues localized in the knob region of Ad5 fiber, such as S408, P409, K417, K420, K442, Y477, L485, Y491, A494 and A503 are directly involved in fiber–CAR interaction. Mutations of these amino acids and of residues 489–492 lead to a drastic reduction in virus transduction [31]. Several studies have successfully combined CAR-binding ablation and targeting ligand insertion for retargeting. For example, Nicklin *et al.* [32] have retargeted AdV to endothelial cells by mutating two amino acid residues (S408E and P409A) in the knob region that blocks CAR interaction, and by incorporating the SIGYPLI peptide, which was demonstrated to bind specifically to endothelial cells. Finally, disruption of α_v integrin at the same time is also critical for fully ablating the natural AdV tropism.

The production of a fully ablated AdV in its endogenous receptor interaction requires: (1) the insertion in the Ad capsid protein of a novel ligand unable to bind to any cells; and (2) the use of novel packaging cells expressing a chimeric receptor that allows entry and replication of the modified AdV [33,34]. Therefore, in addition to the targeting ligand incorporated into the AdV capsid for cell-specific transduction, another pseudoreceptor-binding ligand should also be inserted into the vector for entry into the packaging cells (Fig. 2B). This pair of pseudoreceptor–ligand for virus propagation should preferably be completely artificial, otherwise it might compromise the specificity of the retargeting, since any natural analogue could cause binding of the vector to non-target cells *in vivo*.

4. Gene expression

4.1. High-level expression

Besides transduction efficiency, transgene expression efficiency has also been improved. High expression level of recombinant proteins with low AdV multiplicies of infection (MOI) are desirable to minimize negative effects of high MOI on target cells. Strong promoter-enhancer elements usually derived from viruses have been used to construct expression cassettes. For example, the immediate-early human cytomegalovirus (hCMV) promoter has been utilized extensively because of its high transcriptional activity in a wide range of cells. However, AdV containing standard CMV-based expression cassettes produce proteins at suboptimal levels representing no more than 2–3% of total cell protein (TCP). Optimizing the expression cassette by including the Ad tripartite leader and a small intron as in pAdCMV5 improved performance by 6–12-fold, both at low and high MOI, and achieved up to 35% of TCP in AdV-infected cells at high copy number [5,35].

The expression of some proteins may have deleterious effects on the host cell or on Ad production. Inducible gene expression systems (Chapter 22) have been therefore developed to overcome this problem. Regulation of transgene expression *in vivo* can also improve its safety and efficacy. An ideal inducible system would have low basal expression of transgene without the inducer and very high expression in its presence (reviewed in [36]). The most widely used inducible promoter is the tetracycline-controllable transactivator system. A *trans*-acting factor (tTA) is formed by the fusion of the activation domain of the herpes simplex virus (HSV) protein VP16 with a tetracycline repressor protein from *Escherichia coli*. The tTA transactivator stimulates transcription from a promoter containing the tetracycline operator sequences (tetO), but the interaction of tTA with tetO is prevented in the presence of tetracycline at concentrations not toxic to eukaryotic cells. The use of an optimized tet-regulated promoter, as in pAdTR5, allows the expression of proteins that are cytotoxic or interfere with Ad replication, at 10–15% TCP levels [5,35]. Recently, a new inducible gene switch was developed in our laboratory using the operator sequences and repressor of a bacterial *p-cymeme* operon whose DNA-binding can be regulated by cumate. This cumate-inducible system is comparable to the tet system in many respects. Interestingly, the cumate-inducible promoter achieved higher transgene expression than the tet-regulated promoter at lower MOI, thereby minimizing the interference of AdV at higher MOI. Moreover, combining the tet and cumate regulatory elements reduced leakiness while fully maintaining the expression level of the on state. Several other inducible systems have also been tested in AdV. These include the cre-loxP system from bacteriophage P1, the bacteriophage T7 binary system [37], the insect ecdysone system and the human rapamycine-inducible system. These inducible systems have intrinsic advantages and limitations that should be carefully evaluated in comparative studies in order to select the optimal system for a specific application (reviewed in [38]).

4.2. Specific transgene expression

Many studies have explored the use of tissue-specific promoters mainly to restrict transgene expression to the target tissues in order to avoid unwanted expression and reduce the immune response against the transgene product. Tissue-specific promoters are also used to transduce particular types of tumor cells in cancer therapy, as discussed above.

4.2.1 Tissue-specific promoters
Tissue-specific targeted gene transcription was mainly attempted for expression in the lung, epithelia, liver, pancreas, muscles, neural cells, mammary gland and cardiac cells. For example, the amylase promoter is used to target pancreas cells, truncated muscle creatine kinase (MCK) and the smooth muscle alpha-actin (SMA) promoters for targeting skeletal and smooth muscle cells, ventricle-specific myosin light chain 2, ventricle specific α-myosin heavy chain and troponin T promoters are used for cardiac cells, L7/PCP2 promotor for cerebellar Purkinje cells, and synapsin 1, tubulin alpha 1 and neuron-specific enolase (NSE) promoters are used to target neurons.

Using silencer elements is another approach for specific transgene expression. For example, elements specific to neurons can selectively repress the transcription of genes in non-neuronal cells so that when they are cloned upstream of the ubiquitous phosphoglycerate kinase promoter, transgene expression is restricted to neurons [39].

4.2.2 Tumor-specific promoters
Transcriptional targeting of tumors has been investigated with promoters known to be particularly active in tumor cells. The use of tumor-specific promoters should result in expression of a therapeutic gene primarily in tumor cells, while marginal expression should be observed in normal cells. There are several promoters that meet this objective: the MUC1/Df3 promoter has been used for breast carcinoma, the alpha fetoprotein (AFP) promoter for hepatomas, the cyclooxygenase-2 (cox-2)

promoter for gastrointestinal cancers, the CTP1 promoter for colon cancer, the prostate-specific antigen (PSA), human glandular kallikrein (hk2) and probasin promoters for prostate carcinoma, and the Midkine (MK) promoter for neuroblastoma.

Promoters that are particularly efficient in actively dividing cells are also useful for cancer targeting, since in adults, few cells other than tumors are replicating. The E2F-responsive promoter E2F-1, is one of them. It allows eradication of established gliomas with significantly less toxicity to normal tissue than constitutive promoters [3]. A similar approach made use of the hexokinase type II promoter to preferentially drive the expression of a toxic gene in a variety of cancer cell lines [3]. The human telomerase reverse transcriptase (hTERT) promoter was also shown to be strong and preferentially active in tumor cells. This promoter enabled induction of the *Bax* gene to elicit tumor-specific apoptosis *in vitro*, and *in vivo* it caused the suppression of tumor growth in nude mice while preventing the cytotoxic effects of the *Bax* gene to normal cells [40].

5. *Production and analyses of viral particles*

First-generation AdV are commonly produced in 293 cells that express the E1 genes from a large fragment of the left end of the Ad genome integrated in their chromosomal DNA. However, homologous sequences between AdV and the genome of 293 cells allow for occasional recombination leading to the generation of replication-competent AdV (RCA). This problem has been solved by the development of new E1-complementing cell lines, which do not share homologies with E1-deleted AdV [3,4]. Pharmaceutical and commercial applications of AdV require virus production on a large-scale. This has been facilitated by the development of processes allowing the production of AdV in suspension and in serum-free cultures [41]. Following production, AdV lysates are prepared by three freeze-thaw cycles and removal of cellular debris. These viral crude lysates can readily be used for *in vitro* studies. However, for many applications it is preferable to purify and concentrate the AdV particles. This is most often done by CsCl gradient purification, which relies on the specific buoyancy of AdV. However, this procedure is time-consuming and difficult to scale-up, necessitating the development of other purification protocols based on chromatography [42].

For the production of clinical-grade material, several challenges remain to be addressed to make the process economically viable and fully compliant with regulatory requirements. Thus, the yield, homogeneity, lot consistency, and a reliable characterization of the vector particles and infectivity need to be addressed. Regarding consistency and comparability of AdV production among different laboratories, the establishment of a standard sample and of standard operating procedures (SOPs) was awaited. The Adenovirus Reference Material Working Group (ARMWG) has established such standards. A wild-type Ad5 will be available from the ATCC in 2002, and SOPs for evaluation of viral particles and infectious particles are published at http://www.wilbio.com. The establishment of this standard is crucial to the comparability of AdV quantification across different laboratories. Current titration measures have different efficiency and are all extremely sensitive to absorption conditions (time, volume, agitation). Moreover, the intrinsic sensitivity of each assay varies within one order of magnitude (TCID50, plaque assay, transduction units). A report on the AdV standards has been published [43].

Finally, AdV need to be stored under conditions that ensure their stability over time. Croyle *et al.* [44] have developed a formulation for the long-term storage of lyophilized AdV at room temperature without loss of virus stability. Recently, the ARMWG has proposed a rather simple formulation that performed well for a period of over 6 months. This formulation is composed of 20 mM Tris pH 8.0, 25 mM NaCl, 2.5% glycerol, and is suitable for the stable storage of AdV at a concentration below 5×10^{11} particles [45].

6. Conclusion

The biological features of AdV have contributed to their popularity as tools for functional studies and gene therapy applications as demonstrated by the thousands of reports on AdV in recent years. Because efficacy and specificity is of paramount importance in gene therapy success, considerable efforts have been invested on their improvement. Although several challenges in targeting and reducing the host immune response remain to be addressed, current progress has armed molecular biologists with a wealth of AdV adapted to numerous applications. Moreover, new developments in the construction of Ad-retro, Ad-AAV or Ad-EBV hybrid viruses will further expand their use in therapy that requires long-term transgene expression. Finally, the recent development of Ad-based libraries [8] will lead to new applications yet to be explored.

Acknowledgements

We thank Rénald Gilbert for critical reading of the manuscript. This is a NRC publication #37701.

Abbreviations

AAV	adeno associated virus
Ad	adenovirus
AdV	adenovirus vector(s)
ARMWG	adenovirus reference material working group
CAR	coxsackie virus and adenovirus receptor
CMV	cytomegalovirus
CPE	cytopathic effect
GFP	green fluorescent protein

IRES	internal ribosome entry site
ITR	inverted terminal repeats
MHC	major histocompatibility complex
MOI	multiplicity of infection
ORF	open reading frame
pTP	preterminal protein
PS	adenovirus protease
RCA	replication-competent adenovirus
SOP	standard operating procedure
TCP	total cell protein
tetO	tetracycline operator
tTA	tetracycline trans-activator

References

1. Shenk, T.E. (2001) In: Knipe, D.M. (ed.) Field's Virology, Lippincott Williams & Wilkins, Philadelphia, 4th ed, 2265–2300.
2. Russell, W.C. (2000) J. Gen. Virol. 81, 2573–2604.
3. Oualikene, W. and Massie, B. (2000) In: Al-Rubeai, M. (ed.) Cell Engineering, Kluwer Academic Publishers, London, pp. 80–1542.
4. Fallaux, F.J., Bout, A., van der Velde, I., van den Wollenberg, D.J., Hehir, K.M., Keegan, J., Auger, C., Cramer, S.J., van Ormondt, H., van der Eb, A.J., Valerio, D., Hoeben, R.C. (1998) Hum. Gene Ther. 9, 1909–1917.
5. Massie, B., Mosser, D.D., Koutroumanis, M., Vitté-Monty, I., Lamoureux, L., Couture, F., Paquet, L., Guilbault, C., Dionne, J., Chahla, l., Jolicoeur, P., Langelier, Y. (1998) Cytotechnology 28, 53–64.
6. Danthinne, X. and Imperiale, M.J. (2000) Gene Ther. 7, 1707–1714.
7. Mizuguchi, H., Kay, M.A., and Hayakawa, T. (2001) Adv. Drug Deliv. Rev. 52, 165–176.
8. Elhai, S.M., Oualikene, W., Naghdi, L., O'Connor-McCourt, M., and Massie, B. (2002) Gene Ther. 9, 1238–1246.
9. Hodges, B.L., Serra, D., Hu, H., Begy, C.A., Chamberlain, J.S., and Amalfitano, A. (2000) J. Gene Med. 2, 250–259.
10. Andrews, J.L., Kadan, M.J., Gorziglia, M.I., Kaleko, M., and Connelly, S. (2001) Mol. Ther. 3, 329–336.
11. Armentano, D., Smith, M.P., Sookdeo, C.C., Zabner, J., Perricone, M.A., St George, J.A., Wadsworth, S.C., and Gregory, R.J. (1999) J. Virol. 73, 7031–7034.
12. Grave, L., Dreyer, D., Dieterle, A., Leroy, P., Michou, A.I., Doderer, C., Pavirani, A., Lusky, M., and Mehtali, M. (2000) J. Gene Med. 2, 433–443.
13. Christ, M., Louis, B., Stoeckel, F., Dieterle, A., Grave, L., Dreyer, D., Kintz, J., Ali Hadji, D., Lusky, M., Mehtali, M. (2000) Hum. Gene Ther. 11, 415–427.
14. Kochanek, S. (1999) Hum. Gene Ther. 10, 2451–2459.
15. Morsy, M.A. and Caskey, C.T. (1999) Mol. Med. Today 5, 18–24.
16. Cheshenko, N., Krougliak, N., Eisensmith, R.C., and Krougliak, V.A. (2001) Gene Ther. 8, 846–854.
17. Parks, R.J., Chen, L., Anton, M., Sankar, U., Rudnicki, M.A., and Graham, F.L. (1996) Proc. Natl. Acad. Sci. USA 93, 13565–13570.
18. Umana, P., Gerdes, C.A., Stone, D., Davis, J.R., Ward, D., Castro, M.G., and Lowenstein, P.R. (2001) Nat. Biotechnol. 19, 582–585.
19. Sandig, V., Youil, R., Bett, A.J., Franlin, L.L., Oshima, M., Maione, D., Wang, F., Metzker, M.L., Savino, R., Caskey, C.T. (2000) Proc. Natl. Acad. Sci. USA 97, 1002–1007.
20. Oualikene, W., Lamoureux, L., Weber, J.M., and Massie, B. (2000) Hum. Gene Ther. 11, 1341–1353.

21. Hodges, B.L., Evans, H.K., Everett, R.S., Ding, E.Y., Serra, D., and Amalfitano, A. (2001) J. Virol. 75, 5913–5920.
22. Alemany, R., Balague, C., and Curiel, D.T. (2000) Nat. Biotechnol. 18, 723–727.
23. Arnberg, N., Edlund, K., Kidd, A.H., and Wadell, G. (2000) J. Virol. 74, 42–48.
24. Von Seggern, D.J., Chiu, C.Y., Fleck, S.K., Stewart, P.L., and Nemerow, G.R. (1999) J. Virol. 73, 1601–1608.
25. Dechecchi, M.C., Melotti, P., Bonizzato, A., Santacatterina, M., Chilosi, M., and Cabrini, G. (2001) J. Virol. 75, 8772–8780.
26. Hong, S.S., Karayan, L., Tournier, J., Curiel, D.T., and Boulanger, P.A. (1997) EMBO J. 16, 2294–2306.
27. Krasnykh, V.N., Douglas, J.T., and van Beusechem, V.W. (2000) Mol. Ther. 1, 391–405.
28. Shayakhmetov, D.M., Papayannopoulou, T., Stamatoyannopoulos, G., and Lieber, A. (2000) J. Virol. 74, 2567–2583.
29. Wickham, T.J., Tzeng, E., Shears, L.L., 2nd, Roelvink, P.W., Li, Y., Lee, G.M., Brough, D.E., Lizonova, A., and Kovesdi, I. (1997) J. Virol. 71, 8221–8229.
30. Curiel, D.T. (1999) Ann. N.Y. Acad. Sci. 886, 158–171.
31. Kirby, I., Davison, E., Beavil, A.J., Soh, C.P., Wickham, T.J., Roelvink, P.W., Kovesdi, I., Sutton, B.J., and Santis, G. (2000) J. Virol. 74, 2804–2813.
32. Nicklin, S.A., Von Seggern, D.J., Work, L.M., Pek, D.C., Dominiczak, A.F., Nemerow, G.R., and Baker, A.H. (2001) Mol. Ther. 4, 534–542.
33. Douglas, J.T., Miller, C.R., Kim, M., Dmitriev, I., Mikheeva, G., Krasnykh, V., and Curiel, D.T. (1999) Nat. Biotechnol. 17, 470–475.
34. Einfeld, D.A., Brough, D.E., Roelvink, P.W., Kovesdi, I., and Wickham, T.J. (1999) J. Virol. 73, 9130–9136.
35. Massie, B., Couture, F., Lamoureux, L., Mosser, D.D., Guilbault, C., Jolicoeur, P., Belanger, F., and Langelier, Y. (1998) J. Virol. 72, 2289–2296.
36. Mullick, A. and Massie, B. (2000) In: Speir, R.E. The Encyclopedias of Cell Technology, Wiley Biotechnology Encyclopedias, London, pp. 1140–1164.
37. Tomanin, R., Bett, A.J., Picci, L., Scarpa, M., and Graham, F.L. (1997) Gene 193, 129–140.
38. Fussenegger, M. (2001) Biotechnol. Prog. 17, 1–51.
39. Millecamps, S., Kiefer, H., Navarro, V., Geoffroy, M.C., Robert, J.J., Finiels, F., Mallet, J., and Barkats, M. (1999) Nat. Biotechnol. 17, 865–869.
40. Gu, J., Kagawa, S., Takakura, M., Kyo, S., Inoue, M., Roth, J.A., and Fang, B. (2000) Cancer Res. 60, 5359–5364.
41. Cote, J., Garnier, A., Massie, B., and Kamen, A. (1998) Biotechnol. Bioeng. 59, 567–575.
42. Huyghe, B.G., Liu, X., Sutjipto, S., Sugarman, B.J., Horn, M.T., Shepard, H.M., Scandella, C.J., and Shabram, P. (1995) Hum. Gene Ther. 6, 1403–1416.
43. Hutchins, B. (2002) BioProcessing 1, 25–28.
44. Croyle, M.A., Cheng, X., and Wilson, J.M. (2001) Gene Ther. 8, 1281–1290.
45. Hoganson, D., Ma, J.C., Asato, L., Ong, M., Printz, M.A., Huyghe, B.G., Sosnowski, B.A., and D'Andrea, M.J. (2002) BioProcessing 1, 43–48.

S.C. Makrides (Ed.) *Gene Transfer and Expression in Mammalian Cells*

Virus-based vectors for gene expression in mammalian cells: Vaccinia virus

Miles W. Carroll[1] and Gerald R. Kovacs[2]

[1]*Oxford BioMedica (UK) Ltd., Oxford Science Park, Oxford OX4 4GA, UK; Tel.: +01865 783000, Fax: 01865 783001; E-mail: m.carroll@oxfordbiomedica.co.uk*
[2]*Office of Biodefense Research Affairs, National Institutes of Allergy and Infectious Disease, 6610 Rockledge Drive, Bethesda, MD 20892, USA; Tel.: 301-451-3511; Fax: 301-480-1263; E-mail: gkovacs@niaid.gov*

1. Introduction

In this chapter we aim to present an overview of the basic concepts and applications of poxviruses as recombinant expression vectors. Additional focussed reviews on poxvirus molecular biology [1] construction of recombinant poxviruses [2,3] and their applications as vaccine vectors for infectious diseases and cancer immunotherapy [4–7] are essential reading.

Vaccinia virus (VV), along with smallpoxvirus, is a member of the orthopoxvirus genus, within the family *Poxviridae* [8]. The *Poxviridae* also encompass the avipoxvirus genus, of which fowlpoxvirus (FPV) [9] and canarypoxvirus (CPV) [5] have been developed as candidate human vaccine vectors. Though we will concentrate on the biology and applications of recombinant vaccinia virus (rVV), due to conservation of the replication machinery within the *Poxviridae*, many of the concepts of rVV can be applied to other poxviruses.

The successful eradication of smallpox, in the late 1970s, was aided by several properties of VV, namely [10] its simple and inexpensive manufacture process, stability and antigenicity. Following the completion of the smallpox eradication programme in 1979, research in VV would have probably declined if not for two reports in 1982 illustrating the application of VV as a recombinant expression vector [11,12].

VV vectors provide a degree of versatility that few other vector systems afford. Advantageous properties of the rVV system include: (1) ease of use; (2) ability to accommodate over 25 kilobasepairs (kbp) of foreign DNA [13]; (3) genomic stability; (4) broad host range; (5) cytoplasmic site of gene expression and (6) ability to authentically process eukaryotic proteins. Perhaps it is not surprising that this vector has been used to express hundreds of recombinant genes for a variety of applications including: protein function analysis, antigen processing, reverse genetics of RNA viruses, and recombinant vaccine development.

2. Vaccinia virus molecular biology

The VV genome consists of a single copy of double-stranded DNA approximately 200 kbp in length with terminal covalently closed hairpin loops. Replication of VV occurs exclusively in the cytoplasm; autonomy from the cell nucleus has been achieved by the acquisition of many genes involved in nucleic acid biosynthesis. Expression of the 200 or so genes encoded by VV is regulated primarily at the level of transcription. To date, three gene classes have been classified: early (pre-DNA replication), intermediate and late (post-DNA replication) (reviewed in [1]). Transcriptional regulation is mediated in large part by virus-encoded factors; however, some reports have suggested that viral transcription is not entirely independent from host cell nuclear factors [14,15]. In addition to having one copy of the genome, virions contain a virus-encoded DNA-dependent RNA polymerase and all of the accessory proteins that enable early gene transcription. Early genes are transcribed immediately after infection [16], and encode factors that fall into two broad categories: (1) proteins that block the host's response to infection; and (2) enzymes that are required for viral DNA synthesis. Of the former type, VV encodes factors known as virokines [17] viroceptors [18] and anti-inflammatory proteins [19] that dampen the host's immune response to infection (reviewed in [20]). VV also wards off the intracellular antiviral response by expressing proteins (eg., E3L and K3L) that inhibit RNA-dependent protein kinase (PKR) activation [21–24] and caspase 1 [19] mediated apoptosis (reviewed in [25]). Early gene products also include the DNA polymerase and associated cofactors, enzymes involved in the biosynthesis of deoxynucleotides, and viral intermediate gene transcription factors (VITF). Together, these gene products function to synthesize and process nascent genomes that in turn serve as templates for intermediate and late gene transcription. Intermediate gene products include three unique transcription factors (viral late gene transcription factors or VLTF [26]) required for the final phase of gene expression. Many of the gene products expressed during the late phase are structural and are required for virion formation and cell-to-cell spread [27]. Resolved, nascent viral DNA is packaged along with all of the enzymes and cofactors needed to initiate further rounds of infection (Fig. 1). Intracellular mature viruses (IMV) form in juxtanuclear factories, and are subsequently wrapped in double membranes derived from the trans-golgi and/or endosomal cisternae. The intracellular enveloped virus (IEV) is transported to the cell surface via microtubules where membrane fusion between the virus and host cell leads to release of relatively small quantities of extracellular enveloped virus (EEV). Most of the virus remains associated with the infected cell plasma membrane and is called cell-associated extracellular enveloped virus (CEV). A portion of the CEV associates with actin tails and is propelled from the infected cell surface to neighbouring cells.

2.1. Vaccinia virus promoters

VV promoters are compact and simple, consisting of an initiator region proceeded by a central core, together comprising no more than 30 bp. Detailed mutagenesis studies have clearly defined the optimal sequences for early and late gene transcription initiation [28,29]. Accordingly, VV expression vectors have been developed that comprise optimal regulatory sequences or natural early and/or late promoters. As many features of the viral transcription machinery are conserved

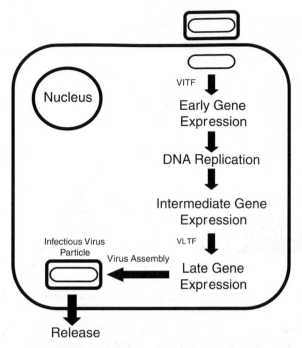

Fig. 1. Vaccinia virus life cycle.

within the *Poxviridae*, most promoters are active when inserted into members of other genera.

In general, early promoters are weaker than late promoters, presumably due to the amplification of the latter proceeding DNA replication (see Table 1). However, depending on the intended use of the rVV, and to avoid issues related to cytotoxicity of the foreign gene and/or the cytopathic effects of viral infection, it is sometimes recommended that weaker promoters be employed. When early promoters are used, it is advised that foreign genes do not contain the sequence TTTTTNT within the open reading frame. This sequence is used by the early viral transcription machinery as a signal for termination [30]. Unlike early mRNAs, intermediate and late mRNAs are not of a defined length, indicating that termination is not sequence-dependent for these gene classes.

3. Construction of recombinant poxvirus vectors

Homologous recombination [10,11], is the most commonly used method for the construction of rVV. A transfer plasmid, that contains the foreign gene adjacent to a suitable VV promoter and flanked by VV-derived DNA, is transfected into cells (Fig. 2). Cells are subsequently infected with wild type VV and, during replication, recombination occurs between the transfer plasmid and homologous regions within the virus genome. The recombinant genome is then packaged and released from the

Table 1

Kinetics and strength of vaccinia virus promoters

VV strain	Promoter	Kinetics	β-galactosidase expression (μg)	
			+ Ara-C	− Ara-C
MVA	$P_{7.5}$	E and L	0.2*	0.5*
MVA	P_{H5}	E and L	1.1*	2.6*
WR	P_{11}	L	0.0	3.9
WR	P_{syn}	E and L	0.3	8.2

Note: This is a modification of the table presented by Wyatt et al. [95] in which recombinant viruses containing the E. coli LacZ gene, under transcriptional control of the various promoters, were used to infect monkey kidney BS-C-1 cells. To differentiate between early and late expression, cells were incubated in the presence or absence of Ara C. Ara C inhibits viral DNA replication, which prevents late viral gene expression. Cells were harvested after 24 h incubation and assessed for β-galactosidase expression.
*Average expression level from two independent recombinant viruses.

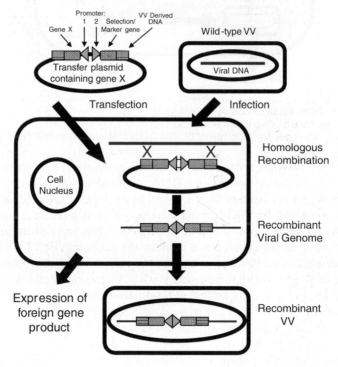

Fig. 2. Vaccinia virus homologous recombination.

cell. An initial site for recombination was the VV thymidine kinase (tk) gene that enabled selection based on a tk-negative phenotype, using BrdU (Fig. 3).

To improve the efficacy of recombinant virus selection and identification, a range of marker genes has been added to the transfer plasmids. These include: the *Escherichia coli* neomycin phosphotransferase enzyme (neo) gene [31],

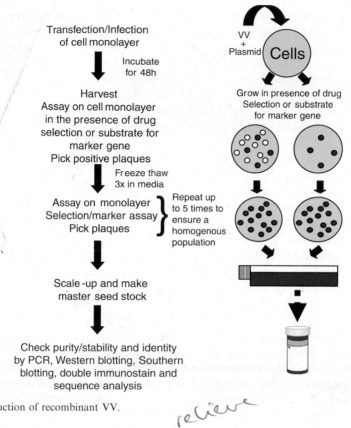

Fig. 3. Production of recombinant VV.

relieve

xanthine–guanine phosphoribosyl transferase (gpt) gene [32] colorimetric markers such as *lac* Z [33] green fluorescent protein (gfp) [34], β-glucuronidase A (gus) gene [35] and fusion proteins such as gpt-gus [36] and neo-gfp genes [37]. Other techniques have been employed that alleviate growth restrictions to certain strains of VV. Genes that restore plaque deficiency and/or host range restrictions are included in the transfer vector and, following recombination, allow replication in the host cell [38–40].

When constructing a rVV for use in clinical trials or if several recombinant genes are to be inserted into one genome, stable integration of selection or marker genes may not be desirable. For these purposes, transfer plasmids have been developed that allow for only transient selection of the rVV [41–45]. Lastly, if the recombinant protein is expressed on the cell surface, and an effective antibody is available, the recombinant virus may be identified by live immunostaining [3] Although, this latter method results in a greater abundance of non-recombinant VV, it does not require a marker or drug selection gene.

Initially, *in vitro* manipulation of the large VV genome (~ 200 kb) was thought to be an impractical approach to constructing rVV. However, it has since been shown that *in vitro* ligation may be used to generate rVV, as an alternative to homologous recombination methodologies [46,47]. VV DNA can be digested with a restriction enzyme that cuts the genome at a unique site, and a recombinant gene, along with a marker gene, is then ligated directly into the viral genome. A replication-restricted helper poxvirus, e.g., a conditionally lethal mutant or psoralin-treated/UV-inactivated VV [48] or fowlpox virus is used to package the non-infectious recombinant genome. Further improvements to these protocols have been made that allow easier identification of rVV [49,50] and should allow insertion of much larger regions of DNA compared to traditional procedures dependent on homologous recombination.

4. Chimeric VV-bacteriophage expression vectors

Although, VV promoter-based vectors are extremely versatile, certain applications have required the development of inducible VV vectors. Chimeric vectors have been developed using several bacteriophage DNA-dependent RNA polymerases. Chimeric vectors have been constructed that express the bacteriophage T7 [51], SP6 [52] or T3 [53] RNA polymerase genes under the transcriptional control of a VV promoter. Trans-gene expression is achieved by transfecting the gene of interest, which is under the control of a T7 promoter, into cells infected with a chimeric vaccinia virus vector. A detailed description of the protocols and vectors required for the T7 system is available [54]. The T7 expression system has now been incorporated into two highly attenuated, host-restricted and safe poxviruses, the attenuated modified VV Ankara (MVA) [55,56], and an avipoxvirus [57]. These systems offer a safe and efficacious small-scale method to produce authentic and potentially toxic recombinant proteins, without the need to construct replication-competent rVV.

For larger scale protein production, two methods may be utilised. In the first configuration, the T7 promoter-driven gene is placed into a second rVV, and is then co-infected with the rVV-T7 vector. Alternatively, the *trans*-gene may be integrated into the rVV-T7, and placed under *lac*I suppression. Modifications to improve stringency in this system have been reported using temperature-sensitive forms of *lac*I [27].

5. Improved safety of VV vectors

Vaccinia virus replicates in human cells, and during its use in the smallpox eradication campaign some VV strains were associated with an unacceptably high incidence of complications [58]. VV is classified in the UK as a category II organism, which requires the use of a class II microbial safety cabinet; in the USA the additional measure of pre-vaccination with VV is required. These factors make the use of replication-defective VV strains an attractive safe alternative. Two replication-defective VV, MVA and NYVAC, are now commonly used as expression vectors and experimental vaccine vectors. More recently, both viruses are being proposed as alternatives to the smallpox vaccine (DryVax) for biodefense purposes.

The replication-competent VV Ankara smallpox vaccine strain, was passaged over 570 times in primary chick embryo fibroblast (CEF) cells to produce the attenuated derivative MVA [59]. During the sequential passaging nearly 30% of the viral genome was deleted [60] and it became non-pathogenic in a number of animal models, and replication-defective in nearly all mammalian cells [61–63]. During the smallpox eradication programme, MVA was used to safely vaccinate over 120000 people who would have been at risk of complications from vaccination with replication-competent vaccine strains [64]. More recently, MVA has been developed as an expression vector [65], and there is now a range of transfer plasmids that allow the construction of rMVA containing several recombinant genes under transcriptional control of a variety of VV promoters [3]. Unlike other replication-defective poxviruses, late gene expression in MVA does occur (Table 2). Blockage of virus replication is evident during a late stage in morphogenesis. MVA has been used to induce Ab and CTL responses to a variety of heterologous viral antigens including those derived from influenza virus [66], parainfluenza virus [67] and SIV [68]. Furthermore, it has been shown to be an effective vector for the induction of anti-tumour immune responses [69,94].

NYVAC is a genetically-engineered VV strain that has 18 genes deleted from its genome which limit its host range and pathogenicity [70]. NYVAC replication is usually blocked at a stage after early gene expression in human cells but is able to replicate and produce infectious progeny in African green monkey kidney (VERO) and chick embryo fibroblast (CEF) cells [71]. NYVAC has also been used extensively in animal vaccines studies and more recently in human clinical trials [5]. Lastly, a conditional-lethal VV expression system has been constructed, which consists of a rVV with a deleted uracil DNA glycosylase gene. The virus is incapable of replication, except in a helper cell line that expresses the VV uracil DNA glycosylase gene [72]. This defective virus is an extremely safe expression vector and has shown much promise as a recombinant vaccine vector [73].

Table 2
Properties of poxvirus expression vectors

Strain	Early expression	Late expression	Cell lysis within 72 h	Inhibition of cell division
VV-WR	Yes	Yes	Yes	Yes
VV-MVA	Yes	Yes	Yes	Yes
VV-NYVAC	Yes	No	Yes	Yes
VV-RDVV	Yes	No	Yes	Yes
FPV	Yes	No	Yes	Yes
ALVAC	Yes	No	Yes	Yes
AmEPV	Yes	No	No	No

Note: The table describes the general properties of selected recombinant poxviruses based on infection of primate cells e.g., CV-1, BS-C-1 or HeLa [96]. It is important to note that there are variations to these general properties in specific cell lines, e.g., ALVAC and NYVAC will carry out early and late gene expression in VERO cells and NYVAC will actually replicate. Though generally all the above poxviruses with the exception of AmEPV cause cell death/lysis within 72 h, the lytic properties do vary considerably, e.g., MVA has significantly less aggressive lytic effects on most mammalian cell lines compared to the replication-competent VV strain Western Reserve (WR).

6. Non-vaccinia poxvirus vectors

The development of non-orthopoxvirus expression vectors for vaccine development has been driven by two general concerns: (1) the safety profile of orthopoxviruses; and (2) pre-existing immunity to VV in the general population vaccinated during the smallpox eradication program. Two members of the genera avipoxvirus, fowlpoxvirus (FPV [9]) and canarypox (commercially known as ALVAC) (reviewed in [5]) have shown promise as safe and effective vectors for use in animals and more importantly humans. Since avian poxviruses are immunologically distinct from VV, pre-existing immunity to the latter does not have a detrimental effect on vaccine efficacy in an animal model [74].

The growth restriction of ALVAC and FPV in mammalian cells is not clearly understood. In human cells ALVAC displays an abortive-early phenotype, significantly shutting down host protein expression by 6 h postinfection ([71], J. Tartaglia personal communication). This type of restriction may limit the use of ALVAC as a vaccine vector in humans. In contrast, the virus displays an abortive-late phenotype with clear production of immature particles in the cytoplasm of infected VERO cells. One other restriction that avipoxviruses encounter in mammalian cells is their inability to counteract the host cell response to viral infection. Premature cell death via activation of the PKR pathway is readily observed in ALVAC-infected mammalian cells. In an attempt to improve ALVAC vectors, the VV E3L and K3L genes have been introduced into their genomes [71,75].

Additional recombinant poxvirus vectors based on ectromelia virus (mouse poxvirus) [76] raccoon poxvirus [77], capripoxvirus [78] and leporipoxvirus [79] have also been developed. The expectations are that these vectors may be used as gene delivery vectors in animals and/or humans.

Lastly, a member of the subfamily entomopoxvirinae, *Amsacti moorei* (AmEPV) has been developed as a recombinant expression system [80]. Though AmEPV can be grown in insect cell lines, it is strictly replication-defective in mammalian cells [81]. Li and colleagues have shown that recombinant AmEPV expressing the *lacZ* gene can enter human liver cells and carry out substantial early gene expression. More importantly, mammalian cells infected with recombinant AmEPV can continue to replicate whilst expressing the recombinant gene product [80]. These properties indicate a possible application of entomopoxviruses as gene therapy vectors.

7. Laboratory and clinical applications

Recombinant poxviruses have been used widely for the analysis of posttranslation modifications, intracellular trafficking, protein–protein interactions, antigen processing and presentation, and the identification of antigens and epitopes for the induction of humoral and cell-mediated immune responses (reviewed in [82]).

Originally the chimeric VV-bacteriophage T7 polymerase vector was developed to obviate the problems associated with expression of cytotoxic foreign gene products; however, several alternative uses have been developed for this vector. Two major scientific breakthroughs were made possible by the rVV-T7 system: (1) the discovery of the HIV-1 T-cell coreceptor [83]; and (2) the rescue of a negative-strand RNA viruses from cDNA clones [84]. More recently, the rVV-T7 system has been used to rescue a positive-strand virus [85], and to package alphavirus replicon vectors [86], two applications that may prove useful in the development of recombinant vaccines. A recent exciting development is the application of the rVV system for the construction of cDNA libraries [87], which enabled the identification of a novel tumour-associated antigen that was recognised by a cytotoxic T cell line. This technology could be used for other discovery areas including identification of genes that induce stem-cell differentiation and those that promote apoptosis.

Studies have shown that rVV vectors can induce both antibody [88] and CTL [89] responses to a variety of recombinant antigens. A comprehensive description of such studies can be found in several detailed reviews [4,6]. Replication-competent rVV vectors first showed great potential as veterinary vaccine vectors [88]. A replication-competent rVV expressing the rabies F glycoprotein has played an important role in the near eradication of rabies in the wild life pool in northern Europe [90] and is currently being applied to a similar task in the USA. Initial human infectious disease vaccine studies using a rVV, based on the Wyeth strain (smallpox vaccine), expressing HIV-1 gp160 was able to induce both HIV antibody and T cell responses in healthy volunteers [91].

Due to environmental concerns regarding release of genetically modified organisms and potential detrimental effects of replication-competent VV in humans, especially in immunocompromised recipients, advancements have been made in the development and evaluation of vectors based on attenuated strains that do not replicate in human cells. Both NYVAC and MVA vectors expressing a variety of recombinant proteins have been assessed in numerous animal models and have demonstrated efficacy [4,5,71]. MVA expressing SIV gag-pol and env genes has been evaluated in the Rhesus macaque model, and has induced protection against challenge with SIV [68,92]. Furthermore, high levels of circulating gag-specific CD8$^+$ cells were identified after inoculation with MVA-gag-pol [93]. Both NYVAC and ALVAC have been evaluated in several clinical trials and have shown the ability to induce antibody and T cell responses to the recombinant proteins [5], [71]. Due to their capacity to induce strong T cell responses, both rVV and ALVAC vectors expressing tumour-associated antigens (TAA) have been evaluated in cancer immunotherapy clinical trials (reviewed in [7]). Initial results have been very encouraging with both antibody and T cell responses induced against the tumour antigen expressed by the recombinant viruses.

Though poxviruses are efficient vectors for the induction of immune responses to recombinant proteins, there are several ways to further improve their efficacy, including manipulation of promoter kinetics, co-expression of immune cofactors and prime boost inoculation regimes (reviewed in [7]).

Recombinant poxviruses are the basis for one of the most popular eukaryotic expression systems due to the extensive range of techniques available for their construction, ease of manipulation, high level expression, authentic posttranslational modifications and broad host range. The development of highly attenuated poxvirus vectors has broadened their application by enhancing their safety profile. The recent application of poxviruses as vectors for expression of cDNA libraries has further enhanced their versatility.

Abbreviations

CEF   chick embryo fibroblast
CPV   canarypox virus
FPV   fowlpox virus
MVA   modified vaccinia virus Ankara
PKR   RNA-dependent protein kinase
rVV   recombinant vaccinia virus
TK   thymidine kinase
VTF   vaccinia transcription factor

References

1. Moss, B. (1996) In: Fields, B.N., Knipe, D.M. and Howley, P.M. Virology. Lippincott-Raven, Phildelphia, pp. 2637–2671.
2. Earl, P., Wyatt, L.S., Moss, B. and Carroll, M.W. (1998) Current Protocols in Molecular Biology Supplement 43 unit 16, 16.17.1–16.17.19.
3. Earl, P., Wyatt, L.S., Cooper, N., Moss, B. and Carroll, M.W. (1998) Current Protocols in Molecular Biology Supplement 43, Unit 16.16.1–16.16.13.
4. Moss, B. (1996) Proc. Natl. Acad. Sci. U.S.A. 93, 11341–11348.
5. Paoletti, E. (1996) Proc. Natl. Acad. Sci. U.S.A. 93, 11349–11353.
6. Flexner, C. and Moss, B. (1997) In: Levine, M.M. Vaccina virus as a live vector for expression of immunogens, Marcel Inc, New York, pp. 297–3147.
7. Carroll, M.W. and Restifo, N.P. (2000) In: Stern, P.L., Beverley, P.L.C. and Carroll, M.W. (eds.) Poxviruses as vectors for cancer immunotherapy. Cambridge University Press, Cambridge, pp. 47–61.
8. Fenner, F. (1996) In: Fields, B.N., Knipe, D.M., Howley, P.M., et al (Ed.), "Virology" Philadelphia: Lippincott-Raven, pp. 2673–2702.
9. Boyle, D.B. and Coupar, B.E. (1988) Virus Res. 10, 343–356.
10. Fenner, F., Henderson, D.A., Arita, I., Jezek, Z. and Ladnyi, I.D. (1988) Smallpox and its Eradication, Geneva.
11. Mackett, M., Smith, G.L., and Moss, B. (1982) Proc. Natl. Acad. Sci. U.S.A. 79, 7415–7419.
12. Panicali, D. and Paoletti, E. (1982) Biotechnology 24, 503–507.
13. Smith, G.L. and Moss, B. (1983) Gene 25, 21–28.
14. Rosales, R., Sutter, G., and Moss, B. (1994) Proc. Natl. Acad. Sci. U.S.A. 91, 3794–3798.
15. Broyles, S.S., Liu, X., Zhu, M., and Kremer, M. (1999) J. Biol. Chem. 274, 35662–35667.
16. Baldick, C.J., Jr. and Moss, B. (1993) J. Virol. 67, 3515–3527.
17. Kotwal, G.J. and Moss, B. (1988) Nature 335, 176–178.
18. Upton, C., Macen, J.L., Schreiber, M., and McFadden, G. (1991) Virology 184, 370–382.
19. Ray, C.A., Black, R.A., Kronheim, S.R., Greenstreet, T.A., Sleath, P.R., Salvesen, G.S., and Pickup, D.J. (1992) Cell 69, 597–604.
20. Smith, G.L., Symons, J.A., Khanna, A., Vanderplasschen, A., and Alcami, A. (1997) Immunol. Rev. 159, 137–154.
21. Beattie, E., Paoletti, E., and Tartaglia, J. (1995) Virology 210, 254–263.
22. Chang, H.W., Watson, J.C., and Jacobs, B.L. (1992) Proc. Natl. Acad. Sci. U.S.A. 89, 4825–4829.

23. Davies, M.V., Elroy-Stein, O., Jagus, R., Moss, B., and Kaufman, R.J. (1992) J. Virol. 66, 1943–1950.
24. Xiang, Y., Condit, R.C., Vijaysri, S., Jacobs, B., Williams, B.R., and Silverman, R.H. (2002) J. Virol. 76, 5251–5259.
25. Alcami, A. and Koszinowski, U.H. (2000) Trends Microbiol. 8, 410–418.
26. Keck, J.G., Baldick, C.J., Jr., and Moss, B. (1990) Cell 61, 801–809.
27. Ward, G.A., Stover, C.K., Moss, B., and Fuerst, T.R. (1995) Proc. Natl. Acad. Sci. U.S.A. 92, 6773–6777.
28. Davison, A.J. and Moss, B. (1989) J. Mol. Biol. 210, 771–784.
29. Davison, A.J. and Moss, B. (1989) J. Mol. Biol. 210, 749–769.
30. Yuen, L. and Moss, B. (1987) Proc. Natl. Acad. Sci. U.S.A. 84, 6417–6421.
31. Franke, C.A., Rice, C.M., Strauss, J.H., and Hruby, D.E. (1985) Mol. Cell. Biol. 5, 1918–1924.
32. Falkner, F.G. and Moss, B. (1988) J. Virol. 62, 1849–1854.
33. Chakrabarti, S., Brechling, K., and Moss, B. (1985) Mol. Cell. Biol. 5, 3403–3409.
34. Wu, G.-Y., Zhou, D.-J., and Cline, H.T. (1995) Neuron 14, 681–684.
35. Carroll, M.W. and Moss, B. (1995) Biotechniques 19, 352–54, 356.
36. Cao, J.X. and Upton, C. (1997) Biotechniques 22, 276–278.
37. Hansen, S.G., Cope, T.A. and Hruby, D.E. (2002) Biotechniques 32, 1178, 1180, 1182–187.
38. Blasco, R. and Moss, B. (1995) Gene 158, 157–162.
39. Perkus, M.E., Limbach, K., and Paoletti, E. (1989) J. Virol. 63, 3829–3836.
40. Staib, C., Drexler, I., Ohlmann, M., Wintersperger, S., Erfle, V. and Sutter, G. (2000) Biotechniques 28, 1137–142, 1144–146, 1148.
41. Falkner, F.G. and Moss, B. (1990) J. Virol. 64, 3108–3111.
42. Isaacs, S.N., Kotwal, G.J., and Moss, B. (1990) Virology 178, 626–630.
43. Spehner, D., Drillien, R., and Lecocq, J.P. (1990) J. Virol. 64, 527–533.
44. Scheiflinger, F., Dorner, F., and Falkner, F.G. (1998) Arch. Virol. 143, 467–474.
45. Kurilla, M.G. (1997) Biotechniques 22, 906–910.
46. Merchlinsky, M. and Moss, B. (1992) Virology 190, 522–526.
47. Scheiflinger, F., Dorner, F., and Falkner, F.G. (1992) Proc. Natl. Acad. Sci. U.S.A. 89, 9977–9981.
48. Timiryasova, T.M., Chen, B., Fodor, N. and Fodor, I. (2001) Biotechniques 31, 534, 536, 538–540.
49. Pfleiderer, M., Falkner, F.G., and Dorner, F. (1995) J. Gen. Virol. 76, 2957–2962.
50. Merchlinsky, M., Eckert, D., Smith, E., and Zauderer, M. (1997) Virology 238, 444–451.
51. Fuerst, T.R., Niles, E.G., Studier, F.W., and Moss, B. (1986) Proc. Natl. Acad. Sci. U.S.A. 83, 8122–8126.
52. Usdin, T.B., Brownstein, M.J., Moss, B., and Isaacs, S.N. (1993) Biotechniques 14, 222–224.
53. Rodriguez, D., Zhou, Y.W., Rodriguez, J.R., Durbin, R.K., Jimenez, V., McAllister, W.T., and Esteban, M. (1990) J. Virol. 64, 4851–4857.
54. Elroy-stein, O. and Moss, B. (1998) Current protocols in Molecular Biology Supplement 43, Unit 16.19.1–16.19.11.
55. Wyatt, L.S., Moss, B., and Rozenblatt, S. (1995) Virology 210, 202–205.
56. Sutter, G., Ohlmann, M., and Erfle, V. (1995) FEBS Lett. 371, 9–12.
57. Britton, P., Green, P., Kottier, S., Mawditt, K.L., Penzes, Z., Cavanagh, D., and Skinner, M.A. (1996) J. Gen. Virol. 77, 963–967.
58. Lane, J.M., Ruben, F.L., Neff, J.M., and Millar, J.D. (1969) N. Engl. J. Med. 281, 1201–1208.
59. Mayr, A., Stickl, H., Muller, H.K., Danner, K., and Singer, H. (1978) Zentralbl. Bakteriol. [B] 167, 375–390.
60. Antoine, G., Scheiflinger, F., Dorner, F., and Falkner, F.G. (1998) Virology 244, 365–396.
61. Meyer, H., Sutter, G., and Mayr, A. (1991) J. Gen. Virol. 72, 1031–1038.
62. Carroll, M.W. and Moss, B. (1997) Virology 238, 198–211.
63. Drexler, I., Heller, K., Wahren, B., Erfle, V., and Sutter, G. (1998) J. Gen. Virol. 79, 347–352.
64. Mahnel, H. and Mayr, A. (1994) Berl. Munch. Tierarztl. Wochenschr. 107, 253–256.
65. Sutter, G. and Moss, B. (1992) Proc. Natl. Acad. Sci. U.S.A. 89, 10847–10851.
66. Sutter, G., Wyatt, L.S., Foley, P.L., Bennink, J.R., and Moss, B. (1994) Vaccine 12, 1032–1040.

136

67. Wyatt, L.S., Shors, S.T., Murphy, B.R., and Moss, B. (1996) Vaccine 14, 1451–1458.
68. Hirsch, V.M., Fuerst, T.R., Sutter, G., Carroll, M.W., Yang, L.C., Goldstein, S., Piatak, M., Jr., Elkins, W.R., Alvord, W.G., Montefiori, D.C., Moss, B., Lifson, J.D. (1996) J. Virol. 70, 3741–3752.
69. Drexler, I., Antunes, E., Schmitz, M., Wolfel, T., Huber, C., Erfle, V., Rieber, P., Theobald, M., and Sutter, G. (1999) Cancer Res. 59, 4955–4963.
70. Tartaglia, J., Perkus, M.E., Taylor, J., Norton, E.K., Audonnet, J.C., Cox, W.I., Davis, S.W., van der Hoeven, J., Meignier, B., Riviere, M. (1992) Virology 188, 217–232.
71. Tartaglia, J., Benson, J., Cornet, B., Cox, W.I., Habib, R.E., -Excler, J.-L., Franchini, G., Goebel, S., Jacobs, B.L., Klein, M., Limbach, K., -Martinez, H., Meignier, B., Pincus, S., Plotkin, S. (1997) In: Girard, M. and Dodet, B. (eds.) Onzieme Colloque des Cent Gardes October 1997. Elsevier, Paris, pp. 187–197.
72. Holzer, G.W. and Falkner, F.G. (1997) J. Virol. 71, 4997–5002.
73. Holzer, G.W., Remp, G., Antoine, G., Pfleiderer, M., Enzersberger, O.M., Emsenhuber, W., Hammerle, T., Gruber, F., Urban, C., Falkner, F.G., Dorner, F. (1999) J. Virol. 73, 4536–4542.
74. Wang, M., Bronte, V., Chen, P.W., Gritz, L., Panicali, D., Rosenberg, S.A., and Restifo, N.P. (1995) J. Immunol. 154, 4685–4692.
75. Fang, Z.Y., Limbach, K., Tartaglia, J., Hammonds, J., Chen, X., and Spearman, P. (2001) Virology 291, 272–284.
76. Kochneva, G.V., Urmanov, I.H., Ryabchikova, E.I., Streltsov, V.V., and Serpinsky, O.I. (1994) Virus Res. 34, 49–61.
77. Hu, L., Esposito, J.J., and Scott, F.W. (1996) Virology 218, 248–252.
78. Romero, C.H., Barrett, T., Chamberlain, R.W., Kitching, R.P., Fleming, M., and Black, D.N. (1994) Virology 204, 425–429.
79. Jackson, R.J., Hall, D.F., and Kerr, P.J. (1996) J. Gen. Virol. 77, 1569–1575.
80. Li, Y., Hall, R.L., and Moyer, R.W. (1997) J. Virol. 71, 9557–9562.
81. Langridge, W.H. (1983) J. Invertebr. Pathol. 42, 77–82.
82. Mackett, M. (1994) Animal Cell Biotechnology 6, 315–371.
83. Feng, Y., Broder, C.C., Kennedy, P.E., and Berger, E.A. (1996) Science 272, 872–877.
84. Schnell, M.J., Mebatsion, T., and Conzelmann, K.K. (1994) EMBO J. 13, 4195–4203.
85. Thumfart, J.O. and Meyers, G. (2002) J. Virol. 76, 6398–6407.
86. Vasilakis, N., Falvey, D., Gangolli, S., Coleman, J., Kowalski, J., Udem, S., Zamb, T., and Kovacs, G.R. (2002) Submitted.
87. Smith, E.S., Mandokhot, A., Evans, E.E., Mueller, L., Borrello, M.A., Sahasrabudhe, D.M., and Zauderer, M. (2001) Nat. Med. 7, 967–972.
88. Mackett, M., Yilma, T., Rose, J.K., and Moss, B. (1985) Science 227, 433–435.
89. Bennink, J.R., Yewdell, J.W., Smith, G.L., Moller, C., and Moss, B. (1984) Nature 311, 578–579.
90. Pastoret, P.P. and Brochier, B. (1999) Vaccine 17, 1750–1754.
91. Cooney, E.L., Collier, A.C., Greenberg, P.D., Coombs, R.W., Zarling, J., Arditti, D.E., Hoffman, M.C., Hu, S.L., and Corey, L. (1991) Lancet 337, 567–572.
92. Ourmanov, I., Brown, C.R., Moss, B., Carroll, M., Wyatt, L., Pletneva, L., Goldstein, S., Venzon, D., and Hirsch, V.M. (2000) J. Virol. 74, 2740–2751.
93. Seth, A., Ourmanov, I., Schmitz, J.E., Kuroda, M.J., Lifton, M.A., Nickerson, C.E., Wyatt, L., Carroll, M., Moss, B., Venzon, D., Letvin, N.L., Hirsch, V.M. (2000) J. Virol. 74, 2502–2509.
94. Carroll, M.W., Overwijk, W.W., Chamberlain, R.S., Rosenberg, S.A., Moss, B., and Restifo, N.P. (1997) Vaccine 4, 387–394.
95. Wyatt, L.S., Shors, S.T., Murphy, B.R., and Moss, B. (1996) Vaccine 15, 1451–1458.
96. Carroll, M.W., Wilkinson, G., and Lundstrom, K. (2001) Mammalian expression systems and vaccination. In: Ring, C. and Blair, E. (eds.) Genetically Engineered Viruses. BIOS Scientific Publishers Ltd, Oxford, UK, pp. 107–158.

S.C. Makrides (Ed.) *Gene Transfer and Expression in Mammalian Cells*

Virus-based vectors for gene expression in mammalian cells: Baculovirus

J. Patrick Condreay and Thomas A. Kost

*Department of Gene Expression and Protein Biochemistry, GlaxoSmithKline Discovery Research,
5 Moore Drive, Research Triangle Park, NC 27709, USA; Tel.: +(919) 483-9942, fax: +(919) 483-0585;
E-mail: patrick.j.condreay@gsk.com*

1. Introduction

The baculoviruses comprise a diverse group of lytic viruses with large double-stranded DNA genomes that infect arthropod species. Baculovirus vectors are, for the most part, derived from the *Autographa californica* nuclear polyhedrosis virus (AcMNPV), which infects cells from lepidopteran species [1]. In the nearly 20 years since the baculovirus expression system was first introduced in 1983 [2,3], these vectors have been used to express hundreds of recombinant proteins in insect cells. Insect cells carry out similar post-translational modifications to proteins as mammalian cells, providing a decided advantage over expression of proteins in bacterial systems. Since their introduction the vectors have been continuously engineered to make virus construction easier and more accessible to a wide spectrum of laboratories. In recent years recombinant baculovirus vectors modified to contain mammalian cell-active promoters have been developed. It has been demonstrated that these viruses can enter a wide variety of mammalian cells and direct the expression of recombinant proteins. The many advantages of this viral gene delivery system, combined with the inherent inability of the virus to replicate in mammalian target cells, make it a good choice for a wide variety of mammalian gene expression applications.

2. Baculoviruses as insect cell expression vectors

Virions of AcMNPV are found in two morphogenetic forms, each of which has a different role in the life cycle of the virus [1]. The occluded virus (OV) form consists of one or more nucleocapsids surrounded by a membrane. These nucleocapsid structures are encased in a protein matrix referred to as a polyhedron. The OV form is responsible for the *de novo* infection of insect larvae in the wild. Polyhedra are ingested by the animal and the polyhedrin matrix is dissolved in the alkaline environment of the midgut releasing nucleocapsids that infect the gut epithelium. Early in the infection cycle viral nucleocapsids are secreted from infected cells, acquiring an envelope upon passage through the cell membrane. This budded virus (BV) is responsible for cell to cell spread within an infected organism. OV is not infectious for cultured insect cells, and conversely, BV is not infectious for the intact

138

insect. Thus, the BV form is commonly used as a gene delivery vector to over-express proteins in insect cell lines.

The most plentiful viral protein synthesized during infection is the product of the polyhedrin gene. The protein is synthesized during the very late phase of infection and the mRNA and protein can constitute as much as 25% of the mRNA or protein in the cell. The polyhedrin gene is nonessential for the growth of virus in cultured cells; when the gene is deleted, only BV is released from cells. The combination of the extremely high transcriptional activity of the polyhedrin gene promoter and the ability to delete the gene from the viral genome without affecting replication formed the basis for the development of baculovirus expression vectors. Recombinant genes are placed under transcriptional control of the polyhedrin gene promoter and the new transcriptional unit is recombined into the polyhedrin gene locus. This results in construction of a recombinant virus that will deliver a recombinant gene to insect cells for high level expression. This is illustrated in Fig. 1, where extracts of insect cells infected with a recombinant virus contain sufficient amounts of recombinant protein to allow visualization on a Coomassie-stained polyacrylamide gel.

There are several additional features of baculovirus that make it highly useful for protein production in insect cells. The virus is advantageous from a biosafety perspective [4]. The BV form used in the laboratory is not infectious for the natural host of the virus, and the virus does not replicate in mammalian cells (discussed below). Thus, concerns about environmental releases are minimal as compared to other viral vector systems. In general, recombinant baculoviruses can be handled at reduced containment levels. The viral genome is able to stably accommodate insertion of large amounts of DNA [5] allowing inclusion of multiple transcriptional units. The virus can be concentrated by ultracentrifugation without a significant loss in titer and remains stable stored at 4°C in the dark [6].

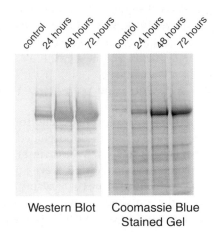

Fig. 1. Over-expression of recombinant protein in baculovirus-infected cells. Time course of expression of a protein from the polyhedrin promoter in infected insect cells. At left detection is by western blot using an antibody directed against the protein, while at right the protein can also be visualized by coomassie blue staining of total protein.

3. Molecular biology of virus vector construction

The first step in construction of a virus is the insertion of a recombinant expression cassette into the viral genome. Three major systems are in general use to derive recombinant baculoviral DNA (Fig. 2). One of these systems relies upon homologous recombination between a plasmid that contains an expression cassette flanked by viral DNA from the polyhedrin gene locus (Fig. 2A). The transfer plasmid is co-transfected into insect cells together with baculovirus DNA. Homologous recombination takes place between these DNAs to incorporate the expression cassette, and virus is produced from the cells. There are a variety of commercially available modified baculovirus nucleic acids that facilitate a high frequency of recombination allowing for efficient isolation of the recombinant viruses.

Another method (Fig. 2B) for generating recombinant baculovirus exploits site-specific recombination between the ends of transposon 7 (Tn7) and its attachment site (Tn7att) to direct the incorporation of an expression construct into viral DNA [7]. Expression cassettes are placed into a shuttle plasmid flanked by the Tn7 ends. The Tn7att site, together with a β-galactosidase selection sequence, has been inserted into a copy of the baculovirus genome that is engineered to exist in a modified *Escherichia coli* host as a single copy plasmid termed a bacmid. This *E. coli* host also contains a plasmid that encodes the Tn7 transposase. Transposition takes place within the bacteria between the transposon-derived DNA elements on the shuttle plasmid and the bacmid to transfer the expression construct into the viral genome. Transfection of the isolated bacmid DNA into insect cells results in recombinant virus production.

The last major theme in virus construction involves direct cloning into the baculovirus genome (Fig. 2C) [8]. Viral DNA has been engineered to contain unique recognition sites for a restriction endonuclease that digests DNA at a non-palindromic site with sequence ambiguity at one position. Digestion at these sites linearizes the DNA just downstream of a promoter and leaves non-complementary ends. The gene to be inserted is flanked with *Eco*R I recognition sites. The ends of these two DNAs can be partially filled in with DNA polymerase to generate complementary ends between the fragments. After ligation of the recombinant gene into the baculovirus DNA, the recombinant genome is transfected into insect cells for virus generation.

4. Baculoviruses as mammalian cell expression vectors

Baculoviruses are lytic viruses and result in death of an infected host. Thus, in the 1980s interest grew in the possibility of using these viruses as agricultural pest control agents. This in turn led to an interest into the consequences of exposure of mammalian cells to baculovirus vectors. Investigators found that viral DNA could be detected in the cytoplasm and nuclei of a variety of mammalian cell lines exposed to baculovirus. However, the viral DNA did not persist in these cultures and no evidence of viral gene expression or replication was detected [9]. Early attempts were

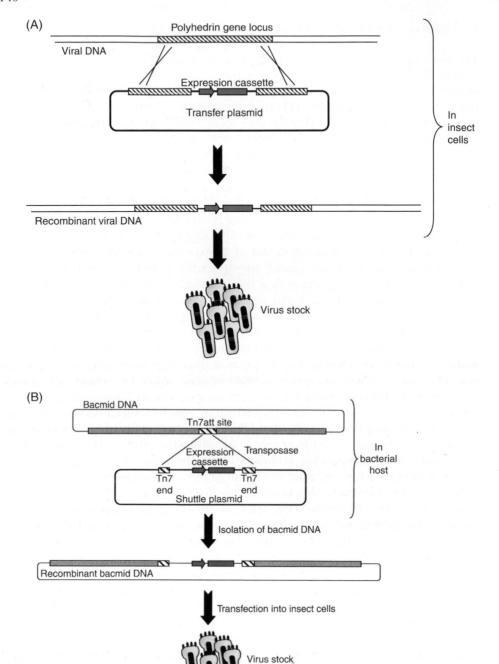

Fig. 2. Methods used for construction of recombinant baculovirus vectors. Details are in the text. (A) Homologous recombination. (B) Site-specific transposition. (C) Direct cloning.

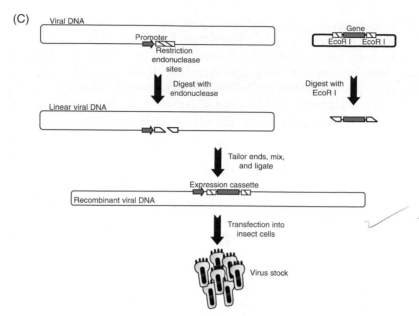

Fig. 2. Continued

made to determine whether baculoviruses containing reporter genes under the control of mammalian promoters would direct gene expression in mammalian cells. Initial findings indicated a low level of reporter gene expression in mouse L929 and human A549 lung carcinoma cells [10]. However, upon further investigation, it was shown that reporter gene expression was detected in the presence of protein synthesis inhibitors. A low level of reporter protein activity was detected associated with the virus particles themselves. These findings led the authors to conclude that the activity seen in cells treated with the modified baculovirus was carried into the cells with the virions and was not due to *de novo* reporter gene expression [11].

In the mid 1990s two groups independently reported the ability of baculoviruses carrying mammalian cell-active transcription cassettes to mediate gene delivery and expression in mammalian cells, primarily of hepatic origin [12,13]. High levels of reporter protein activity were observed after treatment of primary hepatocytes and hepatoma cell lines with viruses containing either the CMV immediate early promoter/enhancer or the RSV LTR to drive expression of reporter proteins. In these initial reports very low levels of reporter activity were observed in COS-1, T-47D, and 293 cells, but no significant expression was obtained in cell lines such as HeLa, CHO, NIH 3T3, or CV-1.

These reports were followed by other investigations that demonstrated baculovirus-mediated transduction of a wide variety of cell types beyond those of hepatic origin. Studies with a baculovirus containing a hybrid CMV-chicken β-actin promoter showed successful transduction of COS-7, HeLa, and porcine kidney cells [14]. A large panel of cell lines was investigated using a baculovirus carrying a CMV promoter-green fluorescent protein (GFP) cassette [15] that allowed analysis not only of gene expression levels, but also of the percentage of cells in a culture that express GFP. This survey greatly expanded the list of susceptible cell types to include many commonly used established cell lines such as 293, CHO, and BHK. Many of the cell lines investigated were transduced at frequencies of 70% or greater. Furthermore, primary cultures of human bone marrow fibroblasts and keratinocytes were efficiently transduced by the virus. Cells of hematopoietic origin, such as THP-1 and U937, were found to be poorly transduced by baculovirus.

5. Characteristics of baculovirus-mediated mammalian gene delivery

The only modification required for adapting baculoviruses for mammalian appli-cations is the addition of an expression cassette controlled by a mammalian cell-active promoter to the recombinant viral DNA. This is necessary since the baculoviral promoters that are generally used for insect cell over-expression are not active in mammalian cells. An example of a modified shuttle vector used for construction of viruses for mammalian gene expression applications is illustrated in Fig. 3. The polyhedrin gene promoter cassette that is present in the shuttle vector used for transposon-mediated recombination (see Fig. 2B) was removed and replaced with expression cassettes from a mammalian expression vector.

Although the initial reports of baculovirus-mediated gene delivery into mammalian cells seemed to imply that the phenomenon was restricted primarily to cells of liver origin, investigators from several laboratories have demonstrated that uptake of the virus by mammalian cells is a general phenomenon. The membrane glycoprotein of the virus, gp64, appears to be important for virus-cell recognition [16]. The cell surface recognition molecule is not well-defined. There is evidence implicating phospholipid [16], heparan sulfate [17], or a protein molecule [12] as the putative "receptor" for the virus.

Whatever the mechanism of virus-cell recognition, it seems to be ubiquitous. As more investigators utilize baculovirus vectors for gene transfer, it is becoming clear

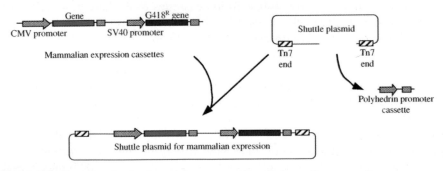

Fig. 3. Modification of a shuttle plasmid for construction of viruses for mammalian gene expression. Conversion of a shuttle plasmid to construct viruses for mammalian cell gene delivery. The polyhedrin promoter expression cassette that is used for expression in insect cells is removed and replaced with expression cassettes containing mammalian cell-active promoters such as the CMV or SV40 promoters.

Table 1

Advantages of baculovirus delivery

• Broad transduced cell profile	• Accessibility
• Versatility	• Ease of use
• Lack of cytotoxicity	• Ease of scale-up
• Modulation of expression	• Virus stability
• Viral multiplicity	• Biosafety profile
• Chemical agents	• Lab form not infectious for natural host
• Transient or stable expression	• Large insert capacity ($\geq 40\,kb$)
• Single copy integration	

that the virus is able to transduce a wide variety of established cell lines and primary cell types (see Table 1 in reference [18]). The major exception appears to be cells of hematopoietic origin [15]. Although, many cell types are transduced by the virus the efficiency of transduction varies among cell lines. This variation may be due in part to transcriptional silencing of the mammalian expression cassettes in the viral DNA by incorporation into nucleosome structures. This can be demonstrated by enhancement of protein expression levels after the addition of inhibitors of histone deacetylase, such as butyrate or trichostatin A, to baculovirus-transduced cells [15]. There is also evidence that pseudotyping the virus envelope with the G glycoprotein from vesicular stomatitis virus (VSV-G) allows more efficient gene delivery by increasing the efficiency of escape of the virus from endosomes [19].

In addition to efficient transient gene expression in mammalian cells, derivatives of transduced cells can be selected that stably maintain portions of the input baculovirus DNA. This was demonstrated by inclusion of two expression cassettes in a recombinant baculovirus. One cassette directs the expression of a reporter protein while the other expresses a dominant selectable marker so that cells that express the protein will be resistant to a particular cytotoxic drug. Chinese hamster ovary cells that were transduced by this virus gave rise to populations of cells that would propagate in the presence of cytotoxic drug and expressed the reporter protein. Analysis of isolated clones from these stable transductants showed that they contain small, single copy regions of the viral DNA (containing the recombinant expression cassettes) randomly integrated into the host cell genome (Fig. 4). These expression cassettes are active in the cell for multiple generations [20].

The manner in which the recombinant baculovirus system can be exploited for mammalian gene expression is illustrated in Fig. 5. As mentioned above, the virus is inherently unable to replicate in the target mammalian cells, thus viral stocks are routinely propagated in insect cells. Virus is added to mammalian cells in a simple liquid addition step when the cells are plated and one of three manipulations is performed. Adding virus alone will result in transient gene expression in the culture, while the addition of histone deacetylase inhibitors can enhance the level of protein expression anywhere from 5- to 40-fold depending on the cell line. Lastly, addition of a selective agent (if the correct selectable marker is present in the virus) will result in the development of stably expressing cells.

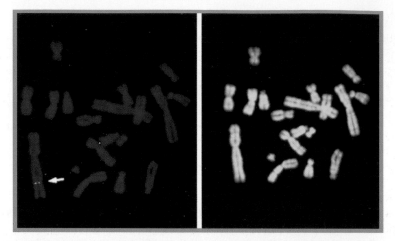

Fig. 4. Incorporation of baculovirus DNA into chromosomal DNA of CHO stable cell line. Fragments of input baculovirus DNA can be detected by fluorescent *in situ* hybridization in chromosomes isolated from a CHO cell line derived from baculovirus-transduced cells (left panel). The right panel shows the chromosomes stained with DAPI. Reprinted from reference [20]; copyright American Society for Microbiology.

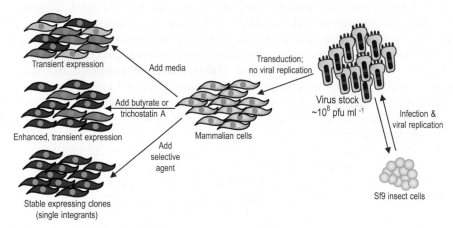

Fig. 5. Use of recombinant baculovirus vectors for gene delivery to mammalian cells. Insect Sf9 cells are used to propagate stocks of recombinant baculoviruses, since the virus is inherently non-replicative in mammalian cells. Mammalian cells can take up virus and will exhibit transient expression of genes under control of mammalian cell-active promoters. Expression levels can be enhanced by the addition of histone deacetylase inhibitors such as butyrate or trichostain A. Inclusion of an expression cassette that expresses a dominant selectable marker allows for the selection of cells that stably maintain the virus-delivered expression cassettes. Adapted from reference 18, used by permission of the publisher (Elsevier).

One further characteristic of the virus that should be mentioned is that it is rapidly inactivated by exposure to human complement [21]. This presents a significant biosafety advantage for *in vitro* mammalian cell gene delivery applications. However, it is a disappointment for those investigators that would like to exploit the inherent replication deficiency of the virus and use it for *in vivo* gene delivery applications. It has been shown that baculovirus is capable of delivering genes in complement deficient animals, or in situations where the virus is protected from exposure to complement. Thus, an active area of investigation in this field is to find methods to modify the physical properties of the virus to render it resistant to complement inactivation, and turn the vector into an *in vivo* gene delivery tool [18,22,23].

6. Applications of the baculovirus mammalian gene delivery vector

Several advantages of the baculovirus system are listed in Table 1. The items listed in the right column are general characteristics of the system that make baculoviruses very desirable for gene expression using either insect or mammalian cells. In the column on the left are characteristics of the virus that are advantageous for its application as a mammalian cell gene delivery vector. The versatility provided by the wide variety of cell types that are transduced by the virus, as well as its utility for transient or stable gene expression, have already been discussed. The level of protein expression obtained by baculovirus transduction can be modulated not only by chemical means (i.e., butyrate or Trichostatin A), but also by the multiplicity of virus used to treat a cell population [12,13,24]. Additionally, treatment with the virus, even at very high multiplicities, does not produce overt cytotoxicity in transduced cultures [18], thus providing a gentle method to introduce recombinant genes into mammalian cells. These advantages provide a unique gene delivery system that has spawned a growing body of literature detailing the application of baculoviruses to overcome some of the problems associated with gene delivery into mammalian cells.

The study of two human pathogenic viruses, hepatitis B virus (HBV) and hepatitis C virus (HCV), is hampered by the inability to propagate the viruses in cultured cell lines. It is unclear precisely where the block to productive infection lies. However, the baculovirus system provides a reliable method to deliver a nucleic acid from either HBV or HCV to the nucleus of a target cell, and launch an infection by these viruses. This concept was demonstrated by placing a greater than full length copy of the HBV genome into a baculovirus vector [25]. Upon delivery to a hepatoma cell line this construct is transcribed to yield the pregenomic RNA which is responsible for starting the HBV infection and virus is produced by the transduced cells. Production of HBV in this way has also been validated for the evaluation of anti-HBV drugs [26,27]. A baculovirus that contains a full-length cDNA copy of the HCV genome under transcriptional control of the CMV promoter can direct the synthesis of HCV proteins when used to transduce hepatoma cells [28]. No evidence was reported to indicate the production of infectious HCV, however, this may have been due to a defect in the HCV clone used in the study. Chimeric viruses such as these make possible the study of the pathogenicity of viruses and the development of cell-based assays to assess the activity of anti-viral drugs.

The production of many viral vectors that are being developed for *in vivo* gene transfer applications is carried out by transfection of one or more plasmids into cells.

This transfection step can be difficult to adapt for large scale production. Recombinant baculoviruses may provide an alternative delivery system for these transcription units. Production of adeno-associated virus (AAV) vectors can be performed using a three baculovirus system, two viruses for production of helper functions and one containing the defective viral genome, similar to the three plasmid system that is usually employed [29]. Transduction of the adenovirus E1A-complementing cell line, 293 cells, with the three baculoviruses results in production of AAV particles. In similar fashion a baculovirus-adenovirus hybrid virus was used to deliver the essential genes for replication and packaging of defective, fully deleted adenovirus vector genomes resulting in manufacture of these vectors [5]. Further modification of these systems, plus adaptation of baculovirus delivery to other viral vector systems, could present viable methods to scale-up the production of vectors for gene therapy applications.

The utility of gene delivery to mammalian cells by baculovirus vectors can be illustrated by some recent publications. Using recombinant baculoviruses that express proteins from herpes simplex virus (HSV) has proven to be an effective complementation system to support growth of viral mutants and aid in investigations into HSV biology [30]. A recombinant baculovirus engineered to express the wild type p53 gene under the control of the CMV promoter was used to exert a therapeutic effect on a p53-deficient osteosarcoma cell line [31]. Viral-mediated p53 expression in the tumor cells induced apoptotic cell death and the effect could be enhanced by addition of the chemotherapeutic agent adriamycin.

The versatility of this system for the development of cell-based functional assays is illustrated by a recent publication using recombinant baculoviruses to deliver the dual subunit K_{ATP} ion channel [24]. Stable cell lines carrying the two subunits ($K_{IR}6.2$ and SUR1) would not yield reproducible functional channel expression, thus making it difficult to develop cell-based assays for identifying modulators of channel function. Following baculovirus delivery of the K_{ATP} channel subunits to CHO and 293 cells, functional channel activity was detected as measured by membrane potential assay, ligand binding, and electrophysiology. Reproducible levels of gene expression could easily be manipulated by varying the multiplicity of the two viruses used to transduce cells. Additionally, different forms of the channel that are controlled by different regulatory subunits (SUR1, 2A, or 2B) were easily generated by mixing different viruses carrying the genes for these proteins. Use of baculovirus gene delivery for the study of this ion channel has made it possible to develop assays that are adaptable to high throughput applications.

7. Points to consider

Within a short period of time since its first demonstration, the use of baculovirus vectors for mammalian gene expression has found utility for a variety of applications. As discussed above there are many advantages of baculoviruses that make them appealing as gene transfer vectors. Also, there are a number of features of the virus that can be modified in attempts to improve the virus for mammalian cell delivery. The virus is enveloped, and can easily be pseudotyped by the addition of a different viral membrane glycoprotein such as the VSV-G protein [19]. Baculoviruses pseudotyped in this manner have been shown to exhibit enhanced transduction of cultured cells and possibly improved gene delivery *in vivo*.

The membrane glycoprotein of baculovirus, gp64, is able to tolerate fusions of proteins to its amino terminus which results in display of these fusion partners on the surface of the virion [32]. This method has also been used to modify the interaction of the virus with mammalian cells. Fusion of proteins involved in destabilizing complement complexes with gp64 render the virus resistant to complement inactivation [33]. These modified viruses are able to deliver genes *in vivo* after systemic delivery. Fusions of protein ligands for cell surface receptors to gp64 have been used in attempts to target the virus to specific cell types [34]. These modified viruses exhibited specific binding to cells, although no improvement in gene expression levels were observed. It is unclear whether the inability of baculoviruses to efficiently transduce certain cell types is due to an inability of the virus to recognize and enter those cells. Quite possibly combinations of gp64 modifications for targeting and the addition of DNA sequences that alter nucleosome structure around expression cassettes might overcome this limitation.

Transduction of mammalian cells with recombinant baculovirus vectors possibly could replace the need to develop cell lines that stably express gene products in order to develop cell-based assays. Derivation of stable cell lines is a multi-step process involving the selection of a stably transfected pool of cells, followed by single cell cloning and screening of the clones for expression and function of the desired protein. This process is made more time-consuming and complex if the active protein is made up of multiple sub-units, or if the sub-units have multiple isoforms. Archiving and maintaining multiple cell lines is also comparatively resource-intensive. In cases where extended expression of a gene product can be deleterious to cells derivation of stable cell lines may be impossible, require the use of inducible expression systems, or culturing of cells in the presence of antagonists of the protein.

The limitations of stable cell lines can be overcome with the use of recombinant baculoviruses [24]. Stocks of virus can be generated in a few weeks and are stable at 4°C for long-term storage. Once a virus stock is obtained it can be used to transduce any of a number of cell lines. Multiple viruses can be used to reconstitute protein function, and viruses that express different isoforms can be mixed to generate functional proteins that are specific for certain cell types or phenotypes. Virus can be added to cells at the time of plating in multi-well formats by simply adding an appropriate volume of virus stock to the cell inoculum, and expression can be detected in as little as 4–6 h. As mentioned above, treatment with the virus has no overt deleterious effects on the cells and expression is sufficiently easy to achieve and reproducible making the process amenable to automation.

As demonstrated by the growing number of recombinant baculovirus applications, there is tremendous potential for these viruses to become the system of choice for cell-based assay development. In addition, the virus also has promise for use as an *in vivo* delivery vehicle for gene therapy applications. However, the use of these viruses for mammalian gene delivery is still relatively uncharacterized with much to learn. Treatment with baculovirus does not cause any overt ill effects on cultures of mammalian cells and the virus does not replicate in these cells. However, it is possible that some of the immediate early gene products of the virus might be expressed in mammalian cells. It has been reported that changes in cellular gene expression may occur in response to the virus [35,36]. These concerns are not limited to baculovirus vectors, and in fact are considerations for any virus-based gene delivery system. Overall, the characteristics of recombinant baculoviruses make them a very attractive choice for mammalian gene delivery.

Acknowledgements

The authors wish to thank their many colleagues at GlaxoSmithKline for their interest and contributions.

Abbreviations

BV	budded virus
CMV	cytomegalovirus
GFP	green fluorescent protein
HBV	hepatitis B virus
HCV	hepatitis C virus
LTR	long terminal repeat
OV	occluded virus
RSV	rous sarcoma virus
Tn	transposon
VSV	vesicular stomatitis virus

References

1. Miller, L.K. (1997) The Baculoviruses. Plenum Press.
2. Smith, G.E. et al. (1983) Mol. Cell. Biol. 3, 2156–2165.
3. Pennock, G.D. et al. (1984) Mol. Cell. Biol. 4, 399–406.
4. Kost, T.A. and Condreay, J.P. (2002) Appl. Biosafety 7, in press.
5. Cheshenko, N. et al. (2001) Gene Ther. 8, 846–854.
6. Jarvis, D.L. and Garcia, A. (1994) Biotechniques 16, 508–513.
7. Luckow, V.A. et al. (1993) J. Virol. 67, 4566–4579.
8. Lu, A. and Miller, L.K. (1996) Biotechniques 21, 63–68.
9. Tjia, S.T. et al. (1983) Virology 125, 107–117.
10. Carbonell, L.F. et al. (1985) J. Virol. 56, 153–160.
11. Carbonell, L.F. and Miller, L.K. (1987) Appl. Environ. Microbiol. 53, 1412–1417.
12. Hofmann, C. et al. (1995) Proc. Natl. Acad. Sci. U.S.A. 92, 10099–10103.
13. Boyce, F.M. and Bucher, N. (1996) Proc. Natl. Acad. Sci. U.S.A. 93, 2348–2352.
14. Shoji, I. et al. (1997) J. Gen. Virol. 78, 2657–2664.

15. Condreay, J.P. *et al.* (1999) Proc. Natl. Acad. Sci. U.S.A. 96, 127–132.
16. Tani, H. *et al.* (2001) Virology 279, 343–353.
17. Duisit, G. *et al.* (1999) J. Gene Med. 1, 93–102.
18. Kost, T.A. and Condreay, J.P. (2002) Trends Biotechnol. 20, 173–180.
19. Barsoum, J. *et al.* (1997) Hum. Gene Ther. 8, 2011–2018.
20. Merrihew, R.V. *et al.* (2001) J. Virol. 75, 903–909.
21. Hofmann, C. and Strauss, M. (1998) Gene Ther. 5, 531–536.
22. Löser, P. *et al.* (2002) Curr. Gene Ther. 2, 161–171.
23. Ghosh, S. *et al.* (2002) Mol. Ther. 6, 5–11.
24. Pfohl, J.L. *et al.* (2002) Receptors Channels 8, 99–111.
25. Delaney, W.E. and Isom, H.C. (1998) Hepatology 28, 1134–1146.
26. Delaney, W.E. *et al.* (1999) Antimicrob. Agents Chemother. 43, 2017–2026.
27. Gaillard, R.K. *et al.* (2002) Antimicrob. Agents Chemother. 46, 1005–1013.
28. Fipaldini, C. *et al.* (1999) Virology 255, 302–311.
29. Sollerbrant, K. *et al.* (2001) J. Gen. Virol. 82, 2051–2060.
30. Ye, G.-J. *et al.* (2000) J. Virol. 74, 1355–1363.
31. Song, S.U. and Boyce, F.M. (2001) Exp. Mol. Med. 33, 46–53.
32. Grabherr, R. *et al.* (2001) Trends Biotechnol. 19, 231–236.
33. Hüser, A. *et al.* (2001) Nat. Biotechnol. 19, 451–455.
34. Ojala, K. *et al.* (2001) Biochem. Biophys. Res. Commun. 284, 777–784.
35. Gronowski, A.M. *et al.* (1999) J. Virol. 73, 9944–9951.
36. Beck, N.B. *et al.* (2000) Gene Ther. 7, 1274–1283.

S.C. Makrides (Ed.) *Gene Transfer and Expression in Mammalian Cells*

Virus-based vectors for gene expression in mammalian cells: Coronavirus

Luis Enjuanes, Fernando Almazán, and Javier Ortego

Department of Molecular and Cell Biology, Centro Nacional de Biotecnología, CSIC, Campus Universidad Autónoma, Cantoblanco, 28049 Madrid, Spain; Tel.: +34 (91) 585-4555; Fax: +34 (91) 585-4915. E-mail: l.enjuanes@cnb.uam.es

1. Introduction

The coronavirus and the torovirus genera form the *Coronaviridae* family, which is closely related to the *Arteriviridae* family. Both families are included in the *Nidovirales* order [1,2]. Recently, a new group of invertebrate viruses, the *Roniviridae*, with a genetic structure and replication strategy similar to those of coronaviruses, has been described [3]. This new virus family has been included within the *Nidovirales* [4].

Coronaviruses have several advantages as vectors over other viral expression systems: (i) coronaviruses are single-stranded RNA viruses that replicate within the cytoplasm without a DNA intermediary, making integration of the virus genome into the host cell chromosome unlikely [5]; (ii) these viruses have the largest RNA virus genome and, in principle, have room for the insertion of large foreign genes [1,6]; (iii) a pleiotropic secretory immune response is best induced by the stimulation of gut associated lymphoid tissues. Since coronaviruses in general infect the mucosal surfaces, both respiratory and enteric, they may be used to target the antigen to the enteric and respiratory areas to induce a strong secretory immune response; (iv) the tropism of coronaviruses may be modified by manipulation of the spike (S) protein allowing engineering of the tropism of the vector [7,8]; (v) non-pathogenic coronavirus strains infecting most species of interest (human, porcine, bovine, canine, feline, and avian) are available to develop expression systems; and (vi) infectious coronavirus cDNA clones are available to design expression systems.

Within the coronavirus two types of expression vectors have been developed (Fig. 1), one requires two components (helper-dependent expression system) (Fig. 1A) and the other a single genome that is modified either by targeted recombination [6] (Fig. 1B.1) or by engineering a cDNA encoding an infectious RNA. Infectious cDNA clones are available for porcine [9,10] (Fig. 1B.2 and B.3), human (Fig. 1B.4) [11], murine [12] and avian (infectious bronchitis virus, IBV) coronavirus [13], and also for the arteriviruses equine infectious anemia virus (EAV) [14], porcine respiratory and reproductive syndrome virus (PRRSV) [15], and simian hemorrhagic fever virus (SHFV) [16]. The availability of these cDNAs and the application of target recombination to coronaviruses [6] have been essential for the development of vectors based on coronaviruses and arteriviruses.

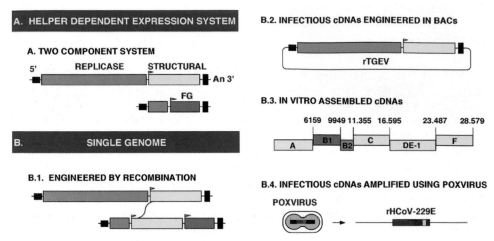

Fig. 1. Coronavirus-derived expression systems. (A) Helper dependent expression system based on two components, the helper virus and a minigenome carrying the foreign gene (FG). An, poly A. (B) Single genome engineered by targeted recombination (B.1), by assembling an infectious cDNA clone derived from TGEV genome in BACs (B.2), by the *in vitro* ligation of six cDNA fragments (B.3), or by using poxviruses as the cloning vehicle (B.4).

This review will focus on the advantages and limitations of these coronavirus expression systems, the attempts to increase their expression levels by studying the transcription-regulating sequences (TRSs), and the proven possibility of modifying their tissue and species-specificity.

2. Coronavirus pathogenicity

Coronaviruses comprise a large family of viruses infecting a broad range of vertebrates, from mammalian to avian species. Coronaviruses are associated mainly with respiratory, enteric, hepatic and central nervous system diseases. In humans and fowl, coronaviruses primarily cause upper respiratory tract infections, while porcine and bovine coronaviruses (BCoVs) establish enteric infections that result in severe economical loss. Human coronaviruses (HCoV) are responsible for 10–20% of all common colds, and have been implicated in gastroenteritis, high and low respiratory tract infections and rare cases of encephalitis. HCoV have also been associated with infant necrotizing enterocolitis and are tentative candidates for multiple sclerosis. In March 2003, a new group of HCoVs has emerged as the ethiological agent of the severe acute pneumonia syndrome (SARS) affecting thousands of people, mostly in China, Singapore, and Toronto. In addition, human infections by coronaviruses seem to be ubiquitous, as coronaviruses have been identified wherever they have been looked for, including North and South America, Europe, and Asia and no other human disease has been clearly associated with them with the exception of respiratory and enteric infections.

3. Molecular biology of coronavirus

3.1. Coronavirus genome

Virions contain a single molecule of linear, positive-sense, single-stranded RNA (Fig. 2). The coronavirus genome with a size ranging from 27.6 to 31.3 kb is the largest viral RNA known. Coronavirus RNA has a 5′ terminal cap followed by a leader sequence of 65–98 nucleotides and an untranslated region of 200–400 nucleotides. At the 3′ end of the genome there is an untranslated region of 200–500 nucleotides followed by a poly(A) tail. The virion RNA, which functions as an mRNA and is infectious, contains approximately 7–10 functional genes, four or five of which encode structural proteins. The genes are arranged in the order 5′-polymerase-(HE)-S-E-M-N-3′, with a variable number of other genes that are believed to be non-structural and largely non-essential, at least in tissue culture [1].

About two-thirds of the entire RNA comprises the Rep1a and Rep1b genes. At the overlap between the Rep1a and 1b regions, there is a specific seven-nucleotide "slippery" sequence and a pseudoknot structure (ribosomal frameshifting signal), which are required for the translation of Rep1b as a single polyprotein (Rep1a/b). The 3′ third of the genome comprises the genes encoding the structural proteins and the other non-structural ones.

Coronavirus transcription occurs via an RNA-dependent RNA synthesis process in which mRNAs are transcribed from negative-stranded templates.

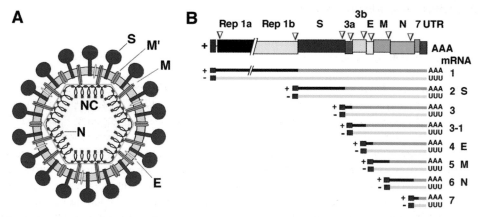

Fig. 2. Coronavirus structure and genome organization. (A) Schematic diagram of coronavirus structure using TGEV as a prototype. The diagram shows the envelope, the core and the nucleoprotein structure. S, spike protein; M and M′, large membrane proteins with the amino-terminus facing the external surface of the virion and the carboxy-terminus towards the inside or the outside surface of the virion, respectively; E, small envelope protein; N, nucleocapsid protein; NC, nucleocapsid. (B) Representation of a prototype TGEV coronavirus genome and subgenomic RNAs. Beneath the top bar a set of positive- and negative-sense mRNA species synthesized in infected cells is shown. Poly(A) and Poly(U) tails are indicated by AAA or UUU. Rep1a and Rep1b, replicase genes; UTR, untranslated region; ∇, indicates presence of a TRS; other acronyms as in A.

Coronavirus mRNAs consist of six to eight types of varying sizes, depending on the coronavirus strain and the host species. The largest mRNA is the genomic RNA that also serves as the mRNA for Rep1a and 1b, the remainder are subgenomic mRNAs (sgmRNAs). The mRNAs have a nested-set structure in relation to the genome structure (Fig. 2B).

3.2. Coronavirus proteins

Coronaviruses are enveloped viruses containing a core that includes the ribonucleoprotein formed by the RNA and nucleoprotein N (Fig. 2A). The core is formed by the genomic RNA, the N protein and the membrane (M) protein carboxy-terminus. Most of the M protein is embedded within the membrane but its carboxy-terminus is integrated within the core and seems essential to maintain the core structure [17,18]. At least in the transmissible gastroenteritis virus (TGEV) the M protein presents two topologies. In one (M′), both the amino and the carboxyl termini face the outside of the virion, while in the other (M) the carboxy-terminus is inside [18]. In addition, the virus envelope contains two or three other proteins, the spike (S) protein that is responsible for cell attachment, the small membrane protein (E) and, in some strains, the hemagglutinin-esterase (HE) [1].

The replicase gene encodes a protein of approximately 740–800 kDa which is co-translationally processed. Several domains within the replicase have predicted functions based on regions of nucleotide homology [19].

4. Helper-dependent expression systems

The coronaviruses have been classified into three groups (1, 2 and 3) based on sequence analysis of a number of coronavirus genes [1]. Helper-dependent expression systems have been developed using members of the three groups of coronaviruses (Fig. 1).

4.1. Group 1 coronaviruses

Group 1 coronaviruses include porcine, canine, feline and HCoVs. Expression systems have been developed for the porcine and HCoVs since minigenomes are only available for these two coronaviruses. One expression system has been developed using TGEV-derived minigenomes [20]. The TGEV-derived RNA minigenomes were successfully expressed *in vitro* using T7 polymerase and amplified after *in vivo* transfection using a helper virus. TGEV-derived minigenomes of 3.3, 3.9 and 5.4 kb were efficiently used for the expression of heterologous genes [21,22]. The smallest minigenome replicated by the helper virus and efficiently packaged was 3.3 kb in length [20].

Using M39 minigenome, a two-step amplification system was developed based on the cloning of a cDNA copy of the minigenome after the immediate-early cytomegalovirus promoter (CMV) [20]. Minigenome RNA is first amplified in the nucleus by the cellular RNA *pol II* and then, the RNAs are translocated into the cytoplasm where they are amplified by the viral replicase of the helper virus. The β-glucuronidase (GUS) and a surface glycoprotein (ORF5), that is the major protective antigen of the PRSSV, have been expressed using this vector [22]. TGEV-derived helper expression systems have a limited stability and minigenomes without the foreign gene replicate about 50-fold more efficiently than those with the heterologous gene [22]. Expression of GUS gene and PRRSV ORF5 with these minigenomes has been demonstrated in the epithelial cells of alveoli and in scattered pneumocytes of swine lungs, which led to the induction of a strong immune response to these antigens [22].

The HCoV-229E has also been used to express new sgmRNAs [23]. It was demonstrated that a synthetic RNA including 646 nt from the 5′ end plus 1465 nt from the 3′ end was amplified by the helper virus.

4.2. Group 2 coronaviruses

Most of the work has been done with mouse hepatitis virus (MHV) defective RNAs (minigenomes) [24,25]. Three heterologous genes have been expressed using the MHV system, chloramphenicol acetyltransferase (CAT), HE, and interferon γ (IFN-γ). Expression of CAT or HE was detected only in the first two passages because the minigenome used lacks the packaging signal [26]. When virus vectors expressing CAT and HE were inoculated intracerebrally into mice, HE- or CAT-specific sgmRNAs were only detected in the brains at days 1 and 2, indicating that the genes in the minigenome were expressed only in the early stage of viral infection [27].

A MHV minigenome RNA was also developed as a vector for expressing IFN-γ. The murine IFN-γ gene was secreted into culture medium as early as 6 h post-transfection and reached a peak level at 12 h post-transfection. No inhibition of virus replication was detected when the cells were treated with IFN-γ produced by the minigenome RNA, but infection of susceptible mice with a minigenome producing IFN-γ caused significantly milder disease, accompanied by less virus replication than that caused by virus containing a control vector [25,28].

4.3. Group 3 coronaviruses

IBV is an avian coronavirus with a single-stranded, positive-sense RNA genome of 27,608 nt. A defective RNA (CD-61) derived from the Beaudette strain of the IBV virus was used as an RNA vector for the expression of two reporter genes, luciferase and CAT [29].

4.4. Heterologous gene expression levels in helper-dependent expression systems

Helper-dependent expression systems have a limited stability probably due to the foreign gene since TGEV minigenomes of 9.7, 3.9 and 3.3 kb, in the absence of the heterologous gene, are amplified and efficiently packaged for at least 30 passages, without generating new dominant subgenomic RNAs [20]. The expression of GUS,

PRSSV ORF5, or CAT using TGEV- or IBV-derived minigenomes in general increases until passages three or four, expression levels are maintained for about four additional passages, and steadily decrease during successive passages [20–22,29]. Using IBV minigenomes CAT expression levels between 1 and 2 µg per 10^6 cells have been described. The highest expression levels (2–8 µg of GUS per 10^6 cells) have been obtained using a two-step amplification system based on TGEV derived mini-genomes with optimized TRSs [20,21]. Using minigenomes derived from TGEV and IBV expression was highly dependent on the nature of the heterologous gene used. Luciferase expression with TGEV and IBV minigenomes was reduced to almost background levels, while expression of GUS or CAT was at least 100–1000-fold higher than background levels, respectively.

5. Single genome coronavirus vectors

5.1. Group 1 coronaviruses

The construction of cDNA clones encoding full-length coronavirus RNAs has considerably improved the genetic manipulation of coronaviruses. The enormous length of the coronavirus genome and the instability of plasmids carrying coronavirus replicase sequences have hampered, until recently, the construction of a full-length cDNA clone. Infectious coronavirus cDNA clones have been described for coronaviruses [9,10,13] and for arteriviruses [14,15].

The strategy used to clone TGEV infectious cDNA was based on three points [9]: (i) the construction was started from a minigenome that was stably and efficiently replicated by the helper virus [20]. During the filling in of minigenome deletions a cDNA fragment that was toxic to the bacterial host was identified. This fragment was reintroduced into the cDNA in the last cloning step; (ii) in order to express the long coronavirus genome and to add the 5′ cap required for TGEV RNA infectivity, a two-step amplification system that couples transcription in the nucleus from the CMV promoter, with a second amplification in the cytoplasm driven by the viral polymerase, was used; and (iii) to increase viral cDNA stability within bacteria, the cDNA was cloned as a bacterial artificial chromosome (BAC), that produces a maximum of two plasmid copies per cell. A fully functional infectious TGEV cDNA clone, leading to a virulent virus infecting both the enteric and respiratory tract of swine was engineered. The stable propagation of a TGEV full-length cDNA in bacteria as a BAC has been considerably improved by the insertion of an intron to disrupt a toxic region identified in the viral genome (Fig. 3) [30]. The viral RNA was expressed in the cell nucleus under the control of the CMV promoter and the intron was efficiently removed during translocation of this RNA to the cytoplasm. Intron insertion in two different positions (nt 9466 and 9596) allowed stable plasmid amplification for at least 200 generations. Infectious TGEV was efficiently recovered from cells transfected with the modified cDNAs. The great advantage of this system is that the performance of coronavirus reverse genetics only involves recombinant DNA technologies carried out within the bacteria.

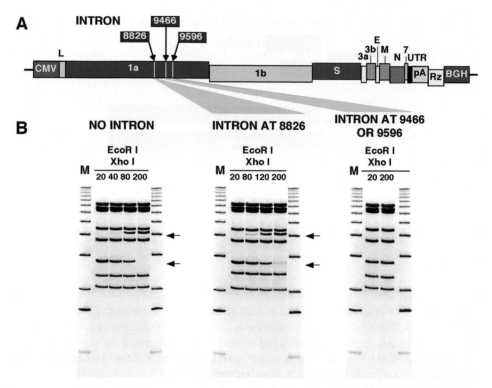

Fig. 3. Intron insertion to stabilize TGEV full-length cDNA. (A) Strategy for the insertion of the 133-nt intron in the indicated positions of the TGEV sequence. (B) Analysis of the three intron-containing TGEV full-length cDNAs in *Escherichia coli* cells. The *Eco*RI–*Xho*I restriction patterns of the three plasmids extracted from *E. coli* cells grown for the indicated number of generations are shown. Arrows indicate disappearance or appearance of a band. M, molecular mass markers. UTR, untranslated region. Rz, hepatitis delta virus ribozyme; BGH, bovine growth hormone termination and polyadenylation sequences. Other acronyms as in Fig. 2.

Using TGEV cDNA the green fluorescent protein (GFP) gene of 0.72 kb was cloned in two positions of the RNA genome: either by replacing the non-essential 3a and 3b genes or between genes N and 7. The engineered genome was very stable (>30 passages in cultured cells) and led to the production of high expression levels ($50\,\mu g/10^6$ cells) when the GFP replaced genes 3a/b but was unstable when cloned between genes N and 7 [31]. In this case, the GFP gene was eliminated by homologous recombination between preexisting TRS sequences and those introduced to express GFP. Using the most stable vector, the acquisition of immunity by newborn piglets breast fed by immunized sows (lactogenic immunity) was demonstrated [31]. GUS expression levels using coronavirus based vectors are similar (Fig. 4) to those described for vectors derived from other positive strand RNA viruses such as Sindbis virus ($50\,\mu g$ per 10^6 cells) [32].

A second procedure to assemble a full-length infectious construct of TGEV was based on the *in vitro* ligation of six adjoining cDNA subclones that span the entire

158

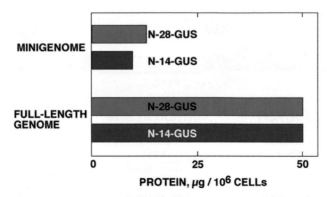

Fig. 4. GUS expression levels using TGEV-based vectors. The amount of protein expressed with the helper dependent expression system (minigenome) is compared with the expression levels from full-length genomes using the same TRSs (N-14 or N-28) from N gene.

TGEV genome [10]. Each clone was engineered with unique flanking interconnecting junctions that determine a precise assembly with only the adjacent cDNA subclones, resulting in a TGEV cDNA. *In vitro* transcripts derived from the full-length TGEV construct were infectious. Using this construct, a recombinant TGEV was assembled that replaced ORF 3a with the GFP gene, leading to the production of a recombinant TGEV that grew with titers of 10^8 pfu/ml and expressed GFP in a high proportion of cells [33].

An infectious cDNA clone has also been constructed for HCoV-229E, another member of group 1 coronaviruses [11]. In this case, the system is based on the *in vitro* transcription of infectious RNA from a cDNA copy of the HCoV-229E genome that has been cloned and propagated in vaccinia virus (Fig. 5). Briefly, the full-length genomic cDNA clone of HCoV-229E was assembled by *in vitro* ligation, and then cloned into the vaccinia virus DNA under control of the T7 promoter. Recombinant vaccinia viruses containing the HCoV-229E genome were recovered after transfection of the recombinant vaccinia virus DNA into cells infected with fowlpox virus. In a second phase, the recombinant vaccinia virus DNA was purified and used as a template for *in vitro* transcription of HCoV-229E genomic RNA that was transfected into susceptible cells for the recovery of infectious recombinant coronavirus (Fig. 5).

A coronavirus replicon has been derived from the HCoV genome using the same procedure described for the full-length genome construction [34]. This replicon included the 5′ and 3′ ends of the HCoV-229E genome, the replicase gene of this virus and a single reporter gene coding for GFP located downstream of a TRS element for coronavirus mRNA transcription. When RNA transcribed from this cDNA was transfected into BHK-21 cells, only 0.1% of the cells showed strong fluorescence. This data shows that the coronavirus replicase gene products suffice for discontinuous sgmRNA transcription, in agreement with the requirements for the arterivirus replicase [35].

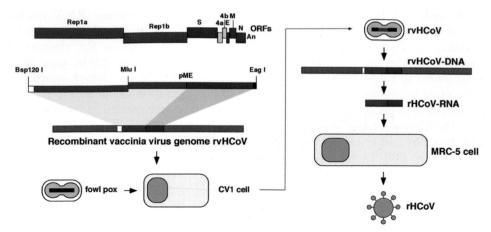

Fig. 5. Infectious HCoV-229E recovery from cDNA. The genetic map of HCoV-229E (top) is shown. ORFs encoding virus replicase proteins and structural or non-structural proteins are colored in light and dark gray, respectively. Vaccinia virus DNA (black box) and the T7 promoter (white box) are also shown [11].

The expression of a heterologous gene (GFP) by a TGEV replicon was increased between 300- and 400-fold when TGEV N protein was *in cis* co-expressed [36]. In addition, expression from a TGEV replicon was also observed when N protein was *in trans* co-expressed using the Venezuelan equine encephalitis virus vector [36]. Furthermore, expression from HCoV based vectors also was significantly increased by co-expression of N gene. Therefore, it seems that N protein either stabilizes coronavirus replicons or increases their replication, transcription or translation.

5.2. Group 2 coronaviruses

Reverse genetics in this coronavirus group has been efficiently performed by targeted recombination between a helper virus and either non-replicative or replicative coronavirus-derived RNAs (Fig. 2B.1). This approach, developed by Masters' group [6,37], was first applied to the engineering of a five-nucleotide insertion into the 3′ untranslated region (3′ UTR) of MHV [37]. This approach was facilitated by the availability of an N gene mutant, designated Alb4, that was both temperature-sensitive and thermolabile. Alb4 forms tiny plaques at restrictive temperature that are easily distinguishable from wild-type plaques. In addition, incubation of Alb4 virions at non-permissive temperature results in a 100-fold greater loss of titer than for wild-type virions. These phenotypic traits allowed the selection of recombinant viruses generated by a single crossover event following cotransfection into mouse cells of Alb4 genomic RNA together with a synthetic copy of the smallest subgenomic RNA (RNA7) tagged with a marker in the 3′ UTR.

An improvement of the recombination frequency was obtained between the helper virus and replicative defective RNAs as the donor species. Whereas between replication competent MHV and non-replicative RNAs a recombination frequency of the order of 10^{-5} was estimated, the use of replicative donor RNA yielded recombinants at a rate of some three orders of magnitude higher [38]. This higher efficiency made it possible to screen for recombinants even in the absence of selection. In this manner, the transfer of silent mutation in Rep1a gene of a minigenome to wild-type MHV at a frequency of about 1% was demonstrated.

Targeted recombination has been applied to the generation of mutants in most of the coronavirus genes. Thus, two silent mutations have been created so far in gene 1a [38]. The S protein has also been modified by targeted recombination. Changes were introduced by one crossover event at the 5′ end of the S gene that modified MHV pathogenicity [39]. Targeted recombination mediated by two cross-overs allowed the replacement of the S gene of a respiratory strain of TGEV by the S gene of enteric TGEV strain PUR-C11 leading to the isolation of viruses with a modified tropism and virulence [7]. In this case the recombinants were selected *in vivo* using their new tropism in piglets. A new strategy for the selection of recombinants within the S gene, after promoting targeting recombination, was based on elimination of the parental replicative TGEV by the simultaneous neutralization with two mAbs (I. Sola and L. Enjuanes, unpublished results). Mutations have also been introduced by targeted mutagenesis within the E and M genes. These mutants provided corroboration for the pivotal role of E protein in coronavirus assembly and identified the carboxyl terminus of the M molecule as crucial to assembly [40].

Targeted recombination was also used to express heterologous genes. For instance, the gene encoding GFP was inserted into MHV between genes S and E, resulting in the creation of the largest known RNA viral genome [41].

An infectious MHV cDNA clone has recently been assembled *in vitro*. A method similar to the one developed to assemble an infectious TGEV cDNA clone based on the *in vitro* ligation of seven contiguous cDNA subclones has been applied to the construction of a cDNA that spanned the 31.5 kb of the MHV A59 strain [12]. The ends of the cDNAs were engineered with unique junctions, which were directed to assembly with only the adjacent cDNAs subclone, resulting in an intact MHV-A59 cDNA construct. The interconnecting restriction site junctions that are located at the ends of each cDNA are systematically removed during the assembly of the complete full-length cDNA product, allowing reassembly without the introduction of nucleotide changes. RNA transcripts derived from the full-length MHV-A59 construct was infectious, although virus recovery was enhanced 10–15-fold in the presence of RNA transcripts encoding the nucleocapsid protein, N.

5.3. Group 3 coronaviruses

The infectious IBV cDNA clone was assembled using the same strategy reported for HCoV-229E with some modifications [13]. Similarly to HCoV-229E, the IBV genomic cDNA was assembled downstream of the T7 promoter by *in vitro* ligation and cloned into the vaccinia virus DNA. However, recovery of recombinant IBV was done after the *in situ* synthesis of infectious IBV RNA by transfection of restricted recombinant vaccinia virus DNA (containing the IBV genome) into primary chick kidney cells previously infected with a recombinant fowlpox expressing T7 RNA polymerase.

Engineered cDNAs are having an important impact on the study of mechanisms of coronavirus replication and transcription and provide an invaluable tool for the experimental investigation of virus–host interactions.

5.4. Replication-competent propagation-deficient coronavirus-derived expression systems

Replication-competent propagation-deficient virus vectors based on TGEV genomes deficient in the essential gene E that are complemented in packaging cell lines have been developed [33,42]. Two types of cell lines expressing TGEV E protein have been established, one with transient expression using the non-cytopathic Sindbis virus replicon pSINrep21 (Fig. 6) and another stably expressing E gene under the CMV promoter. The rescue of recombinant TGEV deficient in the non-essential 3a and 3b genes, and the essential E gene reached high titers ($>6 \times 10^6$ pfu/ml) in cells transiently expressing the TGEV E protein, while this titer was up to 5×10^5 pfu/ml in packaging cell lines stably expressing protein E. Interestingly, virus titers were related to protein E expression levels [42]. Recovered virions showed the same morphology and stability at different pH and temperatures than the wild type virus.

A second strategy for the construction of replication-competent propagation-defective TGEV genomes expressing heterologous genes, involves the assembly of an infectious cDNA from six cDNA fragments that are ligated in vitro [33]. The defective virus with the essential E gene deleted was complemented by the expression of E gene using the Venezuelan equine encephalitis replicon expression vector. However, titers of recombinant TGEV-ΔE expressing the GFP were at least

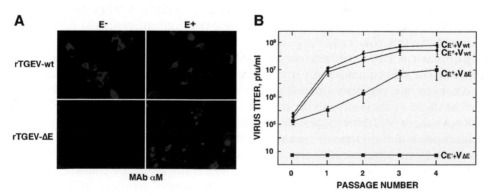

Fig. 6. Rescue of recombinant TGEV-ΔE from cDNA in cells transiently expressing E protein. (A) Analysis by immunofluorescence with M protein specific monoclonal antibodies of the rescue of full-length recombinant TGEV (rTGEV-wt) or TGEV cDNA with the E gene deleted (rTGEV-ΔE) in normal BHK-APN cells (E⁻) or cells expressing TGEV E protein (E⁺). (B) Titers through passage of recombinant TGEV rescued from cDNA in BHK-APN cells (CE⁻) or BHK-APN cells expressing TGEV E protein (CE⁺) transfected either with rTGEV-wt (Vwt) or with rTGEV-ΔE virus (VΔE). Error bars represent standard deviations of the mean from four experiments.

10- or 100-fold lower (around 10^4 pfu/ml) than with the system using stably transformed cells or the SIN vector to complement deletion of the E gene [42].

5.5. Cloning capacity of coronavirus expression vectors

Coronavirus minigenomes have a theoretical cloning capacity close to 27 kb, since their RNA with a size of about 3 kb is efficiently amplified and packaged by the helper virus and the virus genome has about 30 kb. In contrast, the theoretical cloning capacity for an expression system based on a single coronavirus genome like TGEV according to current available knowledge is between 3 and 3.5 kb taking into account that: (i) the non-essential genes 3a (0.2 kb), 3b (0.73 kb), and most of gene 7 [43] have been deleted leading to a viable virus; (ii) the standard S gene can be replaced by the S gene of a porcine respiratory coronavirus (PRCV) mutant with a deletion of 0.67 kb; and (iii) both DNA and RNA viruses may accept genomes with sizes up to 105% of the wild type genome. This cloning capacity will probably be enlarged by deleting non-essential domains of the replicase gene. These domains are being identified by comparing the arterivirus replicase gene (i.e., for EAV 9.7×10^3 nt) and that of coronavirus (i.e., for TGEV 20.3×10^3 nt) [19]. Differences in size between these two replicase genes could correspond to non-essential domains in the coronavirus replicase that may be dispensable.

6. Optimization of transcription levels

To optimize expression levels it is essential to improve virus vector replication levels without increasing virulence, to optimize the accumulation of total mRNA levels, and to improve mRNA translation. These results can only be achieved by determining the mechanism involved in these processes. A brief review of the mechanism of mRNA transcription in coronavirus and arterivirus is described to help achieve this goal.

Coronavirus RNA synthesis occurs in the cytoplasm via a negative-strand RNA intermediate that contains short stretches of oligo(U) at the 5' end. Both genome-size and subgenomic negative-strand RNAs, which correspond in number of species and size to those of the virus-specific mRNAs have been detected [44,45]. Coronavirus mRNAs have a leader sequence at their 5' ends. At the start site of every transcription unit on the viral genomic RNA, there is a TRS that includes a highly conserved core sequence (CS) that is nearly homologous to the 3' end of the leader RNA. This sequence constitutes part of the signal for sgmRNA transcription. The common 5' leader sequence is only found at the very 5' terminus of the genome, which implies that the synthesis of sgmRNAs involves fusion of non-contiguous sequences. The mechanism involved in this process is under debate. Nevertheless, the *discontinuous transcription during negative-strand RNA synthesis* model is compatible with most of the experimental evidence [45–47]. Because the leader–mRNA junction occurs during the synthesis of the negative strand within the sequence complementary to the CS (cCS) the nature of the CS is considered crucial for mRNA synthesis. Transcription levels may be influenced by many factors. The three that we [21] consider most relevant are: (i) potential base pairing between the leader 3' end and sequences complementary to the TRS located at the 5' end of each nidovirus gene (cTRS), that guide the fusion between the nascent negative strand and the leader TRS. A minimum complementarity is needed between the leader-TRS and the cTRSs of each gene. Extension of this complementarity increases mRNA synthesis up to a certain extent, beyond a certain extension addition of 5' or 3' CS flanking sequences does not help transcription [21,48,49]; (ii) proximity of a gene to the 3'. Since the TRSs act as signals to slow down or stop the replicase complex, the smaller mRNAs should be the most abundant. Although this has been shown to be the case in the *Mononegavirales* [50] and in coronaviruses shorter mRNAs are in general more abundant, the relative abundance of coronavirus mRNAs is not strictly related to their proximity to the 3' end [21,51]; and (iii) potential interaction of proteins with the TRSs RNA, and protein–protein interactions that could regulate transcription levels. The reassociation of the nascent RNA chain with the leader TRS is probably mediated by approximation of the leader TRS through RNA–protein and protein–protein interactions.

6.1. Characteristics of the TRS

The three factors implicated in the control of mRNA abundance assume a key role for the TRS. Hence, in order to engineer vectors with high expression levels, it seems relevant to define the characteristics of the TRS, including the size of the 5' and 3' TRS sequences flanking the CS. The CS of coronaviruses belonging to groups I (hexamer 5'-CUAAAC-3') and II (heptamer 5'-UCUAAAC-3') share homology, whereas the CS of coronaviruses belonging to group III, like that of IBV, have the most divergent sequence (5'-CUUAACAA-3'). Also, arterivirus CSs have a sequence (5'-UCAACU-3') that partially resembles that of IBV. Thus, the CSs of different coronaviruses are quite similar though slightly different in length. This CS is essential for mRNA synthesis, and can be considered to be a defined domain in the TRS because it is particularly conserved within a nidovirus family, while the flanking sequences, both at the 5' (5' TRSs) and at the 3' (3' TRSs) have a unique composition for each gene even within the same virus.

The influence of the CS in transcription has been analyzed in detail in the arteriviruses [46,47]. Using an infectious cDNA clone of EAV it has been shown that sgmRNA synthesis requires base pairing interaction between the leader TRS and cTRS in the viral negative strand. The construction of double mutants in which a mutant leader CS was combined with the corresponding mutant RNA7 body CS, resulting in the specific restoration of mRNA7 synthesis, suggested that the sequence of the CS *per se* is not crucial, as long as the possibility for CS base pairing is maintained. Nevertheless, it has been shown that other factors, besides leader–body base pairing, also play a role in sgmRNA synthesis and that the primary sequence (or secondary structure) of TRSs may dictate strong base preferences at certain positions [46]. In addition, detailed analysis of the TRS used in the arteriviruses [47], MHV [52], BCoV [53], and TGEV [31] indicate that non-canonical CS sequences may also be used for the switch during the discontinuous synthesis of the negative strand during transcription in the *Nidovirales*.

The promotion of transcription from a given CS is also a function of the CS flanking sequences. Data from different laboratories working with different nidoviruses have shown that CS flanking sequences can critically influence the strength of a given fusion site [21,48,49,53–55]. Although approximations to the definition of the TRS have been made, the precise length of the TRS requires further work to optimize accumulation of mRNA levels.

6.2. Effect of CS copy number on transcription

Studies on coronavirus transcription were performed using more than one CS to express the same mRNA. The accumulated amounts of sgmRNA remained nearly the same for constructs with one to three CSs, and transcription preferentially occurred at the 3′-most TRS [29,56–58]. This observation is consistent with the coronavirus discontinuous transcription during the negative-strand synthesis model [59].

7. Modification of coronavirus tropism and virulence

Driving vector expression to different tissues may be highly convenient in order to preferentially induce a specific type of immune response, i.e., mucosal immunity by targeting the expression to gut-associated lymph nodes. In addition, it seems useful to change the species specificity of the vector to expand its use. Both tissue- and species-specificity have been modified using coronavirus genomes.

Group 1 coronaviruses attach to host cells through the S glycoprotein by interactions with aminopeptidase N (APN) which is the cellular receptor [60,61]. Group 2 coronaviruses use the carcinoembryonic antigen-related cell adhesion molecules (CEACAM) as receptors. Engineering the S gene can lead to changes both in the tissue- and species-specificity [7,8].

Tropism change in general leads to a change in virulence. Certainly this is the case in porcine coronavirus with a virulence directly related to its ability to grow in the enteric tract [7]. Gene expression among the non-segmented negative-stranded RNA viruses is controlled by the highly conserved order of genes relative to the single transcriptional promoter. Rearrangement of the genes of vesicular stomatitis virus eliminates clinical disease in the natural host and is considered a new strategy for vaccine development [62]. In coronavirus, genes closer to the 3′ end are in general expressed more abundantly than 3′ end distal ones and, in principle, gene order change can also lead to virus attenuation (P. Rottier, personal communication).

8. Expression systems based on arteriviruses

The *Arteriviridae* include four members: EAV, PRRSV, SHFV and lactate dehydrogenase-elevating virus of mice (LDHV). Defective genomes of EAV have been isolated and used to express a reporter gene (CAT) in cell culture [35]. More interestingly, infectious cDNA clones have been obtained for EAV [14], PRRSV [14,63] and SHFV [16] creating the possibility of specifically altering their genomes for vector development and vaccine production. To insert genes in different positions a unique restriction endonuclease site has been introduced between consecutive EAV genes [63]. The viruses recovered expressed epitopes of nine amino acids from MHV within the ectodomain of the membrane (M) protein for at least three passages [35]. Foreign epitopes have also been expressed by using PRRSV vectors [64].

9. Conclusions

Both helper-dependent expression systems, based on two components, and single genomes constructed by targeted recombination, or by using infectious cDNAs, have been developed for coronaviruses. The sequences that regulate transcription have been characterized mainly using helper-dependent expression systems. These expression systems have the advantage of their large cloning capacity, in principle higher than 27 kb, produce reasonable amounts of heterologous antigens ($2–8\,\mu g/10^6$ cells), show a limited stability (synthesis of heterologous gene is maintained for around 10 passages), and elicit strong immune responses. In contrast, coronavirus vectors based on single genomes have at present a limited cloning capacity (3–3.5 kb), expression levels of heterologous genes are 10-fold over those of helper dependent systems ($>50\,\mu g/10^6$ cells) and are very stable (>30 passages). Furthermore, replication-competent propagation-deficient expression systems based on coronavirus genomes have been developed increasing the safety of these vectors. The possibility of expressing different genes under the control of TRSs with programmable strength, and engineering the tissue and species tropism indicate that coronavirus vectors are very flexible. Thus, coronavirus-based vectors are emerging with a high potential for vaccine development and, possibly, for gene therapy.

Acknowledgements

This work has been supported by grants from the Comisión Interministerial de Ciencia y Tecnología (CICYT), La Consejería de Educación y Cultura de la Comunidad de Madrid, and Fort Dodge Veterinaria from Spain, and the European Communities (Life Sciences Program, Key Action 2: Infectious Diseases).

Abbreviations

APN	aminopeptidase N
BAC	bacterial artificial chromosome
BCoV	bovine coronavirus
CAT	chloramphenicol acetyltransferase

cCS	sequence complementary to CS
CEACAM	carcinoembryonic antigen-related cell adhesion molecules
CMV	cytomegalovirus promoter
CS	conserved core sequence
cTRS	sequences complementary to TRS
EAV	equine arteritis virus
GFP	green fluorescent protein
GUS	β-glucuronidase
HCoV	human coronavirus
IBV	infectious bronchitis virus
LDHV	lactate dehydrogenase-elevating virus
MHV	mouse hepatitis virus
PRCV	porcine respiratory coronavirus
PRRSV	porcine respiratory and reproductive syndrome virus
sgmRNA	subgenomic mRNA
SHFV	simian hemorrhagic fever virus
SIN	vector based on Sindbis virus replicon
TGEV	transmissible gastroenteritis coronavirus
TRS	transcription-regulating sequences
UTR	untranslated region

References

1. Enjuanes, L. (2000) In: van Regenmortel, M.H.V. Virus Taxonomy. Classification and Nomenclature of Viruses, Academic Press, New York, pp. 835–849.
2. Enjuanes, L., Spaan, W., Snijder, E., and Cavanagh, D. (2000) In: van Regenmortel, M.H.V. (eds.) Virus Taxonomy. Classification and Nomenclature of Viruses, Academic Press, New York, pp. 827–834.
3. Cowley, J.A., Dimmock, C.M., Spann, K.M., and Walker, P.J. (2000) J. Gen. Virol. 81, 1473–1484.
4. Mayo, M.A. (2002) Arch. Virol. 147, 1655–1656.
5. Lai, M.M.C. and Cavanagh, D. (1997) Adv. Virus Res. 48, 1–100.
6. Masters, P.S. (1999) Adv. Virus Res. 53, 245–264.
7. Sánchez, C.M., Izeta, A., Sánchez-Morgado, J.M., Alonso, S., Sola, I., Balasch, M., Plana-Durán, J., and Enjuanes, L. (1999) J. Virol. 73, 7607–7618.
8. Kuo, L., Godeke, G.-J., Raamsman, M.J.B., Masters, P.S., and Rottier, P.J.M. (2000) J. Virol. 74, 1393–1406.
9. Almazán, F., González, J.M., Pénzes, Z., Izeta, A., Calvo, E., Plana-Durán, J., and Enjuanes, L. (2000) Proc. Natl. Acad. Sci. USA 97, 5516–5521.
10. Yount, B., Curtis, K.M., and Baric, R.S. (2000) J. Virol. 74, 10600–10611.
11. Thiel, V., Herold, J., Schelle, B., and Siddell, S. (2001) J. Gen. Virol. 82, 1273–1281.
12. Yount, B., Denison, M.R., Weiss, S.R., and Baric, R.S. (2002) J. Virol. 76, 11065–11078.
13. Casais, R., Thiel, V., Siddell, S.G., Cavanagh, D., and Britton, P. (2001) J. Virol. 75, 12359–12369.
14. van Dinten, L.C., den Boon, J.A., Wassenaar, A.L.M., Spaan, W.J.M., and Snijder, E.J. (1997) Proc. Natl. Acad. Sci. USA 94, 991–996.
15. Meulenberg, J.J.M., Bos-de-Ruijter, J.N.A., Wensvoort, G., and Moormann, R.J.M. (1998) Adv. Exp. Med. Biol. 440, 199–206.
16. Godeny, E.K., Zeng, L., Smith, S.L., and Brinton, M.A. (1995) J. Virol. 69, 2679–2683.
17. Escors, D., Ortego, J., Laude, H., and Enjuanes, L. (2001) J. Virol. 75, 1312–1324.

18. Escors, D., Camafeita, E., Ortego, J., Laude, H., and Enjuanes, L. (2001) J. Virol. 75, 12228–12240.
19. Gorbalenya, A.E. (2001) In: Lavi, E., Weiss, S. and Hingley, S.T. (eds.) The Nidoviruses (Coronaviruses and Arteriviruses), Vol. 494, Kluwer Academic/Plenum Publishers, New York, pp. 1–17.
20. Izeta, A., Smerdou, C., Alonso, S., Penzes, Z., Méndez, A., Plana-Durán, J., and Enjuanes, L. (1999) J. Virol. 73, 1535–1545.
21. Alonso, S., Izeta, A., Sola, I., and Enjuanes, L. (2002) J. Virol. 76, 1293–1308.
22. Alonso, S., Sola, I., Teifke, J., Reimann, I., Izeta, A., Balach, M., Plana-Durán, J., Moormann, R.J.M., and Enjuanes, L. (2002) J. Gen. Virol. 83, 567–579.
23. Thiel, V., Siddell, S.G., and Herold, J. (1998) Adv. Exp. Med. Biol. 440, 109–114.
24. Liao, C.L., Zhang, X., and Lai, M.M.C. (1995) Virology 208, 319–327.
25. Zhang, X., Hinton, D.R., Cua, D.J., Stohlman, S.A., and Lai, M.M.C. (1997) Virology 233, 327–338.
26. Liao, C.-L. and Lai, M.M.C. (1995) Virology 209, 428–436.
27. Zhang, X., Hinton, D.R., Park, S., Parra, B., Liao, C.-L., and Lai, M.M.C. (1998) Virology 242, 170–183.
28. Lai, M.M.C., Zhang, X., Hinton, D., and Stohlman, S. (1997) J. Neurovirol. 3, S33–S34.
29. Stirrups, K., Shaw, K., Evans, S., Dalton, K., Casais, R., Cavanagh, D., and Britton, P. (2000) J. Gen. Virol. 81, 1687–1698.
30. González, J.M., Penzes, Z., Almazán, F., Calvo, E., and Enjuanes, L. (2002) J. Virol. 76, 4655–4661.
31. Sola, I., Alonso, S., Plana-Durán, J., and Enjuanes, L. (2003) J. Virol. 77, 4357–4369.
32. Frolov, I., Hoffman, T.A., Prágai, B.M., Dryga, S.A., Huang, H.V., Schlesinger, S., and Rice, C.M. (1996) Proc. Natl. Acad. Sci. USA 93, 11371–11377.
33. Curtis, K.M., Yount, B., and Baric, R.S. (2002) J. Virol. 76, 1422–1434.
34. Thiel, V., Herold, J., Schelle, B., and Siddell, S.G. (2001) J. Virol. 75, 6676–6681.
35. Molenkamp, R., van Tol, H., Rozier, B.C., van der Meer, Y., Spaan, W.J., and Snijder, E.J. (2000) J. Gen. Virol. 81, 2491–2496.
36. Curtis, K.M., Yount, B., and Baric, R.S. (2002) J. Virol. 76, 1422–1434.
37. Koetzner, C.A., Parker, M.M., Ricard, C.S., Sturman, L.S., and Masters, P.S. (1992) J. Virol. 66, 1841–1848.
38. van der Most, R.G., Heijnen, L., Spaan, W.J.M., and de Groot, R.J. (1992) Nucleic Acids Res. 20, 3375–3381.
39. Leparc-Goffart, I., Hingley, S.T., Chua, M.M., Phillips, J., Lavi, E., and Weiss, S.R. (1998) J. Virol. 72, 9628–9636.
40. de Haan, C.A.M., Kuo, L., Masters, P.S., Vennema, H., and Rottier, P.J.M. (1998) J. Virol. 72, 6838–6850.
41. Fischer, F., Stegen, C.F., Koetzner, C.A., and Masters, P.S. (1997) J. Virol. 71, 5148–5160.
42. Ortego, J., Escors, D., Laude, H., and Enjuanes, L. (2002) J. Virol. 76, 11518–11529.
43. Ortego, J., Sola, I., Almazan, F., Ceriani, J.E., Riquelme, C., Balach, M., Plana-Durán, J., and Enjuanes, L. (2003) Virology, 308, 13–22.
44. Brian, D.A. (2001) In: Lavi, E., Weiss, S. and Hingley, S.T. (eds.) Nidoviruses, , Vol. 494, Plenum Press, New York, pp. 415–428.
45. Sawicki, D.L., Wang, T., and Sawicki, S.G. (2001) J. Gen. Virol. 82, 386–396.
46. Pasternak, A.O., van den Born, E., Spaan, W.J.M., and Snijder, E.J. (2001) EMBO J. 20, 7220–7228.
47. van Marle, G., Dobbe, J.C., Gultyaev, A.P., Luytjes, W., Spaan, W.J.M., and Snijder, E.J. (1999) Proc. Natl. Acad. Sci. USA 96, 12056–12061.
48. Makino, S. and Joo, M. (1993) J. Virol. 67, 3304–3311.
49. van der Most, R.G., De Groot, R.J., and Spaan, W.J.M. (1994) J. Virol. 68, 3656–3666.
50. Wertz, G.W., Perepelitsa, V.P., and Ball, L.A. (1998) Proc. Natl. Acad. Sci. USA 95, 3501–3506.
51. Penzes, Z., González, J.M., Calvo, E., Izeta, A., Smerdou, C., Mendez, A., Sánchez, C.M., Sola, I., Almazán, F., Enjuanes, L. (2001) Virus Genes 23, 105–118.
52. Zhang, X. and Liu, R. (2000) Virology 278, 75–85.
53. Ozdarendeli, A., Ku, S., Rochat, S., Senanayake, S.D., and Brian, D.A. (2001) J. Virol. 75, 7362–7374.
54. An, S. and Makino, S. (1998) Virology 243, 198–207.

55. Hsue, B. and Masters, P.S. (1999) J. Virol. 73, 6128–6135.
56. Joo, M. and Makino, S. (1995) J. Virol. 69, 272–280.
57. Krishnan, R., Chang, R.Y., and Brian, D.A. (1996) Virology 218, 400–405.
58. van Marle, G., Luytjes, W., Van der Most, R.G., van der Straaten, T., and Spaan, W.J.M. (1995) J. Virol. 69, 7851–7856.
59. Sawicki, S.G. and Sawicki, D.L. (1990) J. Virol. 64, 1050–1056.
60. Delmas, B., Gelfi, J., L'Haridon, R., Vogel, L.K., Norén, O., and Laude, H. (1992) Nature 357, 417–420.
61. Yeager, C.L., Ashmun, R.A., Williams, R.K., Cardellichio, C.B., Shapiro, L.H., Look, A.T., and Holmes, K.V. (1992) Nature 357, 420–422.
62. Flanagan, E.B., Zamparo, J.M., Ball, L.A., Rodriguez, L., and Wertz, G.W. (2001) J. Virol. 75, 6107–6114.
63. de Vries, A.A.F., Glaser, A.L., Raamsman, M.J.B., de Haan, C.A.M., Sarnataro, S., Godeke, G.J., and Rottier, P.J.M. (2000) Virology 270, 84–97.
64. Groot Bramel-Verheije, M.H., Rottier, P.J.M., and Meulenberg, J.J.M. (2000) Virology 278, 380–389.

S.C. Makrides (Ed.) *Gene Transfer and Expression in Mammalian Cells*

Virus-based vectors for gene expression in mammalian cells: Poliovirus

Shane Crotty[*] and Raul Andino

Department of Microbiology and Immunology, University of California, 513 Parnassus Ave., Box 0414, San Francisco, CA 94143, USA; Tel.: +1 (415) 502-6358; Fax: +1 (415) 476-0939

E-mail: andino@itsa.ucsf.edu

1. Introduction

Poliovirus is a member of the picornavirus family (Picornaviridae) of positive strand RNA viruses. Though limited efforts have been made to develop picornaviruses as gene therapy [1] or suicide vectors [2], the vast majority of vector development with picornaviruses has focused on the engineering and use of poliovirus-based vaccine vectors.

Viruses are strong inducers of cellular and humoral immune responses. Thus, activation of the immune responses by vaccination with recombinant viruses expressing relevant antigens derived from other pathogens or tumors is a promising approach for the prevention and treatment of infectious diseases and cellular malignancies. Viral vaccine vectors that have been successfully used in experimental models include poxviruses, adenoviruses, alphaviruses, picornaviruses, and influenza viruses. Vectors based on currently used human vaccine viruses [i.e., poliovirus (the Sabin oral poliovirus vaccine (OPV)) and vaccinia (the smallpox vaccine)] are particularly appealing due to their well characterized safety profiles in millions, if not billions, of vaccinees, as well as the extensively characterized ability of these vaccines to induce strong long-term immune responses in humans.

RNA viruses such as poliovirus have multiple beneficial attributes as well as problems that are likely to be overcome. Herein we will discuss the advantages and disadvantages of the polio system. We will focus on the utilization of poliovirus-based vaccine vectors as mucosal vaccines, particularly with respect to the development of an effective AIDS (acquired immunodeficiency syndrome) vaccine, as the development of a vaccine to prevent AIDS is the most active area of vaccine research in the world today.

2. A poliovirus vector-based HIV vaccine

Sexual transmission is the major route for the spread of the human immunodeficiency virus (HIV) in the AIDS pandemic progression. Almost all attempts to develop a protective HIV vaccine have failed to protect against infection, and in

* Current address: Department of Microbiology and Immunology, Emory University, Rollins Research Center, 1510 Clifton Rd., Rm. G-211, Atlanta, GA 30322, USA; Tel.: (404) 727-9301; fax: (404) 727-3722; E-mail: shanec@alum.mit.edu

most cases candidate vaccines also failed to limit disease despite eliciting strong systemic immune responses. Our basic premise is that an essential requirement for protective immunization against HIV is the induction of a robust mucosal immune response. To address this requirement, we have proposed to use replication-competent live-attenuated Sabin poliovirus recombinants that carry and express antigens derived from the HIV-1 virus. Because there are no good immunological correlates for protection against HIV-1 infection, we elected to express the entire Simian immunodeficiency virus (SIV) genome in defined, discrete overlapping fragments for proof of principle studies that were recently conducted in the SIV model system. With this approach, all of the potentially important antigenic sequences can be effectively expressed at local mucosal sites by the inoculation of a cocktail of recombinant polioviruses that carries a complete set of antigens. The attenuated poliovirus strain Sabin was used as a vector for the insertion of SIV genes to create recombinants expressing SIV antigens. The resulting poliovirus–SIV chimeric recombinant viruses were propagated for multiple generations, and they expressed, at high levels, the SIV antigens. Infection of susceptible mice and cynomolgus monkeys with poliovirus–SIV cocktails elicited serum and secretory humoral responses, as well as cellular immunity to the inserted sequences. Most importantly, vaccination of cynomolgus monkeys with poliovirus–SIV cocktails protected against infection and AIDS after intravaginal challenge with highly pathogenic $SIV_{mac}251$ [3].

We reasoned that by engineering poliovirus to express HIV antigens, the beneficial attributes of the live-attenuated Sabin vaccine can be exploited to develop a cheap, safe, and effective AIDS vaccine. The immunological responses to the inserted proteins in poliovirus vectors should share the same desirable attributes of those elicited against poliovirus proteins: strong humoral and cellular responses, broad mucosal and systemic responses, and long-lasting immunity. Strong mucosal immunity may be required to provide a protective immunological barrier at the most common HIV port of entry: the mucosal surface of the reproductive tract. Since the immunological responses and the antigenic sequences required for protection against AIDS are not well defined, it may be necessary, at least for now, to use complex vaccine formulations to elicit both cellular and humoral responses against most of the HIV proteins. Poliovirus-based libraries expressing a defined collection of HIV antigens can be used to elicit such a broad immunity. Moreover, a prime-boost protocol can be utilized to strengthen and focus the HIV-specific immune response, which can be achieved by priming with Sabin 1-based vectors and boosting with Sabin 2-based vectors.

2.1. Rationale for using poliovirus as a live virus vector for an HIV-1 vaccine

Our choice of poliovirus as a potential AIDS vaccine vector is based on the many advantages of this system:

- Live attenuated poliovirus vectors should elicit a very appropriate and effective immunological response because they mimic the antigen presentation that occurs during natural viral infection. Since virus infection potently induces the innate immune system, the resulting danger signals and cytokine cascades serve as a natural adjuvant to initiate and enhance the adaptive immune response.

- The Sabin OPV is one of the safest and most efficacious human vaccines [4].
- Poliovirus is easy to engineer to create recombinant viruses [5].
- There is a long experience of its use in humans, particularly as a pediatric vaccine, and its safety profile is well established and understood [4].
- It induces both humoral and cellular immunity [4,6].
- It induces both systemic and mucosal immunity [7,8].
- It induces long-lasting immunity [9].
- It is easy to deliver (orally) without the use of needles.
- Manufacturing procedures for poliovirus are well established, and the virus grows to high titers, enabling low cost production [10].
- Poliovirus, like all picornaviruses, lacks a membrane envelope and thus is a very hardy virus and is easy to formulate as a chemically stable vaccine.
- Since poliovirus has a small genome, there is minimal antigenic dilution of the expressed HIV antigens.

Worldwide, it is estimated that over 40 million persons are now living with HIV infections, and in the absence of effective treatment the overwhelming majority of them will die of AIDS [11]. Last year, approximately 3 million people died of HIV-related diseases. Since the emergence of the HIV epidemic, approximately 90% of all HIV infections have occurred in developing countries. A similar or even greater percentage of all new infections also take place in these nations, where over 5 million new HIV infections occur each year, or about 14,000 each day. In sub-Saharan Africa alone, the World Health Organization estimates that AIDS has already produced more than 10 million orphaned or abandoned children as a result of HIV infection of their parents [11]. Only the development of a safe, cheap, and effective vaccine to prevent HIV infection has the potential to contain and ultimately eradicate the AIDS pandemic.

2.1.1. AIDS: a sexually transmitted disease

Of the projected cases of HIV infection, it is believed that 80% will derive from heterosexual viral transmission [11], and most of the rest from homosexual or perinatal transmission. Although the biology of sexual transmission of HIV is poorly understood, it is clear that an essential first step in the process of male transmission involves deposition of infectious virus or HIV-infected cells upon mucosal surfaces in the vagina or rectum of the recipient partner, or for female transmission, the abraded skin of the penis or the urethra. Following introduction into a new host, HIV (or HIV-infected cells) likely soon encounter susceptible host target cells at the mucosal portal of entry where the virus replicates and then infects nearby lymphatic tissues, initiating systemic HIV infection [12]. The SIV-vaginal infection model suggests that the route of dissemination after mucosal exposure to SIV involves infection of $CD4^+$ T cells in the genital mucosa followed by a stepwise progression of the infection from regional lymph nodes to distant lymphoid tissues, presumably by migration of infected antigen-presenting cells (APCs) [13,14]. The human immune system appears unable to clear HIV once infection is established, and the immune system ultimately fails to prevent disease progression in infected individuals [15,16]. As a result, stringent requirements are placed on candidate HIV vaccines, including the induction of immune responses that effectively neutralize the infectivity of cell-free HIV and eliminate virally infected cells before a systemic infection can take hold. To be fully effective, any vaccine must prevent acquisition of the infectious agent or engender sufficient immunity to facilitate clearance of infected cells prior to establishment of persistence and the development of disease. However, many vaccine developers have recently assumed that control of viremia by vaccination may be a more realistic goal than the induction of sterilizing immunity [17].

2.1.2. Criteria for a mucosal HIV vaccine

Considering that the genital mucosal surface is the most common route of acquisition for HIV infection, an immune barrier present at genital mucosal surfaces would have the best chance of interrupting HIV transmission. Unfortunately, few approaches are now available to specifically induce mucosal immune responses. In fact, HIV vaccine development efforts remain severely handicapped by our incomplete understanding of the genital mucosal immune system, as well as by the lack of effective vaccination strategies to raise secretory and cell-mediated antiviral immune responses at mucosal sites.

Parenteral administration of whole inactivated virus, recombinant subunit antigens, or recombinant vaccinia viruses preferentially generates systemic rather than mucosal immune responses [15]. Thus, even if such vaccines proved effective within an experimental paradigm based on intravenous challenge, they may be unable to prevent infection by HIV through the most common routes of exposure [11,15,18]. In fact, it has been shown that the same whole inactivated or attenuated live SIV vaccines that successfully protect rhesus macaques from intravenous challenge with pathogenic SIV do not protect animals from infection or disease following vaginal mucosal challenge with a low dose of cell-free SIV [19]. Thus, systemic immunity alone may be inadequate to interrupt the mucosal transmission of HIV. Given these unique characteristics of HIV infection, an effective vaccine has remained elusive, and most conventional approaches have proven unsuccessful.

Both antibody-dependent and cellular immune responses have been shown to be capable of protecting against a highly virulent SIV. Recently, two groups have clearly demonstrated that passive immunization of macaques with SIV neutralizing antibodies can protect macaques from an intravaginal challenge with SIV [20,21].

The role of $CD8^+$ cytotoxic T lymphocytes (CTLs) in the protective immune response against HIV/SIV is not completely understood. However, because CTLs are important in eliminating a wide variety of intracellular pathogens, stimulating their production may be crucial for vaccination against lentiviruses. Several lines of evidence suggest that CTLs likely play a crucial role in controlling HIV:

(1) In primary HIV infection, HIV-specific CTLs develop prior to the onset of detectable neutralizing antibodies and are temporarily associated with control of viremia [22].

(2) HIV-1-specific CTLs can select for the evolution of escape mutants [23].
(3) HIV-1-specific CTLs are able to inhibit viral replication *in vitro* [24,25].
(4) Depletion of CD8$^+$ T cells from rhesus monkeys shortly before infection with SIV resulted in exacerbated disease [26].
(5) Vaccines that elicit only a strong CD8$^+$ T-cell response can limit disease after infection of monkeys with a pathogenic SHIV (SIV/HIV chimeric virus) [27–30].

Particularly if CTL activity is present locally at the port of entry, i.e., the mucosal surface of the rectum and vagina, more robust protection will be engendered. The fact that SIV-specific CTLs have been demonstrated in the vaginal mucosa of rhesus macaques [31,32] implies that a full range of anti-viral immunity can be brought to bear in the attempt to blunt each stage of virus transmission and dissemination. Furthermore, mucosal immunization can be more effective than systemic immunization against SIV, not only because the natural routes of transmission are primarily through mucosal surfaces, genital or gastrointestinal, but also because the major reservoir of SIV is in the gut mucosa, so that immunization strategies that are more effective at inducing CTLs in this site can be more effective at controlling the virus load. Identifying vectors, viral antigens, and immunization routes which can elicit these immune responses and which can be applied in a clinical setting remains the challenge before us.

2.1.3. The mucosal immune system and vaccine development

Due to the compartmentalization of the secretory and systemic immune systems, parenterally administered antigens do not consistently stimulate mucosal immunity [33]. A particularly relevant example is provided by comparison of the Salk inactivated whole poliovirus vaccine with the Sabin attenuated live poliovirus vaccine. Although the parenterally administered Salk vaccine induces a strong systemic antibody response that blocks viral dissemination to the bloodstream and clinical disease (virus spread to the central nervous system), it does not elicit significant secretory IgA production, block initial viral replication in the gut, or block transmission of virulent poliovirus to other people [4]. In contrast, the orally administered Sabin live-attenuated vaccine, by virtue of its ability to replicate at intestinal sites, induces long-lasting local mucosal immunity that greatly limits poliovirus replication following exposure [4]. As the Sabin attenuated vaccine viruses, like their wild-type parents, replicate extensively in regional and systemic lymph nodes, they also induce levels of polio-reactive serum IgG and IgA antibodies equivalent to those seen following intramuscular administration of the inactivated Salk vaccine [4]. In addition, there is evidence that local immunity can be induced in the genital tract of women. Women with genital herpes simplex virus infection have secretory IgA in their vaginal secretions [34,35]. Intravaginal immunization with *Candida albicans* also induces high titers of specific secretory IgA in vaginal secretions [36]. These findings suggest that it should be possible to produce a specific immune response in the genital tract of women that is capable of preventing sexual transmission of HIV.

2.1.4. Live-attenuated Sabin poliovirus as AIDS vaccine vector

Poliovirus is an attractive live viral vector for multiple reasons. The Sabin live poliovirus vaccine is one of the best human vaccines in the world. It produces long lasting immunity and herd immunity [4]; it is very safe and easy to manipulate experimentally; it has a proven safety and efficacy record in over 1 billion vaccinees [4]; it is cheap to produce and distribute in developing countries [37]; and most importantly, it produces a potent mucosal immune response [8]. Furthermore, the vast experience using OPV as a human vaccine should facilitate rapid transition from the experimental phase of development to the implementation of this approach as a real vaccine. Most importantly, vaccination of cynomolgus monkeys with poliovirus–SIV cocktails protected against infection and AIDS after intravaginal challenge with highly pathogenic $SIV_{mac}251$. This experiment was the first SIV challenge experiment conducted using live poliovirus vaccine vectors, and it demonstrated that our poliovirus-based vaccine candidate is effective at preventing infection and controlling viremia in SIV-infected animals.

The SabRV doses used in our SIV challenge experiment are comparable to normal OPV doses used in infants, children, and adults [38]. It is likely that the Sabin-based vaccine will be substantially more efficacious in humans than it was in monkeys because Sabin viruses are several orders of magnitude more infectious in people ($ID_{50} = 50$ pfu) [39] than in cynomolgus macaques ($ID_{100} = 10^6$ pfu) [40]. Thus, Sabin virus based vectors are also likely to replicate more efficiently in people than in monkeys. The enhanced replication of the vectors is expected to generate a significantly stronger immune response to Sabin-recombinant vaccine antigens in people than in monkeys.

2.1.5. Use of poliovirus vectors in developing countries

For a preventive HIV vaccine to have a significant impact on the HIV pandemic, it must also be appropriate for use in developing countries where an estimated 95% of all new HIV infections occur [11]. In this regard, the abilities to induce long-lasting immunity with a limited number of vaccine doses and to protect against antigenically diverse strains of HIV will be critical. In addition, the vaccine must be easy to store, transport, and deliver affordably in the developing world, where the World Health Organization (WHO) estimates that over 300 million individuals may need to be vaccinated during the first 5 years of a mass immunization campaign.

The live poliovirus vaccine is easy to administer orally, has a low cost for distribution in the developing world, induces both serum antibodies and intestinal mucosal resistance, and confers long-lasting immunity [4]. The success of the polio eradication program is the best evidence that poliovirus-based vectors can be an appropriate vaccine for developing countries.

2.2. Poliovirus and vaccine strains: human clinical experience

2.2.1. Poliovirus vaccines

In 1961, types 1 and 2 monovalent oral poliovirus vaccines (MOPV) were licensed, and in 1962, type 3 MOPV was also licensed. In 1963, trivalent OPV was licensed, which largely replaced inactivated poliovirus vaccine (IPV) use. Trivalent OPV contains live attenuated strains of all three serotypes of poliovirus in a 10:1:3 ratio. The viruses were attenuated by serial passage in monkey kidney, Vero, or human diploid fibroblast cell cultures, allowing them to accumulate mutations. Ultimately, this resulted in attenuated, partially temperature-sensitive viruses that could be given orally to a patient. Following oral administration, OPV replicates in the oropharyngeal and gastrointestinal tracts and in lymph nodes that drain the intestine. These attenuated viruses replicate normally in the intestine, but have tropism-changing mutations that prevent them from invading the central nervous system. OPV strongly induces both systemic and mucosal humoral immune responses, rapidly prompting the production of serum-neutralizing antibodies in the blood and mucosa against all three polio serotypes. This antibody response is long-lasting.

2.2.2. Advantages of oral poliovirus vaccine

OPV is given orally rather than by injection. Its administration does not require a trained health worker or sterile injection equipment. At a current price of 8 US cents a dose, the vaccine is inexpensive, facilitating mass purchasing of the vaccine for National Immunization Days. Short-term shedding of OPV in the stools of recently immunized persons can result in the passive immunization of persons within close contact. OPV is highly effective in producing immunity to poliovirus. A single dose of OPV produces immunity to all three vaccine viruses in about 50% of recipients. Three doses produce immunity to all three poliovirus types in more than 95% of recipients. OPV's ability to stimulate mucosal immunity is responsible for the success of OPV mass campaigns in interrupting wild poliovirus transmission. Therefore, OPV remains the vaccine of choice for the eradication of poliovirus [41]. OPV intestinal immunity reduces the chance that a vaccinated person will become infected with wild virus if he or she is exposed while visiting a polio endemic country. Because attenuated viruses contained in OPV are live viruses that induce the same type of antibody as wild type poliovirus does, and because wild type virus infection is believed to induce lifelong immunity, it has been reasoned by analogy that the immunity induced by OPV is also lifelong [42].

2.2.3. Physical stability of poliovirus vaccine

Temperature instability is often a major problem in the delivery of live virus vaccines. However, OPV has the desirable stability attributes of a non-enveloped picornavirus. It can be stabilized with a solution of magnesium chloride. OPV in $MgCl_2$ is stable for at least 12 months at 4 °C, 6 weeks at 25 °C, and is still sufficiently immunogenic to use after 1 week at 37 °C.

2.2.4. Reversion of OPV to neurovirulent forms

An important concern with the use of live attenuated polio vaccines is their potential for reversion to neurovirulence. All three OPV vaccine strains show some tendency to revert to neurovirulence. Sabin type 3 vaccine viruses revert more easily to neurovirulence phenotypes as compared to types 1 and 2. Single-base mutations that are strongly associated with attenuation and reversion to neurovirulence have been identified. Major attenuating mutations for all three serotypes have been identified in the 5' non-coding region at positions 480 for type 1, 481 for type 2, and 472 for type 3 [43–45]. For type 3, an additional point mutation associated with both attenuation and temperature sensitivity has been identified at position 2034 in theVP3 capsid gene [46]. Vaccination with OPV results in the release of a variety of live mutant viruses in the feces [47]. Despite the excellent safety record of OPV, in approximately 1 in every 3 million doses of the vaccine, the live attenuated vaccine virus in OPV can cause paralysis in the vaccinated child or in a close contact [4]. This extremely low risk of vaccine-associated poliovirus poliomyelitis (VAPP) is accepted by most public health programs in the world because hundreds of thousands of children would be crippled every year without OPV [48]. New methods to stabilize the attenuated phenotype have been proposed using molecular biology [49–51], but vaccine developers have not been convinced to invest in this endeavor. If recombinant Sabin poliovirus vectors prove to be safe and effective for eliciting strong mucosal immunity in humans, it would be possible to incorporate some of the proposed safety changes in the viral vector to produce an even safer and more effective vector. It should be noted that the extremely rare cases of VAPP are predominantly due to genetic agammaglubulinemia immunodeficiencies (lack of antibody development) in the vaccinated individual.

3. Poliovirus vector development and its immunogenic potential

3.1. Poliovirus-derived vaccine vectors

The poliovirus genome consists of a positive-stranded RNA of approximately 7400 nucleotides. The mature poliovirus particle has an icosahedral protein shell (composed of four structural proteins—VP1, VP2, VP3, VP4) surrounding the RNA genome. RNA synthesized *in vitro* from cloned poliovirus DNA yields infectious poliovirus following transfection into susceptible target cells [52].

The poliovirus RNA genome encodes a single open reading frame that, following infection, is translated into a long polypeptide, known as polyprotein (P0). The P0 precursor then is processed proteolytically by the virus-encoded proteases into the mature structural and enzymatic products [53]. Through genetic manipulation of poliovirus molecular clones, we have inserted additional amino acid sequences at several locations in the P0 precursor and placed artificial recognition sequences for the poliovirus protease 3C or 2A between the introduced sequences and the rest of the polyprotein-coding sequences [5]. In this way, a larger than normal precursor is made, which is appropriately cleaved by the viral proteases into the usual array of constituent proteins plus the exogenous antigenic protein. Thus, all of the normal poliovirus protein components are produced accurately, and viral replication proceeds. Importantly, this method yields recombinant viruses that are capable of replicating and propagating within an infected individual without the need for a helper virus [5]. The viruses carry the inserted information as the poliovirus genome spreads from cell to cell in the course of its replicative cycle.

We have demonstrated that recombinant polioviruses express the inserted antigenic protein in readily detectable amounts [5,40]. Inserts encoding up to 200 additional amino acid residues can be successfully accommodated in recombinant polioviruses. As the introduced antigens are expressed within infected cells, stimulation of both cell-mediated and humoral immunity is achieved [3,5,6,40,54]. Importantly, although the recombinants express the foreign sequences during the replicative cycle, these proteins are not included in the mature virus particle. As such, the virion structure, antigenicity, and host range of the recombinant polioviruses is not altered [5]. More recently, coxsackieviruses have been engineered in a similar manner to express model antigens and these recombinants are being employed to study issues related to viral immunology and pathogenesis [55,56] (Fig. 1).

3.2. Other poliovirus and picornavirus vector strategies

Several additional strategies have been reported for engineering poliovirus vectors. In one method, small antigenic epitopes were inserted into one of the capsid proteins, VP1 [57]. In another method, dicistronic poliovirus RNAs were constructed by duplicating the 5' non-coding region of the poliovirus genomic RNA (internal ribosome entry site, or IRES). In this manner, the foreign polypeptide was expressed using one IRES, and essential viral proteins were produced using the other [58]. A third method used poliovirus minireplicon genomes in which poliovirus structural protein genes were replaced by foreign sequences (i.e., HIV-1 gag, env, and pol); in

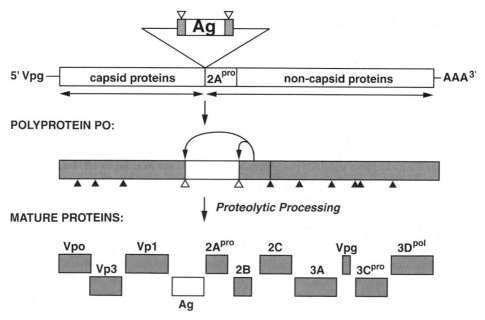

Fig. 1. Schematic representation of replication-competent poliovirus vectors.

this case, the minireplicon genome must be packaged by supplying poliovirus capsid proteins in *trans* [1]. Limitations of these strategies include the small size of the tolerated insert, genetic instability of the inserted sequences, or a requirement for helper virus for viral propagation.

3.3. Restriction of the poliovirus vectors

Two of the more significant restrictions of the live picornavirus vectors are the insert size limit (approximately 200 amino acid residues) and problems with the retention of the insert after several rounds of replication (which leads to the accumulation of viruses lacking the insert after multiple passages) [59,60]. However, strategies have been described to alleviate these two problems [3,40,59].

To overcome insert retention problems, the potential factors governing the genetic stability of foreign inserts were examined by constructing and exploring different recombinant polioviruses, which contain a series of different original or modified sequences of foreign inserts. Importantly, it was demonstrated that the G/C content of the inserted foreign sequence governs the ability of the poliovirus vectors to retain the inserted sequence [61]. Several recombinant viruses carrying exogenous sequences with a high G/C content retain the inserted sequences for more than 20 rounds of replication.

To accommodate the limitation of insert size, we have adopted the strategy of expressing large proteins as a series of defined overlapping fragments. We have described the construction of Sabin poliovirus vectors expressing a defined library of overlapping SIV protein fragments representing the entire sequences of SIV Gag, Pol, Env, and Nef. It was demonstrated that the new Sabin vectors are capable of retaining the full insert; in most cases, they are immunogenic, and they can induce protective immunity in macaques [3].

3.4. Prime-boost approach using different poliovirus serotypes

The construction of recombinant attenuated Sabin polioviruses utilizing the Sabin 1 and 2 vaccine strains carries additional benefit. It has been shown that Sabin 2 poliovirus is not neutralized by antibodies directed against Sabin 1 poliovirus [39]. Thus, by using poliovirus live viral vectors derived from two different serotypes (1 and 2), we hope to avoid vector neutralization problems and to increase the immune response to the booster.

3.5. Testing the immunogenic capacity of poliovirus vaccine vectors

3.5.1. Poliovirus vectors induce CTL responses
It was previously shown that poliovirus recombinants can induce antibody responses against the inserted sequences, but it was not known whether poliovirus or vaccine vectors derived from it elicit effective CTL responses. Therefore, we constructed and produced polio-Ova, expressing the well-characterized model antigen Ova, which was chosen because of the powerful Ova-specific immunological tools and reagents

available. Inoculation of susceptible mice [62,63] with poliovirus vectors expressing Ova elicited a strong insert-specific cytotoxic T-lymphocyte response [6]. Furthermore, vaccination with poliovirus recombinants induced protective immunity against a difficult challenge with lethal doses of a malignant melanoma cell line expressing ovalbumin [6,64]. This virus also has a significant therapeutic effect when used for active immunotherapy to treat solid tumors or experimental pulmonary metastasis [6].

3.5.2. Pre-existing immunity to poliovirus vectors

It has been argued that pre-existing immunity against poliovirus in the human population could restrain the use of poliovirus as vaccine vector. However, 50% of individuals previously immunized with OPV show at least a fourfold increase in antibody titers after administration of OPV booster [7,8,65]. A recent study of children previously immunized with either the enhanced potency inactivated polio vaccine (E-IPV) or OPV indicated that the resistance to infection with a booster dose of type 1 OPV may be relative rather than absolute, and depends on the dose of the booster OPV [66]. In children previously immunized with a complete schedule E-IPV, 46% of those who received a low dose OPV type 1 challenge (500–800 $TCID_{50}$) became infected (as assessed by virus shedding in the stool), while 82% of those who received a higher dose (600,000 $TCID_{50}$) were infected. In children previously immunized with a complete schedule of OPV, 18% of those receiving a low dose challenge became infected, while 31% of those given the high dose challenge became infected [67]. (For reference, the licensed trivalent OPV contains at least 800,000 $TCID_{50}$, 100,000 $TCID_{50}$ and 600,000 $TCID_{50}$ of types 1, 2, and 3 poliovirus, respectively.) These data suggest that poliovirus can replicate in previously vaccinated people.

Therefore, it is likely that higher doses of recombinant polioviruses might be able to establish infection and elicit immune responses against heterologous antigens in individuals previously immunized against polio. Furthermore, it is possible that even limited replication in a relatively immune host might induce immunologic priming to the inserted antigenic proteins at mucosal sites that could then be boosted by subsequent delivery of HIV-1 antigens via another form or route, as previously reported for vaccinia virus recombinants [68,69].

We have recently investigated the effect of pre-existing immunity against poliovirus on the outcome of vaccination with our recombinant poliovirus vaccine vectors in a murine model [64]. Using a recombinant poliovirus expressing a model tumor antigen, we compared antibody titers, T-cell precursor numbers, and efficacy of protection against lethal challenge with malignant melanoma cells. The results showed that the effects of pre-existing immunity to poliovirus could be overcome by a reasonable increase in the vaccine dose [64].

These are encouraging results. However, it is difficult to extrapolate the relevance of these studies to humans. Poliovirus replicates more robustly in humans than in mice, and other species-specific differences (as well as different routes of infection) between mice and humans may influence the outcome of revaccination. It is important to mention that unvaccinated people shed poliovirus vaccine viruses for 4–8 weeks post-vaccination, while previously OPV vaccinated people shed virus for only 1 week after a booster immunization [67]. Thus, it will be important to determine whether Sabin vectors have a good chance to be effective vaccine candidates in humans previously exposed to poliovirus. It is also possible that at higher doses, recombinant polioviruses might be able to establish infection and elicit primary immune responses against foreign antigens in individuals previously immunized against poliovirus. Furthermore, a vaccine that is effective only at immunizing naive or IPV-immunized children would still be extremely worthwhile.

3.6. Immunogenic potential of poliovirus vectors

We constructed recombinant vectors of two different serotypes of poliovirus (types 1 and 2) expressing SIV antigens. To analyze the immunological potential of poliovirus vectors in primates, four cynomolgus macaques were immunized intranasally with recombinant type 1 polioviruses containing SIV p17 and gp41. The animals were boosted with type 2 poliovirus containing SIV p17 and gp41. All macaques generated a mucosal anti-SIV IgA antibody response in rectal secretions. Two of the four macaques generated a mucosal antibody response detectable in vaginal lavages. Strong serum IgG antibody responses lasting for at least 1 year were also detected in two of the four monkeys. SIV-specific T-cell lymphoproliferative responses were detected in three of the four monkeys. SIV-specific CTLs were detected in two of the four monkeys. This study was the first report of poliovirus-elicited vaginal IgA antibodies or CTLs in any naturally infectable primate, including humans. These findings provided further support to the concept that live-poliovirus vectors are a powerful delivery system to elicit humoral and cellular immune responses against exogenous antigens at local mucosal sites [40].

3.7. Induction of protective immunity against a challenge with pathogenic $SIV_{mac}251$

Next, we constructed two libraries, each composed of 20 recombinant polioviruses (using both Sabin 1 and 2 as vaccine strain viruses). These libraries (SabRV1- and SabRV2-SIV, respectively) express the entire SIV genome in overlapping fragments.

3.7.1. Intranasal immunization-induced serum antibody responses
Seven female cynomolgus macaques were immunized intranasally with 5×10^7 pfu of the SabRV1-SIV library. The animals were then boosted by intranasal inoculation with 5×10^6 pfu of the SabRV2-SIV library 20 weeks later. All monkeys seroconverted to SIV after the first SabRV1-SIV library inoculation. The second set of intranasal immunizations using the SabRV2-SIV library boosted the anti-SIV response in all monkeys as evidenced by enhanced levels of serum anti-SIV IgG and IgA. All monkeys maintained a strong long-term anti-SIV IgG response with titers between 1:10,000 and 1:100,000. The anti-SIV serum IgA titers were also enhanced after booster immunization and remained constant thereafter. Western blotting analysis indicated that all seven vaccinated animals seroconverted to multiple SIV

antigens. Immunized monkeys developed anti-SIV antibodies that recognized SIV core, polymerase and envelope antigens [3].

3.7.2. Intranasal immunization-induced rectal and vaginal anti-SIV antibody responses

Following intranasal immunization, every monkey tested positive at some point for vaginal and rectal anti-SIV antibodies. Three monkeys had detectable anti-SIV IgA levels in vaginal lavage after the SabRV1-SIV library intranasal immunization. The vaginal and rectal lavages from all seven monkeys tested positive for anti-SIV IgA after the boost with Sabin 2-SIV library. Seven out of seven vaccinated monkeys produced rectal IgG, and three of seven produced rectal IgA. These results taken together are indicative of robust replication, expression, and immune recognition of individual poliovirus recombinants within the cocktail.

3.7.3. Intranasal immunization-induced CTL responses

Poliovirus vectors can elicit potent CTL responses in both mice [6,54] and primates [40]. In the more recent experiment we were able to detect SIV Env-specific CTLs in three of seven monkeys (25231, 27244, 27250) after SabRV1-SIV vaccination by a standard bulk PBMC cytolytic assay (R.A., unpublished data). After the SabRV2-SIV vaccinations, we detected SIV Gag- and Env-specific CTLs in monkey 25231. The three monkeys (25231, 27244, 27250) that tested positive for SIV-specific CTLs after SabRV-SIV vaccination also tested positive for SIV-specific lymphoproliferative responses (S.I. of 3.3, 4.1, and 2.7, respectively).

3.7.4. Protection from challenge with pathogenic $SIV_{mac}251$

To evaluate our challenge virus and protocol, we initially inoculated six control monkeys with 50 ID_{50} (50% infectious doses, equal to 10^5 $TCID_{50}$) of highly pathogenic $SIV_{mac}251$. One hundred percent of monkeys were infected and demonstrated high levels of viremia. Ten weeks after the final Sabin 2 SIV-library immunization, seven immunized and six additional control cynomolgus monkeys were challenged with an intravaginal inoculation with the highly pathogenic $SIV_{mac}251$ virus. To determine whether inoculated monkeys were infected with SIV, we co-cultivated PBMC with a susceptible cell line and measured production of p27 by ELISA. PBMC of all control monkeys, and five out of seven vaccinated monkeys, were positive for virus isolation 2 weeks post-challenge. Therefore, 100% (12 out of 12) control monkeys were infected. Importantly, one of the vaccinated monkeys was never positive for SIV, and the second protected vaccinated monkey was positive only at week 4.

Virus isolation is a sensitive but non-quantitative assay. Therefore, we determined the SIV viral loads by a sensitive and quantitative PCR assay. SabRV-SIV vaccinated animals 27244 and 27270 had no detectable SIV RNA in plasma at any time point, confirming that these two animals were fully protected (Fig. 2). Compared with the control monkeys, the seven SabRV1/2-SIV vaccinated monkeys had a 3.0 $\log_{10}$ reduction in post-acute geometric mean viral load ($P < 0.01$). Control of post-acute viremia was particularly obvious in two vaccinated monkeys: vaccinated monkey 28508 exhibited stable long-term control of viremia to

$\sim 1 \times 10^3$ copies/ml, and vaccinated monkey 25231 reduced its $SIV_{mac}251$ viremia by more than 10^6-fold during the post-acute phase, implicating a strong vaccine-elicited cellular immune response (Fig. 2).

The clinical outcome of SIV infection was much worse in the control animals compared to the SabRV-SIV immunized animals. Five of six control animals had marked decreases in $CD4^+$ T lymphocyte counts and body weight (Fig. 3A) over the 44-week post-challenge observation period. Three of the six control animals were euthanized at 34, 35, and 44 weeks post-challenge, respectively, due to severe clinical AIDS (Fig. 3B). At necropsy, two of the animals (23414, 26560) had lymphoma and the other animal (28118) had severe non-responsive enteritis and wasting.

In sharp contrast, all seven of the vaccinated monkeys were alive ($P < 0.07$) and healthy, i.e., significantly better body weight ($P < 0.003$) (Fig. 3A) at 44 weeks post-challenge. Although $CD4^+$ T-cell counts declined initially after challenge, the counts stabilized at about 16 weeks post-challenge for five of the seven vaccinated animals [3]. Over the course of the study, the SabRV-SIV vaccinated animals had higher average $CD4^+$ counts than the control animals. At 36 weeks post-challenge, the average $CD4^+$ cell count of vaccinated animals was 840 cells/ml, while five of six control monkeys had $CD4^+$ counts below 150 cells/ml. Two vaccinated animals (27250, 27273) had depressed $CD4^+$ T-cell counts after challenge, consistent with their higher SIV viremia levels, but their body weights remained stable, and they appeared clinically normal. The other five animals gained

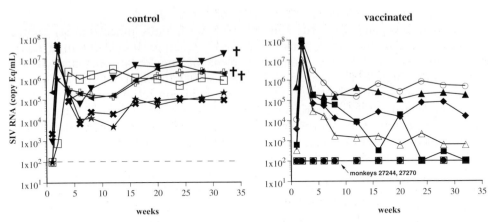

Fig. 2. SIV RNA loads. SIV RNA levels in plasma post-challenge. All seven vaccinated monkeys and six control monkeys were challenged with an intravaginal inoculation of highly virulent $SIV_{mac}251$ at week 30 of the experiment (challenge week 0). Vaccinated monkeys (right panel) are indicated by the symbols: 25231 (black square), 27244 (black circle), 27250 (black triangle), 27253 (black diamond), 27270 (white square), 27273 (white circle), 28508 (white triangle). Control monkeys (left panel) are indicated by the symbols: 26383 (black stars), 26385 (black ×), 23414 (white cross), 26405 (white diamond), 26560 (white triangle), 28118 (black arrow). Vaccinated monkeys 27270 and 27244 were never positive for SIV RNA and appear completely protected. Threshold sensitivity of the assay is 100 RNA copy Eq/ml, indicated by a dashed line. Data points below threshold value are shown at 100. Dagger (†) indicates death of the animal between weeks 32 and 44 post-challenge.

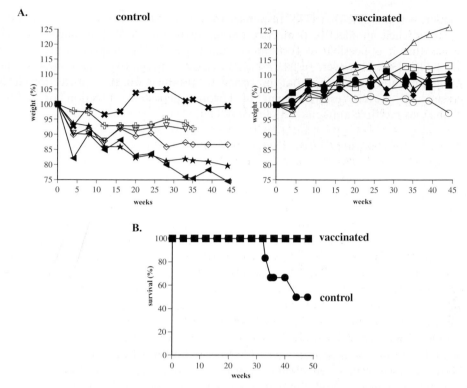

Fig. 3. Clinical outcomes of vaginal challenge with $SIV_{mac}251$. (A) Body weight. Weight of control macaques (left) and SabRV1/2-SIV vaccinated macaques (right), from day of challenge through 44 weeks post-challenge. Weight is indicated as percentage of body weight on the day of challenge. Body weight change of vaccinate animals is significantly better ($P < 0.003$) than the control monkeys. Symbols used are same as in Fig. 2. (B) Mortality curve. SabRV-SIV vaccinated animals (B); control animals (J).

weight steadily since the vaccine challenge. These results indicate that the SabRV-SIV vaccine protects monkeys from SIV-related disease progression.

3.7.5. Using a library of defined SIV antigens expressed by Sabin vectors

It has proven difficult to define HIV antigens that can induce protective immunity in an outbred population. Furthermore, the variability of HIV may render a single antigen vaccine ineffective. Our cocktail approach is designed as an immunological equivalent of a multi drug therapy, whose success is based on the simultaneous targeting of multiple viral proteins. Indeed, the cocktail we have chosen, expressing the entire SIV genome, is an effective immunogen capable of protecting macaques from challenge with a highly virulent SIV strain. Since we did not know which SIV antigens are important for protection, we used Sabin-libraries that express the entire SIV antigenic sequences. Using this approach we have demonstrated that it is possible to protect macaques from a challenge with a very pathogenic SIV strain ($SIV_{mac}251$). We are proposing to use a similar approach to protect against HIV-1.

However, it is possible that CTL responses to only few HIV proteins, for example, Gag, could elicit an effective protective immunity against HIV. Published evidence suggests that it is possible to contain viremia if good CTL responses are elicited against Gag [30]. However, resistant escape viruses emerge over the course of the chronic infection indicating that in order to produce durable protection it is necessary to prevent establishment of chronic persistent infection or induce immunity to multiple antigenic sequences [28,70].

3.8. Immunological complexity and antigen dilution

Immunological complexity is a theoretical and fascinating issue. Large viral vectors, such as vaccinia virus or canary pox, express many "vector" proteins (more than 200 viral proteins for vaccinia), and consequently the immune responses to the inserted HIV sequences may be reduced by competition with the many vector immunodominant antigens. It has been argued that the numerous viral proteins encoded in large DNA virus vectors may "dilute" the foreign sequence and that, for this reason, small RNA viruses could be a better choice as vaccine vectors, *prima facie*. Indeed, the total genetic complexity of the poliovirus recombinants in our Sabin-SIV cocktail, which is comprised of 20 different viruses, used in our pre-clinical "proof-of-concept" experiment, represents only 1/10 of the genome of vaccinia virus. It is also important to point out that complex, effective vaccine formulations are currently used very effectively in humans. For example, there is an effective vaccine with seven components (diphtheria, tetanus, acellular pertussis, *Haemophilus influenza*, and three independent serotypes of IPV), and there are currently efforts to add two additional components (hepatitis B virus and hepatitis A virus vaccines). The most dramatic example is the pneumoccocus vaccine that contains 23 independent components (polysaccharides).

3.9. Poliovirus eradication and vaccine vectors

The WHO poliomyelitis eradication effort has been wonderfully successful, and it is therefore necessary to address the future viability of a vaccine program based on poliovirus live viral vectors. We are optimistic that wild poliovirus infections can be eliminated, and the WHO is well on its way to achieving this goal [37,71]. However, we and others have expressed reservations about the ability to eliminate the Sabin live poliovirus vaccine viruses at any time in the near future [40,72], particularly in the face of the potential use of wild type poliovirus as a bioterrorism agent [73]. The most optimistic view is that the world may be declared free of wild type poliovirus after 2008. And the use of OPV would surely continue for at least several years thereafter. That would give sufficient time to further investigate the potential use of Sabin-based poliovirus as a mucosal vaccine vector. We believe that our work with these recombinant polioviruses demonstrates that the Sabin poliovirus vaccine strains should not be cast aside, as they have strong potential as vectors for creating novel vaccines against major public health threats. The fact that the live poliovirus vaccine is so effective that wild-type poliovirus eradication is conceivable, is a strong argument that we should parlay this knowledge into the development of other vaccines, potentially using the live poliovirus itself as the delivery vector, as we have proposed.

Abbreviations

AIDS	acquired immunodeficiency syndrome
APC	antigen-presenting cell
CTL	cytotoxic T lymphocyte
E-IPV	enhanced inactivated poliovirus vaccine
HIV	human immunodeficiency virus
ID_{50}	infecting dose that causes infection in 50% of the animals or cells
IPV	inactivated poliovirus vaccine
IRES	internal ribosome entry site
MOPV	monovalent oral poliovirus vaccines
OPV	oral poliovirus vaccine
PBMC	peripheral blood monocytes
PFU	plaque forming units
SabRV	Sabin recombinant virus
SHIV	SIV/HIV chimeric virus
SI	syncytial-inducing
SIV	Simian immunodeficiency virus
$TCID_{50}$	tissue culture infectious dose that causes infection in 50% of the cells
VAPP	vaccine associated paralytic poliomyelitis
WHO	World Health Organization

References

1. Bledsoe, A.W., Jackson, C.A., McPherson, S., and Morrow, C.D. (2000) Nat. Biotechnol. 18, 964–969.
2. Gromeier, M., Lachmann, S., Rosenfeld, M.R., Gutin, P.H., and Wimmer, E. (2000) Proc. Natl. Acad. Sci. USA 97, 6803–6808.
3. Crotty, S., Miller, C.J., Lohman, B.L., Neagu, M.R., Compton, L., Lu, D., Lu, F.X., Fritts, L., Lifson, J.D., Andino, R. (2001) J. Virol. 75, 7435–7452.
4. Plotkin, S.A. and Orenstein, W. (eds.) (1999) Vaccines, W.B. Saunders, Philadelphia, PA, 3rd ed.
5. Andino, R., Silvera, D., Suggett, S.D., Achacoso, P.L., Miller, C.J., Baltimore, D., and Feinberg, M.B. (1994) Science 265, 1448–1451.
6. Mandl, S., Sigal, L.J., Rock, K.L., and Andino, R. (1998) Proc. Natl. Acad. Sci. USA 95, 8216–8221.
7. Ogra, P.L. and Karzon, D.T. (1969) J. Immunol. 102, 1423–1430.
8. Ogra, P.L. and Ogra, S.S. (1973) J. Immunol. 110, 1307–1311.
9. Paul, J.R., and Melnick, J.L. (1951) Am. J. Hyg. 54, 275–285.
10. World-Health-Organization (2002) WHO Technical Reports Series 904, 31–84.
11. UNAIDS web site: http://www.UNAIDS.org.
12. Pantaleo, G. and Fauci, A.S. (1994) Curr. Opin. Immunol. 6, 600–604.
13. Miller, C.J., McGhee, J.R., and Gardner, M.B. (1993) Lab. Invest. 68, 129–145.
14. Sewell, A.K. and Price, D.A. (2001) Trends Immunol. 22, 173–175.

186

15. Ada, G.L. (1989) In: Paul, W.E. (ed.) Fundamental Immunology, Raven Press, New York, pp. 985–1032.
16. Weiss, R.A. (1993) Science 260, 1273–1279.
17. Lifson, J.D. and Martin, M.A. (2002) Nature 415, 272–273.
18. Smith, P.D. and Wahl, S.M. (1998) In: Ogra, P. (eds.) Mucosal Immunology, Academic Press, Inc, San Diego, CA, 2nd ed., 985–1032.
19. Sutjipto, S., Pedersen, N.C., Miller, C.J., Gardner, M.B., Hanson, C.V., Gettie, A., Jennings, M., Higgins, J., and Marx, P.A. (1990) J. Virol. 64, 2290–2297.
20. Mascola, J.R., Stiegler, G., VanCott, T.C., Katinger, H., Carpenter, C.B., Hanson, C.E., Beary, H., Hayes, D., Frankel, S.S., Birx, D.L., Lewis, M.G. (2000) Nat. Med. 6, 207–210.
21. Baba, T.W., Liska, V., Hofmann-Lehmann, R., Vlasak, J., Xu, W., Ayehunie, S., Cavacini, L.A., Posner, M.R., Katinger, H., Stiegler, G., Bernacky, B.J., Rizvi, T.A., Schmidt, R., Hill, L.R., Keeling, M.E., Lu, Y., Wright, J.E., Chou, T.C., Ruprecht, R.M. (2000) Nat. Med. 6, 200–206.
22. Koup, R.A., Safrit, J.T., Cao, Y.Z., Andrews, C.A., McLeod, G., Borkowsky, W., Farthing, C., and Ho, D.D. (1994) J. Virol. 68, 4650–4655.
23. Borrow, P., Lewicki, H., Hahn, B.H., Shaw, G.M., and Oldstone, M.B. (1994) J. Virol. 68, 6103–6110.
24. Yang, O.O. and Walker, B.D. (1997) Adv. Immunol. 66, 273–311.
25. Yang, O.O., Kalams, S.A., Trocha, A., Cao, H., Luster, A., Johnson, R.P., and Walker, B.D. (1997) J. Virol. 71, 3120–3128.
26. Schmitz, J.E., Kuroda, M.J., Santra, S., Sasseville, V.G., Simon, M.A., Lifton, M.A., Racz, P., Tenner-Racz, K., Dalesandro, M., Scallon, B.J., Ghrayeb, J., Forman, M.A., Montefiori, D.C., Rieber, E.P., Letvin, N.L., Reimann, K.A. (1999) Science 283, 857–860.
27. Barouch, D.H. and Letvin, N.L. (2001) Curr. Opin. Immunol. 13, 479–482.
28. Barouch, D.H., Kunstman, J., Kuroda, M.J., Schmitz, J.E., Santra, S., Peyerl, F.W., Krivulka, G.R., Beaudry, K., Lifton, M.A., Gorgone, D.A., Montefiori, D.C., Lewis, M.G., Wolinsky, S.M., Letvin, N.L. (2002) Nature 415, 335–339.
29. Amara, R.R., Villinger, F., Altman, J.D., Lydy, S.L., O'Neil, S.P., Staprans, S.I., Montefiori, D.C., Xu, Y., Herndon, J.G., Wyatt, L.S., Candido, M.A., Kozyr, N.L., Earl, P.L., Smith, J.M., Ma, H.L., Grimm, B.D., Hulsey, M.L., Miller, J., McClure, H.M., McNicholl, J.M., Moss, B., and Robinson, H.L. (2001) Science 292, 69–74.
30. Shiver, J.W., Fu, T.M., Chen, L., Casimiro, D.R., Davies, M.E., Evans, R.K., Zhang, Z.Q., Simon, A.J., Trigona, W.L., Dubey, S.A., Huang, L.Y., Harris, V.A., Long, R.S., Liang, X.P., Handt, L., Schleif, W.A., Zhu, L., Freed, D.C., Persaud, N.V., Guan, L., Punt, K.S., Tang, A., Chen, M.C., Wilson, K.A., Collins, K.B., Heidecker, G.J., Fernandez, V.R., Perry, H.C., Joyce, J.G., Grimm, K.M., Cook, J.C., Keller, P.M., Kresock, D.S., Mach, H., Troutman, R.D., Isopi, L.A., Williams, D.M., Xu, Z., Bohannon, K.E., Volkin, D.B., Montefiori, D.C., Miura, A., Krivulka, G.R., Lifton, M.A., Kuroda, M.J., Schmitz, J.E., Letvin, N.L., Caulfield, M.J., Bett, A.J., Youil, R., Kaslow, D.C., and Emini, E.A. (2002) Nature 415, 331–335.
31. Stevceva, L., Kelsall, B., Nacsa, J., Moniuszko, M., Hel, Z., Tryniszewska, E., and Franchini, G. (2002) J. Virol. 76, 9–18.
32. Schmitz, J.E., Veazey, R.S., Kuroda, M.J., Levy, D.B., Seth, A., Mansfield, K.G., Nickerson, C.E., Lifton, M.A., Alvarez, X., Lackner, A.A., Letvin, N.L. (2001) Blood 98, 3757–3761.
33. Mestecky, J., Abraham, R., and Ogra, P.L. (1994) In: Ogra, P.L., Lamm, M.E., McGhee, J.R., Mestecky, J., Strober, W. and Bienenstock, J. (eds.) Handbook of Mucosal Immunity, Academic Press, Inc, San Diego, CA, pp. 357–372.
34. Merriman, H., Woods, S., Winter, C., Fahnlander, A., and Corey, L. (1984) J. Infect. Dis. 149, 505–510.
35. Ashley, R.L., Crisostomo, F.M., Doss, M., Sekulovich, R.E., Burke, R.L., Shaughnessy, M., Corey, L., Polissar, N.L., and Langenberg, A.G. (1998) J. Infect. Dis. 178, 1–7.
36. Oldstone, M.B. (1991) J. Virol. 65, 6381–6386.
37. Hull, H.F., Birmingham, M.E., Melgaard, B., and Lee, J.W. (1997) J. Infect. Dis. 175 Suppl. 1, S4–S9.
38. AHSF Drug Information (1998) American Society of Health-System Pharmacists (http://www.ashp.org). SilverPlatter International, Bethesda, MD.

39. Minor, P.D. (1997) Vaccine 15, 1709.
40. Crotty, S., Lohman, B.L., Lü, F.X., Tang, S., Miller, C.J., and Andino, R. (1999) J. Virol. 73, 9485–9495.
41. Mas Lago, P., Caceres, V.M., Galindo, M.A., Gary, H.E., Jr., Valcarcel, M., Barrios, J., Sarmiento, L., Avalos, I., Bravo, J.A., Palomera, R., Bello, M., Sutter, R.W., Pallansch, M.A., de Quadros, C.A. (2001) Int. J. Epidemiol. 30, 1029–1034.
42. Paul, J.R. (1955) Poliomyelitis: WHO Monograph Series, No. 26, World Health Organization, Geneva, pp. 9–29.
43. Kawamura, N., Kohara, M., Abe, S., Komatsu, T., Tago, K., Arita, M., and Nomoto, A. (1989) J. Virol. 63, 1302–1309.
44. Bouchard, M.J., Lam, D.H., and Racaniello, V.R. (1995) J. Virol. 69, 4972–4978.
45. Macadam, A.J., Pollard, S.R., Ferguson, G., Skuce, R., Wood, D., Almond, J.W., and Minor, P.D. (1993) Virology 192, 18–26.
46. Almond, J.W., Westrop, G.D., Evans, D.M., Dunn, G., Minor, P.D., Magrath, D., and Schild, G.C. (1987) J. Virol. Methods 17, 183–189.
47. Minor, P.D. (1993) Dev. Biol. Stand. 78, 17–26.
48. Esteves, K. (1988) Bull. World Health Org. 66, 739–746.
49. Macadam, A.J., Ferguson, G., Stone, D.M., Meredith, J., Almond, J.W., and Minor, P.D. (2001) Dev. Biol. 105, 179–187.
50. Rowe, A., Burlison, J., Macadam, A.J., and Minor, P.D. (2001) Virology 289, 45–53.
51. Gromeier, M., Alexander, L., and Wimmer, E. (1996) Proc. Natl. Acad. Sci. USA 93, 2370–2375.
52. Racaniello, V.R. (1981) Science 214, 916–919.
53. Krausslich, H.G. and Wimmer, E. (1988) Annu. Rev. Biochem. 57, 701–754.
54. Sigal, L.J., Crotty, S., Andino, R., and Rock, K.L. (1999) Nature 398, 77–80.
55. Slifka, M.K., Pagarigan, R., Mena, I., Feuer, R., and Whitton, J.L. (2001) J. Virol. 75, 2377–2387.
56. Chapman, N.M., Kim, K.S., Tracy, S., Jackson, J., Hofling, K., Leser, J.S., Malone, J., and Kolbeck, P. (2000) J. Virol. 74, 7952–7962.
57. Burke, K.L., Dunn, G., Ferguson, M., Minor, P.D., and Almond, J.W. (1988) Nature 332, 81–82.
58. Alexander, L., Lu, H.H., Gromeier, M., and Wimmer, E. (1994) Aids Res. Hum. Retroviruses 10 Suppl. 2, S57–S60.
59. Tang, S., van Rij, R., Silvera, D., and Andino, R. (1997) J. Virol. 71, 7841–7850.
60. Mueller, S. and Wimmer, E. (1998) J. Virol. 72, 20–31.
61. Lee, S.G., Kim, D.Y., Hyun, B.H., and Bae, Y.S. (2002) J. Virol. 76, 1649–1662.
62. Sigal, L.J., Crotty, S., Andino, R., and Rock, K.L. (1999) Nature 398, 77–80.
63. Crotty, S., Hix, L., Sigal, L.J., and Andino, R. (2002) J. Gen. Virol. 83, 1707–1720.
64. Mandl, S., Hix, L., and Andino, R. (2001) J. Virol. 75, 622–627.
65. Ogra, P.L., Leibovitz, E.E., and Zhao, R.G. (1989) Curr. Top. Microbiol. Immunol. 146, 73–81.
66. Onorato, I.M., Modlin, J.F., McBean, A.M., Thoms, M.L., Losonsky, G.A., and Bernier, R.H. (1991) J. Infect. Dis. 163, 1–6.
67. Modlin, J.F., Halsey, N.A., Thoms, M.L., Meschievitz, C.K., and Patriarca, P.A. (1997) J. Infect. Dis. 175 Suppl. 1, S228–S234.
68. Cooney, E.L., McElrath, M.J., Corey, L., Hu, S.L., Collier, A.C., Arditti, D., Hoffman, M., Coombs, R.W., Smith, G.E., Greenberg, P.D. (1993) Proc. Natl. Acad. Sci. USA 90, 1882–1886.
69. Hu, S.L., Abrams, K., Barber, G.N., Moran, P., Zarling, J.M., Langlois, A.J., Kuller, L., Morton, W.R., and Benveniste, R.E. (1992) Science 255, 456–459.
70. O'Connor, D.H., Allen, T.M., Vogel, T.U., Jing, P., DeSouza, I.P., Dodds, E., Dunphy, E.J., Melsaether, C., Mothe, B., Yamamoto, H., Horton, H., Wilson, N., Hughes, A.L., Watkins, D.I. (2002) Nat. Med. 8, 493–499.
71. Nathanson, N. and Fine, P. (2002) Science 296, 269–270.
72. Racaniello, V.R. (2000) Bull. World Health Org. 78, 359–360.
73. Cello, J., Paul, A.V., and Wimmer, E. (2002) Science 297, 1016–1018.

S.C. Makrides (Ed.) *Gene Transfer and Expression in Mammalian Cells*
© 2003 Elsevier Science B.V. All rights reserved

Virus-based vectors for gene expression in mammalian cells: Sindbis virus

Henry V. Huang and Sondra Schlesinger

*Department of Molecular Microbiology, Washington University School of Medicine, Campus Box 8230, 660
South Euclid Avenue, Saint Louis, Missouri 63110-1093, USA; Tel.: +314-362-2755; Fax: +314-362-7325.
E-mail: huang@borcim.wustl.edu*

1. Introduction

Alphaviruses are relatively simple RNA enveloped viruses that have been valuable models for learning about virus replication and transmission. Their wide host range, small genome length, simple gene organization, and high level of gene expression provided the original motivation to adapt them as gene expression vectors. This review focuses mainly on Sindbis virus, the type species of the Alphavirus genus in the Togaviridae family. In nature, it is transmitted by a number of species of mosquitoes to vertebrates (mammals, birds, reptiles and amphibia), principally birds. Two other members of this genus, Semliki Forest virus (SFV) and Venezuelan equine encephalitis virus (VEEV) are also being developed as vectors. VEEV is a known human pathogen but a vaccine strain exists that is the basis for the VEEV vectors. We include VEEV here; SFV is discussed in Chapter 3.11. The ways in which Sindbis virus vectors are being used cover a wide spectrum. In addition to the illustrations given below, there are many examples in which proteins were expressed to be able to study their cellular localization, modifications, and function. Many of these have been described in previous reviews (see e.g., [1]).

Alphaviruses enter cells through the endosomal pathway (Fig. 1A). Sindbis virus attaches to the cell surface using one of several potential receptors [2]. It is then taken up into an endosome, that, when acidified, triggers the viral surface glycoproteins to fuse the virus membrane with the endosomal membrane, thus depositing the internal nucleocapsid into the cytoplasm. The genomic RNA is capped and polyadenylated, and is the mRNA for the viral nonstructural polyprotein (B). The polyprotein is proteolytically processed to produce the individual subunits, nsP1 through nsP4 by the nsP2 protease. The nonstructural proteins serve as the replicase-transcriptase, that uses the genomic RNA as template to produce a full-length, negative-sense complement, which in turn is template for synthesis of additional genomic RNAs (C). The negative sense complement additionally contains a promoter that is used to produce a subgenomic mRNA for translation of the structural polyprotein (D). The capsid protein located at the amino-terminus is an autoprotease that cleaves itself from the nascent polypeptide, which is then inserted into cellular membranes and is cleaved by host cell proteases to produce the glycoproteins E1 and PE2 (precursor of the virus glycoprotein E2) and a small hydrophobic 6K protein. The capsid protein binds to genomic RNA, facilitated by a packaging signal in the nsP1-encoding

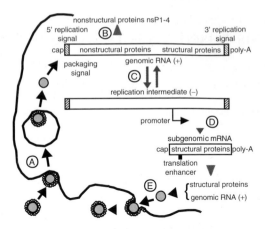

Fig. 1. Sindbis virus lifecycle and gene organization. See text for details.

sequences, to form nucleocapsids (E). (Packaging signals of other alphaviruses map to different parts of the nonstructural proteins coding region). The nucleocapsids interact with the cytoplasmic tail of the E2 glycoprotein, resulting in the budding and release of progeny virions through the host plasma membrane.

2. Viral genetic elements

Many scientists have contributed to the studies that defined the functional significance of the genetic elements required for the virus lifecycle (see [2] for review): The trans-acting factors consist of the viral replicase-transcriptase and the viral structural proteins. The cis-acting signals consist of the replication signals at the termini of the genome, the internal promoter for subgenomic mRNA synthesis [3], the genome packaging signal [4], and the translation-enhancing signal within the capsid gene [5] (Fig. 1). All current vectors consist of combinations of these elements, tailored for the specific applications. Ongoing studies of the elements have provided information to further customize the properties of each, to better suit the particular application.

2.1. Nonstructural proteins

The nonstructural proteins are derived from a polyprotein that is proteolytically processed. The processing intermediates are functionally significant, as they regulate the synthesis of viral minus strands versus positive strands. The nsP1 subunit has guanine-[7]N-methyltransferase and guanylyltransferase activities required for capping of the genomic RNA and subgenomic mRNA, but the 5'-triphosphatase activity appears to be in the nsP2 subunit. nsP2 is additionally an RNA helicase, and its protease activity is required for the processing of the nonstructural polyprotein. The precise function of nsP3 is unclear. The polymerase activity is in the nsP4 subunit. It has the usual polymerase motifs, and a temperature-sensitive mutation in nsP4 results in total cessation of viral RNA synthesis at the nonpermissive temperature. This has been useful for precise regulation of vector gene expression [6,7]. Many other mutations (temperature-sensitive, host-restricted, drug-resistance, virulence) of the nonstructural proteins are available for tailoring the properties of the vectors.

2.2. Structural proteins

The extremely broad host range of Sindbis virus has been a convenience that allows the same vectors to be used in a variety of host species and cell types. The host range is in large part determined by the envelope proteins, so that target cell specificity can be modulated by using variants of the envelope proteins [2]. Despite its broad host range, the virus does have specific tissue tropisms, and it might not infect some cells as well as may be desired. For at least some cell types, this can be overcome by using natural variants of the glycoproteins. For example, infection of mosquito C6/36 cells and mammalian neuronal cells can be substantially improved using the envelope proteins from TE12, a more neurovirulent isolate of Sindbis virus [8,9]. This replacement also conferred preferential infection of neuronal versus accessory cells in explanted rat dorsal root ganglia [9]. Similarly, the glycoproteins of a Malaysian isolate of Sindbis virus, MRE16, give better per os infection of *Aedes aegypti* mosquitoes, due to improved infection of the mosquito's midgut cells [10].

For other applications, e.g., targeting specific cell types, the broad host range of the virus is actually a liability. In these cases, the viral glycoproteins may be modified to obtain exquisite specificities. London *et al.* [11] had mapped sites in E2 that can tolerate insertions, e.g., of foreign epitopes for potential vaccine applications, without substantial effects on infectivity. Insertion of the much larger Fc-binding domain of protein A at one of these sites completely blocked infectivity. Remarkably, infectivity can be restored in the presence of antibodies that bind specific cell-surface antigens [12]. Thus, target cell specificity becomes that of the antibody employed. This same strategy was used to obtain specific targeting of HIV-1 and murine leukemia virus vectors pseudotyped with the modified Sindbis virus glycoproteins [13]. Targeting infection to specific cells *in vivo* is likely to be more complicated, because of the abundance of competing, nonspecific antibodies. In principle, insertion of a single-chain antibody into the same site of E2 might get around this problem.

2.3. Cis-acting signals

2.3.1. Replication signals

A host-derived tRNA^{Asp} sequence can substitute for the normal 5′ end for defective interfering genome replication. Indeed, it confers higher levels of replication. Thus, to minimize packaging of the defective helper genome, the normal 5′ end is preferable (see Section 3). A conserved 19 nt sequence is the replication signal at the 3′ end of the genome, the remainder of the 3′ untranslated sequences appears dispensable. Mutants in the sequence have been studied, but their utility in the vectors has not been explored.

2.3.2. Subgenomic mRNA promoter

The promoter on the minus strand directs the transcription of the subgenomic mRNA. It maps to within −40 nt upstream to 14 nt downstream of the site of transcription initiation. Many mutant promoters with weaker or stronger activities have been characterized [3].

2.3.3. Translation-enhancing signal

The original alphavirus replicons were engineered to have the heterologous gene replace the viral structural protein genes. The level of protein expression was very high, but not as high as the amount of capsid protein produced during normal virus infection. An investigation of this discrepancy led to the identification of a translational enhancer at nt 77–139 of the subgenomic RNA, within the capsid gene [5]. It enhances the level of protein translated by as much as 10-fold.

The enhancer is positioned upstream and in frame with the coding sequences of the gene to be expressed. This amino-terminal appendage can be automatically cleaved from the expressed protein by including sequences that encode the C-terminal part of capsid—the autoprotease—which will then cleave off the capsid fragment. It is also possible to insert an amino acid sequence that can be recognized and cleaved by some other protease. The enhancer is effective only under conditions in which the synthesis of host cell protein is inhibited. It does not increase the expression of heterologous genes embedded in the noncytopathic replicons.

2.3.4. Packaging signal

In cells infected with Sindbis virus (or replicons) the subgenomic RNA is present in excess molar amounts over the genomic RNA, but the latter is preferentially incorporated into nucleocapsids and viral particles. This discrimination is due to a packaging signal in the nt 945–1076 region of the genome. The region binds specifically to the Sindbis viral capsid protein and enhances the packaging of a viral RNA into particles [4]. Although, the packaging signal allows the capsid to discriminate against subgenomic and other RNAs, this discrimination is not absolute.

3. Packaging of replicons

For many studies with alphavirus replicons the goal has been to use particles that only undergo one round of infection and this can be achieved if the defective helper is not packaged. Most of the defective helpers available for packaging replicons lack the packaging signal and contain large deletions in the nsP genes. These helpers package replicons without being packaged themselves or are packaged at very low levels [1,14]. Because co-packaging of the replicon and helper or recombination between these two RNAs would lead to the production of particles able to spread beyond the initially infected cell, several ways have been devised to prevent the formation of such infectious particles. PE2, the precursor of E2 can be mutationally altered so that cleavage does not normally occur. Particles formed are not infectious unless first treated with a specific protease [1,15]. A second way that has proven to be effective is to use split helpers: one expresses the capsid protein and the other expresses the glycoproteins and the 6K protein. Co-packaging of two helpers with a replicon or recombination among the three RNAs has not been observed in laboratory experiments. It will be important to establish that neither co-packaging nor recombination occur during large scale industrial production. Cell lines that contain the Sindbis defective split helpers and are able to package Sindbis replicons have been established in anticipation of the need to prepare large scale amounts of replicon particles [16].

There are times when co-packaging of the replicon and helper is useful, to permit the replicon to spread throughout the culture. There are Sindbis defective helpers that are co-packaged [17] and an example of their utility is described in Section 8.

4. Effects of alphaviruses and replicons on infected cells

Sindbis virus and replicons are cytopathic to vertebrate cells but not insect cells. Cytopathogenicity can be an advantage for vaccine applications, as dying cells will be taken up by macrophages that will then present the resulting antigenic peptides to the immune system. But for most uses, cell survival would be preferred. One approach to preventing cell death has been to use inhibitors of apoptosis (e.g., XIAP, Ref. [18]), as infected cells appear to die from a combination of necrosis and apoptosis [19]. Another approach is to employ pro-survival members of the Bcl-2 family, e.g., Bcl-w [20]. Some of these do appear to slow, but not prevent cell death. Furthermore, in some examples the decrease in cell death may actually be due to a decrease in the number of infected cells. This was what happened when expression of bcl-2 appeared to prevent cell death after infection with Sindbis virus; the expression of bcl-2 in these cells blocked infection by the virus.

A more successful approach has been to obtain viral mutants that are less or not cytopathic. Noncytopathic mutants of Sindbis virus were first obtained from BHK-21 cells that were persistently infected with Sindbis virus. A single change in the nsP2 protein from proline 726 to serine was responsible for this phenotype [21]. By combining this mutation with a temperature-sensitive mutation in nsP4, Boorsma et al. [7] obtained a BHK-21 cell line that contained the Sindbis virus replicon as a cDNA stably integrated into the chromosome and was inducible by shifting the temperature from 37 to 30°C.

Other noncytopathic mutants of Sindbis replicons were isolated by transfecting BHK-21 cells with a replicon that expressed the puromycin-resistance gene and selecting for cells that are puromycin-resistant. Only those cells that contained a noncytopathic replicon could survive in the presence of both replicon and puromycin. Characterization of such a replicon showed that the same proline 726 in nsP2 had been changed, but this time to leucine [22]. Subsequently, several other noncytopathic replicons from both Sindbis virus and SFV were selected using neomycin [23]. One of these Sindbis mutants also had a change from proline 726 to threonine; a second mutant was altered at amino acid 1 in nsP2. The mutations in SFV were also in nsP2.

Replicons with leucine or glutamine at amino acid 726 in nsP2 had 50- to 100-fold decreases in RNA synthesis, and did not inhibit host cell protein synthesis or cause cytopathic effects. Those with serine or threonine at 726 had 10 to 20-fold reduction in RNA levels, and some cell death was observed [24]. Lower levels of replicon RNA synthesis led to significant decreases in inhibition of host cell protein synthesis and cytopathogenicity in essentially all of the different cell lines tested, but did not lead to permanent infection in all of these different cell lines. There appears to be a direct correlation between the ability to establish a persistent infection in which all of the cells are infected and the absence of interferon induction, as cultures in which interferon is induced are cured of infection—a well-established phenomenon with RNA viruses. The expression of MxA or bcl-2 has been shown to inhibit alphavirus replication in some cell lines, but it would be difficult to exploit this means of protecting cells as most of the cells are unable to synthesize any viral products in the presence of bcl-2 or MxA (reviewed in [1]).

It is curious that most of the noncytopathic mutations are in nsP2. The nsP2 protease activity required for polyprotein processing is unaltered in many of the mutants. Normally, a fraction of nsP2 is translocated into the nucleus, specifically to the nucleolus, in infected cells. Mutants of SFV that lack the nuclear localization signal in nsP2 are less virulent than the wild type virus in mice. This suggests that nsP2 nuclear localization may have some role in affecting the response to infection and survival of the host cell.

5. Expression in neurons

One of the strong attractions that alphavirus vectors have for studies in neuro-biology is their ability to infect neurons selectively. A second crucial feature is that infected neurons remain morphologically normal for several days and continue to function as well as their uninfected counterparts [25–28]. Neuronal cells infected with Sindbis virus vectors expressing GFP maintain their integrity for up to three days and are not distinguishable from uninfected cultures based on their resting potential, synaptic transmission and long term potentiation. The use of less cytopathic nsP2 mutants has significantly extended the time in which infected neurons can be analyzed to more than 6 days (Fig. 2). Neurobiologists have now expressed and carried out functional studies for a number of membrane-associated proteins, receptors, and proteins involved in signaling pathways by infection of dissociated neurons and hippocampal slices with Sindbis replicon particles (reviewed in [29]).

Several ways have been used to express more than one protein in neuronal cultures. Co-infection with two different particles seems the most obvious and has been used successfully to express both the AMPA-type glutamate receptor subunit two and PICK1 (protein interacting with C Kinase) [30], but super infection exclusion (virus-specific interference) may prevent both genomes from being

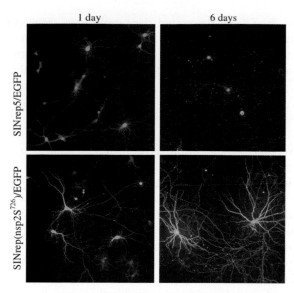

Fig. 2. Hippocampal primary neurons infected by wild type Sindbis replicon or the nsP2 mutant, SINrep(nsP2S726), both expressing EGFP, and observed under living conditions at 1 and 6 days postinfection. (Courtesy of Dr. Pavel Osten, Max Planck Institute for Medical Research, Heidelberg, Germany)

expressed in the same cell [2]. Hayashi *et al.* [31] were able to express both GluR1 and the calcium/calmodulin-dependent protein kinase II from a single Sindbis virus subgenomic RNA by inserting an IRES element upstream of sequences coding for the latter protein. Internal initiation from an IRES element is likely to lead to much lower levels of translation which could be either an advantage or disadvantage. An alternative has been to use vectors with two subgenomic RNA promoters. If both promoters are identical then expression of the two proteins will be similar. Promoter mutants can be used to modulate the level of transcription and thus the level of the translated protein ([3]; P. Osten, personal communication).

6. Insect and crustacea

Some of the most exciting uses of the Sindbis virus vectors have been in studying its alternate host, the mosquito. Female mosquitoes must take a blood meal to lay eggs. The small volume of blood ingested may include infectious viruses if the infected vertebrate host is exhibiting sufficiently high viremia. Sindbis virus in the blood meal first infects midgut cells and is then released into the hemocoel, from whence it spreads to the internal organs, including the fat bodies, neural tissues, and the salivary glands, where it is released into the saliva. The mosquito injects a small volume of virus-laden saliva into the vertebrate host when it takes its next blood meal, thus propagating the infection.

Pioneering work by Olson, Beaty and their colleagues showed that Sindbis virus vectors can express a foreign gene in cultured C6/36 *Aedes albopictus* cells, and *in vivo* in *A. aegypti*, *A. triseriatus* and *Culex pipiens* mosquitoes. Foreign gene expression is found in midgut cells, neural tissues, fat bodies, and salivary glands (see [32] for a review).

6.1. Interrupting mosquito transmission of pathogens

Mosquitoes are vectors of many pathogens of humans and their livestock. A major interest in using the Sindbis virus vectors in mosquito systems is to identify methods to interrupt the infection cycle. One approach to control mosquito-vectored diseases is to kill the mosquito when it takes a blood meal. This is analogous to the use of pesticides for mosquito abatement, but without posing any hazard to bystander species. Higgs *et al.* [33] injected *A. aegypti*, *A. triseriatus* and *C. pipiens* mosquitoes with a double subgenomic RNA Sindbis virus vector expressing the insect-specific toxin, Scotox, from the scorpion *Androctonus australis hector*. The mosquitoes died 1–5 days later. Similarly injected flies (*Musca domesticus*) and ticks (*Dermacentor andersoni*) were unaffected. It is thus conceivable that virus-laden bait, or genetically engineered animals (e.g., sentinel chickens) might be used to deliver a dose of recombinant Sindbis virus vectors to mosquitoes when they feed, to kill them or to immunize them against specific pathogens. Whether killing off just those mosquitoes that feed on the deployed bait will have significant impact on the mosquito population is a key unknown. Nonetheless, this approach is worthy of some consideration. Ideally, the virus used must not be able to propagate nor cause disease in nonmosquito species. This might be achieved by making an E2 derivative bearing a single chain antibody specific for some mosquito midgut cell surface protein, that will serve as a surrogate receptor. The resulting virus should be able to infect mosquito midgut cells, but unable to infect the vertebrate host. Various host-restricted mutations that have been identified may be used as additional virus containment measures.

Another strategy proposed to control mosquito-transmitted diseases is to create genetically modified mosquitoes that are resistant to the particular pathogens of interest, and have these replace the susceptible ones in the wild. This strategy clearly faces many challenges, not the least of which is weighing the ecological impact of releasing genetically modified, fertile mosquitoes into the environment. Nevertheless, significant advances have been made using the Sindbis virus vectors as a tool to identify how mosquitoes may be made resistant to pathogens. Sindbis virus-mediated expression of a scAb that recognizes the circumsporozoite protein provided 99.9% inhibition of *Plasmodium gallinaceum* infection of *A. aegypti* salivary glands [34]. The double subgenomic RNA Sindbis virus vector was also used to express LaCrosse and Dengue virus genome sequences. Replication by LAC, as well as the closely related snowshoe hare and Tahyna viruses, was virtually eliminated when the full-length, antisense S segment of LAC was expressed in C6/36 cells by the Sindbis virus vector. (In contrast, expression of the M segment did not inhibit). The more distantly related trivittatus virus was only partially inhibited. The same recombinant was injected intrathoracically into *A. triseriatus* mosquitoes that had been fed LAC. LAC virus titers were inhibited by 10- to 100-fold relative to animals injected with the nonrecombinant vector. Inhibition of pathogen growth in the mosquito in these studies was impressive and studies with DEN [35] and yellow fever virus [36] further demonstrated that transmission of these viruses could be blocked.

An exciting observation emerging from these studies is that the inhibition is probably RNA-mediated [32]. The precise mechanism of inhibition is yet to be determined, but it could be inhibition of viral processes due to base-pairing between the vector-expressed RNA and the target RNA (all the viruses tested necessarily produce RNA of both polarities during infection, thus potentially providing targets to both sense and antisense RNAs). Another possibility is that the inhibition is mediated by RNAi and active, host-mediated degradation of target virus RNAs [37]. Independent of the mechanism of action, it is clear that the Sindbis virus vector is an excellent tool for identifying means by which to block the transmission of pathogenic viruses.

6.2. Inhibiting gene expression and ectopic gene expression in arthropods

A remarkable observation is that the Sindbis virus vectors can be used to inhibit host gene expression. Prophenoloxidase is considered to be the key enzyme mediating melanization in the insect defense against parasites. A double-subgenomic RNA vector was used to express a 600 nt antisense prophenoloxidase sequence in *Armigeres subalbatus*. Transduced mosquitoes had reduced prophenoloxidase activity in the hemolymph, and were unable to melanize a filarial parasite (*Dirofilaria immitis* mf) when challenged [38].

More classical, ectopic gene expression may also be used to manipulate the host animal. Even though the alternate host of Sindbis virus is normally a mosquito (Diptera), Lewis *et al.* [39] showed that a double subgenomic RNA vector expressing GFP is able to infect a butterfly (*Precis coenia*, Lepidoptera), a flour beetle (*Tribolium castaneum*, Coleoptera), the milkweed bug (*Oncopeltus fasciatus*, Hemiptera), and even a brine shrimp (*Artemia franciscana*), that is a crustacean. Infected cells in the Precis wing disk appeared normal and went on to elaborate their differentiation program normally. They observed no difference in the survival or morphology between the injected animals and mock-injected animals.

The same vector was used to express the Drosophila Ubx-1a protein in the butterfly and the beetle. Ubx is normally expressed in the hindwing but not the forewing of Drosophila and Precis. Ectopic expression of Ubx in the Drosophila forewing results in the homeotic transformation of cell fate to that of the hindwing. This was also true in Precis larvae or pupae injected with the Ubx-expressing vector. Thirty percent of the resulting adults had patches of scales in the forewing that were morphologically hindwing-like. No alteration of the hindwing was detected. The same vector was injected into Tribolium embryos, that were then allowed to complete embryogenesis. Almost one-quarter of the injected animals had morphological defects in the cephalic and thoracic cuticle. No defects were found in parasegments 6 and posterior, where Ubx is normally expressed. This phenotype resembles defects induced in Drosophila when Ubx is expressed in all tissues using a heat shock promoter.

These are truly landmark studies that point to the possibility of facile manipulation of gene expression in diverse arthropod systems, where the well-developed tools available in the Drosophila system are yet to be available.

7. Vaccines for infectious diseases and cancer

Historically, the most effective vaccines against viral diseases have been attenuated versions of the virus that are able to induce long-lasting immunity without causing serious illness. As illustrated with live poliovirus vaccines, however, the potential of reversion or recombination with related viruses leading to virulence can be a serious risk. This is of particular concern in efforts to develop vaccines against HIV or Ebola virus and has been one of the reasons that efforts to engineer other more benign viruses to express specific proteins derived from pathogenic viruses are attracting so much attention.

Three members of the alphavirus family, Sindbis virus, SFV and VEEV are being developed as vaccine vectors, mostly against other virus infections, but also in immunotherapeutic treatments against tumors. Two general types of vaccines are being tested: replicon particles and naked nucleic acids, both DNA and RNA.

Replicon particles are analogous to virus particles, proven as effective vaccines, and replicon genomes packaged into particles infect cells efficiently. The particles can be engineered to target dendritic cells (Section 7.1) and it may be possible to engineer the glycoproteins to target other specific cell types. The most obvious disadvantage of the replicons is that, while they are suicidal vectors that replicate in the infected cell and are incapable of spreading, recombination between the replicon and helper genomes could give rise to infectious virus. Fortunately, neither Sindbis virus nor the vaccine strain of VEEV are considered to be dangerous human pathogens. Furthermore, split helpers can be used, to greatly diminish the probability of co-packing or recombination (Section 3). The potential problem with infectious viruses does not exist with nucleic acid vaccines as they do not include sequences encoding the viral structural proteins.

Naked nucleic acids, DNA or RNA, are receiving considerable attention as vaccines. Both DNA- and RNA-based alphavirus vaccines appear to be effective in mice, but there is still very limited data in primates. Most of the tests of the alphavirus vectors have employed conventional DNA plasmids in which a highly active eukaryotic promoter is used for *in vivo* transcription of the replicon cDNA, which contains the gene coding for the immunogen. Alternately, the RNA replicon can be used directly. It is much more efficient than conventional mRNAs, as it is unlikely that the latter, when injected into animals, would be stable or would be translated to levels sufficient to induce a good immune response.

RNA has several potential advantages over DNA. First, RNA replicon expression in vivo tends to be transient, fading within 10 days to 2 weeks, thus mimicking traditional vaccines. DNA vaccines continue expression for months, but it is currently unclear if the long-term expression by DNA vaccines is advantageous, or if it might cause undesirable complications. Second, DNA itself has immunological effects, one of which is that it has adjuvant activity. The other is more sinister: DNA might activate autoreactive B cells leading to the production of rheumatoid factor [40]. Third, there is the possibility that DNA vaccines might integrate into the host cell genome leading to cell transformation. Fourth, expression of oncogenes and tumor antigens using DNA vaccines could have the potential of cell transformation; this would not be a risk with nonintegrating and transient RNA replicons [41,42]. The main disadvantage of RNA replicons is the fragility of RNA and the relative difficulty in producing it, technical issues that careful formulation and improved technology might overcome.

Replicons, as nucleic acid or as particles, are self-limiting and would not spread from cell to cell. Methods to enhance their vaccine effect have been reported. Sindbis replicon RNA and Sindbis replicon particles were used to express the E7 protein of human papilloma virus type 16 as a vaccine against tumors that express it [43]. The number of cells containing E7 was increased by fusing E7 with VP22, a herpes virus tegument protein that can be transported between cells, bringing with it sequences to which it had been fused. The ability of the Sindbis replicon to protect mice against tumor growth was significantly increased by the presence of VP22 [44,45]. Another method is fusing E7 to the transmembrane and cytoplasmic regions of the lysosome-associated membrane protein (LAMP-1) [46], that is normally targeted to the endosomal/lysosomal compartments, resulting in an increase in T cell mediated immune responses and antitumor activity.

7.1. Replicon particles targeting to dendritic cells

A critical step in the immune response is antigen uptake, processing and presentation to T cells. Dendritic cells (DCs) are specialized for these functions: immature DCs take up and process antigens and are converted to mature DCs able to interact with T cells. MacDonald and Johnston [47] showed that VEEV inoculated into mouse footpads is targeted to Langerhans cells, the dendritic cells resident in the skin. The cells subsequently migrate to the draining lymph node, and are the major infected cells there. This should enhance VEEV vaccine utility due to the more efficient antigen presentation in the lymph node, but the infected Langerhans cells did not show characteristics usually associated with maturation. How this might affect antigen presentation is unknown. Furthermore, targeting of VEEV to DCs in primates and humans remains to be established.

Most strains of Sindbis virus are avirulent in mice and infection does not usually lead to widespread dissemination beyond the site of inoculation. There had been little evidence of infection of DCs until Ryman *et al* infected mice that lacked the alpha/beta interferon receptor and in these mice the virus rapidly spread to the draining lymph nodes and replicated in cells of macrophage lineage [48]. These studies not only suggest that Sindbis virus may also target to DCs, but also that interferon plays an important role in suppressing replication and dissemination of the virus.

Specific amino acids in the alphavirus E2 glycoprotein can determine the ability of the virus or replicon particles to target to dendritic cells. The first example of this came from studies in mice with VEEV. A single mutation in the E2 protein of VEEV destroyed the ability of the virus to target to Langerhans cells and also led to decreased virulence [47]. To understand what determines the ability of Sindbis virus to replicate in dendritic cells Gardner *et al.* turned to human DCs. They passaged Sindbis virus on these cells and isolated a variant that showed an increased efficiency of infection [49]. They found that a change from glutamic acid to glycine at residue 160 of the E2 glycoprotein led to a 10-fold increase in the ability to infect human DCs. (The deletion of the same residue also enhances infection). Differences in the nonstructural proteins also affected the number of cells infected, but to a lesser extent. Infection resulted in dendritic cell activation and maturation. The differences in efficiency of infection observed in human DCs were not as significant in murine DCs and with both viruses, the skin-resident murine dendritic cells that are infected become activated, mature, migrate to the draining lymph node, and induce robust immune responses. In light of the observations by Ryman *et al.* [48,50] on the importance of alpha/beta interferon in the infection of mice by Sindbis virus, it would be interesting to analyze the viruses in animals that are unable to respond to these interferons.

7.2. Vaccination studies in mice and guinea pigs

7.2.1. Replicon particles
Mice serve on the front lines for testing of new vaccines and the alphavirus replicons appear to be effective immunogens for these animals. Extensive studies on the pathogenesis of alphaviruses in these animals serve as a valuable asset for understanding how innate immunity, and in particular the induction by alpha/beta interferon of antiviral pathways, affect the immune response [48,50]. Many of the initial results with replicons have been summarized in reviews [14] and we mention only a few of the more recent studies.

Norwalk-like viruses are positive-strand RNA viruses responsible for many outbreaks of acute nonbacterial gastroenteritis infection. When the capsid protein of the virus is expressed independently of the genome, it forms capsid-like structures that are strongly immunogenic and those obtained from baculovirus expression have been used as an oral vaccine in human trials [51]. VEEV replicon particles have been used to generate Norwalk capsid-like structures in BHK-21 cells. In mice, these particles induce high levels of serum IgG and intestinal IgA levels.

Alphaviruses are among the various vectors that are being designed as vaccines against HIV [14]. Mice were vaccinated with Sindbis virus replicon particles that express HIV-1 gag via several routes of inoculation. Protection against a vaginal challenge with vaccinia virus also expressing HIV-1 gag was observed if the mice were immunized via the vaginal or rectal route, but not when the replicon particles were given intranasally or intramuscularly [52]. Chimeras between the three alphavirus vectors are also being considered for immunization against HIV. They would allow the best features of each of the different alphavirus vectors to be incorporated into one vaccine. In preliminary studies with replicons expressing the HIV p55gag, Sindbis and VEEV replicon particles were compared with chimeric particles in which each replicon was packaged with the structural proteins from the other virus. VEEV replicons packaged with the structural proteins from Sindbis virus were equivalent to pure VEEV particles in their ability to induce $CD8^+$-T cells in BALB/C mice (J. Polo, personal communication). It will be important to learn how these compare in primates.

VEEV replicon particles have been used in immunization studies against Marburg, Ebola and Lassa fever in guinea pig models [53,54]. Vaccinated animals were protected against disease and protection appeared to be due to cell-mediated immunity.

7.2.2. Nucleic acids

Both Sindbis and SFV replicon cDNA plasmids have been used as vaccines in mice. Studies with Sindbis replicon plasmids expressing herpes simplex virus–1 gB (and those with SFV directed against influenza virus HA and nucleoprotein) showed that the replicon cDNA was a much better immunogen than a conventional plasmid in which the gene was expressed directly from a CMV promoter (reviewed in [14]). One explanation for the replicon superiority is that they, as infectious agents, also induce innate immunity, increasing the level of cytokines such as interferon which stimulate the immune response. These results are very promising and further experiments, particularly in nonhuman primates, are essential to determine the true utility of immunizing with replicon nucleic acids.

7.3. Vaccination studies in primates

There are still relatively few published studies in which alphavirus replicons have been used to vaccinate primates. VEEV itself has been used as a vaccine to protect humans from infection by the virus, so it seems that alphaviruses may prove to be effective in humans. In one trial with VEEV replicon particles, cynomolgus macaques were inoculated three times, 28 days apart, with VEEV particles that expressed the Marburg virus GP. The animals were completely protected from viremia and death when subsequently challenged with a lethal dose of Marburg (Musoke) virus [53]. Ebola virus is a filovirus closely related to Marburg virus. In this case, however, the VEEV vector expressing the Ebola virus GP or NP or both did not protect macaques against Ebola (Zaire) virus challenge [55]. The reason for this dramatic difference in outcomes is as yet unclear.

Simian immunodeficiency virus (SIV) infection of rhesus macaques serves as a model for HIV infection of humans. Macaques were vaccinated with a "cocktail" of VEEV replicon particles expressing either the matrix and capsid proteins of SIV or the SIV gp160 [56]. The vaccinated animals had both neutralizing antibodies and a cellular immune response. Upon challenge with SIV, viral loads in the vaccinated animals were several orders of magnitude lower than controls and were below the limit of detection in two of four animals by 6–8 weeks postinfection. The four vaccinated animals remained healthy. In contrast, two of four controls had severe symptoms associated with SIV infection. These data are very promising, but it should be noted that the vaccination protocol consisted of two subcutaneous, two intravenous injections and then two more subcutaneous doses. Further studies will be required to determine if this type of vaccination protocol is required as it will be difficult to implement such a regime for humans even in developed countries.

7.4. Immunotherapy and targeted treatment against tumors

Many of the anti-tumor studies to date have been performed with nucleic acids; both RNA and DNA. Leitner et al. [57] compared replicon plasmids with conventional plasmids in the ability to induce both humoral and cellular immunity against beta-galactosidase which was used as a model 'tumor antigen' in mice. Their results were similar to those found using the replicons as vaccines against infectious agents: replicons were effective at concentrations 100-fold lower than those required to induce a response with conventional plasmids. The level of expression of the antigen from replicons and from conventional plasmids was not correlated with immunogenicity, but expression was measured in BHK-21 cells, not in mice. The authors suggest, however, that several other factors must contribute to the enhanced immunogenicity. Intermediates in replicon RNA synthesis (dsRNA molecules) may serve as inducers of interferon and other cytokines that stimulate the immune system. In addition, replicon-containing cells undergo apoptotic death and these cells would be recognized by phagocytic cells for presentation of immunogenic peptides to the cellular immune system.

Both Sindbis and VEEV replicon particles are being tested as therapeutic anti-tumor agents. The use of Sindbis replicons as vaccines against tumors expressing a protein of human papilloma virus was mentioned at the beginning of Section 7 in describing modifications that enhanced spread of the immunogen. VEEV particles have also been shown to be effective in inducing immunity against such tumors [43]. One of the attractions of particles is based on the premise that they can be specifically targeted to tumor cells [12]. Targeting of Sindbis replicons expressing thymidine kinase (TK) from herpes virus type 1 was demonstrated in vitro [58]. TK causes cells to be sensitive to nucleosides analogs such as ganciclovir and the targeted cells did show a small, but significant increase in sensitivity to ganciclovir. Targeting in vivo will be much more complicated, but holds great promise for therapy.

8. Perspectives

Alphavirus replicons have become an important tool for basic research, for protein production and, potentially, for vaccines against infectious agents and tumors. There are also some rather more novel ways that Sindbis replicons are proving to be of value. They have been used to detect and identify other human viruses (reviewed in [1]). Other examples include the selection of specific protease cleavage sites in vivo using random libraries of sequences inserted into Sindbis replicons [59] and the genera-tion and selection of functional variants of murine leukemia virus receptors [60].

More recently a high-throughput cloning system that uses packageable defective helpers for screening libraries was described [61] and appears to be quite promising for identifying new proteins responsible for a variety of cell functions. These are just a few illustrations and we look forward to future innovations with alphavirus replicons.

Abbreviations

CMV	Cytomegalovirus
DC	Dendritic cell
DEN	Dengue virus
dsRNA	Double-stranded RNA
EGFP	Enhanced green fluorescent protein
GFP	Green fluorescent protein
HIV	Human immunodeficiency virus
IRES	Internal ribosome entry site
LAC	LaCrosse virus
scAb	single chain antibody
SFV	Semliki Forest virus
SIV	Simian immunodeficiency virus
TK	thymidine kinase
VEEV	Venezuelan equine encephalitis virus

References

1. Schlesinger, S. (2001) Expert Opin. Biol. Ther. 1, 177–191.
2. Strauss, E.G. and Strauss, J.H. (1994) Microbiol. Rev. 58, 491–562.
3. Wielgosz, M.M., Raju, R., and Huang, H.V. (2001) J. Virol. 75, 3509–3519.
4. Frolova, E., Frolov, I., and Schlesinger, S. (1997) J. Virol. 71, 248–258.
5. Frolov, I. and Schlesinger, S. (1996) J. Virol. 70, 1182–1190.
6. Xiong, C., Levis, R., Shen, P., Schlesinger, S., Rice, C.M., and Huang, H.V. (1989) Science 243, 1188–1191.
7. Boorsma, M., Nieba, L., Koller, D., Bachmann, M.F., Bailey, J.E., and Renner, W.A. (2000) Nat. Biotechnol. 18, 429–432.
8. Lustig, S., Jackson, A.C., Hahn, C.S., Griffin, D.E., Strauss, E.G., and Strauss, J.H. (1988) J. Virol. 62, 2329–2336.
9. Corsini, J., Traul, D.L., Wilcox, C.L., Gaines, P., and Carlson, J.O. (1996) Biotechniques 21, 492–497.
10. Olson, K.E., Myles, K.M., Seabaugh, R.C., Higgs, S., Carlson, J.O., and Beaty, B.J. (2000) Insect Mol. Biol. 9, 57–65.
11. London, S.D., Schmaljohn, A.L., Dalrymple, J.M., and Rice, C.M. (1992) Proc. Natl. Acad. Sci. USA 89, 207–211.
12. Ohno, K., Sawai, K., Iijima, Y., Levin, B., and Meruelo, D. (1997) Nat. Biotechnol. 15, 763–767.
13. Morizono, K., Bristol, G., Xie, Y.M., Kung, S.K., and Chen, I.S. (2001) J. Virol. 75, 8016–8020.
14. Schlesinger, S. and Dubensky, T.W. (1999) Curr. Opin. Biotechnol. 10, 434–439.
15. Dubuisson, J. and Rice, C.M. (1993) J. Virol. 67, 3363–3374.

204

16. Polo, J.M., Belli, B.A., Driver, D.A., Frolov, I., Sherrill, S., Hariharan, M.J., Townsend, K., Perri, S., Mento, S.J., Jolly, D.J., Chang, S.M., Schlesinger, S., Dubensky, T.W., Jr. (1999) Proc. Natl. Acad. Sci. USA 96, 4598–4603.

17. Bredenbeek, P.J., Frolov, I., Rice, C.M., and Schlesinger, S. (1993) J. Virol. 67, 6439–6446.

18. Sauerwald, T.M., Betenbaugh, M.J., and Oyler, G.A. (2002) Biotechnol. Bioeng. 77, 704–716.

19. Nargi-Aizenman, J.L. and Griffin, D.E. (2001) J. Virol. 75, 7114–7121.

20. Moriishi, K., Koura, M., and Matsuura, Y. (2002) Virology 292, 258–271.

21. Dryga, S.A., Dryga, O.A., and Schlesinger, S. (1997) Virology 228, 74–83.

22. Agapov, E.V., Frolov, I., Lindenbach, B.D., Pragai, B.M., Schlesinger, S., and Rice, C.M. (1998) Proc. Natl. Acad. Sci. USA 95, 12989–12994.

23. Perri, S., Driver, D.A., Gardner, J.P., Sherrill, S., Belli, B.A., Dubensky, T.W., Jr., and Polo, J.M. (2000) J. Virol. 74, 9802–9807.

24. Frolov, I., Agapov, E., Hoffman, T.A., Jr., Pragai, B.M., Lippa, M., Schlesinger, S., and Rice, C.M. (1999) J. Virol. 73, 3854–3865.

25. Ehrengruber, M.U., Lundstrom, K., Schweitzer, C., Heuss, C., Schlesinger, S., and Gahwiler, B.H. (1999) Proc. Natl. Acad. Sci. USA. 96, 7041–7046.

26. Maletic-Savatic, M., Malinow, R., and Svoboda, K. (1999) Science 283, 1923–1927.

27. Osten, P., Khatri, L., Perez, J.L., Kohr, G., Giese, G., Daly, C., Schulz, T.W., Wensky, A., Lee, L.M., Ziff, E.B. (2000) Neuron 27, 313–325.

28. Ehrengruber, M.U., Hennou, S., Bueler, H., Naim, H.Y., Deglon, N., and Lundstrom, K. (2001) Mol. Cell Neurosci. 17, 855–871.

29. Ehrengruber, M.U. (2002) Molecular Neurobiology 26, in press.

30. Perez, J.L., Khatri, L., Chang, C., Srivastava, S., Osten, P., and Ziff, E.B. (2001) J. Neurosci. 21, 5417–5428.

31. Hayashi, Y., Shi, S.H., Esteban, J.A., Piccini, A., Poncer, J.C., and Malinow, R. (2000) Science 287, 2262–2267.

32. Blair, C.D., Adelman, Z.N., and Olson, K.E. (2000) Clin. Microbiol. Rev. 13, 651–661.

33. Higgs, S., Olson, K.E., Klimowski, L., Powers, A.M., Carlson, J.O., Possee, R.D., and Beaty, B.J. (1995) Insect Mol. Biol. 4, 97–103.

34. de Lara Capurro, M., Coleman, J., Beerntsen, B.T., Myles, K.M., Olson, K.E., Rocha, E., Krettli, A.U., and James, A.A. (2000) Am. J. Trop. Med. Hyg. 62, 427–433.

35. Olson, K.E., Higgs, S., Gaines, P.J., Powers, A.M., Davis, B.S., Kamrud, K.I., Carlson, J.O., Blair, C.D., and Beaty, B.J. (1996) Science 272, 884–886.

36. Higgs, S., Rayner, J.O., Olson, K.E., Davis, B.S., Beaty, B.J., and Blair, C.D. (1998) Am. J. Trop. Med. Hyg. 58, 663–670.

37. Adelman, Z.N., Blair, C.D., Carlson, J.O., Beaty, B.J., and Olson, K.E. (2001) Insect Mol. Biol. 10, 265–273.

38. Shiao, S.H., Higgs, S., Adelman, Z., Christensen, B.M., Liu, S.H., and Chen, C.C. (2001) Insect Mol. Biol. 10, 315–321.

39. Lewis, D.L., DeCamillis, M.A., Brunetti, C.R., Halder, G., Kassner, V.A., Selegue, J.E., Higgs, S., and Carroll, S.B. (1999) Curr. Biol. 9, 1279–1287.

40. Leadbetter, E.A., Rifkin, I.R., Hohlbaum, A.M., Beaudette, B.C., Shlomchik, M.J., and Marshak-Rothstein, A. (2002) Nature 416, 603–607.

41. Johanning, F.W., Conry, R.M., LoBuglio, A.F., Wright, M., Sumerel, L.A., Pike, M.J., and Curiel, D.T. (1995) Nucleic Acids Res. 23, 1495–1501.

42. Strong, T.V., Hampton, T.A., Louro, I., Bilbao, G., Conry, R.M., and Curiel, D.T. (1997) Gene Ther. 4, 624–627.

43. Velders, M.P., McElhiney, S., Cassetti, M.C., Eiben, G.L., Higgins, T., Kovacs, G.R., Elmishad, A.G., Kast, W.M., and Smith, L.R. (2001) Cancer Res. 61, 7861–7867.

44. Cheng, W.F., Hung, C.H., Chai, C.Y., Hsu, K.F., He, L., Ling, M., and Wu, T.C. (2001) J. Virol. 75, 2368–2376.

45. Cheng, W.F., Hung, C.F., Hsu, K.F., Chai, C.Y., He, L., Polo, J.M., Slater, L.A., Ling, M., and Wu, T.C. (2002) Hum. Gene. Ther. 13, 553–568.

46. Cheng, W.F., Hung, C.F., Hsu, K.F., Chai, C.Y., He, L., Ling, M., Slater, L.A., Roden, R.B., and Wu, T.C. (2001) Hum. Gene. Ther. 12, 235–252.
47. MacDonald, G.H. and Johnston, R.E. (2000) J. Virol. 74, 914–922.
48. Ryman, K.D., Klimstra, W.B., Nguyen, K.B., Biron, C.A., and Johnston, R.E. (2000) J. Virol. 74, 3366–3378.
49. Gardner, J.P., Frolov, I., Perri, S., Ji, Y., MacKichan, M.L., zur Megede, J., Chen, M., Belli, B.A., Driver, D.A., Sherrill, S., Greer, C.E., Otten, G.R., Barnett, S.W., Liu, M.A., Dubensky, T.W., Polo, J.M. (2000) J. Virol. 74, 11849–11857.
50. Ryman, K.D., White, L.J., Johnston, R.E., and Klimstra, W.B. (2002) Viral. Immunol. 15, 53–76.
51. Harrington, P.R., Yount, B., Johnston, R.E., Davis, N., Moe, C., and Baric, R.S. (2002) J. Virol. 76, 730–742.
52. Vajdy, M., Gardner, J., Neidleman, J., Cuadra, L., Greer, C., Perri, S., O'Hagan, D., and Polo, J.M. (2001) J. Infect. Dis. 184, 1613–1616.
53. Hevey, M., Negley, D., Pushko, P., Smith, J., and Schmaljohn, A. (1998) Virology 251, 28–37.
54. Pushko, P., Geisbert, J., Parker, M., Jahrling, P., and Smith, J. (2001) J. Virol. 75, 11677–11685.
55. Geisbert, T.W., Pushko, P., Anderson, K., Smith, J., Davis, K.J., and Jahrling, P.B. (2002) Emerg. Infect. Dis. 8, 503–507.
56. Davis, N.L., Caley, I.J., Brown, K.W., Betts, M.R., Irlbeck, D.M., McGrath, K.M., Connell, M.J., Montefiori, D.C., Frelinger, J.A., Swanstrom, R., Johnson, P.R., Johnston, R.E. (2000) J. Virol. 74, 371–378.
57. Leitner, W.W., Ying, H., Driver, D.A., Dubensky, T.W., and Restifo, N.P. (2000) Cancer Res. 60, 51–55.
58. Iijima, Y., Ohno, K., Ikeda, H., Sawai, K., Levin, B., and Meruelo, D. (1999) Int. J. Cancer 80, 110–118.
59. Pacini, L., Vitelli, A., Filocamo, G., Bartholomew, L., Brunetti, M., Tramontano, A., Steinkuhler, C., and Migliaccio, G. (2000) J. Virol. 74, 10563–10570.
60. Kazachkov, Y., Long, D., Wang, C., and Silver, J. (2000) Virology 267, 124–132.
61. Koller, D., Ruedl, C., Loetscher, M., Vlach, J., Oehen, S., Oertle, K., Schirinzi, M., Deneuve, E., Moser, R., Kopf, M., Bailey, J.E., Renner, W., Bachmann, M.F. (2001) Nat. Biotechnol. 19, 851–855.

S.C. Makrides (Ed.) *Gene Transfer and Expression in Mammalian Cells*

Virus-based vectors for gene expression in mammalian cells: Semliki Forest virus

Kenneth Lundstrom

Regulon Inc., Biopole Epalinges, Les Croisettes, CH-1066 Epalinges, Lausanne, Switzerland;
Tel.: +41 (76) 418-2167. E-mail: superscientist@hotmail.com

1. Introduction

Among alphaviruses the most popular members developed into expression vectors are Semliki Forest virus (SFV) [1], Sindbis virus (SIN) [2] and Venezuelan equine encephalitis virus (VEE) [3]. The most commonly used SFV expression system is based on two plasmid vectors: the expression vector contains the SFV non-structural and the foreign genes and the helper vector harbors the SFV structural genes (Fig. 1). *In vitro* transcribed RNA molecules from both vectors are co-transfected into BHK-21 cells, where immediate synthesis of non-structural SFV proteins occurs. Expression of nsP1-4 proteins results in formation of the replicase complex, responsible for high-level RNA replication in host cells [4]. Because the generated RNA is of positive polarity, the amplified RNA molecules can be directly translated in the cytoplasm leading to production of large quantities of recombinant proteins. Furthermore, only the recombinant RNA contains a packaging signal, which means that specifically this RNA and not the helper RNA will be packaged into recombinant SFV particles. This feature renders the generated virions

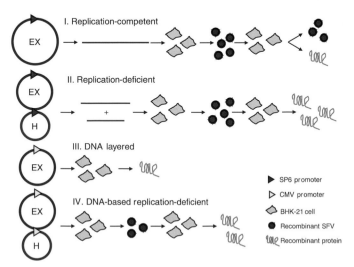

Fig. 1. Schematic presentation of SFV production. (I) Replication-competent vector; (II) Replication-deficient vector system; (III) DNA-layered vector; (IV) DNA-based replication-deficient vector system. EX, SFV expression vector; H, SFV helper vector.

replication-deficient, because they lack all structural genes. The inability to generate virus progeny makes the use of these vectors safe [5]. Alternatively, replication-competent vectors can also be used. In that case, in addition to the full-length genome, a second subgenomic SFV 26S promoter and the foreign gene of interest are introduced. The advantage of this approach is naturally the production of virus progeny capable of infecting neighboring cells and in this way to enhance the spread of infection. The drawback is the increased safety risks associated with the use of such vectors in cell culture and *in vivo*. The third version of SFV vectors is the so called DNA layered vector, where a CMV promoter or another eukaryotic RNA polymerase II type promoter is utilized to drive the transcription of self-amplifying SFV replicon vectors [6]. The advantage of the DNA-based SFV vector system is the high safety standard with no theoretical possibility to generate infectious particles. However, the possibility to infect a broad range of host cells is lost. A fourth type of SFV expression system has been developed where both in the expression vector and the helper vector the SP6 RNA polymerase promoter was replaced by an RNA polymerase II-dependent promoter [7]. Co-transfection of SFV DNA expression and helper vectors generated titers of 10^6 particles/ml, which is approximately 100–1000 lower than using RNA-based SFV vectors.

One of the advantages of using SFV particles is their broad host range [1,8]. A large number of mammalian cell lines are susceptible to high transduction rates of SFV. This makes studies on gene expression feasible in various host cell backgrounds. Introduction of the positive strand RNA into host cells generates more than 200,000 copies of RNA through the efficient replication mechanism and the strong SFV 26S subgenomic promoter is responsible for extreme expression levels of recombinant proteins. Shortly after SFV infection the host cell protein synthesis is dramatically inhibited [4], which means that the recombinant protein is the major protein produced in host cells. In addition, SFV infection causes cell cytotoxicity and induces apoptosis. In general, if the goal is to achieve high overexpression the cytotoxicity and decrease in host cell protein synthesis is not a problem. However, gene expression studies that require an extended cell signal transduction pathway capacity or a longer host cell survival cause serious limitations to the use of SFV vectors. Novel less cytotoxic SFV vectors have therefore been developed [9,10]. Attractive features of SFV vectors are the capability for rapid large-scale recombinant protein production due to the established large-scale technology [11] and the use of SFV vectors for *in vivo* applications without any requirement of virus stock purification or concentration [12].

This review covers primarily the use of replication-deficient vectors for gene expression both *in vitro* and *in vivo*. Expression studies on topologically different recombinant proteins, the use of various mammalian cell lines and the scale-up for large-scale protein production are described. Furthermore, descriptions of applying SFV vectors for primary cells in cultures, for injection of organotypic hippocampal slices and *in vivo* expression are included. Recent development in using SFV vectors for gene therapy is also presented. SFV vectors used for retrovirus particle production and applications of both SFV particles as well as nucleic acid vectors (naked RNA and layered DNA vectors) are summarized. Finally, the engineering of novel less-cytotoxic, temperature-sensitive and down-regulated SFV vectors as well as potential novel applications are discussed.

2. Expression of topologically different proteins

2.1. Intracellular proteins

Topologically different recombinant proteins have successfully been expressed from SFV vectors (Table 1). When the SFV expression system was first established the expression pattern was monitored by the cytoplasmic bacterial reporter gene LacZ to visualize β-galactosidase by X-gal staining and to quantify the expression levels by an enzymatic assay [1]. Metabolic labeling of SFV-LacZ infected BHK-21 cells with [^{35}S] methionine demonstrated high levels of e.g., bacterial β-galactosidase, firefly

Table 1

Examples of recombinant proteins expressed from SFV vectors

Recombinant protein	Host cell	Comments	Reference
Intracellular proteins			
β-Galactosidase	BHK, CHO, HEK293, neurons	Reporter	[1,12]
Dehydrofolate reductase	BHK, CHO	Enzyme	[1]
Green fluorescence protein	BHK, CHO, neurons, etc.	Reporter	[10]
NEDD-2	BHK and neurons	Cysteine protease	[86]
Thyroid peroxidase	CHO	Hemoprotein	[87]
UDP-gluconoryltransferase	BHK, V79	Enzyme	[89]
Aspartylglucosaminidase	Neurons	Enzyme	[13]
Membrane proteins			
Transferrin receptor	BHK	Single transmembrane	[1]
Serotonin 5-HT$_3$ receptor	BHK, CHO, neurons	Ligand-gated ion channel	[8,24]
Purinoreceptor P2X	BHK, CHO	Ligand-gated ion channel	[25]
Neurokinin receptors 1–3	BHK, CHO, HOS	GPCR	[15]
Metabotropic mGluR1, 2, 3, 4, 8	CHO, neurons	GPCR	[20,90,91]
Olfactory receptors	BHK, HEK293, OLF442	GPCR	[22,23]
Dopamine transporter	BHK	Transporter	[29]
Cyclo-oxygenase 1, 2	BHK, CHO	Membrane-associated	[11]
Catechol-o-methyltransferase	BHK, neurons	Membrane-bound	[92]
GABA$_A$ receptor	BHK, neurons	Heteromeric channel	[27]
Potassium channel Kv	CHO	Octomer	[28]
Yeast syntaxin Sso2p	BHK	Membrane insertion	[93]
Secreted proteins			
Secreted alkaline phosphatase	BHK, CHO, HEK293	Reporter	[*]
Interleukin-4	BHK	Cytokine	[*]
Interleukin-12	BHK, tumor cells	Heteromeric p40/p35	[31]
Plasminogen activator inhibitor-1	MEG-01	Inhibitor activity	[14]
Plasminogen activator inhibitor-2	BHK, CHO	Artificial signal sequence	[32]

[*], Lundstrom, unpublished results.

210

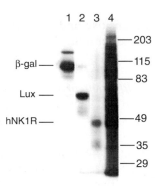

Fig. 2. Metabolic labeling of BHK-21 cells infected with recombinant SFV. Adherent BHK-21 cells were infected with various recombinant SFV particles at an MOI of 10. Cells were metabolically labeled with [^{35}S] methionine followed by SDS-PAGE and autoradiography. Lane 1, SFV-LacZ; lane 2, SFV-Lux (luciferase); lane 3, SFV-hNKR1; lane 4, uninfected BHK-21 cells.

luciferase and human neurokinin-1 receptor (hNK1R) suggesting that the recombinant protein was the major product synthesized in the cells (Fig. 2). Simultaneously, the cellular host cell protein synthesis was severely decreased, which indicated that the efficient replication of SFV RNA almost completely took over the host cell protein synthesis machinery. Moreover, X-gal staining of various host cell lines revealed that efficient infection rates ($>95\%$) were achieved. Enzymatic assays showed extreme β-galactosidase levels in the range of 10^9 molecules per cell and representing up to 25% of total protein yields [1].

Although several cytoplasmic proteins of therapeutic value have been expressed from SFV vectors, two examples are given below. Aspartylglucosaminidase (AGA) was expressed from SFV vectors to study the processing, intracellular localization and endocytosis of this protein, of which mutations result in a neurodegenerative lysosomal storage disease [13]. Expression studies in telencaphilic aspartylglucosaminuria (AGU) mouse neurons in vitro indicated that the processing steps were similar to those in peripheral cells, but both accumulation of inactive precursor and delayed enzyme processing were observed. Additionally, the cell-associated form of plasminogen-activator inhibitor 1 (PAI-1), a key physiological regulator of fibrinolysis, has been expressed from SFV vectors in the megakaryocytic cell line MEG-01 [14]. The SFV-mediated expression enhanced PAI-1 storage by 6–9-fold. Several nuclear proteins including transcription factors and nuclear receptors have been expressed successfully from SFV vectors (Lundstrom, unpublished results).

2.2. Membrane proteins and receptors

Probably the most applied topological groups of proteins to be expressed from SFV vectors are membrane-associated or transmembrane proteins. As membrane proteins have generally been considered to be difficult to express from any vectors, the suitability of SFV vectors has come to an important test. Particularly,

G protein-coupled receptors (GPCRs) have been successfully expressed from SFV vectors [15]. Today more than 50 different GPCRs have been successfully expressed from SFV vectors. Expression levels have been monitored by metabolic labeling experiments (Fig. 2) and binding assays on intact cells as well as on isolated membranes using radioligands [15,16]. In many cases, extremely high B_{max} values, in the range of 50–200 pmol receptor per mg protein were obtained [17]. Binding experiments on whole cells suggested that the receptor density was 3–10 million receptors per cell. The broad host range of SFV vectors allowed easy and fast comparison of expression levels in various cell lines by infection of several host cell lines in parallel with the same SFV stock containing the GPCR of interest. For instance, it was demonstrated that the various subtypes of opioid receptors showed significantly different B_{max} values depending on the host cell line [18]. Overexpression of various GPCRs affected host cells differently. For example, the hNK1R expression seemed to be highly toxic to all host cell lines tested. In contrast, other receptors like the hamster α1B adrenergic receptor and the human histamine H2 receptor, showed binding activity for much longer time periods and the SFV infected cells did not die as rapidly. The functionality of GPCRs can be determined by their coupling to G proteins in several ways. SFV-infected cells have been subjected to measurements of intracellular Ca^{2+}-release [15], inositol phosphate accumulation [19], cAMP stimulation [18] and GTPγS binding [20] as means of functional coupling. Furthermore, the function of metabotropic glutamate receptors was demonstrated by recording of voltage-gated calcium channel inhibition after agonist application [21]. The drawbacks of using SFV for functional studies have been the inhibition of host cell protein synthesis and the extremely high-level receptor expression, which in combination leads to an unfavorable ratio of recombinant receptors and endogenously expressed G proteins. To compensate for the lack of G proteins cells were co-infected with SFV vectors expressing a GPCR and G protein subunits, respectively, which resulted in enhanced functional response as determined by inositol phosphate accumulation [19].

Olfactory receptors, a class of GPCRs, have also been expressed from SFV vectors. Metabolic labeling experiments indicated that high expression levels were obtained in BHK-21 cells and in an olfactory epithelial cell line, OLF442 [22]. Recently, the odorant human OR 17-40 receptor and the rat I7 receptor were expressed as fusion proteins with the N-terminal membrane import sequence of the guinea pig serotonin receptor, and functional response was measured by Ca^{2+}-release in a fura-2 assay after stimulation with the specific agonists helional and octanol [23].

Several ligand-gated ion channels have also been expressed from SFV vectors. The mouse serotonin 5-HT$_3$ receptor showed high binding activity in several mammalian host cell lines and patch-clamp recordings on whole CHO cells generated high amplitude electrophysiological responses [24]. Likewise, several purinoreceptor P2X subtypes showed B_{max} values over 100 pmol/mg protein, electrophysiological responses and functional activity measured by cellular Ca^{2+}-uptake after ATP stimulation [25,26]. The assembly of heteropentameric GABA$_A$ channels was studied by co-expression of α1 and β2 subunits from SFV vectors [27]. Interestingly, homomeric subunits were retained in the endoplasmic reticulum, whereas heteromeric expression resulted in receptor transport to the plasma membrane in both BHK-21 cells and cultured superior cervical ganglia neurons. Furthermore, assembly of voltage-sensitive potassium channel subunits Kv1.1 and 1.2 α and β 1.1 or 1.2 subunits was demonstrated in CHO-K1 cells co-infected with SFV vectors [28]. When membranes from cells co-infected with His-tagged Kv α 1.2 and β 2.1 subunits were solubilized, a 405 kD complex was purified, which suggested that an octomer had been reconstructed.

In addition to GPCRs and ligand-gated ion channels, at least one report indicates that the human dopamine transporter was expressed from SFV vectors in BHK-21 cells [29]. In this case *in vitro* transcribed RNA was electroporated into BHK cells and the expression monitored by [^{3}H] uptake. The molecular mass of the recombinant transporter was 56 kD, which was reduced to 50 kD after tunicamycin treatment, indicating a proper glycosylation pattern. Moreover, the glycosylated transporter was localized to the plasma membrane, whereas the unglycosylated form was retained in the cytoplasm.

2.3. Secreted proteins

Due to the high efficiency of expression of membrane proteins from SFV vectors it was expected that also secreted proteins could be produced in large quantities. Early studies indicated that high expression levels and good secretion efficiency was obtained for chick lysozyme [1]. Expression of secreted alkaline phosphatase (SAP) was achieved in several mammalian host cells [30]. Furthermore, interleukin-4 (IL-4) showed high activity in the supernatant of SFV-infected BHK cells (Blasey and Lundstrom, unpublished results). Typically, the maximum expression was reached much later (at 60 h post-infection) compared to cytoplasmic or membrane proteins (16–24 h post-infection).

The two subunits p40 and p35 of interleukin-12 (IL-12), known to induce tumor regression in animal models, were expressed from the same SFV vector under the control of two SFV 26S subgenomic promoters [31]. This resulted in secretion of high levels of biologically active heterodimeric IL-12. In addition to the cell-associated form of PAI-1, the secreted PAI-1 was enhanced fivefold in MEG-01 cells after SFV infections [14]. To optimize the secretion of PAI-2 an artificial signal sequence was engineered in the coding region, which resulted in more than 90% secretion of PAI-2 [32]. N-Terminal sequencing confirmed that the artificial signal sequence was correctly cleaved off.

3. Host cell range

3.1. Mammalian host cell lines

SFV vectors have been shown to possess an impressively broad host range, which allows for expression studies in a wide variety of mammalian cell lines (Table 2).

Table 2
Cell lines and primary cells in culture susceptible to SFV transduction

Cell type	Transduction rate	Expression capacity	Reference
Mammalian cell lines			
BHK-21, baby hamster kidney	>95%	Very high	[1]
CEF, chicken embryo fibroblast	90%	Very high	[1]
CHO-K1, Chinese hamster ovary	>95%	Very high	[1,8]
C6, rat glioma	>95%	Very high	[18]
COS-7, Green monkey kidney	>95%	Very high	[19]
DU-145, human prostate tumor	>70%	Very high	[47]
HeLa, human epithelial carcinoma	85%	Very high	[1]
HEK293, human embryonic kidney	>95%	Very high	[1]
HOS, human osteogenic sarcoma	>95%	Very high	[1]
Hybridoma 179	30%	Moderate	[30]
MDCK, canine kidney	85%	High	[42]
MME, mouse mammary epithelium	85%	High	[42]
MOLT-4, lymphoblastic leukemia	20%	High	[88]
NIE 115, mouse neuroblastoma	35%	Moderate	[30]
NIH-3T3, mouse embryo	25%	Low	[30]
OLF, olfactory epithelium	90%	Very high	[22]
Raji, human Burkit lymphoma	50%	Moderate	[30]
RPMI 8226, human myeloma	20%	Moderate	[30]
TK6, human lymphoblast	90%	Moderate	[30]
V79, hamster lung fibroblast	80%	High	[89]
Primary cells			
Cortical neurons	70–90%	Very high	[*]
Hippocampal neurons	75–95%	Very high	[42]
Endothelial cells (bovine)	2.5 (60%*)	High	[34]
Fibroblasts	75%	High	[41]
Myotubes	70%	Very high	[*]
Prostate epithelial duct	60%	Very high	[47]
Superior cervical ganglion	70%	High	[92]
Non-mammalian cells			
CHSE-214 (fish, salmon)	70%	High	[36]
C6/36, A. albopictus (insect)	>95%	High	[37]
MOS20, A. aegypti (insect)	>95%	High	[37]
Grasshopper embryos	High	High	[*]
SL3, D. melanogaster	50%	Moderate	[*]
Xenopus oocytes	>50%	High	[*]

*, after PEG treatment; [*], Lundstrom, unpublished results.

Many of the commonly used cell lines, like BHK-21, CHO-K1, COS-7 and HEK293 are showing high transduction rates for SFV vectors (Fig. 3). In general, using multiplicity of infection (MOI) of 5 is enough to achieve expression in almost 100% of the cells. In addition, several less common host cells have shown high susceptibility to SFV [30]. For instance, in adrenal chromaffin cells 40% infection rates were obtained [33]. The membrane capacitance was measured using whole-cell patch-clamp technology, which demonstrated comparable intracellular concentrations, leak currents and cell sizes for SFV-infected cells expressing GFP and control cells.

A   B

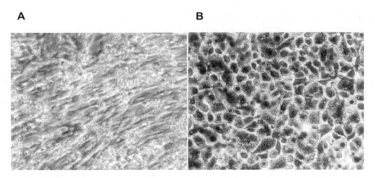

Fig. 3. SFV-LacZ transduction of BHK-21 cells. Adherent BHK-21 cells were infected with SFV-LacZ virus at an MOI of 10 and stained with X-gal at 24 h post-infection: (A) non-infected BHK-21 cells; (B) SFV-LacZ infected BHK-21 cells.

Interestingly, expression levels did not always correlate with the transduction rates. Although human lymphoblast TK-6 cells infected with SFV-LacZ virus showed β-galactosidase expression in almost every cell, the enzymatic activity was only one-tenth of levels obtained in BHK-21 cells. In other cases, cell lines that are less efficiently infected demonstrated high recombinant protein expression in individual cells. The importance of the correct host cell background was demonstrated by studies on localization of HA- and myc-tagged olfactory receptors, where transport to the plasma membrane was only detected in OLF442 cells, but not in BHK-21 cells [22].

Although a wide range of mammalian cell lines is susceptible to SFV infections, there are several examples of cells that tend to be resistant to SFV infection. For instance, endothelial cells exhibited as low as 2.6% infection rates [34]. However, the transduction frequency can be substantially improved by treatment of target cells with polyethylene glycol (PEG). Thus, macrophages and endothelial cells, which ordinarily are almost non-permissive to SFV, showed enhanced infection rates from 5 to 50% and 3 to 65%, respectively, upon PEG treatment [35].

3.2. Non-mammalian cells

SFV vectors are capable of efficient expression of reporter genes, LacZ and firefly luciferase, in fish cell cultures [36]. The expression is cell-type- and temperature-dependent. Maximal expression was obtained in two salmonid-derived cell lines CHSE-214 and F95/9 at 25 and 20 °C, respectively. At temperatures of 15 °C or below, the level of reporter gene expression was reduced 1000-fold, which suggested that the SFV RNA replication was non-functional at low temperatures.

Insect cell lines were also efficiently infected by SFV vectors [37]. SFV-mediated GFP expression indicated that close to 100% infectivity was achieved in *Aedes albopictus* cell lines C6/36 and Aa23T as well as in *Aedes aegypti* cell line MOS20. In another approach, the T7 RNA polymerase sequence was introduced into the SFV vector, which generated high levels of reporter gene expression in mosquito cells [38].

4. Scale-up of protein production

4.1. Drug screening

The ability of SFV vectors to generate robust expression of recombinant proteins in various host cell lines demanded the establishment of SFV-based large-scale expression technology. The critical question was whether SFV infection would be as efficient in suspension cell cultures as it was for adherent cells. Studies on bacterial β-galactosidase revealed that the infection rate of suspension cultures, tested in spinner flasks, was comparable to that of adherent cells. Furthermore, expression of the human cyclo-oxygenase-2 (COX-2) in spinner cultures demonstrated yields in the same range as those obtained for monolayer cultures. Production of COX-2 in five 1-liter spinner cultures resulted in high enzymatic activity and was sufficient to enable compound screening with the material [11]. In contrast, transient transfection of COS-7 cells required production of COX-2 in 20 T150 flasks each week for 6 months to achieve the same amount of material received from 5 L of CHO cells infected with SFV.

The scale-up procedure has facilitated drug-screening procedures significantly. Because GPCRs are considered the most promising drug targets today their large-scale production with SFV vectors has substantially contributed to drug screening programs. Due to the high expression levels obtained for GPCRs it has been possible to provide large quantities of membrane fractions for binding assays of chemical libraries to discover lead compounds. Further compound development can be evaluated in binding assays with the same membrane fractions. Additionally, it is most important that the drugs developed show high selectivity for the targeted GPCR, which can be efficiently tested by expression of other related subtype receptors.

4.2. Structural biology

Another application of large-scale recombinant protein production is to generate sufficient material for structural studies. Overexpression of the mouse serotonin 5-HT$_3$ receptor in BHK-21 cells was performed in a bioreactor with a culture volume of 11.5 L after a C-terminal hexa-histidine tag was engineered for purification purposes [26]. The tagged receptor showed the same pharmacological profile and electrophysiological properties as the wild-type 5-HT$_3$ receptor, and the expression levels were high (B_{max} values of 52 pmol receptor/mg protein, $> 3 \times 10^6$ receptors/cell). The production was highly reproducible with receptor yields of 1–2 mg per liter culture. Solubilization conditions were optimized by testing a panel of detergents. The compound $C_{12}E_9$ showed only minor effect on binding activity [39]. Purification of solubilized 5-HT$_3$ receptor demonstrated a homogenous band with a mobility of 62 kD in SDS-PAGE, which is in good agreement with the size of a glycosylated subunit. According to gel filtration chromatography, the size of the channel complex was 280 kD, the approximate size of a pentameric channel. Preliminary cryo EM data also indicated that the structure included five subunits.

To further improve the yield of recombinant proteins expressed from SFV vectors, a fusion protein construct of the SFV capsid and the hNK1R was engineered. Previous work showed that the N-terminus of the capsid contains a translation enhancement signal that can increase the yield of capsid fusion proteins by 5–10-fold [40]. Moreover, the capsid protein possesses an autocatalytic cleavage property, which cleaves off the recombinant protein from the capsid. Expression of the hNK1R as a capsid fusion protein resulted in 5–10-fold higher expression levels compared to the expression of the hNK1R alone [15]. The yields in CHO-K1 cells cultured in serum-free suspension cultures were 5–10 mg receptor per liter culture. Cell fractionation studies indicated that only 20% of the receptors were located on the plasma membrane. However, hNK1R present in internal membranes demonstrated functional binding activity.

5. Expression in primary cell cultures

5.1. Fibroblasts

Besides their utility in cultured cell lines, SFV vectors can also be used to transfer genes to cultured primary cells and tissue explants. Infection with SFV vectors expressing a mitochondrial precursor protein consisting of the mitochondrial targeting sequence from *Neurospora crassa* ATPase subunit 9 and mouse dihydrofolate reductase resulted in efficient gene delivery in human primary fibroblasts [41]. Processing of the precursor protein to intermediate and mature forms resulted in the translocation of the recombinant protein into the mitochondrial matrix. Immunofluorescence microscopy confirmed the localization of the recombinant protein. In contrast to the rapid processing of the mitochondrial precursor protein in BHK cells, a much slower event was obtained in neurons.

5.2. Neurons

Application of SFV vectors in cultured dissociated primary neurons has demonstrated high efficiency of gene expression [42]. Compared to lipid-based transfection methods (see Chapter 5), which at best generate approximately 25% transfection rates [43], >90% of neurons showed transgene expression after SFV infection [8] (Fig. 4). It also seems that the cytopathic effect is less prominent in non-dividing cells as the neurons stayed intact for at least 3 days post-infection. The functional activity in SFV-infected neurons was monitored by electrophysiological recordings demonstrating viability and signal transduction capacity comparable to non-infected cells [21]. SFV-based expression of the human amyloid precursor protein (APP) and two mutant forms ("Swedish" and "London") in hippocampal neurons implicated that in early onset of Alzheimer disease there is increased production of C-terminally elongated $\beta42$ peptide [44]. Studies on overexpression of SNAP-25 (synaptosomal-associated protein of 25 kD) in hippocampal neurons indicated that the SFV vectors can be efficient tools in analysing the function of proteins involved in neurotransmitter release [45].

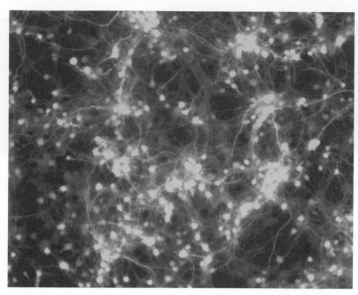

Fig. 4. SFV-GFP infection of hippocampal neurons. Rat primary hippocampal neurons were infected with SFV-GFP virus at an MOI of 10 and visualized by fluorescence at 48 h post-infection.

5.3. Other cell types

Alphavirus vectors have also been employed for expression of heterologous genes in other cell types. For instance, cardiac myocytes showed high infection rates and efficient expression of reporter gene expression when heterologous genes were delivered by SIN vectors [46]. Moreover, cultured primary myotubes were highly susceptible to SFV infections (Lundstrom, unpublished results). Application of tissue explants from prostate biopsies also indicated that high expression levels of SFV-mediated β-galactosidase could be achieved [47].

Certain cell types, however, are only poorly susceptible to SFV vectors. For instance, in addition to endothelial cells, those of a haematopoietic origin are less efficiently infected by SFV. Studies on several B cell types have indicated that the infection rates are modest ($< 5\%$). However, for some B cell lines and Jurkat cells the capacity to deliver SFV RNA is very high. As described earlier, in cases where low or modest infection rates are obtained, improvements can be achieved by addition of PEG to the culture medium of infected cells [35].

6. Expression in hippocampal slice cultures

6.1. Expression in neurons

Organotypic hippocampal slice cultures represent a highly applicable route for fast visualization and functionality of recombinant proteins in a set up that closely

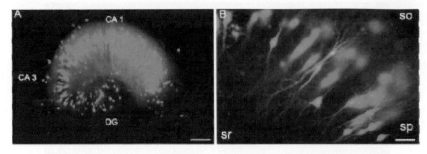

Fig. 5. SFV-GFP infection of organotypic hippocampal slice culture. Rat hippocampal slice cultures were injected with a low concentration (approximately 1000 particles) of SFV-GFP and visualized at 1 day post-infection. (A) Whole slice, bar 300 µm; (B) CA1 pyramidal cells, 75 µm.

resembles an *in vivo* situation. SFV vectors have shown high efficiency of gene delivery in hippocampal slice cultures [48]. SFV-based expression suggested that >90% of the GFP-positive cells were of neuronal origin and very few glial cells, <5% expressed GFP (Fig. 5). SFV vectors were also compared to other viral vectors with respect to gene delivery to hippocampal slice cultures [49]. For instance, the expression levels of GFP were highest and fastest from SFV vectors compared to adenovirus, adeno-associated virus (AAV), measles virus, and lentivirus vectors. Moreover, the neuron-specific expression pattern was highest for SFV and could only be matched by an AAV vector with a tissue-specific promoter. However, the duration of transgene expression lasted for only a short time period (<5 days) for SFV, whereas both AAV and lentivirus demonstrated a long-term expression pattern. On the other hand, neurons in SFV-infected slice cultures demonstrated comparable viability to non-infected cells up to 3 days post-infection measured by electrophysiological responses (resting membrane potential and conductance, action potentials and firing accommodation, and H-current).

6.2. Co-expression of GFP

Two approaches can be used to monitor the expression of transgenes in slice cultures when no antibodies against the target protein are available: one possibility is to generate fusion constructs between the gene of interest and GFP. However, the drawback is that the GFP fusion moiety might interfere with the expression level of the recombinant protein and, more importantly, with its transport and localization. Another approach is to construct an SFV vector with the GFP gene downstream of an internal ribosomal entry site (IRES) (see Chapter 2), which allows for simultaneous expression of the recombinant protein of interest without interfering with protein transport. However, GFP expression will facilitate the identification of SFV-targeted cells. In some cases, the GFP signal expressed from the IRES sequence was weak, a problem that was remedied by the construction of a novel SFV vector with the GFP gene downstream of a second subgenomic SFV 26S promoter (Ehrengruber and Lundstrom, unpublished results).

7. SFV vectors in vivo

7.1. Vaccine production

In addition to infection of cell lines and primary cell cultures SFV vectors can also be used for *in vivo* applications. The first applications included the *in vivo* administration of naked SFV RNA replicons, DNA-based vectors and recombinant SFV particles [50]. SFV vectors expressing many viral antigens have been injected subcutaneously, intravenously and intraperitoneally into various species including mice, guinea pigs and primates. Humoral and CTL responses have been obtained and for several targets protection against lethal viral challenges has been demonstrated. Furthermore, both prophylactic and therapeutic effects could be achieved by vaccination with naked RNA directed against β-galactosidase [51] and a P185 tumor antigen [52]. Mice injected with SFV-LacZ RNA showed protection against tumor challenges and in animals with implanted tumors, regression in tumor size was observed.

7.2. Stereotactic injections

SFV vectors have been applied for stereotactic injections into striatum and amygdala of adult male Wistar rats [12]. Injection of 10^5 SFV-LacZ viral particles generated localized high-level expression of β-galactosidase. The expression showed a strong preference for neurons and the pattern was highly transient. Maximum expression was observed at 1–2 days post-infection. Based on *in situ* hybridization data, LacZ RNA levels decreased rapidly at 4 days post-injection. X-gal staining also showed that β-galactosidase expression was significantly decreased and only minor staining detected at 28 days post-injection. Injection into the sensitive brain regions of striatum and amygdala should generate behavioral responses in the animal if the SFV vectors caused extensive cytotoxicity in the infected neurons. However, no behavioral differences were detected when the SFV-injected rats were compared to animals injected with growth medium from BHK-21 cells and monitored for general health (food intake, body weight and temperature), muscle strength, sensorimotor function and exploratory behavior.

Injection of SFV vectors containing HIV glycoprotein gp120 into rat cortex resulted in high-level expression in neurons [53]. The recombinant protein was detected between days 3 and 9 post-injection. The SFV-mediated expression in rat brain did not cause any detectable brain damage and furthermore, no reactive astrogliosis was observed.

8. Safety of SFV vectors

The expression vectors developed for SFV are based on an attenuated strain and therefore already confer a reduced virulence compared to wild-type SFV [1]. Application of the original SFV helper vector generated some replication-competent particles through homologous recombination during the *in vivo* packaging process in

BHK-21 cells because of the sequence homology in the expression and helper vectors in the nsP1-4 region. Introduction of three point mutations prevented the cleavage of the p62 precursor protein into E2 and E3 envelope proteins, which rendered the SFV particles non-infectious until treated with α-chymotrypsin [5]. This procedure prevented efficiently the amplification of any replication-proficient particles, which could be demonstrated by plaque assays on BHK-21 cells infected with extreme concentrations of virus. To further reduce the theoretical risk of production of SFV particles containing a full-length genome, a split two-helper vector system was engineered. Expression of capsid and envelope proteins from separate vectors resulted in high-titer SFV particles with a theoretical recombination frequency of 4×10^{-17} [54].

The presence of infectious SFV particles in the cell culture medium and associated with infected cells is an important question to address to establish safe procedures for using SFV-infected cells. Studies on BHK-21 cells clearly demonstrated that the infection procedure was remarkably rapid [55]: after a 2-min exposure to SFV, BHK-21 cells showed approximately a 20% infection rate (Fig. 4). Prolongation of the incubation time increased the infection rate resulting in close to 100% infection within 2 h of virus contact with BHK-21 cells. Furthermore, residual recombinant SFV particles could be efficiently removed by a few washes with PBS 2 h post-infection or alternatively prior to harvesting the cells. This procedure ensured the safe use of SFV-infected material outside the cell culture facility. SFV vectors were originally classified for applications only under safety containment level BL2. However, recently the system has been relaxed to containment level BL1 in several European countries (including Germany, Switzerland and UK).

9. SFV vectors in gene therapy

9.1. Intratumoral injections

Because of the capacity of SFV vectors to induce apoptosis in infected host cells [4], which also was shown to hold true for recombinant SFV particles [56], it was potentially attractive to test these vectors on intratumoral injections. Previous studies indicated that human prostate tumor cells were susceptible to SFV vectors [57], and infection with high virus concentrations led to a strong induction of apoptosis [47]. Moreover, infection of prostate epiduct from tissue explants *ex vivo* with SFV-LacZ generated both efficient gene delivery and apoptotic responses. It was therefore not unexpected that intratumoral injections with recombinant SFV particles expressing GFP demonstrated an efficient tumor killing effect in a mouse model with implanted human lung carcinoma [58]. Substantial decrease in tumor volume was observed and in some animals a complete disappearance of tumors was evident. Furthermore, SFV particles expressing the pro-apoptotic *Bax* gene were capable of inhibiting the growth of AT3-Neo and AT3-Bcl-2 tumors in nude mice [59].

In another study, SFV vectors expressing the p40 and p35 subunits of IL-12 were injected intratumorally into a mouse B16 melanoma model [60]. The IL-12 expression induced B16 tumor regression through inhibition of tumor blood vessel formation, which was monitored by Doppler ultrasonography. Interestingly, repeated injections enhanced the tumor-regression and no antiviral response was detected. Another approach described the intratumoral delivery of SFV particles expressing endostatin as an anti-angiogenic therapy for brain tumors [61]. A very significant inhibition of tumor growth and reduction in intratumoral vascularization was observed in mice bearing B16 brain tumors. In comparison to retrovirus-based endostatin expression, the serum levels were after 7 days more than threefold higher after SFV delivery.

9.2. Systemic delivery

Intraperitoneal administration of SFV particles expressing luciferase as a reporter gene showed high expression levels in the peritoneal lining and in tumor cells of mice with ovarian tumors [62]. Only minor expression was detected in spleen, liver and lungs. Additionally, SFV-based expression of granulocyte-macrophage colony-stimulating factor (GM-CSF) activated macrophages to tumor cytotoxicity. Tumor growth was inhibited for 2 weeks in tumor-bearing mice, but did not prolong the survival of animals.

9.3. Targeting

Clearly, the broad host cell range of alphaviruses poses a concern, especially if systemic viral vector delivery approaches were to be considered. It would therefore be advantageous if viral gene delivery could be targeted to the appropriate cell type or tissue. This approach was demonstrated for SIN by the introduction of IgG-binding domains of protein A in the envelope structure of SIN particles [63]. By this procedure, the infection rate of normal host cells was reduced by a factor of 10^5, whereas cells treated with a monoclonal antibody against a surface receptor or marker protein efficiently bound protein A domains and hence targeted the infection to specific cells. Recently, similar constructs have been generated for SFV vectors, which demonstrated that inserts could be introduced only into a relatively small region of the E2 envelope protein to still generate viable recombinant SFV particles (Lundstrom, unpublished results).

10. SFV vectors as tools for virus assembly

The SFV expression system has also been applied for the generation of retrovirus-like particles. This could be achieved by introduction of Moloney murine leukemia virus (MMLV) sequences, gag–pol, env and LTR-ψ^+-neo-LTR into separate SFV expression vectors followed by co-transfection of in vitro transcribed RNA molecules into BHK cells [64]. Retrovirus-like particles with infection properties and reverse transcriptase activity similar to MMLV were packaged. It was also demonstrated that intron-containing retrovirus sequences could be efficiently processed in BHK cells [65]. One advantage of this approach is the rapid and simple generation of helper-free retrovirus particles.

Another approach was to clone retrovirus virion RNA downstream of the SFV 26S subgenomic promoter and to transcribe full-length chimeric SFV-retrovirus RNA for introduction into retrovirus packaging cell lines by electroporation or SFV infection [66]. The generated retrovirus particles were capable of transduction of target cells, showed reverse transcriptase activity and could integrate into the host cell genome. In another study, the SFV envelope genes were replaced with the *env* gene from murine leukemia virus (MuLV), which resulted in minimal virus particles that specifically targeted cells with MuLV receptors [67].

In addition to applying SFV vectors for the generation of retrovirus particles it has also been demonstrated that SFV replicon-based expression of rotavirus structural proteins resulted in viroplasm-like structures in the cytoplasm of BHK-21 cells [68]. In a similar way, expression of human papillomavirus type 16 capsid protein led to self-assembly of virus-like particles in mammalian host cells [69]. Coexpression of HPV16 L1 and L2 generated virus-like particles with incorporation of both L1 and L2 protein.

11. Modifications of SFV vectors

11.1. Non-cytopathogenic vectors

One of the limiting factors in using SFV vectors especially for studies on signal transduction events and expression kinetics has been the strong cytotoxicity that SFV infections induce in mammalian host cells. To overcome this problem, mutant SFV vectors have been engineered. Experience from studies on SIN indicated that mutations in the non-structural genes, preferentially nsP2, either reduced or eliminated the cytopathogenicity of these vectors [70,71]. Sequence analysis of a less virulent replication-proficient SFV strain demonstrated that a single point mutation in the nuclear localization signal (NLS) (Arg649Asp) was responsible for the change in phenotype [72]. Introduction of another point mutation, Arg650Asp, combined with the point mutation Ser259Pro in nsP2, generated a novel SFV vector, SFV-PD, with enhanced transgene expression, lower cytotoxicity and substantially prolonged survival of host cells [9]. Metabolic labeling studies on various mammalian host cell lines and primary neurons showed that the host cell protein synthesis was not as dramatically affected as after infections with the conventional SFV vector. This vector is potentially useful for many new applications of SFV as described below.

Another mutant SFV vector was recently described, which could generate a persistant replication in BHK cells [73]. This vector was also based on a single point mutation in the nsP2 gene, at position 713. Interestingly, the combination of the SFV-PD vector with the Leu713Pro mutation produced a novel triple mutant vector, which showed hardly any cytotoxicity in infected host cells and allowed for expression of GFP in BHK cells for at least 20 days [74].

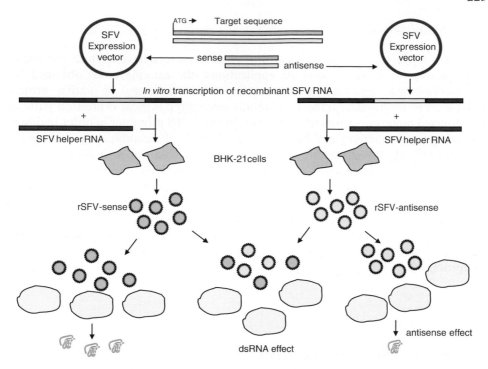

Fig. 6. Schematic presentation of SFV-based RNAi technology. Oligonucleotide primers are designed for the PCR amplification of a 250–500 bp fragment of the target sequence. PCR fragments are introduced in sense and antisense orientation in the SFV-PD vector. Co-infections with rSFV-sense and -antisense particles will generate dsRNA molecules in the target cells, which results in specific inhibition of gene expression.

11.2. Temperature-sensitive mutations

Several temperature-sensitive mutations have been described for alphaviruses [75,76]. Recently, several point mutations in the nsP2 and nsP4 genes of SFV were introduced into the expression vector to study whether a temperature-dependent expression pattern would occur [10]. For instance, a triple mutant rendered the SFV vectors temperature-sensitive, which could be observed in several mammalian cell lines. Moreover, when hippocampal slice cultures were infected with this SFV vector, a predominantly interneuronal GFP expression was observed at 37 °C. In contrast, when the temperature was lowered to 31 °C, mainly pyramidal cells in the CA1 and CA3 regions expressed GFP, as earlier observed for the conventional SFV vector at 37°C. A quadruple SFV mutant vector with three point mutations in nsP2 and one in nsP4 produced a tightly regulated temperature-sensitive phenotype, where no GFP expression was detected at the non-permissive temperature (37°C), but high expression was observed at the permissive temperature (31°C). Furthermore, expression could be induced by a shift in temperature.

11.3. Down-regulated expression

Although less cytotoxic and temperature-sensitive vectors have enabled the use of SFV vectors for a wider range of applications, the extreme levels obtained for alphavirus-based expression do not reflect the conditions of native protein expression in mammalian cells. To obtain a more physiological expression pattern, site-directed mutagenesis was performed on the SFV 26S promoter in order to down-regulate promoter activity and hence reduce the levels of recombinant protein in the host cell [77]. By this procedure, novel vectors were obtained with β-galactosidase and luciferase expression levels of 1, 3 and 30%, respectively, compared to the SFV vector harboring the wild-type promoter. This modification should be particularly useful for establishing functional assays for GPCRs as the SFV-based over-expression decreased the functional coupling to G proteins. Preliminary results indicated that functional activity of adenosine receptors measured by cAMP stimulation could only be established with mutant vectors expressing lower levels of receptors.

12. Novel technologies

12.1. Inducible stable expression vectors

The engineering of temperature-sensitive SFV vectors for which a shift in temperature allows for induction or repression of the heterologous gene expression has opened up the possibility to develop temperature-inducible stable expression vectors as described similarly for SIN [78]. The introduction of a CMV promoter and a selection marker gene into the quadruple vector SFV-PDTE will allow for selection of mammalian clones with the construct of interest integrated into the genomic DNA. The resulting clones can be propagated at the non-permissive temperature and the expression can be simply turned on by a down-shift in temperature. This procedure should be advantageous for clone stability, because of absence of any selection pressure on cells continuously expressing a toxic recombinant protein.

12.2. Antisense and ribozyme applications

Direct antisense applications using the conventional SFV vector have been severely hampered by the strong inhibition of host cell protein synthesis observed shortly after infection with recombinant SFV particles. For this reason, specific gene expression inhibition by introduction of antisense RNA with the aid of SFV has been unfulfilled despite the otherwise advantageous RNA replication typical for alphavirus replicons. For instance, although a CHO cell line stably expressing the orphanin FQ receptor targeted with SFV antisense constructs resulted in decreased receptor levels, a similar effect was obtained after infection with SFV-LacZ particles (Lundstrom, unpublished results).

Despite these observations, an SFV vector was engineered to contain a hammerhead ribozyme targeted to a highly conserved sequence in the U5 region of HIV-1 long terminal repeat [79]. A full-length 8.3 kb chimeric SFV RNA containing the SFV replicon and the ribozyme (SFVRz) was packaged into recombinant particles. Transduction of BHK-21 cells resulted in the production of large amounts of ribozyme-containing RNA with catalytic activity. HIV U5-chloramphenicol acetyl-transferase mRNA expressed in a stable BHK cell line was quantitatively eliminated after infection with SFVRz.

12.3. RNA interference

The use of RNA interference (RNAi) to inhibit gene expression was described earlier for *Caenorhabditis elegans* [80], *Drosophila melanogaster* [81], *Trypanosoma brusei* [82], planarians [83], and plant cells [84]. Recently, RNAi has been demonstrated to function in mammalian cells. Microinjection of synthetic dsRNA molecules into pre-implanted mouse embryos demonstrated specific gene inhibition of E-cadherin and GFP [85]. The novel SFV-PD vector, which does not induce a strong inhibition of host cell protein synthesis, has made it possible to test whether generation of double stranded RNA could be used to affect specific gene inhibition. The approach taken was to introduce a 250–500 bp fragment from the coding region of the target gene into individual SFV-PD vectors in both sense and antisense orientation followed by virus stock generation (Fig. 6). Co-infection of host cells with SFV-PD sense and antisense recombinant particles results in rapid replication of large quantities of sense and antisense RNA molecules, which are capable of forming dsRNA molecules responsible for specific inhibition of target gene expression. Thus, efficient inhibition of aldolase A gene expression was demonstrated in BHK-21 cells, inhibition of cell proliferation by targeting cyclin A and B genes in HEK293 cells, and inhibition of the human dopamine D2 receptor expression in CHO cells (Certa and Lundstrom, unpublished results).

13. Conclusions

As described in this review, SFV vectors are extremely versatile and widely used for many applications in drug discovery, neuroscience and various therapeutic areas. A clear advantage is the rapid production of high-titer recombinant SFV particles, which can be used in a variety of applications. Very importantly, GPCRs that today represent more than 50% of the drug targets are expressed particularly well from SFV vectors. This allows for improvement in drug screening but also the preparation of purified receptors for structural studies.

Moreover, the SFV vectors have turned out to be extremely useful for studies in neurobiology. The strong preference for neuronal infection and expression of heterologous genes in dispersed primary neurons and in organotypic hippocampal slice cultures have made SFV vectors highly useful tools. SFV vectors have been administered successfully into rat brain, demonstrating again a highly neuron-specific expression pattern. Due to the transient nature of the heterologous gene expression, behavioral changes can be monitored during and after the expression phase in the same animals. The short-term expression has also enabled the use of SFV vectors for gene therapy applications, which has been further improved by the targeting or encapsulation of SFV particles. SFV vectors are also potentially attractive as vehicles in vaccine production.

The recent development of novel non-cytopathogenic and temperature-sensitive SFV vectors has further expanded the application range to new areas of signal transduction studies and gene inhibition approaches. It is therefore justified to conclude that SFV vectors have versatile applications from recombinant protein expression to gene therapy trials.

Acknowledgements

I am most thankful to Dr. Markus Ehrengruber (Brain Research Institute, Zurich, Switzerland) for providing Figure 5.

Abbreviations

AAV	adeno-associated virus
AGA	aspartylglucosaminidase
AGU	aspartylglucosaminuria
APP	amyloid precursor protein
β-gal	β-galactosidase
BHK	baby hamster kidney
CHO	Chinese hamster ovary
COX	cyclo-oxygenase
CMV	cytomegalovirus
CTL	cytotoxic T cell response
GABA	gamma-aminobutyric acid
GFP	green fluorescence protein
GM-CSF	granular macrophage-colony stimulating factor
GPCR	G protein-coupled receptor
HA	hemagglutinin
HEK	human embryonic kidney
HIV	human immunodeficiency virus
HPV	human papilloma virus
IL-4	interleukin-4
IL-12	interleukin-12
IRES	internal ribosome entry site
LacZ	gene coding for β-galactosidase
LTR	long terminal repeat
MEG	megakaryotic cell line

MMLV	moloney murine leukemia virus
MOI	multiplicity of infection
MuLV	murine leukemia virus
NK1R	neurokinin-1 receptor
PAI	plasminogen-activator inhibitor
PBS	phosphate buffered saline
PEG	polyethylene glycol
SAP	secreted alkaline phosphate
SDS-PAGE	sodium dodecyl sulfate-polyacrylamide gel elctrophoresis
SFV	Semliki Forest virus
SIN	Sindbis virus
SNAP	synaptosomal-associated protein
VEE	Venezuelan equine encephalitis virus
5-HT	5-hydroxytryptamine

References

1. Liljeström, P. and Garoff, H. (1991) Bio/Technology 9, 1356–1361.
2. Xiong, C., Levis, R., Shen, P., Schlesinger, S., Rice, C.M., and Huang, H.V. (1989) Science 243, 1188–1191.
3. Davis, N.L., Brown, K.W., and Johnston, R.E. (1989) Virology 171, 189–204.
4. Strauss, J.H. and Strauss, E.G. (1994) Microbiol. Rev. 58, 491–562.
5. Berglund, P., Sjöberg, M., Garoff, H., Atkins, G.J., Sheahan, B.J., and Liljeström, P. (1993) Bio/Technology 11, 916–920.
6. Berglund, P., Smerdou, C., Fleeton, M.N., Tubulekas, I., and Liljeström, P. (1998) Nat. Biotechnol. 16, 562–565.
7. Di Ciommo, D.P. and Bremner, R. (1998) J. Biol. Chem. 273, 18060–18066.
8. Lundstrom, K. (1999) J. Recept. Signal Transduct. Res. 19, 673–686.
9. Lundstrom, K., Schweitzer, C., Richards, J.G., Ehrengruber, M.U., Jenck, F., and Mülhardt, C. (1999) Gene Ther. Mol. Biol. 4, 23–31.
10. Lundstrom, K., Rotmann, D., Hermann, D., Schneider, E.M., and Ehrengruber, M.U. (2001) Histochem. Cell Biol. 115, 83–91.
11. Blasey, H.D., Lundstrom, K., Tate, S., and Bernard, A.R. (1997) Cytotechnology 24, 65–72.
12. Lundstrom, K., Richards, J.G., Pink, J.R., and Jenck, F. (1999) Gene Ther. Mol. Biol. 3, 15–23.
13. Kyttälä, A., Heinonen, O., Peltonen, L., and Jalanko, A. (1998) J. Neurosci. 18, 7750–7756.
14. Chuang, J.L. and Schleef, R.R. (2001) J. Cell Biochem. 82, 277–289.
15. Lundstrom, K., Mills, A., Buell, G., Allet, E., Adami, N., and Liljeström, P. (1994) Eur. J. Biochem. 224, 917–921.
16. Lundstrom, K., Schweitzer, C., Rotmann, D., Hermann, D., Schneider, E.M., and Ehrengruber, M.U. (2001) FEBS Lett. 504, 99–103.
17. Lundstrom, K. (2001) In: Kühne, S.R. and de Groot, J.J.M. (eds.) Perspectives on Solid State NMR in Biology, Kluwer Academic Publishers, The Netherlands, pp. 131–139.
18. Lundstrom, K. and Henningsen, R. (1998) J. Neurochem. 71 Suppl. 1, S50D.
19. Scheer, A., Björklöf, K., Cotecchia, S., and Lundstrom, K. (1999) J. Recept. Signal Transduct. Res. 19, 369–378.
20. Monastyrskaia, K., Lundstrom, K., Plahl, D., Acuna, G., Schweitzer, C., Malherbe, P., and Mutel, V. (1999) Br. J. Pharmacol. 128, 1027–1034.
21. Lundstrom, K., Knoflach, F., Goepfert, F., Schaffhauser, H., Pink, J.R., Borer, Y., Ziltener, P., and Mutel, V. (1998) J. Neurochem. 71 Suppl. 1, S86C.

228

22. Monastyrskaia, K., Goepfert, F., Hochstrasser, R., Acuna, G., Leighton, J., Pink, J.R., and Lundstrom, K. (1999) J. Recept. Signal Transduct. Res. 19, 687–701.
23. Hatt, H., Lang, K., and Gisselmann, G. (2001) Biol. Chem. 382, 1207–1214.
24. Werner, P., Kawashima, E., Reid, J., Hussy, N., Lundstrom, K., Buell, G., Humbert, Y., and Jones, K.A. (1994) Mol. Brain Res. 26, 233–241.
25. Michel, A.D., Lundstrom, K., Buell, G.N., Surprenant, A., Valera, S., and Humphrey, P.P.A. (1996) Br. J. Pharmacol. 118, 1806–1812.
26. Lundstrom, K., Michel, A.D., Blasey, H., Bernard, A., Hovius, R., Vogel, H., and Surprenant, A. (1997) J. Recept. Signal Transduct. Res. 17, 115–126.
27. Gorrie, G.H., Vallis, Y., Stephenson, A., Whitfield, J., Browning, B., Smart, T.G., and Moss, S.J. (1997) J. Neurosci. 17, 6587–6596.
28. Shamotienko, O., Akhtar, S., Sidera, C., Meunier, F.A., Ink, B., Weir, M., and Dolly, J.O. (1999) Biochemistry 38, 16766–16776.
29. Lenhard, T., Marheineke, K., Lingen, B., Haase, W., Hammermann, R., Michel, H., and Reiländer, H. (1998) Cell Mol. Neurobiol. 18, 347–360.
30. Blasey, H.D., Brethon, B., Hovius, R., Lundstrom, K., Rey, L., and Bernard, A.R. (1998) In: Merten, O.-W. (ed.) New Developments and New Applications in Animal Cell Technology, Kluwer Academic Publishers, The Netherlands, pp. 449–455.
31. Zhang, J., Asselin-Paturel, C., Bex, F., Bernard, J., Chehimi, J., Willems, F., Caignard, A., Berglund, P., Liljeström, P., Burny, A., Chouaib, S. (1997) Gene Ther. 4, 367–374.
32. Mikus, P., Urano, T., and Liljeström, P. (1993) Eur. J. Biochem. 218, 1071–1082.
33. Ashery, U., Betz, A., Brosc, N., and Rettig, J. (1999) Eur. J. Cell Biol. 78, 525–532.
34. Roks, A.J., Pinto, Y.M., Paul, M., Pries, F., Stula, M., Eschenhagen, T., Orzechowski, H.D., Gschwendt, S., Wilschut, J., van Gilst, W.H. (1997) Cardiovasc. Res. 35, 498–504.
35. Arudchandran, R., Brown, M.J., Song, J.S., Wank, S.A., Haleem-Smith, H., and Rivera, J. (1999) J. Immunol. Methods 222, 197–208.
36. Phenix, K.V., McKenna, B., Fitzpatrick, R., Vaughan, I., Atkins, G., Liljeström, P., and Todd, D. (2000) Mar. Biotechnol. 2, 27–37.
37. Pettigrew, M.M. and O'Neill, S.L. (1999) Insect Mol. Biol. 8, 409–414.
38. Kohl, A., Billecocq, A., Prehaud, C., Yadani, F.Z., and Bouloy, M. (1999) Appl. Microbiol. Biotechnol. 53, 51–56.
39. Hovius, R., Tairi, A.-P., Blasey, H., Bernard, A.R., Lundstrom, K., and Vogel, H. (1998) J. Neurochem. 70, 824–834.
40. Sjöberg, M., Suomalainen, M., and Garoff, H. (1994) Bio/Technology 12, 1127–1131.
41. Huckriede, A., Heikema, A., Wilschut, J., and Agsteribbe, E. (1996) Eur. J. Biochem. 237, 288–294.
42. Olkkonen, V.M., Liljeström, P., Garoff, H., Simons, K., and Dotti, C.G. (1993) J. Neurosci. Res. 35, 445–451.
43. Ohki, E.C., Tilkins, M.L., Ciccarone, V.C., and Price, P.J. (2001) J. Neurosci. Methods 112, 95–99.
44. Tienari, P.J., Ida, N., Ikonen, E., Simons, M., Weidemann, A., Multhaup, G., Masters, C.L., Dotti, C.G., and Beyreuther, K. (1997) Proc. Natl. Acad. Sci. USA 94, 4125–4130.
45. Owe-Larsson, B., Berglund, M., Kristensson, K., Garoff, H., Larhammar, D., Brodin, L., and Low, P. (1999) Eur. J. Neurosci. 11, 1981–1987.
46. Dätwyler, D.A., Eppenberger, H.M., Koller, D., Bailey, J.E., and Magyar, J.P. (1999) J. Mol. Med. 77, 859–864.
47. Hardy, P., Mazzini, M.J., Schweitzer, C., Lundstrom, K., and Glode, L.M. (2000) Int. J. Mol. Med. 5, 241–245.
48. Ehrengruber, M.U., Lundstrom, K., Schweitzer, C., Heuss, C., Schlesinger, S., and Gähwiler, B.H. (1999) Proc. Natl. Acad. Sci. USA 96, 7041–7046.
49. Ehrengruber, M.U., Hennou, S., Büeler, H., Naim, H.Y., Deglon, N., and Lundstrom, K. (2001) Mol. Cell. Neurosci. 17, 855–871.
50. Berglund, P., Smerdou, C., Fleeton, M.N., Tubulekas, I., and Liljeström, P. (1998) Nat. Biotechnol. 16, 562–565.

51. Ying, H., Zaks, T.Z., Wang, R.-F., Irvine, K.R., Kammula, U.S., Marincola, F.M., Leitner, W.W., and Restifo, N.P. (1999) Nat. Med. 5, 823–827.
52. Colmenero, P., Liljeström, P., and Jondal, M. (1999) Gene Ther. 6, 1728–1733.
53. Altmeyer, R., Mordelet, E., Girard, M., and Vidal, C. (1999) Expression and detection of Macrophage-tropic HIV-1 gp120 in the brain using conformation-dependent antibodies. Virology 259, 314–323.
54. Smerdou, C. and Liljeström, P. (1999) J. Virol. 73, 1092–1098.
55. Lundstrom, K., Hermann, D., Rotmann, D., and Schlaeger, E.-J. (2001) Cytotechnology 35, 213–221.
56. Lundstrom, K., Pralong, W., and Martinou, J.-C. (1997) Apoptosis 2, 189–191.
57. Loimas, S., Toppinen, M.-R., Visakorpi, T., Jänne, J., and Wahlfors, J.J. (2001) Cancer Gene Ther. 8, 137–144.
58. Murphy, A.-M., Morris-Downes, M.M., Sheahan, B.J., and Atkins, G.J. (2000) Gene Ther. 7, 1477–1482.
59. Murphy, A.M., Sheahan, B.J., and Atkins, G.J. (2001) Int. J. Cancer 94, 572–578.
60. Asselin-Paturel, C., Lassau, N., Guinebretiere, J.-M., Zhang, J., Gay, F., Bex, F., Hallez, S., Leclere, J., Peronneau, P., Mami-Chouaib, F., Chouaib, S. (1999) Gene Ther. 6, 606–615.
61. Yamanaka, R., Zullo, S.A., Ramsey, J., Onodera, M., Tanaka, R., Blaese, M., and Xanthapoulos, K.G. (2001) Cancer Gene Ther. 8, 796–802.
62. Klimp, A.H., van der Vaart, E., Lansink, P.O., Withoff, S., de Vries, E.G., Scherphof, G.L., Wilschut, J., and Daemen, T. (2001) Gene Ther. 8, 300–307.
63. Ohno, K., Sawai, K., Iijima, Y., Levin, B., and Meruelo, D. (1997) Nat. Biotechnol. 15, 763–767.
64. Li, K.-J. and Garoff, H. (1996) Proc. Natl. Acad. Sci. USA 93, 11658–11663.
65. Li, K.-J. and Garoff, H. (1998) Proc. Natl. Acad. Sci. USA 95, 3650–3654.
66. Wahlfors, J.J., Xanthopoulus, K.G., and Morgan, R.A. (1997) Hum. Gene Ther. 8, 1197–1206.
67. Lebedeva, I., Fujita, K., Nihrane, A., and Silver, J. (1997) J. Virol. 71, 7061–7067.
68. Nilsson, M., von Bonsdorff, C.H., Weclewicz, K., Cohen, J., and Svensson, L. (1998) Virology 242, 255–265.
69. Heino, P., Dillner, J., and Schwartz, S. (1995) Virology 214, 349–359.
70. Agapov, E.V., Frolov, I., Lindenbach, B.D., Pragai, B.M., Schlesinger, S., and Rice, C.M.. Noncytopathogenic Sindbis RNA vectors for heterologous gene expression. (1998) Proc. Natl. Acad. Sci. USA 95, 12989–12994.
71. Dryga, S.A., Dryga, O.A., and Schlesinger, S. (1997) Virology 228, 74–83.
72. Rikkonen, M. (1996) Virology 218, 352–361.
73. Perri, S., Driver, D.A., Gardner, J.P., Sherrill, S., Belli, B.A., Dubensky, T.W., and Polo, J.M. (2000) J. Virol. 74, 9802–9807.
74. Lundstrom, K., Abenavoli, A., Margaroli, A., and Ehrengruber, M.U. (2003) Mol. Ther. 7, 202–209.
75. Hahn, Y.S., Grakoui, S.A., Rice, C.M., Strauss, E.G., and Strauss, J.H. (1989) J. Virol. 63, 1194–1202.
76. Hahn, Y.S., Strauss, E.G., and Strauss, J.H. (1989) J. Virol. 63, 3142–3150.
77. Lundstrom, K., Ziltener, P., Hermann, D., Schweitzer, C., Richards, J.G., and Jenck, F. (2001) J. Recept. Signal Transduct. Res. 21, 55–70.
78. Boorsma, M., Nieba, L., Koller, D., Bachmann, M.F., Bailey, J.E., and Renner, W.A. (2000) Nat. Biotechnol. 18, 429–432.
79. Smith, S.M., Maldarelli, F., and Jeang, K.-T. (1997) J. Virol. 71, 9713–9721.
80. Fire, A., Xu, S., Montgomery, M.K., Kostas, S.A., Driver, S.E., and Mello, C.C. (1998) Nature 391, 806–811.
81. Kennerdell, J.R. and Carthew, R.W. (1998) Cell 95, 1017–1026.
82. Ngo, H., Tschudi, C., Gull, K., and Ullus, E. (1998) Proc. Natl. Acad. Sci. USA 95, 14687–14692.
83. Sanchez Alvarado, A. and Newmark, P.A. (1999) Proc. Natl. Acad. Sci. USA 96, 5049–5054.
84. Waterhouse, P.M., Graham, M.W., and Wang, M.B. (1998) Proc. Natl. Acad. Sci. USA 95, 13959–13964.
85. Wianny, F. and Zernicka-Goetz, M. (1999) Nat. Cell Biol. 2, 70–75.
86. Allet, B., Hochmann, A., Martinou, I., Berger, A., Missotten, M., Antonsson, B., Sadoul, R., Martinou, J.C., and Bernasconi, L. (1996) J. Cell Biol. 135, 479–486.

87. Bikker, H., Bass, F., and de Vijlder, J.-J.M. (1997) J. Clin. Endocrinol. Metab. 82, 649–653.

88. Paul, N.L., Marsh, M., McKeating, J., Schulz, T., Liljeström, P., Garoff, H., and Weiss, R.A. (1993) AIDS Res. Hum. Retroviruses 9, 963–970.

89. Forsman, T., Lautala, P., Lundstrom, K., Monastyrskaia, K., Ouzzine, M., Burchell, B., Taskinen, J., and Ulmanen, I. (2000) Life Sci. 67, 2473–2484.

90. Schweitzer, C., Kratzeisen, C., Adam, G., Lundstrom, K., Ohresser, S., Malherbe, P., Stadler, H., Wichmann, J., Woltering, T., Mutel, V. (2000) Neuropharmacology 39, 1700–1706.

91. Malherbe, P., Kratzeisen, C., Lundstrom, K., Richards, J.G., Faull, R.L.M., and Mutel, V. (1999) Mol. Brain Res. 67, 201–210.

92. Ulmanen, I., Peränen, J., Tenhunen, J., Tilgmann, C., Karhunen, T., Panula, P., Bernasconi, L., Aubry, J.-P., and Lundstrom, K. (1997) Eur. J. Biochem. 243, 452–459.

93. Jäntti, J., Keränen, S., Toikkanen, J., Kuismanen, E., Ehnholm, C., Söderlund, H., and Olkkonen, V.M. (1994) J. Cell Sci. 107, 3623–3633.

S.C. Makrides (Ed.) *Gene Transfer and Expression in Mammalian Cells*

Virus-based vectors for gene expression in mammalian cells: Retrovirus

Cristina Parolin[1] and Giorgio Palù[2]

[1]*Department of Histology, Microbiology and Medical Biotechnologies, University of Padova,*
Via A. Gabelli n.63, 35121 Padova, Italy; Tel.: +049-827-2365; Fax: +049-827-2355;
E-mail: cristina.parolin@unipd.it
[2]*Department of Histology, Microbiology and Medical Biotechnologies, University of Padova,*
Via A. Gabelli n.63, 35121 Padova, Italy; Tel.: +049-827-2350; Fax: +049-827-2355;
E-mail: giorgio.palu@unipd.it

1. Introduction

High-level expression of proteins in animal cells has been very informative in studies of protein and cellular function and in many cases has relied on virus-derived expression vectors. Many viruses have evolved to maximize expression of their proteins in host cells and are therefore a good starting point for construction of efficient expression vectors. Retroviruses are diploid, single-stranded, positive-sense RNA viruses [1]. As an obligate step of the retrovirus life cycle, the RNA genome is converted into DNA and then integrated into the host cell chromosome in the form of provirus. The provirus replicates as the host cell chromosome replicates and is transmitted to all progeny cells. This ability of retroviruses to stably introduce new genetic information into the target cells led to the development of retroviruses as vehicles for the stable transfer of genes [2–4]. Indeed, retroviral vectors have been used for a variety of experimental applications, including insertional mutagenesis, cell lineage studies, the creation of transgenic animals and the expression of foreign genes into mammalian cells, both *in vitro* and *in vivo* as documented by the majority of gene therapy clinical trials. An ideal vector should guarantee not only high efficient gene transfer but also appropriately regulated and stable gene expression from a safely integrated provirus. Currently, efforts are devoted to achieve these goals. This chapter focuses on recent progress on retroviral vector design and applications.

2. Biology of retroviruses

2.1. Virion morphology

Mature particles measure approximately 120 nm in diameter and consist of an inner core that contains two copies of the viral RNA genome complexed with nucleocapsid proteins [1]. Enzymes required for virus particle maturation and for the initial steps of viral replication, i.e., protease, reverse transcriptase and integrase, are also contained in the nucleocapsid, the inner portion of the virion. The outer shell of the virion core is formed by capsid proteins and is surrounded by a continuous lipid bilayer, the envelope, derived from the host cell membrane, including virus-specific glycoproteins. Matrix proteins, which form a layer outside the core, are inserted into the inner surface of the membrane and are responsible for the integrity of the virion.

2.2. Genomic organization

Members of the *Retroviridae* family are classified into seven genera: alpharetroviruses, betaretroviruses, gammaretroviruses, deltaretroviruses, epsilonretroviruses, lentiviruses and spumaviruses. Alpharetroviruses, betaretroviruses, gammaretroviruses and deltaretroviruses, include oncogenic retroviruses of mammals and birds (previously classified as oncoviruses). Epsilonretroviruses include exogenous viruses of fish and reptiles. Lentiviruses are complex viruses which induce slow progressive diseases affecting the immune system. Spumaviruses are a group of apparently non-pathogenic retroviruses that cause a characteristic cytopathic effect in tissue culture [1]. All retroviral genomes are approximately 7–11 kb in length and in the simplest form, as seen with simple retroviruses such as avian leukosis virus (ALV), an alpharetrovirus, and murine leukemia virus (MLV), a gammaretrovirus, contain at least four characteristic genes: *gag, pro, pol* and *env*. The *gag* gene encodes the Gag polyprotein, which, after assembly of the virus particle, is proteolytically cleaved into several structural subunits, the matrix (MA), the capsid (CA) and the nucleocapsid (NC), which constitute the virion. The *pro* gene encodes the viral protease (PR) and is always located between *gag* and *pol* genes. It may be expressed as either a part of the Gag polyprotein or as part of the Gag-Pol polyprotein. The viral protease is required for the translational processing of the Gag and the Gag-Pol polyproteins and the maturation of the viral particles during or after their budding. The *pol* gene encodes the enzymatic activities essential to the process of reverse transcription and integration. The *env* gene encodes the envelope glycoprotein that interacts with the host cell surface receptor. The genome of complex retroviruses, including all lentiviruses, spumaviruses and some oncoviruses, in addition to the *gag, pro, pol* and *env* genes found in simple retroviruses, contains a set of extra regulatory and accessory genes that are involved in modulation of viral gene expression and in viral infectivity.

The full-length linear duplex proviral DNA is synthesized from the genomic RNA shortly after infection. The DNA form of viral genome is called the provirus. The provirus contains a duplication of both regions of long terminal redundancy located at the 5′ and 3′ end of genomic RNA, a consequence of the complex process of reverse transcription (Fig. 1). The long terminal redundancy at either end of the provirus is called the long terminal repeat (LTR). In the integrated DNA provirus, the three structural genes are arranged in the same order (5′-*gag-pol-env*-3′) and are flanked by the characteristic LTRs. The LTRs play diverse and important roles in the virus life cycle. The LTRs are divided into three functionally distinct regions in the following order: unique 3′ (U3), repeat (R) and unique 5′ (U5). The R region forms a direct repeat at both ends of the viral RNA genome and provides the homology sequence necessary for strand transfer during the reverse transcription process. The U5 and U3 regions, while being unique sequences in the viral RNA

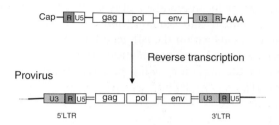

Fig. 1. Structure of the RNA genome and the integrated provirus of a simple retrovirus. The viral RNA genome is reverse-transcribed into a DNA provirus by the viral reverse transcriptase. During this process, the U3 and U5 regions are duplicated to give rise to two identical long terminal repeats (LTRs).

genomes, are duplicated during reverse transcription. Short inverted repeats, the attachment (att) sites, are located at the two ends of the LTR, in particular at the 5′ terminus of U3 in the 5′LTR and in the 3′ terminus of U5 in the 3′LTR. Integrase is thought to interact with the att sites during the integration process.

In the case of MLV and the majority of other oncoviruses, the integrated provirus contains a single splice donor and acceptor combination and encodes only two distinct transcripts. These are the full length unspliced genomic RNA, which also functions as the mRNA for Gag and Pol synthesis and a singly spliced mRNA species that encodes Env. In addition to the singly spliced and unspliced mRNAs observed in the simple retroviruses, complex retroviruses encode a third, multiply spliced class of viral transcripts and are predicted to express multiple distinct mRNA species in a pattern that is temporally regulated. Although, the 5′ and 3′ LTRs are identical, the 5′LTR is used for the initiation and regulation of transcription and the 3′LTR directs the addition of a poly-A tail to the viral mRNA. The U3 region contains the viral promoter, including a TATA box and regulatory elements that bind a complex array of transcription factors. Transcription initiates at the U3/R border and nascent transcripts are capped with 7-methylguanosine by the cellular machinery. Viral transcripts terminate at the R/U5 border in the 3′LTR and sequences in U3 and R are recognized by cellular factors that catalyze the addition of poly-A tail. The untranslated region (leader region) at the 5′ end of the genome contains a primer binding site (PBS) located next to the 3′ end of the U5 that has sequence complementary to a portion of a cellular tRNA. A host tRNA anneals to the PBS on the genomic RNA and serves as a primer for the initiation of reverse transcription. Different tRNAs are used by different viruses as primers for this process [5]. A region between the 5′LTR and the beginning of the gag gene comprises the genome packaging signal, called ψ in MLV and E in avian retroviruses. ψ binds to viral proteins and is required for incorporation of genomic RNA into newly formed virions [6,7]. The MLV ψ is located in the 5′ untranslated region between the PBS and the gag gene. Although, these sequences have been shown to be sufficient even for incorporation of heterologous RNA into viral particles, additional sequences that increase packaging efficiencies have been identified within the gag gene [8]. Signals important for packaging of the human immunodeficiency virus type

234

1 (HIV-1) genome into virion particles have not been completely identified yet, but they have been shown to be located in the 5′ untranslated region and in the 5′ region of the *gag* gene. In human and simian foamy viruses, in addition to the 5′ untranslated region, elements within the *pol* gene have been reported to be involved in viral genome packaging. A polypurine tract (PPT), located upstream of U3, functions as the site of the initiation of the positive strand DNA synthesis during reverse transcription [1].

2.3. The life cycle

The infection process (Fig. 2) begins with the specific interaction of the virion envelope glycoprotein with a host cell surface receptor protein. The binding of the viral glycoprotein to a specific receptor complex on the cell surface leads to fusion of the viral and cellular membranes and release of the viral core in the cytosol. The virus core, once released into the cytoplasm, is partially degraded to form a large nucleoprotein complex. The single-stranded (ss) RNA genome is copied into a linear double-stranded DNA molecule by reverse transcription (preintegration complex) and transported into the nucleus where it integrates into the host genome. Nuclear import of MLV DNA and perhaps that of all oncoviruses requires dissolution of the nuclear envelope with mitosis [9,10]. In contrast, HIV-1 is able to perform this task by exploiting cellular pathways for active nuclear import [11]. Thus, passage through mitosis is required for integration of oncoviruses but not for the human

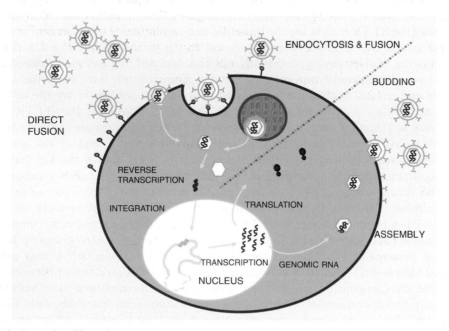

Fig. 2. Retrovirus life cycle.

immunodeficiency virus. The provirus is permanently integrated as part of the host genome. Expression of viral genes is initiated by transcription of provirus into RNA. The latter is transported from the nucleus to the cytoplasm, where it is either used for packaging into progeny particles or as mRNA for translation of viral Gag, Gag-Pol and Env precursor proteins. Gag and Gag-Pol precursors are translated from full length RNA, whereas the Env precursor is made from a subgenomic RNA which is derived from the longer form by RNA splicing. Translation of full-length RNA yields mostly Gag precursor proteins. However, Gag-Pol fusion products are also synthesized either because of suppression of a stop codon between the *gag* and the *pol* genes or because of ribosomal frameshifting. In the spumaviruses, the *pol* gene is expressed by a spliced mRNA. The viral protease is responsible for almost all of the maturation cleavage of the viral precursor proteins. Multiple splicing events occur in complex retroviruses for the generation of the accessory proteins. The packaging signal (ψ) in the viral RNA allows encapsidation of the unspliced RNA only [12–14].

3. Development of recombinant retrovirus vectors

3.1. Replication-competent vectors

3.1.1. Avian retroviruses

Most replication-competent retroviruses are based on the naturally occurring avian leukosis Rous sarcoma virus. Indeed, RSV is a unique example of replication-competent virus expressing a non-essential gene. RSV contains, in addition to a full set of virus genes, the *src* oncogene. For the development of avian replication-competent vectors, this *src* gene has been removed and it has been replaced by foreign DNA. The transgene usually has been inserted in a fashion that permits expression via a spliced, subgenomic transcript initiating from the promoter within the proviral LTR. Viral stocks are prepared by transfecting a plasmid encoding the vector into chicken embryo fibroblasts. High titer virus stocks can be prepared simply by passaging transfected cells and allowing the virus to spread. Replication-competent avian retroviruses have been useful tools for the stable introduction of gene cassettes into avian cells. Avian embryo has been a popular model for studying vertebrate development. In combination with classical techniques of experimental developmental biology, retroviral-competent avian retroviruses have been used to analyze the function of genes involved in cell growth and differentiation of various organs and tissues during avian embryogenesis. Natural avian retroviruses can be classified into several subgroups (A–E and J) on the basis of their *env* genes and therefore are distinguished by virtue of their host range. ALV (or the vectors derived from it) can efficiently infect avian cells while infection of mammalian cells is quite difficult. One approach to overcome this limitation is to generate cell lines or transgenic mice that express the cellular receptor for ALV. Replication-competent avian vectors have been generated that are able to transfer and stably express genes in mammalian cells, as well as in transgenic mice expressing the cellular receptor for subgroup A avian leukosis virus [15–17]. In particular, in transgenic mice the subgroup A receptor has been shown to be expressed in all cells and/or tissues. Following intramuscular injection, the virus can infect any dividing cell it comes in contact with. As an alternative way to infect mammalian cells, replication-competent avian vectors have been initially developed by replacing the avian envelope with the *env*-coding sequence of the amphotropic murine leukemia virus (MLV) [18]. This recombinant virus, once adapted by passage in avian cells, becomes replication-competent and spreads rapidly through an avian cell culture following transfection producing a high-titer viral stock. In addition, it can efficiently transfer genes into cultured mammalian cells. However, while the parental virus caused no detectable cytopathic effect in cultured avian cells, the recombinant virus did so, leading to a reduction of viral titer. This suggested that the murine envelope caused the cytopathic effect. Indeed, the laboratory-adapted avian retrovirus carrying the amphotropic murine envelope contained a point mutation in the extracellular subunit of the envelope glycoprotein. More recently, replication-competent avian derivatives that can infect mammalian cells, but are much less cytopathic, have been developed containing ecotropic MLV envelope or a modified version of the amphotropic MLV envelope, obtained after a second round of selection in tissue culture. As all replication-competent avian retroviruses, these viruses are replication-defective in mammalian cells because of defects in the production of ALV RNAs and proteins. Furthermore, mammalian cells have no endogenous viral sequences that are closely related to ALV, except in the murine amphotropic envelope region. This feature makes recombination with an endogenous mammalian retrovirus unlikely. Replication-competent avian retroviral shuttle vectors have also been developed that contain either the zeocin or blasticidin-resistance gene [19]. The drug-resistance gene was expressed in avian cells as a subgenomic mRNA from the viral promoter within the LTR, since a splice acceptor sequence was located immediately upstream of the cassette. In bacteria, the resistance gene was expressed from a bacterial promoter. It has been demonstrated that these vectors are relatively stable and can be passaged without a substantial loss of the prokaryotic plasmid sequences. An avian retroviral vector containing the amphotropic murine leukemia virus envelope has been used also for gene delivery and insertional mutagenesis in a mammalian system [20]. In this context, expression of a reporter gene lacking promoter sequences and an ATG initiation codon, but containing an appropriate splice acceptor, is dependent on the presence of a chromosomal transcription unit at the level of the integration site. The recombinant ALV-based vector could efficiently infect mammalian cells, but was unable to replicate. The vector contained a green fluorescent protein (GFP)-coding sequence with the 5′ end linked to a splice acceptor and the 3′ end linked to a modified SV40 polyadenylation signal. The proviral form of the vector would express GFP only if integration were to occur in an intron of an actively expressed gene and in the appropriate orientation, adjacent to a splice donor of a cellular gene. This approach allowed the selection of "trapped" cells on the basis of the subcellular localization of GFP.

3.1.2. Other retroviruses

Replication-competent vectors based on MLV have also been described, containing a tRNA suppressor gene, a mutant dihydrofolate reductase gene, the HIV-1 tat gene, sequences encoding puromycin acetyltransferase, hygromycin phosphotransferase or GFP and located at the *env*-3′ untranslated region or in the U3 region of the 3′LTR. While high-titer stocks of replication-competent virus could be obtained, the inserted sequences could be partially or completely lost within passages in cell cultures. Among lentiviruses, replication-competent HIV-1-based vectors that encode the chloramphenicol acetyltransferase or the GFP have been described [21,22]. These recombinant vectors are currently used to study HIV-1 biology and pathogenesis. Replication-competent spumavirus vectors, based on the human foamy virus (HFV) expressing the chloramphenicol acetyltransferase gene or the luciferase gene or the mouse hepatitis virus surface protein have been developed [23]. The genome of HFV contains *gag*, *pol* and *env*, as well as several other genes located between *env* and the 3′LTR. The latter genes, including the *bel-1*, the *bel-2* and the *bel-3* genes, are all derived from singly and/or multiply spliced mRNA. Among them, only the *bel-1* gene, which encodes for a potent transcriptional transactivator, is required for viral replication. In the designed vectors, usually the heterologous gene replaces the *bel-2* and the *bel-3* genes at the 3′ end of the viral genome. With respect to the overall structure and site of insertion of the foreign gene, these recombinant HFV vectors resemble more closely RSV-based recombinant vectors. The HFV vectors represent suitable tools to study different aspects of foamy virus biology.

3.2. Replication-defective vectors

Acutely transforming retroviruses of non human mammals and birds offer a simple example of naturally occurring retrovirus vectors. Indeed, their genome contains a cellular oncogene that during the virus life cycle is transferred to the target cells, resulting in cell transformation and tumour formation. With the exception of some strains of RSV that are replication-competent viruses, since their genome contains the oncogene in addition to a full set of virus genes, most of the transforming viruses are replication-defective. Indeed, these viruses are defective for part of viral coding sequences but contain all the *cis*-acting elements required for reverse-transcription, integration and gene expression. Thus, in the presence of replication-competent helper virus, which supplies the needed viral proteins in *trans*, these transforming viruses can replicate and transfer the oncogene to target cells. However, propagation of retroviral vectors with helper virus would expose the vector preparation to replication-competent retrovirus. As retroviruses are potentially pathogenic, their use as vehicles for gene transfer implies that they have to be engineered to be replication-defective and unable to transfer viral functions by replacing part of the viral genome, the *gag-pol-env* genes, with the desired transgene sequence. Retroviral vectors are able to infect the target cell and to integrate the transgene into the host chromosome, but cannot replicate further and spread to other cells. This process of gene transfer is often referred to as "transduction" rather than "infection" to differentiate this process from productive viral infection by a replication-competent

238

virus. The retroviral vector system is typically separated into two components: the *cis*- and the *trans*-acting viral elements. The retroviral vector contains the transgene and the minimal *cis*-acting elements allowing its efficient incorporation into viral particles, its reverse transcription, integration and expression in target cells; generally, it does not encode viral proteins. The helper function provides in *trans* the viral elements that are not present in the vector and are required for virus replication. Several cell lines for the production of retroviral vectors (packaging cells) have been generated [24]. In these cells, viral proteins needed to propagate retroviral vectors are provided in *trans* by using molecular constructs expressing the viral genes but lacking the specific signals required for packaging. Thus, "empty" particles, unable to replicate, are produced. Upon introduction of the vector into the packaging cell, recombinant viruses are obtained, which contain only the genetic information of the vector (Fig. 3). These particles are infectious but cannot spread because they do not express structural proteins of the virus. The vector carrying the transgene mimics the structure of the viral genome by containing the minimal *cis*-acting sequences necessary for efficient incorporation into viral particles, reverse transcription, integration and expression in target cells.

The possibility of generating replication competent virus (RCV) through genetic recombination raises safety concerns for clinical use of retroviral vectors. RCV can be generated by recombination events taking place between homologous sequences present in the constructs providing the packaging functions and in the vector. In addition, recombination is dependent on residual *cis*-acting sequences in the packaging construct, allowing some level of encapsidation. Therefore, the design of vector and packaging systems with reduced or no potential for the generation of RCV is facilitated by complete segregation of the *cis*- and *trans*-acting elements in

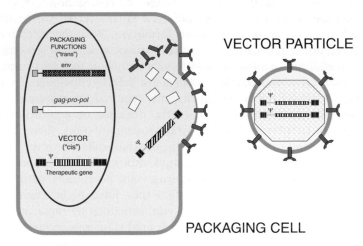

Fig. 3. Schematic diagram of a retroviral vector system. A retroviral vector system is tipically separated into two components: the *trans*- and *cis*-acting elements. Viral functions required for particle formation are provided in *trans* by the packaging component. Only the vector containing the *cis*-acting elements and expressing the transgene is packaged into newly formed recombinant particles.

order to minimize overlapping sequences. RCV can also be generated by recombination events with retroviral elements naturally present in the cells, such as endogenous retroviruses (ERV), and both vector and packaging constructs. Thus, the development of safe packaging cells has been focused on the design of molecular constructs with the minimum chance of recombination, and on the choice of an appropriate cell type.

Many retrovirus species have been employed for the generation of vectors. However, given its ability to efficiently infect a wide range of cells and its relatively well-understood biology, the murine leukemia virus, MLV, has been the most extensively used virus for the generation of recombinant vectors. Three main subgroups of MLV have been identified based on their host range: ecotropic, which infects only rodent cells, xenotropic, which can infect cells from many mammals but not mice, and amphotropic, which can infect a wide range of rodent and non-rodent mammalian cells.

3.2.1. The packaging system

Development of retrovirus packaging cells has focused on reducing the potential of these cell lines to produce replication-competent helper virus while still allowing the production of retroviral vectors at high titers. This goal has been accomplished by generating multiple deletions in the replication-competent retrovirus used to make the packaging cells in a manner that viral proteins are still made but viral RNA encoding these functions cannot be packaged into virions, reverse transcribed or integrated. A key improvement for the establishment of packaging cell lines was the discovery that the *cis*-acting packaging signal ψ, which allows selective incorporation of retroviral RNA genome into particles, mapped near the 5' end of the genome, specifically to the non-coding sequences near the major splice donor and the *gag* start codon. Indeed, the first example of a packaging cell line comes from a retrovirus spontaneous mutant. In particular, the SE21Q1b cell line expressed an avian oncovirus mutant containing a deletion of sequences near the 5' end of the viral RNA that failed to package viral genomic RNA [25,26]. The first generation of packaging cells was produced by providing the packaging functions from one helper construct consisting of a replication-competent retroviral genome from which the packaging signal was deleted from the 5' untranslated region (Fig. 4). These viruses produced all of the retroviral proteins, but RNA was not efficiently encapsidated into virions. Examples of these cells are ψ2 [27] and C3A2 [28]. However, replication-competent virus can easily arise from these cells as one single recombination event between the retroviral vector and the deleted retrovirus contained in the helper cells can restore the packaging signal in the construct expressing the helper functions. A similar recombination event, at a lower efficiency, can also occur between the helper virus and sequences derived from an endogenous virus.

In the second generation of packaging cells further modifications of the helper construct were introduced. In particular, the 3' LTR and the second strand initiation site were replaced with an alternative polyadenylation site such as that from the

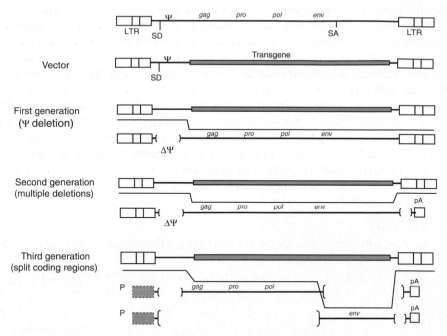

Fig. 4. Strategies for packaging cell lines. Structures of helper provirus and retroviral vector are shown at the top. Three strategies used for the development of packaging cell lines are reported as described in the text. SD and SA: splice donor and splice acceptor sites; ψ, packaging signal. P and pA indicates promoter and polyadenylation signal, respectively. Potential mechanisms for homologous recombination between the packaging cell line and the vector leading to helper virus production are indicated.

SV40. In addition, the 5′ end of the 5′ LTR was deleted. In this context, two recombination events are required to restore a replication-competent virus [29].

A higher level of biosafety was obtained by splitting the retroviral genome into two transcription units, one containing the *gag* and the *pol* genes, and the second containing the *env* gene (third generation packaging cells) (Fig. 4). This strategy enabled the generation of ψ-CRIP, ψ-CRE, GP + *env*AM12, DSN and GP + E-86 cell lines [30–35]. Three recombination events have to occur to generate helper virus. However, production of replication competent virus by a third generation packaging cell line has been described [36,37].

Additional modifications have been introduced in order to further reduce or to completely eliminate the homology between the genomes of the vector and the helper. LTRs from virus species different from those used in the vectors were employed to regulate the expression of the packaging functions. Alternatively, heterologous promoters, such the CMV promoter, were used in the helper constructs. Some packaging cell lines have been designed to express the viral proteins in an inducible manner. The rationale for this approach was to avoid the cytotoxicity induced by high level of protein expression. In an initial attempt, the two helper genomes were introduced by cotransfection with plasmids encoding selectable markers. In this context, no direct selection was applied to the packaging genome itself, and helper functions could be lost during passage of the cells in culture. In a later packaging cell generation, virus protein expression levels were improved by inserting selectable markers downstream of the viral genes in the packaging constructs so that they are translated from the same transcript after ribosomal reinitiation [34]. The pressure of a selective agent can ensure, in these cases, the maximal expression of the viral genes. Transient transfection systems have also been developed for production of retroviral vectors. The vector, along with two different constructs, one expressing the *gag* and the *pol* genes, and the other expressing the *env* gene, are transfected into cells and recombinant viruses are harvested few days later. This approach has been employed to generate high viral titers of MLV-based vectors, HIV-1 vectors and foamy virus vectors [38–48].

3.2.2. *The vector*

Minimal *cis*-acting element sequences allowing efficient encapsidation, reverse transcription, integration and expression of the transgene are incorporated into the retroviral vector (Fig. 5). The elements identified as essential for functional vectors are: (i) a promoter and a polyadenylation signal for expression of the transgene; (ii) the packaging signal (Ψ), necessary for packaging of the retroviral vector into viral particles, usually located between the end of the 5'LTR and the *gag* gene, but sometimes extending into the 5' of the *gag* sequence; (iii) sequences required for reverse transcription, including PBS, PPT and R; and (iv) sequences within U3 and U5 required for integration. The inclusion in the vector of a 5' region extending into *gag* raised the possibility of producing an undesirable viral protein that might elicit immune responses towards the transduced cells. Therefore, a mutation at the ATG start codon for *gag* translation was engineered to prevent any viral protein synthesis [2]. Recently, new safer vectors have been generated, based on evidence that deletion of the entire *gag* sequence does not significantly affect vector packaging or expression [49].

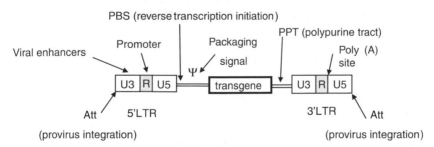

Fig. 5. *Cis*-acting elements in a prototypical retroviral vector. *Cis*-acting sequences required for vector incorporation into viral particle and for its expression in the target cells are shown. These include the packaging signal (ψ), sequences necessary for reverse transcription of the vector RNA (LTR, primer-binding site or PBS, polypurine tract or PPT) and integration of the vector DNA into the host chromosomal DNA (*att* sites). Expression of the transgene can be achieved by the 5'LTR or by an internal heterologous promoter.

4. Expression of the transgene

4.1. Basic design

The simplest vectors contain a single heterologous gene whose expression can be driven by the retroviral promoter within the 5′LTR (Fig. 6). In the 5′ untranslated regions of commonly used vectors several ATG start codons are present. However, the presence of these sequences does not seem to significantly affect transgene expression. Alternatively, a heterologous promoter can be inserted to regulate the expression of the transgene. Interference of the heterologous promoter with the promoter in the 5′ LTR might occur (promoter suppression) [50]. A heterologous promoter can be used to drive the expression of the transgene in the reverse orientation. In this case, the risk of promoter interference is reduced and genetic sequences, including introns, can be correctly maintained. However, it is unclear whether simultaneous expression driven by the LTR and by the heterologous promoter can generate reciprocal antisense transcripts.

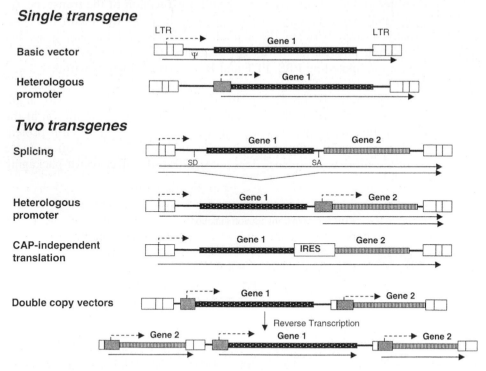

Fig. 6. Strategies for transgene expression from a retroviral vector. Transgene expression can be regulated by the 5′LTR or by an internal heterologous promoter. Dashed arrows indicate promoter activity. Arrows below diagrams indicate pattern of transcription products. Grey boxes indicate heterologous promoter. ψ, packaging signal; SD and SA: splice donor and splice acceptor sites; IRES, internal ribosome entry site. See text for details.

Most retroviral vectors are designed to express not only the gene of interest but also a selectable marker gene. The selectable gene facilitates titration of virus-producing cell lines and allows isolation of the cell population containing the gene of interest. Selectable genes coding for drug resistance or for human cell surface proteins have been used for positive selection of transduced cells *in vitro*. Thus, the presence of two transgenes within the vectors requires designing retroviral vectors able to achieve efficient coexpression of two exogenous genes. Mainly three strategies have been used to allow expression of two or more genes (Fig. 6). In the first strategy, the 5′ LTR can regulate the expression of both transgenes: while the first gene can be expressed as genomic RNA, the second one can be expressed via a spliced, subgenomic transcript. This approach mimics the splicing strategy that retroviruses use to express the *env* gene. The splice donor site that is used to express *env* is located in the 5′ untranslated region of retroviruses. During replication, some full-length viral RNAs are spliced to produce subgenomic viral RNAs that are used to express the envelope glycoprotein. The same strategy is also used to express the *src* gene in RSV. Adopting the same principle, splicing vectors were developed to express two different genes by using the viral splice donor and splice acceptor sites. As the upstream gene flanked by a splice donor and a splice acceptor site is spliced, the downstream gene can be translated from the spliced form of the messenger. In the second strategy, a heterologous promoter can be placed between the two genes to drive the expression of the downstream gene. These vectors allow expression of both genes in the transduced cell population but with a lower efficiency compared to vectors containing only one gene. In addition, selection for expression of one gene may lead to significant reduction in the transcription of the unselected gene, irrespectively of the position of the genes relative to each other, by an epigenetic mechanism.

The third strategy exploits the possibility of using an internal ribosomal entry site (IRES) sequence inserted between the two transgenes to allow cap-independent translation of a downstream gene from the same transcript. IRES sequences were initially identified in the untranslated 5′ ends of picornaviruses where they promote cap-independent translation of viral proteins. In the context of the vector, a single promoter drives expression of one single mRNA that is bicistronic: the first gene is translated in a cap-dependent manner, while a cap-independent mechanism, IRES-regulated, is responsible for translation of the second gene. IRES derived from poliovirus, encephalomyocarditis virus, swine vesicular disease virus and endogenous virus like 30S RNA (VL30) have been already used successfully in retroviral vectors [51]. Bicistronic vectors designed thus far contain the IRES sequence between a 5′ cistron of interest and a 3′ cistron for positive selection. This would ensure that the gene of interest is co-selected with the antibiotic-resistant gene. Indeed, it has been reported that the gene expressed with an IRES is translated less efficiently than the gene located near the 5′ end of the mRNA [52]. IRES and heterologous promoters can be used in the same construct to allow for the efficient expression of up to three genes. Another vector design strategy exploits the property of retroviruses to duplicate in the 5′ LTR the sequences contained in the U3 region of the 3′ LTR at each replication cycle. This approach has been adopted to design vectors expressing two copies of the gene of interest in the resulting provirus (Fig. 6). These are called "double-copy" vectors. A promoter/transgene cassette is thus inserted in the U3 region of the 3′LTR of the retroviral vector upstream of the viral promoter. Alternatively, "double-copy" vectors have also been designed by inserting the transgene in the R region of the 3′LTR.

4.2. Self-inactivating vectors

Retrovirus vectors are currently being used as vehicles for delivering genetic material to humans for gene therapy. These retroviral vectors are replication-defective and lack the coding regions of viral proteins. However, a safety concern is that retrovirus vectors have the potential to activate adjacent cellular proto-oncogenes via their own enhancer-promoter located within the long terminal repeat (LTR). Another safety concern is the potential spread of the therapeutic vector to non-target tissues through generation of replication-competent virus during propagation of the vector. To overcome these problems self-inactivating (SIN) retrovirus vectors have been developed [53]. These vectors contain an intact 5′ LTR and a 3′ LTR with a deletion, which includes sequences encoding the enhancer and the promoter functions. Since sequences contained in the U3 region of the 3′ LTR are duplicated in the 5′ LTR after completion of reverse transcription, removing the U3 sequence from the 3′ LTR in the original retroviral vector results in a final integrated provirus lacking the U3 sequence in both the 5′ and the 3′ LTRs. Since the viral promoter at the 5′ end is deleted, the gene of interest has to be expressed from an internal promoter. This vector design reduces the probability of generation of replication-competent virus and eliminates downstream transcription from the 3′ LTR into genomic DNA. Furthermore, this approach eliminates transcriptional interference resulting from the presence of the strong promoter in the U3 regions of the LTRs. Another type of self-inactivating vector exploits the high frequency of direct repeats deletion during reverse transcription. In this context, vectors are designed to contain direct repeat sequences of the transgene flanking the encapsidation sequence. The vector RNA is efficiently packaged in recombinant particles and reverse-transcribed in target cells. During RNA-dependent DNA synthesis, one copy of the repeat sequence and all of the sequences located between the two repeats can be deleted. Thus, the resulting provirus in the target cell lacks the packaging signal and contains a functional gene. Therefore, retroviral vectors that efficiently delete the packaging signal should reduce the probability of the helper constructs to capture ψ thus suppressing RCV formation.

As an alternative strategy, self-inactivating vectors exploiting the Cre-loxP recombination system have been developed [54]. Bacteriophage P1 is equipped with a site-specific recombination system including the phage-encoded recombinase Cre and a site on the phage genome, loxP, recognized by Cre, where recombination takes place. Cre is both necessary and sufficient for sequence-specific recombination in bacteriophage P1. The recombination occurs between the two loxP sequences. The result of the recombination event depends on the orientation of the interacting loxP sites. Repeats of loxP in the same orientation dictate an excision of intervening

DNA sequences, whereas inverted repeats specify inversion. The ability of the site-specific recombinase Cre to excise retroviral vector sequences positioned between two loxP targets placed in direct orientation has been investigated. *LoxP* sequences were inserted into the 3′ U3 region of a retroviral vector along with a promoter/transgene cassette. LTR-mediated duplication during reverse-transcription would place the viral genome between the two *loxP* sites, thus rendering it susceptible to recombination by the Cre recombinase. The resulting provirus in the host genome corresponds to a single LTR, deleted of the viral enhancers, carrying a single copy of the gene of interest. This strategy has been exploited to delete sequences in the retroviral vectors as well as regions in the helper construct of the packaging component. Moreover, the Cre-*loxP* system has been used to eliminate the selectable marker from the retroviral vector after its integration in the DNA of the target cells.

5. Targeting

5.1. Transcriptional targeting

An important feature of any gene delivery system is the ability to regulate the expression of the delivered gene. This is crucial when examining the product of the delivered gene for its functional role in cell biology and in therapeutic situations when the gene product is toxic or must be maintained at appropriate levels. It is possible to ensure that the transgene is only expressed in the desired target cells by using transcriptional control elements or tissue-specific promoters. Such heterologous regulatory sequences can be inserted in the U3 region of the 3′ LTR to replace the viral promoter and/or the viral enhancer. For example, the insertion of the muscle creatine kinase enhancer into the U3 region of the viral LTR, between the viral enhancer and promoter, resulted in differentiation-specific gene expression in myogenic cells [55]; replacement of the viral enhancer with the tyrosinase enhancer and promoter enabled specific gene expression in melanoma cells [56–58]; muscle creatine kinase enhancer and promoter were similarly used to target vector expression in skeletal muscle tissues [59]; similarly, human pre-pro-endothelin-1 promoter allowed specific gene expression in endothelial cells [60]. As modification of LTRs often compromises vector titre, the success of this technique depends on the ability to obtain functional high titre vector. In addition, in order to achieve complete cell-type/tissue specificity, it is not clear whether it is possible to incorporate all of the control sequences, as regulatory elements can sometimes be located distantly from the structural gene. Alternatively, reversibly inducible vectors can be designed using the tetracycline-inducible (Tet) system (see Chapter 22). The tetracycline-controlled transactivator (tTA)-responsive promoter (Tet system) is a prokaryotic inducible promoter which has been adapted for use in mammalian cells [61]. A Tet system includes a regulator unit and a response unit. The regulator unit encodes a hybrid tTA protein composed of the tetracycline repressor (*tet*R) fused to the herpes simplex virus (HSV) transactivator protein, VP16. The response unit is composed of the *Escherichia coli*-derived tetracycline-resistance operon regulatory element (*tet*O) embedded within a minimal CMV promoter lacking enhancers. In this context, regulation of expression of the delivered gene in an on or off manner can be achieved by addition of tetracycline. Similarly, the Cre-*lox*P recombination system has also been used to either activate or delete transgenes under controlled conditions.

5.2. Cellular targeting

Transgene expression limited to a specific cell type can be achieved by modification of the retroviral envelope gene [62]. As retroviruses can non-specifically incorporate heterologous cell surface proteins in their lipid envelope, recombinant retroviral vectors can be pseudotyped with surface proteins from different viruses. This technique can be useful to target the viral infectivity towards a desired cell type. The glycoprotein most commonly used to pseudotype retroviral vectors is the amphotropic MLV Env from the 4070A strain, which is responsible for the infection of a wide range of human cells. However, glycoproteins from different strains of MLV can also direct gene transfer to human cells: the amphotropic glycoprotein from the 10A1 strain, the xenotropic glycoprotein from the NZB strain of MLV and the polytropic glycoprotein from the mink cell focus-forming virus (MCF). In addition, retroviral vectors can be pseudotyped with surface glyco-proteins from different retrovirus species or different virus families. Pseudotyping with the glycoprotein from gibbon ape leukemia virus (GALV) has come into practice as it enables efficient transduction of cells derived from the hematopoietic lineage. Surface molecules from unrelated viruses have also been successfully incorporated into retrovirus vectors and shown to confer infectivity. Vesicular stomatitis virus G protein (VSV-G) allows efficient infection of a broad range of species [63]. In addition, high titre functional virus pseudotyped by VSV-G can be obtained by ultracentrifugation, due to the monomeric structure of the glycoprotein [64]. However, one main disadvantage of the use of VSV-G is represented by its cellular toxicity, which prevents its constitutive expression in stable packaging cell lines. Among other viral proteins successfully incorporated into MLV particles is the Ebola virus glycoprotein, which was shown to produce enhanced MLV infection of endothelial cells.

In a different approach, retroviral envelope proteins have been manipulated to modify the recombinant virus cell-binding properties, introducing additional receptor binding sites specific for the target cells, like antibody derivatives, single-chain antibodies (scFv) directed against the extracellular domain of cell receptors or natural ligands of surface molecules. However, while viral binding to the targeted cell could be achieved with the modified glycoprotein, the vectors incorporating these chimeric glycoproteins were either not infectious, or very poorly infectious. Thus, the incorporation of modified envelopes into the recombinant virus may not necessarily result in a significant redirection or enhancement of vector particles binding to the target cells.

6. Conclusion and perspectives

Retroviruses have proven to be adaptable to serve as flexible tools to achieve heterologous gene transfer and expression in a number of cells, both *in vitro* and *in vivo*. Moreover, they are the most utilized vectors in clinical trials of gene therapy having a documented safety profile not yet attained by other viral and non-viral vector systems. Knowledge about biology of these viruses has made possible the design of integrating vectors that effect constitutive, regulatable expression of transgenes in the absence of replication-competent helper virus production or harmful activation of cellular genes. Notwithstanding these achievements, retroviral vectors need to be further improved to overcome a number of weaknesses that may limit their *in vivo* applications. Higher titres should be obtained for clinical applications adopting new packaging cells and suitable structural elements. Complement sensitivity, that is innate to viral particles produced from non-primate cells [65,66], needs to be abrogated if vector use is deemed to extend the present *ex vivo* applications. Means are to be sought to confer on retroviral vectors the ability to transduce non-replicating cells. Indeed, transduction of hematopoietic cells can efficiently be achieved *ex vivo* by oncoretroviral vectors, if stem cells are forced to divide. This, however, is detrimental to the staminal potential, which can be lost through differentiation induced by cellular growth *in vitro*. Molecular dissection of the physical interactions occurring between the cell nuclear import system and the lentiviral preintegration complex (see Chapter 3.13) could lead to the design of retroviral vectors engineered with specific elements to permeate the nucleopore in the absence of cell mitosis. Such a newly engineered retroviral vectors would then have access to the nuclear compartment of resting cells (including different types of stem cells or non-cycling tumour cells) that could be genetically treated along with their progeny cells. Certainly the main limitation of all presently available vectors is the low effectiveness of cell-targeting. For this reason, body tissues have to be inoculated directly via a local injection and cannot be reached via the general circulation or other more physiological routes. To improve cell and tissue targeting a new generation of vectors that can be systemically delivered has to be constructed. To this end, recombinant particles should contain complement-resistant outer structures, ligand-fusion components, nuclear-import signals and cell-specific regulatory elements enabling the transgene: (i) to be safely and effectively delivered to specific cell types; (ii) to be integrated into pre-defined sites of the host chromosome; and (iii) to be expressed according to therapeutic needs.

Abbreviations

ALV	avian leukosis virus
att	attachment site
CA	capsid
CMV	cytomegalovirus
E	packaging signal in avian retroviruses

Env	envelope glycoprotein
ERV	endogenous retrovirus
GALV	gibbon ape leukemia virus
GFP	green fluorescent protein
HFV	human foamy virus
HIV-1	human immunodeficiency virus type 1
HSV	herpes simplex virus
IN	integrase
IRES	internal ribosome entry site
LTR	long terminal repeat
MA	matrix
MLV	murine leukemia virus
MoMLV	Moloney murine leukemia virus
NC	nucleocapsid
PBS	primer binding site
PPT	polypurine tract
PR	protease
R	repeat
RCV	replication competent virus
RSV	Rous sarcoma virus
SIN	self-inactivating
SV40	simian virus 40
VSV-G	vesicular stomatitis virus G protein
U3	unique 3'
U5	unique 5'
ψ	packaging signal in MLV

References

1. Goff, S.P. (2001) In: Knipe, D.M. and Howley, P.M. et al. (eds.) Fields Virology, 4th, Vol. 2. Chapter 57. Lippincott-Raven Publishers, Philadelphia, pp. 1871–1939.
2. Miller, A.D. and Rosman, G.J. (1989) BioTechniques 7, 980–982.
3. Anderson, W.F. (1998) Nature 392, 25–30.
4. Palù, G., Parolin, C., Takeuchi, Y., and Pizzato, M. (2000) Rev. Med. Virol. 10, 185–202.
5. Vogt, V.M. (1997) In: Coffin, J.M., Hughes, S.H. and Varmus, H.E. (eds.) Retroviruses, 4th edn. Chapter 2. Cold Spring Harbor Laboratory Press, Cold Spring Harbor, New York, pp. 27–69.
6. Linial, M.L. and Miller, A.D. (1990) Curr. Top. Microbiol. Immunol. 157, 125–152.
7. Berkowitz, R., Fisher, J., and Goff, S.P. (1996) Curr. Top. Microbiol. Immunol. 214, 177–218.
8. Bender, M.A., Palmer, T.D., Gelinas, R.E., and Miller, A.D. (1987) J. Virol. 61, 1639–1646.
9. Roe, T., Reynolds, T.C., Yu, G., and Brown, P.O. (1993) EMBO J. 12, 2099–2108.
10. Miller, D.G., Adam, M.A., and Miller, A.D. (1990) Mol. Cell. Biol. 10, 4239–4242.
11. Lewis, P.F. and Emerman, M. (1994) J. Virol. 68, 510–516.
12. Watanabe, S. and Temin, H.M. (1982) Proc. Natl. Acad. Sci. USA 79, 5986–5990.
13. Mann, R., Mulligan, R.C., and Baltimore, D. (1983) Cell 33, 153–159.
14. Katz, R.A., Terry, R.W., and Skalka, A.M. (1986) J. Virol. 59, 163–167.
15. Bates, P., Young, J.A., and Varmus, H.A. (1993) Cell 74, 1043–1051.

16. Federspiel, M.J., Bates, P., Young, J.A., Varmus, H.A., and Hughes, S.H. (1994) Proc. Natl. Acad. Sci. USA 91, 11241–11245.
17. Federspiel, M.J., Swing, D.A., Eagleson, B., Reid, S.W., and Hughes, S.H. (1996) Proc. Natl. Acad. Sci. USA 93, 4931–4936.
18. Barsov, E.V. and Hughes, S.H. (1996) J. Virol. 70, 3922–3929.
19. Oh, J., Julias, J.G., Ferris, A.L., and Hughes, S.H. (2002) J. Virol. 76, 1762–1768.
20. Zheng, X.H. and Hughes, S.H. (1999) J. Virol. 73, 6946–6952.
21. Terwilliger, E.F., Godin, B., Sodroski, J.G., and Haseltine, W.A. (1989) Proc. Natl. Acad. Sci. USA 86, 3857–3861.
22. Jamieson, B.D. and Zack, J.A. (1998) J. Virol. 72, 6520–6526.
23. Schmidt, M. and Rethwilm, A. (1995) Virology 210, 167–178.
24. Miller, A.D. (1990) Hum. Gene Ther. 1, 5–14.
25. Linial, M., Medeiros, E., and Hayward, W.S. (1978) Cell 15, 1371–1381.
26. Shank, P.R. and Linial, M. (1980) J. Virol. 36, 450–456.
27. Mann, R., Mulligan, R.C., and Baltimore, D. (1983) Cell 33, 153–159.
28. Watanabe, S. and Temin, H.M. (1983) Mol. Cell. Biol. 3, 2241–2249.
29. Miller, A.D. and Buttimore, C. (1986) Mol. Cell. Biol. 6, 2895–2902.
30. Markowitz, D., Goff, S., and Bank, A. (1988) Virology 167, 400–406.
31. Markowitz, D., Goff, S., and Bank, A. (1988) J. Virol. 62, 1120–1124.
32. Danos, O. and Mulligan, R.C. (1988) Proc. Natl. Acad. Sci. USA 85, 6460–6464.
33. Dougherty, J.P., Wisniewski, R., Yang, S., Rhode, B.W., and Temin, H.M. (1988) J. Virol. 63, 3209–3212.
34. Cosset, F.L., Takeuchi, Y., Battini, J.L., Weiss, R.A., and Collins, M.K. (1995) J. Virol. 69, 7430–7436.
35. Rigg, R.J., Chen, J., Dando, J.S., Forestell, S.P., Plavec, I., and Bohnlein, E. (1996) Virology 218, 290–295.
36. Chong, H. and Vile, R.G. (1996) Gene. Ther. 3, 624–629.
37. Chong, H., Starkey, W., and Vile, R.G. (1998) J. Virol. 72, 2663–2670.
38. Parolin, C., Dorfman, T., Palù, G., Gottlinger, H., and Sodroski, J. (1994) J. Virol. 68, 3888–3895.
39. Naldini, L., Blomer, U., Gallay, P., Ory, D., Mulligan, R., Gage, F.H., Verma, I.M., and Trono, D. (1996) Science 272, 263–267.
40. Corbeau, P., Kraus, G., and Wong-Staal, F. (1996) Proc. Natl. Acad. Sci. USA 93, 14070–14075.
41. Russell, D.W. and Miller, A.D. (1996) J. Virol. 70, 217–222.
42. Kafri, T., Blomer, U., Peterson, D.A., Gage, F.H., and Verma, I.M. (1997) Nat. Genet. 17, 314–317.
43. Corbeau, P., Kraus, G., and Wong-Staal, F. (1998) Gene Ther. 5, 99–104.
44. Poeschla, E., Gilbert, J., Li, X., Huang, S., Ho, A., and Wong-Staal, F. (1998) J. Virol. 72, 6527–6536.
45. Heinkelein, M., Schmidt, M., Fischer, N., Moebes, A., Lindemann, D., Enssle, J., and Rethwilm, A. (1998) J. Virol. 72, 6307–6314.
46. Kafri, T., Van Praag, H., Ouyang, L., Gage, F.H., and Verma, I.M. (1999) J. Virol. 73, 576–584.
47. Wu, M. and Mergia, A. (1999) J. Virol. 73, 4498–4501.
48. Vassilopoulos, G., Trobridge, G., Josephson, N.C., and Russell, D.W. (2001) Blood 98, 604–609.
49. Kim, S.H., Yu, S.S., Park, J.S., Robbins, P.D., An, C.S., and Kim, S. (1998) J. Virol. 72, 994–1004.
50. Emerman, M. and Temin, H.M. (1984) Cell 39, 449–467.
51. Pizzato, M., Franchin, E., Calvi, P., Boschetto, R., Colombo, M., Ferrini, S., and Palù, G. (1998) Gene Ther. 5, 1003–1007.
52. Davies, M.V. and Kaufman, R.J. (1992) J. Virol. 66, 1924–1932.
53. Yu, S.F., von Ruden, T., Kantoff, P.W., Garber, C., Seiberg, M., Ruther, U., Anderson, W.F., Wagner, E.F., and Gilboa, E. (1986) Proc. Natl. Acad. Sci. USA 83, 3194–3198.
54. Choulika, A., Guyot, V., and Nicolas, J.F. (1996) J. Virol. 70, 1792–1798.
55. Ferrari, G., Salvatori, G., Rossi, C., Cossu, G., and Mavilio, F. (1995) Hum. Gene Ther. 6, 733–742.
56. Vile, R.G., Miller, N., Chernajovsky, Y., and Hart, I.R. (1994) Gene Ther. 1, 307–316.

57. Vile, R.G., Diaz, R.M., Miller, N., Mitchell, S., Tuszyanski, A., and Russell, S.J. (1995) Virology 214, 307–313.
58. Diaz, R.M., Eisen, T., Hart, I.R., and Vile, R.G. (1998) J. Virol. 72, 789–795.
59. Fassati, A., Bardoni, A., Sironi, M., Wells, D.J., Bresolin, N., Scarlato, G., Hatanaka, M., Yamaoka, S., and Dickson, G. (1998) Hum. Gene Ther. 9, 2459–2468.
60. Jager, U., Zhao, Y., and Porter, C.D. (1999) J. Virol. 73, 9702–9709.
61. Gossen, M., Freundlieb, S., Bender, G., Muller, G., Hillen, W., and Bujard, H. (1992) Science 268, 1766–1769.
62. Lavillette, D., Russell, S.J., and Cosset, F.-L. (2001) Curr. Opin. Biotechnol. 12, 461–466.
63. Yee, J.K., Friedmann, T., and Burns, J.C. (1994) Methods Cell. Biol. 43, 99–112.
64. Burns, J.C., Friedmann, T., Driever, W., Burrascano, M., and Yee, J.K. (1993) Proc. Natl. Acad. Sci. USA 90, 8033–8037.
65. Takeuchi, Y., Cosset, F.L., Lachmann, P.J., Okada, H., Weiss, R.A., and Collins, M.K. (1994) J. Virol. 68, 8001–8007.
66. Takeuchi, Y., Porter, C.D., Strahan, K.M., Preece, A.F., Gustafsson, K., Cosset, F.L., Weiss, R.A., and Collins, M.K. (1996) Nature 379, 85–88.

S.C. Makrides (Ed.) *Gene Transfer and Expression in Mammalian Cells*

Virus-based vectors for gene expression in mammalian cells: Lentiviruses

Mehdi Gasmi[1] and Flossie Wong-Staal[2],*

[1]*Ceregene, Inc., 9381 Judicial Drive #130, San Diego, CA 92121, USA. Tel.: +1 (858) 458-8828;*
Fax: +1 (858) 458-8801; E-mail: mgasmi@ceregene.com
[2]*Department of Medicine MS0665, University of California San Diego, La Jolla, CA 92093, USA*

1. Introduction

Lentiviruses, represented by the human immunodeficiency virus type 1 (HIV-1) are retroviruses that possess complex genomes and a finely regulated mode of replication. These viruses share the common feature of being able to infect post-mitotic cells of the monocyte/macrophage lineage, a phenomenon closely related to their biology and induced pathologies (reviewed in [1]). Gene therapy vectors based on lentiviruses enable the generation of a gene delivery system that combines features of vectors derived from murine oncoretroviruses (i.e., large coding capacity and stable integration of the transgene into host-cell genetic material) and the capability of transducing non-dividing cells. As such, vectors derived from HIV-1 have been designed and the proof of principle of stable transgene delivery in non-dividing cells has been established in a wide variety of systems. The major hurdle that prevents the progression of the HIV-1-based vector system to the clinic for evaluation of therapeutic potential is the association of the parental virus with an incurable and still largely fatal disease in humans. An alternative approach to the construction of HIV-1-derived vectors, is the use of viruses derived from less pathogenic human immunodeficiency virus type 2 (HIV-2) and from simian immunodeficiency virus (SIV). In addition, non-primate lentiviral vector systems derived from feline immunodeficiency virus (FIV), equine infectious anemia virus (EIAV), and visna/maedi virus (VMV) have also been described ([2], for review see [3]). However, the impressive achievements accomplished with HIV-1-based vectors have largely overshadowed the characterization of these other lentiviral vector systems. Therefore, in this chapter we will focus on HIV-1 derived vectors to exemplify the basic design and improvements relevant to the development of a safe and efficient lentivirus-based gene delivery system.

2. Genetic structure and biology of lentiviruses

In this section, we will briefly review those structural and functional characteristics of lentiviruses that are relevant to the development of lentiviral vectors. We refer the

---

*Current address: Immusol, Inc., 10790 Roselle Street, San Diego, CA 92121, USA.

reader to Chapter 3.12 for general retrovirology, and to a review by Tang *et al.* [4] for extensive details on HIV-1 genetic structure and gene expression regulation.

2.1. Genome, structural proteins and enzymes

The HIV-1 proviral genome relative to other lentiviruses is shown in Fig. 1. The HIV-1 proviral genome is flanked by two LTR, which are structurally divided in U3, R and U5 regions where U3 contains the *cis*-acting elements necessary for transcription. As in all retroviruses, HIV-1 structural and enzymatic proteins are encoded in the *gag*, *pol* and *env* genes. The Gag proteins are translated from the full length genomic RNA into a polyprotein precursor that is processed by the viral protease into three major components: matrix (MA), capsid (CA) and nucleocapsid (NC), which constitute the mature viral particle; and minor peptides p1, p2 and p6, which exert more or less defined functions in the HIV-1 replication cycle. The *pol* ORF is translated as a Gag/Pol polyprotein resulting from a ribosomal frameshift at the 3' end of the *gag* gene. Pol proteins contain various enzymatic activities: reverse transcriptase (RT)/RNAseH, protease (PRO) and integrase (IN). In non-primate lentiviruses, the *pol* gene encodes an additional enzymatic activity, deoxyuridine triphosphatase (dUTPase), which prevents misincorporation of deoxyuridine during DNA synthesis. dUTPase is required for efficient replication of FIV and EIAV in non-dividing cells where nucleotide concentration is low.

The env ORF is expressed from singly spliced mRNAs. The Env product is a heavily glycosylated polyprotein that is matured in the Golgi in two polypeptides of 120 and 41 kDa corresponding to the surface (SU) and transmembrane (TM) subunits, respectively. Surface and TM subunits are associated by non-covalent

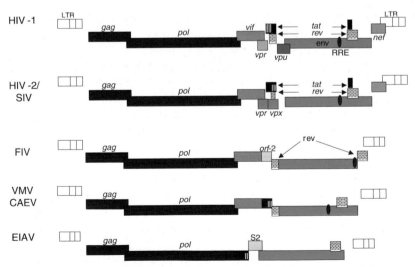

Fig. 1. HIV-1 proviral genome structure relative to other lentiviruses. The relative positions of characterized open reading frames and Rev responsive elements within each genome are shown.

bonds and form oligomers at the surface of HIV-1 virions and infected cells. The *env* gene product is the key determinant in the selection and targeting of infectible cells. Even though the CD4 molecule, long known to be a major cell surface receptor for HIV-1, is present on helper T lymphocytes as well as cells from the monocyte/ macrophage lineage, it is now known that primate lentivirus infection also requires the presence of a co-receptor. The seven-transmembrane chemokine receptors CCR5 and CXCR4 have been identified as coreceptors for infection with macrophage-tropic or lymphotropic HIV-1 virions, respectively. Additional chemokine receptors are also used by HIV-2 and SIV (for review see [5]).

2.2. Regulatory and accessory proteins

Tat and Rev are translated from multiply spliced RNAs and therefore are expressed early in the HIV-1 replication cycle. They participate in the regulation of HIV-1 genome expression at different levels.

The Tat protein activates HIV-1 genome expression by enhancing elongation of nascent viral transcripts. Tat binds to an RNA secondary structure; the Tat responsive element (TAR), which is formed by the first 60 bases of the viral RNA in the R region, and recruits the cyclin T-CDK9 complex that phosphorylates the large subunit of RNA polymerase II necessary to increase enzyme processivity. Other non-primate lentiviruses like the bovine immunodeficiency virus (BIV) and EIAV also encode transactivators, which function through mechanisms similar to that of HIV-1 Tat. On the other hand, the activator proteins of FIV, VMV and that of the caprine arthritis-encephalitis virus (CAEV), behave like conventional transcription factors and interact directly or indirectly with *cis*-elements present in the viral enhancer/ promoter sequences.

The *rev* ORF is found in all lentiviral genomes. Rev triggers the switch between the early and late phases of HIV-1 replication by activating the nuclear export of singly spliced or unspliced viral mRNAs. This process allows expression of the *gag/pol* and *env* genes, and the packaging of the viral genomic RNA. Rev also counteracts the *cis*-acting nuclear retention sequences contained in the viral unspliced mRNA. Rev-responsive RNAs contain a 350 nucleotide-long stem loop structure, the Rev response element (RRE), to which multimers of Rev bind specifically. Structurally, Rev contains both an arginine- and a leucine-rich NES, which allow the shuttling of the protein in and out of the nucleus. The accepted model of Rev nuclear shuttling involves the direct binding of Rev to the karyopherin importin β/Ran-GDP complex enabling its nuclear import. In the nucleus Rev is released from its interaction with importin β and binds to the RRE of viral RNAs. The resulting ribonucleoprotein complexes interact with the nuclear pore complex proteins through exportin-1, triggering their active transport to the cytoplasm (reviewed in [6]).

Although essential for lentiviral *in vivo* infectivity and pathogenicity, the accessory proteins Vif, Vpr, Vpu and Nef of HIV-1 are largely dispensable in virus culture *in vitro* (Table 1; reviewed in [7]) and in vector design.

Table 1

Accessory proteins of HIV-1

Protein	ORF found in:	Function
Vif	All lentiviruses except EIAV	Virion infectivity
Vpr	Primate lentiviruses only	Multifunctional (notably non-dividing cell infection)
Vpu	HIV-1 uniquely	Efficiency of virion release. Downregulation of CD4
Nef	Primate lentiviruses only	Multifunctional (notably CD4 downregulation and apoptosis)

2.3. Redundancy of viral determinants of nuclear import

In HIV-1, the determinants of non-dividing cell infection have been mapped to three viral proteins, namely MA, integrase IN and Vpr, which possess more or less conventional NLS to induce the active nuclear import of HIV-1 PIC. The MA component is dispensable for PIC nuclear import in monocyte-derived macrophages (MDM) [8]. Vpr, which interacts directly with the nuclear pore complex appears to be required for infection of MDM [9]. The only indispensable element for HIV-1 PIC nuclear import in all cell types tested is IN. It has been shown to harbor a non-canonical NLS within its catalytic domain, the mutation of which inhibits nuclear import without impairing the enzyme catalytic function [10]. In addition, this NLS appears to be required both in dividing and non-dividing cells suggesting that translocation of HIV-1 PIC occurs through nucleopores in both cell types. It is not known how these redundant functions are utilized in wild-type HIV-1 infection. In the process of non-dividing cell infection, the leading role of IN is consistent with the fact that non-primate lentiviruses which lack Vpr are nonetheless able to infect macrophages *in vivo* and *in vitro*.

In addition to viral proteins, a nucleotide sequence localized within the IN ORF is implicated in the infection process of non-dividing cells [11]. During reverse transcription, this *cis*-element causes the viral DNA plus strand to be synthesized as two distinct segments. The upstream or U segment originates at the 3' PPT common to all retroviruses, while the other (the downstream or D segment) is primed from a polypurine tract located near the "center" of the viral genome. The U segment synthesis terminates at the CTS located downstream from the central PPT (cPPT), generating a strand displacement of the D segment of 99 nucleotides. The resulting linear 3-stranded DNA structure is referred to as the central DNA flap. Conservative mutations in the cPPT or the CTS sequences severely impair nuclear import of the viral PIC. The central DNA flap is believed to promote the spatial arrangement of the viral PIC into a filamentous shape that would facilitate its penetration through nuclear pores. Sequences more or less homologous to the HIV-1 cPPT/CTS have been found in all lentiviruses, strengthening the hypothesis that the central DNA flap is involved in the capacity of these retroviruses to infect post-mitotic cells. However, apart from HIV-1 the actual central DNA flap synthesis has only been demonstrated in the case of EIAV [12] and FIV [13].

3. HIV-1-derived vector packaging system

Very early versions of HIV-1-based vectors consisted of almost complete viral genomes with partial deletions of either *gag* or *env* genes replaced with reporter gene cassettes (reviewed in [14]). These vectors proved to be valuable tools for the study of HIV-1 infectivity and replication. Conversely, lentiviral vectors designed for gene transfer applications were developed according to the extensive split-genome strategy described for retroviral vectors (Chapter 3.12). The segregation of *cis*- and *trans*-acting elements necessary for vector production offers the advantage of restricting the chances of generating replication-competent viruses by homologous recombination. In addition to addressing safety concerns, the split-genome strategy also offers the advantage of tropism flexibility since heterologous envelopes can be used to pseudotype vector particles (see below). Biosafety is for obvious reasons a particularly sensitive subject in the case of HIV-1-based vectors and therefore, the split-genome strategy had to be enhanced by removing all unnecessary viral sequences for gene transfer that contributed to viral pathogenesis.

Due to the extensive knowledge accumulated on HIV-1 structure and biology, the lentiviral vector technology has been the subject of intensive research in the past few years. These efforts have led to the development of increasingly safer and more potent HIV-1-based vector systems consisting of three or four different constructs. Simultaneously, vectors based on other primate or non-primate lentiviruses have been described, offering potential alternatives to the HIV-1-based system. In the following section we will recapitulate the characteristics of what is now considered the conventional HIV-1 derived lentiviral vector.

3.1. Packaging construct

Similar to the oncoretroviral vector system, the packaging or "helper" construct of lentiviral vectors encodes the *gag* and *pol* genes necessary for the generation of a viral particle. In this construct, gene expression is regulated by a heterologous strong constitutive enhancer/promoter element, usually the human CMV-IE enhancer/promoter, and a heterologous polyadenylation signal such as that of the bovine growth hormone gene. The core-packaging signal and the RNA dimerization domain are removed to avoid packaging interference between the resulting helper transcript and the vector genome in producer cells. The first generation of packaging constructs also encoded the HIV-1 regulatory and accessory proteins. Accessory genes were removed in the second generation of constructs after they were found to be dispensable in a number of applications *in vitro* and *in vivo* [15,16]. Further improvements to the system led to the design of vector constructs whose expression did not rely on the Tat protein. Thus, third generation packaging constructs were built with deletions of the Tat and Rev open reading frames. The Rev protein, still necessary for *gag* and *pol* expression, was supplied from a separate construct, resulting in a conditional packaging system [17].

3.2. Transfer vector

Along with the development of packaging constructs, different versions of HIV-1-derived vector backbones were designed, corresponding to the need for improved biosafety profiles. Today's most widely used HIV-1-derived transfer vector constructs include a minimal amount of HIV-1 derived *cis*-acting sequences flanking the internal expression cassette. These vectors possess a hybrid 5′ LTR where the U3 region has been replaced by a heterologous constitutive enhancer/promoter (RSV U3 region or the CMV-IE promoter). In these hybrid LTRs, the natural position of the HIV-1 TATA box is conserved in order to ensure a proper transcription initiation site. This design allows high levels of vector genomic RNA expression in the producer cells without the need for the Tat protein. The 5′ UTR region is left intact and features the major splice donor and the packaging signal that extends in the 5′ region of the *gag* gene ORF, which is interrupted by a synthetic stop codon for safety purposes. Efficient nuclear export of vector genome in producer cells is ensured by the insertion of the RRE downstream of the *gag* gene sequence. In the third generation construct, the enhancer region in the 3′ LTR has been extensively deleted to generate a self-inactivating (SIN) vector [18,19]. Self-inactivation occurs in the target cells where upon reverse transcription, the inactivated U3 region is copied to the 5′ end of the viral DNA, reconstituting a defective LTR. This is an attractive feature that improves biosafety by further limiting the potential generation of a replication-competent virus by leaving no functional U3 sequence in the vector production system. In addition, the design of a SIN vector obliterates potential cellular gene activation by the viral enhancer following chromosomal integration. SIN retroviruses were previously described for MLV vectors, however, they provided only limited benefit since large deletions in the viral enhancer/promoter resulted in decreased vector titers, or conversely, shorter deletions only partially abolished transcriptional activity. In the case of HIV-1 derived vectors, this modification could be achieved without any major negative effect on vector titer and transduction efficiency. The most recent improvement brought to lentiviral vector system comes from the addition in the backbone of the cPPT/CTS element immediately downstream from the RRE (Fig. 2). Prior to the discovery of the effect of the central DNA flap on the nuclear import of HIV PIC, transduction of non-dividing cells could be obtained with vectors devoid of this structure given the redundancy of viral determinants involved in non-dividing cell infection. There is ample evidence today that these *cis* elements significantly improve transduction efficiency in non-dividing cells [11]. Interestingly cPPT/CTS-containing vectors also showed improved transduction of dividing cells confirming the idea that nuclear import of the PIC occurs through nucleopores in both cell types. Nevertheless, one needs to consider that the insertion of the cPPT/CTS element in HIV-based vector backbones increases the chance for homologous recombination between the packaging and the vector constructs and careful studies are needed to ensure that this does not constitute a potential hazard.

Very similar features are found in the design of other primate and non-primate lentiviral vectors, which are described in detail elsewhere [20–23].

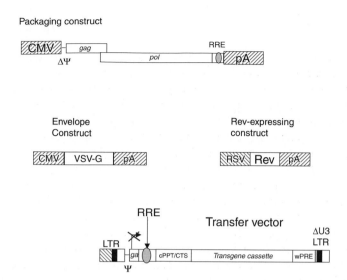

Fig. 2. Diagram showing the most recent 4-construct HIV-1-based vector system (reviewed in [3]). The packaging construct is devoid of all accessory or regulatory protein open reading frames. The Rev protein is encoded on a separate construct. The transfer vector contains the cPPT/CTS sequence element and the wPRE for increased transduction efficiency and transgene expression, respectively (see text for details).

3.3. Requirement for Rev and the RRE

In the quest for an ever-safer vector, the elimination of the Rev/RRE appeared to be an attractive, though challenging goal, since the Rev protein is essential to circumventing negative effects inherent to the gag sequences present in the packaging and vector constructs. Alternatives to the Rev/RRE function exist in simple retroviruses like the Mason-Pfizer monkey virus (MPMV), the simian retrovirus type 1 (SRV-1), the Rous sarcoma virus (RSV) and the spleen necrosis virus (SNV). These viruses possess CTEs to promote the nuclear export of intron-containing viral RNA through interaction with cellular factors in lieu of a Rev-like viral protein. These sequences were found to functionally substitute for RRE/Rev for HIV-1 replication (reviewed in [24]). Unfortunately, the introduction of the MPMV CTE into a HIV-1 vector packaging construct in place of the RRE did not induce enough gag/pol expression and consequently low vector titers were obtained in comparison to the Rev/RRE-dependent systems [25,26].

Early attempts to substitute the RRE by the MPMV CTE in the transfer vector backbone also resulted in decreased vector titers [27]. The same group managed to improve their system by placing the CTE downstream of the transgene ORF in a vector construct devoid of splicing signals [28]. Hybrid vector systems have recently been reported where Rev/RRE and CTE are utilized to promote the nucleocytoplasmic transport of either the packaging or the vector RNAs, respectively [28]. Although still dependent on Rev/RRE, these systems fulfill the purpose of reducing the amount of homologous sequences between the packaging and vector backbones.

Recently, a radical approach to bypass the requirement of Rev/RRE for *gag/pol* expression was described. Investigators synthetically constructed a *gag* gene nucleotide sequence where the majority of codons were optimized for usage in human cells. Codon optimization abolished CRSs and resulted in *gag* expression that was completely independent of any kind of RNA transport element [29]. However, vector production with the modified helper construct still required Rev to be completely efficient given the presence of *gag* sequences in the vector construct that could not be altered. Nonetheless, this approach holds great promise in terms of biosafety since it guards against the generation of replication-competent virus.

3.4. Requirement for accessory proteins

Soon after the first generation of HIV-1 derived vectors was described, the ORFs of HIV-1 accessory proteins Vif, Vpr, Vpu and Nef were readily deleted from the packaging construct in order to increase the biosafety of the vector system [15,16]. Removal of accessory protein ORFs from the system does not hamper HIV-1 vector production. It is rather beneficial since the expression of some of these factors is toxic to producer cells. Moreover, these proteins are not required for efficient transduction of a vast majority of cell types, dividing or growth-arrested, *in vitro* or *in vivo*. Similar observations have been made in non-HIV-1 lentiviral vector systems [21–23]. Consequently, multiply attenuated vectors are being increasingly used in a wide variety of applications with considerable success. Yet, in some instances accessory proteins alone or in combination have been found to enhance transduction efficiency substantially, i.e., Vpr in macrophages, Vif and Vpr in the liver [15,16]. Quiescent peripheral blood lymphocytes are also to various degrees refractory to transduction by vector particles produced in the absence of accessory proteins [30]. Therefore, if HIV-1-derived vectors are to be used for those specific applications, one will have to evaluate the risk–benefit factor of reintroducing one or more of these genes in the vector system in order to obtain satisfactory levels of gene transfer.

4. Lentiviral vector production

Most lentiviral vector preparations are currently made transiently by co-transfection of plasmids encoding vector components (packaging, envelope and vector) into the highly transfectable HEK-293T cell line. This technique allows routine production of 1×10^6 to 1×10^7 TU/ml that can be concentrated 100–1000 times by ultracentrifugation. Although co-transfection is very efficient and relatively easy to handle for bench-scale research and development purposes, it becomes costly and difficult to scale up in order to meet the quantity and quality requirements imposed by clinical applications. Large-scale vector production would be more easily achieved with packaging cell lines (PCLs) and better yet with vector producing cell lines (VPCLs). PCLs are engineered to stably express vector components from the packaging construct, and offer the advantage of avoiding possible recombination between the different plasmids that can occur in multiply transfected cells. Recombinant vector production from PCLs is ensured by transfection of these cells with the vector DNA. VPCLs are generated by stable integration of provector genome into packaging cells. This is optimally achieved by multiple rounds of transduction of PCLs by vector particles as described by Sheridan *et al.* in the case of an oncoretroviral vector encoding the human clotting factor VIII cDNA [31].

The development of an HIV-1 lentiviral PCL has been greatly hampered by the cytotoxic effects of constitutive expression of certain viral proteins such as Vpr, Rev, and PRO. In addition, the pantropic vesicular stomatitis virus G envelope (VSV-G), often used to pseudotype lentiviral vector particles, is also cytotoxic. Limited success has been obtained with HIV-1 PCLs where expression of the viral components was derived from the viral promoter [32]. With the use of inducible expression systems such as the tetracycline-regulated promoter, major improvements have been made in the design of HIV-1 vector PCLs [33,34].

SIN lentiviral vectors preclude the generation of VPCLs by transduction. Taking advantage of the conditional nature of the tetracycline inducible system, Xu *et al.* have generated VPCLs where vector genome expression is placed under the control of a tetracycline inducible promoter, conserving the SIN properties of the vector in target cells [35]. These recent achievements constitute significant advances towards the use of lentiviral vectors in gene therapy.

5. Pseudotyped vectors

As discussed above, the most commonly used heterologous virus envelope is the G protein of VSV. This protein binds to phosphatidylserine molecules of cell membranes of the animal realm, from insect to human cells. Most experiments reported to date describe VSV-G pseudotyped lentiviral vectors that are able to transduce fibroblast cell lines, primary hematopoietic stem cells *in vitro* as well as muscle tissue, liver, and central nervous system *in vivo*. A potential shortcoming of VSV-G is that it was found to be inactivated by human complement *in vitro* [36]. If this holds true *in vivo*, it will pose a serious problem for gene therapy applications that require injection of vectors into the patients' bloodstream.

The amphotropic envelope of MLV, 1040A (Ampho) is another viral glycoprotein that confers broad tropism to pseudotyped vector particles. The receptor for this protein is a phosphate transporter, Pit-2, that is sufficiently homologous among various species to interact with the Ampho protein and mediate virus entry in a wide variety of cell types. However, Ampho-pseudotyped vector titers are routinely found to be 10-fold lower than those obtained with VSV-G pseudotypes either in MLV or HIV-1 vector systems. This observation is related to the limited amount of Pit-2 molecules at the surface of target cells. Ampho does not present the drawbacks of VSV-G regarding pseudo-transduction and sensitivity to human complement [36]. However, the Ampho protein is prone to degradation and therefore precludes vector concentration by ultra-centrifugation as in the case of retroviral vectors [37]. Yet, Reiser recently obtained concentrated (10^8 TU/ml) Ampho pseudotyped-HIV-1 vector preparations by ultra-centrifugation, suggesting that the environment at the surface of a lentiviral particle enhances the stability of the Ampho protein complex [38]. Alternatively, Ampho-pseudotyped lentiviral particles can be concentrated by precipitation with calcium phosphate [39].

Unlike retroviral vectors, lentiviral particles fail to incorporate another amphotropic glycoprotein, the gibbon ape leukemia virus (GaLV) envelope. This protein is of particular interest since its receptor, the phosphate transporter Pit-1, is largely expressed at the surface of hemopoietic stem cells, resulting in efficient transduction of these cells by GaLV-retroviral vectors. Replacement of portions of the GaLV protein C-terminal region by that of the Ampho envelope allows pseudotyping of HIV-1 particles, although inefficiently [40].

Lentiviral particles can be pseudotyped by other envelopes, such as the lymphocytic choriomeningitis virus glycoprotein, the receptor of which is expressed in most tissues [41] and the glycoproteins of rabies and Mokola lyssaviruses that exhibit plciotropic tropisms *in vitro* but are mostly neurotropic *in vivo* [21]. Among tissue specific envelopes, that of HIV-1 naturally comes to mind to restrict transduction to CD4/coreceptor positive cells. Although these cells represent a very interesting target in many gene therapy applications, investigators have avoided using the HIV-1 envelope for fear of increasing the chances of generating replication-competent viruses.

6. Expression from lentiviral vectors

Compared with their oncoretroviral relatives, lentiviral vectors have expanded the number potential applications for gene transfer by permitting transduction of post-mitotic cells and terminally differentiated tissues. However, these vectors face the same challenge as the oncoretroviral or any gene transfer systems, i.e., how much transgene expression can be achieved and for how long?

In most gene transfer applications, gene expression from lentiviral vectors is driven by an internal cassette for two main reasons. First, lentiviral promoters are rather inefficient in human cells or require the presence of a viral transactivator. Second, a SIN vector configuration is necessary to address biosafety concerns. Fortunately, the introduction of heterologous regulatory enhancer/promoter sequences within the lentiviral vector construct does not hamper the capacity to obtain high titer vector preparations, unlike the case in MLV vectors where the viral and internal promoters are found to interfere with one other.

6.1. Promoters

The choice of the internal promoter is obviously driven by the particular gene transfer application. In early studies, expression of the reporter genes *lacZ* and GFP were placed under the control of the constitutive CMV-IE promoter, which functions in various cell types *in vitro* and *in vivo* [42,43]. Examples of other constitutive promoters that have been used in the lentiviral vector context include the phosphoglycerate kinase promoter [17] and the SV40 immediate early promoter [28]. Tissue specific promoters and enhancers have also been used for targeted gene expression [44,45]

6.2. Enhancers of gene expression

Virtually all lentiviral vectors currently used feature the woodchuck hepatitis B virus post-transcriptional regulatory element (wPRE). This element has been found to enhance expression of autologous and heterologous intron-containing mRNAs [46]. The RNA sequence of wPRE forms a secondary structure reminiscent of the lentiviral RREs. The wPRE is believed to increase the post-transcriptional modification of mRNA that in turn increases translation [47]. The use of the wPRE in HIV-1 vector packaging constructs as an alternative to the Rev/RRE system did not induce satisfactory *gag/pol* expression levels [26]. However, when introduced in the vector backbone downstream of the gene of interest, this sequence increases expression three- to fivefold, irrespective of the vector system used [47].

Another way to increase gene expression is to introduce an intron in the expression cassette, as spliced mRNAs are more efficiently expressed than their intronless counterparts due to the fact that the splicing machinery is coupled to the cellular mRNA nucleocytoplasmic export system. Introns have been efficiently introduced into lentiviral vector constructs via intron-containing promoters, such as the human translation elongation factor 1 alpha promoter (EF1α) and the multi-hybrid CAG enhancer/promoter [48]. In vector producer cells, the Rev/RRE system of lentiviruses guarantees the preservation of introns in the vector genomic RNA. Intron-containing cassettes induce stronger levels of transgene expression from lentiviral vectors relative to those obtained with the CMV enhancer promoter notably in human hematopoietic stem cells [48].

6.3. Gene silencing

The success of gene therapy has been hampered by the difficulty of ensuring stable *in vivo* gene expression. Numerous studies using retroviral vectors repeatedly demonstrated epigenetic silencing of gene expression (see Chapter 3.12). To avoid gene silencing, genetic elements such as locus control regions (LCR) (see Chapter 12) have been introduced into these vector backbones. To date, gene expression silencing has not been reported with the use of lentiviral vectors either *in vitro* or *in vivo*. Expression from lentiviral vectors has been observed *in vivo* for up to 8 months in the brain [49] and 12 months in the liver [44]. The discrepancy with oncoretroviral vectors is likely to be due to the difference in the nature of lentiviral vector genome integration sites. Nevertheless, LCRs, which also confer tissue-specific, site-independent and transgene copy number-dependent expression have been used in lentiviral vectors [50,51].

6.4. Regulation of gene expression

The capacity to control timing and level of gene expression in target cells would constitute a definite advantage for certain gene therapy applications. Various gene expression regulatable systems have been described, among which the tetracycline inducible or repressible promoters (see Chapter 22). Lentiviral vectors containing the tetracycline-inducible system have been used to confer tetracycline-dependent expression of GFP *in vitro* and *in vivo* in the rat CNS [52]. Ironically, the HIV-1 Tat and Rev offer another means to achieve regulated expression for the particular application of anti-HIV-1 gene therapy. Genes placed under the control of the viral LTR in Rev/RRE dependent context would be activated when cells express Tat and Rev upon infection with wild type virus [53].

7. Gene transfer applications

It would be impossible to mention herein all the studies achieved to date reporting successful gene transfer with lentiviral vectors. At the time this review was written, no tissue was found to be completely refractory to transduction by these vectors. Pre-clinical studies have shown the potential therapeutic potency of lentiviral vectors in animal models of Parkinson's disease [49] and β-thalassemia [50]. In addition, long-term liver-specific expression of human clotting factor IX from a lentiviral vector has also been reported in SCID mice [44].

8. Conclusion

At the dawn of the new century, labeled by some as the "century of molecular medicine", there is no question that gene therapy of genetic and acquired diseases will involve the use of lentiviral vectors, most likely derived from HIV-1. To be convinced, one just needs to consider the number of laboratories in academia and industry involved in the development of such vectors for therapeutic applications. In addition, the increasing number of successful pre-clinical studies foretells the imminent emergence of lentiviral vectors into the clinic. Nevertheless, achievement of scientific milestones will not diminish the efforts needed to appropriately educate clinicians and lay persons in order to reach the acceptance of HIV-1 derived vectors as therapeutic tools. It is possible that other lentiviral vectors, notably those derived from non-primate lentiviruses, have a role to play in this process. However, it is unclear at present whether they constitute a safer alternative to HIV-1 derived vectors.

Acknowledgements

We thank Dr. Justine Cunningham and Dr. Dominick Vacante for critical review of the manuscript.

Abbreviations

CMV-IE   cytomegalovirus immediate-early promoter
CNS   central nervous system
CTE   constitutive transport element

CRS   *cis*-repressive sequence
CTS   central termination sequence
GFP   green fluorescent protein
LTR   long terminal repeat
NES   nuclear export signal
NLS   nuclear localization signal
ORF   open reading frame
PGK   phosphoglycerate kinase
PIC   pre-integration complex
PPT   polypurine tract
SCID   severe combined immunodeficiency
SIN   self-inactivating
TU   transducing unit

References

1. Joag, S., Stephens, E., and Narayan, O. (1996) In: Fields, B. et al. (eds.) Virology, 3rd ed., Vol. 2. Lippincott-Raven, Philadelphia, PA, USA, pp. 1977–1996.
2. Poeschla, E.M., Wong-Staal, F., and Looney, D.J. (1998) Nat. Med. 4, 354–357.
3. Pfeifer, A. and Verma, I.M. (2001) Annu. Rev. Genomics Hum. Genet. 2, 177–211.
4. Tang, H., Kuhen, K.L., and Wong-Staal, F. (1999) Annu. Rev. Genet. 33, 133–170.
5. Simmons, G., Reeves, J.D., Hibbitts, S., Stine, J.T., Gray, P.W., Proudfoot, A.E., and Clapham, P.R. (2000) Immunol. Rev. 177, 112–126.
6. Hope, T.J. (1999) Arch. Biochem. Biophys. 365, 186–191.
7. Bour, S. and Strebel, K. (2000) Adv. Pharmacol. 48, 75–120.
8. Fouchier, R.A.M., Meyer, B.E., Simon, J.H.M., Fischer, U., and Malim, M.H. (1997) EMBO J. 16, 4531–4539.
9. Vodicka, M.A., Koepp, D.M., Silver, P.A., and Emerman, M. (1998) Genes Dev. 12, 175–185.
10. Bouyac-Bertoia, M., Dvorin, J.D., Fouchier, R.A., Jenkins, Y., Meyer, B.E., Wu, L.I., Emerman, M., and Malim, M.H. (2001) Mol. Cell 7, 1025–1035.
11. Zennou, V., Petit, C., Guetard, D., Nerhbass, U., Montagnier, L., and Charneau, P. (2000) Cell 101, 173–185.
12. Stetor, S.R., Rausch, J.W., Guo, M.J., Burnham, J.P., Boone, L.R., Waring, M.J., and Le Grice, S.F. (1999) Biochemistry 38, 3656–3667.
13. Whitwam, T., Peretz, M., and Poeschla, E. (2001) J. Virol. 75, 9407–9414.
14. Buchschacher, G.L., Jr. and Wong-Staal, F. (2000) Blood 95, 2499–2504.
15. Zufferey, R., Nagy, D., Mandel, R.J., Naldini, L., and Trono, D. (1997) Nat. Biotechnol. 15, 871–875.
16. Kafri, T., Blomer, U., Peterson, D.A., Gage, F.H., and Verma, I.M. (1997) Nat. Genet. 17, 314–317.
17. Dull, T., Zufferey, R., Kelly, M., Mandel, R.J., Nguyen, M., Trono, D., and Naldini, L. (1998) J. Virol. 72, 8463–8471.
18. Miyoshi, H., Blomer, U., Takahashi, M., Gage, F.H., and Verma, I.M. (1998) J. Virol. 72, 8150–8157.
19. Zufferey, R., Dull, T., Mandel, R.J., Bukovsky, A., Quiroz, D., Naldini, L., and Trono, D. (1998) J. Virol. 72, 9873–9880.
20. Poeschla, E., Gilbert, J., Li, X., Huang, S., Ho, A., and Wong-Staal, F. (1998) J. Virol. 72, 6527–6536.
21. Mitrophanous, K., Yoon, S., Rohll, J., Patil, D., Wilkes, F., Kim, V., Kingsman, S., Kingsman, A., and Mazarakis, N. (1999) Gene Ther. 6, 1808–1818.
22. Johnston, J.C., Gasmi, M., Lim, L.E., Elder, J.H., Yee, J.K., Jolly, D.J., Campbell, K.P., Davidson, B.L., and Sauter, S.L. (1999) J. Virol. 73, 4991–5000.

23. Negre, D., Mangeot, P.E., Duisit, G., Blanchard, S., Vidalain, P.O., Leissner, P., Winter, A.J., Rabourdin-Combe, C., Mehtali, M., Moullier, P., Darlix, J.L., Cosset, F.L. (2000) Gene Ther. 7, 1613–1623.

24. Hammarskjold, M.L. (2001) Curr. Top. Microbiol. Immunol. 259, 77–93.

25. Kim, V.N., Mitrophanous, K., Kingsman, S.M., and Kingsman, A.J. (1998) J. Virol. 72, 811–816.

26. Gasmi, M., Glynn, J., Jin, M.J., Jolly, D.J., Yee, J.K., and Chen, S.T. (1999) J. Virol. 73, 1828–1834.

27. Mautino, M.R., Ramsey, W.J., Reiser, J., and Morgan, R.A. (2000) Hum. Gene Ther. 11, 895–908.

28. Mautino, M.R., Keiser, N., and Morgan, R.A. (2000) Gene Ther. 7, 1421–1424.

29. Kotsopoulou, E., Kim, V.N., Kingsman, A.J., Kingsman, S.M., and Mitrophanous, K.A. (2000) J. Virol. 74, 4839–4852.

30. Chinnasamy, D., Chinnasamy, N., Enriquez, M.J., Otsu, M., Morgan, R.A., and Candotti, F. (2000) Blood 96, 1309–1316.

31. Sheridan, P.L., Bodner, M., Lynn, A., Phuong, T.K., DePolo, N.J., de la Vega, D.J. Jr., O'Dea, J., Nguyen, K., McCormack, J.E., Driver, D.A., Townsend, K., Ibañez, C.E., Sajjadi, N.C., Greengard, J.S., Moore, M.D., Respess, J., Chang, S.M.W., Dubensky, T.W. Jr., Jolly, D.J., and Sauter, S.L. (2000) Mol. Ther. 2, 262–275.

32. Corbeau, P., Kraus, G., and Wong-Staal, F. (1996) Proc. Natl. Acad. Sci. USA 93, 14070–14075.

33. Kafri, T., van Praag, H., Ouyang, L., Gage, F.H., and Verma, I.M. (1999) J. Virol. 73, 576–584.

34. Farson, D., Witt, R., McGuinness, R., Dull, T., Kelly, M., Song, J., Radeke, R., Bukovsky, A., Consiglio, A., Naldini, L. (2001) Hum. Gene Ther. 12, 981–997.

35. Xu, K., Ma, H., McCown, T.J., Verma, I.M., and Kafri, T. (2001) Mol. Ther. 3, 97–104.

36. DePolo, N.J., Reed, J.D., Sheridan, P.L., Townsend, K., Sauter, S.L., Jolly, D.J., and Dubensky, T.W., Jr. (2000) Mol. Ther. 2, 218–222.

37. Burns, J., Friedmann, T., Driever, W., Burrascano, M., and Yee, J. (1993) Proc. Natl. Acad. Sci. USA 90, 8033–8037.

38. Reiser, J. (2000) Gene Ther. 7, 910–913.

39. Pham, L., Ye, H., Cosset, F.L., Russell, S.J., and Peng, K.W. (2001) J. Gene Med. 3, 188–194.

40. Stitz, J., Buchholz, C.J., Engelstadter, M., Uckert, W., Bloemer, U., Schmitt, I., and Cichutek, K. (2000) Virology 273, 16–20.

41. Beyer, W.R., Westphal, M., Ostertag, W., and von Laer, D. (2002) J. Virol. 76, 1488–1495.

42. Naldini, L., Blomer, U., Gallay, P., Ory, D., Mulligan, R., Gage, F.H., Verma, I.M., and Trono, D. (1996) Science 272, 263–267.

43. Kafri, T., Blömer, U., Peterson, D.A., Gage, F.H., and Verma, I.M. (1997) Nat. Genet. 17, 314–317.

44. Follenzi, A., Sabatino, G., Lombardo, A., Boccaccio, C., and Naldini, L. (2002) Hum. Gene Ther. 13, 243–260.

45. Yu, D., Chen, D., Chiu, C., Razmazma, B., Chow, Y.H., and Pang, S. (2001) Cancer Gene Ther. 8, 628–635.

46. Donello, J.E., Loeb, J.E., and Hope, T.J. (1998) J. Virol. 72, 5085–5092.

47. Zufferey, R., Donello, J.E., Trono, D., and Hope, T.J. (1999) J. Virol. 73, 2886–2892.

48. Ramezani, A., Hawley, T.S., and Hawley, R.G. (2000) Mol. Ther. 2, 458–469.

49. Kordower, J.H., Emborg, M.E., Bloch, J., Ma, S.Y., Chu, Y., Leventhal, L., McBride, J., Chen, E.Y., Palfi, S., Roitberg, B.Z., Brown, W.D., Holden, J.E., Pyzalski, R., Taylor, M.D., Carvey, P., Ling, Z., Trono, D., Hantraye, P., Deglon, N., and Aebischer, P. (2000) Science 290, 767–773.

50. May, C., Rivella, S., Callegari, J., Heller, G., Gaensler, K.M., Luzzatto, L., and Sadelain, M. (2000) Nature 406, 82–86.

51. Kowolik, C.M., Hu, J., and Yee, J.K. (2001) J. Virol. 75, 4641–4648.

52. Kafri, T., van Praag, H., Gage, F.H., and Verma, I.M. (2000) Mol. Ther. 1, 516–521.

53. Poeschla, E., Corbeau, P., and Wong-Staal, F. (1996) Proc. Natl. Acad. Sci. USA 93, 11395–11399.

S.C. Makrides (Ed.) *Gene Transfer and Expression in Mammalian Cells*

CHAPTER 4

Methods for DNA introduction into mammalian cells

Pamela A. Norton[1] and Catherine J. Pachuk[1,2]

[1]*Department of Biochemistry and Molecular Pharmacology, Jefferson Center for Biomedical Research, Thomas Jefferson University, Doylestown, PA 18901, USA; Tel.: +1-215-489-4903; Fax: +1-215-489-4920; E-mail: pamela.norton@jefferson.edu*
[2]*Nucleonics, Malvern, PA 19355, USA; Tel.: 215-489-4918; Fax: 215-489-4920; E-mail: Cathydsrna@aol.com*

1. Introduction

The development of methods for the introduction of DNA into cultured cells, especially mammalian cells, has proceeded in parallel with advances in molecular cloning techniques. The process of introducing DNA into vertebrate cells is generally referred to as transfection, although some authors have referred to the same process as transformation, by analogy with DNA transfer in prokaryotes. In this chapter, the term transfection will be used to avoid confusion, as transformation can have a distinct meaning with regard to the growth and the morphologic state of mammalian cells.

Ideally, a transfection efficiency of 100% (all target cells acquire and express the introduced DNA) associated with minimal toxicity (all cells survive the procedure) is desired. Although a number of transfection methods have been described, virtually all fall short of these ideals. Increased transfection efficiency often correlates with increased toxicity, necessitating a tradeoff. Despite continual refinements, several different transfection methods remain in common use, as optimal procedures vary from one cell type to another. In addition, certain applications result in a preference for specific procedures. This chapter reviews the most commonly encountered methods, with consideration of their relative merits both in principle and in practice. In addition, the principal barriers to successful transfection are identified and discussed, as further research in this area will likely lead to improved methods in the future.

Transfection methods can be grouped simplistically into three categories: physically mediated delivery, chemically mediated delivery and biological vector-mediated delivery of nucleic acid. Physical methods include electroporation, particle bombardment ("gene guns") and direct microinjection; all permit DNA entry into the cell by temporarily breaching the integrity of the plasma membrane. Chemical facilitation of DNA uptake includes the use of agents such as inorganic calcium phosphate, positively charged polymers such as DEAE dextran and mixtures of lipids, often referred to as liposomes.

Although mammalian cells have been transfected with nucleic acids from a variety of sources, including total genomic DNA and RNA, the vast majority of experiments involve DNA sequences that have been subcloned and propagated in *Escherichia coli*. Before initiating transfection studies, it should be noted that highly purified DNA tends to produce the best results, in terms of both efficiency and reproducibility. In particular, prokaryotic and bacteriophage DNA was noted to be somewhat toxic to many eukaryotic cells [1]. This observation likely is due in part to the carryover of the endotoxin lipopolysaccharide that can contaminate plasmid preparations used for transfection purposes [2], despite the use of standard DNA purification methods including CsCl gradient centrifugation and various types of ion exchange chromatography. Some workers subject the DNA sample to precipitation with ethanol and resuspension under sterile conditions. In our experience this is not necessary, but in the event of bacterial contamination of experiments, such a pretreatment might be considered.

2. Barriers to successful transfection

It is convenient to think about gene transfer in terms of the various obstacles that the DNA must overcome in order to be successfully expressed in the target cell. Although the complexes formed by various cationic and lipid-based reagents may be chemically distinct, certain features of DNA delivery appear to be common. First, the DNA is complexed with the chemical agent, which should condense the DNA, protect it from degradation and facilitate interaction with the negatively charged cell surface. Next, the DNA must cross the plasma membrane and enter the cell; this appears to occur most frequently by endocytosis. Following internalization via endocytosis, the DNA must be released from the vesicles and avoid degradation by lysosomes. In addition, the DNA must escape degradation in the cytoplasm. Finally, the DNA must enter the nucleus, either as naked DNA or in the form of a complex. Most measures of transfection efficiency do not distinguish which of these steps are potentially rate limiting. However, a review of what is known about DNA complex formation and transfection mechanisms offers some insight into these events.

2.1. DNA condensation

DNA condensation is a compaction process by which the inter-atomic distances are minimized between adjacent nucleotides. Condensation of DNA is a prerequisite to the uptake of DNA by cells, as condensed DNA occupies less space, there are more molecules per unit of space, and there are fewer size-related restrictions on cell uptake. Although the major barrier to condensation is electrostatic repulsion, resistance is also contributed by other unfavorable free energies. The relative contributions of the unfavorable free energies vary by several orders of magnitude [3,4]. In supercoiled DNA, the free energy barriers are temporarily overcome for a subset of sequences, yielding loops of single-stranded DNA, with the result that multiple topological forms exist simultaneously.

The potent electrostatic barriers to DNA condensation can be overcome through the use of cations. Charge neutralization of DNA by cations relieves electrostatic repulsion, allowing DNA condensation to occur [5,6]. This free energy change causes a molecular collapse of plasmid DNA resulting in a five to sixfold decrease in the hydrodynamic radii of the DNA molecules, with a concomitant 2- to 3-orders of magnitude decrease in volume occupancy [7,8]. Condensation of DNA is achieved through the use of cationic lipids or cationic amphiphiles, ion-pair reagents that have both lipophilic and hydrophilic properties. Solubility of the lipid in aqueous solution is dependent upon the structure of the specific hydrophilic and lipophilic groups. Depending upon the hydrophobic composition of the molecule, cationic lipids and amphiphiles can adopt a number of different liposomal structures. In polar solvents, a single chain lipid with a cationic headgroup will assemble into micelles containing hydrophobic cores while certain lipids containing two or more hydrophobic chains can form bilayered vesicles [9]. A more complex liposomal structure is formed when a co-lipid is added. The multi-lipid interactions that occur through the addition of co-lipids bring about local rigidity in the otherwise fluid environment of mono-lipid liposomes, further increasing the asymmetry of the complex and also causing a reassortment of bilayers if present and the possible creation of bilayers where originally there were none. Particles with distinctly different biophysical properties from the mono-lipid liposomes are thereby generated.

The structure of DNA–lipid complexes has been investigated, and plausible models for such interactions have been discussed in detail [10]. It has been proposed that a negatively charged plasmid DNA molecule can interact with the surface of one or more positively charged spherical liposomes [11]. However, DNA may also become sandwiched between membrane bilayers [12]. Studies with cationic lipids support the latter configuration for DNA:lipid charge ratios similar to those typically used for DNA transfection, although at lower lipid levels, less compact associations were observed [13]. Indeed, the topology of the lipid/DNA complexes can vary dramatically depending on both the amounts and the composition of the lipids. Differences in these properties may impact the ability of these particles to be taken up by cells [14]. DOPE tends to form a non-lamellar, two dimensional hexagonal phase under physiological conditions [15]. Mixtures of cationic DOTAP and DOPE can condense DNA into a columnar hexagonal array, in which the cationic groups are within the interior of the structure in close association with the DNA [16]. In contrast, the less effective neutral lipid DOPC packages the DNA into a multilamellar structure in which the DNA is trapped between lipid bilayers. These phase transitions have important functional ramifications. DOPE, but not DOPC, can improve transfection efficiency when mixed with polycationic lipids, even if supplied as a separate "helper" liposome [17–19]. The presence of some neutral co-lipids such as DOPE or cholesterol appear to increase transfection capabilities through processes hypothesized to invoke endosomal evasion or escape following cellular uptake [20].

2.2. DNA uptake by cells

The mechanism by which $CaPO_4$ facilitates DNA uptake is understood in part. DNA is entrapped by co-precipitation with the $CaPO_4$, which renders it relatively resistant to nucleases such as those present in serum. The DNA is taken up by the cells in the form of the precipitate, apparently by endocytosis of the aggregates [21].

Similarly, cationic polymers such as PEI interact with the DNA, condensing it and reducing the electrostatic repulsion between the negatively charged DNA and the similarly charged cell surface, permitting endocytosis [22]. Complexes of DNA with polyamidoamine dendrimers also appear to be taken up via endocytosis [23]. The lipid agents, especially cationic lipids, reduce charge repulsion by compacting the DNA [13]. Cationic liposomes bind to the cell surface, apparently via ionic interactions, and are generally taken up by the cells via non-receptor-mediated endocytosis [18,19,24–26]. Thus, although the complexes formed by various cationic and lipid reagents may be distinct, they all appear to promote uptake largely via endocytic processes. These processes appear to be relatively non-specific.

Cellular uptake and nuclear localization of DNA follow the rules of mass action. Maximum bioavailability of DNA can only be realized through active transfection processes, which can be achieved through the use of targeted DNA delivery systems, including receptor-mediated endocytosis. Targeted DNA delivery systems are biochemical approaches involving the use of ligands that interact with and bind to cell surfaces. These ligands can be complexed with DNA through cationic molecules [27,28] or through oligonucleotides that form triplex strands with DNA. DNA can be targeted to specific cell types through the use of a ligand or an antibody specific for that cell type [29–31]. However, DNA uptake does not appear to be the major barrier to optimal transfection efficiency [26], suggesting that the additional steps needed to prepare the ligand-containing complexes are unnecessary for most cells. Nevertheless, receptor-mediated targeting may prove to be valuable for efficient cell-type-specific delivery *in vivo*.

The fate of the DNA once inside the cell also must be considered. Cationic lipids are able to mediate delivery in up to 100% of treated cells, but subsequent barriers prevent most of these cells from actually expressing the transgene [32]. Similarly, most cells transfected using $CaPO_4$-mediated delivery acquire cytoplasmic DNA, suggesting that uptake is not limiting [21]. However, only a small fraction of the intracellular DNA actually enters the nucleus, with most DNA remaining in endosomal vesicles. It is possible that this vesicular DNA can be delivered to the nucleus via fusion with the nuclear envelope [24]. Alternatively, DNA may need to be released from the endosome into the cytoplasm prior to nuclear import. It has been proposed that anionic lipids play a role in the release of DNA from endocytosed cationic liposomes by neutralizing the positive charge of the latter [33]. Several lines of evidence point to a role for the neutral lipid, DOPE, in DNA release from endosomes following cationic lipid-mediated transfection, as discussed in a later section.

2.3. Nuclear entry

Ultimately, DNA must reach the nucleus, whether free or as a complex. Microinjection of naked DNA into the cytoplasm results in a much lower level of transfection efficiency than injection directly into the nucleus [26,34]. Another study suggests that degradation of naked DNA in the cytosol by resident nucleases represents a significant barrier to DNA delivery to the nucleus [35]. Thus, retention of the DNA in a complex with the facilitating polycations or lipids might be advantageous, offering protection from degradation. Consistent with this possibility, injection of PEI–DNA complexes into the cytoplasm resulted in higher levels of reporter gene expression than injection of naked DNA [36]. The direct injection of PEI–DNA complexes into the nucleus demonstrated that the polycation did not interfere with gene expression. In contrast, intranuclear injection of DNA complexed with either the cationic lipid DOTAP or the lipopolyamine DOGS interfered with gene expression [26,36]. Thus, the intracellular pathways that DNA takes after endocytic internalization might not be identical for all transfection facilitators, and cell type-specific differences in transfection efficiency might reflect in part differences in intracellular trafficking patterns and or kinetics.

Dependence upon cell cycle may also be important as has been reported recently for DNA delivered with PEI or by electroporation [37]. Most transfection protocols suggest that cells be in an actively growing state. The breakdown of the nuclear envelope that occurs during mitosis might facilitate DNA entry into the nucleus [26]. Recent studies also support a correlation between cell proliferation and increased efficiency of cationic lipid-stimulated transfection [38]. As a means to improve DNA transport to the nucleus, peptides that promote nuclear localization have been incorporated into lipoplexes [39] (see Chapter 5). Also, nuclear localization signals have been successfully linked to DNA through peptide nucleic acids (PNAs) that enable the formation of triplex structures with plasmid DNA molecules or modified lysines [40,41]. Such procedural refinements likely will continue to improve transfection efficiencies.

In summary, many gaps remain in our understanding of the mechanisms whereby cells take up exogenous DNA. Although the above discussion has focused on the chemical transfection methods, it is obvious that the physical methods are subject to a subset of the same limitations. Interest in gene therapy applications will surely serve to propel further research into the processes involved in intracellular gene delivery, which should benefit all who are interested in using transfection methods.

3. Comparison of available methods

Selection of the optimal transfection method remains an empirical process, as the efficiency of a procedure tends to be highly dependent on the cell type to be used. When planning to transfect a given cell type for the first time, reference to the literature and/or experienced colleagues should be the first step, especially for commonly used cell types. However, the same cell line might behave quite differently from one laboratory to another due to subtle differences in routine culture conditions, necessitating modifications of procedures established in other laboratories. Nevertheless, most cell lines can be transfected successfully, if not optimally, with one or more of the methods described. In the absence of any prior information regarding the optimal transfection method for a given cell type, or if the intended target is a primary cell type, an initial comparison of several procedures is warranted.

Other considerations to weigh include cost and convenience; many commercial reagents and kits are very convenient but can prove expensive over time. Certain procedures (e.g., electroporation) require a specialized apparatus, and accessibility might prove limiting. Another factor is that many cells do poorly in the absence of serum for any length of time, whereas a number of the commercially available lipid reagents are most effective in the absence of serum. The more information and experience a research team has with handling and culturing the cell type of interest, the easier it will be to choose an appropriate transfection method, and the greater the possibility of success. A description of the origin and usage considerations for each of the major methods follows. Step-by step protocols for many of these methods can be found in a recent chapter [42].

3.1. Polycation-mediated DNA uptake

The earliest report of polycation-mediated DNA uptake employed the derivatized polysaccharide DEAE-dextran. Extrapolating from the observation that polycations stimulate infectivity of nucleic acid isolated from RNA viruses, McCutchan and Pagano demonstrated that DEAE-dextran facilitated the uptake of viral DNA by mammalian cells [43]. Efficiency was dependent on both the concentration and the chain length of the dextran. Brief treatment of cells with either DMSO or glycerol after exposure to DNA further enhanced transfection efficiency [44,45]. Other polycations, such as protamine or PEI, a branched polymer, also can be used to introduce DNA into cells [46]. PEI, and probably other polycations, appear to both condense DNA via ionic interactions and promote interaction with the negatively charged cell surface [47]. Following endocytic uptake (see Section 2.2), the excess of protonatable nitrogens likely serves to buffer the lysosomal compartments, decreasing DNA degradation. Although the use of these agents is similar to DEAE dextran (DNA is combined with the polyamine under physiological salt concentrations), the optimal ratio of polyamine to DNA will vary, in part due to the precise charge ratio of the cationic agent to the negatively charged phosphates of the nucleic acid. Some reports suggest that the combination of polycations with lipids may improve delivery when compared with either substance alone (Section 3.3). Specifically, delivery of intact yeast artificial chromosome DNA was improved when the DNA was complexed with PEI prior to lipofection [48].

Polyamidoamine dendrimers are complex polycations consisting of branches with terminal charged amino groups radiating from a central core, forming a rough sphere [23,49]. The DNA is compacted within the dendrimer structure, and the net positive charge is thought to both facilitate association with the plasma membrane as well as to buffer lysosomal pH. These reagents are used in essentially the same manner as the polyamine reagents, and similar principles apply.

Use of polycations entails optimization of cell number plated, amount of DNA used, DNA–polymer ratio, and duration of cell exposure to the complexes. DEAE dextran is generally used at 1.0 mg/ml or lower, with cell toxicity the major reason for reducing the concentration or exposure time. One detailed study has established that a ratio of 9 PEI nitrogens per DNA phosphate is optimal for mouse 3T3 fibroblasts [47], suggesting a starting point.

3.2. Calcium phosphate-mediated DNA uptake

Based on the requirement for divalent cations in bacterial transformation, Graham and Van Der Eb reasoned that a similar strategy might enhance the uptake of DNA by mammalian cells. This prediction was proved to be correct, and a transfection protocol was formulated that since has been used widely [50]. Briefly, the DNA is diluted into a solution buffered within a narrow pH range (typically 6.95–7.1) and containing a low concentration of phosphate ions. Calcium ions, in the form of $CaCl_2$, are then added and the DNA is co-precipitated with the inorganic salts. The precipitates are applied to cells and are apparently endocytosed during subsequent incubation. Further enhancement of transfection efficiency can be obtained by brief treatment of the cells with DMSO after removal of non-adsorbed DNA complexes [51]. A more recent version of this procedure using BES as the buffer in place of HEPES may be more efficient under certain circumstances [52].

Many workers have favored $CaPO_4$–DNA precipitation for the production of stably transfected cells, as co-precipitation of multiple molecules occurs with high frequency, permitting the co-transfection of a selectable marker along with the gene of interest [53]. The reagents are inexpensive, but can require some effort to prepare, due to the need for precise control of pH. In addition, this method tends to require a large mass of DNA in order to drive efficient precipitate formation, necessitating the use of large amounts of the plasmid DNA or the inclusion of carrier DNA in the mixture. Finally, some cell types are quite refractory to $CaPO_4$-mediated transfection. This variability has resulted in a gradual shift toward the use of lipid-based transfection methods, but $CaPO_4$–DNA precipitation remains in use in many laboratories. Optimization of the procedure entails varying cell plating density and time of cell exposure to the precipitated DNA. Incubation of cells under conditions of reduced CO_2 may improve viability.

3.3. Lipid-mediated DNA uptake

The earliest described use of lipids for nucleic acid delivery was for phosphatidylserine-mediated delivery of poliovirus RNA [54]. Although this and other early studies were motivated by a desire to use nucleic acids as a sensitive measure of the intracellular delivery of liposome contents, it was soon recognized that gene delivery was an important goal in itself. Many lipid preparations were tested in an effort to improve the efficiency of delivery [55]. An important development was the inclusion of a cationic lipid, which appears to fulfill two functions: neutralizing the net positive charge of the DNA and complexing with the negatively charged cell surface. The preparation of lipoplexes (Chapter 5), the complex between a cationic lipid and a nucleic acid [18], is fairly straightforward: typically, DNA and lipid are diluted separately, then mixed before application to cells. The large number of reagents commercially available has resulted in a dramatic increase in the popularity of lipid-mediated gene transfer methods over the last 10 years. The use of commercially prepared reagents offers both convenience and reproducibility, although these advantages come at some expense.

Optimization of conditions includes adjusting cell density, DNA:lipid ratio, and duration of cell exposure to the complexes. If a commercial reagent is chosen, then the manufacturer's recommendation is a reasonable starting point in determining the ratio of DNA to lipid. One critical variable is whether the presence of serum in the medium will prove detrimental to gene delivery. Many cells tolerate the removal of serum for the few hours needed for DNA uptake, but if the cells of interest do not fall into this category, a reagent that will perform in the presence of serum should be chosen. In addition, serum-tolerant liposomes will not require supplementation or replacement of medium to restore serum needed for cell growth and viability, further simplifying the experimental procedure.

Evidence suggests that polyamines and liposomes can cooperate to increase transfection efficiency. High molecular weight cations appear to facilitate DNA condensation, reducing particle size [56]. Similarly, encapsulating PEI–DNA complexes into liposomes can augment transfection efficiency [57], suggesting that the combination of the polyamines and the lipids offers benefits that surpass either agent alone. Lipopolyamines are a hybrid between the two classes of molecules, as their name implies. One such compound, DOGS, consists of a spermine head attached to a lipid tail [58]. In theory, the lipopolyamines combine the ability of the polyamine moiety to compact DNA and neutralize its negative charge with the liposomal structures.

3.4. Electroporation and other physical methods

Cells exposed to high voltage electric fields will take up exogenous DNA in a process referred to as electroporation [59]. The electric field induces the formation of transient breaks, or pores, in the plasma membrane, allowing DNA and/or other molecules to pass into the cytoplasm. This relatively non-specific mechanism of action may account for its applicability to a wide variety of cell types [60,61]. A number of electroporation devices are commercially available; those that deliver a pulse that decays exponentially by discharging a capacitor are less expensive and more widely used, but devices that deliver a rectangular or square wave pulse also are available. For mammalian cell transfection, cells and DNA are combined in a small volume in a chamber such as a cuvette with a 0.4 cm gap between the electrodes. The charge is applied, and after a brief incubation to allow the DNA to cross the membrane, the cells are restored to normal culture conditions. Thus, the procedure is quite rapid, especially for cells that normally grow in suspension. On the other hand, treatment of many samples is cumbersome. Another apparent advantage of electroporation over the chemical methods is that even larger DNA molecules such as cosmids appear to be delivered efficiently [60]. The main potential drawback to the method is the need for specialized equipment; note that not all electroporation devices are designed for use with mammalian cells.

Although electroporation is relatively simple in concept, it presents several considerations that are unique to this method. Electric field strength and pulse length are both key parameters that must be optimized to maximize transfection efficiency without excessive cell death. For the more commonly used exponential decay machines, the maximal initial voltage delivered is set, and the duration of the decay is the product of the capacitance setting and the resistance of the sample. Successful transfection of several mammalian cell lines has been performed in HEPES-buffered saline at ca. 500 V/cm and 500–1000 µF, yielding a time constant of ca. 7 ms [62], but higher voltage, low capacitance settings (e.g., 8 kV/cm and 20 mF) also prove effective [61]. Reaction volume also may affect resistance. Another decision is the choice of electroporation buffer; phosphate- or HEPES-buffered salines are good high-salt buffers, but some cells are poorly viable in buffer alone and require serum-free culture medium. Two instrument manufacturers have compiled successful electroporation protocols for numerous cell types (see www.bio-rad.com and www.btxonline.com).

The other physical methods of microinjection and particle bombardment are used less frequently to deliver DNA to cultured mammalian cells. Although the efficiency of DNA transfer by microinjection into the nucleus is high [34], the procedure is laborious and can only be used to deliver DNA to a limited number of cells. Microinjection commonly is used for embryos, however; the larger size of the target cell facilitates injection, and analysis of each gene transfer event on an individual basis is desirable. Similarly, particle bombardment which is not routinely used for cell culture has found a niche for specialized applications such as *in vivo* administration of DNA to animals [63]. As with electroporation, these strategies are limited by the need for specialized equipment.

3.5. Biologically vectored approaches

A number of viral vectors have been used to transduce genes into cells. The first systems to be developed were based on large DNA viruses such as vaccinia virus where foreign genes could be inserted into the viral genome and expressed along with viral genes [64]. Later, other viruses such as adenovirus, herpes viruses, retroviruses and adeno-associated viruses have been used. Recently, much attention has been given to the development of alphavirus-based gene delivery systems for the cytoplasmic expression of transgenes [65]. Of these, the clones of Semliki Forest virus (Chapter 3.11), Sindbis virus (Chapter 3.10) and Venezuela encephalitis virus have been further developed into general expression vectors [66].

The alphavirus genome, a 12 kb single-stranded RNA of positive polarity encodes the viral nonstructural protein replicases within the 5′ terminal two-thirds of the genome while the last one-third encodes the viral structural proteins [67]. In the expression vectors, the region encoding the viral structural proteins is replaced with the transgene of interest, while the viral replicase-coding region and all sequences required in cis for replication and packaging of the full-length recombinant RNA (replicon) are maintained. Upon infection, the alphavirus genomic RNA functions as a mRNA for the translation of the replicases, which subsequently transcribe negative strand RNAs from the genomic RNA templates. The negative strand RNAs then serve as templates for the synthesis of the viral genomic RNAs and the shorter subgenomic mRNAs, which encode the structural proteins. Robust transcription from the internal subgenomic promoter in the negative-strand results in the production of large amounts of subgenomic mRNAs.

Transcription of the full-length replicon RNA can be achieved within a transfected cell by placing an RNA polymerase II-dependent promoter immediately upstream from the replicase genes and the alphavirus regulatory sequences. The replicon RNA is then transported to the cytoplasm, where it acts both as a mRNA for the translation of the viral replicases and as a template for replication. Alternatively, the RNA replicon can be made *in vitro* from a T7 or SP6 promoter, and then directly transfected into cells or injected into animals [68]. When expressed in the presence of the structural proteins, the RNA replicon is packaged into virus particles [65,66]. These particles can be harvested from cells and used to deliver the replicon to cells *in vivo* as well as *in vitro*. The particles have a broad host range and can infect non-dividing cells, which can be difficult to transfect by standard procedures.

4. Applications

The possible experimental applications of DNA transfer techniques are too numerous to mention. However, a certain fundamental distinction can be drawn between two major categories of experimental outcomes: transient versus stable introduction of DNA. In most cases, one is generally concerned with achieving the highest transfection efficiency possible without a need for persistence of the introduced DNA. Some combinations of transfection procedure and cell type can achieve efficiencies in excess of 50% when expression of the introduced DNA is measured 1–3 days post-transfection. The vast amount of this DNA will be lost over the course of several days. However, it is sometimes desirable to have the introduced DNA persist over much longer periods. In this instance, one can generate stably transfected cells, in which the DNA has been integrated into the host genome, so that all daughter cells inherit the foreign gene.

Stable transfection occurs with the integration of the exogenous DNA into the host cell DNA. This process is inefficient, and needs to be accompanied by either the introduced gene conferring a novel phenotype on the cells or, more generally, the simultaneous introduction of a gene that confers a selectable phenotype. Typically, selectable markers confer resistance to a pharmacologic agent normally toxic to the cells. The most commonly used selectable marker is the aminoglycoside phosphotransferase (neo) gene, which confers resistance to the drug geneticin/G418 [69]. Another gene that is useful in mammalian cells is hygromycin-B-phosphotransferase, which confers resistance to hygromycin-B [70]. One to 3 days post-transfection, the appropriate drug is added to the culture medium. Non-transfected cells will be killed, along with those cells that acquired DNA but failed to integrate it, as the episomal plasmid DNA will be lost over the course of several days. Thus, the only cells that remain viable are those that have stably acquired the newly introduced DNA. An alternative strategy is to introduce DNA via a viral vector that can replicate in the cells.

Numerous vectors are available that contain one of these selectable markers as well as cloning sites for insertion of an additional gene. However, it is also possible to co-transfect the selectable marker in trans, as most cells will acquire both plasmids. Plasmid vectors that can be used to establish stable transfectants by either strategy can be purchased from many different commercial vendors; a partial compendium, with links, is found at http://vectordb.atcg.com. Alternatively, some are available from the American Type Culture Collection (http://www.atcc.org).

Finally, although transfection methods have been critical in advancing our understanding of gene expression and function, it is necessary to consider the effect of the transfection procedure itself on the state of the cell. As mentioned, most of the procedures result in some loss of viability. However, other effects can be more subtle. Treatment of cells with $CaPO_4$–DNA precipitates, but not DEAE dextran precipitates, resulted in the induction of several genes that are normally responsive to α- or γ-interferon [71]. Although the mechanism of action was not fully elucidated, there did not seem to be a requirement for the de novo synthesis of interferons. In contrast, a similar induction of interferon-regulated genes by cationic liposomes apparently was mediated by the activation of the β-interferon gene [72]. One can guard against unexpected side effects by performing appropriate controls, such as treating cells with transfection agent in the absence of DNA. If an unwanted effect is observed, another transfection method should be tested. The many options available today for transfecting mammalian cells make success likely.

Acknowledgements

The authors are grateful to Dr C. Satishchandran for helpful discussion. This work was supported by grants from the Commonwealth of Pennsylvania (P.A.N. and C.J.P.) and from the National Institute of Arthritis and Musculoskeletal and Skin Diseases (AR42887) to P.A.N.

Abbreviations

BES	N,N-bis(2-hydroxyethyl)-2-aminoethanesulfonic acid
$CaPO_4$	calcium phosphate
DEAE	diethylaminoethyl-dextran
DMSO	dimethyl sulfoxide
DOGS	dioctadecylamidoglycylspermine
DOPC	dioleoyl phosphatidylcholine
DOPE	L-dioleoyl phosphatidylethanolamine
DOTAP	(dioleoyl trimethylammonium propane)
HEPES	N-(2 hydroxyethyl)piperazine-N'-(2-ethanesulfonic acid)
PEI	polyethyleneimine

276

References

1. Yoder, J.I. and Ganesan, T. (1983) Mol. Cell. Biol. 3, 956–959.
2. Weber, M., Möller, K., Welzeck, M., and Schorr, J. (1995) BioTechniques 19, 930–940.
3. Bloomfield, V.A. (1991) Biopolymers 31, 1471–1481.
4. Bloomfield, V.A., Wilson, R.W., and Rau, D.C. (1980) Biophys. Chem. 11, 339–343.
5. Bloomfield, V.A. (1996) Curr. Opin. Struct. Biol. 6, 334–341.
6. Bloomfield, V.A. (1997) Biopolymers 44, 269–282.
7. Ledley, F.D. (1996) Pharm. Res. 13, 1595–1614.
8. Gosule, L.C. and Schellman, J.A. (1976) Nature 259, 333–335.
9. Cantor, C.R. and Schimmel, P.R. (1980) Biophysical Chemistry Part I: The Conformation of Biological Macromolecules, W.H. Freeman & Co, New York, pp. 207–251.
10. Smith, J.G., Walzem, R.L., and German, J.B. (1993) Biochim. Biophys. Acta 1154, 327–340.
11. Felgner, P.L. and Ringold, G.M. (1989) Nature 337, 387–388.
12. Rädler, J.O., Koltover, I., Salditt, T., and Safinya, C.R. (1997) Science 275, 810–814.
13. Gershon, H., Ghirlando, R., Guttman, S.B., and Minsky, A. (1993) Biochemistry 32, 7143–7151.
14. Allen, T.M. and Austin, G.A. (1991) Biochim. Biophys. Acta 1061, 56–64.
15. Litzinger, D.C. and Huang, L. (1992) Biochim. Biophys. Acta 1113, 210–227.
16. Koltover, I., Salditt, T., Rädler, J.O., and Safinya, C.R. (1998) Science 281, 78–81.
17. Farhood, H., Serbina, N., and Huang, L. (1995) Biochim. Biophys. Acta 1235, 289–295.
18. Felgner, P.L., Barenholz, Y., Behr, J.P., Cheng, S.H., Cullis, P., Huang, L., Jessee, J.A., Seymour, L., Szoka, F., Thierry, A.R., Wagner, E., Wu, G. (1997) Hum. Gene Ther. 8, 511–512.
19. Zhou, X. and Huang, L. (1994) Biochim. Biophys. Acta 1189, 195–203.
20. Maurer, N., Mori, A., Palmer, L., Monck, M.A., Mok, K.W.C., Mui, B., Akhong, Q.F., and Cullis, P.R. (1999) Mol. Memb. Biol. 16, 129–140.
21. Loyter, A., Scangos, G.A., and Ruddle, F.H. (1982) Proc. Natl. Acad. Sci. USA 79, 422–426.
22. Godbey, W.T., Wu, K.K., and Mikos, A.G. (1999) Proc. Natl. Acad. Sci. USA 96, 5177–5181.
23. Kukowska-Latallo, J.F., Bielinska, A.U., Johnson, J., Spindler, R., Tomalia, D.A., and Baker, J.R., Jr. (1996) Proc. Natl. Acad. Sci. USA 93, 4897–4902.
24. Friend, D.S., Papahadjopoulos, D., and Debs, R.J. (1996) Biochim. Biophys. Acta 1278, 41–50.
25. Legendre, J.Y. and Szoka, F.C., Jr. (1992) Pharm. Res. 9, 1235–1242.
26. Zabner, J., Fasbender, A.J., Moninger, T., Poellinger, K.A., and Welsh, M.J. (1995) J. Biol. Chem. 270, 18997–19007.
27. Ferkol, T., Lindberg, G.L., Chen, J., Perales, J.C., Crawford, D.R., Ratnoff, O.D., and Hanson, R.W. (1993) FASEB J. 7, 1081–1090.
28. Michael, S.I. and Curiel, D.T. (1994) Gene Ther. 1, 223–232.
29. Stavridis, J.C., Deliconstantinos, G., Psallidopoulos, M.C., Armenakas, N.A., and Hadjiminas, D.J. (1986) Exp. Cell Res. 164, 568–572.
30. Mohr, L., Schauer, J.I., Boutin, R.H., Moradpour, D., and Wands, J.R. (1999) Hepatol. 29, 82–89.
31. Mizuno, M., Yoshida, J., Sugita, K., Inoue, I., Seo, H., Hayashi, Y., Koshizaka, T., and Yagi, K. (1990) Cancer Res. 50, 7826–7829.
32. Tseng, W.C., Haselton, F.R., and Giorgio, T.D. (1997) J. Biol. Chem. 272, 25641–25647.
33. Xu, Y. and Szoka, F.C. (1996) Biochemistry 35, 5616–5623.
34. Cappechi, M.R. (1980) Cell 22, 479–488.
35. Lechardeur, D., Sohn, K.J., Haardt, M., Joshi, P.B., Monck, M., Graham, R.W., Beatty, B., Squire, J., O'Brodovich, H., Lukacs, G.L. (1999) Gene Ther. 6, 482–497.
36. Pollard, H., Remy, J.-S., Loussouarn, G., Demolombe, S., Behr, J.-P., and Escande, D. (1998) J. Biol. Chem. 273, 7507–7511.
37. Brunner, S., Fürtbauer, E., Sauer, T., Kursa, M., and Wagner, E. (2002) Mol. Ther. 5, 80–86.
38. Mortimer, I., Tam, P., MacLachlan, I., Graham, R.W., Saravolac, E.G., and Joshi, P.B. (1999) Gene Ther. 6, 403–411.
39. Zanta, M.A., Belguise-Valladier, P., and Behr, J.P. (1999) Proc. Natl. Acad. Sci. USA 96, 91–96.
40. Schwarze, S.R. and Dowdy, S.F. (2000) Trends Pharmacol. Sci. 21, 45–48.

41. Branden, L.J., Mohamed, A.J., and Smith, C.I. (1999) Nature Biotechnol. 17, 784–787.
42. Norton, P.A. (2000) In: Norton, P.A. and Steel, L.F. (eds.) Gene Transfer Methods. Eaton Publishing, Natick, MA, pp. 65–91.
43. McCutchan, J.H. and Pagano, J.S. (1968) J. Natl. Cancer Inst. 41, 351–356.
44. Lopata, M.A., Cleveland, D.W., and Sollner-Webb, B. (1984) Nucleic Acids Res. 12, 5707–5717.
45. Sussman, D.J. and Milman, G. (1984) Mol. Cell. Biol. 4, 1641–1643.
46. Godbey, W.T., Wu, K.K., and Mikos, A.G. (1999) J. Control. Release 60, 149–160.
47. Boussif, O., Lezoualc'h, F., Zanta, M.A., Mercny, M.D., Scherman, D., Demeneix, B., and Behr, J.-P. (1995) Proc. Natl. Acad. Sci. USA 92, 7297–7301.
48. Marschall, P., Malik, N., and Larin, Z. (1999) Gene Ther. 6, 1634–1637.
49. Haensler, J. and Szoka, F.C., Jr. (1993) Bioconjug. Chem. 4, 372–372.
50. Graham, F.L. and van der Eb, A.J. (1973) Virology 52, 456–467.
51. Stow, N.D. and Wilkie, N.M. (1976) J. Gen. Virol. 33, 447–458.
52. Chen, C. and Okayama, H. (1988) BioTechniques 6, 632–638.
53. Pellicer, A., Robins, D., Wold, B., Sweet, R., Jackson, J., Lowy, I., Roberts, J.M., Sim, G.K., Silverstein, S., Axel, R. (1980) Science 209, 1414–1422.
54. Wilson, T., Papahadjopoulos, D., and Taber, R. (1977) Proc. Natl. Acad. Sci. USA 74, 3471–3475.
55. Fraley, R., Straubinger, R.M., Rule, G., Springer, E.L., and Papahadjopoulos, D. (1981) Biochemistry 20, 6978–6987.
56. Gao, X. and Huang, L. (1996) Biochemistry 35, 1027–1036.
57. Bandyopadhyay, P., Kren, B.T., Ma, X., and Steer, C.J. (1998) BioTechniques 25, 282–292.
58. Loeffler, J.-P. and Behr, J.-P. (1993) Methods Enzymol. 217, 599–618.
59. Neumann, E., Schaefer-Ridder, M., Wang, Y., and Hofschneider, P.H. (1982) EMBO J. 1, 841–845.
60. Andreason, G.L. and Evans, G.A. (1988) BioTechniques 6, 650–660.
61. Potter, H. (1988) Anal. Biochem. 174, 361–373.
62. Chu, G., Hayakawa, H., and Berg, P. (1987) Nucleic Acids Res. 15, 1311–1326.
63. Nicolet, C.M. and Yang, N.-S. (2000) In: Tymms, M.J. (ed.) Transcription Factor Protocols, Vol. 130 Humana Press, Totowa, NJ, pp. 103–116.
64. Wyatt, L.S., Moss, B., and Rozenblatt, S. (1995) Virology 210, 202–205.
65. Garoff, H. and Li, K.J. (1998) Curr. Opin. Biotechnol. 9, 464–469.
66. Frolov, I., Hoffman, T.A., Prágai, B.M., Dryga, S.A., Huang, H.V., Schlesinger, S., and Rice, C.M. (1996) Proc. Natl. Acad. Sci. USA 93, 11371–11377.
67. Johnston, R.E. and Peters, C.J. (1996) In: Fields, B.N. (ed.) Virology, 3rd ed. Lippincott-Raven, Philadelphia, pp. 843–898.
68. Johanning, F.W., Conry, R.M., LoBuglio, A.F., Wright, M., Sumeral, L.A., Pike, M.J., and Curiel, D.T. (1995) Nucleic Acids Res. 11, 1495–1501.
69. Southern, P.J. and Berg, P. (1982) J. Mol. Appl. Genet. 1, 327–341.
70. Palmer, T.D., Hock, R.A., Osborne, W.R.A., and Miller, A.D. (1987) Proc. Natl. Acad. Sci. USA 84, 1055–1059.
71. Pine, R., Levy, D.E., Reich, N., and Darnell, J.E., Jr. (1988) Nucleic Acids Res. 16, 1371–1378.
72. Li, X.L., Boyanapalli, M., Xiao, W.H., Kalvakolanu, D.V., and Hassel, B.A. (1998) J. Interferon Cytokine Res. 18, 947–952.

S.C. Makrides (Ed.) *Gene Transfer and Expression in Mammalian Cells*

CHAPTER 5

Lipid reagents for DNA transfer into mammalian cells

Christophe Masson, Virginie Escriou, Michel Bessodes, and Daniel Scherman

Université René Descartes Paris 5, Pharmacologie Moléculaire et Structurale, CNRS FRE2463, 4, avenue de l'Observatoire, 75270 Paris Cedex 06, France

1. Introduction

Synthetic DNA delivery agents are of great interest as alternatives to viral vectors since they display potentially fewer risks in terms of immunogenicity and propagation. Essentially two chemical alternatives have been developed: cationic polymers such as poly(ethylene imine) [1] and lipid reagents.

Lipid reagents are among the most frequently used chemical vectors for DNA transfer into mammalian cells: their association with DNA leads to supra-molecular entities, which promote DNA penetration into the cells. As early as 1978 [2], it was discovered that DNA could be encapsulated in lipidic particles such as liposomes. Few years later, it was shown that cationic lipids could strongly interact with DNA and condense it in small ionic particles called lipoplexes [3]. Since this discovery at the end of the 80s [4,5], a large variety of cationic lipids were developed to optimize *in vitro* and *in vivo* DNA transfer. These reagents are yet defined as the most promising systems for DNA transfer, especially because no limitation of the size of transferred DNA occurs with these systems.

Many cationic lipids are commercially available and used *in vitro* due to their high efficiency. Unfortunately, lipoplexes present low efficiency *in vivo* as compared to other systems such as viral vectors. Only 12.8% of clinical trial protocols [6] for gene therapy use "lipofection," i.e., lipid reagents to carry DNA.

2. General structure of cationic lipids

Most cationic lipid structures consist of a cationic head linked to a lipid tail via a chemical spacer [6,7]: the cationic charges induce strong electrostatic interactions with DNA (polyanionic) and the lipid moiety promotes vector auto-association by hydrophobic interactions. Many examples were developed during the last 15 years which can be classified in four families: (1) quaternary ammonium salt lipids such as DOTMA [5], DOTAP [8], DDAB [9], DMRIE [10], DORIE [10]; (2) linear or branched lipopolyamines such as DOGS [11], DPPES [11], RPR209120 [12]; (3) lipids bearing both quaternary ammonium and polyamine like DOSPA,

GAP-DLRIE [13]; (4) lipids with cationic groups different from simple amines such as guanidines [14–17], Imidazole (DOTIM) [18] and aminated carbohydrates [19] (Fig. 1). The lipid moiety of these compounds consists of saturated or unsaturated dialkyl chains of different lengths. However, other hydrophobic tails have been proposed such as cholesterol or other steroids, e.g., BGTC [14], CTAP [20], DC-Chol [21], and cholesteryl spermidine [22]. These lipids were linked to their corresponding cationic heads using a chemical spacer: in most cases, simple linkers such as amino acids or glycerol were introduced, but glycosidic linkers have also been described [23,24]. In addition, the effect of the length of these spacers on transfection activity has been studied [25].

These various cationic lipids were synthesized and evaluated for *in vitro* DNA transfer in order to determine structure-activity relationships. Unfortunately, the *in vivo* efficiency of cationic lipids was often not predictable from the *in vitro* results. The low *in vivo* transfection efficiency with cationic lipid-mediated gene transfer methods resulted from a general lack of knowledge regarding: (i) the structures of cationic lipid–DNA complexes; and (ii) their interactions with cell membranes and the events leading to release of DNA in the cytoplasm for delivery to the nucleus.

3. Formulation and physicochemical properties

Cationic lipids are added to DNA to give supramolecular constructs called lipoplexes. In many cases, cationic lipids were previously mixed with neutral lipid helpers such as dioleyl phosphatidyl ethanolamine (DOPE) or cholesterol: these compounds were used to promote particle formation and were added, respectively, for their fusogenic properties [26] or their ability to stabilize lipid bilayers [27]. Formulation processes were usually inspired from liposome science (Fig. 2): in most cases, lipid reagents are homogenized in organic media to thin films by solvent evaporation. The films are then hydrated in the presence of DNA to form particles. Another approach, only applicable when all lipid reagents are water-soluble, consists of mixing directly all components in water to obtain lipoplexes [28]. Depending on the nature and the relative quantity of neutral and cationic lipids, various structural phases were described such as multilamellar [28–30] or hexagonal [31] supramolecular assemblies. During the condensation step, the cationic lipid tends to neutralize the phosphate groups on DNA, thereby releasing the original counterion in solution. The relative proportion of cationic lipid and DNA determines the properties of the lipoplexes, namely the size, the surface charge monitored by zeta potential determination, the efficacy of complexation, the colloidal stability and the biological activity [28]. In most cases, a large excess of cationic lipid reagent over DNA has been used to build highly positively charged lipoplexes, in which DNA is completely sequestered and condensed. The excess of charges induced electrostatic repulsions between chemical entities leading to stable colloidal suspensions: the size distribution of such cationic lipoplexes is in the range of 80–200 nm, and is not dependent of DNA molecular weight [32]. A similar size distribution is also observed when complexes are prepared with an excess of DNA versus cationic lipids (i.e., negatively

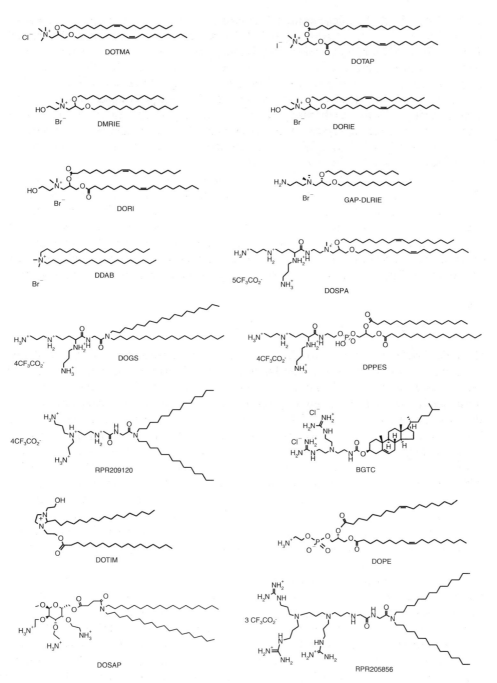

Fig. 1. Various examples of cationic lipids.

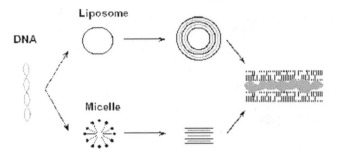

Fig. 2. General scheme of formation of lipoplexes.

charged complexes), although in this case the presence of free DNA is generally observed. On the other hand, complexes prepared using a lipid/DNA charge ratio of approximately 1/1 exhibit a neutral zeta potential, suggesting that DNA neutralizes all the cationic lipid molecules. Such neutral complexes are characterized by a heterogeneous size distribution (mean diameter from 350 to 1200 nm) and usually present a much lower colloidal stability than those exhibiting an excess of net positive or negative charge. This can be attributed to a lack of electrostatic repulsive forces between the lipoplexes, which induces aggregation. The DNA in these formulations is protected from the environment and exhibits increased resistance to nucleases.

4. Mechanisms of in vitro transfection

Schematically, the cellular mechanisms involved in cationic lipid-mediated gene transfer can be divided into five steps (Fig. 3): (1) binding of the cationic lipid–DNA complexes (also termed lipoplexes) to the cell surface; (2) entry of the lipoplexes into the cells; (3) release of the particles into the cytosol where cellular uptake occurs through endocytosis; (4) transport through the cytosol; and (5) entry of DNA into the nucleus followed by transcription. Optimal transfection of the complexes relies on the presence of excess positive charge, which is required for efficient interaction with the negatively charged cell plasma membrane. Membrane-associated proteoglycans are assumed to mediate the binding and delivery of lipoplexes into cells [33]. Several routes of endocytic entry of extracellular material into cells exist: the clathrin-dependent pathway, macropinocytosis, the calveolar pathway, a clathrin- and calveolin-independent pathway and phagocytosis. The former four pathways concern small (<0.2 µm diameter) vesicles transporting fluid and macromolecules while the latter involves uptake of larger particles (>0.5 µm diameter). Although much attention has been focused on elucidating the mechanism by which the DNA is introduced into the cells, a common and generally accepted mechanism has not been identified. The processes that result in the intracellular DNA delivery by cationic lipids might vary depending not only upon the compound and formulation tested but also upon the cell type. Most authors suggest that

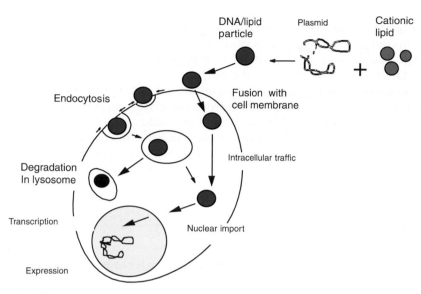

Fig. 3. The different steps involved in non-viral cellular gene transfer.

intracellular delivery of plasmid occurs via endocytosis [34] since lipoplexes have been detected in endosomes and since some endocytosis inhibitors are able to modulate the efficiency of cationic lipid-mediated gene transfer [35]. The prevailing view holds that preventing endosome acidification increases the residence time of DNA within the endosomes and increases the probability of transfer to the cytoplasm. It has also been suggested that the increasing osmotic concentration within the endosome may cause swelling and eventually lysis, releasing the entrapped DNA into the cytoplasm. Phagocytosis has also been proposed to account for the internalization process of the lipoplexes in cultured epithelial cells [36]. We have shown that there was no correlation between the uptake of lipoplexes and gene transfer efficiency [37]. Rather, we found a correlation between the intracellular fluorescent pattern of labeled lipoplexes internalized by the cells and gene transfer efficiency. We suggested that the structural characteristics of a lipoplex interacting with cell membrane might particularly influence the way it will be taken up and determine its intracellular fate. If the primary route of liposome cellular internalisation is the endocytic pathway, a main barrier in lipid-based drug delivery is the escape of hydrolytically sensitive material from degradation in lysosomes. Inclusion of DOPE into lipoplexes has been found to be a key factor for intracellular delivery. In the case of DC-chol, the replacement of the HII-phase lipid DOPE with the bilayer lipid dioleoylphosphatidylcholine (DOPC) either completely inhibits or severely attenuates intracellular delivery [26]. However, it should be noted that most cationic lipids lead to efficient transfection when used alone in the absence of DOPE.

Finally, before transcription can occur, plasmid DNA has to dissociate from the cationic lipid vector and reach the nucleus. The mechanisms by which the plasmid dissociates from the lipid have been studied *in vitro*. Szoka and co-worker have shown that lipoplexes release the associated nucleic acids upon interaction with anionic liposomes [38]. This dissociation probably occurs before transfer of the plasmid to the nucleus, but this point is still unclear. The mechanism by which plasmid DNA is translocated into the nucleus is unknown. In most cell types a fundamental limitation to gene expression in currently used non-viral systems seems to be the inability of DNA to migrate from the cytoplasm into the nucleus. We showed that the efficiency of cationic lipid-mediated gene transfer dramatically increases when cells undergo mitosis, a step in which the nuclear envelope is disrupted [39]. In non-dividing cells, gene transfer is not completely abolished, as the transit of plasmid DNA across the nuclear envelope could still be detected. This, however, is a very rare event.

5. In vivo administration

The effectiveness of *in vivo* gene delivery is poorly predicted by *in vitro* results. This might be because the various biological barriers and the interaction of lipoplexes with blood elements are not reflected in *in vitro* systems. For instance, many formulations leading to efficient *in vitro* gene transfer in the absence of serum are not effective in systems that contain as little as 5–10% serum, an observation that has obvious shortcomings for potential *in vivo* applications. Intravenous administration of lipoplexes can bring about gene expression in many tissues, including those of the heart, lung, liver, spleen and kidney [40]. Although the level of gene expression varies from study to study, the lung invariably shows the highest expression levels. This may be due to a "first-pass effect" because the lung is the first capillary bed the lipoplexe encounters after intravenous administration.

The cationic charges being a major reason for non-specific interactions, one obvious way to overcome this problem would be to prepare neutral lipoplexes using a lower amount of lipid reagents. However attempts to get neutral colloidal suspensions gave large aggregates (> 1 μm) due to the lack of repulsive interactions between lipoplexes. As a result, neutral lipoplexes could not be administered *in vivo* without prior stabilization.

PEGylation of lipoplexes appears to be one of the most promising strategies for overcoming most of these constraints. Poly(Ethylene Glycol) coating was first introduced in the field of liposome science: it was assumed that this polymer played at least a steric role, which allowed stabilization of the particles and prevented the fixation of serum proteins, thus leading to the so-called "stealth" liposomes. In the case of lipoplexes, PEGylation also resulted in colloidal stability, even with neutral lipoplexes, probably due to steric repulsion, which prevents aggregation. PEG was also expected to shield the charges of cationic lipoplexes and more generally making them more stealth *in vivo*: PEGylation could thus prevent lipoplexes interactions with proteins and therefore block their uptake by the reticulo–endothelial system making them more stable and bio-available. Unfortunately, the stability of PEGylated cationic lipoplexes in the blood is often far lower than the conventional stealth behavior obtained with neutral liposomes, suggesting that this polymer

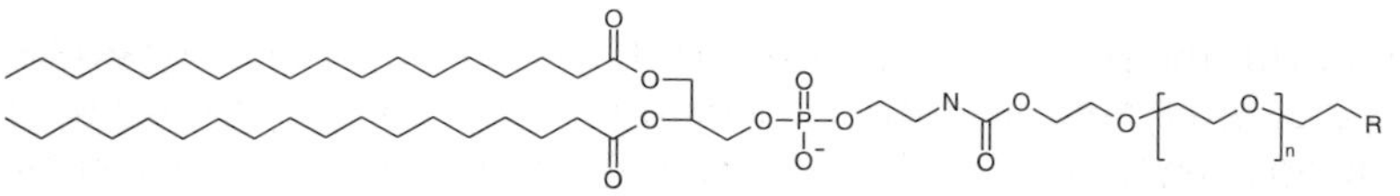

Fig. 4. DSPE–PEG.

cannot totally hide the charges [41]. Nevertheless, the resulting particles lose their transfection efficiency, even *in vitro*, an indication of a relative inhibition of non-specific interactions.

PEGylation of lipoplexes has mainly been obtained from synthetic PEGylated lipids: these lipids were either formulated in combination with the other lipid reagents and DNA to produce directly PEGylated lipoplexes, or were included in the particle composition after the lipoplexes preparation (post-inclusion in the lipid layers of the particles). Many PEGylated lipids have been developed: the size of PEG typically ranges between 2000 and 5000 Da (from about 40 to 110 ethylene glycol monomers). Various lipids have been used such as cholesterol [42] and ceramides [43] but most experiments were performed with commercially available PEGylated phospholipids such as DSPE–PEG [44] (Fig. 4). Other strategies did not use lipid–lipid interactions to insert PEG in the lipophilic layer of the particles but developed PEG post-coating by covalent linkage to previously activated lipoplexes [45] or electrostatic interactions between an anionic derivative of PEG and cationic lipoplexes [46].

Even if PEGylation of particles has mainly been developed for systemic administration, interesting properties of PEGylated particles have been demonstrated in other cases: as far as these lipids can stabilize lipoplexes, they have also shown remarkable properties for DNA local administration and transfer to airways by instillation [47,48].

6. Targeting

Since PEGylated lipoplexes have lost gene transfer efficiency, an intensively investigated strategy to induce specific DNA transfer with PEGylated lipoplexes is to link targeting ligands to these particles: specific recognition of ligands by the cells, which over-express the corresponding membrane receptor could promote selective and active DNA transfer via biological processes such as ligand-mediated endocytosis. Depending on the targeted tissues, various ligands have been envisaged: galactose for hepatoma cells [49], folate for tumor cells [50,51] RGD peptidic sequence [52] for vascular endothelium integrins. In many cases, selective and competitive targeting effects have been measured *in vitro*; only few *in vivo* experiments have been described to date, and the corresponding results still need to be improved for further biological applications.

Another approach was recently developed to promote selective DNA transfer: specifically reversible PEGylation of lipoplexes was introduced by incorporating a chemically labile linker between PEG and its lipid anchor, which is responsible for its interaction with the particles. For instance reduction-sensitive [53] or acid-sensitive bonds [54,55] were used in order to promote PEG escape in reductive or acidic media, respectively. PEG escape leads to classical cationic lipoplexes, which consequently interact strongly with the surrounding membranes in a non-specific manner. If this phenomenon occurs selectively in the targeted tissue, one can hope to restore a specific and efficient *in vivo* DNA transfer. Our group and others recently investigated this strategy by incorporating acid-labile orthoester linkers in the structure of PEGylated lipids to target naturally acidic tumors, giving significant targeting effect after intravenous administration [55].

7. New strategies to improve in vivo transfection

Extracellular traffic and cell penetration are unfortunately not the only barriers to an efficient DNA transfer into mammalian cells: many studies are yet devoted to the optimization of intracellular traffic of DNA after internalization as far as this poly-anion must enter the nucleus to be transcribed. In this context, one major requirement today is to develop new lipid reagents, which can, at least partially, deliver DNA to the cell to allow it to reach the nucleus. Indeed, lipoplexes are small particles where DNA is condensed and protected from proteins responsible for their degradation (nucleases). The final aim being expression of the exogenous DNA in the nucleus, nuclear uptake of accessible DNA is a major requirement to efficient DNA transfer. Several strategies have been proposed to induce intracellular disassembling of lipid reagents/DNA particles. Acid-labile systems were developed to take advantage of acidic compartments such as endosomes to destabilize the liposomes. These pH-sensitive liposomes are composed of lipid reagents, which do not have the same protonation level depending on the pH in a physiological range (from 5 to 8): in acidic media, protonation of the reagents that are involved in the structure of the liposomes modify their conformation and therefore promote drug or DNA delivery by favoring fusion with membranes, for instance to promote endosome escape [56–58]. A related strategy was developed to take advantage of reductive agents *in vivo*: reduction-labile cationic lipids bearing a disulfide bond were synthesized to promote endosome escape by releasing an endosomolytic agent during reduction [59].

8. Conclusion

To conclude on the use of lipid reagents for DNA transfer today, an important point is that most of the lipid systems used for DNA transfer contain at least 2 or 3 lipid reagents, which are complexed with DNA. The main lipid reagent is cationic and therefore permit DNA condensation; a neutral lipid is often used to promote particle formation and to increase its efficiency for gene transfer. In many cases, a PEGylated lipid is also added in order to stabilize and optimize bio-distribution of the particle. The resulting supra-molecular assemblies may allow a selective and efficient *in vivo*

DNA transfer to specific cells or tissues, provided that targeting properties are added. A lot of work was devoted to optimize DNA encapsulation, and condensation of DNA is yet easily obtained from commercially available lipid reagents. Nevertheless, the targeting of specific tissues with lipid reagents has to be enhanced as well as the optimization of intracellular traffic to the nucleus: both extra- and intra-cellular DNA traffic must be improved to enhance DNA transfer into mammalian cells.

Abbreviations

BGTC	bis-guanidinium-TREN-cholesterol
CTAP	N^{15}-cholesteryloxycarbonyl-3,7,12-triazapentadecane-1,15-diamine
DC-Chol	(3β)-[N-(N',N''-dimethylaminoethyl)carbamoyl]cholesterol
DDAB	dimethyldioctadecylammonium bromide
DMRIE	1,2-dimyristyloxypropyl-3-dimethyl-hydroxyethylammonium bromide
DOGS	dioctadecylamidoglycylspermine, 4trifluoroacetic acid
DORIE	1,2-dioleyloxypropyl-3-dimethyl-hydroxyethylammonium bromide
DOSAP	N,N-dioctadecyl-succinamic acid 3,4,5-tris-(2-amino-ethoxy)-6-methoxy-tetrahydro-pyran-2-ylmethyl ester
DOSPA	2,3 dioleyloxy-N-[2(sperminecarboxamido)ethyl]-N,N'-dimethyl-1-propanaminium trifluoroacetate
DOTAP	1,2-dioleoyloxypropyl-3-(trimethylammonio)-propane
DOTIM	1-[2-(oleoyloxy)ethyl]-2-oleyl-3-(2-hydroxyethyl) imidazolinium chloride
DOTMA	N-[1-(2,3-dioleyloxy)propyl]-N,N,N-trimethylammonium chloride
DPPES	dipalmitoylphosphatidylethanolamidospermine, 4trifluoroacetic acid
GAP-DLRIE	($\pm$)-N-(3-aminopropyl)-N,N-dimethyl-2,3-bis (dodecyloxy)-1-propanammonium bromide

References

1. Boussif, O., Lezoualc'h, F., Zanta, M., Maria, A., Mergny, M., Mojgan, D., Scherman, D., Demeneix, B., and Behr, J.-P. (1995) Proc. Natl. Acad. Sci. USA 92, 7297–7301.
2. Szoka, F. and Papahadjopoulos, D. (1978) Proc. Natl. Acad. Sci. USA 75, 4194–4198.
3. Felgner, P.L., Barenholz, Y., Behr, J.P., Cheng, S.H., Cullis, P., Huang, L., Jessee, J.A., Seymour, L., Szoka, F., Thierry, A.R., Wagner, E., Wu, G. (1997) Hum. Gene Ther. 8, 511–512.
4. Behr, J.-P. (1986) Tetrahedron Lett. 27, 5861–5864.
5. Felgner, P., Gadek, T., Holm, M., Roman, R., Chan, H.W., Wenz, M., Northrop, J.P., Ringold, G.M., and Danielsen, M. (1987) Proc. Natl. Acad. Sci. USA 84, 7413–7417.
6. Byk, G. and Scherman, D. (1998) Expert Opin. Ther. Pat. 8, 1125–1141.

288

7. Gao, X.and and Huang, L. (1995) Gene Ther. 2, 710–722.
8. Stamatatos, L., Leventis, R., Zuckermann, M.J., and Silvius, J.R. (1988) Biochemistry 27, 3917–3925.
9. Rose, J.K. and Buonocore, L. (1991) Biotechniques 10, 520–525.
10. Felgner, J.H., Kumar, R., Sridhar, C.N., Wheeler, C.J., Tsai, Y.J., Border, R., Ramsey, P., Martin, M., and Felgner, P.L. (1994) J. Biol. Chem. 269, 2550–2561.
11. Behr, J.-P., Demeneix, B., Loeffler, J.P., and Perez-Mutul, J. (1989) Proc. Natl. Acad. Sci. USA 86, 6982–6986.
12. Byk, G., Scherman, D., Schwartz, B., Dubertret, C. (2001) US Patent, 6171612.
13. Wheeler, C.J., Felgner, P.L., Tsai, Y.J., Marshall, J., Sukhu, L., Doh, S.G., Hartikka, J., Nietupski, J., Manthorpe, M., Nichols, M., Plewe, M., Liang, X., Norman, J., Smith, A., Cheng, S.H. (1996) Proc. Natl. Acad. Sci. USA 93, 11454–11459.
14. Vigneron, J.-P., Oudrhiri, N., Fauquet, M., Vergely, L., Bradley, J.-C., Basseville, M., Lehn, P., and Lehn, J.-M. (1996) Proc. Natl. Acad. Sci. USA 93, 9682–9686.
15. Frederic, M., Scherman, D., and Byk, G. (2000) Tetrahedron Lett. 41, 675–679.
16. Byk, G., Soto, J., Mattler, C., Frederic, M., and Scherman, D. (1998) Biotechnol. Bioeng. 61, 81–87.
17. Byk, G., Frederic, M., Hofland, H., Scherman, D. (1999), PCT Int. Appl. 83 pp. WO 9951581.
18. Solodin, I., Brown, C.S., Bruno, M.S., Chow, C.-Y., Jang, E.-H., Debs, R.J., Heath (1995) Biochemistry 34, 13537–13544.
19. Bessodes, M., Dubertret, C., Jaslin, G., and Scherman, D. (2000) Bioorg. Med. Chem. Lett. 10, 1393–1395.
20. Cooper, R.G., Etheridge, C.J., Stewart, L., Marshall, J., Rudginsky, S., Cheng, S.H., and Miller, A.D. (1998) Chem. Eur. J. 4, 137–151.
21. Gao, X. and Huang, L. (1991) Biochem. Biophys. Res. Commun. 179, 280–285.
22. Moradpour, D., Schauer, J.I., Zurawski, V.R., Jr., Wands, J.R., and Boutin, R.H. (1996) Biochem. Biophys. Res.Commun. 221, 82–88.
23. Banerjee, R., Venkata Mahidhar, Y., Chaudhuri, A., Gopal, V., and Madhusudhana Rao, N. (2001) J. Med. Chem. 44, 4176–4185.
24. Leclercq, F., Dubertret, C., Pitard, B., Scherman, D., and Herscovici, J. (2000) Bioorg. Med. Chem. Lett. 10, 1233–1235.
25. Byk, G., Dubertret, C., Escriou, V., Frédéric, M., Jaslin, G., Rangara, R., Pitard, B., Crouzet, J., Wils, P., Schwartz, B., Scherman, D. (1998) J. Med. Chem. 41, 224–235.
26. Farhood, H., Serbina, N., and Huang, L. (1995) Biochim. Biophys. Acta 1235, 289–295.
27. Kirby, C., Clarke, J., and Gregoriadis, G. (1980) Biochem. J. 186, 591–598.
28. Pitard, B., Aguerre, O., Airiau, M., Lachagès, A.-M., Boukhnikachvili, T., Byk, G., Dubertret, C., Herviou, C., Scherman, D., Mayaux, J-F., Crouzet, J. (1997) Proc. Natl. Acad. Sci. USA 94, 14412–14417.
29. Lasic, D., Strey, H., Stuart, M., Podgornik, R., and Frederik, P.M. (1997) J. Am. Chem. Soc. 119, 832–833.
30. Rädler, J.O., Koltover, I., Salditt, T., and Safinya, C.R. (1997) Science 275, 810–814.
31. Koltover, I., Salditt, T., Rädler, J.O., and Safinya, C.R. (1998) Science 281, 78–81.
32. Kreiss, P., Cameron, B., Rangara, R., Mailhe, P., Aguerre-Charriol, O., Airiau, M., Scherman, D., Crouzet, J., and Pitard, B. (1999) Nucleic Acids Res. 27, 3792–3798.
33. Mislick, K.A. and Baldeschwieler, J.D. (1996) Proc. Natl. Acad. Sci. USA 93, 12349–12354.
34. Zabner, J., Fasbender, A.J., Moninger, T., Poellinger, K.A., and Welsh, M.J. (1995) J. Biol. Chem. 270, 18997–19007.
35. Legendre, J.-Y. and Szoka, F.C. (1992) Pharm. Res. 9, 1235–1242.
36. Matsui, H., Johnson, L.G., Randell, S.H., and Boucher, R.C. (1997) J. Biol. Chem. 272, 1117–1126.
37. Escriou, V., Ciolina, C., Lacroix, F., Byk, G., Scherman, D., and Wils, P. (1998) Biochim. Biophys. Acta 1368, 276–288.
38. Xu, Y.H. and Szoka, F.C. (1996) Biochemistry 35, 5616–5623.
39. Escriou, V., Carrière, M., Bussone, F., Wils, P., and Scherman, D. (2001) J. Gene Med. 3, 179–187.
40. Pouton, C.W. and Seymour, L.W. (2001) Adv. Drug Deliv. Rev. 46, 187–203.
41. Hofland, H., Wils, P. (personal communication).

42. Vertut-Doï, A., Ishiwata, H., and Miyajima, K. (1996) Biochim. Biophys. Acta 1278, 19–28.
43. Wheeler, J.J., Palmer, L., Ossanlou, M., MacLachlan, I., Graham, R.W., Zhang, Y.P., Hope, M.J., Scherrer, P., and Cullis, P.R. (1999) Gene Ther. 6, 271–281.
44. Hong, K., Zheng, W., and Baker, A.D. (1997) FEBS Lett. 400, 233–237.
45. Xu, L. and Chang E.H. (2000) PCT Int. Appl. WO0043043, 41 pp.
46. Finsinger, D., Remy, J.-S., Erbacher, P., Koch, C., and Plank, C. (2000) Gene Ther. 7, 1183–1192.
47. Zhang, Y.P., Sekirov, L., Saravolac, E.G., Wheeler, J.J., Tardi, P., Clow, K., Leng, E., Sun, R., Cullis, P.R., Scherrer, P. (1999) Gene Ther. 6, 1438–1449.
48. Pitard, B., Oudrhiri, N., Lambert, O., Vivien, E., Masson, C., Wetzer, B., Hauchecorne, M., Scherman, D., Rigaud, J.-L., Vigneron, J.-P., Lehn, J.-M., Lehn, P. (2001) J. Gene Med. 3, 478–487.
49. Remy, J.S., Kichler, A., Mordvinov, V., Schuber, F., and Behr, J.P. (1995) Proc. Natl. Acad. Sci. USA 92, 1744–1748.
50. Leamon, C.P. and Low, P.S. (1991) Proc. Natl. Acad. Sci. USA 88, 5572–5576.
51. Hofland, H.E.J., Masson, C., Iginla, S., Osetinsky, I., Leamon, C.P., Scherman, D., Bessodes, M. and Wils, P. (2002) Mol. Ther. in press.
52. Arap, W., Pasqualini, R., and Ruoslahti, E. (1998) Science 279, 377–380.
53. Zalipsky, S., Qazen, M., Walker, J.A., Mullah, N., Quinn, Y.P., and Huang, S.K. (1999) Bioconjug. Chem. 10, 703–707.
54. Guo, X. and Szoka, F.C., Jr. (2001) Bioconjug. Chem. 12, 291–300.
55. Bessodes, M., Masson, C., Scherman, D. and Wetzer B. (2002) PCT Int. Appl. 73 pp. WO 0220510.
56. Gerasimov, O., Qualls, M., and Rui, D.H. (1997) Polym. Mater. Sci. Eng. 76, 499–500.
57. Litzinger, D.C. and Huang, L. (1992) Biochim. Biophys. Acta 1113, 201–227.
58. Budker, V., Gurevich, V., Hagstrom, J.E., Bortzov, F., and Wolff, J.A. (1996) Nat. Biotechnol. 14, 760–764.
59. Byk, G., Wetzer, B., Frederic, M., Dubertret, C., Pitard, B., Jaslin, G., and Scherman, D. (2000) J. Med. Chem. 43, 4377–4387.

S.C. Makrides (Ed.) *Gene Transfer and Expression in Mammalian Cells*

CHAPTER 6

Reporter genes for monitoring gene expression in mammalian cells

Jawed Alam and Julia L. Cook

Ochsner Clinic Foundation, Division of Research, Molecular Genetics,
1516 Jefferson Hwy, New Orleans, LA 70121, USA; Tel.: +504-842-6934;
Fax: 504-842-6956; E-mail: jcook@ochsner.org and jalam@ochsner.org

1. Introduction

Illuminating the mechanisms of many cellular processes has been greatly facilitated by the advent of reporter gene technology. Reporter genes encode proteins with phenotypic properties that are both distinct (from the system being studied) and conveniently monitored. Linkage of the cellular activity being assayed to the phenotypic expression of the reporter is accomplished by fusion of appropriate DNA sequences to the reporter gene. These sequences are either regulatory in nature, typically being responsive to the cellular event under examination, or structural, encoding proteins that mediate such an event. After introduction of the chimera into cells, the qualitative or quantitative activity of the cellular event is extrapolated from the expression of the linked reporter gene product.

Application of reporter gene technology to mammalian cells was introduced in 1982 with the development of plasmid vectors, encoding the bacterial enzymes chloramphenicol acetyltransferase (CAT) [1] or β-galactosidase (β-gal) [2], to study eukaryotic gene regulation. As in these initial studies, reporter genes have been traditionally used to characterize and dissect transcriptionally active regulatory regions. Although characterization of *cis*-elements remains a common usage, reporter gene technology has attained increasing levels of sophistication and is now used in numerous other applications including characterization of transcription factors and associated proteins such as co-activators, delineation of signal transduction pathways, identification of protein–protein interactions, determination of cell fates, visualization of cellular trafficking, high-throughput screening of chemicals for adverse or therapeutic effects, and optimization and monitoring of DNA delivery systems.

Numerous genes, both of prokaryotic and eukaryotic origin, have been proposed for use as reporter genes in mammalian cells. Only a handful of these, however, are widely used in this capacity. Aside from the obvious characteristic that reporter gene products cannot be toxic, the commonly used reporters share two other features [3]: (1) the gene products exhibit phenotypic characteristics that are unique (i.e., foreign to mammalian cells) or can be easily distinguished from any similar endogenous activity; and (2) methods for detection of the reporter are sensitive, generally quantitative, and exhibit a broad linear dynamic range; they are easy to perform, reproducible and reasonably cost-effective. In 1990, we reviewed reporters commonly used for analysis of mammalian gene regulation including CAT, firefly luciferase, human growth hormone, alkaline phosphatase and β-gal [3]. In terms of the reporter genes themselves, the single most significant development since that time is the introduction and evolution of green fluorescent protein (GFP) as a powerful and multi-faceted reporter protein. Naturally, a significant portion of the current article is devoted to this system. An update of some of the other reporter gene systems is also provided.

One obvious trend in the past 12–13 years is movement away from radiochemical-dependent reporter assays to the equally or even more sensitive fluorescent- and chemiluminescent-based detection methods which are also usually more manageable and cost efficient [3,4]. This trend is intrinsically linked to the development of semi-automated systems capable of rapidly monitoring reporter activities in multiple samples (i.e., high throughput analysis) and of methodologies to monitor reporter activities in intact cells or animals. Reporter systems, such as growth hormone, that have not easily adapted to these advancements have largely become obsolete. Another trend, an apparent reflection of the importance of this technology, is the commercialization of the more common reporter gene systems, in particular the reporter vectors themselves. Representative reporter-based plasmids and their sources are described.

2. Reporter gene vectors and fusions

Reporter genes have been incorporated into all common vector formats including those based on viral genomes or bacterial plasmids. Largely because of the commercialization of the technology, the newer generation of plasmid-based reporter vectors is vastly improved over the original constructs. In general, the newer reporter plasmids are more compact, replicate to a higher copy number and contain more extensive multiple cloning sites. Additional general improvements include elimination of potential cis-acting elements from the vector backbone, modification of some reporter genes to remove inconvenient restriction sites and to reflect the codon usage bias observed in mammalian cells, and incorporation of the Kozak consensus sequence for efficient translation. Some vectors also include internal ribosome entry sites (IRES) between the multiple cloning site (MCS) and the reporter gene so that expression of the reporter is dependent on the same transcript that produces the test protein.

Depending on investigational objectives, two general types of experimental reporter constructs are commonly employed: fusion of regulatory sequences that control the expression of the native reporter ("gene fusion") or fusion of coding sequences to generate a chimeric reporter protein ("protein fusion") (Fig. 1). Gene fusions are commonly used to characterize the transcription activity of 5′ flanking regions of genes and to localize active elements. In such cases, the test sequences are typically cloned upstream of the coding portion of the reporter gene in the so-called "promoterless" vectors that lack control sequences for driving expression of the reporter. In specialized situations, such as examination of putative enhancer elements, such sequences are cloned either upstream or downstream of a reporter gene under the control of a heterologous core promoter (i.e., "enhancerless" vector).

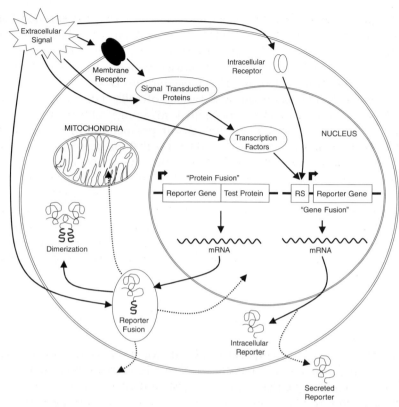

Fig. 1. Schematic representation of cell-based reporter gene applications. Reporter genes can be used in "protein fusion" or "gene fusion" formats. "Gene fusions" are used to test the function of putative transcription regulatory sequences (RS) that may or may not be responsive to extracellular stimuli (e.g., ligands), which can activate signal transduction at one of several steps along the pathway. "Protein fusions" can be used to study protein trafficking (dashed lines) and compartmentalization (e.g., transport into the mitochondria or nucleus) or protein–protein interaction. Extracellular signals may also regulate these activities. Reporters may be intracellular or secretory proteins.

Gene fusions with known, well-characterized regulatory sequences are also often used. For instance, reporter genes under the control of specific response sequences, such as the cAMP- or estrogen-response elements, are used to examine activation of specific signal transduction pathways by extracellular agents or for high-throughput screening of chemical libraries. In addition, gene fusions in which the reporter is under the control of highly active viral promoters are often used for optimization and monitoring of DNA delivery systems. Protein fusions are used to examine sub-cellular trafficking and localization, and protein–protein interactions.

3. Green fluorescent protein

GFP was discovered in the course of investigations of marine bioluminescent organisms [5]. Studies in the early 1960s showed that *Aequoria* jellyfish will emit light upon the addition of Ca^{2+}. No other cofactor is required for the reaction. The substrate, named aequorin, was found to be a precharged, quasi-stable enzyme peroxide intermediate formed by the reaction of apoaequorin with coelenterazine and oxygen. In the presence of Ca^{2+}, it undergoes a bioluminescent reaction to produce a flash of light at 460 nm. The apoaequorin cDNA has been fused to a range of promoters and expressed in a variety of cells. However, the flash-type light emission renders this reporter of limited advantage. Of greater usefulness is GFP itself, a 238 amino acid highly fluorescent polypeptide of molecular weight 27–30 kDa. In the jellyfish, *Aequoria victoria*, GFP fluoresces when light emitted from the aequorin reaction is absorbed and emitted as a 509 nm excitation peak. GFP requires no cofactors or substrates, except oxygen, for generation of green fluorescence.

The gene for *Aequoria* GFP was cloned in 1991; when expressed in cells as a reporter, excitation at 395 nm permits green fluorescence detection at 509 nm. Native GFP is monomeric in dilute solution (<2 mg/ml) and dimeric in concentrated solution (>5 mg/ml). GFP is exceptionally stable with regard to pH, temperature, and proteolysis, at least in part, due to its compact β-can structure [6].

3.1. Modifications and practical applications of GFP

Limited digestion of GFP with papain releases a six amino acid (64–69) chromophore, FSYGVQ, the fluorescence of which requires interactions with other parts of the protein [5]. The organic chemistry and mechanisms behind light absorption and emission by the chromophore are complex and not within the scope of this review. But certain mutations within the chromophore lead to improved properties and have been included in widely utilized "enhanced" vectors. Mutant GFPs exhibit improved fluorescence intensity, folding properties and spectral qualities (reviewed in [7]). Additionally, expression of the human codon-optimized reporter gene is approximately fourfold greater than of the wild-type jellyfish gene in mammalian cells.

The usefulness of GFP as a reporter is, in part, dependent on the time required for post-translational maturation and fluorescence. For the wild-type GFP, expressed in various eukaryotes at room temperature or above, this time is estimated to be 3–5 h [8,9] but may be less (around 0.45 h) for some mutants [10]. In our experience, and that of others [11], this can limit the ability to visualize trafficking such as that characteristic of the secretory pathway. Proteins trafficking through the secretory pathway may move rapidly (up to 1 µm/sec) and therefore, may be secreted before the GFP has assumed the mature folded structure necessary for fluorescence. Temperature reduction can slow movement and improve folding and may permit visualization of GFP fusion proteins targeted for secretion [11,12].

GFP can be used not only to visualize and quantify expression of fusion genes within cells, but also to visualize trafficking, compartmentalization and sophisticated localization mechanisms. In particular, fusion of GFP with a protein of interest permits marking of the relevant protein, generally without affecting biological integrity. For instance, using GFP/5-lipoxygenase (GFP-5LO) fusion constructs, investigators have shown variable, cell-type specific compartmentalization [13]. In HL-60 cells, GFP-5LO is localized to the cytosol, whereas it is distributed throughout the nucleus and cytosol in rat basophilic leukemia cells. Subsequent experiments have shown that differential cellular compartmentalization is controlled by a balance between nuclear localization signal (NLS) and nuclear export signal-dependent mechanisms. The GFP reporter has, thus, augmented the identification and characterization of a complex interplay between two regulatory mechanisms for compartmental sorting.

Coupling of the GFP gene to therapeutically-relevant genes and expression *in vivo* permits monitoring of the biological activity of the protein not only in normal animals but in disease states, following injury, and in senescence. Simultaneous expression of brain-derived neurotrophic factor (BDNF) and GFP via an IRES-dependent bicistronic mRNA has been used to track transgene activity in the substantia nigra following injection of a recombinant adeno-associated virus. Transgene activity was detectable up to 9 months (i.e., through the course of the study) and BDNF was shown to modulate locomotor activity following lesion induction [14]. Similarly, introduction of a recombinant adenoviral vector containing the BDNF/GFP bicistronic transgene into adult rat forebrain ventricular zone induced both striatal and olfactory neuronal recruitment from the endogenous progenitor population [15]. BDNF was, therefore, found to enhance neuronal maturation and survival. Obligate co-expression of GFP with BDNF thus allowed spatial and temporal visualization of the infected cells and characterization of the biological activity.

Creation of proteins with altered spectral properties has allowed not only the localization and tracking of multiple proteins within a cell but also exploration of the interactions between or within proteins (see [16] for review). The latter is possible because of the phenomenon of fluorescence resonance energy transfer (FRET), which can be measured using fast imaging flow cytometry, conventional fluorescence microscopy, and confocal microscopy. FRET involves transfer of energy between fluorescence proteins of different excitation and emission spectra and depends on the use of specific complementary pairs of fluorescence proteins since the distance between the chromophores is critical for energy exchange. This enables energy exchange between fluorescent fusion proteins through the interaction of the fusion partners. As mentioned above, GFP can dimerize at high concentrations, a phenomenon that can complicate FRET studies since one desires that the energy exchange reflect only interactions between the proteins of interest and not the reporters.

FRET was used to demonstrate *in vivo* dimerization of G-protein coupled receptors (GPCR), historically thought to act only as monomers [17]. Co-expression of cyan fluorescent protein- with yellow fluorescent protein-tagged alpha factor receptor, a GPCR from yeast, demonstrated stable dimer association *in vivo*. In this system, oligomerization was further confirmed by rescuing endocytosis-defective receptors with co-expressed wild-type receptors and by inhibition of wild-type receptors with dominant negative mutants. In summary, studies using FRET suggest that oligomerization of GPCRs may be involved in signaling and regulation.

Ligand-mediated receptor redistribution has also been studied using FRET. Measurements using FRET suggest that, while some receptors are dimerized in the absence of ligand, binding of epidermal growth factor (EGF) to the EGF receptor (EGFR) leads to fast and temperature-sensitive microclustering of EGFR [18]. In a related report, FRET was used to show that erb2, a member of the EGF receptor family, self-associates in unstimulated breast tumor lines and that the interaction is enhanced, and the distribution altered, in EGF-treated cells [19]. Similarly, FRET studies show that relative positions of the subunits of the IL-2 receptor complex (α, β, γ) in T lymphoma cells are shifted with the binding of IL-2, IL-7 or IL-15 [20]. IL-2 strengthens the bridges between the subunits while IL-7 and IL-15 weaken the associations. Collectively, these studies show that FRET is immensely useful in reporting dynamic cellular events and that genes encoding fluorescent products are central to this technology.

3.2. Commercially available vectors

The "enhanced" green fluorescent protein (EGFP) marketed by Clontech (BD BioSciences, Palo Alto CA) is a red-shifted variant. EGFP exhibits increased fluorescence intensity, a shorter lag in development of fluorescence after protein synthesis and improved expression in most mammalian cell types. It possesses the double amino acid substitution, F64L and S65T [21] in the chromophore and the coding sequence contains more than 190 silent base changes which correspond to human codon usage preferences [22]. Variants of EGFP are available in yellow (EYFP) and cyan (ECFP) and collectively, the enhanced variants are up to 30-fold brighter than wild-type GFP and more stable for use in mammalian cells. In addition, a gene encoding red fluorescent protein derived from the IndoPacific sea anemone, *Discosoma*, is available in several different vehicles. This protein has a vivid red fluorescence, low background, and high signal-to-noise ratio making it ideal to use alone or in combination with other proteins. For optimum results, Clontech recommends using EGFP with DsRed for dual labeling because of their distinct emission spectra and the combination of DsRed, EYFP and ECFP for triple labeling. The excitation wavelength maxima for DsRed, EGFP, EYFP and ECFP, respectively, are 558, 489, 514 and 434. The emission maxima are 583, 508, 527 and 477.

A partial list of some of the more useful vectors available from Clontech are summarized in Table 1. Plasmids encoding red, green, yellow and cyan fluorescent proteins are available for N- and C-terminal fusion protein design. The basic structure of each of the N- and C-terminal fusion vectors is similar. The multiple cloning site lies downstream of the cytomegalovirus immediate early promoter and upstream (N-terminal) or downstream (C-terminal) of the fluorescent protein gene. They all encode products for kan[r] and neo[r] and possess pUC replication origins for propagation in bacteria and SV40 replication origins for amplification in mammalian cells which express SV40 T-antigen (e.g., COS cells).

The Discosoma Red (pDSRed2) and pECFP, pEGFP and pEYFP vectors have been variably modified by fusing appropriate signal sequences to create controls

Table 1
Select fluorescent vectors from Clontech

Base vector	Extension	Description
pDsRed2	-N1	For fusions to N-terminus of DsRed
	-C1	For fusions to C-terminus of DsRed
	-ER -Mito	Localize to the respective organelle or
	-Nuc -Peroxi	subcompartment (controls)
pEGFP	-N1, -N2, -N3	For fusions to N-terminus of EGFP
	-C1, C2, -C3	For fusions to C-terminus of EGFP
	-Actin -Endo	Localize to the respective organelle or structure
	-Peroxi -Tub	
	-F	
pECFP	-N1	For fusions to N-terminus of ECFP
	-C1	For fusions to C-terminus of ECFP
	-Endo -ER -Golgi	Localize to the respective organelle or structure
	-Mem -Mito -Nuc	
	-Peroxi	
pEYFP	-N1	For fusions to N-terminus of EYFP
	-C1	For fusions to C-terminus of EYFP
	-Actin -Endo, -ER	Localize to the respective organelle or structure
	-Golgi -Mem -Mito	
	-Nuc -Peroxi -Tub	
pIRES2	-EGFP	Bicistronic expression vectors permit expression
pIRES	-EYFP	of gene of interest with fluorescent marker.

which traffic to various subcellular compartments. These compartments and the source of the signal sequence are as follows: the nucleus (3 tandem repeats of the SV40 T-antigen NLS), mitochondria (human cytochrome c oxidase subunit VIII), peroxisomes (peroxisomal targeting signal 1), endoplasmic reticulum (calreticulin ER-targeting and KDEL retrieval sequence), golgi (human β 1,4-galactosyltransferase), endosomes (rhoB), plasma membrane (palmitoylation domain of neuromodulin), actin filaments (human β-actin) and microtubules (human α-tubulin). EGFP is also available as pEGFP-F which possesses the 20 amino-acid farnesylation sequence from c-Ha-Ras and thus localizes to the plasma and internal membranes.

In addition to these, vectors designed to serve as sources of fluorescent protein-encoding genes (EGFP, ECFP, EYFP, DsRed) and promoterless vectors designed to evaluate promoter and enhancer elements are available. Vectors encoding destabilized fluorescent proteins are also available. With a half-life of > 24 h, EGFP accumulates in cells even at low expression levels, which can mask changes in gene expression. PEST sequences, which target proteins for rapid degradation, are added for production of destabilized EGFP, ECFP, and EYFP and corresponding N-terminal fusion protein products. ECFP and EYFP are available with half-lives of 2 h while EGFP is available with half-lives of 1, 2 and 4 h.

In addition, Clontech markets vectors possessing IRES sequences. The incorporation of IRES sequences into vectors permits expression of two proteins from a single mRNA. This, in turn, allows expression of a reporter gene with a gene of interest. The number of gene therapy vehicles that have been designed with this feature reflects the importance and utility of this method. Recombinant disabled herpes vectors (HSV-1) have been developed and used with several reporters (including GFP) as a method of monitoring transfection efficiency, expression efficiency, duration of expression and cell viability over time [23]. Bipromoter vector strategies may not effectively demonstrate these properties for the gene of interest since expression of both genes is not necessarily linked. Retroviral vectors have similarly been constructed from which GFP and drug resistance genes are included as selectable markers [24]. In control lines transduced with vectors encoding bicistronic GFP with a drug resistance marker, up to 99% of infected cells expressed GFP following selection. The GFP co-expressed with a therapeutically relevant protein may be useful in FACs-mediated selection for gene therapy applications, particularly where cells are genetically modified *ex vivo*, selected, and then reintroduced to patients.

Some investigators have found pEGFP to be somewhat toxic, particularly for preparation of stable transfectants (though we have not found toxicity to be a problem in transient assays). Results from one study utilizing four unrelated cell lines and three forms of GFP suggest that GFP can induce apoptosis [25]. Stratagene (La Jolla CA) has responded to these concerns with the development of Vitality^{TM} humanized recombinant (hr) GFP (emission peak at 506 nm) from an undisclosed species of sea pansy. HrGFP is less toxic over time and passage number and may be superior particularly for establishing stable transfectants of fluorescent protein-encoding genes. Stratagene has developed hrGFP counterparts to many of the fluorescent fusion protein vectors marketed by Clontech (see Table 2). These vectors are unique in that they possess *LoxP* sites into which hygromycin, puromycin or neomycin resistance cassettes can be inserted by *in vitro* recombination, thus allowing more flexibility for generating stable transfectants. In the pIRES vectors, the MCS is just downstream of the CMV promoter which permits FLAG (-1) or hemaglutinin (-2) tagging of the inserted gene. The FLAG sequence is followed by an IRES which precedes the hrGFP gene. The Moloney murine leukemia virus-based pFB–hrGFP is a GFP-encoding control for determining transduction and expression efficiency. Corresponding β-gal and firefly luciferase (LUC) counterparts (pFB–Neo–LacZ and pFB–Luc) are also available.

In addition, Stratagene markets a series of *trans*-reporting systems for determining whether a given compound or protein of interest phosphorylates and activates, directly or indirectly, any of six transcription factors (c-Jun, Elk1, CREB, c-Fos, ATF2 or CHOP). The system involves a "transactivator plasmid" possessing a transactivator domain from one of the genes listed above fused to the GAL4 DNA binding domain. If the protein of interest (encoded by an expression vector) or the test compound results in phosphorylation of the activator domain, the GAL4 fusion protein will bind to the GAL4 recognition sequence upstream of the reporter gene and activate transcription of the reporter gene. A "transactivator-less plasmid" is also available into which an activator domain of choice can be ligated and evaluated. The reporter genes available in this system include LUC, β-gal, chloramphenicol

Table 2
Select fluorescent vectors from Stratagene

Base vector	Extension	Description
pHrGFP	-C	For fusions to N-terminus of HrGFP
	-N1	For fusions to C-terminus of HrGFP
	-1	Source vector for hrGFP
	-Golgi -Mito	Localize to the respective organelle (controls)
	-Nucl -Peroxi	
pIRES	-hrGFP-1	FLAG tag ($\times$ 3)
	-hrGFP-2	Hemaglutinin tag
pFB	-hrGFP	Retroviral vector

acetyltransferase, *Renilla* GFP, and secreted alkaline phosphatase. We have found this system to work predictably and effectively with low background reporter expression.

Stratagene also markets a series of *cis*-reporting systems in which gene products and drugs or other compounds of interest can be evaluated for their ability to stimulate, either directly or indirectly, transcriptional activation of gene expression through *cis*-acting elements. The reporters in this series consist of tandem repeats of *cis*-acting elements (AP-1, CRE, SRE, ISRE, GAS, NFAT, NF-kB, p53, SRF, or TARE) upstream of reporters including LUC or GFP. LUC reporter is available for all of these elements; GFP is available only for NF-kB, NFAT, CRE, and AP-1. Again, a cloning plasmid for inserting *cis*-elements of interest is available.

A number of other companies [among them, CPG Inc. (Lincoln Park, NJ), Lab Vision Corp. (Fremont, CA), Qbiogene (Carlsbad, CA), Gene Therapy Systems (San Diego, CA)] offer materials related to fluorescence tagging of gene products, antibodies specific for fluorescence tags, and fluorescence studies of signal transduction pathways. Because of space considerations, these have not been addressed.

4. Luciferases

Luciferase and luciferin are the generic term for the enzyme and substrate, respectively, that upon reaction generate bioluminescence in organisms such as marine bacteria, unicellular algae, coelenterates, beetles and fish. Genes, or cDNAs, encoding luciferases have been isolated from various organisms, including several bacterial species, the North American firefly (*Photinus pyralis*), jellyfish (*Aequoria victoria*), crustacean ostracod (*Vargula hilgendorfii*) and anthozoan sea pansy (*Renilla reniformis*) (a more comprehensive list can be found in 26). Because of the uniqueness of the reaction, many of these luciferases have been successfully employed as reporters in both prokaryotic and eukaryotic cells [26]. Several of the luciferases have been adapted for use as reporters in mammalian cells but only the two most widely used systems, those based on the firefly luciferase (LUC) and *Renilla* luciferase (R-LUC), are described here.

4.1. Firefly luciferase

LUC, encoded by the *luc* gene, catalyzes the oxidation of beetle luciferin to oxyluciferin with emission of green–yellow light (λmax = 562 nm). de Walt *et al.* [27] initially proposed the use of LUC as a reporter enzyme and constructed the corresponding reporter vectors in 1987. Today, LUC is arguably the most widely used reporter gene for quantitative analysis of gene expression in mammalian cells. This should not be surprising as LUC exhibits almost ideal characteristics for a reporter: no equivalent activity is present in mammalian cells; assays for LUC are rapid, simple and inexpensive; the standard luminometer assay is sensitive to as few as $1-3 \times 10^5$ molecules of LUC [3].

Since our last review, the LUC reporter system has expanded in several directions. One area of accelerated growth is the development of sensitive technology to image luciferase activity in living cells and organisms. This topic is beyond the scope of the present article and readers are directed to a recent and extensive review of this subject [26]. Other improvements in the LUC reporter system can be placed in two general categories: modification of the enzyme and modification of the reaction assay.

4.1.1. Modifications of the luc gene

Efforts to study protein structure-function relationships have led to the generation of multiple LUC mutants, most of which have not been exploited for practical applications. One modified *luc*, *luc +*, however, has found wide usage as a reporter gene largely because of its enhanced features and commercial availability. The *luc +* variant was generated by multiple modifications to the wild-type *luc* gene [28]. With respect to reporter performance, however, the most important modification appears to be mutation of the C-terminus, tri-amino-acid peroxisome localization signal. As a consequence, whereas wild-type LUC protein accumulates in peroxisomes of mammalian cells, LUC + is retained in the cytoplasm. This differential compartmentalization appears to lead to a greater stability of the LUC + protein which would also explain the four- to fivefold higher luminescence of LUC + compared to LUC in NIH3T3 cells [28]. When incorporated into new generation vectors (i.e., to generate pGL3, Table 3), *luc +* expression is 20- to 700-fold greater, depending on the cell type examined, than the wild-type *luc* gene cloned into older type vectors [28]. The pGL3 plasmids make luciferase even more useful for studying very weak promoters. Alternatively, lower amounts of DNA and transfection reagents can be used, increasing cost-efficiency.

A different luciferase mutant, LUCODC-DA, generated by fusion of a variant form of the mouse ornithine decarboxylase PEST sequence to the C-terminus of wild-type LUC, displays a functional half-life of 0.8 h compared 3.7 h for wild-type LUC in mammalian cells. The destabilized luciferase provides a more responsive

Table 3
Select, commercially available "gene fusion" reporter vectors

Reporter	Plasmid	Regulatory elements[1]		Source
		Promoter	Enhancer	
GFP	pEGFP-1[2]	−	−	Clontech
hrGFP	pFR-hrGFP[3]	+	+	Stratagene
LUC	pGL3-Basic	−	−	Promega
	pGL3-Promoter	+	−	Promega
	pGL3-Enhancer	−	+	Promega
	pGL3-Control	+	+	Promega
	pLuc-MCS	+	−	Stratagene
	pCRE-Luc[4]	+	+	Stratagene
R-LUC	phRG-B	−	−	Promega
	phRL-CMV[5]	+	+	Promega
SEAP	pSEAP2-Basic	−	−	Clontech
	pSEAP2-Control	+	+	Clontech
CAT	pCAT3-Basic	−	−	Promega
	pCAT3-Promoter	+	−	Promega
	pCAT3-Enhancer	−	+	Promega
	pCAT3-Control	+	+	Promega
β-Gal	pβgal-Basic	−	−	Clontech
	pβgal-Control	+	+	Clontech

[1]Vectors without regulatory elements are used to test putative control sequences. Those containing only core promoter or only enhancer elements are used for analysis of putative enhancers or promoters, respectively. Vectors containing both types of regulatory sequences are used as positive controls.
[2]Also available in ECFP, EYFP and destabilized fluorescent protein formats.
[3]Used in *trans*-reporting systems. Available in other reporter gene formats. See text for details.
[4]Enhancer represents 4 tandem copies of the cAMP response elements. Used in *cis*-reporting systems. Available in other response element and reporter gene formats. See text for details.
[5]Codon optimized R*luc* gene. Also available in other regulatory sequence formats.

reporter system and should be particularly useful in monitoring transient dynamic changes in gene expression that occur in the order of minutes or a few hours [29].

4.1.2. Modifications to the assay

Under standard conditions (i.e., at room temperature in the presence of excess ATP and substrate), the luciferase reaction emits a flash of light that peaks at 0.3 s after initiation of the reaction and then decays in a biphasic manner to approximately 10% of the peak value within 30 s followed by a more gradual decay over a period of several minutes [3]. The flash height is proportional to the concentration of luciferase in the reaction but measurement of peak luminescence requires special care and specialized equipment. For routine analysis, using standard laboratory luminometers, relative luciferase activity is calculated by measuring integrated light output values over a period of time, typically 10 s.

Certain chemicals, such as cytidine nucleotides and coenzyme A, dramatically alter the phenotype of the standard luciferase reaction—from one with flash kinetics to one with "glow" kinetics [30]. These compounds retard the decay phase of light output resulting in a stable luminescent signal with a half-life in the order of 10–15 min [28]. Because luciferase activity is typically measured using time-integrated light output values, glow-type reactions increase assay sensitivity by several fold while also decreasing assay variability at low enzyme concentrations. Some proprietary assay formulations, such as the Steady-Glo^{TM} Reagent (Promega Corp.) and LucLite^{TM} Reagent (Packard Instrument Company, Inc.), produce a luminescent signal that decays with a half-life of several hours [28]. These reagents are used for robotics-controlled, high volume (10,000–100,000 assays/day) analyses of luciferase expression commonly performed in the pharmaceutical industry.

4.2. Renilla luciferase (R-LUC)

R-LUC catalyzes the oxidative decarboxylation of coelenterazine to coelenteramide with emission of blue light ($\lambda_{max} = 480$ nm). The cDNA encoding R-LUC was isolated in 1991, but its use as a reporter gene in mammalian cells was not reported until 1996 [31]. The delayed development of the R-LUC system, coupled to the fact that it does not offer significant advantages over LUC, is probably a major reason for the greater, current popularity of the latter. The R-LUC reporter system also suffers from the higher cost of coelenterazine compared to beetle luciferin and lower sensitivity of the assay arising from non-enzymatic autoluminescence of the substrate and the lower quantum yield of the reaction (5–7% vs. nearly 90% for LUC) [3,28]. For these reasons, R-LUC has been used primarily as the "companion" reporter that functions as an internal control to compensate for experimental variables such as transfection efficiency and sample manipulation.

The use of R-LUC as an internal control reporter has been greatly advanced by the development and promotion of an integrated assay format, called the Dual-Luciferase Reporter (DLR) Assay, that permits rapid quantification of both LUC and R-LUC activities in the same sample and with sub-attomole sensitivity [28]. In this assay system, luciferase activities are measured sequentially with quenching of the LUC signal prior to initiation of the R-LUC reaction. A more economical, non-commercial formulation has also been described [32]. Furthermore, given the large differential between peak light emissions (562 vs. 480 nm) and structurally diverse substrates, with the appropriate instrument it should be possible to measure both LUC and R-LUC activities simultaneously [31].

As the second reporter, R-LUC is not simply limited to the role of an internal control. With the availability of multiple reporter genes and assay formats that are compatible with multiple reporter enzymes, R-LUC (and other reporter proteins) can be used to monitor more than one cellular event within the same system. For instance, in conjunction with LUC, R-LUC has been used to characterize the activity of bi-directional promoters [33], to compare the relative efficiencies of cellular cap-dependent and IRES-dependent translation in a bicistronic message [34], and to

examine the effects of agonists on two separate receptor pathways in the same cell culture [35].

It is necessary to point out that bioluminescence in *Renilla reniformis* resulting from the oxidation of coelenterazine does not correspond to blue light emission ($\lambda_{max} = 480$ nm) but rather green fluorescence with $\lambda_{max} = 509$ nm. This phenomenon results from the quantitative transfer of R-LUC-catalyzed luminescence energy to the *Renilla* GFP, which in turn emits the characteristic fluorescence. Because this bioluminescence resonance energy transfer (BRET), a physiological version of FRET, is strictly dependent on molecular proximity of R-LUC and GFP, this system appears to be ideal for monitoring constitutive and regulated protein–protein interaction in intact cells. Initial studies have provided confirmation for this prediction. For instance, Angers *et al.* [36] have demonstrated constitutive and agonist-stimulated homodimerization of β_2-adrenergic receptor (β_2AR) molecules in HEK293 cells after co-transfection of β_2AR-YFP (a yellow variant of GFP) and β_2AR-R-LUC fusion genes. Interestingly, no BRET signal was detected after co-expression of the unfused YFP and R-LUC proteins, suggesting that background noise due to direct interaction between R-LUC and YFP may not be a significant concern. Much of the future development of the R-LUC system is likely to be directed towards its application to BRET technology.

5. Alkaline phosphatase

Alkaline phosphatase (AP) is a generic term for a class of ubiquitous enzymes that hydrolyze a broad range of monophosphates and exhibit a particularly high pH optimum, around pH 10–10.5. Mammalian APs are glycosylated ectoenzymes attached to the plasma membrane by a carboxy-terminal glycosyl phosphatidylinositol (GPI) anchor. Four distinct mammalian isozymes have been identified [37] and two of the isoforms, the rat tissue non-specific AP (TNAP) and the human placental AP (PLAP), have been proposed as reporter proteins for studying mammalian gene regulation [3]. Their development in this capacity, however, has been severely hampered by the obvious fact that most mammalian cells exhibit abundant endogenous AP activity.

This limitation was circumvented by Cullen and colleagues [38] who generated a secreted form of PLAP, termed SEAP. The enzymatic properties of SEAP are equivalent to that of wild-type PLAP but because activity is assayed using an aliquot of the culture media, interference from endogenous, cellular APs is dramatically reduced. Further reduction in background activity is achieved by prior incubation of samples at 65°C and measuring enzyme activity in the presence of L-homoarginine [3,37,38], conditions that largely inactivate both the TNAP isoform, which accounts for the preponderance of endogenous activity in most cells, and the intestinal AP. SEAP, like PLAP and the germ cell isoform (GCAP), are highly resistant to these manipulations. Being a secreted reporter protein, SEAP offers several advantages over the more conventional intracellular reporter enzymes [3]: (i) the inconvenience of preparing cell lysates is eliminated; (ii) kinetic analysis of gene expression is easily

monitored by periodic collection and examination of the culture medium; and (iii) the transfected cells remain available for measurement of additional phenotypic characteristics such as cell growth rates, expression of specific mRNAs and enzymatic activities.

Aside from the commercial availability of SEAP-based reported vectors, the other major enhancement to this system is the synthesis of chemiluminescent substrates that provide significantly greater sensitivity than the traditional chromogenic substrate *p*-nitrophenyl phosphate [3]. Several chemiluminescent reagents are currently available and one frequently used substrate is a 1,2 dioxetane-based phosphate with the common name of CSPD [4]. CSPD is typically used in conjunction with a proprietary enhancing reagent and in this formulation the assay is capable of detecting as little as 10^4–10^5 molecules of SEAP and is linear over a 10^5-fold range of SEAP concentration.

Although SEAP can be detected with a sensitivity equal to or even greater than LUC, it is still more commonly used as the internal control reporter in promoter analysis experiments. Because of its secretory nature, however, SEAP is the preferred reporter protein in a number of specific applications including development of mammalian and insect cell-based systems for large scale production of recombinant therapeutic proteins, monitoring of *in vivo* tumor growth and anticancer drug efficacy in tumor cell xenografts, and mechanistic characterization of the constitutive and regulated secretory pathways.

Although SEAP is currently the most commonly used secreted reporter protein, it is not without limitations. For instance, endogenous AP isoforms released into the circulation or into the cell culture medium, particularly in neoplastic states or when using transformed cells [3,37] decrease sensitivity of the SEAP reporter system and could be problematic when testing weak promoters. Also, thermal treatment of the samples is an added inconvenience. Because of these limitations, it is not unlikely that SEAP will eventually be replaced by alternative secreted reporter systems. Promising candidates currently under development include the naturally secreted *Vargula* luciferase and a modified *Renilla* luciferase genetically engineered for secretion [39].

6. Chloramphenicol acetyltransferase

CAT was the first reporter utilized for assaying the transcriptional activity of mammalian regulatory sequences. CAT inactivates chloramphenicol, an inhibitor of prokaryotic protein synthesis, by conversion to mono- or di-acetylated molecular species [40]. Consistent with the desired properties of a reporter, mammalian cells do not possess CAT or CAT-like activity, exogenous CAT protein is fairly stable in such cells and assays to measure enzyme activity are reasonably sensitive [40]. These assays, however, require the use of radiolabeled chloramphenicol or acetyl-CoA, and because of the need to separate radiolabeled reactant from product, they are also cumbersome and not easily automated. In spite of these inconveniences, the CAT reporter gene was widely employed in the 1980s and early 1990s to study eukaryotic

gene regulation, largely because of the absence of a viable alternative reporter enzyme. With the introduction of the LUC system, development of chemilumines-cent-based assays for other reporter enzymes, and manufacture of affordable luminometers, there has been a gradual decline in the use of the CAT reporter system. This trend has not been abated even with the development of non-radioactive assays using fluorescent substrates or an ELISA-based CAT detection assay with the potential for automation [40]. Although more manageable, these methodologies still do not provide the sensitivity, linear dynamic range or cost-effectiveness of chemiluminescent-based assays. Today, CAT is still used as a reporter protein to a significant extent, largely because of the prior development of numerous CAT-based reagents (e.g., CAT vectors with and without regulatory sequences, stably transfected cells, transgenic mice) and in those laboratories without access to instruments to quantify chemiluminescent or fluorescent signals. In the absence of some unforeseen development, however, the current decline in the use of the CAT reporter system is likely to continue.

7. β-Galactosidase

The *E. coli lacZ* gene, encoding β-gal, is quite possibly the most recognizable and, cumulatively, the most frequently used reporter gene in science. This stature is due not only to its antiquity—β-gal/*lacZ* was the subject of intense studies throughout the 1950s that led to the operon model of bacterial gene structure and regulation—but also to practical characteristics such as the fact that β-gal is functional and well-tolerated in a wide variety of cell types and activity can be monitored both quantitatively, in cell lysates, and qualititatively, *in situ*, without the necessity of specialized laboratory equipment. In *E. coli*, and presumably in foreign hosts, active β-gal is a homo-tetramer with a subunit size of 1024 amino acids. β-gal hydrolyzes a broad range of natural (e.g., lactose) and synthetic β-D-galactopyranosidase, the latter of which forms the basis of the various detection methods [41].

Although β-gal- and CAT-encoding vectors for the analysis of mammalian gene promoters were developed at essentially the same time [1,2], the latter was used more frequently in this capacity. One reason for this preference was the 10- to 50-fold lower sensitivity of the standard, chromogenic β-gal assay based on the cleavage of ortho-nitrophenyl-β-D-galactoside. More sensitive quantitative assays, based on fluorescent and chemiluminescent substrates (detection limits of 6×10^5 and 4×10^3 molecules, respectively) are now available [41] but because these assays do not provide any advantages over the LUC reporter system, β-gal is primarily used as the internal control reporter in promoter analysis studies. A second disadvantage, recognized even at the time of introduction of the β-gal reporter system, is that mammalian cells possess β-gal enzymes precluding the use of this reporter in cells with high endogenous activity. Fortunately, mammalian β-gal activity can be significantly minimized, although not completely eliminated, by thermal denatura-tion at 50 °C for 1 h without adversely effecting the activity of the bacterial enzyme [4].

Prior to the development of the GFP system, the real advantage that the β-gal system provided over the earlier reporter genes was the ability to conveniently detect reporter activity *in situ* at a single cell resolution, making β-gal the reporter of choice for visualization and tracking of cellular gene activity within a larger population of non-expressing cells. Because of this capability, among other applications, β-gal has been used to characterize the *in vivo* activity of cell-specific, tissue-specific or developmentally regulated gene promoters; to identify regulatory sequences involved in development, pattern formation and organogenesis; to detect patterns of cell movement (e.g., metastases of cancerous cells); to tag and track pre- and post-clinical, therapeutically relevant implanted cells; to identify and isolate lineage- and stage-specific cells; and to gauge the potency of prospective mutagens [41]. *In situ* detection of β-gal activity is most commonly carried out using the substrate 5-bromo-4-chloro-3-indolyl β-D-galactoside (X-Gal), which upon hydrolysis and subsequent reactions yields a blue-indigo precipitate. Positive staining with X-Gal requires fixation of cells and approximately 100–1000 molecules of β-gal per cell [41]. A more sensitive, fluorogenic variant, Fluor-X-Gal, can also be utilized [42]. An alternative *in vivo* detection strategy is based on the substrate fluorescein di-β-D-galactopyranosidase (FdG), which upon hydrolysis releases fluorescein that is retained intracellularly [42]. β-gal dependent, fluorescein-tagged cells can be identified and isolated by fluorescence-activated cell sorting for further analysis and propagation.

A more recent and novel application of the β-gal reporter is for detecting protein–protein associations in live mammalian cells [42]. This application exploits a well-established characteristic of β-gal known as α-complementation, whereby specific β-gal mutants can, to varying degrees, complement one another to reconstitute enzyme activity. Similar to FRET and BRET, the β-gal protein interaction detection system requires fusion of putative interacting proteins to individual reporter proteins, in this case weakly complementary mutants of β-gal. Poorly complementary mutants are necessary so that the β-gal activity detected results from association between the test proteins rather than the β-gal mutants themselves. The incidence of protein–protein interaction, and the regulation of such associations, can be evaluated quantitatively, or qualitatively in intact cells, by any of the β-gal assays described above. Two apparent advantages of the β-gal system over the GFP-based FRET are that: (1) enzymatic amplification of the signal provides high sensitivity; and (2) specialized laboratory equipment is not required.

8. Concluding remarks

Recombinant DNA technology has revolutionized modern biotechnology and medicine. As gene therapy approaches and disease targets continue to proliferate, reporter genes will play an ever more important role in the optimization of gene transduction and transfection, and in *ex vivo* selection of genetically modified cells for introduction *in vivo*. As the menu of new reporters grows, we envision that those which, like the fluorescence markers, permit dynamic real-time visualization of cellular events, will experience tremendous popularity. The discovery of new fluorescent proteins and the improvement of available proteins together with the development of faster and more sensitive camera systems and improved software will all contribute to improved utility of reporter genes.

Acknowledgements

We are grateful to Margaret Overstreet for clerical assistance with this review.

Abbreviations

AP	alkaline phosphatase
β-gal	beta-galactosidase
BRET	bioluminescence resonance energy transfer
CAT	chloramphenicol acetyltransferase
DsRed	*Discosoma* red fluorescent protein
ECFG	enhanced cyan fluorescent protein
EGFP	enhanced green fluorescent protein
EYFP	enhanced yellow fluorescent protein
FRET	fluorescence resonance energy transfer
IRES	internal ribosome entry site
LUC	luciferase
MCS	multiple cloning site
PLAP	placental alkaline phosphatase
R-LUC	*Renilla* luciferase
SEAP	secreted placental alkaline phosphatase
TNAP	tissue non-specific alkaline phosphatase

References

1. Gorman, C.M., Moffat, L.F., and Howard, B.H. (1982) Mol. Cell. Biol. 2, 1044–1051.
2. An, G., Hidaka, K., and Siminovitch, L. (1982) Mol. Cell. Biol. 2, 1628–1632.
3. Alam, J. and Cook, J.L. (1990) Anal. Biochem. 188, 245–254.
4. Olesen, C.E.M., Yan, Y.-X., Liu, B., Martin, D., D'Eon, B., Judware, R., Martin, C., Voyta, J.C., and Bronstein, I. (2000) Methods Enzymol. 326, 175–203.
5. Chalfie, M. and Kain, S. (eds.) (1998) Green Fluorescent Protein: Properties Applications, and Protocols, John Wiley & Sons, Inc., New York, N.Y.
6. Yang, F., Moss, L.G., and Phillips, G.N., Jr. (1996) Nat. Biotechnol. 14, 1246–1251.
7. Welsh, S. and Kay, S.A. (1997) Curr. Opin. Biotechnol. 8, 617–622.
8. Davis, I., Girdham, C.H., and O'Farrell, P.H. (1995) Dev. Biol. 170, 726–729.
9. Brand, A. (1995) Trends Genet. 11, 324–325.
10. Heim, R., Cubitt, A.B., and Tsien, R.Y. (1995) Nature (London) 373, 663–664.
11. Kaether, C. and Gerdes, H.-H. (1995) FEBS Lett. 369, 267–271.

12. Wacker, I., Kaether, C., Kromer, A., Migala, A., Almers, W., and Gerdes, H.-H. (1997) J. Cell Sci. 110, 1453–1463.
13. Hanaka, H., Shimizu, T., and Izumi, T. (2002) Biochem. J. 361, 505–514.
14. Klein, R.L., Lewis, M.H., Muzyczka, N., and Meyer, E.M. (1999) Brain Res. 847, 314–320.
15. Benraiss, A., Chmielnicki, E., Lerner, K., Roh, D., and Goldman, S.A. (2001) J. Neurosci. 21, 6718–6731.
16. Naylor, L.H. (1999) Biochem. Pharmacol. 58, 749–757.
17. Overton, M.C. and Blumer, K.J. (2000) Curr. Biol. 10, 341–344.
18. Gadella, T.W., Jr. and Jovin, T.M. (1995) J. Cell Biol. 129, 1543–1558.
19. Nagy, P., Bene, L., Balazs, M., Hyun, W.C., Lockett, S.J., Chiang, N.Y., Waldman, F., Feuerstein, B.G., Damjanovich, S., Szollosi, J. (1998) Cytometry 32, 120–131.
20. Damjanovich, S., Bene, L., Matko, J., Alileche, A., Goldman, C.K., Sharrow, S., and Waldmann, T.A. (1997) Proc. Natl. Acad. Sci. USA 94, 13134–131139.
21. Cormack, B.P., Valdivia, R.H., and Falkow, S. (1996) Gene 173, 33–38.
22. Haas, J., Park, E.C., and Seed, B. (1996) Curr. Biol. 6, 315–324.
23. Wagstaff, M.J.D., Lilley, C.E., Smith, J., Robinson, M.J., Coffin, R.S., and Latchman, D.S. (1998) Gene Therapy 5, 1566–1570.
24. Levenson, V.V., Transue, E.D.E., and Roninson, I.B. (1998) Hum. Gene Ther. 9, 1233–1236.
25. Liu, H.-S., Jan, M.-S., Chou, C.-K., Chen, P.-H., and Ke, N.-J. (1999) Biochem. Biophys. Res. Comm. 260, 712–717.
26. Greer, L.F., III and Szalay, A.A. (2002) Luminescence 17, 43–74.
27. De Wet, J.R., Wood, K.V., DeLuca, M., Helinski, D.R., and Subramani, S. (1987) Mol. Cell. Biol. 7, 725–737.
28. www.promega.com.
29. Leclerc, G.M., Boockfor, F.R., Faught, W.J., and Frawley, L.S. (2000) Biotechniques 29, 590–601.
30. Ford, S.R. and Leach, F.R. (1998) Methods Mol. Biol. 102, 3–20.
31. Lorenz, W.W., Cormier, M.J., O'Kane, D.J., Hua, D., Escher, A.A., and Szalay, A.A. (1996) J. Biolumin. Chemilum. 11, 31–37.
32. Dyer, B.W., Ferrer, F.A., Klinedinst, D.K., and Rodriguez, R. (2000) Anal. Biochem. 282, 158–161.
33. Igaki, H., Nakagawa, K., Aoki, Y., Ohtomo, K., Kukimoto, I., and Kanda, T. (2001) Biochem. Biophys. Res. Commun. 283, 569–576.
34. Shimazaki, T., Honda, M., Kaneko, S., and Kobayashi, K. (2002) Hepatology 35, 199–208.
35. Stables, J., Scott, S., Brown, S., Roelant, C., Burns, D., Lee, M.G., and Rees, S. (1999) J. Recept. Signal Transduct. Res. 19, 395–410.
36. Angers, S., Salahpour, A., Joly, E., Hilairet, S., Chelsky, D., Dennis, M., and Bouvier, M. (2000) Proc. Natl. Acad. Sci. 97, 3684–3689.
37. Millan, J.L. and Fishman, W.H. (1995) Crit. Rev. Clin. Lab. Sci. 32, 1–39.
38. Berger, J., Hauber, J., Hauber, R., Geiger, R., and Cullen, B.R. (1988) Gene 66, 1–10.
39. Liu, J., O'Kane, D.J., and Escher, A. (1997) Gene 203, 141–148.
40. Bullock, C. and Gorman, C. (2000) Meth. Enzymol. 326, 202–221.
41. Serebriiskii, I.G. and Golemis, E.A. (2000) Anal. Biochem. 285, 1–15.
42. Rossi, F.M., Blakely, B.T., and Blau, H.M. (2000) Trends Cell Biol. 10, 119–122.

S.C. Makrides (Ed.) *Gene Transfer and Expression in Mammalian Cells*

CHAPTER 7

Gene transfer and gene amplification in mammalian cells

Florian M. Wurm and Martin Jordan[1]

*Swiss Federal Institute of Technology Lausanne (EPFL), Faculty of Basic Sciences,
Institute of Chemical and Biological Process Science, Center of Biotechnology, CH-1015
Lausanne, Switzerland; Tel.: +41-21-693-6141; E-mail: Florian.Wurm@epfl.ch*
[1]Tel.: +41-21-693-6142; E-mail: Martin.Jordan@epfl.ch

1. Introduction on the origin of Chinese hamster ovary cells for recombinant protein production

Mammalian genomes have exceptional plasticity. They can exhibit short-term plasticity as evidenced by the fact that some DNA sequences, whether endogenous or introduced by artificial gene transfer, can be induced and/or selected for gene amplification to high copy numbers. In addition, mammalian genomes have an apparently unlimited capacity to take up foreign DNA, the hallmark of this being the presence of retroviral and bacterial DNA sequences that have become an integral part of these genomes. The short-term plasticity and ability to take up DNA makes cell lines useful for the expression and production of recombinant proteins of scientific and pharmaceutical value. Foreign genes can be easily integrated into mammalian cells cultivated *in vitro* and, through selection, populations of cells of clonal origin can be identified that show an increased copy number of the inserted genes. These cell lines express the required genes of interest at elevated levels. This approach has been used widely in both industry and academia, particularly with two immortalized cell lines, Chinese hamster ovary (CHO) cells and a mouse myeloma cell line, NS0. We will primarily concentrate on the *dhfr*/CHO system, since it is the most commonly used system. In general, most of the observations made with this system apply to other immortalized cell lines and thus should give insights into other types of gene transfer and amplification systems in mammalian cells.

CHO cells remain the most popular mammalian host for recombinant protein production since the mid 1980s, when the first product, human recombinant tissue-type plasminogen activator Activase® (rTPA), received regulatory approval for marketing. These cells have also been a preferred choice for studies on gene transfer and gene amplification in mammalian cells. Some of the studies were driven by questions raised by regulatory authorities controlling the clinical evaluation and marketing of pharmaceutical products. Other motivations were the need to maximize gene expression from transfected and amplified plasmid sequences and to establish recombinant cell lines with long-term stability with regard to the genome and to protein productivity. The popularity of CHO cells in recombinant protein production is based on both practical and virological safety reasons. Discussions on virus safety with mammalian production host systems started a half century ago when the first biological products, i.e., vaccines and natural interferons, were developed using primary and transformed mammalian cells. It is an essential goal both for the manufacturers of recombinant protein pharmaceuticals and for regulatory agencies to minimize any potential risks associated with the use of immortalized recombinant mammalian cell hosts.

CHO cells were established for the study of somatic cell genetics in 1957 by Puck [for review see [1]]. These cells are well characterized with respect to karyotype, chromosome structure, gene mapping and culture conditions. Upon the emergence of DNA technology one particular derivative of the original CHO cells (see below) was used in a number of pioneering gene transfer experiments [2,3]. It was possible to sub-cultivate these cells after the identification of genetically altered candidate clones producing the protein(s) of interest under increasingly selective conditions. This resulted in the generation of more productive "amplified" subclones. Another key feature for industrial scale production was the finding that CHO cells can be easily adapted to grow in suspension. In the mid 1980s R. Arathoon, A. Lubiniecki, G. Polastri and others at Genentech Inc. in San Francisco developed a 10,000-liter production process for rTPA, using "deep tank" technology with suspension-adapted CHO cells [4].

CHO cells are considered to be a "safe" host for the production of therapeutic proteins, which will be administered parenterally to human patients. Human pathogenic viruses like Polio, Herpes, Hepatitis B, HIV, Measles, Adenoviruses, Rubella and Influenza do not replicate in these cells. Thus, the risk of a viral adventitious agent being involuntarily carried along with the product of interest is low. Wiebe et al. tested a total of 44 human pathogenic viruses for replication in CHO cells and found that only seven of these (Reo 1,2,3, Mumps, and Parainfluenza 1,2,3) were able to infect these cells [5]. These safety considerations were discussed extensively among scientists from both regulatory agencies and the bio-pharmaceutical sector. Today one can only speculate whether an immortal *human* cell substrate with replicative capacity for a larger number of human viruses would have received approval for the production and marketing of a protein therapeutic. Only very recently (2002) was the marketing approval given for a protein pharmaceutical produced in the immortalized human cell line HEK-293 (Activated Protein C, requiring C-terminal carboxylation, E. Lilly). Due to complex processing and post-translational modifications, many therapeutic proteins must be produced in mammalian cells. At present (2003), at least 50 recombinant proteins made in CHO cells have received approval for marketing. It is difficult to assess the exact number of CHO-based product candidates that are in pre- and clinical evaluation, but it is with certainty above 200.

2. The DHFR/methotrexate/CHO expression system: a multi-layer selection system for high-level expression of recombinant genes

2.1. Gene transfer into immortalized cell lines uses an increasing number of vehicle and selection systems

In 1980, Urlaub and Chasin developed a CHO cell line [6] that was created by mutagenesis of the original K-1 cell line established by Puck and coworkers [7]. As intended by the investigators, this mutant line exhibited a complete deletion of DHFR enzyme activity. Therefore, growth of these cells is dependent on supplementation of the culture medium with hypoxanthine and thymidine (products of the biochemical pathway that includes DHFR). The amino acid glycine also has to be added, since the original K-1 cell line was glycine-dependent. This *dhfr*-deficient cell line quickly became the choice for transfections that employed the use of expression vectors carrying a functional *dhfr* gene and a second gene of interest (either provided on the same or on a separate plasmid). In selective medium (without glycine, hypoxanthine and thymidine; "GHT-minus") *dhfr*-positive clones can be identified after about two weeks usually as circular-shaped colonies of cells growing on the surface of the culture plate (Fig. 1). Many of the recombinant *dhfr*-positive clones also express the gene of interest [8].

Vectors containing the *dhfr* gene have been utilized also in host cells that contain a functional endogenous *dhfr* gene. Such cells require modified selection procedures. For example, a second selectable marker gene [9] or a mutant *dhfr* gene [10] can be used. Furthermore, combinations of several markers have been employed successfully [11]. DHFR selection was also used in insect cells to co-express the Hepatitis B surface protein [12].

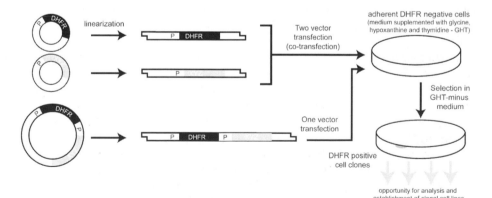

Fig. 1. Transfection and co-transfection of *dhfr*-containing expression vectors into CHO cells deficient in endogenous DHFR activity. Selection of recombinant clones in GHT-minus medium (lacking in glycine, hypoxanthine and thymidine). After transfection most cells die and a few begin to grow, forming visible colonies that can be picked with a cotton swab or with a micropipette tip or using the cloning ring procedure with trypsin to detach cells.

DNA co-precipitation using calcium phosphate was initially the main vehicle system for non-viral DNA transfer into mammalian cells [13]. Improvements to the original protocol by several investigators have resulted in much higher transfection rates and greater reliability [14,15]. More recently, a large number of other vehicle systems have been developed based on lipid formulation, charged polymers, or on more complex mixed formulations. These include polyethyleneimine/DNA complexes (PEI), DEAE dextran, lipofection with lipids such as N-(1(2,3-diolexyloxy)propyl)-N,N,N-trimethylammonium chloride (DOTMA), or cationic polymers that compensate for the negative charge of DNA. A number of companies now offer these vehicles in easy-to-use kits, allowing small-scale gene transfer to immortalized mammalian cells. Comparative studies on efficiency, executed by commercial entities, are difficult to assess. Very few non-commercial comparisons have been published. The availability of many selectable markers and the ease of reproducibly transfecting cells has made the establishment of clonal cell lines less of an art form. Besides *dhfr*, which is used almost exclusively with *dhfr*-negative CHO cell hosts, other selectable markers are in use, such as the glutamine synthetase gene [16]. Antibiotics such as neomycine, hygromycin and puromycin can be used in combination with their respective plasmid vectors encoding these antibiotic-resistance genes.

Electroporation is another widely used technology for transfer of naked DNA to cultured mammalian cells. Electroporation is achieved by supplying an electrical pulse to cells that are suspended in a small volume of buffer, containing a high concentration of DNA. The pulse is calibrated to prevent excessive damage to the plasma membrane, while generating small pores in the membrane through which the DNA-containing solutions can pass [17]. For optimal efficiency with a particular cell population each transfection technique requires some experimentation to identify the most suitable condition. Each transfection procedure results in the delivery of different amounts of DNA since they employ alternative mechanisms. These result in a changed pathway to the nucleus and accordingly differing susceptibilities to loss or degradation of DNA.

We wish to add another comment here on observed variabilities in transfections with mammalian cells. One frequently overlooked phenomenon is that standard laboratory cultures of mammalian cells will have sub-populations that differ in their individual growth rates, which are influenced by media composition. Also, typical cultures are non-synchronized when growing exponentially. We have found that the conditions of growth, degree of cell density and timing of transfection after the most recent feeding with fresh medium have dramatic effects on transfectability. Cell cycle synchronization of cultures prior to transfection and subsequent flow-cytometric analysis indicates that cells have a preferred phase for efficient uptake of DNA, at least with calcium phosphate. Maximum uptake and maximum (transient) expression were observed when cells were transfected in S-phase [18].

2.2. DNA transit to the nucleus and integration into chromosomal DNA is poorly understood

Graham and van der Eb [13] showed that the transfer of DNA into mammalian cells in culture can be achieved by exposing cells to naked DNA complexed into micro-precipitates of calcium phosphate. This simple method remains very popular, particularly after protocol improvements that resulted in high efficiencies of DNA transfer, similar to those seen with commercial transfection vehicles. Osmotic shock is part of some transfection procedures with calcium phosphate. Recent analysis in our laboratory has shown that osmotic shock does not promote a more efficient transfer of DNA across the plasma membrane [19]. We speculate that the rapid modification of the osmotic pressure outside the cell may act indirectly on DNA-containing endosomes, potentially reducing the rate of DNA degradation.

DNA transfer can be almost 100% efficient under optimum conditions. With calcium phosphate, the DNA copy number per cell can reach high values with some cells in a transfected population receiving more than 100,000 plasmid molecules per cell [20]. Tseng et al. [21] used fluorescent ethidium monoacid covalently associated with plasmids to show that cationic liposomes are able to deliver 10^5–10^6 copies of plasmid per cell. However, the transit of DNA from the cytoplasm to the nucleus seems to be a significant barrier to gene transfer, as only a small fraction of intracellular DNA reaches the nucleus [22]. Only 10% of radiolabeled DNA was found in the nuclear fraction, and more than 50% of intracellular radioactivity, representing degraded DNA, was associated with the cytosolic fraction. In spite of a dramatic loss of plasmid copies during transfer to the nucleus, usually a sufficient number of intact plasmids reaches the nucleus, driving transcription, at least during the first few days after gene transfer. The copy number of plasmids reaching the nucleus is in the hundreds of molecules, at least in optimized transfections. This is supported by our observation that short-term expression of green fluorescent protein (GFP) occurred in more than 80% of transfected cells when cocktails for transfection were used in which the GFP-encoding plasmids were present at only a 2% (w/w) ratio [20,23]. This is the technological basis for transient transfection, a method that has become very popular not only for small-scale operation but also for large-scale work [for review see [24].

Transient gene transfer and expression in mammalian cells should not be confused with stable gene transfer. Only a small percentage of transfected cells, usually 0.01–5%, give rise to stably transformed clonal cell lines, depending on the techniques used. The intranuclear processes resulting in the stable chromosonal integration of at least a few functional plasmid molecules are by no means efficient.

Presumably at some time point DNA-vehicle complexes dissociate and nuclear endo- and exonucleases have access to "naked" DNA. It has been shown that circular plasmid DNA becomes linearized within hours after nuclear entry, an essential step for integration into the linear DNA backbone of a chromosome [25]. We have found that linearization of plasmid DNA prior to transfection increases the frequency of stable transformation [15]. However, not all laboratories creating stable cell lines execute transfections with linearized DNA. The mechanism by which DNA is transported to the nucleus for subsequent chromosomal integration remains speculative.

The site of integration into a mammalian genome by foreign DNA is another poorly understood area. It appears that there are no preferred targets for integration. It would be interesting to know whether highly and moderately repetitive DNA sequences or gene-rich areas of the mammalian genome, representing about 3% of the mammalian DNA, are in any way preferred or avoided targets for integration. The variability of expression levels of recombinant proteins in independently established clonal cell lines is indicative of random integration loci. Our own analysis of amplified transfected DNA in CHO cells using fluorescent *in situ* hybridization (FISH) has revealed no preferred chromosome or chromosomal region for integration [26]. It should be noted, however, that loci identified for amplified DNA sequences after selection with drugs that favor gene amplification are most likely not indicative of the primary integration site. We discuss this in more detail in Section 3.

Linearization of closed circular plasmid DNA may be a prerequisite for integration. Single and double strand cleavage of plasmid DNA by host enzymes is expected to occur at random. Depending on the overall size of the plasmid molecule and the size of the *dhfr* gene cassette relative to the vector, a significant number of molecules will be linearized within the *dhfr* sequence jeopardizing plasmid function. Plasmids whose *dhfr* expression cassettes are interrupted or deleted by the linearization process, will not be able to transfer a functional marker gene to the host cells, which subsequently will be eliminated by the selection procedure. For this reason, we recommend restriction enzyme digestion of the circular plasmid molecule prior to transfection for stable transfections (Fig. 1).

Plasmids targeted to specific sites by homologous recombination are linearized within the region having homology to the target site in the genome [27–29]. We have used a similar approach in CHO cells with vectors that contain sequence elements derived from the hamster genome with the expectation that they would facilitate homologous recombination. The target sequences were CHO endogenous retroviral sequences that are widely distributed in the genome [30–32]. Three families of these endogenous retroviral sequences had been observed, two divergent families similar to A-type sequences in the mouse genome, and one family of C-type sequences [33,34]. We found that the presence of retroviral DNA fragments in our transfection cocktails promoted higher transfection efficiencies and improved expression levels when a vector containing a gene of interest (rTPA or a CD4–IgG fusion construct) was co-transfected [30]. Linearization of a DHFR vector within a region of homology to a C-type sequence resulted in a large number of clones expressing the recombinant protein at very high levels. Southern analysis of these clones revealed that the copy number of the integrated plasmid sequences was much lower than the copy numbers in clones derived from control transfections that did not employ retroviral sequences [35]. While a formal proof for homologous recombination using retroviral sequences in transfections has not been established yet, we propose that our data support the notion of an integration into transcriptionally active regions of the CHO genome, since high transcription rates from low-copy number integrants were observed for extended time periods.

Transfected plasmids have also been targeted by homologous recombination to single copy targets whose identities have not been revealed by pharmaceutical companies executing this work. As can be concluded from public presentations about these experiments, a high transcription rate locus has been identified for the targeting of recombinant genes. This approach requires well-designed vectors in combination with plasmids providing enzymes favoring targeted integration such as Bacteriophage P1 Cre recombinase, and lambda phage integrase or yeast Flp recombinase that are capable of exchanging long stretches of DNA that are bordered by regions of homology between the genome and the vector DNA [36]. This recombinase-mediated cassette exchange is reviewed in detail in Chapter 20. Cell lines developed on the basis of this approach have achieved some of the highest productivities reported in the literature (50–100 pg/cell/day).

2.3. Copy number heterogeneity in transfected cell populations

The majority of recombinant cell clones obtained using the calcium phosphate and other transfection methods integrate more than a single copy of the vector DNA [37,38]. Perucho et al. [39] have studied the size of chromosomally integrated concatemeric "package" and found some of them to be up to one million base pairs in length. In our own work, we have verified multi-copy integrations in clonal cell lines selected in GHT-minus medium and have found a wide variation in copy number in those primary clones. The copy number of the rTPA gene in six clones chosen from 12 established and expanded cell lines varied over a 100-fold when analysed by hybridization and computer-assisted densitometry [40]. Insect cells cotransfected with the *dhfr* gene [12] were shown to have integrated between 10 and 240 copies of a vector for the expression of the surface protein of Hepatitis B. The highest copy number was found in clones that had been transfected at a 1:50 ratio of the selection plasmid to the vector containing the HBsAg gene. The phenomenon of co-integration of many copies of transfected plasmid DNA constitutes the basis for the high success rate of co-transfections, in which expression of the gene of interest is to be maximized. Co-integration of co-transfected plasmids (provided in circular or linear form) into the same chromosomal locus is a commonly observed phenomenon. We have analysed 180 random clones from a co-transfection of a TNFR–IgG fusion construct and found a very high co-expression rate (>97%) of *dhfr* and TNFR–IgG [41].

We speculate that the high rate of co-expression is due to the abundant nuclear ligase activity that creates large concatemers, even prior to integration of plasmid DNA into the genome. It has been established that mammalian cells have abundant end-joining capabilities [42] and co-transfected DNA becomes physically and genetically linked [39]. Most of the ligase activity seems to be independent of regions of homology between the ligated partners. Ligation of linearized plasmid molecules generates multimeric chains containing DNA sequences of the original plasmid mixture in a random order and probably proportional in ratio to the initial molar ratios of the plasmids. Indeed, co-transfection protocols usually recommend the use of a plasmid cocktail consisting of a molar ratio of 1:5 to 1:20 of the DHFR plasmid and the plasmid carrying the gene of interest in order to optimize expression of the target gene while providing sufficient expression of the selectable marker gene.

Supportive evidence for the co-integration of co-transfected DNAs is also provided by our own unpublished studies using FISH. We have analyzed 12 clones of co-transfected cell lines expressing *dhfr* and rTPA from two different plasmids. With the exception of two heteroploid cell lines (showing two loci of integration), 10 cell lines had a single integration site in one of the 18–22 chromosomes exhibited by the host. The probe for these hybridizations would have detected each one of the two transfected plasmids and the fact that only one signal was detected indicates that these clones contained both plasmids at one locus. One should be aware that co-integration of multiple plasmids is probably a general phenomenon in eukaryotic cells. Chen *et al.* [43] reported the generation of transgenic rice plants expressing 13 different plasmids out of 14 that were used in the DNA cocktail. Genetic analysis revealed integration into one locus (see also Chapter 8).

Most transfection protocols for *transient* expression succeed in obtaining high efficiencies, thus almost all the cells must have taken up transfected DNA. In contrast, *stably* transfected cell colonies in selective media usually develop from less than one percent of the cells exposed to the DNA. This difference in expression success suggests that stable expression from a genetically maintained DNA represents a rare event. Moreover, the majority of cells that have taken up the DNA are killed by the selection procedure. Thus, integrations in a single cell are highly unlikely to occur at two or more different chromosomal positions. Indeed, Toneguzzo *et al.* reported on restriction analysis of more than 50 clones generated by electroporation of lymphoid cells, and found that more than 90% of these cell lines had a single integration site per cell [44].

The success of establishing stable cell lines is not dependent on the nuclear concentration of exogenously provided DNA. Using calcium phosphate as a delivery vehicle, we showed that cells with the highest intracellular copy number of plasmids had the lowest probability of survival for more than 2 days [19]. While these cells did express the "transiently" delivered DNA, indicating its presence in the nucleus, cell division appeared to be blocked. A viable and proliferating cell line cannot be developed from such cells.

3. Gene amplification is a phenomenon occurring frequently in immortalized mammalian cells

DHFR activity can be inhibited completely by the folate derivative methotrexate (MTX) [45]. By adding it to cell culture media it is taken up readily by cells. Unless individual cells have DHFR enzyme concentrations exceeding the molar concentration of MTX, cells will eventually die due to complete inhibition of DNA synthesis. The exposure of recombinant cells to MTX, therefore, selects for those *individual* cells that show elevated expression of *dhfr*. Typically, protocols are used that apply increasing concentrations of MTX (two- to threefold for each step) from 10 nM to several 100 μM over several weeks of selection.

At each concentration, subpopulations of cells can be identified that have increased DHFR activity as compared to the average expression from the parental generation of cells. In selected cells, a high-level of DHFR activity can be the result of one of two genetic phenomena: (a) high expression level resulting from a few *dhfr*-plasmid molecules integrated into the host genome, probably at a locus that favors high transcription rates; or (b) amplification of *dhfr*-plasmids that express individually at relatively low levels. The level of mRNA molecules transcribed from highly amplified DNA sequences provides the genetic basis for survival under stringent selection. The diagram in Fig. 2 does not distinguish between gene amplification in pools of cells or gene amplification in individually isolated clonal cell lines. We favor single-clone gene amplifications, although this may not necessarily be the preferred mode.

Since chromosomal translocations and gene amplification occur spontaneously at a low frequency in all cells in culture, the phenomena discussed here can be considered Darwinian in nature. They only become apparent when selections are applied favoring survival of the rare occurrences. Surprisingly, gene amplification seems to be a rather frequent phenomenon in immortalized cell lines. In CHO cells and in human tumor cells, where this phenomenon had been recognized initially, gene amplification is associated with the formation of marker chromosomes characterized by long homogeneously staining regions (HSRs). In *dhfr*-transfected CHO cells the HSRs contain plasmid DNA embedded in large areas of endogenous chromosomal DNA [46]. Using a stepwise incremental selection procedure with MTX, a number of cell lines have been developed that have 100s or even 1000s of copies of the transfected plasmid constructs [26,47]. Some gene-amplified cell lines express the desired protein at very high levels, with peak productivities in the several 100s of milligrams per liter in standard cell culture production processes [48–50].

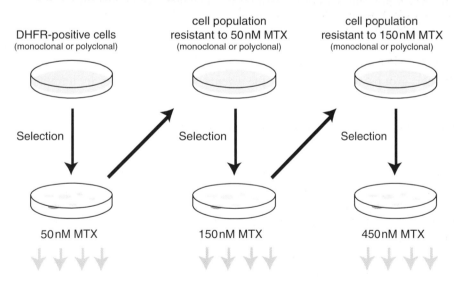

Fig. 2. Incremental increase in MTX concentration in culture and selection of subpopulations of cell clones resistant to high levels of MTX. Most cells die upon elevated exposure to MTX. A few cells with increased *dhfr* expression survive to initiate formation of visible colonies that can be picked and/or continued to be subcultivated as a population and hence be exposed to even higher MTX concentrations.

3.1. Chromosome segments with plasmid sequences are excised and circularized to form replicating episomes, which reintegrate into the host genome

The mechanisms responsible for gene amplification were the subject of intensive research in the 1980s and '90s. These will not be discussed in detail, and we refer the reader to publications cited below. Here, we summarize the current understanding of these phenomena. Most studies have been done with immortalized, i.e., permanent, recombinant and non-recombinant cell lines using antiproliferative agents, including MTX, for selection and subcultivation of mutants. MTX is unique for several reasons. Blocking DHFR activity with MTX lowers the intracellular pool of precursors for the synthesis of adenine and thymidine components because no alternative pathway exists. It has been shown that a low intracellular level of thymidine can result in the misincorporation of uridine monophosphate for thymidine monophosphate during DNA replication, inducing the activity of DNA repair enzymes responsible for excision of mismatched DNA regions. The occurrence of single-stranded gaps within chromosomes has been demonstrated in cells exposed to MTX, and single-stranded gaps in chromosomes have been cytogenetically documented as fragile sites [51,52]. Thus, mismatched DNA regions are assumed to contribute to increased recombinogenic activity of cells treated with MTX. As a consequence, MTX-induced chromosomal plasticity may result in gross rearrangements and amplification [53]. The term "fragile sites" was first coined by Barbara McClintock in her studies with *Zea mays* [54]. Taken together, when added to cells, MTX not only kills those with insufficient DHFR-activity but may also lead to an increased rate in recombinogenic events contributing to amplification of affected chromosomal regions.

Wahl and collaborators [55,56] have provided insights into the early events of integration and amplification of recombinant DNA. Hamlin and coworkers [57,58] and other groups have addressed the events that occur later in amplification. The Wahl group has analyzed CHO cell transformants displaying high-frequency gene amplification upon exposure to high levels of MTX. They use the term "hyperamplifiable" to describe the phenotype of such cells [55]. In these cells, representing about 3% of the initial recombinant transfectants, integration of the transfected DNA into a chromosomal site resulted in the inverted duplication of the plasmid DNA and flanking host DNA within one of the two arms of the chromosomal DNA at the insertion site. The overall size of the duplication (more than 70 kb) was much larger than the plasmid DNA. Convincing evidence was presented that the primary transformants grown *in the absence of MTX* underwent rapid copy number changes. This was most likely due to the generation of large episomal elements (larger than 750 kb but too small to be detected by light microscopy) containing multiple copies (inverted duplication of chromosomal DNA flanking the plasmid DNA). It was proposed that one of the reasons for the almost instantaneous emergence of extrachromosomal elements in transfected cells is chromosomal instability at or around the site of initial integration of the donated plasmid DNA [59]. Since there are no data available at present indicating a preferential site(s) within CHO chromosomes one has to assume that intregration is random. In addition, there is no evidence to indicate a preference for integration of foreign DNA into transcriptionally active regions of the genome. Genetic studies, however, have shown that all regions of the genome are not equally recombinogenic. One hypothesis used to explain such position effects on recombination is to invoke the existence of recombinogenic hot spots equivalent to the *Escherichia coli* chi sites. The rapid emergence of extrachromosomal elements from such sites would provide cells with a hyperamplifiable phenotype and with a high degree of flexibility in response to the environment. If higher copy numbers of such extrachromosomal elements were the reason for a growth advantage over cells with single or low copy numbers, then transient stabilization of this early amplification could occur as long as the selective condition (lack or low level of nucleoside precursors) existed. The exposure of cells with variant copy numbers exhibiting such a wide range of expression of the selectable gene, to low and moderate levels of MTX would result in a further increase of the overall copy number of the transferred DNA within the selected subpopulations.

A FISH cytogenetic study [60] utilizing cells containing a single, stable *dhfr* gene integrated into a CHO chromosome, supported the mechanisms for early gene amplification described above. Exposure of these cells to MTX for only eight to nine cell population doublings resulted in the generation of subclones exhibiting a wide variation in copy number and in the structure of multiple, *extrachromosomal* elements (Fig. 3a). The conclusion from this observation is that MTX addition to cells containing a single copy of the *dhfr* gene in a stable chromosomal site mediated excision of *dhfr*-containing DNA from the chromosome in many cells. This wide range in copy number of extrachromosomal elements fits well into a model for the emergence of "double minutes", small, chromosome-like fragments, visible under light microscopy. Double minute chromosomes occur frequently in mouse cell lines where they may persist for long periods of time, as long as MTX-selection is applied. In CHO cells, they can give rise to HSRs [61]. This is in fact supported by the FISH study of Windle, when the stable cell line with a single *dhfr* gene was exposed for a period of 30–35 days to MTX. A number of resistant clones were found in which the majority of cells (98%) showed *intrachromosomal* hybridization to clustered amplicons (tightly arranged stretches of DNA containing plasmid sequences). The copy number, as measured by fluorescence intensity, varied dramatically from clone to clone (up to 30-fold) and also did the location of these amplicons in chromosomes, but the copy numbers were relatively similar within each individual clone (only a 1.2-fold variation). These phenomena are depicted diagrammatically in Fig. 3.

A different mechanism for gene amplification was proposed by Ma *et al.* [57] who found that the initial event for amplification in CHO-K1 cells was a duplication of the endogenous *dhfr* locus by two types of sister chromatid fusions. One type occurred after symmetric chromosome breaks, the two chromatids of the same chromosome break at the same locus (Fig. 4). The fusion of the two frayed chromatid ends of equal length "healed" the breakpoint prior to the next mitosis. Alternatively, asymmetric breaks of the two chromatids can fuse as well. In either type of event, chromosomes with a duplication of the *dhfr* locus and losses of one telomere and/or two centromeres can emerge. Due to the forces exerted during

320

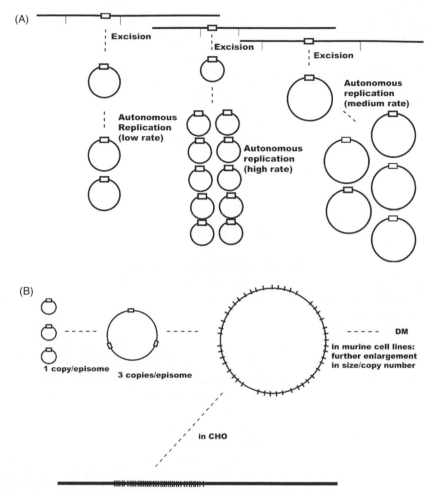

Fig. 3. The two diagrams shown here are based on models proposed by Windle *et al.* [60], Levan *et al.* [61] and others. (a) Three scenarios for the origin of different types in size and copy number of episomal elements that can be generated from daughter cells of a cell line carrying a stable insert of a *dhfr* gene, upon short-term exposure to MTX (5–10 days). (b) Upon long-term exposure to MTX, individual episomes can increase in size through inter- or intramolecular recombination or other mechanisms, and thus increase the copy number of the selective gene (and linked DNA sequences). In CHO cells, reintegration into a chromosomal site eventually stabilizes high copy-number carrying cells and assures long-term stability of the selective gene.

mitotic separation, such chromosomes/chromatids are subject to further breaks that facilitate additional rounds of chromatid fusion and further duplication events in one of the emerging cells after mitosis. This bridge-breakage–fusion mechanism was suggested to explain the high frequency of dicentric chromosomes in cells with amplified *dhfr* copies [57,62]. The model explains why early amplicons are most often located on the same chromosomal arm as the parental single-copy locus, but often quite distal from that site. It also provides an explanation for the clustering of the

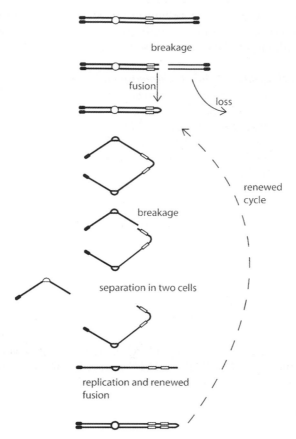

breakage

fusion

loss

renewed
cycle

breakage

separation in two cells

replication and renewed
fusion

Fig. 4. Model of sister chromatid fusion after symmetric breakage, near an endogenous *dhfr* locus. Premitotic fusion of homologous chromatids results during mitosis in breakage so that one cell will receive a chromatid with two *dhfr* loci. The end of the chromatid near the *dhfr* locus does not have a functional telomer. This favors fusion of the two ends after replication of the DNA (during S-phase). Thus, additional rounds of breakage and fusion may occur during subsequent cell cycles, increasing the copy number of the *dhfr* locus. Note: an alternative model, resulting in dicentric chromosomes is not shown here.

amplified sequences, sometimes recognizable as multiple bands following *in situ* hybridization. These two mechanisms discussed here are strikingly different, and both have strong experimental evidence in support of them. We stress, however, that the two models are simplified representations and are not necessarily conclusive.

What happens to *dhfr*-transfected CHO cells that are subsequently exposed to MTX? It is difficult to answer this question confidently. However, it is unlikely that the process of sister chromatid fusion, if in fact it represents a major contributor to gene amplification in CHO cells, would only be restricted to the amplification of endogenous *dhfr* genes. Similar events will likely occur with transfected, chromosomally integrated DNA that was initially amplified via the "excision–episome–expansion–reintegration pathway" suggested by Wahl, Stark and others

[55–57,61,63]. By presenting these mechanisms, we aim to indicate to the reader that many diverse, complex, and only partially understood phenomena may lead to individual cells with elevated expression of *dhfr* during gene amplification.

3.2. CHO cells with highly amplified, chromosomally localized DNA sequences are the result of long-term exposure to incrementally increased MTX concentrations

A survey of the literature suggests that a wide range of MTX concentrations (from nM to μM) are used for the establishment of highly productive CHO cells. Some cell lines have been established that were resistant to more than 100 μM MTX [18]. However, some cell lines achieve a high specific productivity in low level of MTX, depending on the initial site of integration. In comparison to cell lines selected at nM concentrations of MTX, cells resistant to μM concentrations of MTX show even more elongated chromosomal structures containing the transfected DNA [26,37]. Such highly elongated chromosomes are most likely equivalent to the regions described by Biedler and Spengler more than 25 years ago as HSRs [64]. Some of the chromosomes containing a large number of tightly arranged repetitive sequence elements homologous to the transfected DNA, differ dramatically from "normal" CHO chromosomes. While this raises questions about the long-term stability of such unusual chromosomes, conclusions cannot be made based on morphological features alone.

Primary clones isolated at the minimal level of stringency in GHT-minus medium vary in their ability to respond to MTX-induced amplification. The rapidly amplifiable clones may be the ones referred to by Wahl and coworkers as having a hyperamplifiable phenotype. Two groups have isolated sequence elements from mouse and CHO DNA, respectively, characterized by unique AT-rich repetitive elements that facilitate the chromosomal amplification of the transfected DNA when integrated into expression vectors [65,66]. Occasionally, the flanking genomic DNA of the integrated plasmid sequences may have structural and sequence characteristics that facilitate amplification. Evidence for effects mediated by these chromosomal domains on amplification of integrated plasmid sequences has been presented by Carrol *et al.* [67]. These authors demonstrated the involvement of a mammalian replication origin within an episome containing transfected DNA. In addition, other findings [60] provide support for the involvement of host sequences in the facilitation and/or generation of excisional episome formation following plasmid DNA integration. Episomes containing sequences that facilitate replication and amplification may, upon MTX selection, rapidly increase in copy number and eventually reintegrate as multimers into chromosomal DNA. In our own studies, we have found a striking variation of copy numbers of integrated plasmid sequences in primary clones selected in GHT-minus medium. By Southern analysis some clones showed very weak hybridization signals corresponding to one to five copies of plasmid DNA while others showed a very strong signal indicating 50–250 copies of transfected plasmid [40].

Another reason for the observed ranges in expression of different clones from the same transfection, *excluding* copy number effects, might be the influence of the locus into which the primary integration event occurred. Integrated DNA within a chromosomal region of clusters of highly transcribed endogenous genes would likely have a higher transcription rate. In contrast, lower transcription rates are expected when the same DNA becomes inserted into an extended cluster of highly repetitive, non-coding sequence elements. In the first case, progeny cells would be resistant to moderate or high levels of MTX. Only very high levels of MTX would be able to slow down the growth of these cells that express high levels of *dhfr* from a few favorably inserted copies. From such cells, transcribing *dhfr* DNA at high level, subpopulations cannot be easily selected with a further increase in *dhfr* expression, at least when applying moderate concentrations of MTX.

The influence of the original chromosomal environment on transcription and on selectivity would be maintained even when episome formation occurs, since the size of episomes exceeds that of plasmid DNA by several 100-fold, the largest part of the excised episome containing regions of genomic DNA from the original integration locus [67]. The structure of the DNA at the integration locus might have an influence on the rate of episome formation and the extent of copy number amplification within the episomal DNA prior to reintegration.

In summary, it is clear from the above discussion that a multitude of diverse structures are formed during the initial integration and primary amplification events and during periods of stepwise selection in MTX. In order to obtain a reliable source of cells for the stable expression of recombinant proteins, one should consider the genetic stability of these structures both at the level of the individual cell and the level of cell populations generated during selection. To minimize the risk of heterogeneity "young" cell populations generated from a colony on a plate or from a single cell in a well should be evaluated and then expanded to generate a frozen cell bank.

3.3. MTX in culture media affects the copy number of amplified sequences within a cell population

This section addresses studies to evaluate genetic stability of cloned and non-cloned cell populations using FISH. The genetic stability of recombinant sequences in transfected CHO cells is relevant to the study of gene expression and is an essential and required condition for the production of therapeutic proteins in cell lines.

Recombinant cell lines are usually established as clonal progeny of isolated colonies or single cells. Nevertheless, the inherent plasticity of eukaryotic genomes as well as subtle selective pressures applied during continuous cultivation may affect the stability of the transfected and integrated DNA sequences. Hence, verification of the integration and amplification of recombinant sequences is important and may be performed using a variety of techniques, including Southern DNA blotting, cytogenetic analysis of ECRs, and chromosomal *in situ* hybridization. Southern DNA blotting of DNA extracted from a cell line provides little information about subpopulation heterogeneity and the rearrangement of amplified sequences and gives no information on chromosomal location. ECRs are detectable as large, uniformly staining regions (also called HSR), but they are recognized only when they constitute more than 10% of the total length of the chromosome in question. They are also difficult to detect when the cell line under study exhibits a variable karyotype with numerous chromosomal aberrations. This is in fact a typical situation with CHO cells [7]. A more suitable technique in which unique and amplified DNA sequences can be identified and visualized in a convincing manner is FISH, using DNA- or RNA-probes generated with fluorescence-labeled nucleotides. This method offers high sensitivity with markedly improved speed and spatial resolution [68,69].

We and others have used this powerful technique in the past with a number of recombinant CHO cells [70] some of which were established for the production of candidate protein pharmaceuticals. Our goal was to analyze the level of amplification and distribution of recombinant sequences in CHO cells that had been established in moderate to high levels of MTX and cultivated in the presence and absence of MTX [26].

It remains controversial whether the continued presence of MTX in long-term cultures of recombinant CHO cells is required for production stability [71,72]. Over the last 15 years we have obtained in our laboratory ample evidence that many clonally derived cell populations can be cultivated in the absence of MTX without loss of productivity.

Figure 5 shows one example of a cell line that was initially established as a clonal population from a culture containing 500 nM MTX and was grown in the absence of MTX for 8 months without loss or reduction of productivity of a recombinant human immunoglobulin. Continuous subcultivation of a clonal CHO cell population producing an anti-RhD-IgG, cloned after exposure to 500 nM MTX, provided

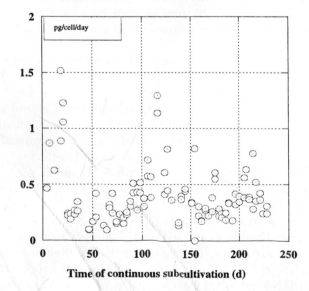

Fig. 5. Continuous subcultivation of a CHO cell line, originally selected in 500 nM MTX, grown in the absence of MTX for almost 8 months. Supernatants from the culture were taken at the time of subcultivation and tested for the quantity of a human recombinant antiRhD–IgG using an ELISA. At the same time, the viable cell concentration in the serum-free culture was determined. The specific productivity was calculated as picogram antibody produced per day per viable cell.

samples for the quantification of secreted IgG. Accumulated IgG was calculated against the cell mass in the vessel and against the time passed since the last subcultivation, usually 3–4 days. Each data point, therefore, indicates approximate specific productivities, expressed as pg/cell/day, obtained at the day of subcultivation of the cells. While a wide fluctuation in productivity can be seen, no general trend towards loss in productivity was observed over the course of 8 months. These data show that studies executed during much shorter periods of time, for example only for a few weeks, may not give sufficient support for general conclusions on stability due to effects on productivity that are not controllable, even in the hands of very experienced operators.

However, the maintenance or loss of productivity as assessed by secreted proteins from a cell population does not provide insight into the synthetic activity of an individual cell. We have therefore recently extended our studies using cell lines that produce cytoplasmic-retained fluorescent proteins allowing microscopic or flow-cytometric studies to provide information on individual cells. Again, we have found that the majority of cell lines established as clonal populations retained recombinant protein production, as assessed by measuring global fluorescence, for many months in the absence of MTX. More importantly, microscopic evaluation revealed that every single cell was expressing the green fluorescent protein over a narrow intensity range at any time point during the passaging of the cell population.

Figure 6 shows a microscopic image of the cell line maintained under non-selective conditions for 4.5 months. It is striking to see that not a single cell can be found that does not exhibit a strong fluorescence, due to synthesized GFP. As can be seen in the two images, taken at 3 (left) and 4.5 months (right) of continuous subcultivation since the time of cloning, every single cell shows significant expression of GFP. Analysis of the cell population at 4.5 months by flow cytometry (data not shown) gave indications that cells not expressing GFP at all (fluorescence 100 times lower than the average fluorescence in the population) were very rare, below the 0.1% range. It is difficult to assess whether these were truly viable cells or whether these few were cells that had died. (Note: a cell culture containing MTX was not carried in parallel, since the purpose of this experiment was to challenge the cell population for the selection of non- or low-producers). We concluded that stable cell lines can be

Fig. 6. GFP-expressing stable CHO cell line, originally cloned from a population of cells selected in 30 nM MTX, but grown in the absence of MTX for 4.5 months. Each cell expresses GFP and the average GFP expression did not decline. (Left panel: cell line after 3 months; right panel: cell line after 4.5 months).

established, expressing significant amounts of recombinant proteins (the GFP-producing cell line shown here produces about 20 pg/cell/day of GFP), that inherit the transgene of interest with high fidelity, showing genetic features of an endogenous gene.

In 1990, we used FISH to analyze the effect of maintaining methotrexate as a selective agent in culture media. For both clonal and non-clonal cell lines, continuous exposure to MTX promotes genetic instability at the chromosomal level resulting in chromosomal rearrangements [26]. Recent elaborate studies by Kim and Lee [72] have verified and extended our conclusions. The first of our FISH analysis studies was done with cell lines expressing an inducible mouse c-myc gene. Under standard growth conditions the non-induced c-myc does not interfere with the host cell, because no c-myc protein is being produced. Observations made under these conditions can be attributed solely to the status (expression level, chromosomal location, degree of amplification) of the *dhfr* marker gene and the inhibitor MTX. Selecting clonal cell lines for growth in 320 μM MTX (the highest MTX concentration reported in the literature) resulted in one clone containing more than 2000 copies of the transfected DNA. The continuous exposure to MTX resulted in a slightly higher copy number of plasmid DNA as compared to a population of cells that had been cultivated in the absence of any selective pressure for at least 4 weeks. The difference was about twofold as indicated by Southern hybridization. The induced level of c-myc mRNA was elevated accordingly [47].

Weidle *et al.* [71] and the group of Piret [73] working independently on the expression of rTPA, have shown a reduction in copy number (but not a complete loss of transfected DNA) and a gradual and sometimes steep decline in productivity when the producer cells were grown without MTX for 40–60 days. Neither group showed any further loss in productivity up to a period of between 100 and 250 days in the absence of MTX. Kim and Lee [72] showed losses in copy number of transfected heavy and light chain plasmid sequences between 90 and 30% in four different subclones (generated from one parental clone) after 21 passages in the absence of MTX. In all cases cited, productivity followed these trends, but stabilized eventually following 60–80 days of culture in non-selective conditions. In the excellent paper by the Piret group two cell lines of four showed stable specific productivities when MTX was kept in the culture medium, while the other two showed trends of weak decline over 90 days.

Thus, neither production stability nor genetic stability can be assured in all cases by the maintenance of the selective agent MTX in the culture medium. We hope that the following sections will help to resolve this dilemma and provide support for our recommendation to eliminate MTX early from selected CHO cultures.

3.4. Unique patterns of chromosomally integrated amplified sequences in clonal recombinant cell lines

The loci of integration and the degree of amplification of recombinant sequences in clonal cells are thought to vary between individual clones and thus may provide the opportunity for identification of recombinant cell lines. FISH analysis with probes complementary to the transfected DNA can be used to label the integrated, amplified DNA sequences and to facilitate cytogenetic identification of recombinant cell lines.

Following our initial observation of a decrease in c-myc copy number in our recombinant CHO cell line cultivated in the absence of MTX, we extended our studies and analyzed the chromosomal integration patterns of transfected DNA in two additional CHO cell lines. The three clonal cell lines in this study expressed human rTPA, a truncated version of the human CD4 receptor [74] or the murine c-myc protein [47]. The three cell lines were established by co-transfecting CHO-DUKX-B11 cells with plasmids carrying one of these genes and a separate plasmid with a dhfr expression construct. Either individual clones or pools of clones were cultivated for extended time periods in media containing stepwise increases in MTX concentration. In each case, a single individual clone expressing the desired level of recombinant protein was expanded for long-term growth and production. FISH analysis was performed using biotinylated probes complementary to the recombinant gene. Prior to hybridization, the cells were grown for at least four subcultivations (14 days) in the absence of MTX. Three different integration patterns (site, degree) of amplified sequences were observed. The FISH signals in the rTPA-expressing cell line were localized to one midsize submetacentric chromosome. A bright band of fluorescence covered most of the short arm and much weaker fluorescence was observed over the center of the chromatids in the long arm. For the cell line expressing the truncated CD4 molecule, a strong signal was found over the telomeric end of a small acrocentric chromosome (possibly a derivative of either chromosome 5, 6, or 7). The hybridization signal covered about one-third of the total length of the chromosome. Finally, cells expressing the mouse c-myc protein exhibited a bright band of hybridization in the center of the long arm of a chromosome believed to be a derivative of chromosome number 2. These unique integration patterns were found in 95–99% of all metaphases from each cell line. Subsequent to the first analysis, we cultivated these cell lines in the absence of methotrexate for extended periods and occasionally performed FISH analysis. We found that the chromosomal structures described above were stable within the observation period (a minimum of 60 days and in one case 160 days). These observations indicate a high degree of genetic stability of chromosomally amplified sequences in the absence of methotrexate. Incidentally, we found that none of the cell lines had lost specific productivity during the observation period.

3.5. MTX induces heterogeneity of amplified sequences

A number of reports describe multiple chromosomal aberrations and a wide variety of diverse structures in individual mouse and human cells that were subcultivated in the presence of MTX [75–77,64]. In an attempt to explain some of the effects observed both in cell culture and in vivo, upon chemotherapy of cancer patients, Goulian and coworkers [53] reported that MTX exposure results in an intracellular depletion of the thymidine pool, possibly increasing the recombinogenic activity in the cells. Our own observations with respect to the dynamics of chromosomally amplified sequences in CHO cells lead us to consider this to be a general phenomenon that is of importance for long-term culture of recombinant cell lines.

Continuous cultivation of cells in media containing high concentrations of MTX is associated with re-arrangement and variable amplification of transfected sequences within one half of the cells of a population expressing *c-myc* [26]. Several types of integration patterns were found in these cells. To simplify a rather confusing picture, we divided the chromosomes containing amplified DNA sequences into two categories termed "variant" and "master" integrations. Variant integrations included: (1) cells with chromosomes with highly extended (long) regions of exogenous DNA; (2) cells containing multiple sites of exogenous DNA on one individual chromosome or (3) on two to four chromosomes; (4) cells containing chromosomes joined at amplified regions; (5) cells containing circular chromosomes entirely consisting of exogenous DNA; and (6) cells with small derivative chromosomes or fragments with exogenous DNA. The second category was termed "master integration" since it was representative of about one-half of the metaphase spreads analyzed. Cells with master integrations had a single "normal" integration within one clearly identifiable chromosome. Similar results were observed with another recombinant CHO cell line producing a chimeric CD4–IgG protein [78]. When grown in the presence of MTX, CD4–IgG-producing cells showed substantial heterogeneity in both chromosomal location and copy number of amplified sequences encoding CD4–IgG. About half of the clonal cell population showed variant integration types of amplified DNA sequences, but the other half was identifiable as a unique, "typical" amplified DNA in one identifiable chromosome.

To explain the discrepancy between the stable, unique structures described in Section 3.4 and the variable, heterogeneic structures outlined here, one has to assume that cell populations go through a transition phase when MTX is removed from the culture medium. Table 1 shows the frequencies of certain types of amplified sequences in cell populations derived from a clonal cell line that were grown in the presence (day 0) or absence of methotrexate for various time periods.

We have verified the loss of variant chromosome types of amplified sequences in four other clonal cell populations and have found it to be of a general nature. Variant integrations decreased in frequency whereas the identifiable master integration increased in frequency in the cell population when grown in the absence of MTX. FISH analysis of three clonal cell lines, originally established at 0.08, 0.32 and 1 µM MTX, but grown in the absence of MTX, showed genetically homogeneous populations with defined and unique hybridization patterns [41]. Thus, it can be concluded that cytogenetic stability in amplified cell lines has two facets. Cell lines with master integrations inherit the amplified sequence as stable genes similar to endogenous gene sequence. In most cases, variant integrations are linked to chromosomes that have a stability problem and thus are lost unless MTX is maintained. In fact, one has to assume that the presence of MTX is probably the reason for a constant recreation of variant integrations that are subsequently lost due to their unstable character.

The variant chromosome structures described above contribute to the overall copy number of transfected DNA within a cell population. Hence, the loss of these

Table 1

Two types of integration of amplified sequences in a clonal CHO cell population expressing the mouse c-myc protein. Metaphase chromosome spreads were prepared from cultured cells in the presence (day 0) or absence of MTX (days 7, 40, 62). Master integrations were characterized by a single hybridization signal over a unique identifiable chromosome. Variant integrations in individual cells were different from master integrations, usually characterized by more than one region of hybridization over various chromosomes.

Type of integration	Day 0 (+MTX) (%)	Day 7 (no MTX) (%)	Day 40 (no MTX) (%)	Day 62 (no MTX) (%)
Master integration	48	58	96	94
Variant integration	52	42	4	6

structures over time may decrease the number of recombinant gene sequences harbored within the cell population. Thus, the phenomena described by Weidle *et al.* [71] and Fann *et al.* [73] are likely to be based on a gradual elimination of cells carrying highly amplified, but cytogenetically unstable chromosomes. This would result in a steady increase in the percentage of cells in the populations that contain a stable master integration but would result in an overall decrease in copy number of amplified genes.

3.6. Genetic instability in immortalized mammalian cells containing amplified DNA sequences is a complex, but solvable problem

Two foremost considerations must be noted when dealing with issues of genetic stability of amplified DNA sequences in immortalized cell lines. The first one deals with the term "clone" which is frequently used in this context. In contrast to most "normal" (primary) cells in culture, immortalized cell populations, even if they have been derived from a single cell (clone), rapidly become heterogeneic with respect to the genetic material they carry. All immortalized cell lines have inherent chromosomal instabilities, and individual cells will vary with respect to chromosome number and chromosome structures [for review see [79]]. The karyotype of immortalized cell populations ("cell lines") usually has a Gaussian distribution of chromosome numbers around one, sometimes two maxima. For example, in cell lines of Chinese hamster origin about 80–85% of cells contain 22 chromosomes (diploid karyotype of Chinese hamsters is 10 pairs of autosomes and 2 sex chromosomes), 2–5% contain 44 chromosomes, and 5–15% of cells contain more or less than 22 chromosomes [80,81].

The second consideration relates to the use of drugs that favor the selection of altered chromosome structures, eventually resulting in gene amplification. As is evident from the preceding discussion, a variety of phenomena, few completely understood, none in any way controlled, affect the chromosomes of the host cells, mostly in a destabilizing manner. Thus the degree of genetic instability is increased further, since most of the selected cells probably survived for a few mitotic cycles due to dramatic rearrangement of major parts of individual chromosomes containing the selective marker gene.

Two additional processes that may complicate the development of permanent cell lines are gene silencing and variegation. Little is known about gene silencing in cell lines containing amplified gene sequences, however, we believe that it could be the basis for some of the expression instability problems observed in immortalized cell lines. Some of the amplified DNA sequences could be localized in chromatin regions that are subject to gene-silencing processes and thus may become inactivated over time. Variegation is also relevant but extremely difficult to study with dynamic and heterogeneic cell populations. The variegation phenomenon is based on gene mobility, which is induced in individual cells. The transposed DNA fragment is subsequently inherited for extended time frames. We can reasonably assume that some integrated DNAs are subject to both gene-silencing and variegation principles, affecting gene expression and thus stability of a given clonal cell population.

An additional reason to eliminate MTX from cell culture fluids, especially when considering large-scale operations, is that MTX is a mutagen, and potential exposure of operators or the environment to large volumes of media and cell culture fluids containing MTX poses unacceptable risks and presents a significant problem with respect to safe disposal of these fluids. Listed below are recommendations for attaining a robust and stable cell line:

(a) A production cell line should be derived from a cloned cell population.
(b) Exposure to selective agents and culture conditions that constitute stress to cells should be minimal.
(c) Culture of cells in medium lacking the selective agent should be executed prior to banking (freezing of cell aliquots in liquid nitrogen) preferably in the medium that will be used in production. More extended periods of subcultivation from a small frozen bank of cells should provide sufficient data and confidence to eventually establish a larger bank of frozen cells from a freshly expanded vial of the first.
(d) Culturing of cells from a freshly thawed vial should follow a well-defined, reproducible protocol.
(e) Cell lines showing signs of decreasing productivity upon removal of selective agents should be avoided.

Thus, experimental approaches exist to assure oneself of the required genetic and production stability of cell populations intended for long-term culture and production of important recombinant proteins.

4. Concluding remarks

In this article, we have summarized some aspects of gene transfer and amplification in *dhfr*-deficient CHO cells, one of the most popular mammalian cell hosts for recombinant protein production. We believe that the observations made and conclusion drawn have value also for work with other mammalian cell hosts for recombinant protein production. Expression vectors containing the target gene are used in combination with a functional *dhfr* expression cassette either on the same or on a separate vector. Generally, there is no risk of separation with two or more co-transfected plasmids. Linkage is probably assured prior to chromosome integration via ligation of linearized plasmids within the host cell nucleus. Optimized vehicle systems or physical techniques deliver plasmids with high efficiency to cells in culture. Most methods are capable of transferring 10^3–10^5 plasmids per cell. Few firmly established insights have been gained about the integration of transfected DNA into the genome of the host cell, but it appears that mitotic activity of the host cell population during the transfection procedure is an important parameter. Selection and identification of recombinant cells is performed using the DHFR phenotype. It is assumed that integration into a chromosomal site within the nucleus is random and rare, usually guaranteeing a singular event. Recombinant cell clones might vary with respect to the stability within the primary locus of integration. Maintenance of selectivity in the culture medium is, however, not always a prerequisite for stable expression of integrated DNA, and clonal cell lines can be identified that inherit the gene of interest with high fidelity in the absence of selectivity.

Profound effects on transcription rates and on the eventual fate of transfected DNA are exerted by the genomic DNA environment at the integration site. The primary integration of plasmid DNA seems to be accompanied by an inverted duplication both of plasmid DNA and large stretches of neighboring chromosomal DNA. Hyperamplifiable phenotypes have been observed, which frequently excise the region of the inverted duplication to create episomes containing both transfected DNA and endogenous chromosomal DNA. Such excised episomes have been found to be circular, self-replicative, submicroscopic elements that undergo rapid fluctuations in copy number. Under stringent MTX-selective conditions, subpopulations of cells can be selected that have a higher copy number of these episomal structures during the initial phase. In CHO cells reintegration of episomal elements into chromosomes occurs frequently, probably at random sites. Double minute chromosomes that are large derivatives of episomes and that have been frequently observed in mouse cell lines, usually do not persist in CHO cells. Integrated episomes in CHO cells can give rise to extended chromosomal regions. These can contain several 100s if not 1000s of copies of the initially transfected DNA. They undergo fluctuations in copy number, chromosomal locus and length when cells are continuously exposed to MTX. These ECR appear to be products of repeated chromatid breakage–fusion mechanisms. In clonal cell lines that have been cultivated for several weeks in the absence of MTX, very specific types of recombinant integration structures are detected in greater than 90% of cells. These "master integrations" are genetically stable and can be used as identifying markers for each clonal cell line. In the presence of MTX, master integrations might be present in 40–50% of the cells, but other, more variable ECRs are observed in the remainder of the cell population. These types of ECRs are mostly unstable since they disappear from the population of cells cultivated in the absence of MTX. However, they contribute to the overall copy number of recombinant genes within a cell population and thus may contribute to overall productivity. In our opinion, removal of MTX may result in an initial drop in productivity due to the elimination of the variant integrations.

Identification of highly productive cell lines requires screening of candidate clonal cell populations. Such screening can occur already at the lowest level of selective stringency (GHT-minus medium for *dhfr*-negative CHO cells), but should also be done at the level of highest selective stringency. Frequently, higher productive cell lines can be identified in populations of cells selected in highest stringency-selections. It is advisable to execute these screenings already in the absence of selective agents, since it will favor the identification of cell lines that have integrated gene sequences and express these sequences from a stable, permanently "open" chromosomal locus.

In spite of many unanswered questions, CHO cells have been extremely successful as a reliable source of large quantities of high-value proteins. No other cell line is available that has been studied to the same extent as CHO cells with respect to a multitude of parameters and characteristics. Though many details remain to be deciphered about gene transfer and amplification in mammalian cells, further study of CHO cells surely will lead to an improved understanding of these complex mechanisms. The knowledge acquired has provided us with improved means to use CHO and other mammalian host cells for the production of valuable pharmaceuticals and for the general study of gene transfer and expression of foreign genes.

Acknowledgements

The authors thank Drs. David Hacker and Tony Mason for carefully reading this manuscript and providing very helpful comments for improvements of the text of this chapter.

Abbreviations

CHO	Chinese hamster ovary
DHFR	Dihydrofolate reductase
DM	Double minute chromosome
ECR	Extended chromosomal region
FISH	Fluorescence *in-situ* hybridization
GFP	Green fluorescent protein
HbsAg	Hepatitis B surface antigen
HEK-293	Human Embryo Kidney – 293,
HSR	Homogenously staining region
IgG	Immunoglobulin G
MTX	Methotrexate
TPA	Tissue-type plasminogen activator

References

1. Puck, T.T. (1995) In: Gottesman, M.M. (ed.) Molecular Cell Genetics, John Wiley and Sons, New York, pp. 37–64.
2. Subramani, S., Mulligan, R., and Berg, P. (1981) Mol. Cell. Biol. 3, 257–266.
3. Crouse, G.F., McEwan, R.N., and Pearson, M.L. (1983) Mol. Cell. Biol. 3, 854–864.
4. Arathoon, W.R. and Birch, J.R. (1986) Science 232, 1390–1395.
5. Wiebe, M.E., Becker, F., Lazar, R., May, L., Casto, B., Semense, M., Fautz, C., Garnick, R., Miller, C., Masover, G., Bergman, D., Lubiniecki, A.S. (1989) In: Spier, R.E, Griffiths, J.B., Stephenne, J. and Crooy, P.J. (eds.) Advances in Animal Cell Biology and Technology for Bioprocesses, Butterworths, London, UK, pp. 68–71.
6. Urlaub, G. and Chasin, L.A. (1980) Proc. Natl. Acad. Sci. USA 77, 4216–4220.
7. Kao, F.-T. and Puck, T.T. (1968) Proc. Natl. Acad. Sci. USA 60, 1275–1281.
8. Kaufman, R.J. and Sharp, P. (1982) J. Mol. Biol. 159, 601–621.
9. Kaufman, R.J., Murtha, P., Ingolia, D.E., Yeung, C.-Y., and Kellems, R.E. (1986) Proc. Natl. Acad. Sci. USA 83, 3136–3140.
10. Simonsen, C.C. and Levinson, A.D. (1983) Proc. Natl. Acad. USA 80, 2495–2499.
11. Wirth, M., Bode, J., Zettlmeissl, G., and Hauser, H. (1988) Gene 73, 419–426.
12. Deml, L., Wolf, H., and Wagner, R. (1999) J. Virol. Meth. 79, 191–203.
13. Graham, F.L. and van der Eb, A.J. (1973) Virology 52, 456–467.
14. Chen, C. and Okayama, H. (1987) Mol. Cell. Biol. 7, 2745–2752.
15. Jordan, M., Schallhorn, A., and Wurm, F.M. (1996) Nucleic Acid Res. 24, 596–601.
16. Bebbington, C.R., Renner, G., Thomson, S., King, D., Abrams, D., and Yarranton, G.T. (1992) Biotechnology 10, 169–175.
17. Neumann, E., Schafer-Ridder, M., Wang, Y., and Hofschneider, P.H. (1982) EMBO J. 1, 841–845.
18. Grosjean, F., Batard, P., Jordan, M., and Wurm, F.M. (2001) In: Lindner-Olsson, E., Chatzissvidou, N. and Lüllau, E. (eds.) Animal Cell Technology: From Market to Target, Kluwer Dordrecht, The Netherlands, pp. 238–240.
19. Batard, P., Jordan, M., Chatellard, P., and Wurm, F.M. (2001) Gene 270, 61–68.
20. Pick, H., Meissner, M.P., Preuss, A.K., Tromba, P., Vogel H. and Wurm F.M. (2002) Biotechnol. Bioeng. (in press).
21. Tseng, W.C., Haselton, F.R., and Giorgio, T.D. (1997) J. Biol. Chem. 272, 25,641–25,647.
22. Orrantia, E. and Chang, P.L. (1990) Exp. Cell Res. 190, 170–174.
23. Girard, Ph., Derouazi, M., Baumgartner, G., Bourgeois, M., Jordan, M., and Wurm, F.M. (2001) In: Lindner-Olsson, E., Chatzissvidou, N. and Lüllau, E. (eds.) Animal Cell Technology: From Target to Market, Kluwer Dordrecht, The Netherlands, pp. 37–42.
24. Wurm, F.M. and Bernard, A. (1999) Curr. Opin. Biotechnol. 10, 156–159.
25. Finn, G.K., Kurz, B.W., Cheng, R.Z., and Shmookler, R.J. (1989) Mol. Cell. Biol. 9, 4009–4017.
26. Pallavicini, M.G., DeTeresa, P.S., Rosette, C., Gray, J.W., and Wurm, F.M. (1990) Mol. Cell. Biol. 10, 401–404.
27. Jasin, M. and Berg, P. (1988) Genes Develop. 2, 1353–1363.
28. Smithies, O., Gregg, R.G., Boggs, S.S., Koralewski, M.A., and Kucherlapati, R.S. (1985) Nature 317, 230–234.
29. Zheng, H. and Wilson, J.H. (1990) Nature 344, 170–173.
30. Wurm, F.M., Johnson, A., Lie, Y., Etcheverry, M.T., and Anderson, K.P. (1992) In: Spier, R.E., Griffiths, J.B. and MacDonald, C. (eds.) Animal Cell Technology Developments, Processes and Products, Butterworth-Heinemannn, Oxford, UK, pp. 34–41.
31. Anderson, K.P., Lie, Y., Low, M.A., Williams, S.R., Fennie, E.H., Nguyen, T.P., and Wurm, F.M. (1990) J. Virol. 64,5, 2021–2032.
32. Anderson, K.P., Low, M.-A., Lie, Y.S., Keller, G.-A., and Dinowitz, M. (1991) Virology 181, 305–311.
33. Ono, M. (1989) Dev. Biol. Stand. 70, 69–81.
34. Ono, M., Toh, H., Miyata, T., and Awaya, T. (1985) J. Virol. 55, 387–394.

334

35. Wurm, F.M., Johnson, A., Etcheverry, T., Lie, Y.S., and Petropoulos, C.J. (1994) In: Spier, R.E., Griffiths, J.B. and Berthold, W. (eds.) Animal Cell Technology: Products for Today and the Future, Butterworth-Heinemann, Oxford, UK, pp. 24–29.
36. Wilson, T.J. and Kola, I. (2001) Methods Mol. Biol. 158, 83–94.
37. Kaufman, R.J. and Sharp, P. (1982) J. Mol. Biol. 159, 601–621.
38. Milbrandt, J.D., Azizkhan, J.C., and Hamlin, J.L. (1983) Mol. Cell. Biol. 3, 1274–1282.
39. Perucho, M., Hanahan, D., and Wigler, M. (1980) Cell 22, 309–317.
40. Wurm, F.M. and Petropoulos, C.J. (1994) Biologicals 22, 95–102.
41. Wurm, F.M., Johnson, A., Ryll, T., Köhne, C., Scherthan, H., Glaab, F., Lie, Y., Petropoulos, C., and Arathoon, W.R. (1996) Ann. NY Acad. Sci. 782, 70–78.
42. Wilson, J.H., Berget, P.B., and Pipas, J.M. (1982) Mol. Cell. Biol. 2, 1258–1269.
43. Chen, L., Marmey, P., Taylor, N.J., Brizard, J.-P., Espinoza, C., D'Cruz, P., Huet, H., Zhang, S., Kochko, A., de Beachy, R.N., Fauquet, C.M. (1998) Nat. Biotechnol. 16, 1060–1064.
44. Toneguzzo, F., Keating, A., Glynn, S., and McDonald, K. (1988) Nucleic Acids Res. 16, 5515–5532.
45. Camargo, M. and Cervenka, J. (1980) Hum. Genet. 54, 47–53.
46. Kaufman, R.J., Wasley, L.C., Spiliotes, A.J., Gossels, S.D., Latt, S.A., Larsen, G.R., and Kay, R.M. (1985) Mol. Cell. Biol. 5, 1750–1759.
47. Wurm, F.M., Gwinn, K.A., and Kingston, R.E. (1986) Proc. Natl. Acad. Sci. USA 83, 5414–5418.
48. Wood, C.R., Dorner, A.J., Morris, G.E., Alderman, E.M., Wilson, D., O'Hara, M., Jr., and Kaufman, R.J. (1990) J. Immunol. 145, 3011–3016.
49. Fouser, A.L., Swanberg, S.L., Lin, B.-Y., Benedict, M., Kelleher, K., Dumming, D.A., and Riedel, G.E. (1992) Biotechnology 10, 1121–1127.
50. Brand, H.N., Froud, S.J., Metcalfe, H.K., Onadipe, A., Shaw, A., and Westlake, A.J. (1994) In: Spier, R.E., Griffiths, J.B. and Berthold, W. (eds.) Animal Cell Technology: Products of Today, Prospects for Tomorrow, Butterworths-Heinemann, Oxford UK, pp. 55–60.
51. Barbi, G., Steinbach, P., and Vogel, W. (1984) Hum. Genet. 68, 290–294.
52. Raffetto, G., Parodi, S., Faggin, P., and Maconi, A. (1979) Mutat. Res. 63, 335–343.
53. Goulian, M., Bleile, B., and Tseng, B.Y. (1980) Proc. Natl. Acad. Sci. USA 77, 1956–1960.
54. McClintock, B. (1950) Proc. Natl. Acad. Sci. USA 36, 344–355.
55. Stark, G.R., Debatisse, M., Giulotto, E., and Wahl, G.M. (1989) Cell 57, 901–908.
56. Ruiz, J.C. and Wahl, G.M. (1988) Mol. Cell. Biol. 8.10, 4302–4313.
57. Ma, C., Martin, S., Trask, S., and Hamlin, J.L. (1993) Genes Develop. 7, 605–620.
58. Hamlin, J.L., Leu, T.-H., Vaughn, J.P., Ma, C., and Dijkwel, P.A. (1991) Prog. Nucleic Acid Res. Mol. Biol. 41, 203–239.
59. Folger, K.R., Wong, E.A., Wahl, G., and Capecchi, M.R. (1982) Mol. Cell. Biol. 2,11, 1372–1387.
60. Windle, B., Draper, B.W., Yin, Y., O'Gorman, S., and Wahl, G.M. (1991) Genes Develop. 5, 160–174.
61. Levan, G. and Levan, A. (1982) In: Schimke R.T. (ed.) Gene Amplification, Cold Spring Harbor Laboratory, Cold Spring Harbor, NY, pp. 91–97.
62. Coquelle, A., Pipiras, E., Toledo, F., Buttin, G., and Debatisse, M. (1997) Cell 89, 215–225.
63. Carrol, S.M., Gaudray, P., DeRose, M.L., Emery, J.F., Meinkoth, J.L., Nakkim, E., Subler, M., Von Hoff, D.D., and Wahl, G.M. (1987) Mol. Cell. Biol. 7,5, 1750.
64. Biedler, J.L. and Spengler, B.A. (1976) J. Natl. Cancer Inst. 57,3, 683–689.
65. Zastrow, G., Koehler, U., Müller, F., Klavinius, A., Wegner, M., Wienberg, J., Weidle, U., and Grummt, F. (1989) Nucleic Acids Res. 17, 1867–1879.
66. McArthur, J.G. and Stanners, C.P. (1991) J. Biol. Chem. 266,9, 6000–6005.
67. Carrol, S.M., Gaudray, P., DeRose, M.L., Emery, J.F., Meinkoth, J.L., Nakkim, E., Subler, M., Von Hoff, D.D., and Wahl, G.M. (1987) Mol. Cell. Biol. 7,5, 1750–1750.
68. Lichter, P., Cremer, T., Chieh-Ju, C.T., Watkins, P.C., Manuelidis, L., and Ward, D.C. (1988) Proc. Natl. Acad. Sci. USA 85, 9664–9668.
69. Pinkel, D., Straume, T., and Gray, J.W. (1986) Proc. Natl. Acad. Sci. USA 83, 2934–2938.
70. Wurm, F.M. and Schiffmann, D. (1999) In: Jenkins, N. (ed.) Animal Cell Biotechnology, Humana Press, Totowa, New Jersey, pp. 49–60.
71. Weidle, U.H., Buckel, P., and Wienberg, J. (1988) Gene 66, 193–203.

72. Kim, S.J. and Lee, G.M. (1999) Biotechnol. Bioeng. 64, 741–749.
73. Fann, C.H., Guirgis, F., Chen, G., Lao, M.S., and Piret, J.M. (2000) Biotechnol. Bioeng. 69, 204–212.
74. Smith, D.H., Byrn, R.A., Marsters, S.A., Gregory, T., Groopman, J.E., and Capon, D.J. (1987) Science 328, 1704–1707.
75. Raggetto, G., Parodi, S., Faggin, P., and Maconi, A. (1979) Mutat. Res. 63, 335–343.
76. Mondello, C., Giorgi, R., and Nuzzo, F. (1984) Mutat. Res. 139, 67–70.
77. Vitek, J.A. (1987) Neoplasma 34,6, 665–670.
78. Capon, D., Chamow, S.M., Mordenti, J, Marsters, S.A., Gregory, T, Mitsuya, H., Byrn, R.A., Lucas, C., Wurm, F.M., Groopman, J.E., Broder, S., Smith, D.H. (1989) Nature 337, 525–531.
79. Hsu, T.C. (1961) Int. Rev. Cytol. 12, 69–161.
80. Puck, T.T. (1985) In: Gottesman, M.M. (ed.) Molecular Cell Genetics, John Wiley and Sons, New York, pp. 37–74.
81. Sager, R. (1985) In: Gottesman, M.M. (ed.) Molecular Cell Genetics, John Wiley and Sons, New York, pp. 75–94.

S.C. Makrides (Ed.) *Gene Transfer and Expression in Mammalian Cells*

CHAPTER 8

Co-transfer of multiple plasmids/viruses as an attractive method to introduce several genes in mammalian cells

Martin Jordan and Florian M. Wurm[1]

Swiss Federal Institute of Technology Lausanne (EPFL), Faculty of Basic Sciences,
Institute of Process Sciences, Center of Biotechnology, CH-1015 Lausanne, Switzerland;
Tel.: +41-21-693-6142; E-mail: Martin.Jordan@epfl.ch
[1]*Tel.: +41-21-693-6141; E-mail: Florian.Wurm@epfl.ch*

1. Introduction

Gene transfer into mammalian cells has become a standard technique with many applications. The transfer may be transient or may serve to establish stable cell lines using viral or non-viral methods, but the general goal in each case is to target the recombinant gene to the nucleus where it can be transcribed. The success rate of transient or stable introduction of genes into the nucleus is highly variable and is dependent on the choice of cell line and DNA transfer method. Despite tremendous progress in this technology, current methods do not allow strict control of the number of copies of the gene that get into the nucleus. This might appear to be a minor issue since success is measured as the amount of recombinant protein expressed, but the gene copy number in the nucleus is relevant to this output. Ignoring this variable may lead to a misinterpretation of experimental results. It is the aim of this article to describe phenomena that depend on the distribution of plasmids or viral genomes following their introduction into cells. Though this question relies heavily on statistical models and analyses, emphasis will not be on calculations but on graphical presentations that attempt to describe DNA levels within the whole cell population. Understanding the principles of such distributions will also help to plan and optimise the co-transfer of multiple plasmids or viral vectors in one step. Although these two different approaches, non-viral and viral, are used for gene transfer and many readers may be more familiar with one than the other, this article presents statistical models that can be applied to either approach.

2. Multiplicity of infection (MOI) and multiplicity of transfection (MOT)

Undoubtedly, viruses are the most efficient vectors for gene transfer. By definition, a single infectious virus is sufficient to generate an expressing cell. By comparison, the initial plasmid copy number that is necessary to obtain an expressing cell by non-viral gene delivery is orders of magnitude higher. For cultured cells, typically up to 10^6 plasmid molecules per cell are added to the culture to obtain reasonable

338

transfection efficiencies and good expression levels. Such a large excess of plasmid suggests a low efficiency as compared to viral systems. Is this conclusion justified? Probably not, since the efficiency in each case depends on how the viral and plasmid vectors are quantified: plaque forming units (PFU) versus μg of DNA. PFUs are determined using a biological assay that estimates infective particles. Thus, for a viral vector the efficiency of gene transfer is always 100% since the assay does not account for "unsuccessful viruses". On the other hand, plasmid DNA is quantified by spectrophotometric methods (UV absorption or fluorescent dyes) that are precise and reproducible if applied to highly pure DNA preparations. Theoretically, plasmid DNA can also be quantified by a biological method similar to the plaque assay. For each preparation of DNA a standardized transfection using low amounts of DNA followed by quantification of the number of positive cells can be performed. The assay would reflect the number of "transfectable plasmid molecules", allowing the "Multiplicity Of Transfection" or MOT to be calculated for each batch of plasmid DNA. This assay is expected to underestimate the DNA quantity as compared to spectrophotometric methods and, thus, to increase the apparent transfection efficiency.

While MOI is an established term and is widely used, MOT does not exist in the literature. However, efforts to quantify DNA uptake by cells began many years ago in order to understand the mechanisms involved [1–3]. Recently, Batard et al. [4] have shown that cells can take up large amounts (>100,000 molecules per cell) of labelled plasmid DNA. Most of these molecules were heterogeneously distributed within the cytoplasm and only a small fraction entered the nucleus. Even though the authors found a good correlation between the overall uptake and the level of expressed reporter protein, they were not able to estimate the MOT (average number of functional plasmids that entered the nucleus) due to the large excess of plasmid DNA present in the cytoplasm.

For stable cell lines, on the other hand, there exist several methods to measure the plasmid copy number in the nucleus [5,6]. This information may be useful for determining the number of plasmids that entered the nucleus and may be related to the MOT for that specific transfection. The integration of 100s to 1000s of copies of plasmid into the host genome has been reported [7–10]. This supports the idea that a large number of plasmid molecules enter the nucleus following transfection. It seems that cells are able to randomly integrate most of the plasmids into a single site within the genome, an event that should not be confused with the process of gene amplification (see Chapter 7). If gene amplification can be ruled out (in such a case large fragments of chromosomal DNA would be amplified as well), a high copy number of integrated plasmid most likely reflects a high initial plasmid copy number. However, the small number of stable clones analysed to date is not sufficient to estimate average values for MOT. In addition, it must be noted that generally less than 1% of the cells within a population that transiently express a protein can be recovered as stable clones. Nevertheless, data about the gene copy number of stable clones indicates that hundreds of plasmids can be transfected into the nucleus. A more accurate conclusion about MOT cannot be made from the data at this point.

3. Random distribution of vector units

Given the large number of vector copies and viable cells (10^4 or more of each) used in a typical gene transfer experiment, each virus or plasmid has an equal chance of getting into one of the cells. In addition, a vector that has been taken up by a cell does not influence the additional uptake of another vector by the same cell. Under these conditions, the law of Poisson describes well the distribution of vector units among the whole cell population. The Poisson distribution, a special case of the binomial distribution, is the best model to describe the probability of rare events, and it was successfully applied to model the dynamics of baculovirus infections [11,12]. The average number of vector copies per cell (**n**) is characteristic for each experiment and is equal to MOI or MOT. **n** can be any positive value including fractions. The number of vector copies in a single cell, however, can only be a whole number, and this number will be variable for each cell. For the examples **n** = 0.75 and **n** = 2 the Poisson distribution is shown in Fig. 1. As can be seen from these examples, **n** affects the frequency of cells with a given number of plasmids. For **n** = 0.75 about 46% of the cells do not get any vector, but when **n** = 2 only 14% do not receive vector.

The distribution, however, is different when the experimental system becomes more complex, e.g., when the cell population is heterogeneous and not all the cells are equally susceptible to gene transfer. This is the case for applications in gene therapy where whole organs containing different cell types are treated. For cultured cell lines, too, the population might be heterogeneous as a result of small genetic variations within the population or other relevant differences including cell size, cell charge or metabolic activity. The vector population may also be heterogeneous. DNA preparations include mixtures of linearized, relaxed and supercoiled plasmids, and virus preparations include intact and defective virus particles. These are critical factors that can affect the statistical analysis. Including such factors would result in a

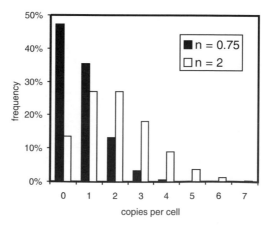

Fig. 1. Distribution of vector in a large homogeneous cell population using two values for **n**. While **n** describes the population and does not characterize individual cells, it affects the frequency at which cells receive a certain number of viruses/plasmids.

distribution that is a combination of several different Poisson distributions known as a negative binomial distribution. A more detailed discussion of such complex heterogeneous cases is clearly beyond the goal of this article.

4. Several copies of one gene

4.1. Frequency of positive cells

As shown in Fig. 1, the frequency of positive cells that is expected following an infection or a transfection can be calculated if the MOI or the MOT is known. For plasmid transfections, however, MOT is not known. Therefore, the frequency of positive cells or "transfection efficiency" is estimated empirically based on expression of a reporter gene. In general, the frequency of positive cells is positively correlated with **n**; as **n** increases, more cells will be positive as predicted by the Poisson distribution. Since **n** represents an average copy number, individual cells may get more than **n** copies while others get less. Thus, at an average copy number of 2, 14% of the cells would still be negative (Fig. 2). In fact, 100% transfection efficiency is only expected for values of **n** equal to or greater than 5 (Fig. 2). When dealing with plasmids, **n** is controlled only indirectly by tuning the parameters that are critical for transfection. Thus, improving the transfection efficiency parallels an increase in **n**, independent of the method used.

Dividing the cells from a transfected/infected population into two categories, positive or negative, can be useful for certain situations. Yet, one should be aware that the positive cell population is highly heterogeneous. Depending on the copy number present within individual cells, the expression level of these cells will be variable over a wide range, as has been observed when cells that transiently express

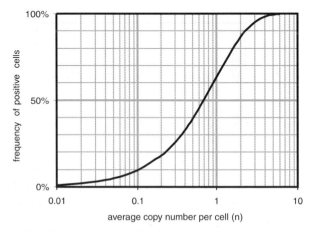

Fig. 2. The frequency of positive cells is related to the average copy number by the Poisson distribution for both transfections and infections. Here, negative cells are defined as those that do not get any genes, while positive cells are those that get at least one copy of the gene.

GFP are analyzed by flow cytometry [13–15]. A similar heterogeneity of β-gal expression has been observed following infection of 293 cells with a recombinant baculovirus [16].

4.2. Detection limits of positive cells

The use of a reporter gene to detect positive cells in transient expression systems is well established. In this case, the definition of a positive cell is based on the expression of a protein. A disadvantage of this approach is that, depending on the sensitivity of the detection system, a specific level of reporter protein is needed to get a positive cell. If the sensitivity is high, then one gene per cell may be sufficient to yield enough protein for detection. Despite the efficacy of current reporter systems such as β-gal or GFP that allow individual expressing cells to be detected, it can not always be taken for granted that the threshold level for detection will be achieved when a cell carries a single copy of the reporter gene. This statement is true regardless of whether the reporter gene is delivered by a plasmid or by a viral vector. In many cases, one copy of the gene will not be sufficient. It is widely known that poor transfection efficiencies, when defined as the percentage of positive cells, can be improved by using a stronger promoter or by exposing the cells to agents such as sodium butyrate [17,18]. In these cases, it is unlikely that the promoter or the treatment changed the average copy number of the gene within the nucleus but rather increased the expression from each copy of the gene.

Convincing quantitative data demonstrating that the detection limit indeed is relevant have recently been published by Condreay et al. [18]. In an elegant approach they used a recombinant baculovirus to express GFP in a wide variety of mammalian cells. They infected several cell lines at a MOI of 200 and found that the frequency of GFP-positive cells was above 50% for most of these. However, for certain "difficult" cell lines the frequency of positive cells was 2% or less. Surprisingly, this low frequency was increased 10-fold or more by adding butyrate, a well-known enhancer of gene expression [19]. This treatment increased the number of positive cells for all the cell lines, but the most dramatic increases were obtained in the "difficult" cell lines. These data support the idea that for some of the cell lines, the gene copy number was not sufficient to yield a detectable level of protein expression.

If several gene copies are needed to detect a positive cell, the relationship between the average copy number and the percentage of positive cells changes. For a given **n**, the frequency of positive cells drops dramatically if the threshold expression can only be achieved by an increased gene copy number. When **n** equals 5, 100% of the cells will be positive if only one copy of the gene is necessary to allow detection of the protein (Fig. 3). However, when at least 10 copies are needed to identify a positive cell, then the frequency of positive cells drops to 3%. If now expression from each plasmid is boosted, then more positive cells will be detected (Fig. 3).

For viral expression systems **n** is usually known since it is defined as the number of PFU/cell. However, **n** may not always be equal to MOI. For baculoviruses that are used as expression vectors in mammalian cell lines, it is not evident that viral

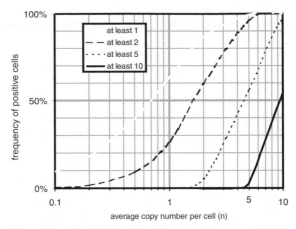

Fig. 3. The frequency of positive cells as a function of the average copy number when at least 1, 2, 5 or 10 copies of the gene are necessary to yield a positive cell.

entry is as efficient in mammalian cells as in insect cells. This means that the MOI for mammalian cells may be lower than the MOI determined in insect cells. Thus, the data obtained by Condreay *et al.* [18] described above do not allow a prediction of the virus copy number per cell that represents the limit of GFP detection. Furthermore, the enhancement of GFP expression with butyrate demonstrated that the sensitivity of detection is also an issue. The observation that certain cells lines are apparently highly resistant to gene transfer may be partially related to the strength of the promoter in these cells. Although these cells appear to be negative, they may contain many copies of the transferred gene. As a summary, it can be stated that positive cells contain at least one functional copy of the gene, but the copy number is significantly higher in most cases. On the other hand, cells judged as negative based on protein expression are not necessarily cells lacking the gene.

4.3. Distribution is important for interpretation

Transient expression systems are frequently used to test the strength of a promoter by quantifying the level of reporter protein expressed. By comparison, measuring the frequency of positive cells is much less reliable. For high frequencies the method will be insensitive to any differences, and for low frequencies misinterpretation will occur. This can be illustrated by an example in which the average gene copy number (**n**) is two. Assuming that at least three copies of the gene are needed to identify a positive cell, then the measured fraction of positive cells would be 32% (Fig. 4A). However, when gene expression is driven by a weaker promoter with 50% less activity, then the detection limit would shift to six gene copies per cell, and consequently the measured number of positive cells would drop from 32% to 1.7% or eventually be reduced 20-fold to 1.7% (Fig. 4B). In this example the 20-fold difference in the frequency of positive cells would overestimate the promoter strength (twofold difference) due to

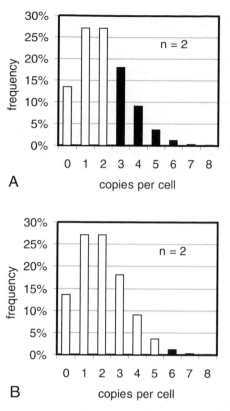

A

B

Fig. 4. The relationship between frequency of positive cells and strength of promoter for an average gene copy number of two. Dark bars highlight positive cell subpopulations. (A) Assuming that the detection limit is achieved with three or more gene copies per cell, then 32% of the cells are positive. (B) A 50% reduction in promoter strength shifts the detection limit to six or more gene copies per cell and reduces the apparent number of positive cells to 1.7%.

the statistical distribution. In a similar way, measuring the frequency of positive cells would also overestimate improvements of transfection methods. Again, small changes in average copy numbers could shift a significant fraction of cells from the negative to the positive pool of cells. These problems would not be relevant if a direct quantification of the level of reporter protein was employed.

5. Co-transfer of multiple genes

In the examples discussed up to this point, all the copies of the transferred gene were identical. Most of the experimental data indicate that several copies of the gene per cell is the rule rather than the exception following transfection. If the methods really transfer several copies of the gene into cells, then the simultaneous transfer of multiple genes on distinct vectors should be possible by simply exposing the cells to a mixture of different vectors.

5.1. Production of stable cell lines

In a large, well-documented study Chen *et al.* tried to co-transfect plant cells with 14 independent plasmid vectors encoding 14 different genes, one of which was a selection marker conferring resistance to hygromycin B [20]. One hundred and twenty five randomly selected R0 plant lines were analysed by PCR for the presence of the 14 vectors, but only three lines containing 13 of the 14 vectors were identified. Each vector, however, was found at a frequency of about 40% in the stable lines and the combinations of integrated vectors observed in the plants were random. It was concluded that each vector had an equal chance of integrating into the host genome (except the selective marker that was present in 100% of the plants). These data are encouraging since they demonstrate that multiple genes can be transferred to cells simultaneously. In principle entire metabolic pathways requiring several genes can be transfected into cells in a single step. Since each gene seems to have an equal chance of integration into the genome, the probability of finding cells with a given set of genes can be calculated by combinatorial statistics. For these types of experiments **n** is a relevant parameter and increasing **n** improves the chances of finding stable cell lines with a complete collection of genes.

In most co-transfections a maximum of two different vectors are used, with one vector containing the gene of interest and the other the selective marker. In such a case it is worth considering a plasmid ratio different from 1:1. By using equivalent amount of plasmid, cells may receive several copies of the marker even though there is no need to have more than one copy per cell. Better expression levels of the gene of interest may result from a high gene to marker ratio. As an example, using a ratio of 50:1 (gene:marker) rather than 1:1 improved the average expression of the gene of interest in a hygromycin-resistant population of S2 cells by a factor of 7 [21]. Two major effects may have contributed to this significant increase in production. Since S2 cells tend to integrate between 10 and 200 copies of the vector [22], a twofold improvement in expression can be expected when the gene of interest represents 98% instead of only 50% of the total plasmid. More importantly, the gene-to-marker ratio may affect the selection procedure. If a cell receives many copies of the marker due to a low gene-to-marker ratio, then the selective pressure for that cell is relatively weak. Even if the marker is integrated at a transcriptionally inactive site, the clone probably would survive selection. On the other hand, if the copy number of the selective marker is close to one as the result of transfection at a high gene-to-marker ratio, then the clone may not survive the selection process. While a 50:1 ratio was the best for the experiment described above, it should not be generalized for other systems where the ratio may need to be different.

5.2. Applications for transient systems

The co-transfer of multiples genes on different vectors also allows high flexibility for transient expression systems. Imhof *et al.* [23] were able to optimize a regulatory network for expression using a transient system. They co-transfected BHK cells with five plasmids including one encoding β-gal as internal standard for transfection

345

efficiency and extract preparation (representing 7% of the mixture), a second encoding luciferase as the principal reporter (3%), a third encoding a transactivator (0.7%), a fourth encoding a tetracycline repressor that regulates both the reporter and transactivator (33%), and pUC18 as carrier DNA. They found a 1000-fold increase in reporter gene expression between the induced and non-induced states. These results demonstrate that multiple plasmids whose genes form a regulatory network can be simultaneously transfected into cells and that the levels of the expressed proteins can be adjusted by varying the ratio of the vectors. This flexibility is an advantage compared to stable systems.

Monoclonal antibodies have been produced in our laboratory by transient transfections of HEK 293 cells expressing Epstein-Barr virus nuclear antigen 1 (EBNA-1) up to a scale of 100 l by co-transfection of two plasmids [24]. In this system the efficiency of expression is directly related to the average plasmid copy number per cell. Since the secretion of an intact monoclonal antibody depends on the expression of heavy and light chains from two different vectors, the cell must have at least one copy of each vector. This additional requirement affects the frequency of expressing cells, which in this case is no longer described by a Poisson distribution (Fig. 5). For an average copy number of 2 and an equivalent amount of each plasmid, one can expect that 60% of cells will not express antibody since 14% of the cells will not receive any vector and 27% of the cells will only receive one plasmid. Furthermore, other non-producers will be observed due to unfavourable combinations of heavy and light chain. About 27% of the cells will have two plasmids, but since there are four possible combinations of heavy and light chain vectors (H + H, H + L, L + H, L + L), only half of these cells will express the

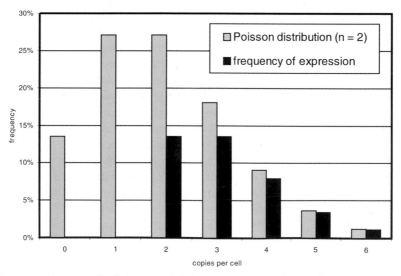

Fig. 5. Co-transfection of antibody heavy and light chain genes. The frequencies of genes per cell and of expressing cells are shown for **n** = 2. The frequency of cells expressing an antibody is lower than the frequency of cells that have taken up genes. This is due to the absence of either the heavy or light chain genes in some cells.

antibody. Among the cells that got exactly three vectors per cell there will still be a small fraction of cells that produce no antibody since the combinations $H + H + H$ and $L + L + L$ are possible. For higher copy numbers, the fraction of non-producing cells shrinks and finally becomes negligible. This example demonstrates that for co-transfections it is no longer sufficient just to get genes into the cells. How the genes are combined with each other becomes important as well. For two genes, combinatorial effects become negligible if n exceeds 10, which is probably the case for the monoclonal antibody expression system.

We routinely include a GFP-encoding vector in our transfections to monitor transfection efficiency. Even if this vector only represents 2% of the DNA in the transfection cocktail, typically its transfection efficiency is still greater than 50%. This indicates that cells get on average at least 0.7 copies of the GFP gene (see Fig. 2). Extrapolating these results to a case in which the GFP vector represents 100% of the plasmid, the average copy number per cell would be 35 or higher. If this copy number can be expected for a single vector, then it is not surprising to see successful co-transfections with two or more plasmids.

Co-infections with two or more viruses are also done routinely and the advantages of rapid transient approaches have already been included in commercial kits (Adeno-X^{TM} adenoviral expression system from Clontech, Palo Alto, CA). In this system, a gene of interest encoded by one virus can be regulated by coinfection of a second virus that expresses the tetracycline repressor. Co-infection at a MOI of 10 leads to a highly regulated system, since it is unlikely that at this MOI a cell gets only one of the two vectors.

6. Conclusion

The co-transfer of multiple genes is an attractive approach in many cases. It is based on the observation that transfected or infected cells usually get several copies of the vector. If the MOI of each virus or the MOT of each plasmid is high enough, then the cells can acquire multiple copies of each. For a non-replicating baculovirus it has been shown that there are no toxic effects with an MOI of 800, and the expression of the reporter gene was found to increase linearly with increasing MOI. Thus, with this system there is a huge capacity to co-infect cells with several viruses carrying different genes (each at a MOI of 10 or above) in one step. Similarly, co-transfections with multiple plasmids can be successful if each plasmid has a high MOT. Unfortunately, MOT is an unknown factor for most of the transfections. Nevertheless, mixtures of different plasmids have been successfully co-transfected [23]. By determining the appropriate concentration of each plasmid within the mixture, the relative expression levels of the proteins can be manipulated [25].

In general, the average vector copy number per cell (**n**) should be sufficiently high when multiple genes are co-transferred. It must be assured that cells get at least one copy of each vector. While the relative ratio of different expressed proteins is not affected by **n** at all [26], the distribution of multiple genes in individual cells strongly depends on the actual copy number as shown in Fig. 5.

For experiments that depend upon the uptake of multiple genes, the number of individual plasmids or viruses that can be used is restricted by the MOI or MOT and also by the ratio of the plasmids or viruses in the mixture. Thus, improving the transfection efficiency or increasing the MOI allows the researcher to increase the number of independent vectors that can be co-transferred at one time. It is already feasible to co-transfer up to 10 different vectors at once with existing methods.

Abbreviations

β-gal   beta-galactosidase
GFP   green fluorescent protein
MOI   multiplicity of infection
MOT   multiplicity of transfection (considering only plasmids within the nucleus)
n   average gene copy number per cell ($\sim$ MOI or MOT)
PFU   plaque forming unit

References

1. Loyter, A., Scangos, G.A., and Ruddle, F.H. (1982) Proc. Natl. Acad. Sci. USA 79, 422–426.
2. Orrantia, E. and Chang, P.L. (1990) Exp. Cell Res. 190, 170–174.
3. Tseng, W., Purvis, N.B., Haselton, F.R, and Giorgio, T.D. (1996) Biotechnol. Bioeng. 50, 548–554.
4. Batard, P., Jordan, M., and Wurm, F. (2001) Gene 270, 61–68.
5. Fu, P., Senior, P., Fernley, R.T., Tregear, G.W., and Aldred, G.P. (1999) J. Biochem. Biophys. Methods 40, 101–112.
6. Fourney, R.M., Aubin, R., Dietrich, K.D., and Paterson, M.C. (1991) In: Murray, E.J. (ed.) Methods in molecular biology: gene transfer and expression protocols. pp. 381–395.
7. Johansen, H., van der Staten, A., Sweet, R., Otto, E., Maroni, G., and Rosenberg, M. (1989) Genes Dev. 3, 882–889.
8. Matsumura, M., Saito, Y., Jackson, M.R., Songs, E.S., and Peterson, P.A. (1992) J. Biol. Chem. 267, 23589–23595.
9. Nellen, W. and Firtel, R.A. (1985) Gene 39, 155–163.
10. Wurm, F.M. and Petropoulos, C.J. (1994) Biologicals 22, 95–102.
11. Licari, P. and Baily, J.E. (1992) Biotechnol. Bioeng. 39, 432–441.
12. Hu, Y.H. and Bentley, W.E. (2000) Chem. Eng. Sci. 55, 3991–4008.
13. Chen, R., Greene, E.L., Collinsworth, G., Grewal, J.S., Houghton, O., Zeng, H., Garnovskaya, M., Paul, R.V., and Raymond, J.R. (1999) Am. J. Physiol. 176, F777–F785.
14. Lybarger, L., Dempsey, D., Franek, K.J., and Chervenak, R. (1996) Cytometry 25, 211–220.
15. Subramanian, S. and Srienc, F. (1996) J. Biotechnol. 49, 137–151.
16. Kost, T.A. and Condreay, J.P. (2002) Trends Biotechnol. 20, 173–180.
17. Walters, M.C., Fiering, S., Eidemiller, J., Magis, W., Groundine, M., and Martin, D.I.K. (1995) Proc. Natl. Acad. Sci. USA 92, 7125–7129.
18. Condreay, J.P., Witherspoon, S.M., Clay, W.C., and Kost, T.A. (1999) Proc. Natl. Acad. Sci. USA 96, 127–132.
19. Hunt, L., Batard, P., Jordan, M., and Wurm, F.M. (2002) Biotechnol. Bioeng. 77, 528–537.
20. Chen, L., Marmey, P., Taylor, N.J., Brizard, J.P., Espinoza, C., D'Cruz, P., Huet, H., Zhang, S., Kochko, A, Beachy, R.N., Fauquet, C.M. (1998) Nat. Biotechnol. 16, 1060–1064.
21. Park, J.H., Kim, J.Y., Han, K.H., and Chung, I.S. (1999) Enzyme Microb. Technol. 25, 558–563.

22. Deml, L., Wolf, H., and Wagner, R. (1999) J. Virol. Methods 79, 191–203.
23. Imhof, M.O., Chatelard, P., and Mermod, N. (2000) J. Gene Med. 2, 107–116.
24. Girard, P., Derouazi, M., Baumgartner, G., Bourgeois, M., Jordan, M., and Wurm, F. (2001) In: Lindner-Olsson, E., Chatzissavidou, N. and Lüllau, E (eds.) Animal cell technology: from target to market, Kluwer, Dordrecht, pp. 37–42.
25. Bramlett, K.S., Dits, N.F.J., Sui, X., Jorge, M.C., Zhu, X., and Jenster, G. (2001) Mol. Cell. Endocrinol. 183, 19–28.
26. Schwarz, D., Kisselev, P., Honeck, H., Cascorbi, I., Schunck, W.H., and Roots, I. (2001) Xenobiotica 31, 345–356.

S.C. Makrides (Ed.) *Gene Transfer and Expression in Mammalian Cells*

CHAPTER 9

Optimization of plasmid backbone for gene expression in mammalian cells

Pascal Bigey, Marie Carrière, and Daniel Scherman

Faculté de Pharmacie Unité INSERM/CNRS de Pharmacologie Chimique et Génétique 4, avenue de l'Observatoire, 75270 Paris Cedex 06, France; E-mail: pascal.bigey@pharmacie.univ-paris5.fr

1. Introduction

Two major groups of vectors are used for gene therapy: viral and non-viral. Viruses are considered to be very efficient vehicles for gene transfer. However, their use is limited by safety concerns, such as immune response, possible mutagenesis and carcinogenesis, and high production costs. Considering these limitations, the delivery of therapeutic genes to target cells upon direct *in vivo* administration of non- viral vectors, i.e., plasmids, is of great value for the development of gene therapy. However, the use of plasmids is plagued by poor transfer efficiency, intracellular penetration and nuclear localization, and low expression level. The success of non-viral gene therapy depends on the development of optimized plasmids. This review will focus on the importance of the plasmid backbone, which is often not considered in gene therapy experiments. Several features will be considered: bacterial DNA sequences, nuclear import, safety. Two examples of vectors will be described.

2. Bacterial DNA sequences

A hallmark of vertebrates genomes is that cytosines within a CpG dinucleotide sequence are modified by a methyl group bound to the 5' position of the cytosine ring (DNA methylation) [1]. Roughly 70–80% of all CpGs are methylated. The pattern of methylation seems to be related to genome defense and transcriptional repression (reviewed in [2]). In general, transcriptionally active genes are undermethylated while silent genes are extensively methylated. Furthermore, CpG dinucleotides are underrepresented in mammalian genomes, a phenomemon referred to as "CpG suppression". In contrast, in bacterial DNA, CpG dinucleotides have the expected frequency of 1:16, and are not methylated. The consequence of these two main differences is that mammalian cells recognize bacterial DNA, including plasmid DNA, as foreign. This results in two major consequences: immune response and

CpG methylation, which might hamper both the efficiency and the safety of the vector.

2.1. Undesirable effects in gene therapy: immune response and toxicity

It is well documented that some of the bacterial non methylated CpG sequences stimulate the mammalian immune system. The most immunostimulatory motif is usually 5′-Pu-Pu-CpG-Pyr-Pyr-3′ [3]. Depending on the DNA backbone and sequence motifs, these non methylated sequences can stimulate B cell division, activate macrophages, NK cells, monocytes and dendritic cells ([4] and references therein, [5]). Although the mechanism is not yet clearly established, it probably involves Toll-like receptor 9 [6]. The activation of the immune system results in the secretion of cytokines, primarily IL-6, IL-12, TNF-α and IFN-γ, which may lead to undesirable immune responses, toxic effects and attenuation of gene expression in therapy.

2.1.1. Immune response
It has been suggested that secretion of proinflammatory cytokines TNF-α and IFN-γ can downregulate gene expression from commonly used viral promoters, for example, the CMV promoter, at the mRNA level [7]. Constitutive mammalian promoters, such as β-actin, were much less sensitive to downregulation by cytokines. One way to avoid cytokine production would be to methylate these immunostimulatory CpG motifs with DNA methyltransferases to make it more mammalian like. Indeed, this results in decreased cytokine levels, but it also dramatically reduces transgene expression [8]. This is not surprising since it appears that DNA methylation represses transcription in vivo [9]. In addition, methylation of episomal plasmids in cultured cells reduces reporter gene expression [10]. This might be a reason for the commonly observed decrease in gene expression in vivo after plasmid DNA transfer. Therefore, methylation of foreign DNA is not a viable solution for long term expression of a transgene, which is often a requirement for gene therapy. An effective approach would be to reduce as much as possible the amount of bacterial CpG sequences in the plasmid backbone and to use mammalian constitutive promoters, such as ubiquitin, muscle creatine kinase or β-actin [11–13]. Constitutive mammalian promoters should not elicit an immune response due to an abnormal CpG content, and they might also be less prone to downregulation by inflammatory cytokines, in contrast to the CMV promoter, as was shown for the β-actin promoter [5]. Furthermore, one problem of gene therapy could be an immune response against the gene product, which is primed by gene expression in professional antigen-presenting transfected cells (APC) (see [13] and references therein). The use of tissue-specific promoters could restrict gene expression to targeted cells, avoid expression in APC cells and thus abrogate the unwanted immune response against the gene product [13]. In addition, restriction of expression to targeted tissues would be a safety advantage in gene therapy.

2.1.2. Toxicity

CpG motifs in plasmid DNA have been shown to be major contributors to induction of proinflammatory cytokines, particularly in combination with cationic lipids [8]. Lung administration of aerosolized lipoplexes (plasmid DNA–cationic lipid complexes) can elicit acute toxicity with mild flu-like symptoms in a clinical trial [14]. Thus, a decrease in the number of CpG motifs is expected to reduce this response and to improve the effectiveness of non-viral gene delivery vectors.

2.2. Vaccination

In contrast to gene expression, CpG motifs might be useful in vaccination: it has been suggested that immunostimulatory CpG sequences are potent adjuvant to the antigen, and the use of optimized sequences elicit a more potent response [15,5], via a Th1-like response. This strategy could also be useful for immunotherapy of cancer (reviewed in [5]). However, an improved knowledge of the mechanisms involved will be necessary in order to increase vaccine efficacy in humans [16].

3. Safety

Regulatory authorities, such as the U.S. Food and Drug Administration (FDA), have made several safety-related recommendations about the plasmid backbone. For example, the use of an antibiotic-resistance gene is allowed only if it is not a member of the penicillin family. Kanamycin or neomycin are suggested. In this case, the final content of antibiotic remaining in the plasmid preparation after purification has to be determined. The use of other means of selection such as suppressor tRNA genes is also considered. In addition, in order to limit the possibility of chromosomal integration by recombination, homology of plasmid DNA to human genome sequences should be avoided (FDA, Docket No. 96N-0400: Points to consider on plasmid DNA vaccines for preventive infectious disease indications. http://www.fda.gov/cber/gdlns/plasmid.pdf).

4. Plasmid size

The size of the plasmid might be an important point to consider: For example, Wells et al. [17] reported that plasmids in the 7 to 16 kb range had no significant effect on in vivo gene expression after intramuscular injection. However, a small increase in gene expression was observed with a 2.8 kb plasmid than with a 5.5 kb plasmid, both in vitro and in vivo after intramuscular injection [18]. Furthermore, large plasmids (> 20 kb) showed less expression than smaller ones after lipofection in cultured cells [19]. This effect remains to be confirmed in vivo.

5. Optimization of plasmids for nuclear localization

5.1. Nuclear import mediated by viral DNA sequences

Nuclear localization of plasmid DNA is a prerequisite for the efficiency of gene therapy [20]. Using a DNA sequence that would facilitate nuclear targeting is an appealing strategy. It has been shown that plasmid DNA containing the origin of replication and early–late promoter domains of SV40 was able to accumulate in TC7 cell nuclei after cytoplasmic microinjection [21]. This nuclear translocation occurred through nuclear pore complexes (NPC), and plasmid DNA appeared to be localized in the areas of active transcription and message processing [21]. This result was reproduced in the model of digitonin-permeabilized HeLa cells, where nuclear import was dependent upon energy and cytoplasmic factors [22].

It was suggested that the 72-bp enhancer domain mediated the import: plasmids encoding this sequence were localized in the nuclei after cytoplasmic microinjection in non-dividing TC7 cells [23]. Two other strong viral promoter and enhancer sequences, e.g. CMV immediate-early promoter/enhancer and RSV LTR did not allow nuclear import of plasmid in the absence of cell division [23]. It should be noted that in our laboratory we have been unable to reproduce these results.

Karyophilic proteins (proteins attracted to the nucleus) bear nuclear localization signals (NLS), which enable them to be recognized by members of the importin family. Transcription factors are among these karyophilic proteins, which bind to DNA. It was hypothesized that newly synthesized transcription factors could bind plasmid DNA in the cytoplasm, and that this DNA-karyophilic protein complex could then be imported into the cell nucleus. It was suggested that the 72-bp enhancer domain of SV40 mediates nuclear import by this mechanism, since it contains binding sites for ubiquitous transcription factors AP1, AP2, AP3, Oct-1 and NFκB.

In experiments using Epstein-Barr virus (EBV)-transformed cells expressing EBNA-1 protein [24], plasmids containing the EBV *oriP* site were up to 100-fold more efficient for gene expression after cytoplasmic microinjection than control plasmids lacking the *oriP* site [24]. Since the EBNA-1 protein is a nuclear protein binding the *oriP* site, it was hypothesized that it mediates the increase in gene expression by stimulating nuclear localization of the *oriP*-containing plasmids.

5.2. Nuclear import mediated by mammalian DNA sequences

The smooth muscle gamma actin (SMGA) promoter was identified as a smooth muscle cell-specific nuclear import sequence, as it is the binding site of various smooth muscle cell-specific transcription factors, like serum response factor (SRF) [25]. After cytoplasmic microinjection, plasmids containing this sequence were localized in the nuclei of smooth muscle cells while they remained cytoplasmic in CV1 cells and fibroblasts.

More recently, DNA vectors containing repetitive binding sites for the inducible transcription factor NFκB showed a 12-fold increase in gene expression after transfection of HeLa cells, and induction of NFκB expression by TNF-α [26].

5.3. Nuclear import mediated by peptide sequences

Nuclear uptake of proteins is a highly selective process directed by NLS peptides. SV40 large T antigen contains a short NLS peptide (PKKKRKV), which has been extensively studied. It was shown that plasmids covalently linked to this NLS peptide were actively imported into the nucleus of digitonin-permeabilized cells [27]. The rate and extent of nuclear import were dependent on the density of NLS peptides coupled to the plasmid. Unfortunately, NLS modification of DNA plasmid totally abolished the expression of the reporter gene, and no nuclear staining could be observed when these NLS-plasmids were microinjected into the cytoplasm of HeLa.

A linearized plasmid encoding the luciferase reporter gene was conjugated to a single NLS peptide. After transfection of various cultured cell lines, a 10- to 1000-fold increase in gene expression was observed [28]. The mechanism suggested here was the following: NLS would promote DNA docking to NPCs, and as soon as the DNA starts entering the nucleus, it would be condensed by histones into chromatin-like structures, which would pull the filamentous molecule into the nucleus. According to this model, multiple NLS coupled to the linearized plasmid would inhibit the nuclear import: the DNA would be stuck on the nuclear envelope if its transfer is initiated at more than one pore.

Regardless of the attractiveness of this approach, the molecule remains difficult to produce, and future pharmaceutical development and production of this chimeric DNA cannot be presently pursued. Moreover, it seems that linearized plasmid is less efficient than supercoiled plasmid for gene expression.

In a parallel effort, we covalently linked SV40 T antigen NLS to circular plasmid DNA [29] in a manner that maintains the structural integrity of the plasmid. In this molecule, the NLS moiety can still interact with its receptors, since the NLS-plasmid conjugate binds an importin-α-GST fusion protein in an NLS-dependent manner. However, expression enhancement of a reporter gene after transfection was not significant when 3 or 8 NLS peptides were coupled to the plasmid, and a 60% decrease in expression was observed when 43 NLS were non site-specifically coupled. In a site-directed ligand-coupling strategy using a triple helix sequence, a single NLS was bound to the plasmid, which remained supercoiled [30]. This approach can be used to produce the plasmid on a large scale, and for any ligand. The expression of the transgene was maintained after transfection on NIH-3T3, but the increase of expression over non-modified plasmid was non significant, both *in vitro* [30] and *in vivo* (Carole Neves, unpublished data).

5.4. Nuclear import mediated by steroids

In a new strategy to improve nuclear import, Rebuffat *et al.* [31] attached a dexamethasone moiety to DNA: a synthetic steroid-spacer-psoralen compound was UV-irradiated in the presence of a reporter plasmid. This caused the psoralen moiety to covalently link to the DNA, leaving the steroid freely accessible. Dexamethasone binds to its cognate glucocorticoid receptor (GR), which actively migrates to the nucleus, once activated by its steroid ligand. This complex is then translocated into the nucleus and accumulates there. A 1.2- to 43-fold increase in gene expression was observed. Notably, even when several dexamethasone moieties were attached to DNA, the docking of DNA to the nuclear envelope described by Zanta *et al.* [28] did not appear to be a problem. This seems a very promising approach, although it remains to be proved if it is a more powerful strategy for enhancing nuclear import than the previously described one [31]. This concept can also be extended to other nuclear receptors such as vitamin D, peroxisome proliferators or retinoic acid.

6. *Examples of vectors: pCOR and minicircles*

Approaches such as minicircle and pCOR were developed to diminish the number of prokaryotic CpG motifs, to avoid the presence of an antibiotic-resistance gene for improved biosafety, and to reduce the size of the plasmid for better bioavailability.

6.1. *pCOR: A plasmid with a conditional origin of replication*

A novel host–vector system with a conditional origin of replication has been designed recently [32]. The vector backbone consists of three bacterial elements: the R6K γ conditional origin of replication, a tRNA suppressor gene, and a *cer* fragment to resolve pCOR oligomers. The size of the pCOR plasmid (2 kb) is thus minimal, and it does not contain immunostimulatory CpG sequences, which should limit the potential inflammatory response to the prokaryotic CpG sequences in human therapy.

Propagation of pCOR plasmid is restricted to a specifically engineered bacterial strain expressing the R6K π initiator protein (*pir* protein). On the other hand, this bacterial strain carries an amber mutation in the *argE* gene, involved in arginine biosynthesis, and thus cannot grow on a minimal medium lacking arginine. This mutation can be corrected by a suppressor tRNA encoded by the *sup* Phe gene on the pCOR plasmid, thus leading to a functionally active argE protein. This allows the strain to grow on minimal medium. With this pCOR construction, no antibiotic resistance gene is required. *In vitro* standard transfection assays and *in vivo* injection into OF1 mouse tibial anterior muscle with *luc* and *luc*+ pCOR led to higher levels of reporter gene expression as compared to a conventional plasmid harboring the same expression cassette. Long-term expression (18 months) was also shown with a secreted alkaline phosphatase (SeAP)-encoding plasmid. The pCOR plasmid is presently being used in a phase II clinical trial using the FGF1 gene for peripheral artery disease.

6.2. *Minicircles*

Another strategy to produce minimal backbone DNA molecules was developed: minicircles, which are supercoiled recombinant DNA molecules containing only the therapeutic gene expression cassette [18]. In contrast to the plasmids currently used in preclinical and clinical trials of gene therapy, minicircles have neither an origin of replication nor an antibiotic selection marker. They cannot self replicate or confer antibiotic resistance to microorganisms. They carry only short bacterial sequences, thus limiting the possible immune response due to immunostimulatory CpG motifs. These minicircles are produced in *Escherichia coli* by site-specific recombination mediated by the phage λ integrase between the *attP* and *attB* recombination sequences, which were introduced upstream and downstream of the expression cassette in the plasmid. This recombination leads to a minicircle (containing the therapeutic gene expression cassette and the recombinant *attR* site) and a miniplasmid (containing the origin of replication, the antibiotic marker and the *attL* site). Unfortunately, the yield of non-recombined plasmid is around 40% of the starting material, and only small quantities of minicircles are recovered because of suboptimal plasmid copy number, incomplete recombination and extensive purification steps. *In vitro* experiments were performed with a minicircle encoding the luciferase gene: the minicircle led to a 2- to 10-fold higher luciferase activity in various cell lines compared to the same copy number of non-recombined plasmid. *In vivo* injection of minicircle into mouse skeletal muscle and experimental tumors gave strongly higher reporter gene expression compared to non recombined plasmid. The percentage of transfected myofibers per muscle was also higher, with minimal levels of muscle damage [18].

The purification of minicircles is currently the major limitation to their use. Chromatographic purification methods have been tested with satisfactory results. Triple helix-mediated affinity chromatography of plasmids [33] represents a promising technology to efficiently separate minicircle from miniplasmid after successful recombination.

7. Conclusions

Both safety and efficiency considerations tend to the development of optimized plasmids containing minimal bacterial DNA sequences. Foreign CpG DNA sequences could be a threat to gene therapy: they can induce an immune response resulting in either elimination of transfected cells, and/or in production of proinflammatory cytokines that can downregulate promoters such as the CMV promoter. Furthermore, in combination with cationic lipids, this immune response due to bacterial CpG can also elicit an acute toxicity. Second, CpG sequences from the plasmid might be prone to methylation, which can result in gene silencing in the transfected cells, without loss of the plasmid. On the other hand, the CpG motifs can be useful in vaccination and immunotherapy of cancer since they enhance the immune response against the antigen.

As nuclear localization of plasmid DNA is essential for efficient gene therapy, several strategies have been used to improve it:

(i) Inserting in the plasmid backbone various DNA sequences (either viral or mammalian) that should allow active nuclear import mediated by karyophilic proteins. This resulted in 12- to 100-fold increase in gene expression.

(ii) Grafting NLS peptides on the plasmid backbone to facilitate nuclear import through the importin family pathway. This is an attractive strategy, which resulted in 10- to 1000-fold increase in gene expression. However, production of these molecules remains a difficult and expensive process.

(iii) Grafting steroids on the plasmid backbone to mediate nuclear import through the glucocorticoid receptor pathway. This promising approach needs to be investigated further.

None of these strategies is currently sufficiently satisfactory to effect a major improvement in gene expression. In conclusion, although more efficient non-viral vectors have been constructed (for example, the above described pCOR and minicircle), backbone optimization or functionalization have yet to solve one of the major problems in gene therapy, which is gene delivery to quiescent non-dividing cells.

Abbreviations

APC	antigen-presenting cells
CMV	cytomegalovirus
EBNA-1	Epstein-Barr nuclear antigen 1
EBV	Epstein-Barr virus
FGF	fibroblast growth factor
GR	glucocorticoid receptor
GST	glutathione-S-transferase
IFN-γ	interferon gamma
IL	interleukin
LTR	long terminal repeat
NFκB	transcription factor NF Kappa B
NLS	nuclear localization signal
NPC	nuclear pore complex
NK	natural killer
pCOR	plasmid with a conditional origin of replication
Pu	purine
Pyr	pyrimidine
RSV	respiratory syncitial virus
SeAP	secreted alkaline phosphatase
SMGA	smooth muscle gamma actin
SRF	serum response factor
Th-1	type 1 helper
TNF-α	tumor necrosis factor alpha

References

1. Razin, A. and Riggs, A.D. (1980) Science 210, 604–610.
2. Robertson, K.D. and Wolffe, A.P. (2000) Nat. Rev. Genet. 1, 11–19.
3. Krieg, A.M., Yi, A.-K., Matson, S., Waldschmidt, T.J., Bishop, G.A., Teasdale, R., Koretzky, G., and Klinmman, D. (1995) Nature 374, 546–549.

4. Krieg, A.M. (2002) Annu. Rev. Immunol. 20, 709–760.
5. Scheule, R.K. (2000) Adv. Drug Deliv. Rev. 44, 119–134.
6. Aderem, A. and Ulevitch, R.J. (2000) Nature 406, 782–787.
7. Qin, L., Ding, Y., Pahud, D.R., Chang, E., Imperiale, M.J., and Bromberg, J.S. (1997) Hum. Gene Ther. 8, 2019–2029.
8. Yew, N.S., Wang, K.X., Przybylska, M., Bagley, R.G., Stedman, M., Marshall, J., Scheule, R.K., and Cheng, S.H. (1999) Hum. Gene Ther. 10, 223–234.
9. Siegfried, Z., Eden, S., Mendelsohn, M., Feng, X., Tsuberi, B.-Z., and Cedar, H. (1999) Nat. Genet. 22, 203–206.
10. Hong, K., Sherley, J., and Lauffenburger, D.A. (2001) Biomol. Eng. 18, 185–192.
11. Yew, N.S., Zhao, H., Wu, I.-H., Song, A., Tousignant, J.D., Przybylska, M., and Cheng, S.H. (2000) Mol. Ther. 1, 255–262.
12. Yew, N.S., Przybylska, M., Ziegler, R.J., Liu, D., and Cheng, S.H. (2001) Mol. Ther. 4, 75–82.
13. Weeratna, R.D., Wu, T., Efler, S.M., Zhang, L., and Davis, H.L. (2001) Gene Ther. 8, 1872–1878.
14. Alton, E.W., Stern, M., Farley, R., Jaffe, A., Chadwick, S.L., Phillips, J., Davies, J., Smith, S.N., Browning, J., Davies, M.G., Hodson, M.E., Durham, S.R., Li, D., Jeffery, P.K., Scallan, M., Balfour, R., Eastman, S.J., Cheng, S.H., Smith, A.E., Meeker, D., and Geddes, D.M. (1999) Lancet 53, 947–954.
15. Sato, Y., Roman, M., Tighe, H., Lee, D., Corr, M., Nguyen, M.-D., Silverman, G.J., Lotz, M., Carson, D.A., Raz, E. (1996) Science 273, 352–354.
16. Manders, P. and Thomas, R. (2000) Inflamm. Res. 49, 199–205.
17. Wells, D.J., Maule, J., McMahon, J., Mitchell, R., Damien, E., Poole, A., and Wells, K.E. (1998) J. Pharm. Sci. 87, 763–768.
18. Darquet, A.M., Rangara, R., Kreiss, P., Schwartz, B., Naimi, S., Delaère, P., Crouzet, J., and Scherman, D. (1999) Gene Ther. 6, 209–218.
19. Kreiss, P., Cameron, B., Rangara, R., Mailhe, P., Aguerre-Charriol, O., Airiau, M., Scherman, D., Crouzet, J., and Pitard, B. (1999) Nucleic Acids Res. 27, 3792–3798.
20. Escriou, V., Carrière, M., Bussone, F., Wils, P., and Scherman, D. (2001) J. Gene Med. 3, 179–187.
21. Dean, D.A. (1997) Exp. Cell Res. 230, 293–302.
22. Wilson, G.L., Dean, B.S., Wang, G., and Dean, D.A. (1999) J. Biol. Chem. 274, 22025–22032.
23. Dean, D.A., Dean, B.S., Muller, S., and Smith, L.C. (1999) Exp. Cell Res. 253, 713–722.
24. Längle-Rouault, F., Patzel, V., Benavente, A., Taillez, M., Silvestre, N., Bompard, A., Sczakiel, G., Jacobs, E., and Rittner, K. (1998) J. Virol. 72, 6181–6185.
25. Vacik, J., Dean, B.S., Zimmer, W.E., and Dean, D.A. (1999) Gene Ther. 6, 1006–1014.
26. Mesika, A., Grigoreva, I., Zohar, M., and Reich, Z. (2001) Mol. Ther. 3, 653–657.
27. Sebestyén, M.G., Ludtke, J.J., Bassik, M.C., Zhang, G., Budker, V., Lukhtanov, E.A., Hagstrom, J.E., and Wolff, J.A. (1998) Nat. Biotechnol. 16, 80–85.
28. Zanta, M.-A., Belguise-Valladier, P., and Behr, J.-P. (1999) Proc. Natl. Acad. Sci. USA 96, 91–96.
29. Ciolina, C., Byk, G., Blanche, F., Thuillier, V., Scherman, D., and Wils, P. (1999) Bioconjug. Chem. 10, 49–55.
30. Neves, C., Byk, G., Scherman, D., and Wils, P. (1999) FEBS Lett. 453, 41–45.
31. Rebuffat, A., Bernasconi, A., Ceppi, M., Wehrli, H., Brenz Verca, S., Ibrahim, M., Frey, B.M., Frey, F.J., and Rusconi, S (2001) Nat. Biotechnol. 19, 1155–1161.
32. Soubrier, F., Cameron, B., Manse, B., Somarriba, S., Dubertret, C., Jaslin, G., Jung, G., LeCaer, C., Dang, D., Mouvault, J.M., Scherman, D., Mayaux, J.F., Crouzet, J. (1999) Gene Ther. 6, 1482–1488.
33. Wils, P., Escriou, V., Warnery, A., Lacroix, F., Lagneau, D., Ollivier, M., Crouzet, J., Mayaux, J.F., and Scherman, D. (1997) Gene Ther. 4, 323–330.

S.C. Makrides (Ed.) *Gene Transfer and Expression in Mammalian Cells*

CHAPTER 10

Use of scaffold/matrix-attachment regions for protein production

Pierre-Alain Girod and Nicolas Mermod

*Laboratory of Molecular Biotechnology, Center for Biotechnology UNIL-EPFL and
Institute of Biotechnology, University of Lausanne, 1015 Lausanne, Switzerland;
Tel.: +41-21-693-6151; Fax: +41-21-693-7610; E-mail: nicolas.mermod@ibt.unil.ch*

Abstract

MARs (Matrix Attachment Regions), also called SARs (Scaffold Attachment Regions), are 300–3000 bp long DNA elements proposed to play a role in nuclear and chromosomal architecture. They were proposed to attach chromatin to proteins of the nuclear matrix and thereby partition the eukaryotic genomes into independent chromatin loops. Because of their co-localization with transcription units and regulatory elements in genomes, MARs have been implicated in the regulation of gene expression. For instance, MARs were shown to control gene expression by facilitating interactions between DNA activating complexes and genes, and by controlling chromatin accessibility. The ability of MARs to protect transgenes against transcriptional silencing effects has been used to augment expression of heterologous genes. This review provides a possible explanation on the function of MARs and presents recent data on their use to increase protein production.

1. Introduction

The applications of gene transfer technology in mammalian cells are diverse and include (i) *in vitro expression*, in cells engineered for production of proteins of pharmaceutical/industrial interest or with altered metabolism (e.g., for use in biotransformation), and (ii) *in vivo expression*, with gene transfer in explanted cells that will be reimplanted or gene transfer *in situ* within the organism (e.g., gene therapy). The main requirements of gene transfer are usually either high level expression, for instance in the case of *in vitro* protein production, or temporal and/or spatial regulation in the case of biotransformation and gene therapy. However, the levels and pattern of transgene expression often differ among different transformant cells and/or tend to decline with time. This variability of expression and rapid extinction are manifestations of (i) structural effects such as gene elimination or epigenetic silencing and (ii) the influence of dominant regulatory elements flanking the sites of integration of the transgene (the so-called position effect). These effects are thought to result from chromatin structure effects, which has led to intense studies to identify genetic elements that may experimentally define or control transgene expression at the chromatin structure level.

2. Chromatin structure and control of gene expression

In the early days of the analysis of transgene expression in eukaryotic cells, an activator protein was presumed to access a promoter DNA sequence and recruit proteins to direct the assembly of a transcription pre-initiation complex composed of RNA polymerase II and other general transcription factors. This view is considered nowadays too simplistic, as it is clear that another major function of transcription activating proteins is to recruit chromatin remodelling enzymes that govern chromatin structure, to create a permissive environment for the transcription machinery at specific phases of the cell cycle. Thus, transgenes may be embedded in relatively open chromatin structure that is permissive for transcription, or to the contrary, in the closed chromatin structure typical of fully silenced genes. The chromatin status of the transgene and its evolution are thus governed by the interplay of regulatory elements on the transgene and on the neighbouring chromosomal sequences. Our laboratory has long been interested in investigating the relationship between chromatin structure and gene expression, and we have been looking for DNA sequences that could be inserted in transgene constructs to minimize silencing by regulatory elements at the site of integration into the genome. One type of DNA elements that have been used successfully in animal and plant cells and organisms [1] are the matrix attachment regions (MARs), also known as scaffold attachment region (SARs), which define domains of independent gene activity.

2.1. MARs and loop domain organization

The nuclear matrix was originally described as a framework of the nucleus that remains insoluble after selective extraction of histones and DNA in the chromatin loops [2]. However, the very existence of a nuclear matrix that would organize chromosomes has been questioned. The complex procedure used for isolation of nuclear matrix may be fraught with artefacts. It has been proposed that aggregation of most of the nuclear proteins may occur upon extraction [3]. Consistent with this, although electron microscopy has discerned granules and fibrils in the eukaryotic interphase nucleus, with a skeleton substructure that resembles a network of filaments, no rigid framework proteins similar to those found in the cytoplasm have been isolated from the nuclear matrix [4,5]. Instead, all the matrix proteins identified thus far appear to be regulatory as opposed to structural proteins. Notably, it is natural to assume that transcription factors and RNA processing proteins form temporary sites of DNA attachment to the nuclear matrix [3]. For instance, heterogenous nuclear ribonucleoproteins (hnRNP) are major protein components of the nuclear matrix. Because multiple hnRNP factors form complexes that assemble onto pre-messenger RNA transcripts, the nuclear matrix may be composed of the protein machineries that mediate nascent transcript production and RNA maturation [6].

MARs are defined as the genomic DNA sequences at which the chromatin is anchored to the nuclear matrix proteins during interphase [7]. The organization of chromatin into discrete looped domains is believed to contribute structurally to the packaging of DNA in the nucleus and functionally to the regulation of expression and replication of the genome. The metabolic activities of the matrix proteins acting at MARs constrain the chromatin fiber and bring about the morphological changes of chromosomes by close packing of pre-existing or de novo formed loops. Thus, within individual chromosome territories, the chromatin fiber appears highly contorted, looping back and forth between the interior and periphery bringing together distant loop domains into close proximity (Fig. 1A). By means of *in situ* hybridization it has been found that the actively transcribed genes are associated with the nuclear matrix, and that they are localized preferentially at the periphery of the chromosome where they are assembled into large macromolecular transcription and splicing complexes [4,8–10]. Consequently, the chromosome substructures arise transiently as a consequence of the nuclear metabolic activities such as transcription and replication. In contrast, chromatin domains that belong to closed and silent

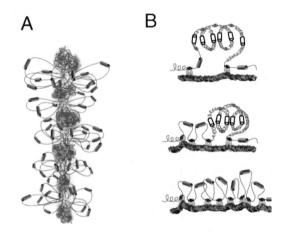

Fig. 1. (A) Comprehensive chromosome model. A portion of a chromosome is shown, along with its interspersed inactive and active regions segregated from one another within distinct chromosome sub-regions. Inactive genes are compacted into tightly folded dense regions. Active genes are found in more unfolded regions characterized by loop bodies. Active genes (solid tubes) tend to be preferentially localized at the periphery of these regions. RNA transcripts are formed at the surface of the loop region within large assemblies of complexes involved in transcription and splicing [9,83]. Transcripts are then shed into the interchromosomal domains for further processing and transport. (B) Establishment and maintenance of spatial patterns of gene expression. In this cartoon, a coiled structure indicates a closed chromatin configuration, and an open cylinder a silenced gene. In contrast closed cylinders and linear portions of chrosmosomes indicate active loci. The MAR elements are schematically indicated by crosses along the DNA fiber. In the One gene-One MAR hypothesis, each gene is regulated by a MAR element that dictates its chromosomal expression status. In some cases, such as Hox gene loci, genes can be colinear on the chromosome and expressed in a precise sequential order. Sequential activation may involve the progressive increase of the accessibility of genes to the transcription machinery through a directional release of a repressive state. This transition may result from the relocalization of the gene as mediated by the attachment of the MAR element to the matrix. The sequential binding of putative MAR sequences that regulate each gene individually, could help to compartmentalize the gene array by creating transcriptionally active subloops.

chromatin are topologically arranged and compacted in a way that precludes the accessibility of genes to the transcription complexes [11].

2.2. MAR-mediated transcriptional regulation

Evidence that MARs structurally separate chromatin into independent chromatin domains first came from a number of studies demonstrating that MARs augmented the expression of a reporter gene in stably transformed cell lines but not in transient transfection experiments [12–14]. This is in contrast to enhancers which also function in transient transfection assays. Consequently, the reporter gene randomly integrated into the genome was proposed to be structurally isolated from the chromosomal environment by MARs. It was hypothesised that this separation from the chromosomal environment was required for the appropriate expression of the transgene, without interference from chromosomal elements.

Later work focusing on position-effect protection showed that cell lines and transgenic mice carrying multiple copies of the reporter gene flanked by MARs expressed the gene at levels proportional to copy number [15,16]. This led to the proposal that multiple copies of genes each bracketed by two MAR elements were all competent for transcription. Thus, a transgenic chromatin domain located between MAR elements might function as an independent regulatory unit that would form a chromatin loop extending from the nuclear scaffold.

Whether MARs are similar to other chromatin-acting genetic elements, such as the boundary elements (BE) that prevent the propagation of silent chromatin, or insulators that block enhancer action on promoters (Chapter 11), is still a matter of debate [17]. Interestingly, the ability of MARs to prevent integration site effects could be blocked by a G + C rich element called a CpG island when the latter was positioned between the MAR and the enhancer [18]. Thus, the action of MAR may itself be subjected to insulator-like effects like that mediated by the CpG island, or, in an alternative model, it may be incapable of preventing the action of silencing elements that are located within the MAR-bracketed domain.

An additional feature of MARs is their function as origins of replication in combination with other genetic elements. MAR AT-rich sequences were reported to facilitate dissociation of the two DNA strands, and may thereby open chromatin and allow interaction with factors of the DNA replication machinery [19]. This has allowed the construction of episomally replicating expression vectors for mammalian cells [20]. MAR association with elements of the nuclear matrix was found to be essential for stable plasmid maintenance during mitosis [21].

2.2.1. The One gene-One MAR hypothesis
Generally, it is agreed that the average size of chromatin loops is around 86 kb, with sizes ranging between 5 and 200 kb [22]. The original chromatin loop model predicted that multiple genes within the same MAR-bordered chromatin domain would exhibit coordinated regulation, and that loops are more or less static or architectural structures. But this view is not fully consistent with some aspects of

current models of gene regulation. Firstly, it was found that highly transcribed genes are assigned to smaller loop sizes (3–14 kb) [23]. Secondly, MARs appear to be free from interactions with the nuclear matrix in cells that do not express adjacent genes [8]. Together, these studies indicated that the interaction of MARs with the matrix is gene- and cell-type specific, and that this interaction is somehow linked with the expression status of adjacent genes. Furthermore, examination of the distribution of MARs in long genomic DNA sequences of plants and animals revealed that it is similar to the average gene density, implying that each gene has its own MAR [24–26]. Another important feature is that MARs were generally found to lie close to enhancers and promoters or in the first intron. Furthermore, the MAR-mediated transcriptional effect was found to depend on its distance from the promoter and the direction of transcription, with the gene proximal to the MAR being strongly expressed whereas those located more distally being weakly expressed [14,27]. These data led to the emerging view that MARs are parts of genetic units like promoters and enhancers, and that they have an active regulatory role rather than the simple structural role of anchoring DNA to the nuclear matrix that is traditionally assigned to them.

2.2.2. MARs as regulatory switches

Gene mapping has revealed that coordinately or sequentially regulated genes belonging to a common developmental program are often clustered along the chromosome, and may form large loops comprising an inactive chromatin locus. During sequential activation, genes within the same loop domain may associate with the nuclear matrix in a cell cycle and developmental stage-specific pattern (Fig. 1B). Indeed, analysis of HOX gene loci in mice where genes are sequentially activated along time and position on the chromosome implied the occurrence of genetic elements with properties resembling those of MAR elements, although matrix-attachment properties were not directly demonstrated [28].

Thus, chromatin loops are not immobile or fixed to a certain position, but undergo significant diffusive motion. Switching of gene expression status involves changing the attachment points of chromatin loops, implying that the association of MARs with the matrix is regulated and depends on the cell type [8]. This view is supported by the findings that MARs are not necessarily active in all cells but control gene expression in a tissue-specific manner [29–32]. For example, we found that the use of chicken lysozyme MAR (cLysMAR) elements cotransfected with the erythropoietin gene allowed regulated and long term expression in undifferentiated C2C12 myoblasts. However, the erythropoietin gene became silenced in most C2C12 cells when they were committed to myotube differentiation ([30]; M. Imhof and N. Mermod, unpublished data).

A role of MARs in gene regulation is further supported by the observation that MARs can also act as gene silencers to downregulate transcription of reporter genes, depending on the context [31,33,34]. For instance, when dissecting the elements responsible for the cLysMAR activity, we found some multimerized portions of the MAR will increase transgene expression while others will inhibit the transgene

(P.-A. Girod and N. Mermod, unpublished data). Therefore, these data indicate that *cLysMAR* is a complex element with separate activating and silencing activities.

2.3. DNA sequence features of MARs

2.3.1. Structural motifs
The genome may be viewed as a mosaic of DNA segments of varying GC content. Interestingly, MARs have a higher AT-content ($>70\%$ A/T) and in this regard they differ from promoters and coding sequences. However, comparative analysis of MAR's AT-rich tracts failed to reveal a detectable organization of A/T motifs or consensus sequences [35], despite the fact that MARs are evolutionarily conserved [36]. Nevertheless, a number of reports have described motifs embedded along MAR sequences [37]. Most MARs appear to contain a MAR-specific sequence called "MAR recognition signature," which is a bipartite sequence that consists of two individual sequences AATAAYAA and AWWRTAANNWWGNNNC within 200 bp. Other sequences, proposed to be indicative of MAR sequences, are the DNA-unwinding motif (AATATATTAATATT), replication initiator protein sites (ATTA and ATTTA), homo-oligonucleotide repeats (e.g., the A-box AATAAAYAAA and the T-box TTWTWTTWTT), DNase I-hypersensitive sites, potential nucleosome-free stretches, polypurine/polypyrimidine tracks, and sequences that may adopt non-B-DNA or triple-helical conformations under conditions of negative supercoiling [38]. However, a periodic pattern of motifs has not yet been detected, and it remains unknown whether particular combinations of these motifs mediate MAR activity.

MARs exhibit various lengths, and the shorter elements, like the intronic MARs that combine enhancer- and position-effect activities, appear to be structurally relatively simple. In contrast, MARs embedded in long regions of condensed chromatin characterized by high levels of CpG DNA methylation and low levels of histone acetylation are generally larger and more complex. Whether, this reflects distinct activities or regulatory properties is an interesting but mostly unexplored possibility.

2.3.2. MAR functional elements
MAR activity is unlikely to arise uniquely from the intrinsic properties of its DNA motifs. Rather, the ability to protect against position effect and to regulate transcription may depend on the contribution of the protein factors that bind these motifs [35]. These elements can be classified in sites which are recognized by (i) abundant multifunctional matrix proteins (i.e. high mobility group HMG-I/Y protein TATTATATAA binding sites, or topoisomerase II RNYNNCNNGYNGKTNYNY elements) and by (ii) transcription factors such as H-box (A/T_{25}), MTAATA, Y-box (CCAAT) and CTAT repeats-binding proteins [3,39]. Some of the latter appear to activate transcription, as exemplified by the Bright protein. Remarkably, there also exists a class of transcription factors that interfere with MAR-matrix productive effects on transcription. These include the methyl-CpG binding protein ARBP/MeCP2, SATB1 (special AT-rich binding protein 1), and CDP (CCAAT

displacement protein) that may participate in the gene switch or gene inhibition properties of MARs.

Recently, an algorithm was designed to predict the potential of DNA sequences to form MAR elements, with the goal of deducing the motif pattern that confers MAR-forming ability (http://www.futuresoft.org/modules/MarFinder/index.html; [37]). An illustration of the arrangement of these motifs on the *cLysMAR* is shown in Fig. 2. However, because of the absence of an easily detectable MAR code comparable to other periodic elements carried by DNA (i.e., open reading frames and the chromatin code), the contribution from each individual element is still unclear. Nevertheless, deletion experiments have suggested interdependence between the elements. Consistently, the mere presence of binding sites for activator proteins does not constitute a MAR. For instance, Bright is unable to transactivate expression from a concatamerized binding site [40]. Finally, the identification of DNA element(s) that mediate MAR activity is complicated by the cell type-specific expression or activity of MAR-binding matrix proteins [32]. For instance, the deletion of CDP-binding sites in the MAR activates the expression of the nearby immunoglobulin heavy chain gene in immature B-cells

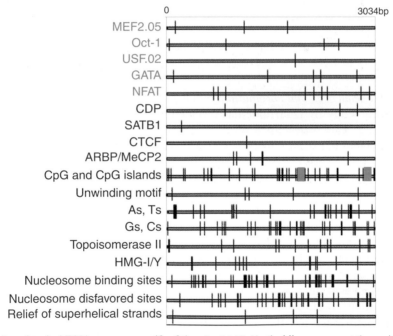

Fig. 2. Map of typical DNA sequence motifs of the *cLysMAR*. Vertical lines represent the position of the computer-predicted sites or sequence motifs along the 3034 base pairs of the *cLysMAR*. The putative transcription factor sites were identified using TRANSFAC, and CpG islands were identifed with CPGPLOT. The motifs shown here include CpG dinucleotides and CpG islands, unwinding motifs, poly As and Ts, poly Gs and Cs, topoisomerase II and High mobility group I (HMG-I/Y) protein binding sites, nucleosome-binding and nucleosome disfavouring sites [48], and a motif thought to relieve the superhelical strand of DNA.

where the gene is normally silent. In contrast, deletion of the full MAR sequence decreased transcription in expressing cells [41,42]. Thus, MARs may function to either activate or repress the expression of adjacent genes, depending on the cell type.

2.4. MAR mode of action

2.4.1. Chromatin acetylation and accessibility

Expression of genes in higher organisms is acutely dependent on DNA accessibility. To date, the most characterised structural modification linked to transcriptional activation is histone acetylation [43,44], but other histone modifications such as phosphorylation and hypomethylation also play a role in gene activation. Conversely, transcriptional repression is usually accompanied by the deacetylation of histones which will lead to DNA methylation at CpG islands adjacent to, or overlapping, the transgene [45]. DNA methylation can be viewed as a locking mechanism that ensures maintenance of the inactive state. Histone H1, which associates preferentially with methylated DNA, is thought to compact nucleosomes in heterochromatin [46], forming 30–400 nm thick compact fibers that pose a barrier to transcription by blocking access to promoter DNA. Such closed chromatin structures are thought to propagate along the chromosome, originating either from chromosomal loci that act as silencers or from telomeric and centromeric sequences.

2.4.2. MAR sequences and chromatin dynamics

An intriguing study by Izaurralde has revealed that histone H1 binds preferentially to MARs [47]. This provocative result suggested that MARs might operate as *cis*-acting sequences that would nucleate co-operative H1 assembly along the MAR into the flanking non-MAR DNA.

Consistently, DNA sequences that favour nucleosome formation and positioning [48] are significantly more frequent in MARs as well as at the 3'-end of transcriptional units as compared to promoter and coding regions. Thus, MAR distinctive structural features, such as A/T-tracts that bend DNA and direct nucleosome positioning [38,40,48,49], could regulate nucleosome organization. The cooperative binding of the high mobility group protein HMG-I/Y, which associates with histone acetyltransferases such as the general co-activator p300 [50], may lead to the acetylation of nucleosome core histones and to the displacement of histone H1 [51,52]. In this respect, it is interesting to note that the *cLysMAR* has nucleosome positioning sequences and that it is bordered by HMG-I/Y potential binding sites (Fig. 2). In this model, negative supercoiling would induce the A/T-tracts to locally unwind the DNA strands, further decompacting the nucleosomal array [53,54]. Consistently, MAR-mediated unwinding was shown to be important for the attachment to the nuclear matrix proteins [19]. The perturbation of the nucleosomal array over MAR elements could potentially disrupt heterochromatin locally, and/or block the propagation of silent chromatin structures. This mechanism is consistent with the appearance of DNAse I-hypersensitive sites within MARs, which is one of the hallmarks of active regulatory regions of DNA [54].

An additional feature of MARs such as *cLysMAR* is that they contain CpG islands (Fig. 2). Recently, Antequera and Bird [55] speculated that CpG islands are associated with regulatory regions that are transcriptionally active at totipotent stages of development, but may later become methylated as gene silencing progresses along cell differentiation pathways. Thus, CpG islands may mediate the gene switch function of MARs during developmental regulation.

2.4.3. MARs as entry sites for chromatin remodelling activities

Gene activation requires chromatin remodelling enzymes that acetylate histones. Comparison of the distribution of MARs and histone acetylation patterns of the β-globin locus has revealed a correlation between the location of histone hyperacetylation and MAR position, consistent with the patterns of expression of neighbouring genes [24,56,57]. Furthermore, high levels of acetylation inhibit the spread of DNA methylation. Thus, MARs may indirectly regulate the recruitment of DNA demethylase [58,59]. Indeed, MARs from various origins have been shown to be required for efficient initiation of transcription from CpG methylation-silenced genes [60,61].

As detailed above, particular MAR elements can also have gene silencing effects. The silencing activity of MARs may similarly involve the histone acetylation status. Indeed, MAR-binding repressors like SATB1 [53], ARBP/MeCP2 [62], CDP [33,63–66], and CRBP [31] mediate repression of transcription through recruitment of components of the histone deacetylating HDAC pathway. This contrasts with the action of MAR-binding activators such as Bright and SAF-A, which recruit SWI or p300 [40,67,68] and effect the acetylation and remodelling of nucleosomes [69].

In addition to their chromatin remodelling activities, MAR-binding transcription factors may also contribute to transgene regulation by directly interacting with components of the general transcription machinery. For instance, the ubiquitous SAF-B and SATB1 associate with RNA-polymerase II and/or with RNA processing factors [70,71]. Thus, MAR-binding proteins could serve as a molecular base to assemble the transcription apparatus and chromatin remodelling network. Together, these results suggest that MARs may act as permanent entry gates into a silent heterochromatin environment, allowing DNA–protein and/or protein–protein interactions, thereby increasing the probability of establishing an active locus. This property may be the basis for the ability of MARs to protect against position effects.

2.5. MAR-associated chromatin remodelling network

Activation of the correct gene, in the proper cell and at the right time, requires that binding of chromatin remodelling enzymes is a strictly controlled event. Thus, only the genes that have recruited the appropriate combination of remodelling enzymes will be expressed. This is an important feature of regulation as it generates selectivity at promoters and enhancers. MARs and promoters were long considered to be controlled by separate pathways. However, because MARs contain numerous potential binding sites for transcription factors, and transcription factors are known to be associated with the nuclear matrix, it is an intriguing possibility that these regulators could have a dual role on promoters and MARs to fine-tune the expression of different sets of genes [3,39,72]. Thus, gene regulation would require the concerted action of MARs, enhancers and promoters.

The composition of the nuclear matrix has been found to change in relation to the biological function of the cell and with transformation to cancer [73,74]. Specifically, proteins that mediate MAR functions (i.e. SATB1, CDP, Bright) appear to control expression of numerous genes expressed in differentiated cells. SATB1 is expressed predominantly in thymus but also in brain and several other organs, while CDP expression occurs in all but terminally differentiated cells. Binding of SATB1 and CDP to regulatory elements has been associated with transcriptional repression of numerous genes expressed in differentiated cells. For instance, SATB1 regulates gene expression both spatially and temporally during T-cell development. Recently, SATB1-null mice were generated, and genome-wide expression profiling analysis indicated that about 2% of all genes become significantly derepressed in thymocytes [34].

In contrast, the Bright activator is present in differentiated cells, where it may compete with repressors for MAR-binding sites. Thus, there would be a dynamic process of activation/repression forming temporary sites of DNA attachment. Bright overexpression was found to enhance transgene expression, indicating that the normal activator protein level is limiting. In contrast, cotransfection of CDP abrogated Bright transactivation and reduced the basal expression level [66], indicating that CDP is able to override the DNA binding and/or transactivation capacity of Bright [75]. A recent study by Liu et al. [65] presented additional evidence that the balance between these factors is critical for the regulation of MAR activity. It was shown that SATB1 binds to CDP, abrogating the MAR-binding ability of both proteins, and resulting in reactivation of the silent heterochromatic gene. Different degrees of repression could be mediated by competition between repressors and activators for MAR binding (Fig. 3).

Together, these observations suggest that the different MAR-binding proteins associate with diverse sets of proteins involved in transcriptional regulation and pre-mRNA processing that constitute the nuclear matrix. Thus, the MAR-binding matrix proteins could serve as a molecular base to assemble a 'transcriptosome complex' in the vicinity of actively transcribed genes [52,71]. By bringing together genetic loci and the components of the gene expression machineries, the MAR-organizing structures would regulate the efficiency of the transcription processes. Consistently, the MFP1 MAR-binding proteins were reported to cluster in speckle-like structures at the nuclear periphery, which may correspond to actively transcribing areas of the cell nucleus [76–78]. Assemblies of dedicated factories at specific sites in the nucleus would in turn impose constraints on the location of transcription, and RNA processing events.

Overall, a unified picture emerges from these and other studies. An early event in gene activation is the relocation of a silent gene embedded in heterochromatin and

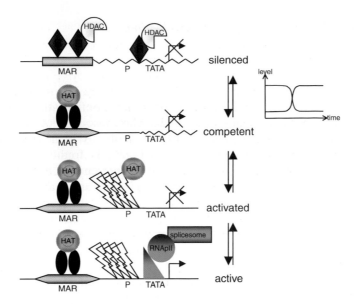

Fig. 3. Schematic model showing how the binding of specific proteins to MARs may establish an active chromatin domain. This cartoon summarizes that the binding of specific matrix proteins to MARs is one of the mechanisms whereby chromatin boundaries can be established. The co-operative binding of regulators could destabilize higher order folding of heterochromatin region surrounding the MAR, thereby propagating modifications to the neighbouring genes. Thus, access of transcriptional regulators at the promoter-enhancer region would be facilitated. HDAC, HAT indicate histone deacetylases and histone acetyltransferases, respectively, as recruited by inhibiting (diamond) and activating (ellipses) DNA-binding regulatory proteins. Activation of the transgene results from its association with structural proteins and components of the basal transcription and RNA processing machineries that may compose the nuclear matrix.

located at the nuclear periphery to a euchromatic, nucleocentral structure. This would be accompanied by the attachment of a gene-proximal MAR to the nuclear matrix. The precise structure of the nuclear matrix or scaffold is still unclear, but it has been found to be enriched in transcription and RNA processing activities. Thus, the MAR-matrix interaction would localize genes close to the activities required for gene expression. In turn, this would favour occupancy of the locus control regions and enhancer by sequence specific transcription factors. The continuous action of multiple transcription factors and remodelling complexes would further propagate an open chromatin structure towards the promoter and coding sequences. Promoter occupancy would culminate in the association with the basal transcription machinery and RNA polymerase II. These steps have in common the requirement for chromatin packaging and remodelling prior to protein–DNA interactions, consistent with the fact that histone- and DNA-modifying enzymes are involved in the functioning of each regulatory element. Each of these steps would be regulated by the availability of particular DNA-binding proteins, allowing for several layers of gene control. In this complex interplay, MAR would act as a very early event and as a master switch upon which all subsequent steps would depend.

3. Use of MAR elements for protein production

3.1. Copy number dependence of transgene expression

It has long been recognized that transgenes are generally not integrated into the genome as single copies, but rather in multiple repeat configurations of intact and sometimes also rearranged genes at a single chromosomal locus. Most often, multiple transgenes are not expressed proportionally to their copy number in independent cell clones, but the expression per gene is found to decrease with increasing copy number. Is has been argued that the presence of multiple integrated copies of the transgene favours silencing, while single-copy integrants are more likely to be stably expressed, thus contradicting the assumption that an increase in copy number of a transgene always leads to higher expression. In addition to such effects, gene silencing may also stem from the site of integration of the transgenes in the chromosome, for endogenous genetic elements nearby on the chromosomal locus may negatively affect transgene expression. Because of such silencing effects, transgene expression is not necessarily uniform in every cell of a clonal population, thereby leading to a variegated expression within the population. Overall, the combination of these effects is that transgene expression will vary greatly in various cell clones having integrated the same expression construct in distinct copy number and chromosomal integration sites. Silencing of introduced transgenes therefore constitutes a major bottleneck that demands lengthy cell clone selection, limits protein production, and thereby hampers the exploitation of eukaryotic cultured cells as protein factories.

The mechanism by which multiple gene copies are silenced is unclear. It may result from the propensity of repeated elements to pair and assemble in heterochromatin, by analogy to natural repeated or satellite elements that constitute constitutive heterochromatin. Alternatively, gene heterochromatization may stem from the expression of double-stranded and/or small interfering RNAs from repeated genes or gene parts that lead to epigenetic silencing. Other potential causes for transgene silencing may also include high GC content of the vector sequences and the inability of bacterial sequences to be packed properly in a chromatin structure. The incidence of silencing has been shown to increase with high transcriptional state of the transgene, possibly because transcription may facilitate the pairing of tandem repeats. Alternatively, chromatin condensation may also stem from a previous promoter inactivation, because the assembly of a closed chromatin structure would not be perturbed by the interaction of the transcription machinery with DNA. However, some chromosomal loci must have evolved a mechanism to avoid such silencing effects. Particularly the rRNA, tRNA and histone genes, which are often organized as high-copy concatemers of repeated units, are expressed at high levels indicating that multiple-copy arrays need not always be subjected to repeat-induced silencing. The intergenic spacers of these genes may play an important role in maintaining gene expression, but whether this results from nucleosome positioning sequences or from other DNA features, remains unclear.

3.2. Use of MAR elements to increase protein production by CHO cell lines

With regard to the use of mammalian cell systems for industrial protein production processes, one of the time-consuming steps is the selection and analysis of transformed cells to obtain highly expressing stable cell lines that produce recombinant proteins of the desired quality. Numerous approaches have been attempted to bypass this bottleneck, including vector engineering to improve gene expression and decrease clonal variations, and work on processes accelerating the selection of high producer clones. However, the current world limitations in the GMP production of proteins for clinical applications still acutely drives the need for higher producing clones.

While numerous studies suggested that genetic elements such as MARs may control chromatin structure and may be used to improve protein expression in plant and animal cells, there are few reports of their use in cell lines suitable to industrial scale protein production processes [30,79]. Therefore, we set up to systematically analyze the functioning of MAR and other such genetic elements in CHO cells, and to determine if they might be used in long-term production processes.

We selected the well characterized 3-kb MAR elements bordering the 5′-end of the chicken lysozyme (*cLys*) locus (reviewed in [80]), as well as the *Drosophila scs* boundary element and *hspSAP* MAR, and the mouse T-cell receptor *TCRα* and the rat *LAP* locus control regions, as all of these genetic elements have been reported to mediate permissive chromatin structures in a variety of systems [30]. These elements were introduced in expression vectors and the resulting transgene expression was assayed by stable transfection of CHO cells. Most of these elements showed modest effects on the expression levels of the transgene in CHO cell pools. In contrast, *cLysMAR* was six-fold more effective than the second best of these elements. Moreover, a further four-fold increase in expression level was seen when two *cLysMAR* elements flanked the expression cassette, as shown previously in other cell types [49].

Multiple copies of *cLysMAR* are large (6 kb), which thereby restricts the general use of such chromatin elements taking into account the size of the expression vectors. One approach to improve MAR versatility consists of co-transfecting the transgene expression cassette and various amounts of *cLysMAR* in *trans*, on a separate plasmid. This was shown to enhance expression to levels higher than those obtained with plasmids bearing just one *cLysMAR* element. Thus, MAR-bearing plasmids can be simply added to current expression vectors, in co-transfections, to significantly increase expression levels [30]. In another approach, short functional elements of *cLysMAR* were defined by deletion mutagenesis. These portions of the MAR, when multimerized, were found to be equally active as the full-length element, although of much smaller size (P.-A. Girod and N. Mermod, unpublished data).

While first assays made use of easily assayable reporter genes, we next used transgenes of clinical relevance coding for full-size IgG antibodies composed of two "heavy" and two "light" polypeptide linked together by disulfide bridge. Multimeric protein production faces one major problem: to have efficient production, all subunits need to be co-ordinately synthesized at stoichiometric levels. Therefore, identical expression signals, such as promoters and 3′-regions were used in the different expression cassettes. Two linearized plasmids encoding the heavy- and light-chain expression cassette and harbouring adjacent chromatin elements of various kinds were transfected simultaneously into CHO cells, in presence or absence of MAR elements. Once again, addition of *cLysMAR* elements increased the average and maximal productivities of isolated CHO cell clones by 5- to 10-fold [30].

One of the optimal settings consisted of adding the MAR elements both in *cis*, on each of the heavy and light chain expression vectors, and in *trans*, by further adding to the transfection mix an additional MAR-containing plasmid. Fig. 4 shows that co-transfection of *cLysMAR* in *cis* and *trans* increased the mean expression level in a pool of stably co-transfected cells. Moreover, it reduced the expression variability in different clones thus allowing the isolation of clones exhibiting high secretion levels at a higher frequency.

3.3. MAR elements and copy number-dependent expression of the transgene

We then addressed whether copy number-dependent expression was obtained using MAR elements in CHO cell lines. Fig. 5 shows that increasing number of copies integrated in the genome correlates with an enhanced production of IgGs up to a

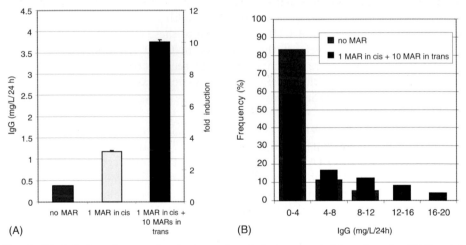

(A)　　　　　　　　　　　　(B)

Fig. 4. (A) IgG titers from pools of primary transfectants generated by using *cLysMAR*-containing vectors and control vectors. CHO cell pools of primary transfectants were selected in antibiotic containing medium for a period of two weeks and seeded in 6-well plates (6×10^5 cells/well) in 3 ml of medium. The amount of IgG secreted in the medium after 24 h was assayed with an ELISA and are expressed as arbitrary units. Different combinations of plasmids were used during transfection, with *cLysMAR* inserted upstream of the SV40 promoter (MAR in *cis*) or further added on a separate plasmid to the *cLysMAR*-containing expression vectors (MAR in *cis* + *trans*). (B) Selection and analysis of individual clones in IgG producing pools of primary transfectants containing no MAR or *cLysMAR* in *cis* + *trans*. Randomly selected clones were isolated by picking and the IgG titers were determined as in part (A). The clones were classified according to productivity, as indicated.

certain point. However, a decrease in production rate per gene copy was noticed in clones containing the higher gene copy numbers.

Several explanations may account for the latter effect. High densities of MAR elements in a single chromosomal locus may saturate MAR-binding activities in particular areas of the nucleus, and all MAR elements may not be able to bind to the nuclear matrix simultaneously. To test this possibility, nuclear matrix was isolated by selectively extracting histones from chromatin preparations. Using restriction enzyme hypersensitivity assays on the MAR and IgG gene sequences and quantitative PCR, we determined that at least 40% of the immunoglobulin light chain gene are physically linked to the matrix in a clone containing a total of ~ 70 copies of heavy and light chains genes, representing a locus size estimated to > 1 Mbp (P.-A. Girod and N. Mermod, unpublished data). These results suggested that not all *cLysMARs* are matrix associated at this locus and that a MAR-binding activity may be limiting in CHO cells.

Since multiple copies of MAR elements may not all associate with the matrix and may consequently not fully prevent transgene silencing, we next determined the effect of gene silencing inhibitors such as sodium butyrate. It has been reported previously that the expression of certain genes responds positively to this HDAC inhibitor, and that this effect may be larger if the MAR elements are located next to the transgene [81]. Although, butyrate increased expression levels in several CHO cell clones, as expected from a general increase of histone acetylation, no selective enhancement could be seen in high copy number clones *vs* low copy number ones (Fig. 5). This

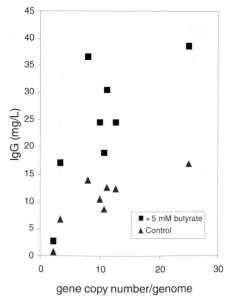

Fig. 5. Productivities of randomly picked CHO cell clones as a function of transgene copy number. CHO cell transfections and clone selection were performed as in Fig. 4A, using SV40 promoter-based expression vectors and *cLysMAR* elements added in *cis* and in *trans*. IgG secretion titers were determined as in Fig. 4, from cells treated or not with butyrate, and were plotted against gene copy numbers.

indicated that chromatin-mediated repression effects are not more prominent in clones with the higher copy number. Thus, we conclude that the saturation of protein production seen with high number of copies of both trangenes and MAR elements does not result from a specific silencing effect in the high copy number clones.

3.4. The effect of MAR elements on transgene expression level and stability

There have been contradictory reports on whether MAR elements may prevent transgenes from the gradual decline in expression observed with some clones [1]. As post-transcriptional silencing may ultimately result in DNA methylation of the chromosomal locus, we treated several cell lines with the drug 5-aza-deoxycytidine, which removes methyl groups from CpG dinucleotides. While the drug did not affect protein production in cell lines expressing IgGs under the control of the SV40 promoter, we found a strong enhancement of production for all clones containing the CMV promoter. These and other data suggested that epigenetic silencing is common in CHO cells for some of the CMV promoter-based vectors, and that this occurs irrespective of the presence of MAR elements (P.-A. Girod and N. Mermod, unpublished results). Thus, although expression levels are increased by MAR elements using both SV40 and CMV promoters, MAR elements cannot fully oppose the intrinsic propensity to silence some CMV promoters. These results are consistent with a mechanism whereby MAR would shield transgenes from silencing effects arising from the chromosomal elements near the site of integration, but may not prevent silencing arising from elements within the transgene expression cassette itself.

These results indicate that the MAR-mediated inhibition of gene silencing can be influenced by properties of the expression cassette. In the case of the SV40 promoter constructs, expression of the transgene increases with copy number, implying a relative independence of the transgene from the silencing effects of chromosomal elements adjacent to the insertion locus. The distinct behaviour of the SV40 promoter may be explained by the fact that it binds the Sp1 transcription activator previously proposed to inhibit DNA methylation. Furthermore, the SV40 promoter has relatively few methylation sites, with just one cluster meeting the criteria of a typical CpG island. In contrast, the CMV promoter has three islands that overlap the entire promoter-enhancer region, making a continuous array towards the cLysMAR (Fig. 6), and may be intrinsically more susceptible to silencing. The fact that the cLysMAR does not always block gene silencing is consistent with a recent report showing that it cannot fully insulate the lysozyme gene against X chromosome condensation-mediated gene inactivation when introduced by homologous recombination within an X-linked gene in female mice, while multicopy transgenic mice with randomly inserted autosomal copies of this locus demonstrated expression dependent on the transgene copy number [82]. Thus, MARs may not have evolved as simple genetic elements that insulate genes, but most probably regulate chromatin accessibility in concert with other activities (see also Section 2).

To ascertain the long term stability of the transgene expression, CHO clones expressing IgGs from SV40 promoters were transferred to spinner cultures. High levels of expression were maintained stably during 228 days of culture in serum-free

375

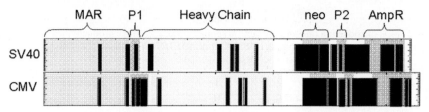

Fig. 6. Maps of IgG heavy-chain encoding plasmids under the control of the SV40 (top) and CMV (bottom) promoter/enhancer regions. The positions of the CpG islands were determined using bioinformatics tools as in Fig. 2 and are indicated by black rectangles. Neo denotes the neomycin-resistance gene. P1 and P2 indicate the SV40 or CMV and the neomycin-resistance gene promoters, respectively. AmpR denotes the bacterial ampicillin-resistance gene. The IgG heavy chain expression cassette and the selection cassette were cloned in a tail-to-tail orientation.

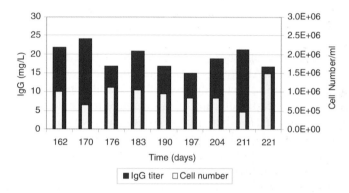

Fig. 7. Growth kinetics and IgG titer in spinner suspension culture. Spinners were inoculated at 200,000 cells/ml. Cells were passaged every 3 to 4 days. Cell titers and secreted IgGs were maintained for multiple passages over a period of 8 months (courtesy of Michaela Bourgeois and Florian Wurm).

medium in the absence of selection (Fig. 7). In contrast, expression was variable among lines that did not contain MAR elements (data not shown). The MAR was thus capable of protecting the transgene significantly against long term silencing.

An approximately 15-fold increase in IgG specific production rate with the clones generated with *cLysMAR*, as compared to the best clone obtained without MAR (Table I). Consistently, sequence determination of more than 100 randomly selected cDNAs from MAR-containing CHO cell clones demonstrated that the level of transcripts encoding the transgenes was high, amounting 10–15% of the total mRNAs of the cell, a ratio equivalent to that found for the transcripts of various ribosomal proteins. It is noteworthy that high expression rates were allowed by MAR elements using non-optimized expression vectors and transgenes containing the SV40 early promoter and human IgG polyadenylation sequences. The amount of total protein produced per day per CHO cell was approximately 200 pg, consistent with IgG productivities of about 25 pg/cell/day. In bioreactors, the simple addition of MAR elements in *trans* to commonly available vectors raised productivity from 10–20 to 100–200 mg/l of IgGs when using standard synthetic media and processes,

Table I

Production rates of seedtrains in spinner culture. Average productivities of seedtrains containing no MAR or *cLysMAR* in *trans* or in *cis + trans*, cultured in spinner flasks in synthetic medium (ProCHO5)

MAR	Productivity (pg/cell/day)
cis + trans	8.7 ± 2.9
trans	1.6 ± 0.8
no	0.6 ± 0.3

Courtesy of Michaela Bourgeois and Florian Wurm.

with potential for production in the gram per liter range when using MAR elements in *cis* and *trans* combinations.

4. Conclusion

Earlier work in a number of laboratories had demonstrated the usefulness of the 3-kb chicken lysozyme MAR (*cLysMAR*) element in protecting reporter genes from chromosomal position effect in diverse cell lines and in transgenic animals. Our recent studies demonstrated the efficacy of this element for protein production using CHO cell lines. The *cLysMAR*, when used in conjunction with appropriate promoter elements, appears to insulate the transgenes from chromosomal integration site effects, conferring gene copy number-dependent expression. Thus, high producer cell lines can be obtained at a higher frequency, with consistent increases in specific productivities of around 10-fold.

The *cLysMAR* element can be conveniently used in *trans* in co-transfection experiments, thus obviating the need for expression vector re-engineering, as well as in *cis* or using combinations of the two approaches. As such, stable cell lines exhibiting elevated transgene expression can be obtained without gene amplification systems (i.e., dihydrofolate reductase gene selection), although expression levels can be further increased when combining both approaches. High-level expression is stably maintained in protein production conditions. The MAR elements may therefore prove to be versatile and generally applicable tools for the large-scale production of proteins.

Acknowledgements

We thank Michaela Bourgeois, Maria de Jesus, Florian Wurm and other members of the Swiss Priority Program in Biotechnology for sharing data and helpful comments, and gratefully acknowledge support by the Swiss National Science Foundation.

Abbreviations

BE	Boundary element
CDP	CCAAT displacement protein

H1	histone H1
HMG	High-mobility group
hnRNP	heterogeneous nuclear ribonucleoproteins
IgG	Immunoglobulin G
LCR	Locus control region
MAR	Matrix attachment region
SAR	Scaffold attachment region
SATB1	special AT-rich binding protein 1

References

1. Allen, G.C., Spiker, S., and Thompson, W.F. (2000) Plant Mol. Biol. 43, 361–376.
2. Berezney, R. and Coffey, D.S. (1974) Biochem. Biophys. Res. Commun. 60, 1410–1417.
3. Sjakste, N.I. and Sjakste, T.G. (2001) Molecular Biology 35, 627–635.
4. Lamond, A.I. and Earnshaw, W.C. (1998) Science 280, 547–553.
5. Cmarko, D., Verschure, P.J., Rothblum, L.I., Hernandez-Verdun, D., Amalric, F., van Driel, R., and Fakan, S. (2000) Histochem. Cell Biol. 113, 181–187.
6. Stratling, W.H. and Yu, F. (1999) Crit. Rev. Eukaryot. Gene Expr. 9, 311–318.
7. Mirkovitch, J., Mirault, M.E., and Laemmli, U.K. (1984) Cell 39, 223–232.
8. de Belle, I., Cai, S., and Kohwi-Shigematsu, T. (1998) J. Cell. Biol. 141, 335–348.
9. Cmarko, D., Verschure, P.J., Martin, T.E., Dahmus, M.E., Krause, S., Fu, X.D., van Driel, R., and Fakan, S. (1999) Mol. Biol. Cell 10, 211–223.
10. Maniatis, T. and Reed, R. (2002) Nature 416, 499–506.
11. Cremer, T., Kreth, G., Koester, H., Fink, R.H., Heintzmann, R., Cremer, M., Solovei, I., Zink, D., and Cremer, C. (2000) Crit. Rev. Eukaryot. Gene Expr. 10, 179–212.
12. Stief, A., Winter, D.M., Stratling, W.H., and Sippel, A.E. (1989) Nature 341, 343–345.
13. Phi-Van, L., von Kries, J.P., Ostertag, W., and Stratling, W.H. (1990) Mol. Cell. Biol. 10, 2302–2307.
14. Klehr, D., Maass, K., and Bode, J. (1991) Biochemistry 30, 1264–1270.
15. Bonifer, C., Vidal, M., Grosveld, F., and Sippel, A.E. (1990) EMBO J. 9, 2843–2848.
16. Bonifer, C., Yannoutsos, N., Kruger, G., Grosveld, F., and Sippel, A.E. (1994) Nucleic Acids Res. 22, 4202–4210.
17. Recillas-Targa, F., Pikaart, M.J., Burgess-Beusse, B., Bell, A.C., Litt, M.D., West, A.G., Gaszner, M., and Felsenfeld, G. (2002) Proc. Natl. Acad. Sci. USA 99, 6883–6888.
18. Poljak, L., Seum, C., Mattioni, T., and Laemmli, U.K. (1994) Nucleic Acids Res. 22, 4386–4394.
19. Bode, J., Kohwi, Y., Dickinson, L., Joh, T., Klehr, D., Mielke, C., and Kohwi-Shigematsu, T. (1992) Science 255, 195–197.
20. Piechaczek, C., Fetzer, C., Baiker, A., Bode, J., and Lipps, H.J. (1999) Nucleic Acids Res. 27, 426–428.
21. Baiker, A., Maercker, C., Piechaczek, C., Schmidt, S.B., Bode, J., Benham, C., and Lipps, H.J. (2000) Nat. Cell Biol. 2, 182–184.
22. Jackson, D.A., Dickinson, P., and Cook, P.R. (1990) EMBO J. 9, 567–571.
23. Strissel, P.L., Dann, H.A., Pomykala, H.M., Diaz, M.O., Rowley, J.D., and Olopade, O.I. (1998) Genomics 47, 217–229.
24. van Drunen, C.M., Sewalt, R.G., Oosterling, R.W., Weisbeek, P.J., Smeekens, S.C., and van Driel, R. (1999) Nucleic Acids Res. 27, 2924–2930.
25. Chernov, I.P., Akopov, S.B., Nikolaev, L.G., and Sverdlov, E.D. (2002) J. Cell Biochem. 84, 590–600.
26. Frisch, M., Frech, K., Klingenhoff, A., Cartharius, K., Liebich, I., and Werner, T. (2002) Genome Res. 12, 349–354.
27. Bode, J., Benham, C., Knopp, A., and Mielke, C. (2000) Crit. Rev. Eukaryot. Gene Expr. 10, 73–90.
28. Kondo, T. and Duboule, D. (1999) Cell 97, 407–417.

ubeler, D., Mielke, C., Maass, K., and Bode, J. (1996) Biochemistry 35, 11160–11169.

ahn-Zabal, M., Kobr, M., Girod, P.A., Imhof, M., Chatellard, P., de Jesus, M., Wurm, F., and Mermod, N. (2001) J. Biotechnol. 87, 29–42.

Youn, B.S., Lim, C.L., Shin, M.K., Hill, J.M., and Kwon, B.S. (2002) Mol. Cells 13, 61–68.

. Lenartowski, R. and Goc, A. (2002) Neurosci. Lett. 330, 151.

3. Stunkel, W., Huang, Z., Tan, S.H., O'Connor, M.J., and Bernard, H.U. (2000) J. Virol. 74, 2489–2501.

34. Alvarez, J.D., Yasui, D.H., Niida, H., Joh, T., Loh, D.Y., and Kohwi-Shigematsu, T. (2000) Genes Dev. 14, 521–535.

35. Liebich, I., Bode, J., Reuter, I., and Wingender, E. (2002) Nucleic Acids Res. 30, 3433–3442.

36. Cockerill, P.N. and Garrard, W.T. (1986) FEBS Lett. 204, 5–7.

37. Singh, G.B., Kramer, J.A., and Krawetz, S.A. (1997) Nucleic Acids Res. 25, 1419–1425.

38. Boulikas, T. (1995) Int. Rev. Cytol. 162A, 279–388.

39. van Wijnen, A.J., Bidwell, J.P., Fey, E.G., Penman, S., Lian, J.B., Stein, J.L., and Stein, G.S. (1993) Biochemistry 32, 8397–8402.

40. Kaplan, M.H., Zong, R.T., Herrscher, R.F., Scheuermann, R.H., and Tucker, P.W. (2001) J. Biol. Chem. 276, 21325–21330.

41. Blasquez, V.C., Xu, M., Moses, S.C., and Garrard, W.T. (1989) J. Biol. Chem. 264, 21183–21189.

42. Scheuermann, R.H. and Chen, U. (1989) Genes Dev. 3, 1255–1266.

43. Turner, B.M. (2000) Bioessays 22, 836–845.

44. Fry, C.J. and Peterson, C.L. (2002) Science 295, 1847–1848.

45. Burgers, W.A., Fuks, F., and Kouzarides, T. (2002) Trends Genet. 18, 275–277.

46. Herrera, J.E., West, K.L., Schiltz, R.L., Nakatani, Y., and Bustin, M. (2000) Mol. Cell Biol. 20, 523–529.

47. Izaurralde, E., Kas, E., and Laemmli, U.K. (1989) J. Mol. Biol. 210, 573–585.

48. Thastrom, A., Lowary, P.T., Widlund, H.R., Cao, H., Kubista, M., and Widom, J. (1999) J. Mol. Biol. 288, 213–229.

49. Phi-Van, L. and Stratling, W.H. (1996) Biochemistry 35, 10735–10742.

50. Hoffmeister, A., Ropolo, A., Vasseur, S., Mallo, G.V., Bodeker, H., Ritz-Laser, B., Dressler, G.R., Vaccaro, M.I., Dagorn, J.C., Moreno, S., Iovanna, J.L. (2002) J. Biol. Chem. 277, 22314–22319.

51. Zhao, K., Kas, E., Gonzalez, E., and Laemmli, U.K. (1993) EMBO J. 12, 3237–3247.

52. Jenuwein, T., Forrester, W.C., Fernandez-Herrero, L.A., Laible, G., Dull, M., and Grosschedl, R. (1997) Nature 385, 269–272.

53. Dickinson, L.A., Joh, T., Kohwi, Y., and Kohwi-Shigematsu, T. (1992) Cell 70, 631–645.

54. Ramakrishnan, M., Liu, W.M., DiCroce, P.A., Posner, A., Zheng, J., Kohwi-Shigematsu, T., and Krontiris, T.G. (2000) Mol. Cell Biol. 20, 868–877.

55. Antequera, F. and Bird, A. (1999) Curr. Biol. 9, R661–667.

56. Gribnau, J., Diderich, K., Pruzina, S., Calzolari, R., and Fraser, P. (2000) Mol. Cell 5, 377–386.

57. Recillas-Targa, F., Pikaart, M.J., Burgess-Beusse, B., Bell, A.C., Litt, M.D., West, A.G., Gaszner, M., and Felsenfeld, G. (2002) Proc. Natl. Acad. Sci. USA 99, 6883–6888.

58. Jost, J.P., Oakeley, E.J., Zhu, B., Benjamin, D., Thiry, S., Siegmann, M., and Jost, Y.C. (2001) Nucleic Acids Res. 29, 4452–4461.

59. Zhu, B., Benjamin, D., Zheng, Y., Angliker, H., Thiry, S., Siegmann, M., and Jost, J.P. (2001) Proc. Natl. Acad. Sci. USA 98, 5031–5036.

60. Kirillov, A., Kistler, B., Mostoslavsky, R., Cedar, H., Wirth, T., and Bergman, Y. (1996) Nat. Genet. 13, 435–441.

61. Forrester, W.C., Fernandez, L.A., and Grosschedl, R. (1999) Genes Dev. 13, 3003–3014.

62. Weitzel, J.M., Buhrmester, H., and Stratling, W.H. (1997) Mol. Cell. Biol. 17, 5656–5666.

63. Ai, W., Toussaint, E., and Roman, A. (1999) J. Virol. 73, 4220–4229.

64. Li, S., Moy, L., Pittman, N., Shue, G., Aufiero, B., Neufeld, E.J., LeLeiko, N.S., and Walsh, M.J. (1999) J. Biol. Chem. 274, 7803–7815.

65. Liu, J., Barnett, A., Neufeld, E.J., and Dudley, J.P. (1999) Mol. Cell. Biol. 19, 4918–4926.

66. Wang, Z., Goldstein, A., Zong, R.T., Lin, D., Neufeld, E.J., Scheuermann, R.H., and Tucker, P.W. (1999) Mol. Cell. Biol. 19, 284–295.

67. Fernandez, L.A., Winkler, M., and Grosschedl, R. (2001) Mol. Cell. Biol. 21, 196–208.
68. Martens, J.H., Verlaan, M., Kalkhoven, E., Dorsman, J.C., and Zantema, A. (2002) Mol. Cell. Biol. 22, 2598–2606.
69. Davie, J.R. (1997) Mol. Biol. Rep. 24, 197–207.
70. Durrin, L.K. and Krontiris, T.G. (2002) Genomics 79, 809–817.
71. Nayler, O., Stratling, W., Bourquin, J.P., Stagljar, I., Lindemann, L., Jasper, H., Hartmann, A.M., Fackelmayer, F.O., Ullrich, A., Stamm, S. (1998) Nucleic Acids Res. 26, 3542–3549.
72. Forrester, W.C., van Genderen, C., Jenuwein, T., and Grosschedl, R. (1994) Science 265, 1221–1225.
73. Getzenberg, R.H., Pienta, K.J., Huang, E.Y., and Coffey, D.S. (1991) Cancer Res. 51, 6514–6520.
74. Galande, S. and Kohwi-Shigematsu, T. (2000) J. Cell. Biochem. Suppl. 35, 36–45.
75. Herrscher, R.F., Kaplan, M.H., Lelsz, D.L., Das, C., Scheuermann, R., and Tucker, P.W. (1995) Genes Dev. 9, 3067–3082.
76. Gindullis, F. and Meier, I. (1999) Plant. Cell. 11, 1117–1128.
77. Gindullis, F., Peffer, N.J., and Meier, I. (1999) Plant Cell. 11, 1755–1768.
78. Ishii, K., Arib, G., Lin, C., Van Houwe, G., and Laemmli, U.K. (2002) Cell 109, 551–562.
79. Koduri, R.K., Miller, J.T., and Thammana, P. (2001) Gene 280, 87–95.
80. Bonifer, C., Jagle, U., and Huber, M.C. (1997) J. Biol. Chem. 272, 26075–26078.
81. McCaffrey, P.G., Newsome, D.A., Fibach, E., Yoshida, M., and Su, M.S. (1997) Blood 90, 2075–2083.
82. Chong, S., Kontaraki, J., Bonifer, C., and Riggs, A.D. (2002) Mol. Cell. Biol. 22, 4667–4676.
83. Verschure, P.J., van Der Kraan, I., Manders, E.M., and van Driel, R. (1999) J. Cell. Biol. 147, 13–24.

S.C. Makrides (Ed.) *Gene Transfer and Expression in Mammalian Cells*

CHAPTER 11

Chromatin insulators and position effects

David W. Emery, Mari Aker, and George Stamatoyannopoulos

*Department of Medicine, Division of Medical Genetics, Box 357720, University of Washington,
Seattle, WA 98195, USA; Tel.: +206-543-3526; Fa: +206-543-3050.
E-mail: gstam@u.washington.edu*

1. Introduction

Most mechanisms of stable gene transfer in mammalian cells, including methods of physical transfection and virus transduction, require chromosomal integration. This allows for the stable and faithful transmission of the transferred gene to all of the progeny of the targeted cell. However, once integrated, expression of the transferred gene is subject to the influence of surrounding chromatin, a phenomenon known as *position effects*. Since chromosomal integration is generally random and the majority of the mammalian genome is often in the form of silent and condensed heterochromatin, position effects generally manifest as the partial or complete loss of expression. It is important to note that the loss of expression due to position effects is distinct from the loss of expression due to the lack of essential *cis*-regulatory elements or the presence of elements which may function to actively silence expression.

There are several potential approaches to overcoming the negative influence of position effects on the expression of randomly integrated genes. As described in other chapters, such effects can be mitigated through the use of appropriate transcriptional promoters or other *cis*-regulatory elements such as enhancers, matrix-attachment regions, and locus control regions (LCRs). However, the actions of these elements are generally gene- and tissue-specific, and may not allow for the desired level or precise regulation of gene expression. Another approach involves the use of homologous recombination to integrate the transferred gene into a specific site where the surrounding chromatin can support the desired pattern of expression. However, current methods for targeted integration are limited and suffer from exceedingly low frequencies [1]. In gene transfer studies with mammalian cell lines a selection strategy is often used to simply circumvent the problem of silencing position effects. In such cases a reporter gene is linked to the gene of interest, and a selection strategy is employed to pull out the rare clones where the transferred gene has integrated at transcriptionaly favorable locations. However, in most cases it is necessary to maintain the selective pressure, and this approach has met with mixed success when applied *in vivo*.

A growing body of literature has emerged on the use of an alternative class of *cis*-regulatory elements, known as chromatin insulators, as a means of overcoming the influence of position effects on the expression of genes transferred in mammalian cells. These are naturally occurring DNA elements found in a diverse range of species which function as boundary elements to separate differentially regulated chromosomal loci. This chapter outlines the problems of position effects as they apply to gene transfer and expression in mammalian cells, the properties of chromatin insulators which make them useful for abrogating these position effects, and several examples where chromatin insulators have been used successfully for this and other purposes.

2. Chromosomal position effects

Chromosomal position effects arise from two central facets of mammalian cell biology: most methods used for stable gene transfer in mammalian cells lead to random chromosomal integration; and mammalian chromosomes are partitioned into active and inactive domains. Below we summarize current information on both of these topics. We also discuss the various patterns of expression which can result from position effects, and comment on the importance of experimental design in the detection of these patterns. We then conclude by discussing some specific sources of silencing position effects.

2.1. Random integration and silencing

Position effects arise in part from the fact that most forms of stable gene transfer result in essentially random integration in the target cell genome. This is true for mechanical means of transfection with plasmid DNA as well as transduction with oncoretrovirus vectors [2,3]. This pattern is thought to reflect the fact that integration occurs during cell division before the chromatin has attained the higher level organization characteristic of G1 phase. In the case of recombinant adeno-associated virus vectors, initial studies suggest that integration events may favor a particular gene-rich region of chromosome 19, although the impact of this property on expression has yet to be elucidated [4]. It has been suggested that the ability of lentivirus vectors to enter the nucleus through nuclear pores in the absence of cell division may be associated with preferential integration at more accessible chromosomal locations [5,6]. However, careful analysis following single-copy transduction indicate that the expression of lentivirus vectors can vary widely depending on the site of integration [7,8]. Vectors based on recombinant foamy virus, which are also able to transduce cells in stationary phase, appear to integrate randomly as well [9]. A significant effort is underway to catalog the sites of integration for several classes of gene transfer vectors in pre-clinical and clinical gene therapy studies [10]. Such studies should allow for a more precise analysis of integration site preferences.

In practice position effects typically manifest as a reduction or loss in expression. This probably reflects the fact that the majority of the mammalian genome is transcriptionaly silent throughout much of development. Data from the human genome project indicates that only about 1.5% of the human genome consists of coding sequences [11]. However, because of the relatively large size of introns and the fact that thousands of genes produce noncoding RNA as their ultimate product, these studies also estimate that as much as one-third of the human genome is transcribed in genes. This in turn suggests that two-thirds or more of the genome does not contain transcribed genes. Further, it is worth noting that this estimate relies in part on expressed sequence tag analysis from essentially all stages of development and all tissues. It is likely that the fraction of the genome which is actively transcribed within any one tissue or stage of development is substantially lower than one-third. This supposition is borne out by data arising from the increasing number of expression array studies which focus on specific tissues. For example, in the erythroid lineage the number of highly expressed genes declines from perhaps a thousand [12] to only a handful during the terminal stages of differentiation when the chromatin is sequentially condensed and the nucleus is eventually lost.

2.2. Chromatin structure

The chromosomes of mammalian cells are arranged in higher order chromatin structures, which play a critical role in the regulation of gene expression. As extensively reviewed [13–15] and diagramed in Fig. 1, DNA strands wrap around histones to form nucleosomes, which in turn combine with linker histones and other molecules to compact into inaccessible condensed chromatin. At a larger scale mammalian chromosomes are divided into discrete functional domains which either support or repress transcriptional activation. In the case of active euchromatin, the

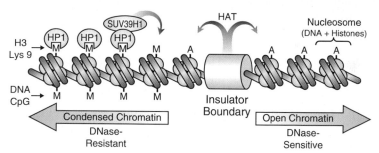

Fig. 1. Chromatin structure. Mammalian chromatin is comprised of a highly ordered arrangement of DNA and proteins. At the primary level of condensation chromosomal DNA is wrapped around core histones to form nucleosomes. In the case of open euchromatin the histones are hyperacetylated (A) and the DNA is relatively sensitive to digestion with DNase I. In the case of transcriptionally repressive condensed heterochromatin the histones are hypoacetylated and methylated (M) and the DNA is relatively resistant to DNase I digestion. This state reflects the secondary level of compaction. The frequency of DNA methylation at CpG dinucleotides is also frequently associated with silent condensed chromatin. Emerging evidence suggests that competing reactions to methylate and acetylate the Lys 9 moiety of the core histone H3 may serve to regulate these two states. Methylated H3 Lys 9 recruits heterochromatin protein 1 (HP1), which in turn recruits the histone methyltransferase SUV39H1 which serves to methylate H3 Lys 9 in adjoining nucleosomes, effectively propagating the heterochromatin signal. Emerging evidence also suggest that chromatin insulators may block this propagating signal by recruiting histone acetyltransferases (HAT) which serve to actively acetylate H3 Lys 9.

core histones H3 and H4 are hyperacetylated and the DNA is relatively sensitive to endonucleases such as DNase I, indicating an open conformation. In the case of inactive heterochromatin, the core histones are hypoacetylated and histone H1 is present and acts to link neighboring nucleosomes together. As a result inactive chromatin is relatively resistant to endonuclease cleavage. Condensed chromatin is also associated with methylation (rather than acetylation) of H3 Lys 9. Together these characteristics constitute part of the histone code of a chromosomal region. The difference between active and inactive chromatin is also frequently correlated with the level of methylation at CpG dinucleotides in the DNA sequence. The acetylation and methylation status of the nucleosomal code and the methylation status of the DNA code are tightly correlated with transcriptional activity.

As recently reviewed [16], the roles of the histone code and DNA methylation in gene regulation are complex. Current data suggest that genes which are either actively targeted for silencing or simply not undergoing active transcription become sensitive to *de novo* DNA methylation through the methyltransferases DNMT3A and DNMT3B. This state of methylation is typically maintained by another methyltransferase, DNMT1, which reestablishes symmetric CpG methylation on newly synthesized hemimethylated substrates during DNA replication. It is worth noting that DNMT3A and 3B are highly expressed in early embryonic cells where most programmed *de novo* methylation takes place. DNA methylation can interfere with transcriptional expression either by directly blocking the attachment of critical transcription factors, or by recruiting a repressor complex such as Mi-2/NuRD through direct interaction with the component methyl-CpG-binding protein MBD3. In the latter scenario, this repressor complex functions to alter the nucleosome code through a combination of nucleosome remodeling and histone deacetylase components. Recent studies suggest however that transcriptional silencing is often preceded by changes in the histone code. Examples include the silencing of retro-virus vectors during embryogenesis, and silencing of X-linked genes through X-chromosome inactivation [13,17]. There are several sequence-specific mechanisms that have been proposed for the initiation of a repressive histone code which typically involve nucleosome remodeling and histone deacetylation. Recent studies also suggest that the repressive histone code may be propagated from adjoining areas through a mechanism mediated by members of the heterochromatin protein 1 (HP1) family of proteins [14]. As diagrammed in Fig. 1, HP1 proteins are recruited to heterochromatin by their ability to bind the methylated Lys 9 of histone H3. In turn, HP1 recruits both transcriptional repressors and the histone methytransferase SUV39H1 which serves to methylate H3 Lys 9 in adjoining nucleosomes, effectively propagating the heterochromatin signal.

2.3. Manifestation of position effects

Effects of chromosomal position on the expression of genes transferred in mammalian cells can be manifested in several ways. As seen in the left panel of Fig. 2, the most straightforward manifestation is a variation in the level (or presence) of expression between independently isolated clones. In its simplest form this variation

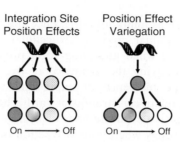

Fig. 2. Manifestations of position effects. In the case labeled "Integration Site Position Effects" (left), position effects are manifested immediately after gene transfer leading to a heritable variation in the level of expression between independent clones due to the effects of the surrounding chromatin. In the case labeled "Position Effect Variegation" (right), there can also be a variation in the level of expression within the progeny of a single clone due to the stochastic or developmental changes in the surrounding chromatin.

in the level of expression is heritable so that all of the progeny of a particular clone continue to express the introduced gene at the same level. As seen in the right panel of Fig. 2, variation in expression can also be observed between the progeny of individual clones following gene transfer in mammalian cells. This phenomenon is known as position effect variegation. This form of position effect often arises from the fact that the silencing effects of the surrounding chromatin may not act immediately after gene transfer or may be subject to stochastic mechanisms so that the same site of integration may succumb to the silencing effects of surrounding chromatin at different times and to varying degrees. Position effect variegation can also be used to explain the temporal loss of expression within the progeny of a genetically modified clone by simply invoking a mechanism whereby an inactive state is fixed once it is reached. Finally, it is important to note that the problem of position effects can be particularly severe when genetically modified cells undergo programmed differentiation [17,18]. Such programs often involve the remodeling of chromatin and the specific activation and repression of distinct genetic loci.

2.4. Issues of experimental design

There are several experimental parameters which can obscure the presence or degree of position effects. For example, the ability to detect expression variation between clones is lost when expression is analyzed in polyclonal pools rather than individual clones. Likewise, it is difficult to assess the expression of individual integration events if the method of gene transfer favors the integration of several copies of the transferred gene per cell. This is often the case with conventional methods of plasmid transfection and microinjection, and when virus vectors are used at high multiplicities of infection. It is important to realize that the use of drug resistance or other reporter systems to select for gene transfer biases the repertoire of integration events because only those clones with the transferred gene integrated at transcriptional favorable sites will be available for analysis. When possible it is preferable to measure the presence and level of gene expression using a cell-autonomous assay (such as immunofluorescence and flow cytometry) in order to distinguish between pan-cellular and hetero-cellular expression profiles; such distinctions are not possible when analyzing total protein or mRNA from bulk cell preparations. When studying the incidence of overt silencing it is especially important to correlate the incidence of expression with the presence of the transferred gene. This is relatively straightforward when working with a clonal population of cells, such as tissue culture lines or transgenic animals. However, this task is more difficult in settings where only a fraction of the cells may contain the transferred gene. This is often the case in gene transfer studies using virus vectors and primary tissues such as hematopoietic stem cells or whole organs such as liver, muscle, or lung [19]. Finally, it is important to take into consideration the role of differentiation on expression: transferred genes can silence at different stages of differentiation because of a lack of tissue-appropriate promoters or enhancers as well as by the negative influence of surrounding chromatin.

2.5. Sources for silencing position effects

Different cellular mechanisms, though not mutually exclusive, operate to silence genes transferred in mammalian cells. As discussed above, one frequent cause of silencing arises when transgenes integrate into repressive heterochromatin. Telomeric and centromeric heterochromatin in particular can exert strong repression or confer instability on transgene expression. For example, inclusion of telomeric sequences in a luciferase reporter plasmid reduced expression 10-fold in telomerase-positive HeLa cells [20]. The silencing effects of centromeric integration are even dominant over the enhancing effects of the β-globin locus control region (LCR) in transgenic mice carrying a 150 kb human β-globin transgene [21], even though such constructs resist position effects when integrated at other sites. Other heterochromatic regions such as satellite and LINE-rich regions and facultative heterochromatin such as the inactive X chromosome may have similar repressive effects. This type of silencing often displays variegation, due to the variable spread of heterochromatin into the gene. The heterochromatin spread may be orientation specific, in that one orientation of the construct allows spread, but the other does not [22].

Gene silencing can also arise when expression cassettes are integrated in tandem arrays, a common event with methods of gene transfer such as plasmid transfection and pronuclear injection. This type of silencing may result from RNA polymerase II loading difficulties due to the close spacing of the promoters (termed transcriptional interference), or from a cellular mechanism that has been termed repeat-induced silencing. In transgenic mice, transcriptional interference can lead to silencing of a downstream promoter, regardless of orientation [23]. A good example of repeat-induced silencing also comes from studies in transgenic mice, where a 1000-fold increase in expression was observed when a tandem array of lacZ reporter genes were reduced from 100 copies down to one or a few copies using a cre/lox system [24].

Finally, silencing position effects can be significantly increased due to the presence of specific sequences within the transferred construct which either attract or serve as targets for silencing complexes. Such negative regulators can be developmental stage- or tissue-specific, or they can be ubiquitous. For example, the long terminal repeats (LTRs) of first generation oncoretrovirus vectors based on the Moloney Murine Leukemia Virus contain inhibitory sequences that bind ubiquitously present negative regulators. Efforts have been made to either remove or replace the offending elements. [25,26]. However, these vectors are still sensitive to more conventional position effects [17,19,27,28]. As discussed earlier, some studies suggest that recombinant lentivirus vectors may completely avoid silencing position effects [29]. However, careful analysis following single-copy transduction indicates that the expression of lentivirus vectors can vary widely depending on the site of integration [7,8]. Sequences of prokaryotic origin can also serve to increase the sensitivity of transferred genes to silencing position effects in mammalian cells [30]. This may reflect a cellular defense mechanism against parasitic sequences with reading frames that are not codon-optimized or sequences that have an unusual base composition [31].

3. Chromatin insulators

As discussed above, the mammalian genome is organized into domains which support or repress gene expression. This is critical for proper gene regulation and coordination during development and differentiation. Growing evidence indicates that the boundaries between these differentially regulated domains are often associated with a class of regulatory elements known as chromatin insulators. As diagrammed in Fig. 3, these elements function to block the effects of regulatory signals in adjacent domains. This includes the ability to block the activating effects of enhancers and, in some cases, the negative effects of encroaching heterochromatin or functional repressors. Unlike more conventional *cis*-regulatory elements, chromatin insulators do not exhibit inherent transcriptional enhancing or repressing activities on their own. As such, they allow such *cis*-regulatory elements to function without interference from the effects of surrounding chromatin. The identification and functional characterization of chromatin insulators have been the topic of several recent reviews [32–34]. We summarize below some of the salient features of chromatin insulators as they apply to the use of these elements to protect the expression of gene cassettes following transfer in mammalian cells.

3.1. Insulator activities

Most chromatin insulators have been characterized by their ability to block the activating effects of enhancers on promoters. This reflects a critical property of boundary elements which serve to prevent the illegitimate activation by potent enhancers from adjoining loci. In addition to blocking the positive effects of enhancers, some chromatin insulators also have the ability to act as barriers to silencing position effects. This property also reflects a critical property of boundary elements which serve to prevent the processive encroachment of repressive heterochromatin from surrounding loci. A recent survey indicates that this barrier activity has only been documented in a minority of chromatin insulators [32],

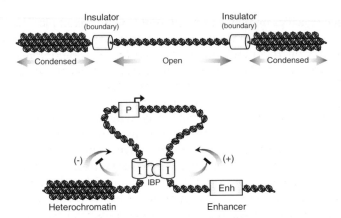

Fig. 3. Role and function of chromatin insulators. As seen in the top diagram, chromatin insulators are naturally occurring elements which serve as boundaries to separate regions of repressive (condensed) heterochromatin and supportive (open) euchromatin. As seen in the bottom diagram, chromatin insulators (I) can function to block the supportive effects of enhancers (+) and the repressive effects of heterochromatin (−) when placed between the *cis*-effector and gene promoter (P). They appear to function best when used in pairs to flank the gene of interest, and function through interactions with specific insulator-binding proteins (IBP). One model suggests that insulator activity is mediated through the formation of a loop structure as indicated.

Table 1

Insulators known to function in mammalian cells

Name	Block[a]	Species	Gene/locus	References
scs/scs′	E & R	Fly	Flank *87A7 hsp70* locus	[59,60]
gypsy	R	Fly	Retrotransposable element	[60]
sns	E	Sea Urchin	3′ boundary to H2A enhancer	[61]
URI	R	Sea Urchin	5′ boundary to *Ars* gene	[62]
Lys 5′ A	E & R	Chicken	5′ boundary to lysozyme gene	[63]
HS4	E & R	Chicken	5′ boundary to β-globin locus	[38,40,19,32]
3′HS	E	Chicken	3′ boundary to β-globin locus	[64]
BEAD-1	E	Mouse	Separate the TCR α/β loci	[59]
HS2-6	E & R	Mouse	External boundary for TCR α locus	[65]
DMD/ICR	E	Mouse	Between imprinted *Igf2* & *H19* genes	[66,67]
DMD/ICR	E	Human	same as for mouse	[66,67]
5′HS5	E & R	Human	5′ boundary to β-globin locus	[44,45]
apoB (−57kb)	E	Human	5′ boundary to *apoB* gene domain	[68]
apoB (+43kb)	R	Human	3′ boundary to *apoB* gene domain	[47]
DM1	E	Human	Separate the *DMPK* and *SIX5* genes	[69]

[a]E, enhancer-blocking; R, repressor-blocking.

although it is important to note that most of these elements were initially identified using enhancer-blocking assays. A list of chromatin insulators which have been shown to have either enhancer- or repressor-blocking activity in human cells is presented in Table 1. Of these elements, the chicken β-globin HS4 elements (from here on referred to as cHS4) has been most widely studied and employed in mammalian gene transfer studies. The dual functionality of this element makes it useful in the setting of mammalian gene transfer studies where position effects frequently lead to silencing of expression, or where a fine degree of gene regulation is either the subject or a critical requirement of the study.

3.2. Topological considerations

As diagramed in the bottom of Fig. 3, chromatin insulators must be physically located between the promoter of interest and any undesired *cis*-effectors. This is true both for the ability to block the inducing effects of enhancers or the repressive effects of heterochromatin (or other silencing elements). Chromatin insulators have little or no effect when located outside of the sequence between the promoter and *cis*-effectors. Although this property has some times been referred to as a direction- or orientation-dependency, these terms refer to the topological placement of the insulator rather than the relative 5'–3' orientation. Several lines of evidence suggest that chromatin insulators function, at least most efficiently, when used in pairs to flank the gene of interest. It is not clear whether such an arrangement simply provides insulation from *cis*-acting sequences on both sides of the integration site, or whether flanking copies are a mechanistic requirement. As diagramed in the bottom of Fig. 3, one proposed model of insulator function involves the formation of a loop structure between flanking copies of a chromatin insulator (discussed below). The question of whether chromatin insulators function best when used in flanking pairs is difficult to address from the literature because most studies to date have either employed flanking copies exclusively, or have used plasmid transfection where integration of tandem arrays automatically leads to a flanking arrangement. It is also possible that isolated insulators may form functional interactions with naturally occurring insulator elements in the neighboring chromatin, although direct evidence for such interactions is lacking. Given our current state of knowledge, it is prudent to use chromatin insulators in a flanking arrangement whenever possible, especially when they are used to protect a transferred gene from position effects where repressive heterochromatin can encroach from both sides. This can either be done directly in the case of engineered plasmid transfer cassettes, or by using a "double-copy" configuration in the case of recombinant retrovirus vectors (see below). The fact that proper chromatin insulator function requires a flanking arrangement does not necessarily indicate that more copies are always better. Although there is some evidence that the core of the cHS4 element may function more efficiently when used as a dimer, there is also evidence that deployment of the *Drosophila gypsy* insulator in a tandem array actually negates its function [35].

3.3. Mechanisms of action

The ability of chromatin insulators to block promoter-enhancer interactions appears to be mediated by specific *trans*-acting insulator-binding proteins. Although several such proteins have been identified in *Drosophila*, only one has been identified to date in mammals; a zinc-finger DNA binding protein called CTCF (CCCTC-binding factor). A recent survey indicates that essentially all of the chromatin insulator elements with enhancer-blocking activity identified in vertebrates to date bind CTCF, and that mutations which disrupt CTCF binding result in a loss of enhancer-blocking activity [32]. As diagramed in the bottom of Fig. 3, one proposed model of insulator function involves a formation of a loop structure between flanking copies of a chromatin insulator mediated by insulator-binding proteins such as CTCF. Such a structure presumably functions as a physical block to the ability of the enhancer to reach the promoter by a tracking or looping mechanism.

The mechanism by which a subclass of chromatin insulators function as barriers to silencing position effects is more poorly understood. Studies with the cHS4 element have demonstrated that the enhancer-blocking and repressor-barrier functions can be mapped to distinct subdomains of the DNase-I hypersensitive core, and that the repressor-barrier function is not dependent on binding of CTCF [36]. However, the subdomain with repressor-barrier activity does exhibit a distinct DNase I footprint indicating the involvement of DNA binding proteins in the barrier activity. As summarized in Fig. 1, recent studies also indicate that cHS4 barrier function couples high histone acetylation levels with specific protection of promoter DNA methylation [37]. These studies suggest that the cHS4 insulator prevents silencing by either inhibiting histone deacetylases and/or recruiting histone acetyl transferases. In one current model presented in Fig. 1, specific acetylation of H3 Lys9 prevents the processive chromatin condensation mediated by methylation of this moiety and the subsequent recruitment of HP1 and SUV39H1 as described above.

4. Specific uses of chromatin insulators

The ability of chromatin insulators to block the activating effects of enhancers and the repressive effects of heterochromatin and functional repressors make these elements ideal for many mammalian gene transfer applications. Beside their obvious use for abrogating the effects of surrounding chromatin, these elements can also be used to isolate regulatory elements within engineered expression constructs. Following are some specific examples which serve to illustrate these uses.

4.1. Preventing silencing in transgenic animals

The most obvious application of chromatin insulators in the context of gene transfer and expression in mammalian cells involves the use of these elements as barriers to silencing position effects. The ability of the cHS4 element to prevent transcriptional silencing in mammalian cells was first demonstrated by plasmid transfection in cell lines [38]. This approach was also used to study in detail the effects of the cHS4 element on chromatin structure and transcriptional activity at the molecular level [39]. Subsequent studies demonstrated that this element also functioned efficiently in transgenic mice to block the suppressive effects of surrounding chromatin and thus allow for integration site-independent expression of a transgene [40]. These studies involved the generation of transgenic mice with a ligand-inducible expression cassette for a chimeric transcriptional activator. In the absence of insulation, this cassette was transcriptionaly silent in 10 of 10 founder lines. However, when the same cassette was flanked with the cHS4 chromatin insulator, the transgene was found to be expressed in 3 of 4 founder lines. The ability of the cHS4 chromatin insulator to reduce the variability in expression between transgenic founder mice has been confirmed in several settings. In one system, flanking with the cHS4 element significantly reduced the variability in expression of a tyrosinase minigene between transgenic lines, and resulted in a more uniform distribution of expression in different tissues [41]. The ability of the cHS4 element to increase the uniformity of expression in various cell types has been confirmed in another transgenic mouse model utilizing a luciferase reporter gene [42]. The insulating activity of the cHS4 element has also been confirmed in a transgenic rabbit system, where inclusion of this element allowed for the ubiquitous and high-level expression of cassettes for the human inhibitors of complement activation CD55 and CD59 transcribed from a human $EF1\alpha$ gene promoter [43].

Two other chromatin insulators have also been found to reduce the impact of silencing position effects in the setting of transgenic mice: the HS5 element from the human β-globin LCR; and the $+43$ kb element from the human $apoB$ loci. The first evidence that the HS5 element from the human β-globin LCR could function to protect a transferred gene from silencing position effects came from plasmid transfection studies in cell lines [44]. Recent studies from our laboratory demonstrated that a 3.2 kb fragment containing HS5 could protect a position-sensitive expression cassette for human γ-globin against silencing position effects in transgenic mice. This activity required that the expression cassette be bracketed with the HS5 element [45]. The first evidence that the $+43$ kb element from the human $apoB$ loci also could function to protect a transferred gene from silencing position effects also came from plasmid transfection studies in cell lines [46]. Subsequent studies in transgenic mice (as well as transgenic *Drosophila*) confirmed the repressor-barrier function of this element [47].

4.2. Preventing silencing of oncoretrovirus vectors

Gene transfer vectors based on recombinant oncoretroviruses are increasingly used for both basic research and clinical gene therapy applications [48]. Early on, it was noticed that transcription from the vector LTR is often subject to silencing *in vitro* and *in vivo*. The mechanism of retrovirus silencing involves DNA methylation within the LTR itself, as well as transcriptional inhibition by a repressive histone code [17,49,50]. Attempts have been made to overcome these problems by either genetically modifying the virus LTR to remove sequences thought to be involved in temporal quiescence or by using LTR sequences from alternative viruses [25,50]. However, even with these changes the expression of oncoretrovirus vectors are still sensitive to chromosomal position effects, leading to striking variations in expression both *in vitro* and *in vivo*, especially following transfer into primitive embryonic [17] and hematopoietic stem cells [18]. Oncoretrovirus vectors for globin genes in particular seem to be profoundly sensitive to position effects, with a high incidence of integrated provirus which are transcriptionally silent observed *in vitro* and *in vivo* [18,52].

In one approach to overcoming this problem we and others have turned to the use of chromatin insulators. As diagramed in Fig. 4, this approach entails flanking an oncoretrovirus vector with a 1.2 kb fragment containing the cHS4 chromatin insulator using a "double-copy" arrangement. In this arrangement, the chromatin insulator is inserted into the 3′ LTR, and from there it is copied into the 5′ LTR during formation of provirus. In cell lines, the cHS4 insulator element has been shown to prevent methylation of sequences in the oncoretrovirus vector 5′ LTR [53]. Building on these studies, we showed that the use of this element could protect an oncoretrovirus reporter vector from such position effects in a mouse bone marrow transduction and transplantation model [19]. For this purpose we used a reporter vector which has two expression cassettes: a GFP gene transcribed from the virus LTR promoter (LTR → GFP) and a *neo* gene transcribed from a *pgk* gene promoter (Pgk → Neo). We found that flanking this vector with the cHS4-containing fragment in one particular orientation insulated the LTR → GFP cassette 29% of the time in white blood cells, and the Pgk → Neo cassette 73% of the time in myeloid progenitors. In order to more thoroughly characterize this activity, the promoter-gene combinations in the reporter vector were switched and expression of these vectors flanked with the cHS4 fragment in both orientations was again analyzed following bone marrow transduction and transplantation in mice [54]. The results of this study indicate that the cHS4 fragment can function in both orientations and can insulate both the virus LTR promoter and an internal *pgk* promoter. However, insulation of the LTR promoter diminished when the orientation of the cHS4 fragment placed the core element immediately proximal to the U3 region, suggesting a minimal distance requirement. Further, placement of the cHS4 fragment in the U3 region of the 3′ LTR dramatically decreased the level of expression from an internal *pgk* promoter, presumably by blocking interaction between this promoter and enhancer in the 3′ LTR.

In more recent studies, we tested whether flanking with the cHS4 element could also reduce the impact of silencing position effects on expression of an

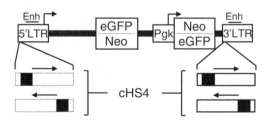

Fig. 4. Oncoretrovirus vectors used to test the cHS4 chromatin insulator. As previously reported [19,54], a 1.2 kb fragment containing the cHS4 chromatin insulator at one end (black box) was used to flank reporter vectors using a "double-copy" configuration. The vectors were based on the oncoretrovirus vector MSCV [25]. One set of vectors used the promoter in the virus LTR to transcribe eGFP (LTR → eGFP) and a promoter from the housekeeping gene *pgk* to transcribe *neo* (Pgk → Neo). In the other set of vectors the genes were switched to generate an LTR → Neo cassette and a Pgk → eGFP cassette. The locations of the enhancers in the LTRs are indicated.

oncoretrovirus vector for human γ-globin, again using a mouse bone marrow transduction and transplantation model [52]. When present, the γ-globin cassettes from the uninsulated vectors were expressed in only 2–5% of peripheral red blood cells (RBC) long-term, indicating they are highly sensitive to silencing position effects. In contrast, when present the γ-globin cassette from the insulated vector was expressed in about 50% of RBC long-term. RNase protection analysis indicated that the insulated γ-globin cassette was expressed at 6% of total mouse α-globin in transduced RBC. These results demonstrate that flanking an oncoretrovirus vector for γ-globin with the cHS4 insulator increased the likelihood of expression nearly 10-fold, which in turn allows for γ-globin expression approaching the therapeutic range for sickle cell anemia and β thalassemia in transplanted mice.

4.3. Improving gene regulation

In addition to preventing the repressive effects of surrounding heterochromatin, chromatin insulators can also be used to improve the degree of gene regulation within complex expression cassettes transferred in mammalian cells. As described above, we have found that flanking an oncoretrovirus vector with the cHS4 element using a conventional "double-copy" configuration can interfere with the interaction between an internal promoter and the enhancer in the virus 3' LTR [54]. Other studies using transfected embryonic stem cells and transgenic mice have shown that flanking with the cHS4 element can also prevent the sequence-specific silencing mediated by the LTRs of both oncoretrovirus vectors and lentivirus vectors [17]. It is important to note, however, that inclusion of duplicated sequences within retrovirus vector constructs typically result in vector rearrangements [55]. The enhancer-blocking ability of the cHS4 chromatin insulator has also been employed to prevent the inappropriate activation of inducible expression cassettes. In one example with an adenovirus vector, flanking copies of a 1.2 kb fragment containing the cHS4 core were used to prevent the transactivation by virus enhancers of a metal-inducible promoter linked to an expression cassette for human α_1-antitrypsin [56]. In another example with an adeno-associated virus vector, the 42-bp enhancer-blocking core of the cHS4 chromatin insulator was again used to prevent the transactivation by virus enhancers of various doxycycline-responsive expression cassettes [57]. Finally, the cHS4 chromatin insulator also appears to prevent transcriptional interference, which often occurs when two expression cassettes are located in close proximity by placement of this element between the two expression cassettes [23,58].

Abbreviations

cHS4	DNase I hypersensitive site 4 from the chicken β-globin LCR
CpG	DNA dinucleotide with a cytosine followed by a guanine base
CTCF	CCCTC-binding factor
H3 Lys 9	histone H3 lysine 9
HAT	histone acetyltransferase

394

HP1   heterochromatin protein 1
HS5   DNase I hypersensitive site 5 from the human β-globin LCR
LCR   locus control region
LTR   long terminal repeat
RBC   red blood cell

References

1. Hirata, R., Chamberlain, J., Dong, R., and Russell, D.W. (2002) Nat. Biotechnol. 20, 735–738.
2. Holmes-Son, M.L., Appa, R.S., and Chow, S.A. (2001) Adv. Genet. 43, 33–69.
3. Bushman, F.D. (2002) Curr. Top. Microbiol. Immunol. 261, 165–177.
4. Miller, D.G., Rutledge, E.A., and Russell, D.W. (2002) Nat. Genet. 30, 147–148.
5. Depienne, C., Mousnier, A., Leh, H., Le Rouzic, E., Dormont, D., Benichou, S., and Dargemont, C. (2001) J. Biol. Chem. 276, 18,102–18,107.
6. Elleder, D., Pavlicek, A., Paces, J., and Hejnar, J. (2002) FEBS Lett. 517, 285–286.
7. Jordan, A., Defechereux, P., and Verdin, E. (2001) J. 20, 1726–1738.
8. Pawliuk, R., Westerman, K.A., Fabry, M.E., Payen, E., Tighe, R., Bouhassira, E.E., Acharya, S.A., Ellis, J., London, I.M., Eaves, C.J., Humphries, R.K., Beuzard, Y., Nagel, R.L., Leboulch, P. (2001) Science 294, 2368–2371.
9. Josephson, N.C., Vassilopoulos, G., Trobridge, G.D., Priestley, G.V., Wood, B.L., Papayannopoulou, T., and Russell, D.W. (2002) Proc. Natl. Acad. Sci. USA 99, 8295–8300.
10. Schmidt, M., Hoffmann, G., Wissler, M., Lemke, N., Mussig, A., Glimm, H., Williams, D.A., Ragg, S., Hesemann, C.U., von Kalle, C. (2001) Hum. Gene. Ther. 12, 743–749.
11. Lander, E.S. et al. (2001) International Human Genome Sequencing Consortium. Nature 409, 860–921.
12. Gubin, A.N., Njoroge, J.M., Bouffard, G.G., and Miller, J.L. (1999) Genomics 59, 168–177.
13. Bird, A. (2002) Genes. Dev. 6, 6–21.
14. Lachner, M. and Jenuwein, T. (2002) Opin. Cell. Biol. 14, 286–298.
15. Dillon, N. and Festenstein, R. (2002) Trends Genet. 18, 252–258.
16. Bird, A.P. and Wolffe, A.P. (1999) Cell 99, 451–454.
17. Pannell, D. and Ellis, J. (2001) Med. Virol. 11, 205–217.
18. Rivella, S. and Sadelain, M. (1998) Semin. Hematol. 35, 112–125.
19. Emery, D.W., Yannaki, E., Tubb, J., and Stamatoyannopoulos, G. (2000) Proc. Natl. Acad. Sci. USA 97, 9150–9155.
20. Baur, J.A., Zou, Y., Shay, J.W. and Wright, W.E. (2001) Science 292, 2075–2077.
21. Alami, R., Greally, J.M., Tanimoto, K., Hwang, S., Feng, Y.Q., Engel, J.D., Fiering, S., and Bouhassira, E.E. (2000) Hum. Mol. Genet. 9, 631–636.
22. Feng, Y.Q., Lorincz, M.C., Fiering, S., Greally, J.M., and Bouhassira, E.E. (2001) Mol. Cell. Biol. 21, 298–309.
23. Hasegawa, K. and Nakatsuji, N. (2002) FEBS Lett. 520, 47–52.
24. Garrick, D., Fiering, S., Martin, D.I., and Whitelaw, E. (1998) Genetics 18, 56–59.
25. Hawley, R.G., Lieu, F.H., Fong, A.Z., and Hawley, T.S. (1994) Gene. Ther. 1, 136–138.
26. Challita, P.M., Skelton, D., el-Khoueiry, A., Yu, X.J., Weinberg, K., and Kohn, D.B. (1995) J. Virol. 69, 748–755.
27. Cherry, S.R., Biniszkiewicz, D., van Parijs, L., Baltimore, D., and Jaenisch, R. (2000) Mol. Cell. Biol. 20, 7419–7426.
28. Kurre, P., Morris, J., Andrews, R.G., Kohn, D.B., and Kiem, H.P. (2002) Mol. Ther. 6, 83–90.
29. Pfeifer, A., Ikawa, M., Dayn, Y., and Verma, I.M. (2002) Proc. Natl. Acad. Sci. USA 99, 2140–2145.
30. Artelt, P., Grannemann, R., Stocking, C., Friel, J., Bartsch, J., and Hauser, H. (1991) Gene 99, 249–254.
31. Walsh, C.P. and Bestor, T.H. (1999) Dev. 13, 26–34.
32. West, A.G., Gaszner, M., and Felsenfeld, G. (2002) Genes Dev. 16, 271–288.

33. Bell, A.C., West, A.G., and Felsenfeld, G. (2001) Science 291, 447–450.
34. Gerasimova, T.I. and Corces, V.G. (2001) Annu. Rev. Genet. 35, 193–208.
35. Mongelard, F. and Corces, V.G. (2001) Nat. Struct. Biol. 8, 192–194.
36. Recillas-Targa, F., Pikaart, M.J., Burgess-Beusse, B., Bell, A.C., Litt, M.D., West, A.G., Gaszner, M., and Felsenfeld, G. (2002) Proc. Natl. Acad. Sci. USA 99, 6883–6888.
37. Mutskov, V.J., Farrell, C.M., Wade, P.A., Wolffe, A.P., and Felsenfeld, G. (2002) Genes Dev. 16, 1540–1554.
38. Chung, J.H., Whiteley, M., and Felsenfeld, G. (1993) Cell 74, 505–514.
39. Pikaart, M.J., Recillas-Targa, F., and Felsenfeld, G. (1998) Genes Dev. 12, 2852–2862.
40. Wang, Y., DeMayo, F.J., Tsai, S.Y., and O'Malley, B.W. (1997) Nat. Biotechnol. 15, 239–243.
41. Potts, W., Tucker, D., Wood, H., and Martin, C. (2000) Biochem. Biophys. Res. Commun. 273, 1015–1018.
42. Ciana, P., Di Luccio, G., Belcredito, S., Pollio, G., Vegeto, E., Tatangelo, L., Tiveron, C., and Maggi, A. (2001) Mol. Endocrinol. 15, 1104–1113.
43. Taboit-Dameron, F., Malassagne, B., Viglietta, C., Puissant, C., Leroux-Coyau, M., Chereau, C., Attal, J., Weill, B., and Houdebine, L.M. (1999) Transgenic. Res. 8, 223–235.
44. Li, Q. and Stamatoyannopoulos, G. (1994) Blood 84, 1399–1401.
45. Li, Q., Zhang, M., Han, H., Rohde, A., and Stamatoyannopoulos, G. (2002) Nucleic Acids Res. 30, 2484–2491.
46. Kalos, M. and Fournier, R.E. (1995) Mol. Cell. Biol. 15, 198–207.
47. Wang, D.M., Taylor, S., and Levy-Wilson, B. (1996) J. Lipid Res. 37, 2117–2124.
48. Emery, D.W., Nishino, T., Murata, K., Fragkos, M., and Stamatoyannopoulos, G. (2002) Int. J. Hematol. 75, 228–236.
49. Challita, P.M. and Kohn, D.B. (1994) Proc. Natl. Acad. Sci. USA 91, 2567–2571.
50. Lorincz, M.C., Schubeler, D., and Groudine, M. (2001) Mol. Cell. Biol. 21, 7913–7922.
51. Halene, S., Wang, L., Cooper, R.M., Bockstoce, D.C., Robbins, P.B., and Kohn, D.B. (1999) Blood 94, 3349–3357.
52. Emery, D.W., Yannaki, E., Tubb, J., Nishino, T., Li, Q. and Stamatoyannopoulos, G. (2002) Blood 100, 2012–2019.
53. Rivella, S., Callegari, J.A., May, C., Tan, C.W., and Sadelain, M. (2000) J. Virol. 74, 4679–4687.
54. Yannaki, E., Tubb, J., Aker, M., Stamatoyannopoulos, G., and Emery, D.W. (2002) Mol. Ther. 5, 589–598.
55. Pathak, V.K. and Temin, H.M. (1990) Proc. Natl. Acad. Sci. USA 87, 6024–6028.
56. Steinwaerder, D.S. and Lieber, A. (2002) Gene Ther. 7, 556–567.
57. Fitzsimons, H.L., Mckenzie, J.M., and During, M.J. (2001) Gene Ther. 8, 1675–1681.
58. Villemure, J.F., Savard, N., and Belmaaza, A. (2001) J. Mol. Biol. 312, 963–974.
59. Zhong, X.P. and Krangel, M.S. (1997) Proc. Natl. Acad. Sci. USA 94, 5219–5224.
60. van der Vlag, J., den Blaauwen, J.L., Sewalt, R.G., van Driel, R., and Otte, A.P. (2000) J. Biol. Chem. 275, 697–704.
61. Di Simone, P., Di Leonardo, A., Costanzo, G., Melfi, R., and Spinelli, G. (2001) Biochem. Biophys. Res. Commun. 284, 987–992.
62. Akasaka, K., Nishimura, A., Takata, K., Mitsunaga, K., Mibuka, F., Ueda, H., Hirose, S., Tsutsui, K., and Shimada, H. (1999) Cell. Mol. Biol. 45, 555–565.
63. Bell, A.C., West, A.G., and Felsenfeld, G. (1999) Cell 98, 387–396.
64. Saitoh, N., Bell, A.C., Recillas-Targa, F., West, A.G., Simpson, M., Pikaart, M., and Felsenfeld, G. (2000) EMBO J. 19, 2315–2322.
65. Zhong, X.P. and Krangel, M.S. (1999) J. Immunol. 163, 295–300.
66. Hark, A.T., Schoenherr, C.J., Katz, D.J., Ingram, R.S., Levorse, J.M., and Tilghman, S.M. (2000) Nature 405, 486–489.
67. Bell, A.C. and Felsenfeld, G. (2000) Nature 405, 482–485.
68. Antes, T.J., Namciu, S.J., Fournier, R.E., and Levy-Wilson, B. (2001) Biochemistry 40, 6731–6742.
69. Filippova, G.N., Thienes, C.P., Penn, B.H., Cho, D.H., Hu, Y.J., Moore, J.M., Klesert, T.R., Lobanenkov, V.V., and Tapscott, S.J. (2001) Nat. Genet. 28, 335–343.

CHAPTER 12

Locus control regions

Xiangdong Fang[1], Kenneth R. Peterson[2,3], Qiliang Li[1], and George Stamatoyannopoulos[1,4]

[1]*Division of Medical Genetics, and* [4]*Department of Genome Sciences, University of Washington, Seattle, WA 98195, USA*
[2]*Department of Biochemistry & Molecular Biology, and* [3]*Department of Anatomy & Cell Biology, University of Kansas Medical Center, Kansas City, KS 66160, USA*

1. Introduction

Locus control regions (LCRs) are regulatory elements which are operationally defined by their ability to enhance the expression of linked genes in a tissue-specific and copy number-dependent manner at ectopic chromatin sites. The LCRs are composed of a series of DNase I-hypersensitive (HS) sites usually located near the genes they control. Each HS consists of a core element and flanking sequences. The core elements usually are approximately 200 nucleotides long and are composed of arrays of multiple ubiquitous and lineage-specific transcription factor binding sites.

The first LCR to be discovered was the human β-globin locus LCR (reviewed in [1–3]). Initial clues about the existence of a regulatory element upstream of the β-globin locus came from *in vivo* observations in β-thalassemia mutants. Certain thalassemia mutants had the characteristic phenotype of absence of β-globin gene expression in *cis*, despite the fact that the β-like globin genes of the β-globin locus were intact. Characteristic of these thalassemias were deletions of upstream sequences [4,5] that were associated with a closed chromatin conformation spanning the entire β-globin locus [5,6]. These observations raised the possibility that the deleted DNA segment contained *cis*-acting regulatory elements required for β-globin gene expression *in vivo*. Another important clue for the existence of an upstream regulatory sequence was the discovery of a series of DNase I-hypersensitive sites (HSs) [7,8] located 6 to 20 kb upstream of the ε-globin gene of the β-globin locus. Four of these HSs, 5'HS1 to 5'HS4, were found to be erythroid-specific, whereas one, 5'HS5, was considered to be ubiquitous (subsequently it was demonstrated that 5'HS5 is not ubiquitous, but has limited lineage specificity). These HSs were developmentally stable [7–9]. The appearance of these sites before the initiation of transcription was considered evidence that they played a role in the activation of the β-globin locus [9]. Hence, this DNase I-HSs region was initially called the Locus Activation Region (LAR) [9]. Definitive proof for the existence of a very powerful regulatory element in this region came from studies in transgenic mice [10]. Early studies showed that a 5-kb β-globin gene segment, including a 1.5-kb promoter region, was expressed in erythroleukemia cell lines, implying that it contains all the regulatory elements necessary for proper expression [11,12]. However, when the same construct was used for production of transgenic mice [13–15], expression was extremely variable between lines. Only a small proportion of transgenic mice expressed the gene, and in these mice expression was far below physiologically significant levels. Grosveld *et al.* [10] demonstrated that linkage of the β-globin gene to the 6- to 22-kb upstream region containing the DNase I-hypersensitive sites resulted in expression of the gene at a level comparable to the endogenous mouse β-globin genes in a position-independent, copy number-dependent manner. Because of these properties, they called this region the Dominant Control Region (DCR) [10,16]. These nomenclatures, LAR and DCR are found in the literature until 1990 when it was decided to call this region the Locus Control Region (LCR) [17]. The term LCR is used exclusively to describe regulatory elements with functional properties similar to those of the β-globin locus LCR.

2. The β-globin locus LCR

A diagram showing the location of the LCR in the β-globin locus is presented in Fig. 1. The DNase I-hypersensitive sites 5'HS1 to 5'HS5, are numbered from 3' to 5'. Two new sites, 5'HS6 and 5'HS7 have been described and are located 6 and 12 kb upstream of the 5'HS5, respectively [18], but it is unclear whether they constitute components of the β-globin locus LCR. The HSs of the LCR are much stronger than the HSs located in the globin gene promoters, and frequently they are referred to as "super-hypersensitive sites." Notice that another HS, 3'HS1, is found approximately 20 kb downstream from the β-globin gene. Its function is unclear, although recent evidence suggests that it may function as a chromatin insulator [19].

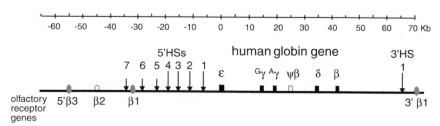

Fig. 1. Human β-globin gene cluster. The human locus consists of five functional genes, indicated as filled boxes, arrayed in their order of developmental expression, 5'-ε-^{G}γ-^{A}γ-δ-β-3'. There are two developmental switches in globin chain synthesis coincident with changes in site and type of erythropoiesis. During primitive erythropoiesis, the ε-globin gene is expressed in the embryonic yolk sac. The first switch occurs at approximately eight weeks gestation; the ε-globin gene is silenced and the ^{G}γ- and ^{A}γ-globin genes are expressed during definitive erythropoiesis in the fetal liver. The second switch occurs shortly after birth; the γ-globin genes are silenced and the β-globin gene and, to a lesser extent, the δ-globin gene, are activated in the bone marrow. 5'HS1-7 are located -6, -11, -15, -18, -22, -28, and -35 kb relative to the β-globin gene, respectively, and are indicated by arrows. 5'HS1-4 are erythroid-specific, but 5'HS5-7 are not. Another HS is located 20 kb downstream of the β-globin gene (3'HS1). 3'HS1 is found only in erythroid cells. Boxes are globin genes and ovals are olfactor receptor genes (HOR) with filled ones representing the productive genes and shaded ones being pseudogenes. (Modified from [104])

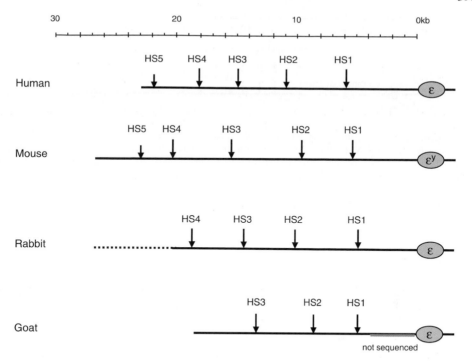

Fig. 2. Human, mouse, rabbit, and goat LCR. The dotted line for the rabbit LCR indicates that the specified sequence does not match the human sequence.

The β-globin locus LCRs of several species have been characterized and compared to the human LCR [20–26]. The general organization and spatial array of the individual HSs has been conserved (Fig. 2). The goat and human LCR share 6.5 kb that are roughly 60% conserved. The homology is 80–90% within and adjacent to the individual HSs. The mouse and human LCRs have an identical organization, although insertion of repetitive sequences within this region during evolution has altered the distances between the HSs. There are extended regions of significant homology, with the highest conservation within and adjacent to the individual HSs. Overall, evolutionary comparisons suggest that the functionally important sequences extend beyond the core elements of the HSs and are required for LCR activity in cultured cells and transgenic mice [20].

The LCR of the chicken β-globin gene cluster has been extensively characterized [27,28] (Fig. 3). Several insights have been gained with studies of the chicken LCR. Organization of the chicken LCR is similar to its human counterpart except that one of the LCR elements is located between the adult β^A and embryonic β genes (the β^A/ε enhancer) [29]. This enhancer is able to confer site-independent expression to the chicken β^A-globin gene in transgenic mice [30]. Chromatin unfolding of the chicken β-locus requires the presence of both the LCR and the promoter [31]. Chicken HS4 demarcates the 5' border of the locus, which functions as a powerful chromatin insulator [32]. The insulating function of chicken HS4 is manifested by

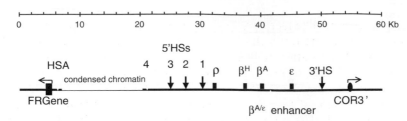

Fig. 3. Chicken β-globin gene cluster. At the 5′ end is a folate receptor gene (FR). HSA is a hypersensitive site associated with FR. A second hypersensitive site, HS4, marks the insulator element of the β-globin locus. The 16 kb sequence between HSA and HS4 is occupied by a region of condensed chromatin. Beyond HS4 are found the hypersensitive sites HS3 to HS1, part of the globin LCR. A strong enhancer, βA/ε, lies between the βA and ε-globin genes. The 3′ end of the locus has another hypersensitive site (3′HS); beyond that is an olfactory receptor gene (COR3′)

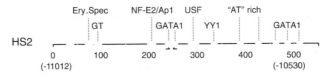

Fig. 4. Position of binding motifs for various proteins within individual hypersensitive sites. (Modified from [1])

enhancer blocking activity and position-effect protection. These two activities are separable [33]: the former is mediated by the transacting factor CTCF [34] and the latter function may be achieved by highly efficient recruitment of histone acetyltransferase by the HS4 element [35]. Chicken 5′HS4 also acts as an insulator in retroviral vectors [36]. In addition, the 3′HS1 of the chicken LCR functions as an insulator.

Mapping of individual HSs has shown that 250–500 bp of DNA is exposed to nuclease action at each site [37–41]. Formation of these sites is thought to represent displacement of one or two nucleosomes, the number usually associated with this length of DNA. Each HS contains one or more binding motifs for the two erythroid-restricted transcriptional activators, GATA-1 and NF-E2, and for other ubiquitous DNA-binding proteins, most notably those associated with the CACC/GTGG class of binding motifs (Fig. 4).

The HSs have been extensively characterized with respect to their factor-binding properties *in vitro* and *in vivo* [37,38,40–48]. The proteins bound by the different HSs include tissue-restricted and ubiquitously expressed transcriptional factors. One or more GATA-1 binding motifs are found within each HS. Overall, the data suggest critical roles for GATA proteins in LCR function pertaining both to chromatin structure and gene activity. Although, these roles are generally ascribed to GATA-1, it should be noted that GATA-2 is also expressed in primitive hematopoietic cells. The combination of GATA-1 binding sites and a GGTGG motif occurs repeatedly and appears to be associated with erythroid specificity. Each characterized HS has binding motifs of this type. Experiments with 5′HS3 in transgenic mice showed that the erythroid Krüppel-like factor (EKLF) interacts with the GT site *in vivo* and that the binding of EKLF induces a change in chromatin structure as detected by DNase I sensitivity [49]. These data correlate well with the fact that EKLF binds to members of the SWI/SNF family of proteins, which modify chromatin structure [50,51]. A number of other factors have been detected that bind these sequences, but it is unknown whether any of them are functional *in vivo*. Another important binding motif is AP1/NF-E2. The binding motif for NF-E2 is also found in each of the characterized HSs [37] (see below). In the β-globin LCR 5'HS2, tandem motifs are phased at a 10 bp intervals, establishing binding sites on successive turns of the DNA double helix [38,52–54]. Other widely expressed proteins, such as USF and YY1, have been shown to interact with the LCR, but the functional relevance of such interactions has not been defined.

3. Properties of LCRs

3.1. Enhancer activity

As mentioned earlier, the LCR was first identified because of its ability to confer consistent physiological levels of expression upon a linked β-globin gene in transgenic mice, indicating that it functions, in part, as an enhancer [10]. Deletion of the LCR *in situ*, in knock-out mice or in erythroleukemia cell lines, results in a striking decline of β-globin locus gene expression activity in *cis* [55–57], indicating that the LCR is also acting as a very strong enhancer at its endogenous location on the chromosome. The enhancer activity of the LCR resides in 5'HS2, 3 and 4, but not in 5'HS1 or 5'HS 5 (reviewed in [1–3,20]). The enhancer activity of 5'HS2 can be demonstrated in transient transfection assays and therefore it behaves as a classical enhancer. Enhancer activity in 5'HS3 or 5'HS4 can be detected only when they are integrated into chromatin (reviewed in [20]), suggesting that alteration of chromatin structure may be involved in propagating the enhancer activity of these two HSs.

LCR activities may be orientation-dependent [58] and distance-dependent [59–61]. A combination of HSs can deliver enhancer activity over longer distances than individual HSs [62]. The enhancer activity of the LCR is tissue-specific; that is, the expression of globin genes or heterogeneous non-globin genes is confined to erythroid cells when linked to the β-globin LCR [10,16], although ectopic expression was observed for some non-globin genes in transgenic mice [63]. These data suggest that tissue specificity is not really an intrinsic property of the LCR, but depends upon both the LCR and the promoter that it interacts with.

The enhancer activity of the LCR resides in the HS core elements that contain binding sites for ubiquitous and erythroid-specific trans-acting factors. In 5'HS2, a conserved sequence, TGCTGA(C/G)TCA(T/C), is critical for strong enhancer activity [38,52]. This Maf recognition element (MARE) is bound by multiple homodimeric and heterodimeric transcription factors *in vitro* [64], including Maf homodimers, heterodimers containing a Maf subunit and another bZIP protein (NF-E2, Nrf1, Nrf2, Bach1, Bach2), and heterodimers lacking a Maf subunit

(AP1) [65–70]. The NF-E2 heterodimer binds the tandem MAREs of 5′HS2, and globin gene expression closely parallels the level of NF-E2 binding activity [71]. The p45 subunit of NF-E2 is restricted to erythroid and megakaryocytic cells in contrast to most other MARE-binding proteins, which are expressed in wide range of cells. It is likely that the LCR is occupied by small Maf proteins, and during erythroid maturation, the NF-E2 complex is recruited to the LCR, as well as to the active globin promoters, even though the promoters do not contain MAREs [72,73]. This recruitment of the NF-E2 complex correlates with a greater than 100-fold increase in mouse β^{maj}-globin transcription [73]. However, inactivation of the p45 gene does not inhibit globin gene expression [74] in transgenic mice. In addition, the absence of a phenotype in the p45 knockout mice is not due to compensation by Nrf2 (a factor closely related with p45) since double knockout p45 Nrf2 mice express the α- and β-globin genes at normal levels [75]. These observations can be explained if there are interchangeable functions between members of the cap'n collar (CNC) subfamily of bZIP transcription factors. Formation of hypersensitivity is a result of interaction of multiple ubiquitous and erythroid-specific *trans* factors in the HS regions [76]. NF-E2 plays an essential role in disruption of chromatin structure of HS2 of the LCR [77,78]. Nucleosome disruption by NF-E2 is an ATP-dependent process, suggesting the involvement of energy dependent nucleosome remodeling factors [77]. The modulation of DNA structure by HMY I/Y is a critical regulator of long-range enhancer function [79,80].

3.2. Copy number-dependent gene expression

Characteristic of the LCR is its ability to confer copy number-dependent, position-independent expression on linked genes [10]. Copy number-dependent expression is considered to be indicative of open chromatin structure; that is, a structure in which DNA is accessible to transcription factors. In transgenic mice carrying single copies of small, recombinant globin gene constructs linked to the HSs of the LCR, only 5′HS3 is able to confer copy number-dependent gene expression, suggesting that 5′HS3 possesses the dominant chromatin-opening activity of the LCR [81]. These results indicate that only the intact LCR can provide position-independent chromatin-opening activity in single-copy transgenic mice. Single copy transgenic mice carrying the entire human β-globin locus, showed that deletion of any one of the HSs resulted in β-globin gene expression that was sensitive to the position of integration [82].

3.3. Timing and origin of DNA replication

The human β-globin locus replicates late in most cell types, but replicates early in erythroid cells [83,84]. The β-globin LCR region (5′HSs 1–5) is sufficient to direct replication timing in a developmentally specific manner *in vivo* [85]. LCR-mediated chromatin opening activity may be intertwined with the ability to direct early replication timing [86]. Targeted deletions of LCR 5′HSs 2–5 has shown that long-distance control of origin choice and replication timing in the human β-globin locus are independent of the LCR [87]. Therefore, an as yet undefined class of *cis*-acting elements may play a role in mediating control of replication timing, independent of transcription. One of the major roles for replication timing control within the globin locus may be to set up late replication in non-erythroid cells. In this manner, repression of background transcription may be achieved [85].

3.4. Histone modification

The effects of LCRs on chromatin acetylation have been studied in different model systems. Function of the human growth hormone (hGH) LCR has been linked to specific patterns of core histone acetylation. In the pituitary, the hGH locus is encompassed in a somatotrope-specific domain of hyperacetylated chromatin that extends from the most 5'LCR component to the hGH-N gene promoter. Further analysis demonstrated that the deletion of HSI of the hGH LCR, located 14.5 kb upstream from the hGH-N gene promoter, resulted in hypoacetylation of a 32-kb region spanning the entire hGH LCR through the hGH-N gene promoter. These data suggested that LCR long-range activity is mediated by targeting, and extensive spreading, of core histone acetylation [88].

Deletion of 5'HS2-5 of the human β-globin LCR did not affect the general pattern of histone H4 acetylation of a β-globin locus transgene [89]. Although, deletion of the murine β-globin LCR decreased the rate of β-globin transcription, it did not alter the acetylation status of histone H3 or H4 within the promoter region [90]. NF-E2 is required for histone hyperacetylation at the adult β-globin promoter, but not at the LCR [72]. In addition, the acetylation pattern of the LCR and promoters varied at different developmental stages [91]. Dynamic histone acetylation and deacetylation activities may play an important role in the developmental control of β-globin gene expression.

LCRs may play a role in tissue-specific DNA demethylation. Studies of the methylation status of the LCR for the mouse T-cell receptor α/δ locus support such a role. The elements of an LCR may control tissue-specific DNA methylation patterns both in transgenes and in native loci [92].

4. LCR knockout mice

Generally, it is assumed that the role of the LCR *in vivo* is to open and/or maintain a permissive chromatin conformation of the β-globin locus within erythroid cells, although enhancement of transcription is also an essential function. Elements of the LCR also function as chromatin insulators, as clearly demonstrated with the chicken β-globin LCR, where both 5'HS4 and the 3'HS1 insulate the locus (see Chapter 11). Therefore, it was a surprise to find that when the entire mouse LCR (5'HS1-6) was deleted by homologous recombination, transcription from the locus was strikingly reduced [55,93], but the formation of the general DNase I sensitivity associated with the β-globin locus domain was not affected [93]. This phenotype differed from the LCR deletions that produce β-thalassemia in humans because the latter results not

404

only in total absence of globin gene expression, but also in total inactivation of the β-globin locus domain which becomes DNase I-resistant and hypermethylated [6]. An explanation for the discrepancy between ΔLCR mice and ΔLCR human mutants is not clear. Perhaps the LCR is simply another enhancer within the β-globin locus [55], although this explanation does not account for the observations in transgenic mice where chromatin-opening activity is a consistent function associated with the LCR.

5. Mechanisms of globin gene activation by the LCR

The molecular mechanism by which the LCR functions over a long distance is still speculative. Three models have been proposed, looping, tracking, and linking. Existing data do not refute any of these.

The looping model (Fig. 5) suggests that the HSs of the LCR fold to form a "holocomplex". It is postulated that the core elements of the HSs bind transcription factors and form the active site of the holocomplex while the core-flanking sequences constrain the holocomplex in the proper conformation. This structure physically "loops" so that the LCR comes in close proximity to the appropriate promoter. This close association of the holocomplex with promoter and enhancer elements allows the interaction between LCR-bound transcription proteins and other co-activators with the basal transcription apparatus already bound at the promoter, resulting in formation of a stable transcription complex and enhancement of globin gene expression [82,94–98].

Results from transgenic mouse studies support the looping model. While deletion of the 5'HS2 core abolishes expression of the ε-, γ-, and β-globin genes [95], deletion of the entire 5'HS2 region (core and flanking sequences) has only minor effect on ε-, γ-, and β-globin gene expression [97]. The removal of the 5'HS2 core in effect destroys the active site of the holocomplex, resulting in a dominant negative mutation that abolishes LCR function. In contrast, when the entire 5'HS2 region is deleted [97], the remaining 5'HSs may adapt an alternate holocomplex conformation. Similarly, deletion of the core element of 5'HS3 produces striking, functionally

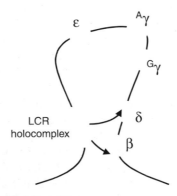

Fig. 5. The looping model. (Modified from [1])

abnormal expression of the embryonic and fetal globin genes [99,100], while the effects of deleting the entire 5′HS3 region (core plus flanking sequence) are minor [97]. Additional evidence supporting the looping model has been provided from studies showing that the LCR interacts with only one globin gene promoter at any given time and during switching "flip-flops" between two or more promoters [82,98]. The LCR holocomplex is thus free to move from gene to gene. Two parameters, the availability of specific transcription factors and distance of a gene from the LCR determine the frequency of LCR-gene interactions during development [59–61].

According to the tracking, or scanning, model, an activator complex is formed in the LCR by the binding of erythroid-specific and ubiquitous transcription factors and co-factors with the HS motifs. The activation complex subsequently migrates, or tracks, linearly along the DNA helix of the locus [101,102]. When this activation complex encounters the basal transcription machinery bound to a globin gene promoter, the complete transcriptional apparatus is assembled and transcription of that gene is initiated.

The linking model proposes that chromatin facilitator proteins bound throughout the locus define the domain to be transcribed and mediate the sequential stage-specific binding of transcription factors [103]. A chain of Chip-like protein containing complexes bound to chromatin via interaction with homeodomain transcription factors or other proteins is nucleated at the LCR by higher-order protein complexes and extends along the chromatin fiber between the LCR and the gene to be transcribed. When the γ-globin genes are transcribed, the promoter-bound complex acts as a boundary to the further spread of the chain to more distal genes. When the γ-globin genes are inactivated and β-globin gene is transcribed, the γ-globin gene promoters form another link in the chain [103].

6. LCRs of other genes

Several LCR or LCR-like elements have been described in mammals, including human, mouse, rat, chicken, rabbit, sheep, and goat [104]. All of these LCRs are comprised of varying numbers of tissue-specific DNase I-hypersensitive sites. The HSs that have been extensively characterized consist of a 150–300 bp central core containing a high density of transcription factor binding sites [105–107]. The HSs composing these LCRs are not necessarily located together; they may be upstream of, downstream of, or within the genes they control. Some LCRs are a collection of elements, with different numbers of HSs spread over large distances. Functionally, all LCRs exhibit some or all of the properties associated with the β-globin LCR, most commonly the hallmark of copy number-dependent, site-of-integration-independent expression of their cognate loci or linked transgenes.

Most of the data regarding LCR function have been obtained from studies of β-globin LCRs. Studies of other genes regulated by LCRs largely corroborate LCR properties first characterized for β-globin locus LCRs. However, some novel data regarding LCR function are derived from studies of non-globin LCRs. For example, the human CD2 LCR was shown to be essential for establishing an open chromatin configuration, even in the absence of enhancer activity [108,109]. The T cell receptor α/δ (TCRα) LCR has been implicated in tissue-specific DNA demethylation, an important role for LCRs since DNA methylation may cause chromatin closure and gene silencing. Studies of the human growth hormone (hGH) LCR suggest that LCRs may increase histone acetylation by targeted recruitment and subsequent spreading of histone acetyltransferase activity to encompass and activate remote target genes [110].

Acknowledgements

This work was supported by National Institutes of Health grants DK53510, HL67336, DK61804 and a Faculty Scholar Award from the Madison and Lila Self Graduate Fellowship awarded to KRP.

Abbreviations

EKLF   erythroid Krüppel-like factor
hGH   human growth hormone
HSs   hypersensitive site
LCR   locus control region
MARE   Maf recognition element

References

1. Stamatoyannopoulos, G. and Grosveld, F. (2001) In: Varmus, H. (ed.) Molecular Basis of Blood Diseases, 3rd Ed., W.B. Saunders Company, Philadelphia, PA, pp. 135–182.
2. Li, Q., Harju, S., and Peterson, K.R. (1999) Trends Genet. 15, 403–408.
3. Fraser, P. and Grosveld, F. (1998) Curr. Opin. Cell. Biol. 10, 361–365.
4. Driscoll, M.C., Dobkin, C.S., and Alter, B.P. (1989) Proc. Natl. Acad. Sci. USA 86, 7470–7474.
5. Kioussis, D., Vanin, E., deLange, T., Flavell, R.A., and Grosveld, F.G. (1983) Nature 306, 662–666.
6. Forrester, W.C., Epner, E., Driscoll, M.C., Enver, T., Brice, M., Papayannopoulou, T., and Groudine, M. (1990) Genes Dev. 4, 1637–1649.
7. Forrester, W.C., Thompson, C., Elder, J.T., and Groudine, M. (1986) Proc. Natl. Acad. Sci. USA 83, 1359–1363.
8. Tuan, D., Solomon, W., Li, Q., and London, I.M. (1985) Proc. Natl. Acad. Sci. USA 82, 6384–6388.
9. Forrester, W.C., Takegawa, S., Papayannopoulou, T., Stamatoyannopoulos, G., and Groudine, M. (1987) Nucleic Acids Res. 15, 10159–10177.
10. Grosveld, F., van Assendelft, G.B., Greaves, D.R., and Kollias, G. (1987) Cell 51, 975–985.
11. Antoniou, M., deBoer, E., Habets, G., and Grosveld, F. (1988) EMBO J. 7, 377–384.
12. Wright, S., deBoer, E., Grosveld, F.G., and Flavell, R.A. (1983) Nature 305, 333–336.
13. Magram, J., Chada, K., and Costantini, F. (1985) Nature 315, 338–340.
14. Kollias, G., Wrighton, N., Hurst, J., and Grosveld, F. (1986) Cell 46, 89–94.
15. Townes, T.M., Lingrel, J.B., Chen, H.Y., Brinster, R.L., and Palmiter, R.D. (1985) EMBO J. 4, 1715–1723.
16. Blom van Assendelft, G., Hanscombe, O., Grosveld, F., and Greaves, D.R. (1989) Cell 56, 969–977.

17. Stamatoyannopoulos, G. (1991) In: Stamatoyannopoulos, G. and Nienhius, A.W. (eds.) The Regulation of Hemoglobin Switching, 3rd, The Johns Hopkins University Press, Baltimore, MD, pp. xiii.

18. Bulger, M., van Doorninck, J.H., Saitoh, N., Telling, A., Farrell, C., Bender, M.A., Felsenfeld, G., Axel, R., Groudine, M., von Doorninck, J.H. (1999) Proc. Natl. Acad. Sci. USA 96, 5129–5134.

19. Farrell, C.M., West, A.G., and Felsenfeld, G. (2002) Mol. Cell. Biol. 22, 3820–3831.

20. Hardison, R., Slightom, J.L., Gumucio, D.L., Goodman, M., Stojanovic, N., and Miller, W. (1997) Gene 205, 73–94.

21. Li, Q., Zhou, B., Powers, P., Enver, T., and Stamatoyannopoulos, G. (1991) Genomics 9, 488–499.

22. Li, Q.L., Zhou, B., Powers, P., Enver, T., and Stamatoyannopoulos, G. (1990) Proc. Natl. Acad. Sci. USA 87, 8207–8211.

23. Moon, A.M. and Ley, T.J. (1990) Proc. Natl. Acad. Sci. USA 87, 7693–7697.

24. Hug, B.A., Moon, A.M., and Ley, T.J. (1992) Nucleic Acids Res. 20, 5771–5778.

25. Jimenez, G., Gale, K.B., and Enver, T. (1992) Nucleic Acids Res. 20, 5797–5803.

26. Slightom, J.L., Bock, J.H., Tagle, D.A., Gumucio, D.L., Goodman, M., Stojanovic, N., Jackson, J., Miller, W., and Hardison, R. (1997) Genomics 39, 90–94.

27. Evans, T., Felsenfeld, G., and Reitman, M. (1990) Annu. Rev. Cell. Biol. 6, 95–124.

28. Mason, M.M., Abruzzo, L.V., and Reitman, M. (1995) In: Stamatoyannopoulos, G. (ed.) Molecular Biology of Hemoglobin Switching, Andover, U.K, Intercept, pp. 13–22.

29. Felsenfeld, G. (1993) Gene 135, 119–124.

30. Reitman, M., Lee, E., Westphal, H., and Felsenfeld, G. (1990) Nature 348, 749–752.

31. Reitman, M., Lee, E., Westphal, H., and Felsenfeld, G. (1993) Mol. Cell. Biol. 13, 3990–3998.

32. Chung, J.H., Whiteley, M., and Felsenfeld, G. (1993) Cell 74, 505–514.

33. Recillas-Targa, F., Pikaart, M.J., Burgess-Beusse, B., Bell, A.C., Litt, M.D., West, A.G., Gaszner, M., and Felsenfeld, G. (2002) Proc. Natl. Acad. Sci. USA 99, 6883–6888.

34. Bell, A.C., West, A.G., and Felsenfeld, G. (1999) Cell 98, 387–396.

35. Litt, M.D., Simpson, M., Gaszner, M., Allis, C.D., and Felsenfeld, G. (2001) Science 293, 2453–2455.

36. Emery, D.W., Yannaki, E., Tubb, J., and Stamatoyannopoulos, G. (2000) Proc. Natl. Acad. Sci. USA 97, 9150–9155.

37. Stamatoyannopoulos, J.A., Goodwin, A.J., Joyce, T.M., and Lowrey, C.H. (1995) In: Stamatoyannopoulos, G. (ed.) Molecular Biology of Hemoglobin Switching. Andover, U.K, Intercept, pp. 71–86.

38. Talbot, D. and Grosveld, F. (1991) EMBO J. 10, 1391–1398.

39. Philipsen, S., Talbot, D., Fraser, P., and Grosveld, F. (1990) EMBO J. 9, 2159–2167.

40. Pruzina, S., Hanscombe, O., Whyatt, D., Grosveld, F., and Philipsen, S. (1991) Nucleic Acids Res. 19, 1413–1419.

41. Lowrey, C.H., Bodine, D.M., and Nienhuis, A.W. (1992) Proc. Natl. Acad. Sci. USA 89, 1143–1147.

42. Talbot, D., Philipsen, S., Fraser, P., and Grosveld, F. (1990) EMBO J. 9, 2169–2177.

43. Walters, M., Kim, C., and Gelinas, R. (1991) Nucleic Acids Res. 19, 5385–5393.

44. Strauss, E.C. and Orkin, S.H. (1992) Proc. Natl. Acad. Sci. USA 89, 5809–5813.

45. Ikuta, T. and Kan, Y.W. (1991) Proc. Natl. Acad. Sci. USA 88, 10188–10192.

46. Reddy, P.M. and Shen, C.K. (1991) Proc. Natl. Acad. Sci. USA 88, 8676–8680.

47. Stamatoyannopoulos, J.A., Goodwin, A., Joyce, T., and Lowrey, C.H. (1995) EMBO J. 14, 106–116.

48. Pomerantz, O., Goodwin, A.J., Joyce, T., and Lowrey, C.H. (1998) Nucleic Acids Res. 26, 5684–5691.

49. Gillemans, N., Tewari, R., Lindeboom, F., Rottier, R., de Wit, T., Wijgerde, M., Grosveld, F., and Philipsen, S. (1998) Genes Dev. 12, 2863–2873.

50. Zhang, W., Kadam, S., Emerson, B.M., and Bieker, J.J. (2001) Mol. Cell. Biol. 21, 2413–2422.

51. Armstrong, J.A., Bieker, J.J., and Emerson, B.M. (1998) Cell 95, 93–104.

52. Ney, P.A., Sorrentino, B.P., McDonagh, K.T., and Nienhuis, A.W. (1990) Genes Dev. 4, 993–1006.

53. Caterina, J.J., Ryan, T.M., Pawlik, K.M., Palmiter, R.D., Brinster, R.L., Behringer, R.R., and Townes, T.M. (1991) Proc. Natl. Acad. Sci. USA 88, 1626–1630.

54. Moi, P. and Kan, Y.W. (1990) Proc. Natl. Acad. Sci. USA 87, 9000–9004.

408

55. Epner, E., Reik, A., Cimbora, D., Telling, A., Bender, M.A., Fiering, S., Enver, T., Martin, D.I., Kennedy, M., Keller, G. (1998) Mol. Cell. 2, 447–455.

56. Reik, A., Telling, A., Zitnik, G., Cimbora, D., Epner, E., and Groudine, M. (1998) Mol. Cell. Biol. 18, 5992–6000.

57. Bender, M.A., Bulger, M., Close, J., and Groudine, M. (2000) Mol. Cell. 5, 387–393.

58. Tanimoto, K., Liu, Q., Bungert, J., and Engel, J.D. (1999) Nature 398, 344–348.

59. Hanscombe, O., Whyatt, D., Fraser, P., Yannoutsos, N., Greaves, D., Dillon, N., and Grosveld, F. (1991) Genes Dev. 5, 1387–1394.

60. Peterson, K.R. and Stamatoyannopoulos, G. (1993) Mol. Cell. Biol. 13, 4836–4843.

61. Dillon, N., Trimborn, T., Strouboulis, J., Fraser, P., and Grosveld, F. (1997) Mol. Cell. 1, 131–139.

62. Bresnick, E.H. and Tze, L. (1997) Proc. Natl. Acad. Sci. USA 94, 4566–4571.

63. Tewari, R., Gillemans, N., Harper, A., Wijgerde, M., Zafarana, G., Drabek, D., Grosveld, F., and Philipsen, S. (1996) Development 122, 3991–3999.

64. Motohashi, H., Shavit, J.A., Igarashi, K., Yamamoto, M., and Engel, J.D. (1997) Nucleic Acids Res. 25, 2953–2959.

65. Mignotte, V., Wall, L., deBoer, E., Grosveld, F., and Romeo, P.H. (1989) Nucleic Acids Res. 17, 37–54.

66. Andrews, N.C., Erdjument-Bromage, H., Davidson, M.B., Tempst, P., and Orkin, S.H. (1993) Nature 362, 722–728.

67. Caterina, J.J., Donze, D., Sun, C.W., Ciavatta, D.J., and Townes, T.M. (1994) Nucleic Acids Res. 22, 2383–2391.

68. Chan, J.Y., Han, X.L., and Kan, Y.W. (1993) Proc. Natl. Acad. Sci. USA 90, 11371–11375.

69. Moi, P., Chan, K., Asunis, I., Cao, A., and Kan, Y.W. (1994) Proc. Natl. Acad. Sci. USA 91, 9926–9930.

70. Oyake, T., Itoh, K., Motohashi, H., Hayashi, N., Hoshino, H., Nishizawa, M., Yamamoto, M., and Igarashi, K. (1996) Mol. Cell. Biol. 16, 6083–6095.

71. Kotkow, K.J. and Orkin, S.H. (1995) Mol. Cell. Biol. 15, 4640–4647.

72. Johnson, K.D., Christensen, H.M., Zhao, B., and Bresnick, E.H. (2001) Mol. Cell. 8, 465–471.

73. Sawado, T., Igarashi, K., and Groudine, M. (2001) Proc. Natl. Acad. Sci. USA 98, 10226–10231.

74. Shivdasani, R.A. and Orkin, S.H. (1995) Proc. Natl. Acad. Sci. USA 92, 8690–8694.

75. Martin, F., van Deursen, J.M., Shivdasani, R.A., Jackson, C.W., Troutman, A.G., and Ney, P.A. (1998) Blood 91, 3459–3466.

76. Goodwin, A.J., McInerney, J.M., Glander, M.A., Pomerantz, O., and Lowrey, C.H. (2001) J. Biol. Chem. 276, 26883–26892.

77. Armstrong, J.A. and Emerson, B.M. (1996) Mol. Cell. Biol. 16, 5634–5644.

78. Gong, Q.H., McDowell, J.C., and Dean, A. (1996) Mol. Cell. Biol. 16, 6055–6064.

79. Bagga, R., Michalowski, S., Sabnis, R., Griffith, J.D., and Emerson, B.M. (2000) Nucleic Acids Res. 28, 2541–2550.

80. Bagga, R. and Emerson, B.M. (1997) Genes Dev. 11, 629–639.

81. Ellis, J., Tan-Un, K.C., Harper, A., Michalovich, D., Yannoutsos, N., Philipsen, S., and Grosveld, F. (1996) EMBO J. 15, 562–568.

82. Milot, E., Strouboulis, J., Trimborn, T., Wijgerde, M., de Boer, E., Langeveld, A., Tan-Un, K., Vergeer, W., Yannoutsos, N., Grosveld, F. et al. (1996) Cell 87, 105–114.

83. Epner, E., Forrester, W.C., and Groudine, M. (1988) Proc. Natl. Acad. Sci. USA 85, 8081–8085.

84. Dhar, V., Skoultchi, A.I., and Schildkraut, C.L. (1989) Mol. Cell. Biol. 9, 3524–3532.

85. Simon, I., Tenzen, T., Mostoslavsky, R., Fibach, E., Lande, L., Milot, E., Gribnau, J., Grosveld, F., Fraser, P., Cedar, H. (2001) EMBO J. 20, 6150–6157.

86. Allshire, R. and Bickmore, W. (2000) Cell 102, 705–708.

87. Cimbora, D.M., Schubeler, D., Reik, A., Hamilton, J., Francastel, C., Epner, E.M., and Groudine, M. (2000) Mol. Cell. Biol. 20, 5581–5591.

88. Ho, Y., Elefant, F., Cooke, N., and Liebhaber, S. (2002) Mol. Cell. 9, 291–302.

89. Schubeler, D., Francastel, C., Cimbora, D.M., Reik, A., Martin, D.I., and Groudine, M. (2000) Genes Dev. 14, 940–950.

90. Schubeler, D., Groudine, M., and Bender, M.A. (2001) Proc. Natl. Acad. Sci. USA 98, 11432–11437.
91. Forsberg, E.C., Downs, K.M., Christensen, H.M., Im, H., Nuzzi, P.A., and Bresnick, E.H. (2000) Proc. Natl. Acad. Sci. USA 97, 14494–14499.
92. Santoso, B., Ortiz, B.D., and Winoto, A. (2000) J. Biol. Chem. 275, 1952–1958.
93. Bender, M.A., Mehaffey, M.G., Telling, A., Hug, B., Ley, T.J., Groudine, M., and Fiering, S. (2000) Blood 95, 3600–3604.
94. Gribnau, J., de Boer, E., Trimborn, T., Wijgerde, M., Milot, E., Grosveld, F., and Fraser, P. (1998) EMBO J. 17, 6020–6027.
95. Bungert, J., Tanimoto, K., Patel, S., Liu, Q., Fear, M., and Engel, J.D. (1999) Mol. Cell. Biol. 19, 3062–3072.
96. Grosveld, F. (1999) Curr. Opin. Genet. Dev. 9, 152–157.
97. Peterson, K.R., Clegg, C.H., Navas, P.A., Norton, E.J., Kimbrough, T.G., and Stamatoyannopoulos, G. (1996) Proc. Natl. Acad. Sci. USA 93, 6605–6609.
98. Wijgerde, M., Grosveld, F., and Fraser, P. (1995) Nature 377, 209–213.
99. Navas, P.A., Peterson, K.R., Li, Q., Skarpidi, E., Rohde, A., Shaw, S.E., Clegg, C.H., Asano, H., and Stamatoyannopoulos, G. (1998) Mol. Cell. Biol. 18, 4188–4196.
100. Bungert, J., Dave, U., Lim, K.C., Lieuw, K.H., Shavit, J.A., Liu, Q., and Engel, J.D. (1995) Genes Dev. 9, 3083–3096.
101. Blackwood, E.M. and Kadonaga, J.T. (1998) Science 281, 61–63.
102. Tuan, D., Kong, S., and Hu, K. (1992) Proc. Natl. Acad. Sci. USA 89, 11219–11223.
103. Bulger, M. and Groudine, M. (1999) Genes Dev. 13, 2465–2477.
104. Li, Q., Peterson, K.R., Fang, X. and Stamatoyannopoulos, G. (2002) *Blood*, 100, 3077–3086.
105. Bonifer, C., Vidal, M., Grosveld, F., and Sippel, A.E. (1990) EMBO J. 9, 2843–2848.
106. Aronow, B.J., Silbiger, R.N., Dusing, M.R., Stock, J.L., Yager, K.L., Potter, S.S., Hutton, J.J., and Wiginton, D.A. (1992) Mol. Cell. Biol. 12, 4170–4185.
107. Kushida, M.M., Dey, A., Zhang, X.L., Campbell, J., Heeney, M., Carlyle, J., Ganguly, S., Ozato, K., Vasavada, H., Chamberlain, J.W. (1997) J. Immunol. 159, 4913–4929.
108. Festenstein, R., Tolaini, M., Corbella, P., Mamalaki, C., Parrington, J., Fox, M., Miliou, A., Jones, M., and Kioussis, D. (1996) Science 271, 1123–1125.
109. Greaves, D.R., Wilson, F.D., Lang, G., and Kioussis, D. (1989) Cell 56, 979–986.
110. Elefant, F., Cooke, N.E., and Liebhaber, S.A. (2000) J. Biol. Chem. 275, 13827–13834.

S.C. Makrides (Ed.) *Gene Transfer and Expression in Mammalian Cells*
© 2003 Elsevier Science B.V. All rights reserved

CHAPT

Protein synthesis, folding, modification, a. secretion in mammalian cel.

Stacey M. Arnold[2] and Randal J. Kaufman[1,2]

[1]*Howard Hughes Medical Institute, and* [2]*Department of Biological Chemistry, University of Michigan Medical School, 4570 MSRB II, 1150 W. Medical Center Dr., Ann Arbor, MI 48109-0650, USA; Tel.: +1-734-763-9037; Fax: 1-734-763-9323; E-mail: kaufmanr@umich.edu*

1. Introduction

Protein synthesis in eukaryotic cells is a complex process dependent upon numerous mechanisms to ensure the successful production and targeting of proteins. Most protein synthesis begins in the cytoplasm, with the exception of a small number of mitochondrial-encoded proteins that are synthesized within the mitochondria. Proteins destined for the mitochondria, peroxisomes, and nucleus are fully translated in the cytosol and delivered post-translationally to their final destinations. Translation of resident proteins of the endoplasmic reticulum (ER), Golgi apparatus, and lysosome, as well as of proteins secreted into the extracellular environment, begins in the cytosol and is completed on the membrane of the ER.

Protein translocation into and transport through the secretory pathway in eukaryotic cells is accompanied by a multitude of covalent modifications to the polypeptide backbone. Processing events are carefully regulated both temporally and spatially, working in concert to promote productive folding, correct localization, and appropriate activation of synthesized proteins. Post-translational modifications can affect the half-life of a protein in the extracellular environment and may be required for recognition by receptors, cofactors, and/or substrates. Most of the pivotal processing events are localized to the ER, including signal peptide cleavage, asparagine-linked oligosaccharide addition and modification, and post-translational modification of specific amino acid residues. These modifications promote the correct formation of final tertiary or quaternary structures before transit of proteins through the remainder of the secretory pathway. The machinery that directs post-translational modifications recognizes specific structural and/or sequence determinants within the polypeptide backbone. The efficiency of these reactions is determined by the host cell enzymatic repertoire, the availability of cofactors, and the character of the polypeptide to be modified. Investigators have employed numerous techniques to evaluate the role of post-translational modification in protein function. These include treatment with chemicals or enzymes to remove modifications, analysis of proteins synthesized in the presence of inhibitors of specific modification reactions, expression of proteins in either different cell types or in cell mutants with defects in specific modification machinery, or expression of recombinant proteins containing sequence mutations that prevent modifications. This chapter will provide a basic overview of the mechanisms involved in the biosynthesis,

...ranslational modification, and transport of proteins through the secretory
...way.

2. Modification of proteins in the early secretory pathway

2.1. Translocation

In mammalian cells, translocation of proteins into the ER for entrance into the secretory pathway most often occurs cotranslationally, while in yeast, translocation into the ER can occur co- or post-translationally [1]. For transport across the ER membrane, most proteins contain, within their first 30–60 amino acids, a signal sequence. Signal sequences have the following properties: an amino terminal segment with net positive charge, a central region of 6–15 hydrophobic residues, and a subsequent region containing a helix-breaking amino acid such as proline, glycine, or serine [2]. This 'signal peptide' is recognized by the signal recognition particle (SRP), a complex composed of six heterogeneous polypeptides and one 7S RNA molecule [3,4]. The 54-kDa polypeptide of SRP, SRP54, is the component that binds the signal peptide [5,6]. When it does, elongation is halted [1] and the complex is directed to the ER membrane SRP receptor (SR), composed of an α and a β subunit [7,8]. SRP54, SRα, and SRβ are GTPases [9], and they utilize the energy from GTP hydrolysis to promote docking of the ribosome to the translocon [10], transfer of the signal peptide to the translocon [11], and release of SRP from SR [12]. Once SRP is released, translation resumes and the growing peptide is translocated into the ER. See Fig. 1 for an overview.

The mammalian ER translocon is composed of an aqueous pore formed by the Sec61p complex (Sec61α, β, and γ) [13], which in most cases recruits TRAM (translocating chain-associating membrane protein) to aid in the translocation of proteins [14]. The soluble, ER-resident, peptide-dependent ATPase BiP, a member of the hsp70 family, is also associated with the translocon. It seals the lumenal side of the aqueous pore while the cytosolic opening is exposed to the cytoplasm [15]. It has also been suggested that BiP might work as a molecular motor to pull the elongating polypeptide into the ER lumen [16,17].

2.2. Asparagine-linked glycosylation

Most of the proteins that pass through the ER membrane contain consensus sites (Asn-X-Ser/Thr, where X is any amino acid except proline [18]) for asparagine-linked glycosylation (Fig. 2A), which involves the oligosaccharyltransferase complex-mediated transfer of the structure $Glc_3Man_9GlcNAc_2$ from a dolichylpyrophosphate precursor. This oligosaccharide is well conserved through evolution, as evidenced by its role as the core N-linked structure in mammalian, plant, and fungal cells [19]. Transfer occurs when the acceptor Asn is 12–14 amino acids from the lumenal side of the ER membrane [20]. The likelihood that a consensus site will become N-glycosylated depends, in part, on two factors: (1) the identities of, and

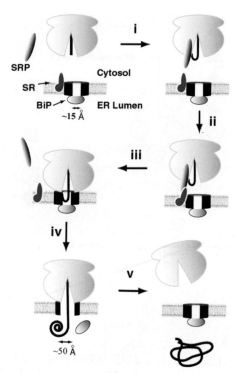

Fig. 1. Protein translocation into the ER. Chain elongation proceeds until the signal peptide is accessible to the signal recognition particle (SRP). Upon signal peptide binding to SRP, translation arrests and the translation complex (i) is targeted to the SRP receptor (SR, ii). SRP interaction with SR promotes ribosome binding to the Sec61 translocon concomitant with release of the ribosome from the SRP/SR complex (iii). Translation elongation continues and the growing polypeptide chain is translocated into the ER lumen. At the same time, BiP is released from the lumenal side of the translocon to permit the growing polypeptide chain to enter the ER lumen (iv). Completion of translation is concomitant with BiP binding to the Sec61 channel (blocks leakage between lumen and cytoplasm) and dissociation of the ribosome (v). Figure reproduced with permission [148].

possible steric hindrances induced by, amino acids flanking the consensus site [18]; and (2) the rate at which the region containing the consensus site begins to fold. The faster a peptide attains its secondary structure, the less likely its consensus N-glycosylation sequences will be exposed for glycosylation [21]. Interference with N-glycosylation, e.g. treatment of cells with tunicamycin, a drug that prevents formation of the dolichol-linked core structure, results in the accumulation of unfolded proteins in the ER. Global glycosylation is therefore vital to the folding process; the hydrophilic nature of glycans promotes protein solubility and affects the conformation of residues adjacent to linked asparagines, resulting in the formation of more compact structures [22].

Immediately after the oligosaccharide linkage is formed, the terminal glucose and mannose residues are removed by glucosidase and mannosidase activities [23] (Fig. 2A). Glucosidase I (GI), an $\alpha(1,2)$ glucosidase that resides in the ER membrane, removes the terminal glucose to generate $Glc_2Man_9GlcNAc_2$ [19].

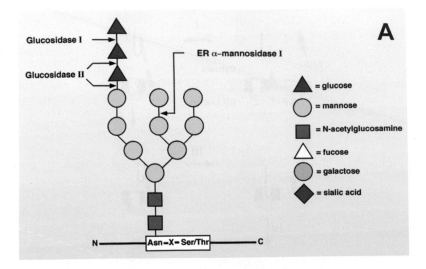

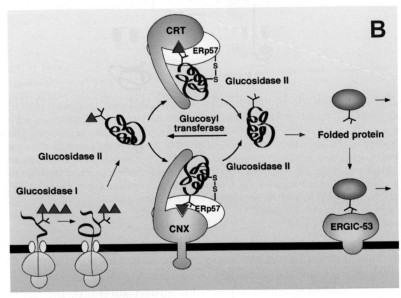

Fig. 2. The role of N-linked glycosylation in protein folding. Panel A, The structure of the N-linked oligosaccharide core. Glucosidase and mannosidase activities remove sugar residues where indicated. Removal of the terminal mannose residue from the middle branch generates the Man8B isomer. Panel B, The calnexin (CNX)/calreticulin (CRT) cycle. Trimming of glucose residues from the N-linked oligosaccharide core attached to newly synthesized glycoproteins leads to the exposure of a monoglucosylated structure that binds to the ER chaperones CNX and CRT. Both CNX and CRT interact with ERp57, a protein disulphide isomerase that promotes protein folding through rearrangement of disulphide bonds. Release of the folded glycoprotein is concomitant with removal of the final glucose by glucosidase II so the glycoprotein can exit the ER (in some cases via a transporter like the mannose lectin ERGIC53). Upon release of unfolded glycoproteins from CNX or CRT, the glucose is also trimmed, but the unfolded glycoprotein is recognized by UDP-glc:glycoprotein glucosyltransferase to reestablish the monoglucosylated structure that promotes reentry into the folding cycle by interaction with CNX or CRT. Figures reproduced with permission [22].

Glucosidase II (GII), a soluble $\alpha(1,3)$ glucosidase composed of two subunits, one catalytic and the other conferring ER retrieval to the heterodimer [24], removes the remaining two glucose residues. ER mannosidase I is an $\alpha(1,2)$-mannosidase that removes a single mannose from the middle, $\alpha(1,3)$-linked branch of the N-linked structure to generate $Glc_{0-3}Man_8GlcNAc_2$, the Man8B isomer [25]. ER mannosidase II is also an $\alpha(1,2)$-mannosidase, but it removes a single mannose from the outer, $\alpha(1,6)$-linked branch of the N-linked structure to produce the Man8C isomer [25]. Neither of these mannosidase activities disrupts the mannose linkage from which the glucose residues stem, and remaining mannosidase activities are localized to the Golgi complex. As will be discussed below, the presence of N-linked glycans and the characteristics of their trimmed forms dictate the manner in which proteins interact with the protein folding machinery of the ER.

2.3. ER chaperones

The ER has a relatively low proportion of folding substrates compared to its high concentration of proteins responsible for aiding in the folding process [26]. Among this arsenal are proteins that take advantage of the ER's oxidizing environment to introduce disulphide bonds: protein disulphide isomerase (PDI, or ERp59), and its family members ERp72 and ERp57 [27,28]. PDI homologues display thiol oxidoreductase and isomerase activities. Also instrumental in the ER milieu are chaperones responsible for facilitating folding by sequestering unfolded substrates to prevent their aggregation. This group includes BiP (GRP78), GRP94, GRP170, and the lectin-like homologues calnexin (CNX) and calreticulin (CRT) [29–33]. GRP78 and GRP94 earned their designation as **G**lucose-**R**egulated **P**roteins when their induction was detected in sarcoma virus-transformed cells as a consequence of glucose deprivation [34]. This nutrient stress was later conjectured to elicit its effect on the production of ER chaperones by promoting the accumulation of unfolded proteins [35]. This theory stemmed in part from the work of Haas and Bole, in which GRP78 was found to bind nonsecreted immunoglobulin (Ig) heavy chains, a means of preventing transport of incompletely assembled Ig intermediates from the ER [29,36]. In proof of the theory, over-expression of a folding-deficient variant of influenza hemagglutinin (HA) lead to induction of BiP and GRP94 [35].

CNX and CRT are unique in that they mediate their chaperone functions through specific interactions with monoglucosylated, N-linked structures of the form $Glc_1Man_{6-9}GlcNAc_2$ [37–39] (Fig. 2B). In addition, both facilitate protein folding through association with ERp57 [40]. Interestingly, ERp57 appears to require this association for its function, as it accelerated *in vitro* refolding of monoglucosylated RnaseB only in the presence of CNX or CRT [41], and it was prevented from forming mixed disulphides with substrate glycoproteins *in vivo* when their association with CNX or CRT was prohibited by exposure to castanospermine (a GI/GII inhibitor) [42]. Mammalian CNX is a type I transmembrane protein for which the lumenal domain structure has recently been solved [43]. Extending from a globular core is a finger-like projection comprised of the CNX P-domain, a region known to contain two contiguous motifs each repeated in tandem four times [44].

The globular portion is proposed to contain the lectin-binding determinant, while the P-domain is conjectured to loop around unfolded sections of substrates to provide a protective environment. It is thought that the P-domain might additionally bind ERp57, bringing it into the proximity of the folding glycoprotein [45]. CRT is homologous to CNX, though slightly smaller and soluble. Its P-domain structure was recently determined [46], and, though comprised of three rather than four tandem repeats of motifs 1 and 2 [44], displays a flexibility consistent with the role proposed for the analogous portion of CNX [45].

Elongating chains containing N-linked oligosaccharides within the first ~ 50 amino acids of the N-terminus interact preferentially with CNX and CRT; interaction with BiP is delayed until later in the folding process. Conversely, chains containing N-linked glycans more remote from the N-terminus interact first with BiP, and association with CNX and CRT is post-translational [47]. Folding peptides are further differentiated by the fact that CNX and CRT interact with only a partially overlapping set of substrates [48]; furthermore, when CNX and CRT bind the same protein, their glycan affinities are often distinct. An example of the latter is the processing of HA; found in ternary complexes with CNX and CRT during processing, its rapidly folding top-hinge domain appears to be the preferential target of CRT-mediated glycan association, while its stem domain, a region closer to the ER membrane, is the target of choice for CNX-mediated glycan association [49]. This disparity is thought to be attributable to the fixed vs. soluble localizations of CNX and CRT, respectively, and is supported by the fact that modification of CNX to soluble and of CRT to membrane-bound form results in an essential reversal of roles [50,51]. Folding transmembrane proteins might interact not only with ER-resident chaperones but with cytosolic chaperones as well. For example, the cystic fibrosis transmembrane conductance regulator, CFTR, binds Hsp90 and Hdj-2/Hsc70 on the cytosolic side of the ER membrane [52].

2.4. The CNX/CRT cycle

CNX and CRT further distinguish themselves from other ER chaperones by their participation in a cycle designed to impart quality control (Fig. 2B). Association of glycoproteins with CNX or CRT is eventually terminated by removal of the final glucose residue by GII [53]. The next step for a glycoprotein depends upon its folding status. If it has become properly folded, the protein can leave the ER for further processing in the Golgi, or, if ER-resident, take up its role as a member of this organelle. However, if the protein has not attained its native fold, it is recognized by UDP-Glc:Glycoprotein Glucosyltransferase (UGT), an enzyme that transfers a single glucose residue from UDP-Glc to $Man_{7-9}GlcNAc_2$ structures, to reform the monoglucosylated glycan and promote another round of interaction with the CNX/ERp57 or CRT/ERp57 machinery. Evidence suggests that UGT specifically recognizes unfolded, and not native, substrates by recognition of two determinants: hydrophobicity associated with tertiary structures characteristic of late stages of folding [54,55], and the innermost GlcNAc unit of the N-linked oligosaccharide [56]. The innermost GlcNAc residue should serve as an indicator of folding status because it is expected to be accessible to recognition by UGT (i.e., separated from interactions with neighboring amino acid residues) only when the amino acids in its vicinity are in an unfolded conformation. This theory is predicated by the fact that in many native proteins the innermost GlcNAc is not accessible to agents such as Endo H, an N-acetylglucosaminidase that cleaves the bond joining the two GlcNAc residues of the N-linked core, unless the protein is first denatured [57]. The two recognition components must be covalently linked, i.e., present on the same molecule [56]. In addition, UGT requires Ca^{++} ions in millimolar concentrations for its function [58].

2.5. ER-resident proteins

Glycoproteins cycle through this interaction with CNX/ERp57 or CRT/ERp57, mediated by GII and UGT activities, until their proper folds are obtained, or, barring successful completion of folding, they are removed from the secretory pathway by ER-associated degradation (described below). Soluble ER-resident proteins contain a retrieval mechanism at their C-termini known as the KDEL signal (its name describes the consensus sequence); interaction with the KDEL receptor in the compartment intermediate between the ER and Golgi (ERGIC) or in the cis-Golgi brings these proteins back to their proper location through incorporation into COPI coated vesicles (described below). Type I (carboxy terminus in cytosol) transmembrane ER-resident proteins have a C-terminal signal comprised of the consensus sequence KKXX (where X is any amino acid) [59], that mediates an interaction with the COPI component of secretory vesicles for retrieval from the ERGIC and Golgi to the ER [60–62]. ER retention mechanisms might exist as well [63,64], perhaps involving a dynamic network of weak interactions stabilized by the high ER calcium concentration [65,66].

2.6. Gamma-carboxylation of glutamic acid residues

The vitamin K-dependent coagulation factors contain the post-translationally modified amino acid γ-carboxy-glutamic acid (Gla). The Gla residues are essential for these proteins to attain a calcium-dependent conformation and for their ability to bind phospholipid surfaces, an essential interaction for their function [67]. The precursor of proteins subject to γ-carboxylation contains a propeptide that directs γ-carboxylation of glutamic acid residues immediately downstream of the propeptide [68]. The propeptides of these factors share amino acid similarity by conservation of the γ-carboxylase recognition site and the site for cleavage of the propeptide. NMR structural analysis identified that the propeptide is an amphipathic α-helix with the carboxylase recognition site N-terminal to the helix [69].

2.7. Beta-hydroxylation of amino acids: lysine, proline, aspartic acid, and asparagine

Proline and lysine residues are hydroxylated in procollagen by prolyl-3-hydroxylase and prolyl-4-hydroxylase. This modification increases the stability of the collagen triple helix. Prolyl-4-hydroxylase acts on prolines in the context of G-X-P, where X is any amino acid. This reaction requires molecular oxygen, ferrous ion, ascorbic acid, and α-ketoglutarate. Vitamin C deficiency results in scurvy because of reduced proline hydroxylation to form a stable collagen triple helix.

Post-translational β-hydroxylation of aspartic acid and asparagine occurs within the ER. Hydroxylation of both aspartic acid and asparagine is catalyzed by aspartyl β-hydroxylase, which requires 2-ketoglutarate and Fe^{+2} [70,71]. The β-hydroxylase was molecularly cloned and characterized [72,73]. A consensus β-hydroxylation site within EGF domains (Cys-X-Asp/Asn-X-X-X-X-Phe/Tyr-X-Cys-X-Cys) was proposed [70,71,74], and is inhibited by agents that inhibit 2-ketoglutarate-dependent dioxygenases [75].

3. Golgi modifications

3.1. Golgi glycosylation events

Proteins that transit to the Golgi complex receive further modifications of their N-linked oligosaccharides. Golgi-resident carbohydrate-modifying enzymes share the following features in common: a short cytoplasmic N-terminal domain, a membrane-spanning α helix, and a large C-terminal portion that contains the catalytic domain and faces the Golgi lumen. Retention of these Golgi-resident proteins seems to be solely dependent upon the presence of the transmembrane α helix. Though the mechanism for this retention is not fully understood, the possibility exists that adjacent Golgi residents bind via their helical transmembrane domains, forming complexes that may in turn associate with cytoskeletal proteins like actin or tubulin and preventing incorporation into transport vesicles. Upon transit through the Golgi apparatus a series of additional carbohydrate modifications occur that are separated spatially and temporally and involve the removal of mannose residues by Golgi mannosidases I and II and the addition of N-acetylglucosamine, fucose, galactose, and sialic acid residues. These reactions occur by specific glycosyltransferases that modify the high mannose carbohydrate to complex forms.

Also within the Golgi apparatus, O-linked oligosaccharides are attached to the hydroxyl group of serine or threonine residues through an O-glycosidic bond to N-acetylgalactosamine. Serine and threonine residues subject to glycosylation are frequently clustered together and contain an increased frequency of proline residues in the region, especially at positions -1 and $+3$, relative to the glycosylated residue [76]. Galactose, fucose, and sialic acid are frequently attached to the serine/threonine-linked N-acetylgalactosamine. O-Glycosylation occurs in the Golgi complex concomitant with complex processing of N-linked oligosaccharides.

3.2. Tyrosine sulfation

Sulfate addition to tyrosine as an O4-sulfate ester is a common post-translational modification of secretory proteins that occurs in the *trans*-Golgi apparatus [77] and is mediated by tyrosylprotein sulfotransferases that utilize the activated sulfate donor 3'-phosphoadenosine 5'-phosphosulfate (PAPS) [78,79]. Two closely related and ubiquitously expressed Golgi isoenzymes, tyrosylprotein sulfotransferases −1 and −2, have been characterized and molecularly cloned [80,81]. Gene deletion studies in the mouse indicate that these two isozymes have distinct biological roles that may reflect differences in the substrate specificity [82]. Tyrosine sulfation occurs on many secretory and transmembrane proteins that play important roles in inflammation, hemostasis, immunity, and growth. The proteins that contain this modification include hormones, P-selectin glycoprotein ligand-1 [83], platelet glycoprotein Ibα, G-protein coupled receptors, extracellular matrix proteins, and a number of proteins that interact with thrombin, such as hirudin [84], coagulation factors VIII and V [85–87], fibrinogen [88], heparin cofactor II [89], α2-antiplasmin [90], vitronectin [91], and factor X [92]. Comparison of all known tyrosine sulfation sites yielded a consensus sequence that has three aspartic acid and/or glutamic acid residues near the sulfated tyrosine (± 5 amino acids), a turn-inducing residue within $+7/-7$, and the absence of cysteine or N-linked glycosylation sites within seven residues of the sulfated tyrosine [77]. Although tyrosine sulfation is found on many proteins that transit the secretory apparatus, there are only a few examples where this modification is required for secretion or functional activity of the molecule. For example, tyrosine sulfation in the hormone cholecystokinin is required for its biological activity [93]. However, tyrosine sulfation can modulate the biological activity, binding affinities, and secretion of specific proteins [77]. For example, tyrosine sulfation at the carboxy terminus of hirudin increases its binding affinity to the anion-binding exosite of thrombin [94,95].

3.3. Proteolytic processing

Processing of the proproteins of zymogens, peptide hormones, complement proteins, and clotting factors, to name a few, is essential for their activation and can take place either pre- or post-secretion (for review, see Refs. [96,97]). Cleavage sites are typically composed of single (Arg) or paired (Arg-Arg, Lys-Arg, Arg-Lys, Lys-Lys) basic amino acids. The major processing proteases, the subtilisin-like proproteins convertases (SPCs), prefer to cleave after Arg residues, though cleavage after Lys residues is possible. After cleavage, the N-terminal fragment is further processed by removal of the basic residues by carboxypeptidases. Carboxypeptidase E fulfills this role for components of regulated secretion within the endocrine system, while Carboxypeptidase D processes N-terminal fragments in the constitutive secretory pathway [98]. These activities might be necessary to prevent product inhibition of the SPCs [99]. One further modification is possible for N-terminal fragments exposing a glycine residue following carboxypeptidase-mediated trimming. Often, this glycine is amidated by peptidyl-glycine-α-amidating mono-oxygenase, a modification that might increase the stability of the N-terminal peptide [96].

SPCs are calcium-dependent serine proteases that specialize in cleavage after the motif (Arg/Lys)-$(X)_n$-Arg, where $n = 0, 2, 4$, or 6 and X is any amino acid except Cys, and, with few exceptions, Pro [97]. While cleavage of proproteins following non-basic amino acids has been documented, only one protease has been discovered so far that is capable of this type of cleavage. SKI-1 appears to cleave after the motif (Arg/Lys)-X-(Hydrophobic)-(Leu, Tyr) [100].

There are numerous known SPCs, often capable of redundant function but specialized by restriction to specific tissue types. No two SPCs have an identical pattern of tissue expression. Mouse knockout models for individual SPCs often demonstrate the existence of partially compensatory mechanisms [96].

SPCs share structural and functional characteristics in their N-termini. The N-terminus is composed of a signal peptide, pro-segment, catalytic domain, and P domain. The pro-segment is 80–90 amino acids in length and acts both as an intramolecular chaperone to promote proper folding and as a competitive inhibitor of SPC function. By these means, it prevents the convertase from becoming activated during its synthesis. The pro-segment is cleaved by an autocatalytic mechanism in the ER (with the exception of that of SPC2, discussed below), an event that proceeds correct sorting of the enzymes upon exit from this organelle. Cleavage is after the motif Arg-X-Arg/Lys-Arg, and the released pro-segment most likely binds the active site to exert its inhibitory function. It remains tightly associated until the convertase reaches its final destination [96]. Cleavage after a second set of basic amino acid residues in the pro-segment releases it from the catalytic domain for activation of the SPC [101]. The catalytic domain contains the catalytic triad characteristic of serine proteases, comprised of Asp, His, and Ser residues, and its overall structure determines substrate specificity. Negatively charged residues in the catalytic domain interact in a binding pocket with positively charged residues of the substrate. The P domain makes strong hydrophobic contacts with the catalytic domain and is therefore integral in promoting correct folding of the SPC [96]. It also mediates the calcium dependence of the SPC [102].

It is in the C-termini that SPCs differ by presenting a variety of domain structures, including cysteine-rich regions (SPC1, SPC4, and SPC6), transmembrane (SPC1 and SPC7) and cytosolic domains (SPC1) important for cell sorting, and amphipathic helices (SPC2 and SPC3). In addition, SPC7 has a Ser/Thr-rich region of unknown function [96].

There are seven well-characterized members of the SPC family. SPC1, or furin, is considered a 'housekeeping' SPC since it processes constitutively secreted proteins in all cell types [96]. It is essential for normal development of the mouse embryo [103]. SPC7 also operates in the constitutive secretory pathway [97]. SPC2 and SPC3 process proteins destined for regulated secretion. SPC2 cleavage is extensive, resulting in the generation of low molecular weight species, while SPC3 cleavage is more limited and results in the production of higher molecular weight proteins. These convertases are largely restricted to the neuroendocrine system [96]. SPC2, as mentioned above, is not cleaved of its pro-segment in the ER. This processing event is restricted to the *trans* Golgi network [97]. Removal of the pro-segment is followed by truncation of the C-terminal domain and glycosylation [104]. SPC2 is additionally unique in that it is the only SPC that requires a binding protein, 7B2, to improve the efficiency of pro-segment cleavage. 7B2 coexpression ensures the proper folding and sorting of SPC2 [97]. The N-terminal domain of 7B2 has a chaperone-like function, while its C-terminus is a specific SPC2 inhibitor [96]. An SPC1-like activity mediates the dissociation of 7B2 from SPC2 [105]. SPC4 and SPC6 appear to process components of both the constitutive and regulated secretory pathways [97]. SPC5 is predominantly localized to testicular germ cells. While no substrates have been identified for SPC5 [96], knockout results indicate that it is essential for successful male fertility and for early embryonic development [97]. SPC2, SPC3, and SPC6 appear to be exclusively localized to neuronal cells, while SPC1, SPC4, and SPC7 are present both in neuronal and glial cells [96].

4. Quality control

In the secretory pathway, there are four strategies proposed to control the quality of the proteins synthesized (Fig. 3): (1) ER retention; (2) retrieval of proteins from downstream organelles back to the ER; (3) rerouting of proteins from the Golgi complex to the lysosomes or to endosomes; and (4) ERAD [106].

4.1. ER retention

ER retention primarily involves the recognition and sequestration of nonnative proteins exposing hydrophobic patches and structural mobility indicative of a lack of folded compactness. Classical chaperones, like the GRPs, recognize these characteristics directly, while the nonclassical chaperones, CNX and CRT, require UGT to detect and flag unfolded substrates for their interaction. PDIs also get into the retention game; for example, immunoglobulin subunits that have not yet oligomerized on the basis of interchain disulphide linkages are not only detained by interactions with BiP, as described earlier, but by formation of intermolecular disulphide bonds with the ER-resident thiol oxidoreductases [107]. Unfolded proteins can even associate with one another to form large ER aggregates; their size makes them transport-incompetent, so escape from the ER is hindered [106]. For proteins such as thyroglobulin, MHC Class II, and procollagen, transient aggregate formation appears to be part of the folding process [108–110].

ER retention is vital, however, in those cases where mutations prevent the protein from obtaining a functional native conformation. Mutations often leading to ER retention are ones affecting the following processes: disulphide bond formation (cysteine residues), signal peptide cleavage, GPI anchor addition, glycosylation, ligand binding, and metal ion coordination [106]. In certain cases, however, sequence changes with no effect on protein function can have a minor effect on protein folding that leads to ER retention, and accumulation of the variant causes the development of an ER storage disease like α1-antitrypsin deficiency [111].

4.2. Retrieval of proteins from downstream organelles

Examples of the second means of quality control, retrieval of proteins from downstream organelles, occur for folding transmembrane proteins. They, in addition

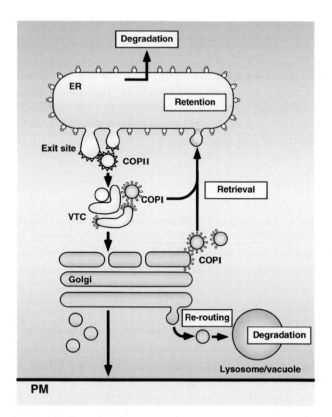

Fig. 3. Mechanisms for quality control in the ER. Numerous mechanisms exist to ensure the quality of proteins that transit the secretory pathway. ER retention is mediated by the CNX/CRT cycle, specific association of unfolded proteins with BiP and PDI family members, and by formation of large aggregates. ER retrieval involves masking and unmasking of protein sequence signals to ensure the correct assembly of oligomeric complexes or the correct folding of individual transmembrane proteins. These signals often work in concert with the cargo carriers of the COPI- and COPII-coated vesicles that mediate retrograde transport from the Golgi to the ER and anterograde transport from the ER to the Golgi, respectively. Targeting of proteins to the lysosome occurs in the Golgi. Finally, ERAD promotes retrograde translocation of terminally misfolded proteins from the ER to the cytosol where they are degraded by the proteasome. Figure reproduced with permission [106].

to a final means of ER retention, involve masking or unmasking of signals to ensure subunit assembly or to prevent misfolded proteins from reaching the cell surface. These factors can be contained in transmembrane domains: the immunoglobulin μ heavy chain contains two hydrophilic residues in its membrane-spanning region that prevent transport until assembly with light chain has occurred [112]. Other signals reside within the cytoplasmic domain. The α subunit of the human high-affinity receptor for IgE contains the retrieval signal KKXX, which is masked once oligomerization with the γ subunit has occurred [113]. Subunits of the ATP-sensitive K^+ channel appear to contain a different retrieval signal, RKR, which prevents transport of incompletely assembled channels to the cell surface [114]. This motif can also be found in CFTR and is responsible for preventing a misfolded mutant, CFTR ΔF508, from transiting to the cell surface [115].

4.3. Rerouting from Golgi to the endosome

The third category of quality control, rerouting of proteins from the Golgi to the endosome, is more a yeast than a mammalian phenomenon. *S. cerevisiae*, for example, lacks homologues of GRP94 and CRT and appears to be missing a CNX cycle [116]. Thus, it lacks the stringency of ER quality control seen in mammals, and must rely upon a backup mechanism to rid itself of undesirable components. A transmembrane protein, Vps10p, has been implicated as the shuttle responsible for rerouting, and it transports not only unfolded proteins but also native vacuolar proteins like proteinase A and carboxypeptidase Y (CPY) [117–120].

4.4. ER-Associated degradation

The final means of quality control, ERAD, is a mechanism for which great strides in identifying recognition components have recently been made. It is an important part of cell processing under normal conditions, and it attains even greater significance under conditions of cell stress, when processing of folded proteins is greatly impaired and the accumulation of incomplete products must be attenuated. ERAD appears to involve retrograde transport of unfolded or misfolded proteins through the Sec61 complex out into the cytosol [121], where they are deglycosylated by an N-glycanase activity [122,123], ubiquitinated, and degraded by the 26S proteasome [124]. In addition, the cell must have a sensor(s) capable of identifying proteins that cannot be folded and a means of delivering these proteins to the translocon.

Initial indications as to the character of the sensing apparatus came from the finding that degradation of overexpressed prepro-α factor is arrested in transfected rat epithelial cells in response to inhibition of $\alpha(1,2)$ mannosidase activity [125]. Support for the role of an oligosaccharide as marker came from work demonstrating that N-linked glycosylation, more precisely, a $Man_8GlcNAc_2$ glycan of the Man8B form, is necessary for degradation of a mutant CPY in yeast [126,127]. Proteasome inhibition in mammalian cells expressing a misfolded variant of α1-antitrypsin containing $Glc_1Man_8GlcNAc_2$ glycans leads to ER retention and a stable association with CNX [128]. Thus, ER mannosidase I appears to be playing a role in ERAD, perhaps in conjunction with CNX. However, since proteins capable of folding correctly are also subject to trimming by ER mannosidase I, a specific role for ER mannosidase I in ERAD would necessitate the acquisition, by nonproblematic proteins, of their native forms before their N-linked glycans are cleaved to the Man8B isoform. This timing contingent might involve a property of ER mannosidase I itself: the yeast homologue contains a deep substrate-binding pocket that presents a structural hindrance to access and therefore greatly slows the catalytic rate [129,130].

Three models have been proposed to describe how modification of N-linked glycans might promote recruitment of misfolded proteins to ERAD [131]. The first is designed to explain the retention of α1-antitrypsin by CNX in the presence of proteasome inhibitors. After numerous cycles of interaction with CNX/ERp57, a misfolded protein would finally be cleaved by ER mannosidase I. It would re-associate with CNX, only to become stuck because its new structure, $Glc_1Man_8GlcNAc_2$, is a very poor substrate for glucosidase II [132]. The malformed glycoprotein would thus be captured and shuttled off to degradation. The second model is based upon the finding that CRT and Golgi endomannosidase (cleaves the glucose-linked mannose from the outer α(1,3)-linked branch of the N-linked structure) have been shown to co-purify [38]. CRT would play a role analogous to that of CNX in the first model, in that its dissociation from substrate would be impeded. However, in a processing event not described for the CNX scenario, endomannosidase would cleave the N-linked glycan to produce Glc-Man and promote release from CRT for degradation. The third model seeks to dismiss the role of CNX or CRT for cases in which interactions with other chaperone systems predominate. In this scenario, a Man8B-binding lectin would shuttle unfolded proteins directly to degradation. The identification of an inactive ER mannosidase I homologue in yeast (Mnl1p/Htm1p) [133,134] and mammals (EDEM) [135] (see Fig. 4) has buoyed this third hypothesis: overexpression of Mnl1p or EDEM leads to increased degradation of misfolded glycoproteins, and deletion of Htm1p from yeast leads to the stabilization of glycosylated, but not non-glycosylated, proteins. The first two models share in common a requirement for UGT as sensor of nonnative structure. The third model could still involve a role for UGT in determining folding status, leading to the specific recognition by Mnl1p/Htm1p/EDEM of $Glc_1Man_8GlcNAc_2$ glycans. Alternatively, an unidentified protein could play the UGT-like role, or Mnl1p/Htm1p/EDEM could itself recognize folding status.

The existence of ERAD mechanisms distinct from the above scenarios is suggested by recent evidence. Specifically, the PIZ variant of α1-antitrypsin is degraded in mouse hepatoma cells in a non-proteasome-dependent manner [136], even though its degradation in extrahepatic cell lines is proteasome-dependent [137,138]. This differential sorting may be attributable to the fact that in hepatoma cells, most of the PIZ variant produced undergoes a process of loop sheet polymerization [139], an extended conformation that leaves its oligosaccharides more accessible. These glycans are therefore subject to cleavage by ER mannosidase II [136], in addition to cleavage by ER mannosidase I, resulting in the generation of $Man_7GlcNAc_2$ isomers that are poor substrates for UGT [140]. Removal of PIZ variant is therefore independent of CNX association. The likelihood, overall, is that ER-associated degradation is a program of action that will differ among species and cell types, depending upon the ER components available to conduct its operation.

5. Transport of proteins through the secretory pathway

COPII-coated vesicles mediate transport of proteins from the ER to the Golgi apparatus (Fig. 5). Only properly folded proteins can be packaged [141], though

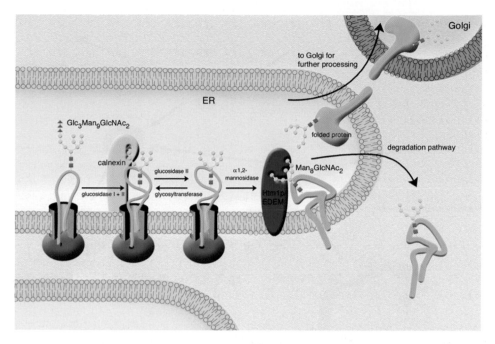

Fig. 4. Targeting of misfolded and/or unfolded proteins for proteasomal degradation. N-linked core oligosaccharides are added to newly synthesized glycoproteins and glucose residues are immediately removed by glucosidase activities. Removal of the terminal mannose residue of the middle branch by ER mannosidase I (α1,2-mannosidase) leads to one of two fates. If the glycoprotein is folded correctly, it transits to the Golgi for further processing. If the glycoprotein remains unfolded after several rounds of interaction with CNX/CRT, the Man8B isomer may act as a signal for interaction with EDEM/Htm1p to target the glycoprotein for retrograde transport through the translocon and cytosolic degradation. Figure reproduced with permission [149].

specific motifs involved in the recognition of these proteins as cargo have yet to be identified [142]. Some cargo adapters, on the other hand, have defined motifs that promote their accumulation in cargo-rich areas. For example, ERGIC-53, an adapter protein that recognizes high mannose oligosaccharides for cargo transport, contains a C-terminal di-phenylalanine signal that promotes its inclusion in the budding vesicle [143]. Exit from the ER takes place at specific sites termed transitional elements [144] wherein cargo selected for transport becomes concentrated and from which ER-resident proteins are excluded.

Formation of COPII-coated vesicles begins with the recruitment, by membrane-bound Sec12p, of Sar1p-GDP to the ER. Sec12p is the guanine exchange factor (GEF) for Sar1p-GDP, and activation of Sar1p initiates coat formation. Specifically, Sar1p-GTP promotes the association of Sec23p:Sec24p with cargo. Sar1p-GTP activates Sec23p to bind SNARE proteins. These, in turn, provide targeting specificity by promoting the fusion of COPII vesicles with acceptor membranes [145]. In addition, SNAREs may provide scaffolding that promotes the polymerization of the COPII coat at ER exit sites [146]. Following recruitment

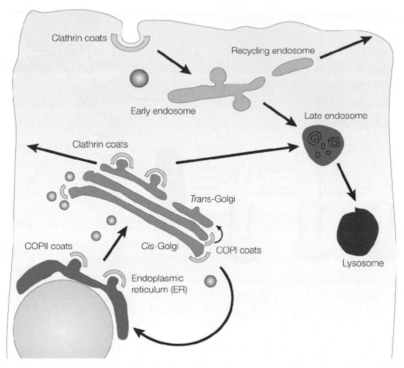

Fig. 5. Protein transport through the secretory pathway. Proteins transiting the secretory pathway are directed from the ER to the Golgi by inclusion into COPII-coated vesicles, while transport through the Golgi relies upon the formation of COPI vesicles. COPI-coated vesicles also mediate the retrograde transport of proteins that reside in the ER. Clathrin-coated vesicles emerge from the *trans*-Golgi network and mediate stimulus-induced regulated secretion. Formation of clathrin-coated vesicles at the plasma membrane is the first step in the endosomal pathway. Late endosomes fuse with clathrin-coated vesicles containing lysosomal hydrolases, promoting their delivery to the lysosome. Figure reproduced with permission [145].

of the Sec23p:Sec24p complex, Sec13p:Sec31p is recruited. Polymerization of the latter on the coat probably drives membrane deformation for the completion of vesicle budding [145]. Incorporation of the Sec13p:Sec31p complex promotes the GTPase activating (GAP) function of Sec23p with respect to Sar1p-GTP; Sar1p-GTP hydrolysis promotes uncoating, a necessary precursor to fusion of the vesicle with the target membrane [143]. The function of a final COPII coat component, Sec16p, is uncertain. This protein interacts with Sec23p:Sec24p, as well as the C-terminus of Sec31p, and is conjectured to act either as an additional scaffold for COPII coat assembly or as a factor to accelerate the GAP activity of Sec23p. COPII-derived vesicles form or fuse with vesicular tubular clusters (VTCs), transitory structures lying between the ER and Golgi that either fuse with or become the *cis*-Golgi [142].

COPI vesicles participate in both anterograde and retrograde protein transport, shuttling proteins through the Golgi and from the Golgi to the ER, respectively (Fig. 5). ARF1GEF recruits ARF1-GDP to the Golgi membrane.

Unlike Sar1p-GTP, which forms no direct association with the ER membrane, ARF1-GTP displays a myristoylation that facilitates membrane binding. In the GDP-bound state, this myristoylation is hidden in a binding pocket: activation by ARF1GEF promotes the conformational change that exposes ARF1's myristoylated tail. In parallel to COPII coat formation, ARF1-GTP activates coat assembly. Assembly of the COPI-coated vesicle involves incorporation of COPI coatamer and recruitment of cargo. COPI is made up of seven proteins, α, β, β', γ, δ, ε, and ζ. The γ subunit recognizes dilysine motifs and therefore appears to direct cargo recognition. p24 family members are also a part of the coatomer makeup, possibly playing a role in coatomer recruitment and vesicle assembly. ARF1GAP stimulates ARF1 GTP hydrolysis and thereby promotes uncoating, which again permits fusion of the vesicle with the target membrane [145].

Clathrin-coated vesicles are involved in transport from the *trans*-Golgi network (TGN, see Fig. 5). As with COPI-coated vesicles, ARF1 is recruited to the Golgi membrane by ARFGEF [145]. AP-1 is an adapter composed of γ, $\beta 1$, $\mu 1$ and $\sigma 1$ adaptins that is recruited to the membrane by ARF1 [147]. It binds cargo via sorting signals including the sequence FDNPVY, tyrosine-based motifs of the sequence YXXØ (where X is any amino acid and Ø is an amino acid with a bulky hydrophobic side chain), and dileucine motifs. AP-1 is responsible for recruiting clathrin triskelions to the forming vesicle. These triskelions polymerize to form a cage, the generation of which leads to membrane deformation and budding. The GTPase dynamin is recruited to the membrane, as is a dynamin-associating protein, endophilin. In the presence of dynamin, endophilin, an acyltransferase, generates lysophosphatidic acid from arachidonic acid or palmitic acid, a reaction that produces a negative curvature at the neck of the vesicle. Hydrolysis of dynamin-GTP then pinches off the vesicle from the Golgi. Hsc70 and auxilin, a J-domain-containing protein, work in concert to promote uncoating in an ATP-dependent event, and the vesicle is ready to fuse with its target membrane [145].

Sorting of lysosomal hydrolases is dependent upon the formation of clathrin-coated vesicles at the TGN and fusion of these vesicles with the endosomal pathway, which has its genesis in the formation of clathrin-coated vesicles at the plasma membrane (Fig. 5). Clathrin-mediated endocytosis uses a separate adapter, AP-2. Lysosomal enzymes are synthesized in the secretory pathway and are exposed to unique glycosylation events, in that certain oligosaccharide moieties are modified in the *cis*-Golgi by the phosphorylation of mannose residues to produce mannose-6-phosphate (M6P) groups. This specific feature allows recognition of the hydrolases by M6P receptors in the TGN, and the hydrolase:receptor pairs are incorporated into clathrin-coated vesicles. Upon fusion of the vesicles with late endosomes, exposure to the acidic environment of the latter promotes dissociation of the hydrolase from its receptor. The receptors are recycled to the TGN, while the enzymes are delivered to lysosomes [147].

Regulated secretion also involves clathrin-coated vesicles, although the role of their involvement is presently unclear. Two models have been postulated. One hypothesizes that the contents of immature secretory granules (ISGs) are selectively aggregated in the TGN by components of the clathrin machinery. The second suggests that the formation of ISGs is relatively non-specific, and that the function of clathrin-coated vesicles is the removal of non-specific cargo before the maturation of ISGs is complete [147].

6. Conclusion

Since the first demonstration that cultured animal cells have the capacity to synthesize, secrete, and produce in a functional form a variety of complex proteins for therapeutic use, a tremendous knowledge base has been generated concerning the steps that limit the production of proteins at high levels in mammalian cells. It is remarkable that most mammalian cells have all the machinery to synthesize, fold, and modify a protein so that it is secreted in a biologically active form. Therefore, the information that directs all these reactions is contained within the amino acid sequence of the polypeptide.

Analysis of the individual rate-limiting steps in protein biosynthesis, modification, folding, assembly, and transport has identified individual enzymes that are rate-limiting in the high level production of biologically active polypeptides. It is now possible to engineer mammalian cells so that they can optimally produce a desired polypeptide in a form that is fully modified for best biological activity. Future studies should further elucidate the specific folding requirements for individual proteins so it will be possible to engineer cells to optimize the production of any particular secreted protein. This area of research has spawned a new era in bioengineering that will see tremendous advances in the coming years.

Abbreviations

BiP	immunoglobulin binding protein
CFTR	cystic fibrosis transmembrane conductance regulator
CNX	calnexin
CPY	carboxypeptidase Y
CRT	calreticulin
ER	endoplasmic reticulum
ERAD	ER-associated degradation
ERGIC	ER-Golgi intermediate compartment
GI	glucosidase I
GII	glucosidase II
GRP	glucose-regulated protein
HA	hemagglutinin
ISG	immature secretory granule
M6P	mannose-6-phosphate
PDI	protein disulphide isomerase
SPC	subtilisin-like proprotein convertase
SR	SRP receptor
SRP	signal recognition particle

TGN   *trans*-Golgi network
TRAM   translocating chain-associating membrane protein
UGT   UDP-Glc:glycoprotein glucosyltransferase
VTC   vesicular tubular cluster

References

1. Kaufman, R.J. (1999) In: Flickinger, M.C. and Drew, S.W. (eds.) The Encyclopedia of Bioprocess Technology: Fermentation, Biocatalysis, and Bioseparation, John Wiley & Sons, New York, pp. 2366–2378.
2. Nielsen, H., Engelbrecht, J., Brunak, S., and von Heijne, G. (1997) Protein Eng. 10, 1–6.
3. Walter, P. and Blobel, G. (1980) Proc. Natl. Acad. Sci. USA 77, 7112–7116.
4. Walter, P. and Blobel, G. (1982) Nature 299, 691–698.
5. Krieg, U.C., Walter, P., and Johnson, A.E. (1986) Proc. Natl. Acad. Sci. USA 83, 8604–8608.
6. Kurzchalia, T.V., Wiedmann, M., Girshovich, A.S., Bochkareva, E.S., Bielka, H., and Rapoport, T.A. (1986) Nature 320, 634–636.
7. Gilmore, R., Walter, P., and Blobel, G. (1982) J. Cell. Biol. 95, 470–477.
8. Tajima, S., Lauffer, L., Rath, V.L., and Walter, P. (1986) J. Cell. Biol. 103, 1167–1178.
9. Keenan, R.J., Freymann, D.M., Stroud, R.M., and Walter, P. (2001) Annu. Rev. Biochem. 70, 755–775.
10. Rapiejko, P.J. and Gilmore, R. (1992) J. Cell. Biol. 117, 493–503.
11. Connolly, T. and Gilmore, R. (1989) Cell 57, 599–610.
12. Connolly, T., Rapiejko, P.J., and Gilmore, R. (1991) Science 252, 1171–1173.
13. Hartmann, E., Sommer, T., Prehn, S., Gorlich, D., Jentsch, S., and Rapoport, T.A. (1994) Nature 367, 654–657.
14. Voigt, S., Jungnickel, B., Hartmann, E., and Rapoport, T.A. (1996) J. Cell. Biol. 134, 25–35.
15. Hamman, B.D., Hendershot, L.M., and Johnson, A.E. (1998) Cell 92, 747–758.
16. Lyman, S.K. and Schekman, R. (1997) Cell 88, 85–96.
17. Matlack, K.E., Misselwitz, B., Plath, K., and Rapoport, T.A. (1999) Cell 97, 553–564.
18. Bause, E. (1983) Biochem. J. 209, 331–336.
19. Kornfeld, R. and Kornfeld, S. (1985) Annu. Rev. Biochem. 54, 631–664.
20. Nilsson, I.M. and von Heijne, G. (1993) J. Biol. Chem. 268, 5798–5801.
21. Holst, B., Bruun, A.W., Kielland-Brandt, M.C., and Winther, J.R. (1996) EMBO J. 15, 3538–3546.
22. Helenius, A. and Aebi, M. (2001) Science 291, 2364–2369.
23. Moremen, K.W., Trimble, R.B., and Herscovics, A. (1994) Glycobiology 4, 113–125.
24. Trombetta, E.S., Simons, J.F., and Helenius, A. (1996) J. Biol. Chem. 271, 27509–27516.
25. Weng, S. and Spiro, R.G. (1993) J. Biol. Chem. 268, 25656–25663.
26. Quinn, P., Griffiths, G., and Warren, G. (1984) J. Cell. Biol. 98, 2142–2147.
27. Hirano, N., Shibasaki, F., Kato, H., Sakai, R., Tanaka, T., Nishida, J., Yazaki, Y., Takenawa, T., and Hirai, H. (1994) Biochem. Biophys. Res. Commun. 204, 375–382.
28. Mazzarella, R.A., Srinivasan, M., Haugejorden, S.M., and Green, M. (1990) J. Biol. Chem. 265, 1094–1101.
29. Haas, I.G. and Wabl, M. (1983) Nature 306, 387–389.
30. Lin, H.Y., Masso-Welch, P., Di, Y.P., Cai, J.W., Shen, J.W., and Subjeck, J.R. (1993) Mol. Biol. Cell. 4, 1109–1119.
31. Ostwald, T.J. and MacLennan, D.H. (1974) J. Biol. Chem. 249, 974–979.
32. Sorger, P.K. and Pelham, H.R. (1987) J. Mol. Biol. 194, 341–344.
33. Wada, I., Rindress, D., Cameron, P.H., Ou, W.-J., Doherty, J.J., II, Louvard, D., Bell, A.W., Dignard, D., Thomas, D.Y., Bergeron, J.J.M. (1991) J. Biol. Chem. 266, 19599–19610.
34. Shiu, R.P., Pouyssegur, J., and Pastan, I. (1977) Proc. Natl. Acad. Sci. USA 74, 3840–3844.
35. Gething, M.J. and Sambrook, J. (1992) Nature 355, 33–45.
36. Bole, D.G., Hendershot, L.M., and Kearney, J.F. (1986) J. Cell. Biol. 102, 1558–1566.

430

37. Hammond, C., Braakman, I., and Helenius, A. (1994) Proc. Natl. Acad. Sci. USA 91, 913–917.
38. Spiro, R.G., Zhu, Q., Bhoyroo, V., and Soling, H.D. (1996) J. Biol. Chem. 271, 11588–11594.
39. Ware, F.E., Vassilakos, A., Peterson, P.A., Jackson, M.R., Lehrman, M.A., and Williams, D.B. (1995) J. Biol. Chem. 270, 4697–4704.
40. Oliver, J.D., van der Wal, F.J., Bulleid, N.J., and High, S. (1997) Science 275, 86–88.
41. Zapun, A., Darby, N.J., Tessier, D.C., Michalak, M., Bergeron, J.J., and Thomas, D.Y. (1998) J. Biol. Chem. 273, 6009–6012.
42. Molinari, M. and Helenius, A. (1999) Nature 402, 90–93.
43. Schrag, J.D., Bergeron, J.J., Li, Y., Borisova, S., Hahn, M., Thomas, D.Y., and Cygler, M. (2001) Mol. Cell. 8, 633–644.
44. Parodi, A.J. (2000) Annu. Rev. Biochem. 69, 69–93.
45. Ellgaard, L. and Helenius, A. (2001) Curr. Opin. Cell. Biol. 13, 431–437.
46. Ellgaard, L., Riek, R., Herrmann, T., Guntert, P., Braun, D., Helenius, A., and Wuthrich, K. (2001) Proc. Natl. Acad. Sci. USA 98, 3133–3138.
47. Molinari, M. and Helenius, A. (2000) Science 288, 331–333.
48. Peterson, J.R., Ora, A., Van, P.N., and Helenius, A. (1995) Mol. Biol. Cell. 6, 1173–1184.
49. Hebert, D.N., Zhang, J.X., Chen, W., Foellmer, B., and Helenius, A. (1997) J. Cell. Biol. 139, 613–623.
50. Danilczyk, U.G., Cohen-Doyle, M.F., and Williams, D.B. (2000) J. Biol. Chem. 275, 13089–13097.
51. Wada, I., Imai, S., Kai, M., Sakane, F., and Kanoh, H. (1995) J. Biol. Chem. 270, 20298–20304.
52. Meacham, G.C., Lu, Z., King, S., Sorscher, E., Tousson, A., and Cyr, D.M. (1999) EMBO J. 18, 1492–1505.
53. Trombetta, S.E., Bosch, M., and Parodi, A.J. (1989) Biochemistry 28, 8108–8116.
54. Labriola, C., Cazzulo, J.J., and Parodi, A.J. (1999) Mol. Biol. Cell. 10, 1381–1394.
55. Trombetta, E.S. and Helenius, A. (2000) J. Cell. Biol. 148, 1123–1129.
56. Sousa, M. and Parodi, A.J. (1995) EMBO J. 14, 4196–4203.
57. Trimble, R.B. and Maley, R. (1984) Anal. Biochem. 141, 515–522.
58. Trombetta, S.E. and Parodi, A.J. (1992) J. Biol. Chem. 267, 9236–9240.
59. Nilsson, T. and Warren, G. (1994) Curr. Opin. Cell. Biol. 6, 517–521.
60. Cosson, P., Demolliere, C., Hennecke, S., Duden, R., and Letourneur, F. (1996) EMBO J. 15, 1792–1798.
61. Cosson, P. and Letourneur, F. (1994) Science 263, 1629–1631.
62. Letourneur, F., Gaynor, E.C., Hennecke, S., Demolliere, C., Duden, R., Emr, S.D., Riezman, H., and Cosson, P. (1994) Cell 79, 1199–1207.
63. Andersson, H., Kappeler, F., and Hauri, H.P. (1999) J. Biol. Chem. 274, 15080–15084.
64. Sonnichsen, B., Fullekrug, J., Nguyen Van, P., Diekmann, W., Robinson, D.G., and Mieskes, G. (1994) J. Cell. Sci. 107, 2705–2717.
65. Booth, C. and Koch, G.L.E. (1989) Cell 59, 729–737.
66. Sambrook, J.F. (1990) Cell 61, 197–199.
67. Ware, J., Diuguid, D.L., Liebman, H.A., Rabiet, M.J., Kasper, C.K., Furie, B.C., Furie, B., and Stafford, D.W. (1989) J. Biol. Chem. 264, 11401–11406.
68. Jorgensen, M.J., Cantor, A.B., Furie, B.C., Brown, C.L., Shoemaker, C.B., and Furie, B. (1987) Cell 48, 185–191.
69. Sanford, D.G., Kanagy, C., Sudmeier, J.L., Furie, B.C., Furie, B., and Bachovchin, W.W. (1991) Biochemistry 30, 9835–9841.
70. Gronke, R.S., VanDusen, W.J., Garsky, V.M., Jacobs, J.W., Sardana, M.K., Stern, A.M., and Friedman, P.A. (1989) Proc. Natl. Acad. Sci. USA 86, 3609–3613.
71. Stenflo, J., Holme, E., Lindstedt, S., Chandramouli, N., Huang, L.H., Tam, J.P., and Merrifield, R.B. (1989) Proc. Natl. Acad. Sci. USA 86, 4440–4447.
72. McGinnis, K., Ku, G.M., VanDusen, W.J., Fu, J., Garsky, V., Stern, A.M., and Friedman, P.A. (1996) Biochemistry 35, 3957–3962.
73. Jia, S., VanDusen, W.J., Diehl, R.E., Kohl, N.E., Dixon, R.A., Elliston, K.O., Stern, A.M., and Friedman, P.A. (1992) J. Biol. Chem. 267, 14322–14327.

74. Stenflo, J., Lundwall, A., and Dahlback, B. (1987) Proc. Natl. Acad. Sci. USA 84, 368–372.
75. Derian, C.K., VanDusen, W., Przysiecki, C.T., Walsh, P.N., Berkner, K.L., Kaufman, R.J., and Friedman, P.A. (1989) J. Biol. Chem. 264, 6615–6618.
76. Wilson, I.B., Gavel, Y., and von Heijne, G. (1991) Biochem. J. 275, 529–534.
77. Huttner, W.B. (1988) Annu. Rev. Physiol. 50, 363–376.
78. Baeuerle, P.A. and Huttner, W.B. (1987) J. Cell. Biol. 105, 2655–2664.
79. Niehrs, C. and Huttner, W.B. (1990) EMBO J. 9, 35–42.
80. Beisswanger, R., Corbeil, D., Vannier, C., Thiele, C., Dohrmann, U., Kellner, R., Ashman, K., Niehrs, C., and Huttner, W.B. (1998) Proc. Natl. Acad. Sci. USA 95, 11134–11139.
81. Ouyang, Y., Lane, W.S., and Moore, K.L. (1998) Proc. Natl. Acad. Sci. USA 95, 2896–2901.
82. Ouyang, Y.B., Crawley, J.T., Aston, C.E., and Moore, K.L. (2002) J. Biol. Chem. 277, 23781–23787.
83. Wilkins, P.P., Moore, K.L., McEver, R.P., and Cummings, R.D. (1995) J. Biol. Chem. 270, 22677–22680.
84. Braun, P.J., Dennis, S., Hofsteenge, J., and Stone, S.R. (1988) Biochemistry 27, 6517–6522.
85. Pittman, D.D., Wang, J.H., and Kaufman, R.J. (1992) Biochemistry 31, 3315–3323.
86. Michnick, D.A., Pittman, D.D., Wise, R.J., and Kaufman, R.J. (1994) J. Biol. Chem. 269, 20095–20102.
87. Pittman, D.D., Tomkinson, K.N., Michnick, D., Selighsohn, U., and Kaufman, R.J. (1994) Biochemistry 33, 6952–6959.
88. Farrell, D.H., Mulvihill, E.R., Huang, S.M., Chung, D.W., and Davie, E.W. (1991) Biochemistry 30, 9414–9420.
89. Hortin, G., Tollefsen, D.M., and Strauss, A.W. (1986) J. Biol. Chem. 261, 15827–15830.
90. Hortin, G., Fok, K.F., Toren, P.C., and Strauss, A.W. (1987) J. Biol. Chem. 262, 3082–3085.
91. Jenne, D., Hille, A., Stanley, K.K., and Huttner, W.B. (1989) Eur. J. Biochem. 185, 391–395.
92. Morita, T. and Jackson, C.M. (1986) J. Biol. Chem. 261, 4008–4014.
93. Mutt, V. (1980) In: Glass, G.B.J. (ed.) Gastrointestinal Hormones, Raven Press, New York, pp. 169–221.
94. Niehrs, C., Huttner, W.B., Carvallo, D., and Degryse, E. (1990) J. Biol. Chem. 265, 9314–9318.
95. Rydel, T.J., Ravichandran, K.G., Tulinsky, A., Bode, W., Huber, R., Roitsch, C., and Fenton, J.W. (1990) Science 249, 277–280.
96. Bergeron, F., Leduc, R., and Day, R. (2000) J. Mol. Endocrinol. 24, 1–22.
97. Seidah, N.G. and Chretien, M. (1999) Brain Res. 848, 45–62.
98. Dong, W., Fricker, L.D., and Day, R. (1999) Neuroscience 89, 1301–1317.
99. Day, R., Lazure, C., Basak, A., Boudreault, A., Limperis, P., Dong, W., and Lindberg, I. (1998) J. Biol. Chem. 273, 829–836.
100. Seidah, N.G., Mowla, S.J., Hamelin, J., Mamarbachi, A.M., Benjannet, S., Toure, B.B., Basak, A., Munzer, J.S., Marcinkiewicz, J., Zhong, M., Barale, J.C., Lazure, C., Murphy, R.A., Chretien, M., Marcinkiewicz, M. (1999) Proc. Natl. Acad. Sci. USA 96, 1321–1326.
101. Anderson, E.D., VanSlyke, J.K., Thulin, C.D., Jean, F., and Thomas, G. (1997) EMBO J. 16, 1508–1518.
102. Zhou, A., Martin, S., Lipkind, G., LaMendola, J., and Steiner, D.F. (1998) J. Biol. Chem. 273, 11107–11114.
103. Roebroek, A.J., Umans, L., Pauli, I.G., Robertson, E.J., van Leuven, F., Van de Ven, W.J., and Constam, D.B. (1998) Development 125, 4863–4876.
104. Shen, F.S., Seidah, N.G., and Lindberg, I. (1993) J. Biol. Chem. 268, 24910–24915.
105. Paquet, L., Bergeron, F., Boudreault, A., Seidah, N.G., Chretien, M., Mbikay, M., and Lazure, C. (1994) J. Biol. Chem. 269, 19279–19285.
106. Ellgaard, L., Molinari, M., and Helenius, A. (1999) Science 286, 1882–1888.
107. Reddy, P.S. and Corley, R.B. (1998) Bioessays 20, 546–554.
108. Kellokumpu, S., Suokas, M., Risteli, L., and Myllyla, R. (1997) J. Biol. Chem. 272, 2770–2777.
109. Kim, P.S., Bole, D., and Arvan, P. (1992) J. Cell. Biol. 118, 541–549.
110. Marks, M.S., Germain, R.N., and Bonifacino, J.S. (1995) J. Biol. Chem. 270, 10475–10481.

432

111. Perlmutter, D.H. (1996) Ann. Med. 28, 385–394.
112. Stevens, T.L., Blum, J.H., Foy, S.P., Matsuuchi, L., and DeFranco, A.L. (1994) J. Immunol. 152, 4397–4406.
113. Letourneur, F., Hennecke, S., Demolliere, C., and Cosson, P. (1995) J. Cell. Biol. 129, 971–978.
114. Zerangue, N., Schwappach, B., Jan, Y.N., and Jan, L.Y. (1999) Neuron 22, 537–548.
115. Chang, X.B., Cui, L., Hou, Y.X., Jensen, T.J., Aleksandrov, A.A., Mengos, A., and Riordan, J.R. (1999) Mol. Cell. 4, 137–142.
116. Zapun, A., Jakob, C.A., Thomas, D.Y., and Bergeron, J.J. (1999) Structure Fold. Des. 7, R173–R182.
117. Cooper, A.A. and Stevens, T.H. (1996) J. Cell. Biol. 133, 529–541.
118. Hong, E., Davidson, A.R., and Kaiser, C.A. (1996) J. Cell. Biol. 135, 623–633.
119. Jorgensen, M.U., Emr, S.D., and Winther, J.R. (1999) Eur. J. Biochem. 260, 461–469.
120. Marcusson, E.G., Horazdovsky, B.F., Cereghino, J.L., Gharakhanian, E., and Emr, S.D. (1994) Cell 77, 579–586.
121. Wiertz, E.J., Tortorella, D., Bogyo, M., Yu, J., Mothes, W., Jones, T.R., Rapoport, T.A., and Ploegh, H.L. (1996) Nature 384, 432–438.
122. Suzuki, T., Park, H., Kitajima, K., and Lennarz, W.J. (1998) J. Biol. Chem. 273, 21526–21530.
123. Wiertz, E.J., Jones, T.R., Sun, L., Bogyo, M., Geuze, H.J., and Ploegh, H.L. (1996) Cell 84, 769–779.
124. Hiller, M.M., Finger, A., Schweiger, M., and Wolf, D.H. (1996) Science 273, 1725–1728.
125. Su, K., Stoller, T., Rocco, J., Zemsky, J., and Green, R. (1993) J. Biol. Chem. 268, 14301–14309.
126. Jakob, C.A., Burda, P., Roth, J., and Aebi, M. (1998) J. Cell. Biol. 142, 1223–1233.
127. Knop, M., Hauser, N., and Wolf, D.H. (1996) Yeast 12, 1229–1238.
128. Liu, Y., Choudhury, P., Cabral, C.M., and Sifers, R.N. (1999) J. Biol. Chem. 274, 5861–5867.
129. Dole, K., Lipari, F., Herscovics, A., and Howell, P.L. (1997) J. Struct. Biol. 120, 69–72.
130. Romero, P.A., Vallee, F., Howell, P.L., and Herscovics, A. (2000) J. Biol. Chem. 275, 11071–11074.
131. Cabral, C.M., Liu, Y., and Sifers, R.N. (2001) Trends Biochem. Sci. 26, 619–624.
132. Grinna, L.S. and Robbins, P.W. (1980) J. Biol. Chem. 255, 2255–2258.
133. Jakob, C.A., Bodmer, D., Spirig, U., Battig, P., Marcil, A., Dignard, D., Bergeron, J.J., Thomas, D.Y., and Aebi, M. (2001) EMBO Rep. 2, 423–430.
134. Nakatsukasa, K., Nishikawa, S., Hosokawa, N., Nagata, K., and Endo, T. (2001) J. Biol. Chem. 276, 8635–8638.
135. Hosokawa, N., Wada, I., Hasegawa, K., Yorihuzi, T., Tremblay, L.O., Herscovics, A., and Nagata, K. (2001) EMBO Rep. 2, 415–422.
136. Cabral, C.M., Choudhury, P., Liu, Y., and Sifers, R.N. (2000) J. Biol. Chem. 275, 25015–25022.
137. Novoradovskaya, N., Lee, J., Yu, Z.X., Ferrans, V.J., and Brantly, M. (1998) J. Clin. Invest. 101, 2693–2701.
138. Qu, D., Teckman, J.H., Omura, S., and Perlmutter, D.H. (1996) J. Biol. Chem. 271, 22791–22795.
139. Lomas, D.A., Evans, D.L., Finch, J.T., and Carrell, R.W. (1992) Nature 357, 605–607.
140. Sousa, M.C., Ferrero-Garcia, M.A., and Parodi, A.J. (1992) Biochemistry 31, 97–105.
141. Antonny, B. and Schekman, R. (2001) Curr. Opin. Cell. Biol. 13, 438–443.
142. Barlowe, C. (2000) Traffic 1, 371–377.
143. Gorelick, F.S. and Shugrue, C. (2001) Mol. Cell. Endocrinol. 177, 13–18.
144. Jamieson, J.D. and Palade, G.E. (1967) J. Cell. Biol. 34, 577–596.
145. Kirchhausen, T. (2000) Nat. Rev. Mol. Cell. Biol. 1, 187–198.
146. Springer, S., Spang, A., and Schekman, R. (1999) Cell 97, 145–148.
147. Gu, F., Crump, C.M., and Thomas, G. (2001) Cell. Mol. Life Sci. 58, 1067–1084.
148. Johnson, A.E. and van Waes, M.A. (1999) Annu. Rev. Cell. Dev. Biol. 15, 799–842.
149. Braakman, I. (2001) EMBO Rep. 2, 666–668.

S.C. Makrides (Ed.) *Gene Transfer and Expression in Mammalian Cells*

CHAPTER 14

Pathways and functions of mammalian protein glycosylation

Dale A. Cumming

Acton Biotech Consulting, 57 Hammond Street, Acton, MA 01720-3204, USA;
Tel.: + 1-978-263-2876, Fax. + 1-978-635-9124. E-mail: actonbiotech@mac.com

1. Introduction

Protein glycosylation is a prevalent post-translational modification in eukaryotic cells, occurring on more than half of all proteins [1]. The structural features and functional roles of protein-linked glycans (oligosaccharides) remain today the subject of intensive investigation. Simply put, this effort is underway because of the emerging picture of protein glycosylation as a sophisticated, non-linear mechanism of genetic information flow (e.g. [2]) complementing that described by the central dogma. Thus, the study of protein-linked glycans is a crucial adjunct to current efforts in proteomics and functional genomics.

In the broadest terms, there are three families of protein-linked glycans categorized on the basis of the type of amino acid side chains through which protein attachment occurs: those linked through asparagine side chains (the "N-linked" oligosaccharides), those linked through hydroxylated amino acid side chains (for example, the "O-linked" oligosaccharides linked through the side chains of serine or threonine), and finally those linked through the C-terminal amino acid of a protein (the glycosylphosphatidyl-inositols; GPIs). In addition to these major types of protein-linked glycans, a number of other structural motifs have been recently identified [3]. However, this simple description is deceptive since mammalian protein glycosylation is a sophisticated biological process composed of elaborate biosynthetic pathways whose essential features are highly conserved and that are tightly regulated in terms of time, sub-cellular localization, and tissue specificity.

A considerable investment of cellular resources is ultimately responsible to support these elaborate pathways of protein glycosylation. This investment includes not only the enzymes that process the oligosaccharides of nascent peptide chains but also the cellular systems responsible for, as examples, the synthesis and transport of donor substrates (nucleotide sugars and lipid-linked monosaccharides) or information-decoding lectins to specific subcellular locales. Thus, the "glycosylation apparatus" of the cell extends well beyond the set glycosyltransferases and hydrolases that execute biosynthesis.

In the following pages, we will cover the biosynthetic pathways for each of the three major types of protein-linked glycans and review the factors that specify what glycan structures are ultimately attached to a given protein. We will then conclude with a brief discussion of the emerging functional roles for protein-linked glycans that ultimately justify the large cellular investment in protein glycosylation.

2. Glycosylation of asparagine residues: N-glycosylation

2.1. Characteristics of N-linked glycosylation

The N-glycosylation of proteins is a (nearly) co-translational event that occurs on nascent polypeptide chains as they enter the lumen of the endoplasmic reticulum (ER). N-glycosylation requires as a necessary, but not sufficient, condition a tripeptide consensus sequon defined as Asn-X-Ser/Thr/Cys where X can be any amino acid save Pro. The biosynthetic elements that comprise the N-glycosylation pathway are dispersed and segregated along various components of the endomembrane system of eukaryotic cells. This system is composed of the ER (both rough and smooth), the pre-Golgi compartments such as the ER-Golgi Intermediate compartment (ERGIC), the Golgi apparatus, and a variety of transport vesicles (for a comprehensive review, see [4]).

2.1.1. Types of N-linked glycans
All N-linked glycans contain an invariant inner core (Fig. 1) although this pentasaccharide is rarely found attached to protein without further monosaccharide additions. The monosaccharides which populate the extra-core region can be various, including mannose (Man), galactose (Gal), N-acetylglucosamine (GlcNAc), and sialic acid (NANA). The monosaccharide composition of the "outer region" allows further definition of types of N-linked glycans: "high-mannose" type, "hybrid" type, and "complex" type (Fig. 1). Further structural classification of the "complex" type glycans is made on the basis of the number and position of "antennae" that emanate from the invariant core, always through the addition of GlcNAc residues to the three Man residues of the invariant core (Fig. 1). The number of antennae can range from two to seven giving rise to biantennary through hepta-antennary complex type N-glycans. Each antenna can be differentially elongated with other monosaccharides and thereby a large number of distinct N-glycan structures can be generated at any given site of N-glycosylation.

Given the existence of an invariant core structure, it is perhaps surprising that the biosynthesis of N-glycans does not proceed by simply adding the invariant core to the appropriate Asn residues and adding additional monosaccharides as available. Instead, the biosynthesis of N-glycans begins with the co-translational attachment of a large, preformed, lipid-linked oligosaccharide to the nascent peptide chain. An intricate pathway then follows that is "constructively destructive"; the glycan chains are trimmed down and, if needed for a hybrid or complex-type glycan, built up again to their final form.

2.2. Assembly of the lipid-linked precursor

The process of protein N-glycosylation begins with the transfer, either co-translationally or nearly so, of a preformed oligosaccharide from a lipid-linked precursor to the nascent polypeptide chain in the lumen of the RER. The lipid component of the precursor is a partially saturated polyisoprenoid termed

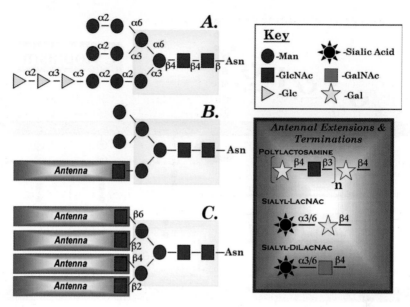

Fig. 1. Structural classes of N-glycans. All N-glycans contain an invariant pentasaccharide core, highlighted in yellow. The three principle classes of N-glycans are illustrated. (A) High-mannose-type N-glycan. The example shown is the glucosylated high-mannose glycan typically transferred to nascent protein chains in the ER. (B) Hybrid-type N-glycan. Failure to complete the processing of the upper Man(α1,6) arm can lead to further elaboration of the Man(α1,3) (lower) arm, creating an "antenna" (purple box). (C) Complex-type glycan. Successful processing leads to multiple antennae emanating from the invariant core, in this case a tetra-antennary glycan. The monosaccharide composition of each antenna can be different but are most often composed of Gal, NANA (sialic acid), GalNAc, GlcNAc, or fucose. A few examples of observed antennal structures are shown in the box at the lower right.

dolichol. When completed under normal conditions, the oligosaccharide is composed of 14 monosaccharide residues (Fig. 1A) attached through a pyrophosphoryl linkage to dolichol and yielding a composition of $Glc_3Man_9GlcNAc_2$. The essential features of this biosynthetic pathway are shown in Fig. 2. The genesis of the lipid-linked precursor begins on the cytoplasmic surface of the ER where a series of enzymes mediate the synthesis of a specific Man_5-PP-dolichol utilizing nucleotide sugars as donor substrates. The partially assembled precursor is then translocated to the lumenal side of the ER by a "flippase", an element of a complex system mediating bilateral molecular trafficking at the ER [5]. The yeast *RFT1* gene product has been reported to be a likely candidate for this "flippase" and homologs of *RFT1* are present in all but one eukaryotic genome sequenced to date [6].

Once present on the lumenal surface of the ER, further elongation of the precursor oligosaccharide occurs but now the donor substrates are Dol-P-Man and Dol-P-Glc instead of the corresponding nucleotide sugars. Indeed, these dolicholphosphoryl-monosaccharides are themselves synthesized on the cytoplasmic leaflet of the ER [7] and translocated to the lumen by a "flippase" apparently distinct from the *RFT1* gene product and its homologues [6,8]. These observations suggest that there is a

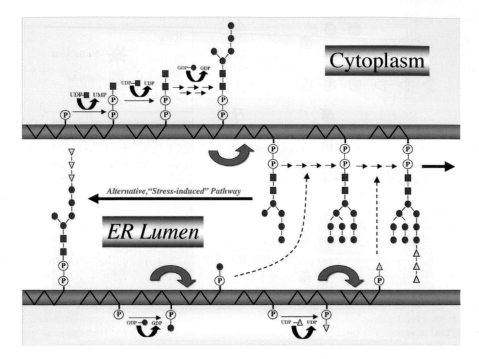

Fig. 2. Biosynthetic pathway for mammalian dolichol-linked oligosaccharides. Biosynthesis of the large lipid-linked oligosaccharide employed in N-glycosylation begins on the cytoplasmic surface of the ER by the generation of Dol-PP-GlcNAc from UDP-GlcNAc and dolichylphosphate (upper left). The dolichol-linked mannose is synthesized by the Dol-P-Man synthase, a heterotrimer consisting of the catalytic, regulatory, and stabilization subunits DPM1, DPM2, and DPM3, respectively (lower left; [7]). Dol-P-Glc is similarly synthesized on the cytoplasmic surface of the ER and translocates to the lumen prior to participating in the final assembly of the oligosaccharide precursor. Note that the failure to extend structures generated on the cytoplasmic surface gives rise to an "alternative" oligosaccharide precursor for transfer to nascent protein chains. The extent and state of glucosylation of this "alternative" Man$_5$ structure varies considerably, though they are greatly preferred over their non-glucosylated counterparts for transfer to protein.

biological necessity to support a distinct set of donor substrates for final assembly of the lipid-linked oligosaccharide precursor in the lumen of the ER. This is apparently the case as certain genetic disorders, viral infections, or cellular stress (e.g. glucose starvation) lead to the generation of an "alternative" lipid-linked precursor that can be used for protein glycosylation (Fig. 2).

2.3. Transfer of precursor to polypeptide backbone

The preformed dolichol-linked Glc$_3$Man$_9$GlcNAc$_2$ moiety is transferred *en bloc* to the side chains of asparagine in the nascent polypeptide chain, in the context of the Asn-X-Thr/Ser/Cys canonical sequon, on the luminal face of the ER. Oligosaccharide transfer is mediated enzymatically through the action of an oligosaccharyltransferase (OST48) [9] but actually accomplished through the coordinated action of several molecular assemblies: the ribosome nascent chain complex (RNC), the oligosaccharyltransferase (OT) complex, the ER resident conducting channel formed by oligomers of Sec61 (the translocon), the nascent polypeptide associated complex (NAC) and the signal recognition particle (SRP).

RNCs have a site (the M-site) that itself possesses the intrinsic ability to bind to the ER resident translocon [10]. To prevent indiscriminate binding of any RNC to the RER, NAC binds to the ribosomal M-site regardless of whether the nascent peptide chain contains a signal sequence. However, if a signal sequence is present, the ribonucleoprotein SRP binds to the ribosomal M-site, displacing NAC. After binding of the SRP-RNC complex to the SRP receptor, the SRP is displaced from the M-site allowing ribosomal binding to the translocon, releases the signal peptide in a GTP-dependent process, and thus enables cotranslational translocation of nascent chains to the lumen of the ER.

In mammals, the oligosaccharyltransferase complex is composed of at least four proteins: ribophorins I and II, OT, and DAD1 while in yeast there are nine protein components in the complex [11]. OT contains a dilysyl motif that helps keep the complex localized within the ER and the entire complex interacts with Sec61 to catalyze the attachment of the preformed oligosaccharide to the nascent polypeptide chain.

2.4. Processing of N-linked glycans

An elaborate biochemical process that degrades the newly attached N-glycan chain begins immediately after addition to the polypeptide chain in the lumen of the RER. As the nascent protein makes its way through the endomembrane system, a series of glycosidases can sequentially remove specific monosaccharide residues from the N-glycan. Once in the *medial*-Golgi, a set of glycosyltransferases can catalyze the branching and terminal elaboration of N-glycan chains. Features of this "processing" pathway are illustrated in Fig. 3 and Fig. 4. It is important to note that the following description applies to a site of N-glycosylation destined to be a highly branched, complex type structure. At any given site in any given protein, the process can terminate as determined by any one of several factors as discussed in Section 5.

The initial step in the processing of N-glycans is the removal of the three terminal glucose residues. This is accomplished by two ER-resident glucosidases: glucosidase I which catalyzes the removal of the terminal (α1-2)-linked Glc residue and glucosidase II that removes the internal Glc(α1-3)-linked residues. Of particular note is the existence of a system for the dynamic reglucosylation/de-glucosylation of nascent glycoproteins, an important component of the ER "Quality Control" system for the proper folding of proteins. This subject is discussed fully elsewhere in this volume (see also [12]), but here we will note that the dynamic cycling of the glucosylated state of a nascent protein, mediated by glucosidase II and an ER resident glucosyltransferase, serve as a "folding sensor" that allow the nascent chain to associate with chaperones such as calnexin or calreticulin. When properly folded, the nascent chain disengages from this dynamic recycling through the action of ER mannosidase I.

438

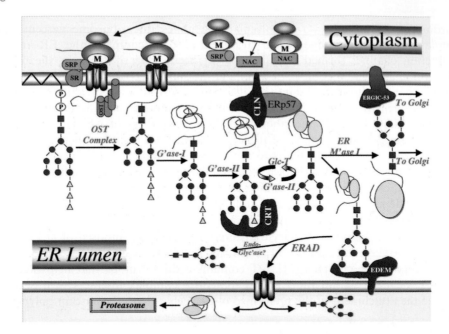

Fig. 3. N-glycosylation of nascent proteins and glycan processing in the ER.Ribosomes synthesizing polypeptide chains destined for the endomembrane/secretory pathway are first encountered in the cytosol, precluded from high affinity M-site binding to the ER translocon by virtue of bound NAC (top, middle). In the presence of a nascent chain signal sequence, SRP displaces NAC and can bind to the SRP receptor (SR; top, left) enabling binding of the ribosome to the translocon and, as the nascent chain is extruded into the lumen of the ER, subsequent association with the oligosaccharyl-transferase complex. Once in the luminal compartment, glucosidases (G'ases) and mannosidases (M'ases) execute an ordered series of cleavages as the polypeptide chain folds. Note the four ER-resident lectins (purple; calnexin [CLN], calreticulin [CRT], ERGIC-53, and EDEM) that in part mediate the ER QC (CRT and CLN) and export to the Golgi (ERGIC-53) or to ERAD via the Sec61 translocon (EDEM). Erp57 is a homolog of protein disulfide isomerase, involved in proper folding of nascent proteins. It is likely that all four of these ER-resident lectins interact not only with bound glycans but the nascent chain as well, though no attempt has been made in the figure to illustrate both interactions.

ER mannosidase I is a member of the class 1 group of (α1,2)-mannosidases. Class I mannosidases can be divided into two distinct subgroups. The first of these includes ER mannosidase I while subgroup 2 includes the Golgi-resident (α1,2)-mannosidases of mammalian cells. The ER mannosidase I displays a very narrow substrate specificity acting on $Man_9GlcNAc_2$ oligosaccharides to generate a specific isomer of $Man_8GlcNAc_2$, termed the B isomer (Fig. 3). In contrast, the Golgi resident class I mannosidases can act virtually on all Man(α1, 2) residues in the $Man_9GlcNAc$ structure, including the ability to generate $Man_8GlcNAc_2$ isomers A and C. All members of the class I group share a similar 3D topology and the structural basis for the substrate specificity differences between the two subgroups has recently been attributed to a single amino acid difference that creates a more spacious active site in the Golgi-resident subgroup [13].

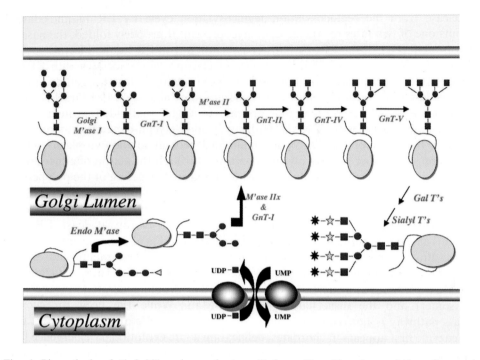

Fig. 4. Biosynthesis of "hybrid" and complex-type N-glycans.**Top**: The glycan chains of properly folded proteins entering from the ERGIC (left) are further cleaved by sequential exo-mannosidase activity of Golgi mannosidases to yield a specific $Man_5GlcNAc_2$ structure that is the preferred acceptor substrate for the medial-Golgi enzyme GlcNAc-TI. The resulting $GlcNAc_1Man_5GlcNAc_2$ glycan represents a key point of commitment in the processing of N-glycan chains. Further processing by Golgi mannosidase II (or mannosidase IIx) allows the action of other GlcNAc transferases to form multiple antennae on the N-glycan, thus defining a commitment to "complex-type" N-glycans. Note that mannosidase II activity is dependent upon the prior action of GlcNAc-TI. The pathway illustrated shows the generation of a "tetra-antennary" N-glycan through the sequential action of GlcNAc-TII, -TIV, and –TV. Failure of mannosidase II to act commits the processing pathway to the "hybrid-type" of N-glycan where the newly created GlcNAc "antenna" is further extended by trans-Golgi enzymes.*Bottom*: Glucosylated N-glycans can also enter the cis-Golgi after protein folding due to the failure of ER-glucosidase II to act prior to ER mannosidase-I. The "alternative" $Glc_1Man_5GlcNAc_2$ structure is illustrated here, though other structures, such as $Glc_1Man_{8-9}GlcNAc_2$ can also enter. Because the terminal Glc residue blocks the action of Golgi mannosidases on that branch, these N-glycans are restored to the normal pathway through the action of an endo-mannosidase that liberates a Glc-Man disaccharide and enables subsequent action by mannosidases (including mannosidase II or IIx if the substrate is $Glc_1Man_{8-9}GlcNAc$) and eventually GlcNAc-TI. The fate of the "alternative" $Man_5GlcNAc_2$ structures is less clear. After endo-mannosidase action, the resulting $Man_4GlcNAc_2$ structures could be substrates for mannosidase cleavage to a $Man_3GlcNAc_2$ structure that is also a substrate (though a poor one) for GlcNAc-TI. Also shown in the bottom middle of the figure is a typical sugar nucleotide antiporter, in this case the UDP-GlcNAc antiporter, responsible for bringing glycosyltransferase donor substrates within the endomembrane system from the cytoplasm (where they are synthesized). After transferase action, the resulting nucleotide diphosphate is cleaved by a luminal pyrophosphatase to yield the nucleotide monophosphate that is transported to the cytoplasm by the antiporter in exchange for the sugar nucleotide.

Release from the reglucosylation/deglucosylation cycle by ER mannosidase I permits one of two fates for the nascent glycoprotein. If properly folded, the nascent glycoprotein can exit the ER, perhaps in association with ER luminal lectins such as ERGIC-53. If improperly folded, the nascent glycoprotein associates with EDEM (ER degradation enhancing α-mannosidase-like) protein, a close homologue of ER mannosidase I but without demonstrable mannosidase activity [14]. EDEM is a type II ER transmembrane protein that appears to target malfolded protein to the ER-associated degradation (ERAD) pathway in an N-glycan-dependent manner. On or about the point of transfer of the malfolded protein to the cytoplasm via the translocon, the $Man_8GlcNAc_2$ glycan is removed through the action of either an ER or cytosolic *endo*-oligosaccharidase [15] with subsequent targeting of the polypeptide chain to the proteasome.

An important "bypass" mechanism exists in higher eukaryotes comprised of the enzyme endomannosidase [16]. This Golgi-resident (or at least ERGIC-resident) enzyme acts on glucosylated N-glycans, regardless of previous mannosidase action, thus providing substrates suitable for Golgi-resident mannosidases and eventual hybrid- and complex-type N-glycan formation [17]. Thus, utilizing the $Glc_1Man_9GlcNAc_2$ as a substrate, endomannosidase would generate $Man_8GlcNAc_2$ (A isomer) and the disaccharide $Glc(\alpha 1,3)$-Man. While the need for such a bypass pathway is not completely understood, phylogenetic analysis indicates that this enzyme first appears in chordates, delineating a late evolutionary emergence [18].

Folded glycoproteins entering the *cis*-Golgi undergo further processing of their N-glycan chains by the aforementioned Golgi mannosidases to generate the key glycan intermediate $Man_5GlcNAc_2$ (Fig. 4). The next processing step marks the point of commitment to hybrid or complex-type N-glycans and is mediated by the enzyme N-acetylglucosaminyltransferase-I (GlcNAc-TI). The addition of this $GlcNAc(\beta 1-2)$ residue to the $Man(\alpha 1,3)$-$Man(\beta)$ moiety (the Man-3 arm) is a necessary step for removal of two additional Man residues through the action of the *medial*-Golgi enzymes mannosidase II or mannosidase IIx [19]. At this point, another N-acetylglucosaminyltransferase, GlcNAc-TII, can add a $GlcNAc(\beta_{1,2})$ to the newly exposed $Man(\alpha_{1-6})$-$Man(\beta_1-)$ [the Man-6 arm] non-reducing terminus to create the most rudimentary "biantennary" complex-type glycan. Failure to act further on the Man-6 arm after GlcNAc-TI action can yield "hybrid-type" N-glycans through substitution of the $GlcNAc(\beta 1-2)Man(\alpha 1-3)$-moiety with galactose and sialic acid. A set of distinct N-acetylglucosaminyl-transferases (GlcNAc-TIII through GlcNAc-TVII) can add further GlcNAc residues to create additional antennae (Fig. 4). Each of these antennae can be separately elongated by a variety of *trans*-Golgi glycosyltransferases (e.g. galactosyltransferases and sialyltransferases) to yield numerous mature glycan structures.

3. Glycosylation of serine/threonine residues: O-glycosylation

The modification of hydroxylated amino acid side chains represents a more diverse set of post-translational modifications than do N-linked glycans. This type of substitution covers historically distinct groups of protein-linked glycans including the "mucin type" glycans, proteoglycans (glycosaminoglycans), the cytosolic/ nuclear O-GlcNAc substitutions, and the O-fucosylation/glucosylation of proteins.

3.1. The "mucin type" of O-glycans: GalNAc substitution of Ser/Thr

The biosynthesis of mucin type O-glycans initiates in the *cis* Golgi (or at least the ERGIC) with the transfer of GalNAc from UDP-GalNAc to a protein acceptor by the enzyme UDP-GalNAc::polypeptide N-acetylgalactosaminyl transferase (GalNAc-T). Unlike N-glycans or GPIs, there is no preformed oligosaccharide that is transferred to the polypeptide. Rather, the initial step, and all subsequent elaborations, occurs through the stepwise addition of monosaccharides from nucleotide sugar donors, and most accessible Ser/Thr residues are sites of potential O-glycosylation. However, detailed analysis of known O-glycosylation sites has yielded some general rules summarized in [1].

There is not a single gene responsible for coding the GalNAc-T activity but rather a large family numbering twelve gene products in mammals [20]. Detailed investigations have demonstrated distinct substrate specificities and different kinetic parameters for members of the family, as well as differential tissue and temporal regulation of expression. These observations have suggested a mechanism to control O-glycan site occupancy and density. It has been observed that GalNAc-T's possess two major domains: an N-terminal catalytic domain and a C-terminal lectin-like domain. Further, the lectin domains in select members of the GalNAc-T family (e.g. GalNAc-T4) appear to be operative and enable selective modification of Ser/Thr residues only after modification of other O-glycosylation sites [21]. Thus a "hierarchical" process of initial addition of GalNAc residues to a protein by GalNAc-T1 or T3 could be followed by the GalNAc-dependent initiation of new site of O-glycosylation through the action of GalNAc-T2, -T4, or -T7 [22].

Further elaboration from the initial GalNAc residue leads to eight distinct "core structures" as illustrated in Fig. 5. Further elongation and elaboration of these core structures can occur through the action of a battery of transferases and can yield complex structures bearing polylactosaminyl, fucosyl, and sialyl modifications and can be further sulphated or acetylated. A key enzyme in mucin O-glycan biosynthesis is "Core 2 transferase" which adds GlcNAc(β1-6) "branches" to the reducing GalNAc of core 1 and core 3 structures, generating core 2 and core 4 O-glycans. Core 2 transferases are a family of three enzymes: Core 2L (leukocytic form), Core 2M (expressed predominantly in the alimentary canal; also called Core 2b), and Core 2c (expressed in the thymus). Core 2L and Core 2c prefer core 1 substrates and Core 2M acts on both core 1 and core 3 structures.

3.2. Xylose substitution of Ser/Thr

Attachment of xylose residue to Ser (and occasionally Thr) is the initial step in the biosynthesis of proteoglycans, a large and diverse set of glycoconjugates that are major components of the extracellular matrix. The protein-linked carbohydrate

442

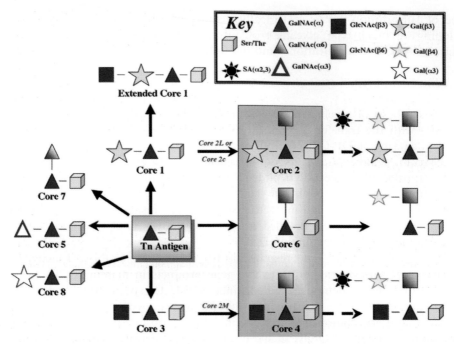

Fig. 5. Early biosynthetic pathways of mucin-type O-glycansThe biosynthesis of all mucin-type glycans begins with the formation of the Tn antigen through the action of an N-acetylgalactosaminyltransferase. Further elaboration of this structure can give rise to eight distinct "core structures" as shown. Perhaps the best known of these is the core 1 (T-antigen, Thomsen-Friedenreich antigen) structure. The addition of a GlcNAc(β1,6) residue to the GalNAc-Ser/Thr moiety generates core 2, core 4, and core 6 structures from core 1, core 3, and the Tn antigen, respectively. Note that core structures such as core 1, core 2, core 3, and core 4 can be further elaborated by the addition of Gal, GlcNAc, and NANA residues to yield more complex structures, including select blood group antigens or polylactosaminyl repeats . The elaborated core 6 structure is also known as the F1α antigen. The Tn antigen can also be (α2,6)-sialylated to form the sialyl-Tn antigen.

chains of proteoglycans are classified into four main groups based upon their composition and internal linkages. These four groups are chondroitin sulfate, dermatan sulfate, heparin, and heparan sulfate. The glycans of all four groups are attached through a "linkage region" composed of a common tetrasaccharide. Further elaborations from the linkage region are found as a set of repeat disaccharides that differ between each group. These elemental architectural features of proteoglycans are summarized in Fig. 6.

The biosynthesis of the glycan moieties initiates in the Golgi or ERGIC through the action of xylosyltransferase (XT). This transferase utilizes UDP-Xyl as a donor substrate creating a linkage to specific Ser residues within the context of the sequence x-x-x-x-G-S-G-x-y-x (where x = E or D and y = E, D,or G) [23]. Human, rat and Drosophila XT have been cloned and shown to be members of the Core 2/I GlcNAc-T family (e.g. [24]). The linkage region is next extended through the action of Gal-TI and Gal-TII and completed through the action of glucuronyltransferase I.

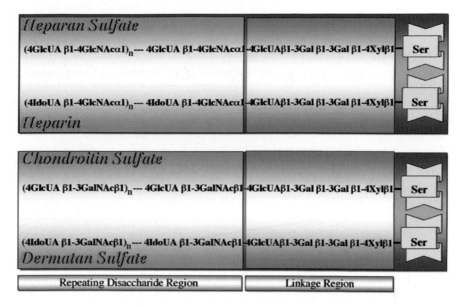

Fig. 6. Summary of the four major glycosaminoglycans of proteoglycans.Biosynthesis of the glycosaminoglycan of proteoglycans begins with the sequential synthesis of the tetrasaccharide "linkage" that modifies Ser residues at consensus sites. All four of the glycosyltransferases involved in linkage region biosynthesis utilize sugar nucleotides as donor substrates. The linkage region can be further modified by 2-O-phosphorylation of the Xyl residue and 4,6-sulfation of the Gal residues, though no clear function has been ascribed to these modifications. Competition between GlcNAc- and GalNAc-T's determine subsequent commitment to the heparin/heparan sulfate (GlcNAcα1,4 addition) pathways or chondroitin/dermatan sulfate (GalNAcβ1,4 addition) pathways. The competition is controlled by protein determinants either near the site of attachment or in more distal domains, with the GalNAc pathway being the "default". The glycosaminoglycan chains then elongate by the alternating addition of monosaccharide residues as per their respective class. As elongation of the repeat disaccharide region proceeds, there is extensive remodeling and modification of the residues in this region. In the case of heparin/heparan sulfate, these remodeling/modification events include epimerization of glucuronic acid (GlcUA) to iduronic acid (IdoUA), de-N-acetylation, subsequent N-sulfation of GlcNAc residues, and variable O-sulfation of GlcNAc, N-sulfo-glucosamine, and uronic acid (IdoUA and GluUA) residues. For a more thorough discussion, see [65] and references therein.

All three of these enzymes are Golgi enzymes acting sequentially using sugar nucleotide donors and their precise roles in generating the linkage region have been defined using a variety of techniques, including CHO mutant cells lines (e.g. [25]). Interestingly, the activity of Gal-TI has recently been reported to be activated in the presence of ATP, presumably through the action of a kinase [26].

The repeat disaccharides that emanate from the linkage region are composed of alternating uronic acids (glucuronic or iduronic acid) with amino sugars (glucosamine or galactosamine), the identity of which defines one of the four groups of proteoglycans defined above (Fig. 6). Additional, non-carbohydrate modifications of both the disaccharide repeat and linkage regions, specifically sulfation and phosphorylation, can occur thus further extending the structural diversity of proteoglycans (Fig. 6).

3.3. O-fucosylation/glucosylation

Epidermal growth factor (EGF) domains are a common component of extracellular, multidomain proteins. A unique modification of EGF domains is O-fucosylation or O-glucosylation (for a review, see [27]), the latter being further elaborated by the addition of one or two xylose residues linked (α1,3). The generation of glucosylated EGF domains requires the consensus sequence –Cys-Xaa-Ser-Xaa-Pro-Cys- in proteins such as tPA and the blood coagulation factors VII and IX.

O-fucosylation of EGF domains also requires a consensus sequence, –Cys-Xaa-Xaa-Gly-Gly-Ser/Thr-Cys-, as found in proteins such as human factor IX, tPA, and urokinase. In the case of human factor IX, the O-fucose residue was further elongated to produce a sialylated tetrasaccharide [28].

3.4. O-GlcNAc substitution of cytosolic/nuclear proteins

The modification of Ser and Thr residues of cytosolic and nuclear proteins by O-linked GlcNAc has emerged as an important mechanism of cellular regulation [29]. Three features underscore a potential role in cellular regulation: (1) this modification competes for the same substrate sites as Ser/Thr kinases; (2) O-GlcNAc substitutions are dynamic and responsive to changes in cellular state; (3) for certain proteins such as RNA Pol II and c-Myc, O-GlcNAc substitution is "reciprocal" with Ser/Thr phosphorylation [30].

The state of O-GlcNAc substitution of cytosolic and nuclear proteins is maintained through the actions of two enzymes: a UDP-GlcNAc:polypeptide O-β-N-acetylglucosaminyltransferases (OGT) to add the monosaccharide to the protein and an O-β-N-acetylglucosaminidase (O-GlcNAcase) to remove it. OGT is a highly conserved protein and has been cloned independently by two groups [31,32]. OGT is typically a homotrimer with each monomer composed of two domains: an N terminal series of 9–12 tetratricopeptide repeats (TPR) and a C-terminal catalytic domain [31]. The former is known to mediate protein-protein interactions and is responsible for multimerization of OGT [33]. OGT is different from other glycosyltransferases in that is has a type I membrane topology and exhibits the curious property of having multiple K_m values for its donor substrate (UDP-GlcNAc) while donor substrate concentration directly affects the extent of activity towards acceptor substrates [33]. The latter properties would facilitate fine biological control over enzymatic activity. The O-GlcNAcase is widely expressed in many tissues as one of two splice variants. Only the larger of the two variants is active as an O-GlcNAcase and this form is localized to the cytoplasm [30].

4. Glycosyl phosphatidyl inositol (GPI)-anchored proteins

GPI is a distinct type of protein glycosylation that is employed to anchor proteins to membranes. Abundant genetic evidence demonstrates that GPIs are essential for appropriate cell surface expression of select proteins (termed glypicans), defects of which result in developmental abnormalities [34] and human diseases such as paroxysmal nocturnal hemoglobinuria [35]. Attachment of GPIs occurs at the C-terminus of target proteins, displacing a C-terminal peptide fragment. While there is not a clear consensus sequence for attachment of GPIs, some general characteristics of the GPI signal sequence have been elucidated [36].

4.1. Structural classes of eukaryotic GPIs

GPIs are frequently distinguished from the total set of PI glycolipids by defining an essential structural motif of Man(α1-4)GlcNH$_2$(α1-6)-PI, thus omitting the mannosylated-PIs of lower eukaryotes [37]. With this definition, a structurally conserved core has been delineated: Man(α1-2)-Man(α1-X)-Man(α1-4)- GlcNH2(α1-6)-PI. Three subtypes of GPIs have been identified. The subtypes are designated one to three: for type 1 the "X" linkage is (α1-6) while in type 2 it is (α1-3). In the hybrid type, both substitutions of the Man(α1-4) residue are present. The vast majority of protein-linked GPIs are of type 1.

4.2. Biosynthesis of GPIs

The biosynthetic pathway for GPIs shares several features in common with that for N-glycans. GPI biosynthesis can be topologically dispersed from the ER to the Golgi, generates a pre-assembled lipid-linked precursor that is transferred to the protein backbone, and utilizes both dolichol and nucleotide sugar donors within the pathway. The GPI biosynthetic pathway for mammalian cells is illustrated in Fig. 7.

The first step in the biosynthesis of GPIs occurs on the cytosolic face of the ER and is catalyzed by a complex of four proteins (Pig-a, Pig-c, Pig-h, and hGPI-1) that mediate the transfer of GlcNAc from the sugar nucleotide to PI. This complex is further associated with two other proteins, Pig-p and DPM-2 [38]. The association of DPM-2, one of three subunits of Dol-P-Man synthase [7] is not essential for GlcNAc transfer, as Lec15 cells still can synthesize GlcNAc-PI while cells defective in Pig-p are unable to mediate this process. In the next step, the GlcNAc residue is de-N-acetylated to generate glucosamine (GlcNH$_2$) by the PIG-L gene product. Further elongation of the GPI requires ER-luminal enzymes [39], although the topological disposition of GPIs is unclear as they can readily flip back and forth between cytosolic and luminal leaflets until attachment to proteins (see [40]). In mammalian cells, 2-O-acylation of the inositol moiety then occurs as a necessary prelude to mannosylation. Dolichol-phosphorylmannose serves as the sugar donor for the addition of all three mannose residues. The initial Man(α1-4) residue is transferred by Pig-m while the terminal (α1-2) mannose residue being catalyzed by the PIG-B gene product.

Mannosylated GPIs can be further modified by the addition of ethanolamine phosphate on potentially all three mannose residues, a process that involves multiple gene products including Pig-f, Mcd4, Yll031c, and Gpi7, Gpi11 and Gpi13. Homologous proteins between yeast and human appear to mediate slightly different

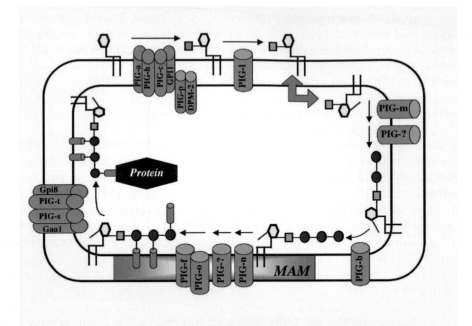

Fig. 7. Biosynthetic pathway for mammalian GPIs.Beginning with phosphatidylinositol (PI; upper left), the first step in the pathway is the addition of a residue of GlcNAc to PI on the cytoplasmic leaflet of the ER, an event mediated by a multiprotein complex as described in the text. Following de-N-acetylation of the GlcNAc residue and subsequent acylation (blue line) of the inositol moiety (yellow hexagon), all subsequent biosynthetic events occur on the luminal leaflet of the ER although GPI precursors can apparently freely translocate between ER membrane leaflets. These events include the sequential addition of three mannose residues (red circles) to generate a Man(α1,2)Man(α1,6)Man(α1,4)GlcNH$_2$-PI. 2-O-acylation of the inositol moiety is a prerequisite for mannosyltransferase action with Dol-P-Man acting as the donor substrate. Ethanolamine substituents are added from phosphatidylethanolamine donors to one or more of the mannose residues in a specialized region of the ER termed MAM (see text). Such substitution of the terminal Man(α1,2) residue is a prerequisite for transfer of the GPI precursor to the C-terminal consensus sequence of the target protein, a reaction carried out by the GPI transamidase complex as shown. After transfer to the target protein, the inositol moiety is deacylated (not shown). The gene products mediating the transfer of the Man(α1,6) and its ethanolamine substituent are not known.

functions but the Pig-f and Pig-o proteins appear responsible in mammalian cells for modification of the Man(α1-2) residue while Pig-n mediates EtNH$_2$ addition to the Man(α1-4) residue [41,42]. The addition of ethanolamine phosphate to nascent GPI chains may occur in a specialized region of the ER associated with mitochondria termed MAM (for a thorough discussion, see [40]). MAM is enriched in the enzymes and substrates required for select lipid biosynthesis.

The addition of ethanolamine phosphate to the Man(α1-2) residue is necessary for attachment to the protein by GPI transamidase complex . It is generally believed that the transamidase complex binds to the C-terminal signal peptide and attacks the carbonyl group of ω site amino acid, releasing the signal peptide and generating a carbonyl intermediate with the target protein. The amino group in the terminal ethanolamine of the preformed GPI would then attack the intermediate to complete the reaction [36]. The transamidase complex is composed of four proteins: Pig-s, Pig-t, Gaa1, and the catalytic Gpi8 [43]. The relative contributions of the four components of the transamidase complex are still emerging, but Gaa1 and Pig-s mediate stable carbonyl intermediate formation, and Pig-t stabilizes the rest of the complex. Gpi8 mediates the cleavage of the GPI consensus sequence and has homology to cysteine proteases.

5. What specifies glycan structure?

Having reviewed the pathways of mammalian protein glycosylation, one might reasonably ask if any typical characteristics exist for the mature glycoproteins emerging from the pathways' end. Global characteristics of protein glycosylation include the following: (1) Protein glycosylation is a non-random process. The action of biosynthetic enzymes is determined by the previous action of other enzymes and, in turn, can affect subsequent action by "downstream" enzymes; (2) Site occupancy where a protein can have multiple sites of glycan attachment. However, a given site of glycan attachment can be variably occupied giving rise to a specific form of heterogeneity on proteins; and (3) Site-specific microheterogeneity, where a given site of glycosylation frequently bears multiple glycan structures creating, when summed over all sites of glycosylation, an ensemble of protein molecules that bear distinct combinations of attached glycans (glycoforms).

The major factors that determine the extent and type of protein glycosylation can be summarized as: (1) the polypeptide chain itself, (2) the cellular complement of processing enzymes, and (3) the "environmental" state of the secreting cell/tissue. In a stable cellular environment, a given cell-type will secrete a glycoprotein with a consistent and reproducible set of glycan structures at each site of modification (e.g. [44]).

5.1. The polypeptide chain

The significant role that the polypeptide chain plays in determining the structure of covalently bound glycans has been evident for some time and has been the subject of several previous reviews (see [45,46]). While all levels of protein structure (i.e. primary, secondary, tertiary, and quaternary structure) have been deemed relevant in specifying attached glycan structures, especially in terms of the class(es) of structures present at a given site, only recently have integrated theories been put forward to explain these observations. The key set of observations cited include:

- Expression of recombinant glycoproteins in a number of heterologous cell systems has enabled direct evaluation of structures present at individual glycosylation sites. The principal finding from these studies is that the site-specific glycan structure class observed at a given glycosylation site is preserved regardless of the cell type of expression, as long as each cell-type has the "machinery" to produce those structures.

- Expression of a series of parental and mutant forms of a protein in the same cell type has shown that changes to the protein structure can alter the site-specific glycan structural class.
- Protein glycosylation is intimately associated with protein folding. Numerous studies have demonstrated that improper glycosylation leads to abnormal protein folding and that conditions that give rise to aberrant protein folding also give rise to abnormal glycosylation. As we have already seen, there exists a complex "editing" apparatus within the endoplasmic reticulum that monitors the integrity of nascent proteins.

We have previously suggested [45] that the underlying mechanism by which the polypeptide chain directs the oligosaccharide structures is by modulation of the "local" three-dimensional topology at a glycosylation site and we would term this hypothesis "site-specific topological modulation". In this scheme protein–protein, protein–oligosaccharide, and oligosaccharide–oligosaccharide interactions would provide the mechanisms for topological alterations about a glycosylation site. In turn, the precise topology at a nascent glycosylation site would influence, for example, the effective K_m and/or V_{max} of a processing enzyme.

In addition, several other features of protein glycosylation are also consistent with this hypothesis. It has been demonstrated that several glycosyltransferases exhibit a catalytic preference for specific antennae of N-linked oligosaccharides (reviewed in [47]). For example, the glycosyltransferase that catalyzes the α1,3-linked galactosylation of the non-reducing termini of isolated biantennary glycans shows a marked preference for galactose addition to the α1,6-Man antenna [48]. Topological modulation about protein-linked oligosaccharides could likely affect these arm preferences, perhaps by altering conformational equilibria of glycan antennae. Similar observations can be made for the "ordered addition" observed for many glycosyltransferases, especially where the action of one transferase precludes or enables the subsequent action of a different transferase [47]. For example, mannosidase II cannot act until GlcNAc-TI does while GlcNAc-TIII, -IV, and -V all require that GlcNAc-TII act first. These are examples of "GO" signals while "STOP" signals can be exemplified by the action of GlcNAc-TIII, which precludes subsequent action by mannosidase II and GlcNAc-TV. Thus, the hypothesis of site-specific topological modulation should provide a useful conceptual framework for further studies into the means by which a polypeptide chain directs its own glycosylation.

5.2. The glycosylation phenotype of the cell

The glycosylation phenotype of a cell, defined by the complement of expressed glycosylation processing enzymes (glycohydrolases and glycosyltransferases), is essential in specifying the structural class and terminal elaborations of the resulting glycan chains. It is well known that different cell types possess distinct sets of processing enzymes and that the same polypeptide chain synthesized in different cells can bear glycans with distinct primary structures. Thus, CHO cells do not normally attach sialic acid to N-linked glycoconjugates in an (α2,6)-linkage, since the necessary sialyltransferase(s) is not expressed, while BHK cells are capable of forming this linkage [49].

5.3. Environmental effects

The cellular glycosylation phenotype can be altered by environmental (physiological) factors, such as hormones, pathogens, metabolites, and growth factors (for a review, see [50]. The sensitivity of protein glycosylation to environmental factors may provide an efficient cellular mechanism for modulating the biological attributes of a protein in response to an altered environment, or at a minimum, serve as a convenient and sensitive indicator of changes in cellular phenotype.

6. Functions of protein-linked glycans

Our understanding of the functional roles played by protein-linked oligosaccharides has progressed considerably over the past decade and the reader is referred to a number of outstanding reviews on the subject that have appeared recently [46,51,52]. In a very general sense, the functional role for all protein-linked glycans is to carry genetic information, and this carrier role manifests itself in two ways: (1) as specific binding determinants or (2) as modulators of the attached protein's functional attributes. For the former, the information contained within protein-linked glycans is "decoded" by a diverse set of proteins that bind to specific carbohydrate structures, lectins. Lectins are known to mediate important physiological events and the reader is referred to a number of reviews on the subject (e.g. [53,54]). For the latter functional mechanism, the attached glycan chains can modulate various biochemical and biophysical attributes of the attached protein, including substrate/ ligand binding affinity, catalytic efficiency, and thermostability.

The accelerated discovery of the functional roles for protein-linked glycans over the past decade has been facilitated by observations from several areas of study: (1) the generation of genetically engineered strains of mice that either have targeted deletions or overexpression of glycosyltransferases or glycohydrolases; (2) the human genetics of the glycome (that portion of the genome ascribable to the glycosylation apparatus); and (3) genetic and molecular identification of proteins key to (patho)physiological processes. Transgenic and "knockout" mice have been generated for a number of glycosyltransferases and evaluation of the resulting phenotype has provided insight into the functional roles of protein-linked glycans (e.g., [55]). For example, an apparent reciprocity has been observed between O-GlcNAc glycosylation and O-phosphorylation of Ser/Thr residues on nuclear and cytosolic proteins. This "yin-yang" relationship has led to the hypothesis that one function of this saccharide is to transiently block phosphorylation [29]. In this context, O-GlcNAc modification of cytoplasmic proteins is viewed as an intracellular signaling mechanism that modulates the pool of sites available for phosphorylation. The consequences of such modulation could either be through directly blocking specific signaling pathways or affecting kinase-dependent protein–protein interactions [29]. And once again, environmental signals are an important driver of this mechanism since changes in intracellular UDP-GlcNAc concentration affect the activity of OGT. In this case, intracellular levels of UDP-GlcNAc are sensitive to extra-cellular glucose concentration allowing the sugar nucleotide to function as a glucose sensor. Thus, elevated blood glucose levels, such as those that occur in type II diabetes, would increase intracellular UDP-GlcNAc levels, affecting OGT and increasing the extent of O-GlcNAc substitution of cytosolic proteins. Indeed, transgenic overexpression of OGT under the control of a GLUT4 promoter produces by itself the type II diabetic phenotype: insulin resistance and hyperleptinemia [56].

The use of genetically engineered mice to probe the function of the protein-linked glycans is not without it limitations, notably that strain/species differences can preclude generalization of conclusions drawn from transgenic or knockout mice. This is a recurrent issue for biotechnology, where the glycobiology of the mouse can be different than that of humans. Nonetheless, mice genetically engineered with an altered protein glycosylation apparatus are a powerful tool in elucidating the functional roles of protein-linked glycans, particularly when dealing with multigene families of related glycosyltransferases or hydrolases.

The clinical evaluation of human genetic defects in the protein glycosylation apparatus also provides powerful insights into protein-linked glycan function. The Congenital Disorders of Glycosylation (CDG) are a set of multisystemic disorders caused by genetic defects in the glycosylation apparatus (reviewed in [57]). There are two major subtypes of CDGs: Type I disorders involve defects in the synthesis of lipid-linked precursors for N-glycosylation while Type II disorders involve defects in the processing of protein-linked glycans. There are currently seven groups of Type I disorders (CDG Ia-Ig) characterized by central and peripheral nervous system deficits as well as by hypoglycosylation (including lowered sequon occupancy) of serum proteins. Some of these diseases (e.g., CDG 1a and 1b) can be compensated by dietary supplementation of monosaccharides. In contrast, there are currently four known Type II CDGs (CDG IIa–IId) where, in contrast to CDGS I individuals, patients show more severe psychomotor retardation, no peripheral neuropathy and a normal cerebellum. Perhaps the best known CDG II is Leukocyte Adhesion Deficiency Type II (LAD-II; CDG IIc). LAD-II patients suffer from recurrent infections, marked leukocytosis, and severe growth and psychomotor retardation. The molecular defect in LAD-II has recently been identified in the GDP-Fuc transporter and these patients exhibit the near absence of fucosylated glycoconjugates and, hence, the Bombay phenotype. Other genetic defects, for example in proteoglycan or GPI-anchor biosynthesis [58,59], also have a significant impact on the well being of affected individuals. Thus, the CDGs and other genetic defects in protein glycosylation demonstrate the importance of this process in development and homeostasis.

A third avenue for elucidating the functional roles of protein-linked glycans is from detailed studies of important molecular mediators of biological processes including disease. For example, the congenital muscular dystrophies (CMDs) are a group of disease characterized by progressive loss of muscle strength that becomes apparent (unlike Duchenne's muscular dystrophy) around the time of birth. Genetic mutations have been identified in two of these CMDs: muscle-eye-brain disease (MEB) and Fukuyama CMD (FCMD). The genes for both MEB and FCMD have similarity to known glycosyltransferases and recent work has shown that hypoglycosylation of dystroglycan, a component of the dystrophin glycoprotein complex that bridges intracellular F-actin-dystrophin to extracellular laminin, is common to both MEB and FCMD [60]. These observations in humans are consistent with the neuromuscular defects observed in the *myd* mouse where again a genetic mutation is found in a gene homologous to glycosyltransferases [61]. A conditional neuronal dystroglycan knockout mouse also reproduces the neuronal defects observed in MEB, FCMD, as well as another CMD, Walker-Warburg syndrome [62]. Thus, the molecular biology and genetics of (patho)physiological processes can provide insight into the functional roles of protein-linked glycosylation.

6.1. Evolutionary considerations

Current thoughts around glycan function are rooted in evolutionary considerations where the increased complexity of eukaryotic cells and the organisms they form has demanded sequentially expanded utility of constituent biopolymers, including protein-linked glycans. Helenius and colleagues (e.g., [12]) first suggested that the crucial role played by protein-linked glycans in the "quality control" of folding nascent peptide chains would justify the highly conserved generation of an elaborate, preformed oligosaccharide, its transfer to nascent peptide chains, and its tightly regulated initial deconstruction (Fig. 3). Subsequently, Drickamer and Taylor [63] conjoined this and other observations to suggest that the mammalian pathway of glycoprotein synthesis reflects the sequential development of distinct functions and that the biosynthetic apparatus that supports these functions are segregated along the endomembrane system (Fig. 8). For example, the addition of terminal elements in the medial and *trans*-Golgi, which can serve as binding determinants for intra- and extra-cellular lectins, would have arisen after the role for the core element of N-glycans in the ER QC function. Similarly, as noted in [63], comparison of the N- and O-linked glycosylation pathways suggests that the latter are more like the terminal elaborations of N-glycans: construction of the glycan occurs by the stepwise addition from sugar nucleotide donors largely, if not exclusively, in the Golgi. While O-glycans are known to exhibit a number of functional roles, a well known role for O-glycans is as a selectin ligand [64]. Here again, a process occurring late in the endomembrane system has a functional role associated with higher eukaryotes, cellular adhesion events for immune cell trafficking. Similar observations can be made for GPIs and proteoglycans. Therefore, the functional roles for protein-linked glycans can be viewed as the progressive exploitation of pathways to meet new demands for physiological systems in increasingly complex organisms. Such a

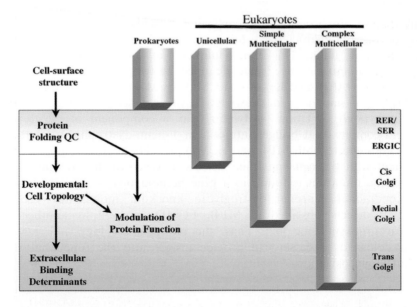

Fig. 8. Sequential progression of protein-linked glycan function.The evolutionary progression of protein-linked glycan function is illustrated from prokaryotes, where the major functional role is structural, to higher eukaryotes where glycans can serve major functional roles as ligands for proteins (lectins) or as modulators of attached protein function. Note that this progression is tracked by a parallel progression in the use of the endomembrane system, with glycan determinants supporting the "latest" functions synthesized in the trans-Golgi, despite the fact that the most primitive eukaryotic members have essentially the same elaborate endomembrane system as higher eukaryotes. The figure is redrawn from [63] and modified to include a distinction between "simple" and "complex" multicellular organisms. This distinction is predicated upon the observation that knockout mice with deletions in the Gn-TI step of N-glycan are embryonic lethal, suggesting an important role for cis/medial-Golgi elements of protein-linked glycan (e.g., antenna formation in complex-type glycans) in development. Moreover, knockout animals with targeted deletions in genes responsible for the terminal elaborations of antennae are often viable though with impaired abilities to mediate certain specific physiological processes (e.g. cell adhesion). A similar set of conclusions can be drawn from the CDG genetic disorders.

perspective provides a rationale for the impressive cellular investment in the maintenance and regulation of the protein glycosylation apparatus.

Abbreviations

CHO	Chinese hamster ovary
Dol	dolichol
EGF	epidermal growth factor
ER	endoplasmic reticulum
ERAD	ER associated degradation
ERGIC	endoplasmic reticulum Golgi intermediate compartment
Fuc	fucose

GalNAc	N-Acetylgalactosamine
Glc	glucose
GlcNAc	N-Acetylglucosamine
GlucUA	glucuronic acid
GPI	glycosyl phosphatidyl inositol
IdoUA	iduronic acid
MAM	mitochondria associated ER membrane
Man	mannose
NAC	nascent polypeptide associated complex
NANA	N-acetylneuraminic acid
OT	oligosaccharyltransferase complex
RNC	ribosome nascent chain complex
SRP	signal recognition particle
Xyl	xylose

References

1. Van den Steen, P., Rudd, P.M., Dwek, R.A., and Opdenakker, G. (1998) Crit. Rev. Biochem. Molec. Biol. 33, 151–208.
2. Lehrman, M.A. (2001) J. Biol. Chem. 276, 8623–8626.
3. Vliegenthart, J.F.G. and Casset, F. (1998) Curr. Opin. Struct. Biol. 8, 565–571.
4. Roth, J. (2002) Chem. Rev. 102, 285–303.
5. Suzuki, T., Yan, Q., and Lennarz, W.J. (1998) J. Biol. Chem. 273, 10083–10086.
6. Helenius, J., Ng, D.T.W., Marolda, C.L., Walter, P., Valvano, M.A., and Aebi, M. (2002) Nature 415, 447–450.
7. Maeda, Y., Tanaka, S., Hino, J., Kangawa, K., and Kinoshita, T. (2000) EMBO J. 19, 2475–2482.
8. Bossuyt, X. and Blanckaert, N. (1993) Biochem. J. 296, 627–632.
9. Silberstein, S. and Gilmore, R. (1996) FASEB J. 10, 849–858.
10. Moller, I., Jung, M., Beatrix, B., Levy, R., Kreibich, G., Zimmerman, R., Wiedmann, M., and Lauring, B. (1998) Proc. Natl. Acad. Sci. USA 95, 13425–13430.
11. Sanjay, A., Fu, J., and Kreibich, G. (1998) J. Biol. Chem. 273, 26094–26099.
12. Helenius, J. and Aebi, M. (2001) Science 291, 2364–2369.
13. Lobsanov, Y.D., Vallee, F., Imberty, A., Yoshida, T., Yip, P., Herscovics, A., and Howell, P.L. (2002) J. Biol. Chem. 277, 5620–5630.
14. Hosokawa, N., Wada, I., Hasegawa, K., Yorihuzi, T., Tremblay, L.O., Herscovics, A., and Nagata, K. (2001) EMBO Reports 2, 415–422.
15. Durrant, C. and Moore, S.E. (2002) Biochem. J. 365, 239–247.
16. Lubas, W.A. and Sprio, R.G. (1988) J. Biol. Chem. 263, 3990–3998.
17. Zuber, C., Spiro, M.J., Guhl, B., Spiro, R.G., and Roth, J. (2000) Mol. Biol. Cell 11, 4227–4240.
18. Dairaku, K. and Spiro, R.G. (1997) Glycobiology 7, 579–586.
19. Oh-eda, M., Nakagawa, H., Akama, T.O., Lowitz, K., Misago, M., Moremen, K.W., and Fukuda, M.N. (2001) Eur. J. Biochem. 268, 1280–1288.
20. Schwientek, T., Bennett, E.P., Flores, C., Thacker, J., Hollmann, M., Reis, C.A., Behrens, J., Mandel, U., Keck, B., Schafer, M.A., Haselmann, K., Zubarev, R., Roepstorff, P., Burchell, J.M., Taylor-Papadimitriou, J., Hollingsworth, M.A., Clausen, H. (2002) J. Biol. Chem. 277, 22623–22638.
21. Hassan, H., Reis, C.A., Bennett, E.P., Mirgorodskaya, E., Roepstorff, P., Hollinngsworth, M.A., Burchell, J., Taylor-Papadimitriou, J., and Clausen, H. (2000) J. Biol. Chem. 275, 38197–38205.
22. Hanisch, F.G., Reis, C.A., Clausen, H., and Paulsen, H. (2001) Glycobiology 11, 731–740.
23. Brinkmann, T., Weilke, C., and Kleesiek, K. (1997) J. Biol. Chem. 272, 11171–11175.
24. Wilson, I.B. (2002) J. Biol. Chem. 277, 21207–21212.

25. Bai, X., Wei, G., Sinha, A., and Esko, J.D. (1999) J. Biol. Chem. 274, 13017–13024.

26. Higuchi, T., Tamura, S., Tanaka, K., Takagaki, K., Saito, Y., and Endo, M. (2001) Biochem. Cell Biol. 79, 159–164.

27. Harris, R.J. and Spellman, M.W. (1993) Glycobiology 3, 219–224.

28. Harris, R.J., van Halbeek, H., Glushka, J., Basa, L.J., Ling, V.T., Smith, K.J., and Spellman, M.W. (1993) Biochemistry 32, 6539–6547.

29. Wells, L., Vosseller, K., and Hart, G.W. (2001) Science 291, 2376–2378.

30. Zachara, N.E. and Hart, G.W. (2002) Chem. Rev. 102, 431–438.

31. Kreppel, L.K., Blomberg, M.A., and Hart, G.W. (1997) J. Biol. Chem. 272, 9308–9315.

32. Lubas, W.A., Frank, D.W., Krause, M., and Hanover, J. (1997) J. Biol. Chem. 272, 9316–9324.

33. Kreppel, L.K. and Hart, G.W. (1999) J. Biol. Chem. 274, 32015–32022.

34. Kawagoe, K., Kitamura, D., Okabe, M., Taniuchi, I., Ikawa, M., Watanabe, T., Kinoshita, T., and Takeda, J. (1996) Blood 87, 3600–3606.

35. Takeda, J., Miyata, J., Kawagoe, K., Iida, Y., Endo, Y., Fujita, T., Takahashi, M., Kitani, T., and Kinoshita, T. (1993) Cell 73, 703–711.

36. Udenfriend, S. and Kodukula, K. (1995) Annu. Rev. Biochem. 64, 563–591.

37. Ferguson, M.A.J. (1999) J. Cell Sci. 112, 2799–2809.

38. Watanabe, R., Murakami, Y., Marmor, M.D., Inoue, N., Maeda, Y., Hino, J., Kangawa, K., Julius, M., and Kinoshita, T. (2000) EMBO J. 19, 4402–4411.

39. Maeda, Y., Watanabe, R., Harris, C.L., Hong, Y., Ohishi, K., Kinoshita, K., and Kinoshita, T. (2001) EMBO J. 20, 250–261.

40. McConville, M.J. and Menon, A.K. (2000) Mol. Memb. Biol. 17, 1–16.

41. Taron, C.H., Wiedman, J.M., Grimme, S.J., and Orlean, P. (2000) Mol. Biol. Cell 11, 1611–1630.

42. Hong, Y., Maeda, Y., Watanabe, R., Inoue, N., Ohishi, K., and Kinoshita, T. (2000) J. Biol. Chem. 275, 20911–20919.

43. Ohishi, K., Inoue, N., and Kinoshita, T. (2001) EMBO J. 20, 4088–4098.

44. Cruz, H.J., Conradt, H.S., Dunker, R., Peixoto, C.M., Cunha, A.E., Thomaz, M., Burger, C., Dias, E.M., Clemente, J., Moreira, J.L., Rieke, E., Carrondo, M.J. (2002) J. Biotechnol. 96, 169–183.

45. Camphausen, R.T., Yu, H.-A., and Cumming, D.A.The Role of Polypeptide in the Biosynthesis of Protein-Linked Oligosaccharides (1995) In: Montreuil, J., Schachter, H. and Vliegenthart, J.F.G. (eds.) Glycoproteins in New ComprehensiveBiochemistry. Amsterdam, Elsevier Science.

46. Dwek, R.A. (1995) Biochem. Soc. Trans. 270, 3261–3267.

47. Schachter, H. (1986) Biochem. Cell Biol. 64, 163–181.

48. Joziasse, D.H., Schiphorst, W.E.C.M., van den Eijden, D.H., van Kuik, J.A., van Halbeek, H., and Vliegenthart, J.F.G. (1987) J. Biol. Chem. 252, 2025–2033.

49. Tsuda, E., Goto, M., Murakami, A., Akai, K., Ueda, M., and Kawanishi, G. (1988) Biochemistry 27, 5646–5654.

50. Andersen, D.C. and Goochee, C.F. (1994) Curr. Opin. Biotechnol. 5, 546–549.

51. Kim, Y.J. and Varki, A. (1997) Glyconj. J. 14, 569–576.

52. Dwek, R.A. (1998) Dev. Biol. Stand. 96, 43–47.

53. Fujita, T. (2002) Nat. Rev. Immunol. 2, 346–353.

54. Dodd, R.B. and Drickamer, K. (2001) Glycobiology 11, 71R–79R.

55. Daniels, M.A., Devine, L., Miller, J.D., Moser, J.M., Lukacher, A.E., Altman, J.D., Kavathas, P., Hogquist, K.A., and Jameson, S.C. (2001) Immunity 15, 1051–1061.

56. McClain, D.A., Lubas, W.A., Cooksey, R.C., Hazel, M., Parker, G.J., Love, D.C., and Hanover, J.A. (2002) Proc. Natl. Acad. Sci. (USA) 99, 10695–10699.

57. Schachter, H. (2001) Cell Mol. Life Sci. 58, 1085–1104.

58. Kim, B.T., Kitagawa, H., Tamura, J., Saito, T., Kusche-Gullberg, M., Lindahl, U., and Sugahara, K. (2001) Proc. Natl. Acad. Sci. USA 98, 7176–7181.

59. Lyakisheva, A., Felda, O., Ganser, A., Schmidt, R.E., and Schubert, J. (2002) Exp. Hematol. 30, 18–25.

60. Michele, D.E., Barresi, R., Kanagawa, M., Saito, F., Cohn, R.D., Satz, J.S., Dollar, J., Nishino, I., Kelley, R.I., Somer, H., Straub, V., Mathews, K.D., Moore, S.A., Campbell, K.P. (2002) Nature 418, 417–422.

61. Yoshida, A., Kobayashi, K., Manya, H., Taniguchi, K., Kano, H., Mizuno, M., Inazu, T., Mitsuhashi, H., Takahashi, S., Takeuchi, M., Herrmann, R., Straub, V., Talim, B., Voit, T., Topaloglu, H., Toda, T., Endo, T. (2001) Dev. Cell 1, 717–724.

62. Moore, S.A., Saito, F., Chen, J., Michele, D.E., Henry, M.D., Messing, A., Cohn, R.D., Ross-Barta, S.E., Westra, S., Williamson, R.A., Hoshi, T., Campbell, K.P. (2002) Nature 418, 422–425.

63. Drickamer, K. and Taylor, M.E. (1998) TIBS 23, 321–324.

64. Sperandio, M., Thatte, A., Foy, D., Ellies, L.G., Marth, J.D., and Ley, K. (2001) Blood 97, 3812–3819.

65. Esko, J.D. and Lindahl, U. (2001) J. Clin. Invest. 108, 169–173.

S.C. Makrides (Ed.) *Gene Transfer and Expression in Mammalian Cells*

CHAPTER 15

Metabolic engineering of mammalian cells for higher protein yield

Hitto Kaufmann[1] and Martin Fussenegger[2]

[1]*Walter and Eliza Institute of Medical Research, Post Office,*
Royal Melbourne Hospital Victoria, 3050, Australia
[2]*Institute of Biotechnology, Swiss Federal Institute of Technology, ETH Zurich,*
CH-8093 Zurich, Switzerland; Tel.: +41 1 633 34 48; Fax: +41 1 633 10 51.
E-mail: fussenegger@biotech.biol.ethz.ch

1. Therapeutic protein production from mammalian cells

The advent of recombinant DNA technology enabled heterologous gene expression in living cells and thereby large-scale production of protein therapeutics. Many protein pharmaceuticals are produced using bacterial and yeast cells because of their rapid growth and high expression levels. However, many of the most promising protein therapeutics require post-translational modifications such as glycosylation for full therapeutic efficacy. Therefore, mammalian cells are rapidly becoming the standard for industrial production of pharmaceutical proteins. Chinese hamster ovary (CHO) cells have emerged as the most widely used industrial production cell line since it is: (i) well characterized; (ii) known for the stability of chromosomal transgene integrations; (iii) easy-to-cultivate at large scale; and (iv) simple to adapt for suspension growth in serum- and protein-free media [1].

The expanding list of therapeutic proteins produced by mammalian cell culture processes includes blood clotting factors, monoclonal antibodies and thrombolytics. A prominent recent example is Herceptin (Trastuzumab), the first humanized monoclonal antibody licensed for the treatment of metastatic HER2 (human epidermal growth factor receptor 2) expressing breast cancer [2].

Currently, there is a worldwide shortage in mammalian cell-based production capacities. As of the beginning of the new millennium more then 100 biologicals were either in phase III clinical trials or awaiting FDA approval [3]. An annual approval rate of up to 10 therapeutic monoclonal antibodies is expected in the years to come [3]. Large-scale production of monoclonal antibodies is particularly challenging because the average dosing regimes are typically up to an order of magnitude higher compared to other protein therapeutics. Considering these developments in the biotechnology industry as well as the high cost/low yield characteristics of mammalian cell culture processes compared to bacterial or yeast fermentations, there is increasing demand for optimization of various production parameters including cell-specific productivity and product quality.

Early approaches to optimize mammalian cell culture performance were entirely focused on improving process strategies and reactor design. The concept of metabolic engineering introduced by James E. Bailey in 1991 marked the start of a new era in biopharmaceutical manufacturing aiming at improving production of high-quality heterologous protein therapeutics by engineering of enzymatic, transport and regulatory functions of cells using recombinant DNA technology [4]. Here we review current metabolic engineering strategies to improve the productivity of mammalian cell lines.

2. Initial metabolic engineering approaches

Initial metabolic engineering strategies to improve mammalian production cell lines and reduce production costs focused on adaptation of mammalian cells for growth in suspension and serum- or even protein-free media. Examples for non-engineering-based approaches to produce anchorage- and serum-independent cell lines included cultivation of CHO-DUKX cells in decreasing concentrations of serum. However, this strategy is time-consuming, unpredictable and difficult to standardize [5]. At that time, two research groups took a metabolic engineering approach to reprogram mammalian cells for growth in suspension and protein-free media. Stable expression of transferrin and insulin-like growth factor I (IGF-I) in CHO-K1 cells resulted in a cell line able to proliferate under protein-free conditions [6]. Similar results were obtained when the cell-cycle regulating protein cyclin E was expressed in CHO-K1 cells [7]. However, even in the early days of mammalian cell metabolic engineering increasing the product yield was a major focus. For example, Pendse and Bailey could show that dexamethasone-responsive expression of *Vitreoscilla* hemoglobin in CHO cells increased the production of human tissue plasminogen activator (tPA) by 100% [8]. However, at that stage metabolic engineering of mammalian cells consisted of a collection of examples rather than an integrated long-term strategy for rational genetic improvement of key cellular production characteristics.

3. Transcriptional hotspot targeting, chromosomal locus amplification and episomal expression strategies

Most cell lines currently used for biopharmaceutical manufacturing contain a constitutive expression unit stably integrated in the target chromosome. However, the transcriptional activity varies depending on the cellular levels of the relevant transcription factors and on the chromatin structure at the integration site. The dominance of illegitimate recombination (IR) in immortalized cell lines results in rare transgene integrations randomly distributed on the target chromosome. This requires intensive clonal screening programs to identify the desired high-level producers. In order to reduce the expensive and time-consuming screening procedures, a variety of strategies have been designed to enable single-step development of a high-level producer: (i) Complementation-based screening and amplification of desired chromosomal loci using physiological markers such as dihydrofolate reductase (*dhfr*) and glutamine synthetase (*gs*) [9–11]. Coamplification of the product gene and the selection marker results from a step-wise increase of relevant drug concentrations (methotrexate for *dhfr*; methionine sulphoximine for *gs*) (see Chapter 7). (ii) Screening for high-level expression of a reporter gene that is linked to expression of the product gene [12,13]. For example, FACS-mediated screening for high-level expression of green fluorescent protein will determine high level producing cell clones. (iii) Recombinase-mediated cassette exchange procedure (RMCE) (see Chapter 20) represents an advanced technology based on site-specific recombinases such as Cre or Flp [14]. RMCE enables expression of the product gene in the absence of any coexpressed selection marker but requires prescreening for transcriptional hotspots, which can then be used as expression loci for any desired transgene. (iv) Chromatin structures such as boundary elements (BE) (see Chapter 11), matrix attachment regions (MAR) (see Chapters 10 and 20) and locus control regions (LCR) (see Chapter 12), have been shown to boost transgene expression levels. Work by Mermod and coworkers showed that MARs flanking the expression cassette increase the frequency of obtaining high producing CHO cell clones [15] (see Chapter 10).

Another strategy to avoid 'positional effects' due to random chromosomal integrations is expression of product genes from episomally replicating vectors. Most currently used artificial episomes are based on virus-encoded transacting factors such as the SV40 large T antigen or the Epstein-Barr virus nuclear antigen (EBNA) which mediate time-limited episomal amplification of plasmids encoding the respective viral origins of replication (ori_{SV40} or *oriP*; see Ref. [16] for a review). Unfortunately, these episomal expression systems are transient in nature unless ori_{SV40} is combined with the scaffold/matrix attachment region (S/MAR) on the same expression vector which mediates stable selection-independent maintenance and replication for more than 100 generations [17].

4. Biphasic production processes—controlled proliferation in biotechnology

4.1. The concept

In biopharmaceutical manufacturing a high proliferation rate of production cell lines was long considered desirable since improved process yields often resulted from an increase in cell number. However, current research interests are focused on studying production under proliferation-controlled conditions [18–22]. The predominant biopharmaceutical manufacturing set-up is batch production which is based on cultivating cell lines adapted for growth in suspension and protein-free media in a stirred tank bioreactor. Typically, production cell lines are seeded into the vessel at a low density and grow for 1–2 weeks while producing the desired product protein. A classical batch cultivation process consists of four sequential phases: (i) lag phase; (ii) exponential growth phase; (iii) stationary phase; and (iv) cell death or decline phase. In a standard production process cells continuously proliferate/produce at high rates until cell death resulting from medium consumption or accumulation of waste products terminates the process. Several findings pointed to a positive correlation between reduced growth rates

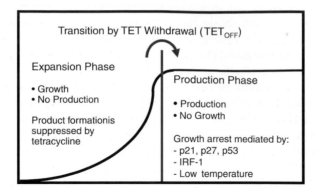

Fig. 1. The biphasic production process. This novel bioprocess consists of two distinct phases, an expansion phase where product formation is suppressed while cells proliferate followed by a production phase in which production is initiated concomitant with growth inhibition. Induction of the growth arrest genes may, for example be initiated by the tetracycline-responsive expression system.

and specific productivity. For example, hyperosmolarity-induced growth rate reduction of hybridoma cultures resulted in increased specific antibody productivity [23,24]. Based on these initial results metabolic engineering strategies to control proliferation of mammalian production cell lines have been developed to establish biphasic bioprocesses with the following characteristics (Fig. 1): (i) In the first phase production cell lines proliferate at high rates until they reach desired cell densities. During this expansion phase no product proteins are produced (non-producing expansion phase). (ii) At the desired cell density proliferation is arrested and transgene expression is induced (proliferation-arrested production phase). This biphasic controlled proliferation process combines high cell density cultivation, required for achieving elevated product yields, with production under proliferation-controlled conditions.

4.2. Separating the phases: regulated gene expression in mammalian cells

Restriction of heterologous protein formation to the second, growth-arrested production phase is a major challenge for controlled proliferation technology. A variety of heterologous mammalian gene control systems have been established. For an overview on heterologous gene regulation strategies we refer to Chapter 22 in this volume as well as to [25]. The most widely used adjustable gene control systems are the "Tet systems" which are responsive to antibiotics of the tetracycline (tet) class. Tet systems enable tight repression as well as high-level expression of genes driven by a chimeric tetracycline-responsive promoter ($P_{hCMV*-1}$). The initial configuration of the Tet systems was based on a chimeric tetracycline-dependent transactivator (tTA), a fusion of the bacterial regulator of the tetracycline-resistance regulon (TetR) and the eukaryotic transactivation domain derived from *Herpes simplex* protein VP16 [26]. In the absence of tet, tTA (TetR-VP16) binds to $P_{hCMV*-1}$, consisting of seven tandem tet operators (*tetO₇*) cloned adjacent to a minimal eukaryotic promoter, and activates the desired transgene [26]. Expression levels can be adjusted by addition of tet since binding of tTA to P_{hCMV^*-1} is modulated in a tet-dependent manner. Tetracycline is a light- and temperature-sensitive agent and degrades under cell culture conditions in a self-sufficient manner following simple first-order kinetics that are independent of the cell line and cell culture medium used. An elegant approach involved coexpression of product and growth-arresting genes under the control of the tet-regulatable promoter. Tet added at the beginning of the process at concentrations sufficient to repress product and growth arrest genes degrades until its concentration reaches a threshold level below which product and proliferation-controlling genes are concomitantly induced. This results in an automatic transition from the non-proliferating expansion to the proliferation-controlled production phase. Recent experimental data demonstrate that this transition can be timed to occur at any desired time point only depending on the initial tet concentration [27].

4.3. Proliferation control of mammalian cells by overexpression of growth inhibitory proteins

Human cancers develop as a result of accumulation of functional mutations in key cell growth and death regulators. For example, tumor suppressor genes either inhibit proliferation, apoptosis, induce/maintain terminally differentiated cells states, or ensure the genomic integrity of cells. One of the first controlled proliferation technologies was based on the growth-arresting potential of the interferon regulatory factor 1 (IRF-1) [28]. Baby hamster kidney (BHK) cells were stably transfected with a fusion between the genes encoding IRF-1 and the hormone-binding domain of the human estrogen receptor [29]. In this configuration, estradiol is required to activate IRF-1 and elicit sustained growth arrest. Unfortunately, prolonged periods of hormone exposure resulted in cell death and lower specific productivities [29,30]. Recently, an improved culture regime has been described for IRF-1-induced controlled proliferation technology in BHK cells. In a microcarrier-based perfusion reactor hormone exposure of the production cell line was limited to short time intervals to avoid the negative impact of estradiol-induced IRF-1 expression on culture viability [20]. Furthermore, induction of protein production at the onset of growth arrest was achieved by placing the product gene (human factor VII) under the control of an IRF-1-responsive promoter thereby creating an autoregulated expression system that results in high level expression of the recombinant blood clotting factor [20].

Over the past years, dramatic progress has been made towards understanding how core regulatory proteins are integrated in a cellular network to control the mammalian cell cycle [31]. The identification of cyclin-dependent kinase inhibitors (CKIs) and other proteins that inhibit cell growth at specific points of the cell cycle enabled the development of novel metabolic engineering strategies to control proliferation of mammalian cells by conditional overexpression of cell-cycle modulating genes.

Pioneering experiments focused on transient expression of p21, p27 and an apoptosis-deficient p53 mutant to evaluate the potential of these cell-cycle regulators to arrest CHO cells in the G1-phase. Interestingly, in all cases G1-specific growth arrest correlated with enhanced specific productivity of the cocistronically encoded human model glycoprotein SEAP (human secreted alkaline phosphatase) [18]. Clonal cell lines engineered for tetracycline-responsive expression of the CKI p27 and SEAP also showed a sustained G1-specific growth arrest upon withdrawal of the regulating agent tetracycline, which was accompanied by up to 10-fold higher specific SEAP productivities compared to proliferation-competent control cells (producing about 8 ng SEAP cell^{-1} day^{-1}) [32]. In a parallel approach SEAP was expressed in CHO-derived cells concomitant with the CKI p21 and the differentiating factor CCAAT/enhancer-binding protein in a tricistronic expression configuration [19]. Again, efficient proliferation control was achieved and was accompanied by 10- to 15-fold increased specific productivity [19].

Another strategy to establish biphasic proliferation-controlled production processes is based on low temperature-induced growth arrest. In mammalian cells, a mild temperature reduction to 27–30°C was shown to result in a sustained G1-specific growth arrest associated with increased culture viability due to delayed onset of apoptosis [21]. Following a downshift in temperature from 37 to 30°C CHO batch cultivations demonstrated a significant increase of specific SEAP productivity [21]. In addition, low temperature allowed extended cultivation times resulting in a 3.4-fold higher overall product yield [21].

Initial studies on novel metabolic engineering strategies are often performed in small scale cultures using adherent cells which produce a quantifiable model glycoprotein. Although such small-scale strategies enable fast and reliable proof-of-concept studies evaluation of the full impact of a new process strategy on biopharmaceutical manufacturing requires global consideration of the following issues: (i) Does the engineering strategy influence product quality? (ii) Can improved cell characteristics be transferred to established processes using different cell lines and product proteins? (iii) Can the metabolic engineering concept be applied to large-scale production using cells which grow in suspension and serum- or preferably, protein-free media?

Getting the glycosylation right is a crucial requirement regarding product quality and pharmacokinetics. The glycosylation pattern of a product protein represents the entity of its various glycoforms and directly influences product quality (see Chapter 14). The glycosylation pattern may vary in an unpredictable way with any subtle change in bioprocess parameters. Therefore, the impact of controlled proliferation technologies on protein glycosylation has been analyzed: (i) for SEAP produced in low temperature or p27-arrested CHO cells; and (ii) for erythropoietin secreted by IRF-1 growth-controlled BHK cells [33,34]. The data suggest that product quality of p27-arrested CHO cells remains unchanged compared to proliferating production cell lines whereas a temperature shift to 30°C or IRF-1-mediated growth control increase protein sialylation, a parameter correlating with improved pharmacokinetics [33]. Overall, there is no evidence so far that controlled proliferation technologies interfere negatively with current product quality standards of biopharmaceutical manufacturing.

Although predicted by transient transfections in CHO cells, it has only recently been reported that inducible expression of p21 alone results in a sustained G1-specific growth arrest in an antibody-producing NSO myeloma cell line without showing any undesired impact on cell viability [22]. The specific antibody productivity increased more than fourfold in these proliferation-controlled NSO cultures compared to the same cell line grown in the absence of p21 expression. Furthermore, controlled proliferation technology was engineered by muristerone-inducible p27 expression in *dhfr*-deficient, anchorage- and serum-independent CHO-DUKX B11-derived cells constitutively producing the common cold therapeutic soluble intercellular adhesion molecule 1 (sICAM) [35]. In this configuration, the G1-specific growth arrest resulted in up to fivefold higher specific sICAM productivity compared to the same cell line assessed during proliferation (absence of muristerone, p27 repressed) [35]. It is interesting to note that in both the NSO and the CHO-DUKX B11 cell lines, the product gene is expressed constitutively thus enabling direct analysis of enhanced productivity by CKI-mediated proliferation control over a true isogenic control [22,35]. Therefore, any production cell line that constitutively produces a desired protein therapeutic may be engineered for controlled proliferation to achieve increased productivities.

Decreased cell viability upon prolonged exposure to estradiol was found to be a major challenge in adapting IRF-1-based controlled proliferation technologies for biopharmaceutical manufacturing realities (see above). However, perfusion technology enabled fine-tuning of estradiol concentrations and managed the detrimental effects of extended IRF-1 expression. This strategy enabled successful scale up of BHK-based IRF-1-arrested production set-ups to produce hetero-chimeric antibodies in up to 4-liter reactors [20] and the blood coagulation Factor VII in 2-liter fermentors [30]. Both studies report efficient proliferation management concomitant with high cell viability over a cultivation period of several weeks [20,30]. The accumulating data on proliferation control of various mammalian production cell lines point to a high potential of this engineering strategy for large scale industrial protein production in the near future.

5. Counteracting suicidal tendencies: preventing cell death in mammalian cell cultures

Cells from higher organisms are capable of committing suicide by initiating a highly conserved molecular program known as apoptosis. Over the past decade basic research has elucidated the fundamentals of the genetic control of apoptosis [36]. All signaling pathways that trigger apoptosis within the cell ultimately lead to activation of caspases, aspartate-specific, cysteine-dependent proteases. Proteins such as IAPs or Bcl-2 family members have been shown to regulate this process within the cell (Fig. 2). Several studies suggest that apoptosis is the predominant form of cell death in mammalian cell batch cultivations [37]. High percentages of apoptotic cells have been described starting from late exponential growth phase in hybridoma, myeloma, CHO and BHK cultures. Up to 80% of cell death in standard serum-free CHO batch

464

Core Pathways of Programmed Cell Death

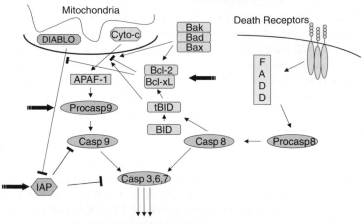

Programmed Cell Death

Fig. 2. Central components of the mammalian cell death machinery. Programmed cell death, or apoptosis is an evolutionary conserved process by which organisms remove damaged or superfluous cells. At the core of apoptosis execution are caspases, a family of aspartate-specific, cysteine-dependent proteases. Caspases function in a cascade in which procaspases are activated by upstream death signals. These signals include cytokine binding to receptors such as members of the TNF receptor family and release of cytochrome C or DIABLO/Smac from the mitochondria. Antiapoptotic members of the Bcl-2 family of proteins can protect cells from various death stimuli acting on upstream death signaling molecules. Inhibitor of apoptosis proteins (IAPs) such as XIAP bind and antagonize processed caspase 3 and 9. Arrows indicate protein targets for antiapoptosis engineering approaches.

cultivations has been associated with apoptosis [38]. Elimination of serum from most biopharmaceutical production processes is thought to contribute significantly to the sensitivity of cultured cells to programmed cell death. In fact, cells adapted to growth in serum/protein-free media could be protected from programmed cell death caused by nutrient deprivation by re-adding serum to the culture medium [39]. The loss in production yield related to apoptosis has initiated a new field in mammalian cell culture engineering, so called "antiapoptosis engineering", aimed at reducing programmed cell death in production cultures by targeted molecular interventions in key apoptosis regulatory pathways. Most antiapoptosis engineering strategies are focused on overexpression of anti-apoptotic genes of the bcl-2 family of proteins (for example, bcl-2 and bcl-x_L) in various mammalian cell lines [19,37,40–46] (Fig. 2).

Several groups have shown that apoptosis in hybridoma cell cultures can be reduced or its onset delayed by overexpression of bcl-2 [47]. Similarly, bcl-2 could suppress apoptosis in NSO cells producing a chimeric antibody [45]. Although bcl-2-based antiapoptosis engineering resulted in 20% increased viability and extended cultivation times of NSO cell cultures the antibody titer remained unchanged [45]. However, when the cultivation mode was changed from batch to fed batch, bcl-2-expressing cells reached 60% higher maximum cell numbers correlating with a 40% increased product titer compared to relevant control cells [45]. Several studies investigated the possible benefits of *bcl-2* overexpression in CHO cells: (i) A γ-interferon-producing CHO cell line engineered for overexpression of *bcl-2* could be protected from cell death following withdrawal of insulin and transferrin from the culture medium [38]. (ii) The effect of overexpression of *bcl-2* and *bcl-x$_L$* on the survival of BHK and CHO cells has been investigated in various configurations. Both proteins could mediate protection in response to culture insults such as glucose deprivation or the presence of millimolar quantities of ammonium [43]. (iii) Fixed-bed cultivation of a CHO cell line expressing *bcl-2* and SEAP in a dicistronic tetracycline-responsive manner resulted in a twofold higher viable cell density [41]. A recent study on antiapoptosis engineering of mammalian production cell lines grown in protein-free media demonstrated that *bcl-2* family members did not increase the viability of CHO-DG44 cell cultures under conditions mimicking biopharmaceutical manufacturing. Overexpression of *bcl-2* even resulted in a dramatic decrease in the product yield [48]. However, coexpression of *bcl-x$_L$* increased the product yield for various intracellular and secreted product proteins. Advantages in apoptosis protection could only be observed following *dhfr*-based amplification of *bcl-2* and *bcl-x$_L$* [48]. This result may be linked to the increased numbers of mitochondria present in CHO cell derivatives growing under serum- or protein-free conditions [48].

Since antiapoptosis engineering strategies have been successful in many configurations, genetically optimized *bcl-2* family members have recently been designed to improve their capacity to increase viability. Determination of the 3D structure of Bcl-x$_L$ revealed a large unstructured loop region [49]. Structure modeling suggests a similar domain for Bcl-2. CHO- and BHK-derived cell lines expressing a Bcl-2 loop deletion mutant demonstrated a higher degree of protection against culture insults such as serum starvation and sindbis virus infection compared to cells expressing the wild type protein [40]. Another interesting addition to the antiapoptosis engineering strategies was the overexpression of a dominant negative caspase-9 mutant that was shown to inhibit apoptosis in various cell lines and bioreactor configurations [50]. However, the caspase-9 dominant negative mutant was unable to completely suppress apoptosis. The caspase mutant rather delayed the onset of apoptosis by approximately 2 days in standard bioreactor operations (van de Goor, personal communication). Besides Bcl-2 family members and dominant negative caspase-9 mutants, caspase inhibitors such as XIAP have been successfully used to delay the onset of apoptosis in biotechnologically relevant cell lines [51].

6. Third generation metabolic engineering—multiregulated multigene metabolic engineering

The classic metabolic engineering approach consists of overexpression of a single transgene followed by detailed analysis of desired cell phenotypes including cell viability, product yield and quality. The metabolic engineering determinants may be expressed in sense or antisense orientation and/or placed under control of an adjustable or constitutive promoter. Yet, most approaches are based on constitutive expression of the full-length wild type gene. This is true for example for the aforementioned antiapoptosis engineering strategies based on overexpression of *bcl-2* family members. However, advanced future metabolic engineering strategies may require simultaneous expression of several transgenes to achieve increasingly complex phenotypes, in particular as basic research reveals more and more attractive target molecules that could be used to modulate biotechnologically relevant cell characteristics [52,53].

Pioneering work on cocistronic expression of p27 with *bcl-x$_L$* suggests that efficient genetic engineering of production cells may require adjusting the level of more than one transgene [19]. Cell clones overexpressing both genes displayed higher specific productivities compared to clones that only express the cyclin-dependent kinase inhibitor p27. However, in this early multigene metabolic engineering approach expression of both genes was driven by the same tetracycline-inducible promoter. Latest-generation multigene metabolic interventions are based on independent control of different transgenes, so-called multiregulated multigene metabolic engineering. Such technology requires at least two compatible heterologous gene regulation systems, an option that became available since the development of the novel streptogramin- and macrolide-responsive gene control systems [54] (see Chapter 22). Functional compatibility with the tetracycline-regulatable expression system enabled independent regulation of two genes (dual-regulated expression technology) within the same cell by titration of pristinamycin and tetracycline concentrations in the culture medium. Dual-regulated expression technology based on sense and antisense expression of p27 enables complete proliferation management of mammalian cells. Antisense expression of this cyclin-dependent kinase inhibitor led to increased proliferation of engineered CHO cells, whereas expression of p27 in the sense orientation resulted in a sustained G1-specific growth arrest [55]. Such proliferation management could be seen as a pilot metabolic engineering strategy towards complex expansion (induction of growth in grafted cells) and reimplantation (induction of growth arrest in treated cells) scenarios urgently required for advanced gene therapy and tissue engineering. There, growth arrest is required to prevent any carcinogenic characteristics of treated cells following reimplantation into the patient. Also, ongoing *in vivo* selection of the growth-arrested phenotype may require sophisticated surface selection recently developed based on multidomain surface markers [56].

7. Outlook

The goal to improve mammalian cell culture processes can be achieved by diverse approaches including: (i) new bioreactor design; (ii) altered physical or chemical cultivation parameters; or (iii) metabolic engineering of production cell lines. Although metabolic engineering is certainly the most powerful strategy to improve cell phenotypes and product characteristics, integrated approaches combining (i), (ii) and (iii) will likely lead to the best biopharmaceutical manufacturing results. The experience in metabolic engineering accumulated over the past decades suggests that any perturbation of environment or genome will result in a multigene/multitranscript/multiprotein response [57]. The development of novel technologies now provides the platform for obtaining a global picture of host cell responses to distinct genetic or environmental changes. The emerging science of functional genomics with its core element transcriptomics, the analysis of mRNA expression profiles, and proteomics, the study of protein abundance patterns, promises to provide a vast amount of valuable information for mammalian cell culture process development.

As bioengineers become increasingly competent in rational reprogramming of mammalian cells they will have a powerful technology at their hands enabling the development of strategies to reduce production costs and to overcome the worldwide shortage in biopharmaceutical manufacturing capacity. The metabolic engineering community may also expand to produce therapeutic proteins at the place where they are required—in the human body. In this respect, metabolic engineering strategies may gain impact in the gene therapy and tissue engineering arena in the not-too-distant future.

Acknowledgements

We thank Stefan Schlatter and Monilola Olayioye for critical comments on the manuscript. Work in the laboratory of M.F. is supported by the Bundesamt für Bildung und Wissenschaft (BBW) within the 5th Framework Programme of the European Commission.

Abbreviations

BHK	baby hamster kidney
CHO	Chinese hamster ovary
CKI	cyclin-dependent kinase inhibitor
IRF-1	interferon regulatory factor 1
SEAP	secreted alkaline phosphatase
sICAM	soluble intercellular adhesion molecule
Tet	tetracycline

References

1. Zang, M., Trautmann, H., Gandor, C., Messi, F., Asselbergs, F., Leist, C., Fiechter, A., and Reiser, J. (1995) Biotechnology (NY) 13, 389–392.
2. Brenner, T.L. and Adams, V.R. (1999) J. Am. Pharm. Assoc. (Wash) 39, 236–238.
3. Garber, K. (2001) Nat. Biotechnol. 19, 184–185.
4. Bailey, J.E. (1991) Science 252, 1668–1675.
5. Sinacore, M.S., Charlebois, T.S., Harrison, S., Brennan, S., Richards, T., Hamilton, M., Scott, S., Brodeur, S., Oakes, P., Leonard, M., Switzer, M., Anagnostopoulos, A., Foster, B., Harris, A., Jankowski, M., Bond, M., Martin, S., Adamson, S.R. (1996) Biotechnol. Bioeng. 52, 518–528.

6. Pak, S.C.O., Hunt, S.M.N., Bridges, M.W., Sleigh, M.J., and Gray, P.P. (1996) Cytotechnology 22, 139–146.
7. Renner, W.A., Lee, K.H., Hstzimanikatis, V., Bailey, J.E., and Eppenberger, H.M. (1995) Biotechnol. Bioeng. 47, 476–482.
8. Pendse, G.J. and Bailey, J.E. (1994) Biotechnol. Bioeng. 44, 1367–1370.
9. Barnes, L.M., Bentley, C.M., and Dickson, A.J. (2001) Biotechnol. Bioeng. 73, 261–270.
10. Bebbington, C.R., Renner, G., Thomson, S., King, D., Abrams, D., and Yarranton, G.T. (1992) Biotechnology (NY) 10, 169–175.
11. Kaufman, R.J. (1990) Methods Enzymol. 185, 537–566.
12. Kaufman, R.J., Davies, M.V., Wasley, L.C., and Michnick, D. (1991) Nucl. Acids Res. 19, 4485–4490.
13. Hunt, L., Batard, P., Jordan, M., and Wurm, F.M. (2002) Biotechnol. Bioeng. 77, 528–537.
14. Baer, A. and Bode, J. (2001) Curr. Opin. Biotechnol. 12, 473–480.
15. Zahn-Zabal, M., Kobr, M., Girod, P.A., Imhof, M., Chatellard, P., De Jesus, M., Wurm, F., and Mermod, N. (2001) J. Biotechnol. 87, 29–42.
16. Calos, M.P. (1996) Trends Genet. 12, 463–466.
17. Piechaczek, C., Fetzer, C., Baiker, A., Bode, J., and Lipps, H.J. (1999) Nucl. Acids Res. 27, 426–428.
18. Fussenegger, M., Mazur, X., and Bailey, J.E. (1997) Biotechnol. Bioeng. 55, 927–939.
19. Fussenegger, M., Schlatter, S., Datwyler, D., Mazur, X., and Bailey, J.E. (1998) Nat. Biotechnol. 16, 468–472.
20. Geserick, C., Bonarius, H.P., Kongerslev, L., Hauser, H., and Mueller, P.P. (2000) Biotechnol. Bioeng. 69, 266–274.
21. Kaufmann, H., Mazur, X., Fussenegger, M., and Bailey, J.E. (1999) Biotechnol. Bioeng. 63, 573–582.
22. Watanabe, S., Shuttleworth, J., and Al-Rubeai, M. (2002) Biotechnol. Bioeng. 77, 1–7.
23. Yang, X.H., Oehlert, G.W., and Flickinger, M.C. (1996) Biotechnol. Bioeng. 5, 184–196.
24. Bibila, T.A., Ranucci, C.S., Glazsomitsky, K., Buckland, B.C., and Aunins, J.G. (1994) Biotechnol. Prog. 10, 87–96.
25. Fussenegger, M. (2001) Biotechnol. Prog. 17, 1–51.
26. Gossen, M. and Bujard, H. (1992) Proc. Natl. Acad. Sci. USA 89, 5547–5551.
27. Mazur, X., Eppenberger, H.M., Bailey, J.E., and Fussenegger, M. (1999) Biotechnol. Bioeng. 65, 144–150.
28. Kirchhoff, S., Schaper, F., and Hauser, H. (1993) Nucl. Acids Res 21, 2881–2889.
29. Kirchhoff, S., Kroger, A., Cruz, H., Tummler, M., Schaper, F., Koster, M., and Hauser, H. (1996) Cytotechnology 22, 147–156.
30. Carvalhal, A.V., Moreira, J.L., and Carrondo, M.J. (2001) J. Biotechnol. 92, 47–59.
31. Fussenegger, M. and Bailey, J.E. (1998) Biotechnol. Prog. 14, 807–833.
32. Mazur, X., Fussenegger, M., Renner, W.A., and Bailey, J.E. (1998) Biotechnol. Prog. 14, 705–713.
33. Kaufmann, H., Mazur, X., Marone, R., Bailey, J.E., and Fussenegger, M. (2001) Biotechnol. Bioeng. 72, 592–602.
34. Mueller, P.P., Schlenke, P., Nimtz, M., Conradt, H.S., and Hauser, H. (1999) Biotechnol. Bioeng. 65, 529–536.
35. Meents, H., Enenkel, B., Werner, R.G., and Fussenegger, M. (2002) Biotechnol. Bioeng. 79, 619–627.
36. Vaux, D.L. and Strasser, A. (1996) Proc. Natl. Acad. Sci. USA 93, 2239–2244.
37. Al-Rubeai, M. and Singh, R.P. (1998) Curr. Opin. Biotechnol. 9, 152–156.
38. Goswami, J., Sinskey, A.J., Steller, H., Stephanopoulos, G.N., and Wang, D.I. (1999) Biotechnol. Bioeng. 62, 632–640.
39. Zanghi, J.A., Fussenegger, M., and Bailey, J.E. (1999) Biotechnol. Bioeng. 64, 108–119.
40. Figueroa, B., Jr., Sauerwald, T.M., Mastrangelo, A.J., Hardwick, J.M., and Betenbaugh, M.J. (2001) Biotechnol. Bioeng. 73, 211–222.
41. Fussenegger, M., Fassnacht, D., Schwartz, R., Zanghi, J.A., Graf, M., Bailey, J.E., and Portner, R. (2000) Cytotechnology 32, 45–61.
42. Kim, N.S. and Lee, G.M. (2000) Biotechnol. Bioeng. 71, 184–193.
43. Mastrangelo, A.J., Hardwick, J.M., Zou, S., and Betenbaugh, M.J. (2000) Biotechnol. Bioeng. 67, 555–564.

44. Mastrangelo, A.J., Hardwick, J.M., Bex, F., and Betenbaugh, M.J. (2000) Biotechnol. Bioeng. 67, 544–554.
45. Tey, B.T., Singh, R.P., Piredda, L., Piacentini, M., and Al-Rubeai, M. (2000) J. Biotechnol. 79, 147–159.
46. Tey, B.T., Singh, R.P., Piredda, L., Piacentini, M., and Al-Rubeai, M. (2000) Biotechnol. Bioeng. 68, 31–43.
47. Simpson, N.H., Singh, R.P., Perani, A., Goldenzon, C., and Al-Rubeai, M. (1998) Biotechnol. Bioeng. 59, 90–98.
48. Meents, H., Enenkel, B., Eppenberger, H.M., Werner, R.G., and Fussenegger, M. (2002) Biotechnol. Bioeng. 80, 706–716.
49. Muchmore, S.W., Sattler, M., Liang, H., Meadows, R.P., Harlan, J.E., Yoon, H.S., Nettesheim, D., Chang, B.S., Thompson, C.B., Wong, S.L., Ng, S.L., Fesik, S.W. (1996) Nature 381, 335–341.
50. Angelastro, J.M., Moon, N.Y., Liu, D.X., Yang, A.S., Greene, L.A., and Franke, T.F. (2001) J. Biol. Chem. 276, 12,190–12,200.
51. Sauerwald, T.M., Betenbaugh, M.J., and Oyler, G.A. (2002) Biotechnol. Bioeng. 77, 704–716.
52. Moon, N.S., Premdas, P., Truscott, M., Leduy, L., Berube, G., and Nepveu, A. (2001) Mol. Cell. Biol. 21, 6332–6345.
53. Kaufmann, H., Marone, R., Olayioye, M.A., Bailey, J.E., and Fussenegger, M. (2001) J. Biol. Chem. 276, 29,987–29,993.
54. Fussenegger, M., Morris, R.P., Fux, C., Rimann, M., Von Stockar, B., Thompson, C.J., and Bailey, J.E. (2000) Nat. Biotechnol. 18, 1203–1208.
55. Fux, C., Moser, S., Schlatter, S., Rimann, M., Bailey, J.E., and Fussenegger, M. (2001) Nucl. Acids Res. 29, E19.
56. Schlatter, S., Bailey, J.E., and Fussenegger, M. (2001) Biotechnol. Bioeng. 75, 597–606.
57. Bailey, J.E. (1999) Nat. Biotechnol. 17, 616–618.

S.C. Makrides (Ed.) *Gene Transfer and Expression in Mammalian Cells*

CHAPTER 16

Translational regulation in mammalian cells

Marilyn Kozak

Department of Biochemistry, Robert Wood Johnson Medical School, University of Medicine and Dentistry of New Jersey, 675 Hoes Lane, Piscataway, NJ 08854, USA; Tel.: +1-732-235-5355; Fax: +1-732-235-5356. E-mail: kozakma@umdnj.edu

1. Introduction

Translational regulation is important in determining when and where a protein is produced, how much is made and, in some cases, the actual structure of the protein. Because most control mechanisms operate during the early steps wherein the ribosome/factor complex assembles on the mRNA and selects the AUG start codon, the bulk of this review focuses on the initiation phase of translation. Near the end (Section 6), I describe some regulatory mechanisms that operate during the elongation and termination phases. The simplest indication that translation might be regulated, namely a discrepancy between mRNA and protein levels, turns out sometimes to result from post-translational control mechanisms. These and other complications are discussed briefly in the closing section. Translational perturbations that underlie human diseases are the subject of another review [1].

2. Overview of the initiation phase of translation in eukaryotes

Because of constraints imposed at the level of translation, the usual pattern of transcription in eukaryotes is that each gene has its own promoter, generating a monocistronic mRNA. The few exceptions only underscore the rule; i.e., some animal viruses produce mRNAs that are structurally polycistronic but when put to the test, only the 5′ proximal cistron is translated [2]. Thus all eukaryotic mRNAs are *functionally monocistronic*, except under the special conditions described below, and even those escape mechanisms (leaky scanning and reinitiation) are limited to the vicinity of the 5′ end. The reason for this constraint is that eukaryotic ribosomes initiate translation via a scanning mechanism wherein the small (40S) ribosomal subunit enters at the 5′ end of the mRNA and migrates linearly until the first AUG codon is encountered (reviewed in [2,3]).

The 5′ end of all cellular and most viral mRNAs is capped, and the 7-methylguanosine (m7G) cap strongly promotes ribosome entry. The 5′ restriction on initiation still holds, however, with the uncapped mRNAs produced by certain plant viruses as well as with synthetic uncapped mRNAs constructed for laboratory studies [2–4]. This tells us that, although the cap increases the efficiency of ribosome entry, a more fundamental process—possibly threading of the ribosome onto the mRNA—defines the initiation mechanism.

Eukaryotic mRNAs differ from prokaryotic mRNAs in overall structure—monocistronic versus polycistronic—and in the frequency with which a codon other than AUG serves as the start site. Whereas about 10% of bacterial genes employ a GUG or UUG start codon, use of a codon other than AUG as the primary initiation site is extremely rare in eukaryotes [2,3]. As mentioned below, however, an upstream non-AUG codon is used occasionally in addition to the first AUG codon. In such cases, ACG and CUG are the usual supplementary start codons [5]. Curiously, ACG and CUG are not able to function as alternative start codons in prokaryotes.

To summarize: the scanning mechanism postulates that the 40S ribosomal subunit carrying Met-tRNA$_i$ enters at the (usually capped) $5'$ end of the mRNA and migrates linearly, stopping when it encounters the first AUG codon, whereupon the 60S ribosomal subunit joins. The resulting 80S ribosome, with Met-tRNA$_i$ bound in the P-site, is ready for the next aminoacyl-tRNA to enter the A-site and thus to begin the elongation phase. The AUG^{START} codon is recognized by base-pairing with the anticodon in Met-tRNA$_i$ [6]. Because an ACG or CUG codon which might precede the first AUG pairs less well with Met-tRNA$_i$, most (not always all) ribosomes bypass those alternative start codons. The stop-scanning step is modulated by sequences flanking the initiator codon, as explained below.

2.1. Initiation factors

The numerous protein factors that participate in the initiation reaction (Table 1 in [7]) can be divided into three functional groups. The following summary is based on biochemical and genetic studies in yeast and mammals. Most but not all points have been confirmed in both systems.

(i) The GTP form of *eIF2* binds and escorts Met-tRNA$_i$ onto the 40S ribosomal subunit, forming a 43S initiation complex which subsequently binds mRNA. Although *eIF1A* and *eIF3* do not interact directly with Met-tRNA$_i$, these factors stabilize eIF2-mediated binding of the initiator tRNA to the 40S subunit. eIF2 is ultimately released from the ribosome as a stable complex with GDP, necessitating a guanine-nucleotide exchange factor (*eIF2B*) to regenerate eIF2 · GTP.

(ii) Factors that somehow promote ribosome/mRNA interactions include *eIF4E*, which binds the m7G cap; *eIF4G* which binds to eIF4E; and *eIF4A*, an RNA-dependent ATPase. A popular hypothesis is that eIF4A might unwind mRNA secondary structure in preparation for ribosome entry [7], but the evidence for helicase activity is weak: a 200-fold molar excess of the protein is required to disrupt a short duplex structure [8] and, unlike well characterized helicases, eIF4A has the anomalous ability to unwind blunt-ended substrates. An alternative hypothesis is that ATP hydrolysis by eIF4A might drive the 40S ribosome/factor complex during the scanning phase of initiation. The multisubunit factor eIF3, which associates with the 40S ribosomal subunit before or during binding of Met-tRNA$_i$, is likely to play a major role in subsequent steps (mRNA binding, scanning), but this remains to be elucidated. eIF1 and eIF1A have discernible effects on the formation of ribosome/mRNA complexes *in vitro* [7] but nothing is known about the underlying mechanism. The postulated involvement of other factors (eIF4B, eIF4H) requires further study.

(iii) When the 40S subunit reaches the AUG^{START} codon, *eIF5* activates hydrolysis of the GTP moiety carried by eIF2 [9]. The eIF5-mediated reaction is a timing device that helps to identify authentic start codons: only at authentic start sites is the pause, caused by pairing with Met-tRNA$_i$, long enough for GTP to be hydrolyzed [10]. Release of eIF2 · GDP is followed by entry of *eIF5B* · GTP which promotes joining of the 60S subunit. (eIF5B was initially identified by its structural homology to the prokaryotic factor IF2, which escorts Met-tRNA onto the small ribosomal subunit and then promotes joining of the large ribosomal subunit; only the latter activity is retained by eIF5B [11].) Joining of the 60S subunit triggers hydrolysis of the GTP moiety carried by eIF5B, which then dissociates from the ribosome. This marks the end of the initiation phase of translation.

The best studied substeps are those involving eIF2. That key factor is the focal point of regulatory mechanisms which are described in Section 4. The least well understood substeps are those involving the eIF4 set of proteins. Various interactions—e.g. between eIF4E and eIF4G, between eIF4G and eIF4A—have been demonstrated *in vitro* or deduced via genetic tests [7], but we do not know the natural sequence of events. We know neither when and how these factors assemble on mRNA (whether before or in conjunction with the 40S ribosomal subunit) nor when and how the factors are released.

Without some understanding of the normal assembly/disassembly cycle, it is hard to make sense of putative regulatory proteins such as an eIF4E binding protein (4E-BP1) which, when dephosphorylated, is purported to block the formation of an eIF4E · eIF4G complex [7]. Although the effects of dephosphorylation can be seen easily when protein interactions are monitored in an artificial *in vitro* system, the *in vivo* picture is unclear: in one study, cell proliferation was not impaired under conditions which maintained 4E-BP1 in the dephosphorylated (supposedly inhibitory) state [12]; a different study contends that prevention of 4E-BP1 phosphorylation promotes apoptosis [13]. When the function was probed by disrupting the 4E-BP1 gene in mice, there were no gross abnormalities in development; the major finding was a reduction in adipose tissue mass [14]. That result does not reveal a clear role for 4E-BP1 in regulating translation.

2.2. Selection of translational start sites

The position effect on start codon selection is the strongest evidence for a linear scanning mechanism. The position effect is evident when a silent internal AUG codon becomes activated upon removal of the upstream start codon. This happens with many genes that produce, along with the full length protein, an N-terminally truncated isoform: the shorter form of the protein is translated via a second form of mRNA from which the upstream start codon has been eliminated by alternative splicing or use of an internal promoter. Fig. 1A depicts one such example; many others are described elsewhere [2].

Whereas initiation from an internal start codon, via a second mRNA, is part of the normal pattern of gene expression in the case of LEF1 (Fig. 1A), in other cases

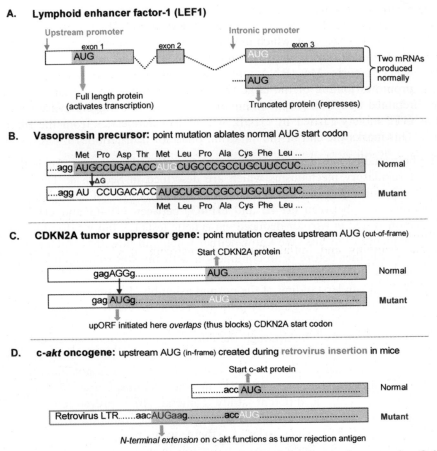

A. Lymphoid enhancer factor-1 (LEF1)

Upstream promoter   Intronic promoter

exon 1   exon 2   exon 3
AUG   AUG   Two mRNAs
                                                  produced
                           AUG   normally

Full length protein   Truncated protein (represses)
(activates transcription)

B. Vasopressin precursor: point mutation ablates normal AUG start codon

Met Pro Asp Thr Met Leu Pro Ala Cys Phe Leu ...
...agg AUGCCUGACACC AUG CUGCCCGCCUGCUUCCUC...............   Normal
         ↓AG
...agg AU CCUGACACC AUGCUGCCCGCCUGCUUCCUC...............   Mutant
         Met Leu Pro Ala Cys Phe Leu ...

C. CDKN2A tumor suppressor gene: point mutation creates upstream AUG (out-of-frame)

                    Start CDKN2A protein
        gagAGGg.........................AUG........................   Normal

        gag AUGg.........................AUG........................   Mutant

upORF initiated here *overlaps* (thus blocks) CDKN2A start codon

D. c-*akt* oncogene: upstream AUG (in-frame) created during retrovirus insertion in mice

                    Start c-akt protein
        ....acc AUG.....................   Normal

Retrovirus LTR.......aacAUGaag.............acc AUG....................   Mutant

N-terminal extension on c-akt functions as tumor rejection antigen

Fig. 1. Proximity to the 5′ end of the mRNA dictates which AUG functions as the start codon. Only the portion of each mRNA relevant to understanding initiation of translation is depicted. The coding domain is shaded gray. Silent internal AUG codons are white. Mutations are highlighted in red. The first two examples [15,16] show how an internal AUG codon becomes activated when the first AUG codon is removed by restructuring the mRNA (A) or by a point mutation (B). The position effect on start codon selection is seen also in the third and fourth examples, where an AUG codon is introduced upstream from the normal site of initiation. The shift in initiation thereby effected can have important biological consequences. (C) In some families predisposed to melanoma, translation of CDKN2A is prevented by a G → T mutation in the 5′ UTR that creates a strong upstream out-of-frame AUG codon [17]. (D) In one type of mouse leukemia caused by insertion of a retrovirus into the c-akt proto-oncogene, the creation of an upstream in-frame AUG codon adds 33 amino acids to the N-terminus of the c-akt protein. This novel peptide was found to mediate tumor rejection by cytotoxic T lymphocytes [18].

an internal AUG codon is activated only when a mutation destroys the normal AUG^{START} site. The gene that encodes vasopressin is an interesting example (Fig. 1B). The second AUG codon, which takes over as the initiation site upon ablation of the first AUG, produces a truncated signal peptide. This slight change in structure prevents proper folding and processing of the vasopressin precursor protein, and the resulting failure to secrete vasopressin causes neurohypophyseal diabetes insipidus [16].

The position effect can be seen also in the opposite direction; i.e., when a mutation creates an AUG codon upstream from the normal AUG^{START} codon. The examples in Fig. 1C and D show that a shift in the site of initiation brought about in this way can have important biological consequences.

Initiation at the first AUG codon—the so-called position rule—depends to some extent on the flanking sequence. By testing the effects of point mutations (reviewed in Refs. [2,3]), the optimal context for initiation was shown to be GCCRCCaugG. The purine in position −3 (three nt upstream from the AUG codon, which is numbered +1 to +3) and the G in position +4 make the strongest contributions; as long as those two positions conform, deviations from the rest of the consensus sequence have little consequence. In position −3, A is somewhat more effective than G. These rules define two extremes: (i) when the first AUG codon occurs in a strong context (ANNaugN or GNNaugG), all or almost all ribosomes stop and initiate at that point; (ii) when the first AUG resides in a very weak context, lacking both R in position −3 and G in position +4, some ribosomes initiate at that point but most continue scanning and initiate farther downstream. mRNAs that support dual initiation via this mechanism are discussed in the next section under the heading "leaky scanning".

It is harder to predict what happens at start sites that fall between the extremes; i.e., mRNAs in which the first potential start codon has the sequence YNNaugG, GNNaugY or GNNaugA. Leaky scanning is seen in some but not all such cases. With test transcripts, the leaky scanning normally engendered by a suboptimal context can be suppressed by downstream secondary structure, which presumably slows scanning and thus provides more time for codon/anticodon pairing; that explanation might apply to natural mRNAs in which translation initiates uniquely at the aforementioned sites despite deviation from the consensus sequence.

The validity of these context rules is evident from mutagenesis studies with test transcripts [2,3] and from analysis of natural mRNAs that display leaky scanning (see below). Hatzigeorgiou [19] has shown that use of the context/scanning rules enables the start site in cDNA sequences to be predicted with high accuracy. Other investigators have taken a contrary approach, however: from a statistical analysis of cDNA sequences they deduce "rules for translation" which they believe argue against the scanning mechanism. The following clarifications are offered in response to one such challenge [20]: (i) The assertion that "most authentic AUGs contain three or more mismatches from the consensus sequence" means nothing because it rests on giving equal weight to all positions from −5 to +4, which is not how the consensus sequence is defined. The authors unaccountably failed to point out that, in the critical −3 position, a purine was present in 84% of the cDNAs they examined

[20]. In other cDNA surveys [21], conformity to the −3 rule was ≥ 90%. (ii) Because many mammalian genes produce transcripts with alternative 5′ leader sequences, the number and significance of upstream AUG codons is misrepresented when a cDNA survey scores only the longest 5′ untranslated region (UTR). (iii) Small upstream open reading frames (upORFs) *are translated*, as shown by detecting the encoded peptide or by fusing a reporter gene to the upORF. Thus, the repeated reference to "unused" upstream AUG codons [20] is inappropriate and misleading. (iv) Since small upORFs are translated, there is no reason why upstream AUG codons should not reside in a good context. (v) The presence of upstream AUG codons does not contradict the scanning model. On the contrary, the presence of small upORFs in mRNAs *which are thereby translated inefficiently* (see Section 3.2) constitutes evidence in support of the scanning mechanism.

3. Regulation of initiation via mRNA structure

While the scanning model predicts that translation should initiate at the AUG codon nearest the 5′ end of the mRNA, the scanning model does not predict that translation is invariably limited to the first AUG. Two well defined mechanisms allow access to AUG codons farther downstream.

3.1. Context-dependent leaky scanning

When the 5′ proximal AUG codon resides in a suboptimal context, some ribosomes will initiate there but many ribosomes will bypass the first AUG and initiate downstream. More than 30 mRNAs (cellular and viral) that produce two proteins via this mechanism have been documented [2]. The number of bifunctional mRNAs is much larger if one counts instances in which translation initiates to a limited extent at an upstream CUG or ACG codon in addition to the first AUG [1–3].

Figure 2 depicts a few examples of context-dependent leaky scanning. The first three cases, chosen to illustrate the variety of biological problems that can be solved by this device, are typical in that the context flanking the first AUG deviates from the consensus motif in position −3 and/or position +4. The significance of this was proved by experimentally improving the context; e.g., a mutation that substituted G for C in position −3 of the human glucocorticoid receptor gene (Fig. 2A) abolished leaky scanning [22]. A similar verification was carried out in many other cases [2]. The glucocorticoid receptor gene is representative of a number of mammalian genes that use leaky scanning to produce long and short protein isoforms which have distinguishable functions as transcriptional activators or repressors. Whereas the first and second AUG^{START} codons are in the same reading frame in those cases, in some other genes the alternative start codons reside in different reading frames. That arrangement enables one mRNA to produce two completely unrelated proteins, as illustrated by the influenza A virus RNA diagrammed in Fig. 2B. The apoptosis-promoting protein initiated from the downstream AUG codon in this mRNA is postulated to kill host immune cells that

A. Glucocorticoid receptor (two isoforms from same reading frame)

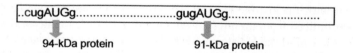

B. Influenza A virus RNA 2 (two proteins from *different reading frames*)

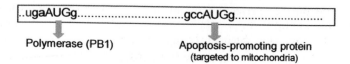

C. Macrophage colony stimulating factor (*different reading frames*)

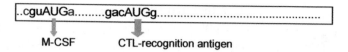

D. Histone H4 + osteogenic growth peptide (same reading frame)

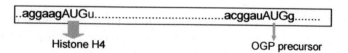

E. c-myc (three isoforms from same reading frame)

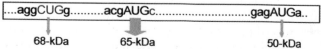

Fig. 2. Two or more proteins can be produced from one mRNA via context-dependent leaky scanning. Highlighted in red are sequences around the 5′ proximal start codon that promote leaky scanning; i.e., these deviations from the optimal ANNaugG motif enable some ribosomes to bypass the first start site and advance to a second (or occasionally a third) downstream start site. In the mRNAs shown here [22–26], the distance between upstream and downstream start codons ranges from 38- to 300-nt. Whereas most instances of leaky scanning are straightforward in that ribosomes initiate at the first (weak) and second AUG codons [2], with influenza A virus mRNA the proteins identified in panel B derive from the first and fourth AUG codons. The two intervening AUGs initiate small ORFs, 2–8 codons in length. Here and in a few other cases, including c-myc, a combination of leaky scanning and reinitiation presumably provides access to the far downstream start codon.

respond to the viral infection [23]. A downstream-initiated peptide produced from the human macrophage colony-stimulating factor gene (Fig. 2C) has the opposite function; i.e., this 25-amino acid antigenic peptide stimulates the cytotoxic T lymphocyte response which defends against tumor cells and viruses [24].

Low level leaky scanning is seen occasionally with mRNAs in which the first AUG codon resides in a good *but not perfect* context. For example, studies carried out with rat histone H4 mRNA showed that an osteogenic growth peptide (OGP) is produced from an internal AUG codon despite the favorable context (A in position −3) around the H4 start codon (Fig. 2D). The occurrence of low level leaky scanning in this case was demonstrated in two ways: (i) production of the downstream-initiated peptide increased when the upstream H4 start codon was eliminated; and (ii) production of the downstream-initiated peptide was abolished when the good-but-not-perfect upstream start site (AGGAAGaugU) was changed to GCCACCaugG [25].

The final example (Fig. 2E) shows that the scanning/context rules prevail even in rare cases where translation initiates from three sites in one mRNA. In mice, the longest c-myc protein (68-kDa) derives from a CUG start codon which, despite a strong context, is leaky because the codon is not AUG. The middle-sized c-myc protein initiates from the first AUG codon. The presence of A in position −3 makes this a strong start site (indeed the predominant translation product is the 65-kDa isoform) but absence of the rest of the consensus sequence allows a small number of ribosomes to advance to a downstream AUG codon, producing the third (50-kDa) form of c-myc [26]. That all this is due to leaky scanning was deduced from the effects of mutations: strengthening the context around the first AUG codon (ACGaugC → ACCaugG) suppressed production of the 50-kDa protein, while changing the upstream CUG codon to AUG suppressed production of both the 65- and 50-kDa proteins.

3.2. Modulation of translation via small upstream open reading frames

A subset of mammalian mRNAs—including many that encode cytokines, growth factors, kinases, transcription factors, and other regulatory proteins [27]—contain small open reading frames (ORFs) upstream from the major coding domain. This arrangement forces translation to occur via a reinitiation mechanism. There are significant gaps in our understanding of how reinitiation occurs in eukaryotes but the basic idea is that, following translation of the first small ORF, the 60S ribosomal subunit dissociates while at least some 40S subunits remain bound to the mRNA, resume scanning and—after reacquiring Met-tRNA$_i$—reinitiate at an AUG codon farther downstream. The need to reacquire Met-tRNA$_i$ (without which the downstream AUG codon could not be recognized) creates the potential for regulation: *an AUG codon positioned too close to the termination site of the upORF might be bypassed if the ribosome has not yet regained Met-tRNA$_i$*. The rate at which Met-tRNA$_i$ is reacquired depends in part on the concentration of eIF2 which, as discussed in the next section, is modulated in response to many stimuli. Thus, one consequence of a 5′ leader sequence that has small upORFs is the possibility of switching between alternative reinitiation sites, depending on the availability of eIF2. The best studied example of this qualitative regulation is the yeast *GCN4* gene [7,28]. A few candidate mammalian genes have been identified [2,7] but not yet studied thoroughly.

Quantitative regulation of translation by small upORFs is a well established phenomenon in mammals, however. Fig. 3 depicts a few examples; many others are compiled elsewhere [2]. The presence of small upORFs causes a quantitative reduction in translation because reinitiation is usually less efficient than straightforward initiation at the first AUG codon. (We do not know if this is because some 40S ribosomal subunits dissociate from the mRNA after translating the upORF or if all 40S subunits resume scanning but only some are competent to reinitiate.) The length of the upORF is not usually a significant variable—upORFs must be short because reinitiation cannot occur following translation of a full-length ORF—but the inhibitory effect of small upORFs is modulated in other ways.

The first example (Fig. 3A) illustrates a common phenomenon in mammals wherein a gene produces mRNAs with alternative 5′ leader sequences, only some of which have inhibitory upORFs. Whereas the form of mdm2 mRNA that has two small upORFs is translated inefficiently via reinitiation, the upORFs are eliminated by a promoter switch when elevated production of the protein is required [29]. Here and in many other mRNAs where small upORFs terminate before the start of the major cistron, their function is simply to down-modulate translation [1,2].

Imposition of an upORF that overlaps the start of the major cistron has more drastic consequences. Because ribosomes cannot backup to reinitiate (scanning and hence reinitiation occur strictly in the 5′ to 3′ direction), the overlapping upORF in the major form of thrombopoietin (TPO) mRNA almost completely precludes translation (Fig. 3B). A small amount of TPO is produced from a minor alternative mRNA (not depicted) which has a different, less inhibitory arrangement of upORFs [30]. The poor translatability of the major form of TPO mRNA is deliberate: mutations that restructure the upORF and thus elevate translation of this potent cytokine cause hereditary thrombocythemia.

Translation of the ADAR2 enzyme (adenosine deaminase) is turned on and off via an RNA editing mechanism that causes a shift in the reading frame, as a result of which the normal AUG^{START} codon becomes the start of an inhibitory upORF [31]. As illustrated in Fig. 3C, conversion of adenosine to inosine by ADAR2—this is an *auto*regulatory mechanism—creates a new splice junction motif. Translation of the alternatively spliced mRNA begins at the normal AUG^{START} codon and then (after 9 codons) shifts to a new reading frame that terminates prematurely; the product is an inactive protein fragment. Because the next potential in-frame start codon (Met25) resides within what is effectively an "upORF" (see diagram), translation cannot initiate or reinitiate from Met25. Studies with experimental constructs show that initiation at Met25 would produce an 80-kDa form of ADAR2 that has enzymatic activity, but this protein isoform is not detected in rat brain extracts wherein the edited transcript is the predominant mRNA.

In the foregoing examples, as with most other mRNAs that have small upORFs, the reduction in translation is caused simply by the presence of an upstream minicistron without regard to what it encodes. Only in rare cases has the upORF-encoded peptide itself been implicated in the inhibition of translation [32]. This has been claimed with a few additional mRNAs, but caution is needed in interpreting experiments wherein mutations that change the coding capacity of an

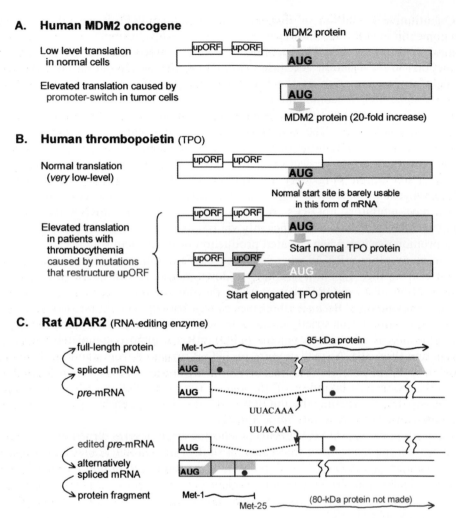

Fig. 3. Small upstream ORFs regulate translation in various ways. (A) upORFs that terminate before the start of the MDM2 coding domain (shaded gray) reduce translational efficiency. The constraint is relieved by switching to a downstream promoter in tumor cells [29]. (B) In normal individuals, the major (most abundant) transcript encoding TPO is virtually untranslatable due to an overlapping upORF. This upORF has a strong start site (GCCGCCUCCaugG), which precludes leaky scanning, and the overlapping arrangement precludes reinitiation. Translation of TPO is elevated in some individuals by a mutation that truncates the upORF and thus makes reinitiation possible, or by a mutation that shifts the upORF into the same reading frame as TPO. In the latter case, the elongated form of TPO appears to function normally; the pathological effects are due to overproduction of the cytokine. These and other aspects of TPO translation are reviewed in Ref. [30]. (C) Editing (A → I) of ADAR2 mRNA creates the equivalent of an inhibitory upORF. Due to the change in splicing, translation that initiates from the normal AUG[START] codon generates only an inactive protein fragment; the arrangement of this 82-codon "first ORF" precludes translation of the rest of the ADAR2 domain [31]. The dotted line depicts the intron in ADAR2 pre-mRNA that undergoes alternative splicing. A large blue dot indicates the position of the (silent) AUG codon for Met25, which is discussed in the text.

upORF diminish its inhibitory effect. While that result could indicate a functional role for the encoded peptide, it could also be explained by depletion of some critical tRNA due to slow recycling of peptidyl-tRNA. The latter mechanism, documented with prokaryotic mRNAs [33], merits consideration in eukaryotes especially when the experimental protocol involves overexpressing the upORF-containing mRNA.

One notable way in which upstream minicistrons are *not* used in eukaryotes is to effect a change in conformation of the mRNA. In prokaryotes, translation of a small upORF is used in clever ways to unmask a downstream initiation site [34,35]. Because eukaryotic ribosomes enter at the 5′ end of the mRNA, however, disrupting secondary structure around an internal AUG codon is not sufficient to activate translation.

3.3. Effects of mRNA secondary structure

The postulated linear migration of 40S ribosomal subunits en route to the AUG^{START} codon explains the strong inhibition imposed by secondary structure within the 5′ UTR [36]. If the translation of structured mRNAs were to improve in response to mitogens, for example, it would be an important regulatory mechanism, given that many mammalian mRNAs carry GC-rich—hence highly structured— leader sequences [27]. But a change in the translational machinery that makes it easier for ribosomes to penetrate base-paired structures is only a hypothetical possibility. What we know for a fact is that the expression of many genes is regulated by restructuring the mRNA; i.e., when elevated translation is required, a switch in the transcriptional start site replaces the long GC-rich leader sequence with a less encumbered form [reviewed in Refs. [2,27].

Experimental probing of the effects of structured leader sequences is not simple, in part because the GC-rich sequences that restrict translation often harbor elements that promote transcription. This might explain occasional reports in which a GC-rich 5′ UTR inhibits translation *in vitro* but not *in vivo*. Base-paired structures near the 5′ end may also have effects, positive and negative, on mRNA stability [37].

4. Regulation via initiation factors

As mentioned in Section 2.1 and outlined in Fig. 4A, the eIF2 cycle depends on a helper factor, eIF2B, which dislodges GDP and thus regenerates eIF2·GTP, the form that binds Met-tRNA$_i$. This regeneration is an important point of regulation. Mammalian cells contain four kinases, each of which phosphorylates Ser51 in the smallest (α) subunit of eIF2. This phosphorylation increases the affinity of eIF2B for eIF2, but the interaction is abortive; GDP is not released (Fig. 4B). Phosphorylation of even a portion of the eIF2 pool can shut down protein synthesis by diverting the relatively small pool of eIF2B to this nonproductive pathway. In experimental tests, the inhibition of translation attributed to phosphorylation of eIF2(α) is prevented by mutation of Ser51 to Ala; mutation of Ser51 to Asp, on the other hand, mimics the

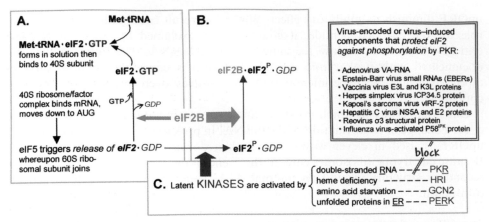

Fig. 4. The guanine nucleotide exchange reaction on eIF2 is catalyzed by eIF2B and regulated by phosphorylation. (A) depicts the normal pathway for regenerating eIF2·GTP. (B) depicts the abortive pathway caused by phosphorylation of eIF2. (C) summarizes the eIF2 kinases that are activated under various conditions in mammalian cells. Activation of PKR by double stranded RNA involves dimerization of the enzyme which is prevented by some of the viral components listed as blocking agents. Other agents protect by sequestering double stranded RNA. Herpes simplex virus protein ICP34.5 is postulated to protect by targeting a phosphatase to eIF2. The protective protein P58[IPK] activated by influenza virus is encoded not by the virus but by the host cell. In addition to preserving enough eIF2 in the active state to support synthesis of viral proteins, these PKR-blocking agents also prevent apoptosis and thus promote the establishment of persistent viral infections [38].

charge and hence the deleterious effects of phosphorylation. These tests and a wealth of other evidence underlying the eIF2 story are documented elsewhere [7,38,39]. eIF2B can also be inactivated by phosphorylation [40] but, for simplicity, I will focus here on modification of eIF2.

The eIF2 kinases exist as latent enzymes, each of which is regulated by a specific ligand (Fig. 4C).

- PERK, a kinase that resides in the endoplasmic reticulum (ER), is held in the latent state by the chaperonin BiP until unfolded proteins begin to accumulate in the ER [41]. The release of BiP activates PERK, which inactivates eIF2, and protein production is thus prevented from overwhelming the protein-maturation machinery.
- In erythroid cells, the eIF2 kinase called HRI (heme-regulated inhibitor) shuts down protein synthesis when heme becomes limiting, thus coordinating the synthesis of globin polypeptide chains with availability of the prosthetic group [42].
- In yeast, starvation for amino acids generates uncharged tRNAs which activate an eIF2 kinase designated GCN2 [28]. A mammalian enzyme that resembles GCN2 in possessing a synthetase-like, tRNA-binding domain has been identified but not yet shown to be activated in response to amino acid deprivation. In mammals, phosphorylation of eIF2 is not likely to be the sole mechanism underlying nutrition-induced changes in translation [43].

• The best studied eIF2 kinase in mammals is PKR, which is activated when double-stranded RNA binds to two sites, causing the enzyme to dimerize and undergo autophosphorylation. The activation is inhibited by 2-aminopurine, which is thus a useful experimental tool. PKR is often activated in virus infected cells as a result of double-stranded RNA produced in the course of viral transcription or replication. The fact that many viruses go to the trouble of encoding a special protein or RNA (identified in Fig. 4C and Ref. [38]) that functions to *block* PKR-induced phosphorylation of eIF2 is evidence of how potent this shutoff mechanism is.

Further evidence of the real-life importance of modulating eIF2 function comes from the recent discovery of human hereditary diseases caused by mutations in eIF2B [44] or in PERK [45]. The serious biological effects—pancreatic dysfunction, erythroid cell death—produced in transgenic mice when one of the eIF2 kinases is eliminated [42,46] leave no doubt that cells need to down-modulate translation under certain circumstances and this is accomplished by blocking the regeneration of eIF2 · GTP.

The above description is provided only as an introduction. Activation of eIF2 kinases is more complicated than outlined above (e.g., HRI is activated without regard to heme availability when cells are heat shocked) and eIF2 might not be the only target. Study of the phosphatases—the other half of the equation—is just beginning. Whereas the effect of eIF2 depletion on most mRNAs is quantitative (a simple reduction in protein yield), mRNAs that have small upORFs might show a qualitative change in translation; i.e., a switch in the site of reinitiation, as discussed in Section 3.2. mRNAs removed from the actively translated pool due to phosphorylation of eIF2 form ribonucleoprotein particles called stress granules, the function of which is under study [47].

I emphasized the "real life importance" of regulation via eIF2 because the literature is bogged down with papers claiming translational regulation via other factors—claims based in part on artificial manipulations (e.g., forced overexpression of eIF4E) and in part on looking only at the factor of interest in situations where translation is seen to increase or decrease. In no case does the quality of evidence come anywhere near that described for eIF2.

The postulated role of eIF4E in regulating translation is one of the much discussed topics for which the evidence is ambiguous. eIF4E is said to be the rate-limiting component in initiation [48], although that basic premise has been challenged [49]. The function of eIF4E is said to be augmented by phosphorylation, which initially was thought to occur on Ser53 but later was mapped to Ser209. The kinase responsible for phosphorylating eIF4E is MNK1 [48]. The hypothesis that phosphorylation of eIF4E up-regulates translation is contradicted by the results of one study wherein MNK1 activity was experimentally increased or decreased [50]. Other recent studies contradict the claim that the phosphorylated form of eIF4E has a higher affinity for the m7G cap [51]. An investigation carried out in *Drosophila* did show that transgenic organisms are delayed in development when Ser251 (the equivalent of Ser209 in mammalian eIF4E) is changed to Ala [52]. That result is interesting but not definitive. (In the early days, when Ser53 was thought to be the site of phosphorylation, a mutated form of eIF4E in which Ser53 was replaced by Ala was tested and found to be impaired [53]. The moral is that substitution of Ala for Ser sometimes perturbs the function of a protein for reasons that have nothing to do with phosphorylation.) The biological effects of phosphorylating eIF4E could conceivably involve changes in its intracellular distribution, since eIF4E resides not only in the cytoplasm but also in the nucleus. The apparent involvement of eIF4E in nucleocytoplasmic transport of certain mRNAs [54] is a warning against thinking only in terms of translation when eIF4E levels are perturbed.

Proteolysis of eIF4G was first detected in poliovirus-infected cells, where loss of eIF4G function was postulated to underlie the shutoff of host protein synthesis [38]. That hypothesis continues to color much of the thinking in the field despite inconsistencies in the supporting data (e.g., cleavage of IF4G does not abolish its function [55]) and the belated recognition that many other cellular proteins undergo proteolysis in the course of infection by poliovirus. Cleavage of eIF4G has been postulated to cause inhibition of protein synthesis during apoptosis; but again, given that so many other proteins (including other translation factors) are cleaved [56], I see no evidence that proteolysis of eIF4G is of primary importance.

Although ribosomal protein S6 (rpS6) is not an initiation factor, it bears mention here by way of illustrating how hard it can be to determine the functional consequences of modifications in the translational machinery. A correlation between mitogen-induced activation of translation and phosphorylation of rpS6 has been documented in countless reports over the course of 25 years [57]. The kinase (S6K1, $p70^{S6K}$) that targets rpS6 was identified recently, as was a set of mRNAs— ribosomal protein mRNAs and others that carry a 5'-terminal oligopyrimidine tract designated "TOP"—whose translational efficiency was said to depend on phosphorylation of rpS6 [57,58]. But the most recent studies reveal that "TOP" mRNAs undergo normal translational shifts in the absence of rpS6 phosphorylation [59,60]. Thus, the functional significance of this ubiquitous modification, detected in organisms ranging from fruit flies to humans, remains unclear.

5. Regulation of initiation via mRNA-specific binding proteins

The most straightforward example in this category is the repression of ferritin mRNA translation by iron regulatory proteins (IRPs) [30,61]. When intracellular iron levels are low, the iron-free form of IRP-1 binds tightly to a stem-loop structure, called the iron response element or IRE, near the 5' end of ferritin mRNA. Binding of the repressor protein thus blocks ribosome entry.[1] When iron levels are elevated,

---

[1] When ferritin mRNA is experimentally restructured in a way that moves the IRE farther from the 5' cap, IRP-1 still binds but no longer inhibits translation, presumably because the 40S ribosomal subunit, which now has room to enter, dislodges the protein in the course of scanning.

IRP-1 assumes a different structure due to assembly of an iron-sulfur cluster. This form of the protein dissociates from the mRNA, thereby turning on production of ferritin which sequesters and thus detoxifies the excess iron. A backup repressor protein (IRP-2) undergoes oxidation when iron levels are elevated, and the oxidized form of IRP-2 is rapidly degraded [62]. The physiological significance of this regulatory loop is evidenced by the discovery of naturally occurring mutations in the IRE that allow ferritin production in the absence of iron overload. This loss of regulation causes hereditary hyperferritinemia/cataract syndrome, the severity of which correlates with the affinity of IRPs for the mutated IRE [30].

There are many other reports of mRNA binding proteins that inhibit, or occasionally appear to augment, translation. What is missing in most cases is: (i) a mechanism to achieve *reversible* binding, such as the iron-induced changes in IRPs; and (ii) clear evidence of *biological significance*, such as the naturally occurring mutations in the IRE. Sometimes an RNA-binding protein that strongly augments translation *in vivo* has only a small effect on translation *in vitro* [63], which might be a clue that the vivo augmentation is indirect.

Evidence for the biological significance of fragile X mental retardation protein (FMRP) is overwhelming—mutations that prevent production of this RNA-binding protein cause X-linked mental retardation—but it is not yet clear whether or how FMRP regulates translation. Recent reports identified some of the targeted mRNAs in mouse brain and even pinpointed the recognition motif [64]. The problem—the complication—is that some of these mRNAs show decreased association with polysomes in the absence of FMRP while other targeted mRNAs show increased polysome association. Thus, despite remarkable progress, we cannot yet say how FMRP affects translation or even that translation, rather than nucleocytoplasmic transport, is the primary level of action of this important protein.

RNA binding proteins, many of which bind to the 3′ UTR, are major players in the temporal and position-dependent expression of developmentally-regulated mRNAs. Rather than skim the surface of this important topic, I refer the reader to a detailed review of translational activation of stored mRNAs in oocytes and embryos [65]. Attempts to figure out how translation might be regulated by 3′ UTR-binding proteins are complicated by the frequent involvement of these proteins in mRNA stability and localization; this is taken up in Section 7. Some stories that appear to involve translational regulation via a 3′ UTR-binding protein start out strong, but discrepancies develop when one tries to pin down the mechanism. In the case of reticulocyte 15-lipoxygenase (LOX), early studies revealed that, although production of the enzyme is detectable only in mature erythrocytes, LOX mRNA is present in abundance in immature precursor cells [66]. (The mRNA has to be produced ahead of time because the nucleus is ejected from reticulocytes in the course of maturation.) The current working hypothesis is that, in immature erythroid cells, translation of LOX mRNA is blocked by binding of hnRNP-K protein to the 3′ UTR [67]. In contrast to the clean off/on expression seen in the initial studies with natural LOX mRNA [66], however, a reporter gene bearing the LOX 3′ UTR was inhibited only ~twofold upon overexpression of hnRNP-K in HeLa cells [68]. Quantitatively, that is a rather weak result. Qualitatively, the experimental results are flawed because binding of hnRNP-K to the 3′ UTR is postulated to prevent joining of the 60S ribosomal subunit [67] but there was no accumulation of the expected 40S subunit/mRNA intermediate. The putative inhibition of LOX translation by hnRNP-K requires direct testing in developing erythroid cells.

6. Regulation of translation subsequent to the initiation step

Programmed frameshifting is the best understood regulatory mechanism that operates during the elongation phase of protein synthesis. Most examples of frameshifting by eukaryotic ribosomes involve re-pairing (slipping) of tRNAs during the translation of viral mRNAs (reviewed in Ref. [69]). Antizyme mRNA is the only cellular mRNA known to require frameshifting, and the mechanism is unusual. The +1 frameshift required to produce the full-length antizyme protein occurs only when polyamine levels are elevated [69]. (Antizyme itself inhibits the synthesis and uptake of polyamines; therefore the production of antizyme only when polyamines are elevated is a homeostatic device.) The working hypothesis is that a high concentration of polyamines distorts the ribosome and/or the tRNA bound in the P site and thus blocks the adjacent base in the mRNA. A direct test of this occlusion mechanism is difficult, but mutational studies do appear to rule out a conventional mechanism of frameshifting via tRNA slipping [70].

An unprecedented mode of regulation was discovered recently in connection with translation of the yeast *HAC1* gene [71]. Some but not all aspects of this mechanism apply also to the mammalian gene that encodes transcription factor *XBP1*; I will describe the more thoroughly studied example from yeast. HAC1 mRNA encodes a transcription factor that turns on expression of a large set of genes required to process and dispose of unfolded proteins in the ER. Thus, this regulatory mechanism is part of what is called the unfolded protein response. Although HAC1 mRNA is produced constitutively, an intron near the C-terminal end of the coding domain prevents complete translation. The intron is removed via a nonconventional cytoplasmic splicing mechanism catalyzed by Ire1p [71]. Here the relevant point is that the unspliced mRNA is associated with "frozen" polysomes that generate no polypeptide product [72]. Mutational probing revealed that the translational block involves base pairing between the intron and the 5′ UTR—an interaction that presumably forms after the initiation step and is relieved when Ire1p (activated upon accumulation of unfolded proteins) removes the intron.

A complete elongational block such as occurs during translation of HAC1 mRNA is unusual, but transient pausing at a particular point during elongation is seen with some other mRNAs. A pause in elongation, caused by a codon for which the complementary tRNA is scarce, is postulated to play a role in promoting degradation of c-myc mRNA [73]. During the synthesis of membrane-targeted proteins, there is a transient delay in elongation when the signal peptide emerges from the ribosome and is bound by the signal recognition particle [74]. This pause allows transfer of the ribosome/nascent-chain complex to the ER membrane before polypeptide extension resumes. A pause during polypeptide elongation is seen also with some viral mRNAs that require frameshifting: a pseudoknot downstream from the frameshift site slows elongation and thus gives the tRNAs time to slip. (The mechanism is not that simple, however, as not every structure-induced delay in elongation promotes frameshifting [75].)

Spatial constraints permit discussion of but one example of regulation during the termination phase of protein synthesis. In a small but important set of mRNAs, a UGA terminator codon positioned within the coding domain is translated as selenocysteine [76]. The most extreme example is the mRNA for liver selenoprotein P, where suppression (i.e., translation) of UGA is required at 10 or more positions. The insertion of selenocysteine is mediated by a special tRNA in which the anticodon is complementary to UGA; this tRNA is recognized by a special elongation factor. In prokaryotic mRNAs, a stem-loop structure recognized by the selenocysteine-specific elongation factor is positioned immediately after the UGA codon that requires suppression. The eukaryotic mechanism is more complicated in that the identifying stem-loop structure resides in the 3' UTR and is recognized by a protein distinct from (but bound to) the special elongation factor.

7. Complicated issues and closing notes

7.1. Regulation via 3' UTR sequences

Selenocysteine insertion is but one of many instances in which the 3' UTR is implicated in some aspect of translation. In no case, however, is the mechanism clear. A popular idea is that poly(A)-binding protein (PABP), positioned at the 3' end of the mRNA, binds eIF4G and transfers it back to the 5' end; but this interpretation—supported by static pictures of a "bridge" [77]—awaits verification via a chase experiment which shows direct transfer of eIF4G without dilution through the soluble pool of factors. An alternative interpretation is that interactions between PABP and other protein factors might augment translation simply by counteracting diffusion; i.e., maintaining a pool of initiation factors close to where they are needed. The importance of the eIF4G/PABP interaction is called into question by recent studies wherein the preferential translation of yeast poly(A)$^+$ mRNAs was found to persist even when mutations in eIF4G eliminated binding to PABP [78].

One concern about these and other experiments in which the mRNA is introduced into cells by electroporation is that the (possibly small) fraction of input mRNA that actually supports translation might be unusually susceptible to degradation, which could be counteracted by 3' UTR-binding proteins or by 3'–5' interactions. In other words, answers obtained with this unnatural assay system might not apply to endogenous mRNAs. An effect on mRNA stability could explain why 3' UTR elements sometimes have a more pronounced effect on expression *in vivo* than when translation is measured directly *in vitro* [4].

In Section 5, I skipped lightly over examples that are hard to dissect because a protein targeted to the 3' UTR regulates the localization and/or stability of the mRNA in addition to apparently regulating translation. I say "apparently" because

surprises keep turning up. Recent studies of repressed mRNAs in lower eukaryotes—
e.g., the classical *nanos* in *Drosophila* and *lin-28* in *Caenorhabditis elegans*—revealed
that the "repressed" mRNAs are actually associated with polysomes [79,80]. These
polysomes are not frozen; i.e., down-regulation does not result from an elongational
block of the sort described for HAC1 mRNA (Section 6). The most likely, albeit not
yet proven, explanation is that factors bound to the 3′ UTR of *nanos* and *lin-28*
mRNAs promote rapid proteolysis of the nascent polypeptides. Along the same
lines, the mechanism whereby a point mutation in the 3′ UTR of mouse Ran mRNA
causes a 5- to 10-fold change in accumulation of Ran protein appears to involve rapid
degradation when the protein lingers in the cytoplasm [81]. Movement of Ran protein
into the nucleus protects against proteolysis, presumably due to interactions with
other proteins. The mutation in the 3′ UTR is postulated to affect localization of Ran
mRNA which in turn affects the intracellular distribution—and hence stability—of
the encoded protein. These are but a few examples wherein what appears to be
translational control via the "wrong end of the mRNA" is actually post-translational
control. The estimate that some 50% of polypeptides are subject to co-translational
proteolysis [82] warns against jumping to the conclusion that *translation* is regulated,
every time a discrepancy between mRNA and protein levels is detected.

Sometimes a noncoding sequence (3′ or 5′) augments translation indirectly, as was
shown with a retrovirus element that promotes translation by directing mRNAs to
exit the nucleus via a specific pathway [83]. A possible mechanistic explanation is
that the export pathway links the emerging mRNAs to the cytoskeleton, but the
mechanism whereby the cytoskeleton augments translation [84] remains unclear.
Linkage to the cytoskeleton can obviously localize mRNAs to certain areas of the
cell; what we do not understand is how translational efficiency is elevated in some
areas.

7.2. Internal initiation as an alternative to scanning

The scanning mechanism described in Section 2 is sometimes said to be inadequate—
an alternative mechanism said to be needed [7,38]—when an mRNA leader
sequence is burdened with upstream AUG codons or secondary structure. That
simplistic view misses the point of the TPO story (Fig. 3B), which teaches that some
mRNAs are deliberately burdened with upstream AUG codons in order to impede
scanning and thus limit the translation of proteins that would be harmful if
overproduced.

The postulated alternative to the scanning mechanism is direct entry of ribosomes
at an internal position (internal ribosome entry sequence, IRES) located close to the
AUG^{START} codon. The experimental evidence for IRES elements in cellular mRNAs
has serious flaws, however [85]. The proffered rationale for a cap-independent
internal initiation mechanism is that it would enable certain mRNAs to be translated
when eIF4E levels decline (e.g., when phosphorylation of 4E-BP1 is blocked), but
recent experiments presented in support of that idea [13] used the 5′ UTR from
poliovirus rather than 5′ UTRs from the cellular mRNAs that are supposedly
regulated via "a dynamic interplay between cap-dependent and cap-independent

processes". *The absence of any common sequence or structure among candidate cellular IRES elements is a major stumbling block*, raising concern that the dicistronic constructs[2] used to test for internal initiation might generate false positive results. Indeed, the synthetic dicistronic test gives erratic results even with the paradigmatic IRES elements derived from picornaviruses [86]. Advocates of the internal initiation hypothesis press the point that the phenomenon shows sequence specificity (i.e., translation of the 3′ cistron is inactivated by certain mutations in the putative IRES), but the required sequence might function in unanticipated ways—e.g., as a splicing signal, or a transcriptional promoter [87], or a target for ribonucleases [88]. These unanticipated mechanisms could produce additional forms of mRNA that would undermine the design of the dicistronic test.

Unlike transcription, where specialized factors operate with different genes, the translational machinery employs generic protein factors. The special elongation factor that mediates selenoprotein biosynthesis (Section 6) is a rare exception. At the level of initiation, there is only the standard set of factors; alternative isoforms directed to specific mRNAs have not been found.[3] The proteins postulated to bind to, and thus promote the use of, putative IRES elements are either generic initiation factors (eIF4G, eIF3) or general RNA binding proteins (La, pyrimidine tract binding protein, nucleolin, hnRNP-E2, etc.; reviewed in Ref. [89]). Because the latter proteins have no recognizable function in translation, the proffered explanation is that they stabilize some undefined structure in the mRNA; how this promotes ribosome entry is not explained. While some (not all) viral IRES elements have been shown to bind eIF4G or a subunit of eIF3 [7], this has not been demonstrated for any candidate cellular IRES element. Even the viral story is incomplete, inasmuch as the required chase experiment has not been carried out to show that pre-bound eIF4G or eIF3 can serve as a functional intermediate in initiation.

7.3. Some unanswered questions

Many of the topics discussed herein, albeit prominently represented in the literature, are premature. One cannot really understand the regulation of translation until the basic mechanism is established. We do not yet know the basic initiation pathway, i.e., the point at which each protein factor enters and leaves. We do not even know whether the full set of initiation factors is in hand. We do not know what drives migration of the 40S ribosomal subunit during the scanning phase or how the GCCRCC motif promotes recognition of the AUG "stop" codon. The first round of translation might be found to differ from subsequent rounds vis-à-vis which

[2]The customary procedure is to transpose a suspected IRES element from its natural location (the 5′ UTR of one or another cellular mRNA) to the midpoint of a dicistronic vector; translation of the 3′ cistron is then taken as evidence for internal initiation. The puzzling *absence of natural dicistronic mRNAs* is ignored by advocates of the internal initiation hypothesis.
[3]Influenza virus NS1 protein has been postulated to function as a "virus-specific initiation factor" by virtue of its ability to bind eIF4G [90]. The only clear point, however, is that NS1 and eIF4G can interact. There is no proof that translation is augmented by pre-binding of eIF4G, and binding of NS1 (which has many functions unrelated to translation) is not limited to viral mRNAs.

initiation factors are required. If translation begins as the mRNA emerges from the nucleus, the act of translation might promote mRNA export; but that idea, and the possibility that initiation factor requirements might differ depending on the export pathway, awaits testing.

My list of unanswered questions includes some problems that others consider to be solved. The fact that the immunosuppressive drug rapamycin blocks phosphorylation of rpS6 and 4E-BP1 is intriguing;[4] but if those modifications are central to the control of cell growth, it is not clear how. The speculation that translation of key regulatory proteins is turned on and off by (de)phosphorylation of 4E-BP1 or rpS6 is only speculation. The popular idea that phosphorylation of rpS6 controls translation of mRNAs that contain a "TOP" sequence appears to be wrong (Section 4); and the idea that translation of IRES-containing mRNAs might persist when phosphorylation of 4E-BP1 is blocked [13] awaits credible evidence that IRES elements indeed exist in cellular mRNAs.[5] The rapamycin/TOR pathway inhibits cell growth in yeast as well as in mammals, but 4E-BP1 is absent from yeast[6] and mutations that prevent phosphorylation of the yeast equivalent of rpS6 are without consequence [93]. This and other discrepancies,[7] including the fact that mTOR is found in abundance in the mitochondrial and nuclear compartments of mammalian cells [95,96], suggest a need for caution in ascribing the growth inhibitory effects of rapamycin primarily to a translational block caused by under-phosphorylation of rpS6 and 4E-BP1. It is interesting that the rapamycin-induced inhibition of cell growth can be partially reversed by restoring rpS6 phosphorylation (via introduction of a rapamycin-resistant form of S6K1) or by overexpressing eIF4E [97], but one wonders whether rescue might also be achieved by overexpressing other components. Only S6K1 and eIF4E were tested.[8]

---

[4]mTOR (mammalian target of rapamycin, also called FRAP), the kinase inhibited by the protein (FKBP12) to which rapamycin binds, is the first of several enzymes that phosphorylate sites in 4E-BP1. mTOR is not directly responsible for phosphorylating rpS6 but lies upstream in the pathway [58].

[5]The existence of an IRES must be shown directly (e.g., via the dicistronic test discussed in Section 7.2); it cannot be deduced simply from the fact that a given cellular mRNA shows reduced dependence on the cap binding protein. With certain (capped) viral mRNAs that show reduced dependence on eIF4E, the possibility of direct internal initiation was tested and ruled out [91], and some other mRNAs that are naturally uncapped were shown nevertheless to utilize the standard 5'-end-dependent initiation mechanism [2–4]. Thus, it is fallacious to equate "cap independence" with "internal initiation".

[6]A yeast protein designated Eap1p, unrelated to 4E-BP1, was postulated to bind eIF4E and regulate translation in response to rapamycin [92]; but that story was put out in advance of determining basic points, such as whether Eap1p resides in the cytoplasm and whether it is phosphorylated by TOR. Disruption of the EAP1 gene has no effect on yeast growth at the normal (30°C) temperature.

[7]In mammalian cells, the kinetics of events following rapamycin treatment are not reassuring: whereas 4E-BP1 is completely dephosphorylated after ~1 h, translation is inhibited only slowly and incompletely (~50% inhibition after 20 h) [94].

[8]Indeed, even if eIF4E and S6K1 proved to be the only components capable of effecting rescue, it would not prove that rapamycin inhibits cell growth primarily by inhibiting translation. S6K1 has other targets, including nuclear proteins that participate in splicing [98]. A reduction in eIF4E levels was shown to impair nucleocytoplasmic transport of cyclin D1 mRNA [54], demonstrating that eIF4E also has non-translational functions.

Other complicated questions include how higher-order structures, such as the assembly of tRNA synthetases into macromolecular complexes [99], affect function; how the machinery for protein synthesis in integrated with quality-control mechanisms that degrade aberrant mRNAs and proteins; and how the translational machinery is affected by contact with the cytoskeleton.

Acknowledgements

Research in the author's laboratory is supported by grant GM33915 from the National Institutes of Health.

Abbreviations

eIF	eukaryotic initiation factor
ER	endoplasmic reticulum
4E-BP1	eIF4E-binding protein 1
FMRP	fragile X mental retardation protein
HRI	heme-regulated inhibitor
IRE	iron response element
IRES	internal ribosome entry sequence
IRP	iron regulatory protein
LOX	15-lipoxygenase
m7G	7-methylguanosine
mTOR	mammalian target of rapamycin
OGP	osteogenic growth peptide
ORF(s)	open reading frame(s)
PABP	poly(A) binding protein
PERK	ER-resident protein kinase
PKR	double-stranded RNA activated protein kinase
rpS6	ribosomal protein S6
S6K1	kinase targeted to rpS6
TPO	thrombopoietin
upORF(s)	upstream open reading frame(s)
UTR	untranslated region

References

1. Kozak, M. (2002) Mamm. Genome 13, 401–410.
2. Kozak, M. (2002) Gene 299, 1–34.
3. Kozak, M. (1999) Gene 234, 187–208.
4. Guo, L., Allen, E.M., and Miller, W.A. (2001) Mol. Cell 7, 1103–1109.
5. Kozak, M. (1991) J. Biol. Chem. 266, 19,867–19,870.
6. Cigan, A.M., Feng, L., and Donahue, T.F. (1988) Science 242, 93–97.

492

7. Dever, T.E. (2002) Cell 108, 545–556.
8. Rogers, G.W., Jr., Lima, W.F., and Merrick, W.C. (2001) J. Biol. Chem. 276, 12,598–12,608.
9. Das, S. and Maitra, U. (2001) Prog. Nucl. Acid Res. Mol. Biol. 70, 207–231.
10. Huang, H.-K., Yoon, H., Hannig, E.M., and Donahue, T.F. (1997) Genes Devel. 11, 2396–2413.
11. Roll-Mecak, A., Cao, C., Dever, T.E., and Burley, S.K. (2000) Cell 103, 781–792.
12. Marx, S.O. and Marks, A.R. (1999) Mol. Cell. Biol. 19, 6041–6047.
13. Li, S., Sonenberg, N., Gingras, A.-C., Peterson, M., Avdulov, S., Polunovsky, V.A., and Bitterman, P.B. (2002) Mol. Cell. Biol. 22, 2853–2861.
14. Tsukiyama-Kohara, K., Poulin, F., Kohara, M., DeMaria, C.T., Cheng, A., Wu, Z., Gingras, A.-C., Katsume, A., Elchebly, M., Spiegelman, B.M., Harper, M.E., Tremblay, M.L., Sonenberg, N. (2001) Nature Med. 7, 1128–1132.
15. Hovanes, K., Li, T.W.H., Munguia, J.E., Truong, T., Milovanovic, T., Marsh, J.L., Holcombe, R.F., and Waterman, M.L. (2001) Nature Genet. 28, 53–57.
16. Beuret, N., Rutishauser, J., Bider, M.D., and Spiess, M. (1999) J. Biol. Chem. 274, 18,965–18,972.
17. Liu, L., Dilworth, D., Gao, L., Monzon, J., Summers, A., Lassam, N., and Hogg, D. (1999) Nature Genet. 21, 128–132.
18. Wada, H., Matsuo, M., Uenaka, A., Shimbara, N., Shimizu, K., and Nakayama, E. (1995) Cancer Res. 55, 4780–4783.
19. Hatzigeorgiou, A.G. (2002) Bioinformatics 18, 343–350.
20. Peri, S. and Pandey, A. (2001) Trends Genet. 17, 685–687.
21. Pesole, G., Gissi, C., Grillo, G., Licciulli, F., Liuni, S., and Saccone, C. (2000) Gene 261, 85–91.
22. Yudt, M.R. and Cidlowski, J.A. (2001) Mol. Endocrinol. 15, 1093–1103.
23. Chen, W., Calvo, P.A., Malide, D., Gibbs, J., Schubert, U., Bacik, I., Basta, S., O'Neill, R., Schickli, J., Palese, P., Henklein, P., Bennick, J.R., Yewdell, J.W. (2001) Nature Med. 7, 1306–1312.
24. Probst-Kepper, M., Stroobant, V., Kridel, R., Gaugler, B., Landry, C., Brasseur, F., Cosyns, J.-P., Weynand, B., Boon, T., Van den Eynde, B.J. (2001) J. Exp. Med. 193, 1189–1198.
25. Bab, I., Smith, E., Gavish, H., Attar-Namdar, M., Chorev, M., Chen, Y.-C., Muhlrad, A., Birnbaum, M.J., Stein, G., Frenkel, B. (1999) J. Biol. Chem. 274, 14,474–14,481.
26. Spotts, G.D., Patel, S.V., Xiao, Q., and Hann, S.R. (1997) Mol. Cell. Biol. 17, 1459–1468.
27. Kozak, M. (1991) J. Cell Biol. 115, 887–903.
28. Hinnebusch, A.G. (1997) J. Biol. Chem. 272, 21,661–21,664.
29. Landers, J.E., Cassel, S.L., and George, D.L. (1997) Cancer Res. 57, 3562–3568.
30. Cazzola, M. and Skoda, R.C. (2000) Blood 95, 3280–3288.
31. Rueter, S.M., Dawson, T.R., and Emeson, R.B. (1999) Nature 399, 75–79.
32. Morris, D.R. and Geballe, A.P. (2000) Mol. Cell. Biol. 20, 8635–8642.
33. Heurgué-Hamard, V., Dincbas, V., Buckingham, R.H., and Ehrenberg, M. (2000) EMBO J. 19, 2701–2709.
34. Decatur, A., McMurry, M.T., Kunkel, B.N., and Losick, R. (1997) J. Bacteriol. 179, 1324–1328.
35. Mayford, M. and Weisblum, B. (1989) EMBO J. 8, 4307–4314.
36. Kozak, M. (1986) Proc. Natl. Acad. Sci. USA 83, 2850–2854.
37. Stefanovic, B., Hellerbrand, C., and Brenner, D.A. (1999) Mol. Cell. Biol. 19, 4334–4342.
38. Gale, M., Jr., Tan, S.-L., and Katze, M.G. (2000) Microbiol. Mol. Biol. Rev. 64, 239–280.
39. Clemens, M.J. (2001) Prog. Mol. Subcell. Biol. 27, 57–89.
40. Pap, M. and Cooper, G.M. (2002) Mol. Cell. Biol. 22, 578–586.
41. Bertolotti, A., Zhang, Y., Hendershot, L.M., Harding, H.P., and Ron, D. (2000) Nature Cell Biol. 2, 326–332.
42. Han, A.-P., Yu, C., Lu, L., Fujiwara, Y., Browne, C., Chin, G., Fleming, M., Leboulch, P., Orkin, S.H., Chen, J.-J. (2001) EMBO J. 20, 6909–6918.
43. Kimball, S.R. (2001) Prog. Mol. Subcell. Biol. 26, 155–184.
44. Van der Knaap, M.S., Leegwater, P.A.J., Könst, A.A.M., Visser, A., Naidu, S., Oudejans, C.B.M., Schutgens, R.B.H., and Pronk, J.C. (2002) Ann. Neurol. 51, 264–270.
45. Delépine, M., Nicolino, M., Barrett, T., Golamaully, M., Lathrop, G.M., and Julier, C. (2000) Nature Genet. 25, 406–409.

46. Harding, H.P., Zeng, H., Zhang, Y., Jungries, R., Chung, P., Plesken, H., Sabatini, D.D., and Ron, D. (2001) Mol. Cell 7, 1153–1163.

47. Kedersha, N., Cho, M.R., Li, W., Yacono, P.W., Chen, S., Gilks, N., Golan, D.E., and Anderson, P. (2000) J. Cell Biol. 151, 1257–1268.

48. Pyronnet, S., Imataka, H., Gingras, A.-C., Fukunaga, R., Hunter, T., and Sonenberg, N. (1999) EMBO J. 18, 270–279.

49. Rau, M., Ohlmann, T., Morley, S.J., and Pain, V.M. (1996) J. Biol. Chem. 271, 8983–8990.

50. Knauf, U., Tschopp, C., and Gram, H. (2001) Mol. Cell. Biol. 21, 5500–5511.

51. Scheper, G.C., van Kollenburg, B., Hu, J., Luo, Y., Goss, D.J., and Proud, C.G. (2002) J. Biol. Chem. 277, 3303–3309.

52. Lachance, P.E.D., Miron, M., Raught, B., Sonenberg, N., and Lasko, P. (2002) Mol. Cell. Biol. 22, 1656–1663.

53. Lazaris-Karatzas, A., Montine, K.S., and Sonenberg, N. (1990) Nature 345, 544–547.

54. Cohen, N., Sharma, M., Kentsis, A., Perez, J.M., Strudwick, S., and Borden, K.L.B. (2001) EMBO J. 20, 4547–4559.

55. Ali, I.K., McKendrick, L., Morley, S.J., and Jackson, R.J. (2001) EMBO J. 20, 4233–4242.

56. Clemens, M.J., Bushell, M., Jeffrey, I.W., Pain, V.M., and Morley, S.J. (2000) Cell Death Differ. 7, 603–615.

57. Volarevic, S. and Thomas, G. (2000) Prog. Nucl. Acid Res. Mol. Biol. 65, 101–127.

58. Meyuhas, O. (2000) Eur. J. Biochem. 267, 6321–6330.

59. Tang, H., Hornstein, E., Stolovich, M., Levy, G., Livingstone, M., Templeton, D., Avruch, J., and Meyuhas, O. (2001) Mol. Cell. Biol. 21, 8671–8683.

60. Barth-Baus, D., Stratton, C.A., Parrott, L., Myerson, H., Meyuhas, O., Templeton, D.J., Landreth, G.E., and Hensold, J.O. (2002) Nucl. Acids Res. 30, 1919–1928.

61. Muckenthaler, M., Gray, N.K., and Hentze, M.W. (1998) Mol. Cell 2, 383–388.

62. Iwai, K., Drake, S.K., Wehr, N.B., Weissman, A.M., LaVaute, T., Minato, N., Klausner, R.D., Levine, R.L., and Rouault, T.A. (1998) Proc. Natl. Acad. Sci. USA 95, 4924–4928.

63. Timchenko, N.A., Iakova, P., Cai, Z.-J., Smith, J.R., and Timchenko, L.T. (2001) Mol. Cell. Biol. 21, 6927–6938.

64. Brown, V., Jin, P., Ceman, S., Darnell, J.C., O'Donnell, W.T., Tenenbaum, S.A., Jin, X., Feng, Y., Wilkinson, K.D., Keene, J.D., Darnell, R.B., Warren, S.T. (2001) Cell 107, 477–487.

65. Wickens, M., Goodwin, E.B., Kimble, J., Strickland, S., and Hentze, M.W. (2000) In: Sonenberg, N., Hershey, J.W.B. and Mathews, M.B. (eds.) Translational Control of Gene Expression, Cold Spring Harbor Press, New York, pp. 295–370.

66. Thiele, B.J., Andree, H., Höhne, M., and Rapoport, S.M. (1982) Eur. J. Biochem. 129, 133–141.

67. Ostareck, D.H., Ostareck-Lederer, A., Shatsky, I.N., and Hentze, M.W. (2001) Cell 104, 281–290.

68. Ostareck-Lederer, A., Ostareck, D.H., Cans, C., Neubauer, G., Bomsztyk, K., Superti-Furga, G., and Hentze, M.W. (2002) Mol. Cell. Biol. 22, 4535–4543.

69. Ivanov, I.P., Gesteland, R.F., and Atkins, J.F. (2000) Nucl. Acids Res. 28, 3185–3196.

70. Matsufuji, S., Matsufuji, T., Miyazaki, Y., Murakami, Y., Atkins, J.F., Gesteland, R.F., and Hayashi, S. (1995) Cell 80, 51–60.

71. Ma, Y. and Hendershot, L.M. (2001) Cell 107, 827–830.

72. Rüegsegger, U., Leber, J.H., and Walter, P. (2001) Cell 107, 103–114.

73. Lemm, I. and Ross, J. (2002) Mol. Cell. Biol. 22, 3959–3969.

74. Mason, N., Ciufo, L.F., and Brown, J.D. (2000) EMBO J. 19, 4164–4174.

75. Kontos, H., Napthine, S., and Brierley, I. (2001) Mol. Cell. Biol. 21, 8657–8670.

76. Hatfield, D.L. and Gladyshev, V.N. (2002) Mol. Cell. Biol. 22, 3565–3576.

77. Wells, S.E., Hillner, P.E., Vale, R.D., and Sachs, A.B. (1998) Mol. Cell 2, 135–140.

78. Searfoss, A., Dever, T.E., and Wickner, R. (2001) Mol. Cell. Biol. 21, 4900–4908.

79. Clark, I.E., Wyckoff, D., and Gavis, E.R. (2000) Curr. Biol. 10, 1311–1314.

80. Seggerson, K., Tang, L., and Moss, E.G. (2002) Devel. Biol. 243, 215–225.

81. Wong, P.M.C., Yuan, Q., Chen, H., Sultzer, B.M., and Chung, S.-W. (2001) J. Biol. Chem. 276, 33,129–33,138.

82. Turner, G.C. and Varshavsky, A. (2000) Science 289, 2117–2120.
83. Dangel, A.W., Hull, S., Roberts, T.M., and Boris-Lawrie, K. (2002) J. Virol. 76, 3292–3300.
84. Stapulionis, R., Kolli, S., and Deutscher, M.P. (1997) J. Biol. Chem. 272, 24,980–24,986.
85. Kozak, M. (2001) Mol. Cell. Biol. 21, 1899–1907.
86. Hennecke, M., Kwissa, M., Metzger, K., Oumard, A., Kröger, A., Schirmbeck, R., Reimann, J., and Hauser, H. (2001) Nucl. Acids Res. 29, 3327–3334.
87. Han, B. and Zhang, J.-T. (2002) Mol. Cell. Biol. 22, 7372–7384.
88. Elgadi, M.M. and Smiley, J.R. (1999) J. Virol. 73, 9222–9231.
89. Vagner, S., Galy, B., and Pyronnet, S. (2001) EMBO Reports 21, 893–898.
90. Aragón, T., de la Luna, S., Novoa, I., Carrasco, L., Ortín, J., and Nieto, A. (2000) Mol. Cell. Biol. 20, 6259–6268.
91. Dolph, P.J., Huang, J., and Schneider, R.J. (1990) J. Virol. 64, 2669–2677.
92. Cosentino, G.P., Schmelzle, T., Haghighat, A., Helliwell, S.B., Hall, M.N., and Sonenberg, N. (2000) Mol. Cell. Biol. 20, 4604–4613.
93. Johnson, S.P. and Warner, J.R. (1987) Mol. Cell. Biol. 7, 1338–1345.
94. Beretta, L., Gingras, A.C., Svitkin, Y.V., Hall, M.N., and Sonenberg, N. (1996) EMBO J. 15, 658–664.
95. Desai, B.N., Myers, B.R., and Schreiber, S.L. (2002) Proc. Natl. Acad. Sci. USA 99, 4319–4324.
96. Zhang, X., Shu, L., Hosoi, H., Murti, G., and Houghton, P.J. (2002) J. Biol. Chem. 277, 28,127–28,134.
97. Fingar, D.C., Salama, S., Tsou, C., Harlow, E., and Blenis, J. (2002) Genes Dev. 16, 1472–1487.
98. Wilson, K.F., Wu, W.J., and Cerione, R.A. (2000) J. Biol. Chem. 275, 37,307–37,310.
99. Kim, J.Y., Kang, Y.-S., Lee, J.-W., Kim, H.J., Ahn, Y.H., Park, H., Ko, Y.-G., and Kim, S. (2002) Proc. Natl. Acad. Sci. USA 99, 7912–7916.

S.C. Makrides (Ed.) *Gene Transfer and Expression in Mammalian Cells*

CHAPTER 17

Pathways of mammalian messenger RNA degradation

Angela Inácio[1] and Stephen A. Liebhaber[2]

[1]*Instituto Nacional de Saúde Dr. Ricardo Jorge, Lisboa, Portugal*
[2]*Departments of Genetics and Medicine, University of Pennsylvania, Room 428 CRB, 415 Curie Boulevard, Philadelphia, PA 19104, USA. Fax: 215-573-5157. E-mail: Liebhabe@mail.med.upenn.edu*

1. Summary

Degradation of messenger RNAs plays a central role in mammalian gene expression. mRNAs can differ dramatically in their intrinsic stabilities and numerous pathways have evolved to alter this property in response to developmental, physiologic, and environmental stimuli. The rates and modes of decay are heavily impacted by structures at the 5′ and 3′ termini of a mRNA as well as by its translational state. The co-ordinated action of 5′ and 3′ terminal structures and the linkage of decay with translation may be explained by the "closed loop" model of mRNA structure. Both exonuclease and endonuclease activities are involved in mRNA decay. An emerging concept is that many of the essential activities in mRNA decay are encompassed in the multiprotein exosome complex. Specific mRNA decay pathways may be dedicated to mRNA surveillance, assuring that information flowing from the nucleus to the cytoplasm is of high quality prior to translation. Finally, studies are currently defining specific sequence determinants and corresponding *trans*-acting proteins that function to co-ordinate decay rates of subsets of mRNAs. The multiple pathways of mRNA decay are thus complex and multifaceted and exert dynamic control over mammalian gene expression.

2. mRNA decay comprises a major control in gene expression

Intrinsic stabilities of mRNA can differ by over a 100-fold. Short-lived mRNAs tend to encode proteins that must undergo dramatic changes in expression levels within a brief time frame. Such proteins include cell cycle control factors, protooncogenes, and cytokines. At the opposite end of the mRNA stability spectrum are highly stable mRNAs. These mRNAs tend to encode bulk proteins expressed at high levels in terminally differentiated cells. Examples of such highly stable mRNAs include globins, collagens, crystallins, and immunoglobulins.

mRNA half-lives can be dramatically and selectively altered by a broad spectrum of developmental cues as well as by alterations in nutritional, environmental, hormonal, and pharmacological variables. Most often these changes in mRNA stability complement parallel changes in transcriptional rates to effect tight control over protein synthesis.

Since stable and unstable mRNAs co-exist in the same cell, there must be ways to selectively target specific mRNAs for degradation and/or stabilization. *Cis*-acting determinants involved in these processes have been identified in a number of cellular and viral mRNAs. Corresponding sets of RNA-binding proteins that mediate stability control are being identified, as are the enzymes involved in the decay processes.

In addition to their roles in controlling patterns of gene expression, specific RNA decay pathways are involved in quality control. Genetic mutations and defects in transcript processing result in mRNAs that encode mutant, truncated, and/or potentially deleterious proteins. Thus, mature mRNAs must be screened for structural defects and degraded prior to extensive translation [1]. Distinct surveillance mechanisms appear to be targeted to specific subsets of defective mRNAs. These pathways have revealed fascinating and complex interactions between the nuclear and cytoplasmic compartments [2].

3. General pathways of mRNA decay

The combined effects of shared and mRNA-specific structures control decay rates of mRNAs. The 5′ and 3′ terminal structures common to almost all eucaryotic mRNAs are essential determinants of the major decay pathways in eucaryotic cells. These determinants and their corresponding binding proteins are discussed below, to be followed by a discussion of more specialized structures singular to specific mRNAs or subgroups of mRNAs.

3.1. The closed loop model

A 5′-terminal m^7G cap and a 3′-terminal polyA tail are present in almost all eucaryotic mRNAs. These modifications are added co-transcriptionally and are specific to RNA polymerase II (RNAP II) transcripts. In the cytoplasm, the m^7G cap is tightly bound by the cap-binding protein, eIF4E (Fig. 1). This protein anchors the two additional components of the eIF4F cap-binding complex, eIF4A and eIF4G, to the 5′ terminus of the mRNA. At the 3′ terminus, the polyA tail is bound by the polyA-binding protein, PABP. Each PABP protects approximately 15–20 As and assembles on the polyA tail in a "beads-on-a-string" configuration. The 5′ and 3′ RNP complexes provide a general protection from exonuclease attack. The 'closed loop' model proposes functional and physical interactions between the 3′ and 5′ termini of a mRNA [3]. The 'closed loop' is mediated by simultaneous interactions of eIF4G with both eIF4E + eIF4A and with PABP (Fig. 1). While these interactions have been most clearly defined in yeast (*Saccharomyces cereviciae*), evidence suggests that they also occur in mammalian systems [4]. Two likely functions of the 'closed loop' are protection of the 5′ and 3′ termini from attack by ubiquitous exonucleases and the enhancement of ribosome re-initiation during translation. The linkage of these two functions has certain logic as it selectively facilitates the translation of physically intact mRNAs.

General Pathways of mRNA Decay

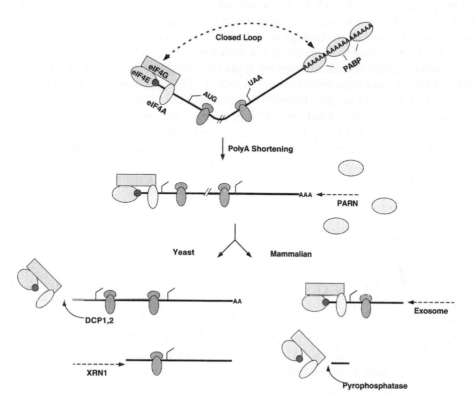

Fig. 1. General pathways of mRNA decay. The 5′ terminal ^{m7}G cap and a 3′ terminal polyA tail interact to form a 'closed loop'. The ^{m7}G cap is bound by eIF4E (cap-binding protein) that anchors eIF4A and eIF4G to the 5′ terminus of the mRNA. Together those proteins form eIF4F, the cap-binding complex. At the 3′ terminus, the polyA tail is bound by the polyA-binding protein (PABP). Interaction between the termini is mediated by simultaneous interaction of eIF4G with eIF4E + eIF4A and with PABP. The first step in the general decay pathway is the shortening of the polyA tail. Once shortened to a critical size, PABP can no longer remain bound. Following this critical polyA shortening the major yeast and mammalian pathways appear to diverge. In yeast, the next step is decaping by Dcp1p and Dcp2p and subsequent 5′ to 3′ decay by Xrn1p. In mammalian cells the deadenylation is followed by continued 3′ to 5′ exonuclease activity mediated by the multiprotein exosome complex. The exosome digestion generates 5′ terminal mRNA fragments that are decapped by a scavenger pyrophosphatase activity.

3.2. 3′ Terminal deadenylation as a rate-limiting step in mRNA decay

The major mRNA decay pathways are controlled by rate-limiting steps involved in disruption of the protective 5′ and 3′ terminal structures (Fig. 1). In most cases the initial step involves shortening of the polyA tail. A polyA tail of approximately 200 bases is added during the initial 3′ processing of RNAP II transcripts in the nucleus. This tail is gradually shortened as the mRNA "ages" in the cytoplasm. This shortening comprises a "molecular clock" for an mRNA. Rates of deadenylation differ among mRNA species and to a certain extent this rate dictates their respective cytoplasmic half-lives. PolyA decay characteristically occurs in increments of ~ 20–30 As. This pattern of decay may reflect rapid removal of polyA segments that are transiently exposed by 'breathing' between the polyA tail and PABP [5]. The role of PABP in protecting the polyA tail has been most clearly documented by the use of conditional PABP mutants in yeast [6]. Once the polyA tail is shortened to a critical size, PABP can no longer bind to the 3′ terminus. This minimal length, less than 20–25 As, correlates with the minimal polyA size sufficient to bind a single PABP [7,8]. It is interesting to consider that the stabilizing effect of the polyA tail may not relate to polyA *per se* but instead may entirely reflect the presence of the bound PABP. This model is supported by studies in which the stabilizing function of the polyA tail was fully replaced by artificially tethering PABP to the 3′ end in the absence of polyA [9].

The enzymatic activities involved in polyA shortening are being defined. PolyA degrading activity, first reported over a decade ago [10], has recently been purified and cloned. The major polyA nuclease in yeast is encoded at the PAN locus [11]. In mammalian cells, a polyA-specific 3′ exonuclease was initially described as the deadenylating nuclease (DAN) and subsequently renamed polyA-specific ribonuclease, PARN [5]. As might be predicted, PARN activity is inhibited by PABP, presumably owing to competition for binding and access to the polyA tail [12]. Remarkably, PARN has cap-binding activity and m^7G stimulates PARN-mediated deadenylation [13]. In a reciprocal fashion PARN activity is inhibited by the eIF4E and the decapping enzyme (Dcp1p). The apparent competition between PARN and eIF4E for the cap is consistent with the observation that PARN is excluded from polysomes. Thus analysis of PARN activity supports a functional communication between the 5′ and 3′ termini of a mRNA and supports the linkage between the translational activity and decay of an mRNA.

3.3. 5′ Terminal decapping

The steps in mRNA decay that follow critical polyA shortening may differ substantially between yeast and mammalian systems (Fig. 1). In yeast deadenylation is ordinarily followed by 5′ decapping and subsequent 5′ to 3′ decay [5,8,12]. Decapping must precede the onset of 5′ → 3′ exonuclease activity because the unique 5′–5′ phosphodiester bond within the cap structure is resistant to ribonucleases [6]. The enzymes in yeast that carry out the 5′ decapping and subsequent 5′ → 3′ decay are Dcp1p and Xrn1p, respectively. A corresponding set of enzymes capable of cap hydrolysis (Dcp1p-like decapping activity [14]) and subsequent 5′ → 3′ degradation (Xrn1p-like activity [5]) have been described in mammalian cells [8]. As in the yeast system, interaction of PARN with the 5′ cap structure inhibits mammalian decapping activity [14].

In addition to the yeast correlates, an additional pyrophosphatase activity has been identified in mammalian cells. This enzyme, which can hydrolyze the 5′cap and release m^7GMP, may scavenge caps from 5′ terminal mRNA fragments generated by 3′ → 5 exonuclease decay of the mRNA (see below) [15]. The functional relationship between the Dcp1p-like and the pyrophosphatase decapping activities in mammalian cells is presently unclear. These two activities appear to represent two distinct proteins, suggesting the existence of at least two distinct decapping proteins in eucaryotes.

3.4. 3′ → 5′ exosome-mediated decay of mammalian mRNAs

In distinction to yeast, the major mRNA decay pathway in mammalian cells appears to involve 3′ → 5′ exonuclease decay after the critical 3′ polyA shortening step (Fig. 1). This decay process appears to be independent of 5′ decapping; mRNA degradation is finalized by hydrolysis of the cap from 5′ terminal fragments by a scavenger pyrophosphatase activity [15].

A 3′ → 5′ exonuclease activity was initially identified in mammalian cells in 1987 [10]. This polysome-associated 33 kDa exonuclease was purified but not further defined. More recently, a multiprotein 'exosome' complex found in both the nucleus and the cytoplasm has been linked to the 3′ → 5′ exoribonucleolytic activity [16]. In addition to its putative role in mRNA decay, this exosome may be involved in 3′ processing of rRNA and snoRNAs [17]. The role of the exosome in mRNA decay is supported by studies in yeast that document accumulation of polyadenylated transcripts in exosome mutants. However, it is not clear if these yeast mutants are defective in mRNA processing, transport, or degradation. It is also unclear whether the exosome can mediate the initial polyA shortening reaction seen in the general mRNA turnover pathways. In addition to its role in general mRNA turnover, the exosome has been specifically linked to the decay of hyperunstable mRNAs (see Section 6.2) [18] and to an mRNA surveillance pathway (see Section 5.2) [19].

Although 3′ → 5′ exosome-mediated degradation may comprise the dominant mRNA decay pathway in mammalian cells, data suggest that a yeast-like pathway of 5′ decapping followed by 5′ → 3′ exonuclease activity also exists in mammalian cells [5,16,20]. A 5′ → 3′ exonuclease activity was observed and partially purified in cytoplasm extracts from mouse sarcoma 180 ascites cells [5] and has been detected in HeLa extracts [16]. Furthermore, the homolog of the yeast 5′ → 3′ exonuclease (Xrn 1p-like) has been identified in mouse cells [5]. Finally, mapping of mammalian mRNA decay intermediates establishes a link in *cis* between deadenylation and decapping [20]. These data suggest that the relative roles of the 3′ → 5′ and 5′ → 3′ pathways in mammalian cells should remain open to further study. Determining why yeast cells and mammalian cells might differ so dramatically in their major pathways of mRNA decay will be of particular interest.

4. Regulation of mRNA decay by rate-limiting endonuclease cleavage

The major pathways of mRNA decay involve the removal of 3′ and 5′ terminal barriers followed by exonuclease decay. In contrast to these general pathways, stability of specific mRNA species may be selectively controlled by decay pathways that circumvent 3′ or 5′ terminal events. One general model for selective stability control involves endonuclease cleavages. The sites of endonuclease attack can be specified by unique sequences and structures in the target mRNA. The rate of mRNA decay can be modulated by controlling endonuclease activity or by controlling endonuclease access to the mRNA target site. In general endonuclease-controlled decay pathways appear to be independent of polyA tail shortening and thus can control RNA levels acutely and without waiting for mRNA 'aging' to occur.

Endonuclease cleavage sites utilized in these pathways have been identified in the coding region (e.g., c-*myc*), and 3′UTR (e.g., transferrin receptor (TfR) and insulin like growth factor II mRNAs) [8,10]. In general, the endonuclease(s) that mediate rate-limiting cleavages in selective mRNA targets are not well defined and modifications that modulate their activities have not been described. The rate limiting cleavages described to date are regulated by RNA-binding proteins that recognize and bind the endonuclease-sensitive site and protect it from hydrolysis. Examples of *cis–trans* interactions that regulate endonuclease controlled decay pathway are detailed below.

Endonucleotic cleavage also plays a central role in the RNA interference pathway (RNAi). In this pathway, double-stranded RNAs (dsRNAs), often generated in the process of a viral infection, target the destruction of homologous mRNAs. In the first stage of this process a dsRNA is bound by cellular factors and processed into 21–23 nucleotide fragments. These cleavages are mediated by Dicer, an RNase III-related enzyme with helicase and dsRNA-binding motifs [21]. In the second stage, the small dsRNA fragments serve as guides to recruit a degradation complex (RNA induced silencing complex; RISC) to the targeted RNA [21]. The native function of this pathway in mammalian cells, aside from proposed role in viral defence, is open to speculation.

5. Decay pathways involved in mRNA surveillance

mRNA surveillance pathways comprise an important component of the quality control network that filters information flow in eucaryotic cells [22]. These pathways target and destroy categories of mRNAs that are unable to synthesize functional proteins. Two independent mRNA surveillance pathways have been described. Both of these pathways are conserved between yeast and mammalian cells. The first pathway, referred to as nonsense-mediated mRNA decay (NMD), eliminates mRNAs containing premature termination codons (PTCs). The second pathway targets mRNAs lacking termination codons ('nonstop' mRNAs).

5.1. Nonsense-mediated mRNA decay

NMD targets mRNAs contain PTCs. PTCs can be generated by nonsense and frameshift mutations, abnormalities in transcript splicing, and out-of-frame translation due to leaky scanning or to the utilization of aberrant AUG initiation sites (see Chapter 16). NMD has been characterized in detail in yeast and a corresponding pathway has been identified in mammalian cells [1,23,24]. In yeast,

NMD is initiated by 5′ decapping of a PTC-containing mRNA (Fig. 2). This decapping is independent of prior polyA shortening. Such circumvention of polyA shortening may be logical as it allows for rapid destruction of newly synthesized PTC-containing mRNAs prior to their 'aging' and translation of detrimental proteins. The 5′ decapping, mediated by Dcp1p, is followed by 5′ → 3′ exonuclease degradation by the Xrn1p. The NMD pathway is dependent on the UPF gene products Upf1p, Upf2p, and Upf3p. The unique dependence of the NMD pathway on these proteins serves to distinguish it from the other major decay pathways in the cell.

While NMD in yeast appears to be a cytoplasmic event, the site of the corresponding decay process in mammalian cells is less clear. A major question is whether PTC recognition and NMD occur in the nucleus, at the nuclear/cytoplasmic pore during mRNA transport, or in the cytoplasm. Recognition of PTCs implies that a codon reading mechanism must be in place to target accelerated decay of nonsense mRNAs. According to conventional thinking this would rule out a direct nuclear

Surveillance Pathways of mRNA Decay

Nonsense-mediated decay

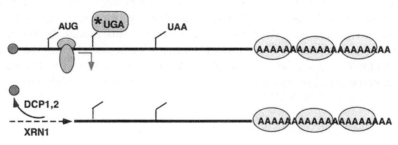

Nonstop decay

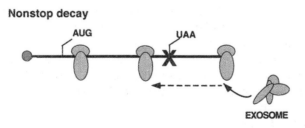

Fig. 2. Surveillance pathways of mRNA decay. Two mRNA surveillance pathways are described. Nonsense-mediated decay (NMD) eliminates mRNAs containing premature termination codons. In yeast this pathway is triggered by direct 5′ decapping independent of prior polyA shortening. The decapping, mediated by Dcp1,2, is followed by 5′ to 3′ Xrn1p exonuclease degradation. In mammalian cells the sequence of steps remains less well defined. Nonstop decay targets mRNAs lacking stop codons; the decay appears to be mediated by 3′ to 5′ exosome activity and to be quite independent from the components of the NMD pathway.

contribution. However, NMD is also dependent on the spatial relationship of the translation termination codon to the most 3' exon–exon junction to distinguish a normal from a premature termination event [2]. The vast majority of normal termination codons are located in the last exon or in close proximity to the last intron. If a PTC is introduced more than 50 to 55 bases 5' to the last exon–exon splice junction it is recognized as a PTC and triggers NMD. Since splicing is a nuclear event and translation occurs in the cytoplasm, how can these two events serve to identify the same mRNA?

Several models have been proposed to link nuclear splicing patterns and translation. The first model is based on the finding that the splicing apparatus deposits an RNP complex approximately 20–24 nt upstream of each exon–exon junction [2]. Proteins in this complex include Y14 (an RNA-binding protein), Aly/Ref (an RNA binding and export factor), RNSP1 (an RNA binding protein associated with splicing), SRm160 (a protein associated with splicing but lacking RNA binding activity), DEK (a phosphoprotein that binds to SRm160), and magoh (a protein involved in the mRNA export to the cytoplasm which binds to Y14 and TAP) [24]. Both Y14 and Aly/Ref are deposited selectively on spliced mRNAs. Aly/Ref transfer from the spliceosome to the spliced RNA during processing may constitute the primary event in the complex deposition. Once splicing is complete, a processed mRNA with the associated RNP splice-marker complexes is transported out of the nucleus [24]. Thus cytoplasmic mRNA available for translation contains a physical record of its splicing history. According to this model, NMD pathway is triggered if (cytoplasmic) translation terminates sufficiently 5' to the last splice-marker complex.

Two alternative, but not necessarily mutually exclusive, models have also been proposed. One is that NMD may be directly triggered in the nucleus by nuclear translation. The unconventional notion that translation occurs in the nucleus is supported by recent studies [24]. A third model proposes that the NMD pathway is triggered in a functional compartment that straddles the nuclear and cytoplasmic compartments. This model proposes that an mRNA undergoes an initial or "pioneer" round of translation as it exits the nuclear pore [25]. Pioneer translation allows for detection of PTCs by cytoplasmic translation while the mRNA is still nucleus-associated and potentially interacting with the splicing apparatus [24].

Homologs of the yeast NMD components have been identified in mammalian cells. hUpf1 has been cloned and implicated in the mammalian NMD pathway [23]. Its role may be to link exon–exon boundary complexes, the translation termination complex, and the mRNA cap complex. hSMG1 a kinase component of the mammalian NMD surveillance complex activates hUPF1 by phosphorylation [26]. hUpf3 binds to Y14 within the splice-marker complex. hUpf2 and hUpf1 appear to reside at the periphery of the nucleus and in the cytoplasm, respectively. Upf2p may serve as a bridge between hUpf1p and the hUpf3p in the splice-marker complex. Thus the Upf proteins may serve as critical linkages among the structural determinants that identify nonsense mRNAs and trigger the NMD pathway.

How does the translating ribosome recognize a nonsense codon as abnormal? As noted above, the positioning of the codon relative to the last exon–exon junction appears to be important in this process. A prevalent model proposes that splice-marker complexes are stripped from mRNAs by the elongating ribosome during its initial ('pioneer') passage [25]. If a PTC causes translation to halt more than 50 to 55 bases 5' to the terminal exon–exon junction, the last splice-marker complex remains on the mRNA. Upf1p present in the retained splice-marker complex then interacts with the translation termination complex and recruits Dcp2p. This protein activates Dcp1p-mediated decapping with subsequent mRNA decay [24]. Alternative models have also been proposed [1]. One model suggests that one or more elements 3' to the termination codon are searched for after translation termination by a scanning 'surveillance complex'. A related model suggests that structures 3' to the termination codon positively influence the events within the termination complex via direct interactions. In this model, recognition of a mRNA as normal requires that a terminating ribosome encounters a 3' terminal mRNP complex, which signals to the ribosome that it has approached the 3' end of the normal coding region. Translation termination that occurs during or after this encounter would be considered proper. In contrast, translation termination prior to interaction with this terminal mRNP complex would be considered improper and lead to NMD. The structure of this putative recognition complex remains to be defined.

It should be noted that the models linking transcript splicing to NMD are not all encompassing. NMD can be triggered in the β-globin mRNA in the absence of the last intron [27] and NMD of human HEXA mRNA can be triggered on an mRNA encoded by an intronless gene [28]. Also, the rules regarding the relationship of nonsense mutations to introns can be breached; nonsense mutations within the first exon of the immunoglobulin μ heavy chain gene [29], TPI gene [30] and β-globin gene [31] apparently circumvent NMD surveillance. Finally, boundary-independent nonsense-mediated decay was recently reported in a T-cell receptor gene; rather than a definitive boundary position, nonsense codons had a polar effect, such that nonsense codons distant from the terminal downstream intron triggered robust NMD and proximal nonsense codon caused modest NMD [32]. Thus, while much is now understood regarding the biochemistry of NMD, there are numerous aspects of this pathway that invite further study.

5.2. Other surveillance pathways

A second surveillance decay pathway exists in yeast and mammalian cells (Fig. 2). This pathway targets mRNAs that lack termination codons [19]. Such 'nonstop' mRNAs may be generated by aberrant 3' processing and/or utilization of cryptic polyA sites within exons. Alternatively, such mRNAs could be generated by partial $3' \rightarrow 5'$ decay of actively translating mRNAs. Genetic studies in yeast demonstrate that this 'nonstop' decay pathway is distinct and independent from NMD as well as from the pathways involved in the decay of normal cellular mRNAs; nonstop mRNA decay can occur in the absence of Upf1p as well as in the absence of Dcp1p, Xrn1p, and the major 3' deadenylation activity. Nonstop mRNAs are destroyed by

$3' \rightarrow 5'$ exosome; this conclusion is supported by the dependence of this pathway on the essential exosome protein Ski7p. A model has been proposed in which Ski7p binds to a stalled ribosome at the terminus of the 3'-truncated mRNA and recruits the exosome to this site with subsequent and rapid decay.

6. Examples of decay pathways controlled by defined cis–trans interactions

Determinants of mRNA stability can either accelerate decay or stabilize an mRNA. The great majority of stability determinants are located in the 3'UTR. This positioning is logical as it shelters critical RNA–protein complexes (RNPs) from scanning 40S ribosomal subunits and/or translating 80S ribosomes. In addition, RNP complexes located in the 5'UTR would be likely to interfere with translation initiation by either blocking 40S ribosome binding or blocking subsequent scanning. Consistent with this reasoning, it is interesting to note that the stability determinant in polio viral mRNA comprises a 5'UTR RNP complex that is located upstream (5') to the site of 40S ribosome binding in the internal ribosome entry site (IRES) [33]. The structures of a number of stability determinants have been defined in some detail. Representative examples, detailed below (Fig. 3), highlight the variety of RNA structures, RNP complexes, and mechanisms that can be employed to control mRNA decay.

6.1. Cell cycle control of mRNA stability

Histone mRNAs are unique among metazoan mRNAs in that they lack polyA tails. The polyA tail has been replaced by a 3' terminal stem-loop structure (a 6-bp double-stranded stem and 4-bp loop). Stabilization of the histone mRNA appears to rely on formation of a complex between this 3' terminal RNA structure and a 31 kD 'stem-loop-binding protein' (SLBP) [5] (Fig. 3). Histone mRNAs are also unusual in that their translation is linked to S-phase of the cell cycle; following S-phase the mRNAs are rapidly degraded. The unique timing of histone mRNA stabilization (S-phase) and destabilization (post S-phase) parallels the demand for histone synthesis during DNA replication and chromatin assembly. The post S-phase decay pathway appears to be mediated by a $3' \rightarrow 5'$ exonuclease. How the 3' terminal RNP complex monitors the cell cycle and coordinates mRNA decay with these events is a topic of significant interest. This control may reflect fluctuations in the level or activity of SLBP during the cell cycle [34]. The signal transduction pathway that mediates these putative alterations in SLBP function remains to be defined.

6.2. Destabilization of mRNAs by the 3' AU-rich element

A well-characterized AU-rich element (ARE) is present within the 3'UTR of a large subclass of highly unstable mammalian mRNAs. These mRNAs encode proteins that must be rapidly induced and subsequently silenced within a narrow time frame to serve their biologic functions. Examples include immediate early proteins, cell cycle control proteins, lymphokines, cytokines, and proto-oncogenes [35]. ARE control appears to be exerted by rapid shortening of the polyA tail (Fig. 3). Distinct structural features of the AREs and corresponding polyA decay patterns and kinetics suggest that ARE-containing mRNAs can be subgrouped [36]. Class I AREs, found in c-*fos* and c-*myc* mRNAs, contain 1–3 scattered copies of the pentanucleotide AUUUA embedded within U-rich regions. Class II AREs, present in cytokine mRNAs such as granulocyte-macrophage colony stimulating factor (GM-CSF) and tumor necrosis factor alpha (TNF-α), contain multiple overlapping copies (5–8 copies) of the AUUUA motif. Class III AREs, such as that in the c-*jun* mRNA, lack the hallmark AUUUA pentanucleotide but require a U-rich sequence and possibly other unknown features for its destabilizing function. The overall importance of these sub-classifications is at this point unclear.

AREs control mRNA stability via binding one or more ARE binding proteins (ARE-BPs). At least four distinct ARE-BPs have been reported. These proteins, once bound to the ARE, can mediate rapid decay or, alternatively, can block the action of the ARE and stabilize the mRNA. ARE-BPs that appear to trigger hyperinstability include TTP, GAPDH, and AUF1 (isoforms p37, p40, p42 and 45) [5]. In contrast, binding of HuR to the ARE appears to mediate mRNA stabilization [37]. Thus it may be misleading to consider the ARE-containing mRNAs as intrinsically unstable; instead one might view them as mRNAs that have a wide range of potential stabilities with the unusual capacity for rapid decay when bound by the appropriate control protein(s).

The initial step in ARE-mediated mRNA decay is rapid shortening of the polyA tail. This is followed by $3' \to 5'$ exosome-mediated degradation [16,18] (Fig. 3). Certain of the destabilizing ARE-BPs (possibly TTP) fulfil their function by recruiting the $3' \to 5'$ exosome [18]. According to this model, HuR, which is not able to interact with the exosome, may stabilize ARE-containing mRNAs by binding to the ARE and blocking exosome recruitment [18]. AREs may also function via additional mechanisms such as direct stimulation of decapping activity [12]. As current evidence does not link ARE-BPs to PARN, and the exosome is not presently thought to be involved in polyA shortening, it will be important in future studies to determine how the ARE-mediated polyA shortening is linked to exosome recruitment.

6.3. Control of mRNA stability in response to intracellular iron concentration

The iron response element (IRE) is one of the earliest and best described RNA stability determinants. The prototype mRNA controlled by this pathway is the transferrin receptor (TfR). The abundance of the TfR mRNA is regulated by intracellular iron levels [38]. Under conditions of reduced iron concentration, TfR mRNA is stabilized resulting in increased TfR protein expression and increased uptake of iron (Fig. 3). Reciprocally, in the presence of high iron levels, TfR mRNA is destabilized with an appropriate drop in TfR expression. This switch in TfR mRNA stability is regulated by the binding of two related iron responsive proteins (IRP) to the IRE with consequent protection of the 3′UTR from endonuclease

506

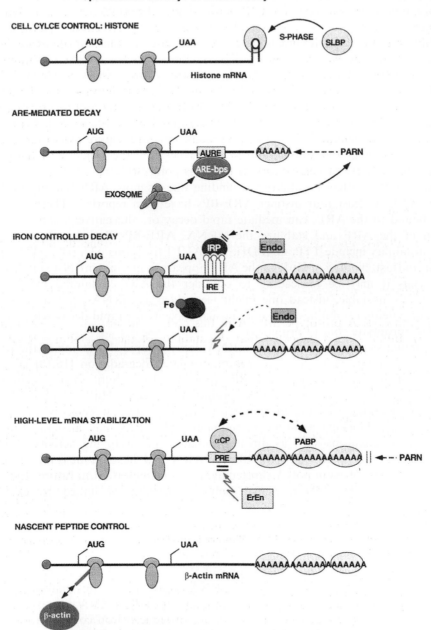

Fig. 3. Examples of specialized mRNA decay pathways. *Cell cycle control: histone mRNA*. Histone mRNAs lack polyA tail; instead they contain at their 3′ termini a highly conserved stem-loop that is bound by a specific SLPB (stem-loop-binding protein). This complex stabilizes mRNA during S phase of the cell cycle. After S-phase the SLBP is released (mechanism unknown) and the histone mRNA is rapidly

attack. IRP structure and its binding affinity for the IRE is highly sensitive to iron levels; in the iron-poor state the IRP binds tightly to the IRE while in a high iron environment the IRP dissociates from the IRE and exposes an endoribonuclease-sensitive site [38]. IREs are present in a well-defined set of genes whose expression is co-ordinately controlled by iron [38,39]. In a subset of these genes, including the TfR, the IRE is present in the 3'UTR and controls expression via alterations in stability. In a second set of iron-responsive mRNAs, best described for the ferritin mRNA, the IRE is located within the 5'UTR. In this case the reversible binding of the IRP mediates a highly efficient translational control via blocking 40S ribosome attachment and translation initiation [39]. Thus the reciprocal control of ferritin and TfR expression by translational and mRNA stability mechanisms is mediated and integrated by IRE/IRP interactions in the 5' and 3'UTRs, respectively.

6.4. Determinants of highly stable mRNAs

Certain mRNAs are highly stable. These mRNAs tend to encode bulk proteins expressed in terminally differentiated cells. This relationship is logical as these mRNAs do not need to respond to rapid changes in gene expression profiles and their stability affords the cell an economy of scale. A number of highly stable mRNAs have been analysed to determine the basis for their relative resistance to decay pathways. A prototype mRNA in this class is the human α-globin mRNA (Fig. 3). Initial studies indicated that the stability of the hα-globin mRNA was contributed to by a determinant in its 3'UTR. This determinant comprises three non-contiguous C-rich sites (pyrimidine-rich element; PRE) and binds a 39 kD RNA binding protein [40,41]. The binding protein, α-globin polyC-binding protein (αCP), also known as polyC-binding protein (PCBP or hnRNP E), contains 3 KH domain RNA-binding motifs. The 'α-complex' formed between the PRE and αCP stabilizes the hα-globin mRNA in a translation-independent manner [42]. The α-complex

---

degraded. *ARE-mediated decay.* AREs (AU-rich elements) are present within 3'UTR of a large subclass of highly unstable mRNAs. ARE-bps (ARE-binding proteins) once bound to the ARE, can mediate rapid decay via rapid shortening of the polyA tail, followed by 3' to 5' exosome-mediated degradation. *Iron controlled decay: transferrin receptor mRNA.* The decay of mRNAs carrying an IRE (iron response element) in the 3'UTR is regulated by intracellular iron levels. If there is no iron, IRPs (iron responsive protein) bind the IRE and stabilizes mRNA. When the intracellular iron levels are high, iron complexes with the IRP and causes a major allosteric shift with resultant loss in affinity for the IRE. The IRE is now exposed to sequence-specific endonucleolytic cleavage. *High-level mRNA stabilization; human α-globin mRNA.* The human α-globin mRNA, a highly stable mRNA, has a C-rich, pyrimidine-rich (PRE) element within 3'UTR. This determinant binds αCP (α-globin PolyC-binding protein), forming an RNP complex that stabilizes mRNA. This stabilization may be mediated via a direct interaction of the αCP protein with the PABP. Alternatively, release of αCP from the PRE (mechanisms unknown) may expose the PRE to cleavage by an erythroid-enriched endonuclease (ErEN). Whether the endonuclease cleavage and/or the polyA shortening constitute the rate-limiting event in turnover of this highly stable mRNA is unknown. *Nascent peptide control: β-tubulin mRNA.* This decay pathway is controlled by the amount of free intracellular tubulin. When uncomplexed tubulin is present (i.e., in excess), it interacts with the nascent N-terminus of the β-tubulin protein and promotes decay of the mRNA *in cis*. This mechanism remains undefined.

appears to inhibit polyA shortening [43,44]. This protection may be mediated via a direct interaction of the αCP protein with PABP [44]. An erythroid-enriched endoribonuclease (ErEN) has been described that can cleave the α-globin mRNA within the PRE [44,45]. This ErEN cleaving activity appears to be controlled via interaction of αCP with the polyA-associated PABP and may function subsequently to polyA tail shortening. Thus this model of α-globin mRNA decay is novel in that it links polyA shortening with endonuclease cleavage.

The α-complex or a closely related complex may constitute a general determinant of mRNA stabilization [46]. Subsequent to the definition of the PRE stability determinant in hα-globin mRNA a series of additional highly stable cellular mRNAs including αI(1) collagen and tyrosine hydroxylase, were found to contain pyrimidine (usually C-rich) elements in their 3′UTRs and bind αCP. In several of these cases the α-complexes has been functionally linked to stabilization [46]. Stabilization of the polio viral mRNA also has been linked to formation of the α-complex. In this case the complex forms at the 5′ terminal cloverleaf. This binding site is located 5′ to the site of 40S ribosome binding (IRES) thus avoiding interference with translational controls (see above).

6.5. Coding region stability determinants

Coding region determinants of stability remain poorly defined and often work in conjunction with cis-acting elements elsewhere in the mRNA [10]. As an example, c-myc mRNA has an ARE within its 3′UTR that appears to work in conjunction with a coding determinant [5]. This coding region site appears to contain an endonuclease-sensitive site that is exposed during active translation by the elongating ribosome [6]. Proteins that bind on this region and protect it from cleavage [6,10] have been identified. A similar situation may exist for the c-fos mRNA. This mRNA contains an ARE in the 3′UTR as well as two purine-rich destabilizing elements in the coding region [7,10]. As in the case of the c-myc mRNA, this decay pathway is translation-dependent [10]. Two proteins have been shown to interact with this coding region determinant, one is associated with the polysome and the other is located in the post-ribosomal fraction [10]. Based on the spacing constraints between the coding determinant and the polyA tail a model was proposed in which a bridging complex formed between these two determinants is disrupted during translation leading to mRNA decay [47]. The presence of multiple, and apparently redundant destabilization determinants in mRNAs such as myc and fos may reflect the importance of rapid mRNA decay and extinction of the corresponding protein signals to cell function.

6.6. Triggering mRNA destabilization by its protein product

An unusual mode of stability control has been described in the case of the β-tubulin mRNA [3,10]. The stability of this mRNA is controlled by translation of its first 13 nucleotides (Fig. 3). The N-terminal tetrapeptide encoded by the β-tubulin mRNA provides a signal to target rapid decay of the mRNA in cis. This decay appears to function as part of a feedback loop; free β-tubulin subunits interact either directly or indirectly with the nascent N-terminus of the β-tubulin protein as it emerges from the ribosome. The mechanism that links the N-terminal peptide synthesis with accelerated mRNA decay and the pathway involved in this process remain to be deciphered.

7. The role of translation in mRNA decay pathways

Translation is often a key element in mRNA decay pathways. The decay rates of many mRNAs are correlated with their translational state [10,47]. Pharmacologic inhibition of translation and/or mutations that alter translational efficiency or accuracy can impact on mRNA stability [3]. This linkage between mRNA translation and decay may reflect one or more effects of the elongating ribosome on mRNA structures and/or mRNP complex assembly. Translational activity may impact on the structure and function of the 'closed loop' and association of ribonucleases and decay factors with the translational apparatus may facilitate their access to critical regions [3]. As detailed above, NMD is blocked by inhibition of translation and a number of the Upf factors are associated with translational initiation and termination complexes [12].

A clear example of how translation can impact on mRNA stability comes from analysis of a naturally occurring mutation in human α-globin mRNA. The 'Constant Spring' mutation converts the α-globin termination codon to a sense codon (UAA → CAA). This mutation represents the most prevalent form of nondeletional α-thalassemia and is carried by millions of individuals in Southeast Asia. The translating ribosomes extend into the 3′UTR of the α-globin mRNA for additional 31 codons prior to terminating at an in-frame UAA embedded within the normal polyadenylation signal (AAUAAA). This mutation results in a dramatic loss of mRNA stability [48]. While this translation-dependent destabilization may be contributed to by displacement of the 3′UTR α-complex, it is also possible that this mutant mRNA may be recognized by a novel surveillance pathway.

8. Degradation of mRNA by external factors

The decay pathways discussed in the preceding text are all intrinsic to the cell. mRNA stability can also be impacted by environmental alterations, hormonal fluctuations, and viral infections. An example of environmental influence is the stability regulation of cholesterol 7-α-hydroxylase (CYP7A1) mRNA in response to diurnal variation. This pathway is dependent on the AREs in the 3′UTR of the CYP7A1 [5]. Estrogen and glucocorticoids destabilize IL-1, interferon, and collagen mRNAs, as a component of their anti-inflammatory effect [10]. A number of viral infections trigger substantial and general decay of cellular mRNA, freeing up the cellular translational apparatus for expression of viral proteins. Some viruses, including influenza, contain specific nucleases that cleave cell mRNAs to generate oligonucleotides containing capped 5′ ends that serve as primers for their mRNA synthesis [34]. The Herpesvirus infections result in the expression of a virally encoded mRNA degradation factor, Vhs, that targets cellular mRNAs. A number of studies suggest that Vhs is either an endonuclease or activates a cellular endonuclease. Recent studies show that Vhs interacts with the cellular translation initiation factors suggesting a mechanism for targeting Vhs to mRNA and to regions of translation initiation [49].

9. Conclusions

In conclusion, mRNA decay pathways in eukaryotic cells are varied and flexible. These pathways contribute to the ability of a cell to respond rapidly to changes in intrinsic and extrinsic stimuli as well as to eliminate potential deleterious transcripts through multiple surveillance mechanisms. Structures at the 5′ and 3′ termini of a mRNA have a major role in stabilizing a mRNA and controlling its rate of decay. Our understanding of how these termini physically and functionally communicate with each other and with additional *cis* elements has undergone a substantial recent advance and has led to more refined models of stability control. Specific sequence determinants and corresponding *trans*-acting proteins are also involved in the targeted control over the decay of subsets of mRNA. The mechanisms involved in these targeted pathways often utilize rate-limiting steps that circumvent the general decay pathways. The recent definition of a multiprotein exosome has expanded our understanding of many of these pathways but many details of its targeting and mode of action remain to be explored. The pathways of mRNA decay thus represents a field of study that offers novel insights into eukaryotic gene expression and poses many interesting challenges for future study.

Acknowledgements

Angela Inácio was supported by the Portuguese Fundação para a Ciência e Tecnologia. Work in Dr. Liebhaber's laboratory is supported by grants from the NIH (HL65449, HL38632, CA7276) and by generous support of the Doris Duke Foundation. We apologize to those whose work was not cited directly; due to lack of space a large number of primary citations were encompassed in the cited reviews.

Abbreviations

5′UTR	5′ untranslated region
3′UTR	3′ untranslated region
ARE	AU-rich element
DAN	deadenylating nuclease
DCP	5′ decapping enzyme

NMD   nonsense-mediated RNA decay
ORF   open reading frame
PABP   polyA-binding protein
PAN   polyA nuclease
PARN   polyA ribonuclease
PTC   premature termination codon
RNAi   RNA interference
RNAPII   DNA-dependent RNA polymerase II
RNP   RNA-protein complexes
rRNA   ribosome RNA
snoRNA   small nucleolar RNA
XRN1   5′ exoribunuclease

References

1. Hilleren, P. and Parker, R. (1999) Ann. Rev. Genet. 33, 229–260.
2. Maquat, L.E. and Carmichael, G.G. (2001) Cell 104, 173–176.
3. Jacobson, A. and Peltz, S. (1996) Ann. Rev. Biochem. 65, 693–739.
4. Imataka, H., Gradi, A., and Sonenberg, N. (1998) EMBO J. 15, 7168–7177.
5. Guhaniyogi, J. and Brewer, G. (2001) Gene 265, 11–23.
6. Sachs, A. (1993) Cell 74, 413–421.
7. Shyu, A., Belasco, J., and Greenberg, M. (1991) Genes Dev. 5, 221–231.
8. Beelman, C. and Parker, R. (1995) Cell 81, 179–183.
9. Grey, N.K., Coller, J.M., Dickson, K.S., and Wickens, M. (2000) EMBO J. 19, 4723–4733.
10. Ross, J. (1995) Microbiol. Rev. 59, 423–450.
11. Brown, C. and Sachs, A. (1998) Mol. Cell Biol. 18, 6548–6559.
12. Mitchell, P. and Tollervey, D. (2001) Curr. Opin. Cell Biol. 13, 320–325.
13. Dehin, E., Wormington, M., . Korner, C., and Wahle, E. (2000) EMBO J. 19, 1079–1086.
14. Gao, M., Wilusz, C., and Peltz, S. (2001) EMBO J. 20, 1134–1143.
15. Wang, Z. and Kiledjian, M. (2001) Cell 107, 751–762.
16. Mukherjee, D., Gao, M., O'Connor, J., Raijmakers, S., Pruijin, G., Lutz, C., and Wilusz, J. (2002) EMBO J. 21, 165–174.
17. van Hoof, A. and Parker, R. (1999) Cell 99, 347–350.
18. Chen, C., Gherzi, R., Ong, S., Chan, E., Raijmakers, R., Pruijin, G., Stoecklin, G., Moroni, C., Mann, M., Karin, M. (2001) Cell 107, 451–464.
19. Frischmeyer, P., van Hoof, A., O'Donnell, K., Guerreiro, A., Parker, R., and Dietz, H. (2002) Science 295, 2258–2262.
20. Couttet, P., Fromont-Racine, M., Steel, D., Pictet, R., and Grange, T. (1997) Proc. Natl. Acad. Sci. USA 94, 5629–5633.
21. Mango, S. (2001) Trends Genet. 17, 646–5325. Ishigaki, Y., Li, X., Serin, G. and Maquat, L.E. (2001) Cell 106, 607–617.
22. Culberston, M. (1999) Trends Genet. 15, 74–80.
23. Sun, X., Perlick, H., Dietz, H., and Maquat, L. (1998) Proc. Natl. Acad. Sci. USA 95, 10,009–10,014.
24. Byers, P. (2002) J. Clin. Invest. 109, 3–6.
25. Ishigaki, Y., Li, X., Serin, G., and Maquat, L.E. (2001) Cell 106, 607–617.
26. Yamashita, A., Ohnishi, T., Kashima, I., Taya, Y., and Ohno, S. (2001) Genes Dev. 15, 2215–2228.
27. Zhang, J., Sun, X., Qian, Y., Laduca, J., and Maquat, L. (1998) Mol. Cell. Biol. 18, 5272–5283.
28. Rajavel, K. and Neufeld, E. (2001) Mol. Cell. Biol. 21, 5512–5519.
29. Buzina, A. and Shulman, M. (1999) Mol. Biol. Cell 10, 515–524.

512

30. Zhang, J., Sun, X., Qian, Y., Laduca, J., and Maquat, L. (1998) RNA 4, 801–815.
31. Romão, L., Inácio, A., Santos, S., Ávila, M., Faustino, P., Pacheco, P., and Lavinha, J. (2000) Blood 96, 2895–2901.
32. Wang, J., Gudikote, J., Olivas, G., and Wilkinson, M. (2002) Embo Reports 3, 274–279.
33. Murray, K.E., Roberts, A.W., and Barton, D.J. (2001) RNA 7, 1126–1141.
34. Atwater, A., Wisdom, R., and Verma, I. (1990) Ann. Rev. Genet. 24, 519–541.
35. Ross, J. (1996) TIG 12, 171–175.
36. Xu, N., Chen, C., and Shyu, A. (2001) Mol. Cell. Biol. 21, 6960–6971.
37. Brennan, C.M. and Steitz, J.A. (2001) Cell. Mol. Life Sci. 58, 266–277.
38. Rouault, T. and Klausner, R. (1997) Curr. Topics Cell. Regul. 35, 1–19.
39. Theil, E. (1993) Biofactors 4, 87–93.
40. Wang, X., Kiledjian, M., Weiss, I., and Liebhaber, S. (1995) Mol. Cell. Biol. 15, 1769–1777.
41. Kiledjian, M., Wang, W., and Liebhaber, S. (1995) EMBO J. 17, 4357–4364.
42. Weiss, I. and Liebhaber, A.S. (1995) Mol. Cell. Biol. 15, 2457–2465.
43. Morales, J., Russel, J.E., and Liebhaber, S. (1997) J. Biol. Chem. 272, 6607–6613.
44. Wang, Z. and Kiledjian, M. (2000) EMBO J. 19, 295–305.
45. Rodgers, N., Wang, Z., and Kiledjian, M. (2002) J. Biol. Chem. 277, 2597–2604.
46. Holcik, M. and Liebhaber, S. (1997) Proc. Natl. Acad. Sci. USA 94, 2410–2414.
47. Grosset, C., Chen, C., Xu, N., Sonenberg, N., Sablon, H., and Shuy, A.-B. (2000) Cell 103, 29–40.
48. Weiss, I. and Liebhaber, S. (1994) Mol. Cell. Biol. 14, 8123–8132.
49. Feng, P., Everly, D., and Read, G. (2001) J. Virol. 75, 10,272–10,280.

S.C. Makrides (Ed.) *Gene Transfer and Expression in Mammalian Cells*
© 2003 Elsevier Science B.V. All rights reserved

CHAPTER 18

Pathways of mammalian protein degradation

William A. Dunn, Jr.

*Department of Anatomy and Cell Biology, University of Florida, College of Medicine,
Gainesville, FL 32610, USA; Tel.: + 1-352-392-1872; fax: + 1-352-392-3305.
E-mail: dunn@anatomy.med.ufl.ed*

1. Introduction

Cellular proteins are continually synthesized and degraded at different rates. Many regulatory proteins (e.g., cell cycle proteins, metabolic enzymes, and transcription factors) have short half-lives of minutes to hours while a majority of the house keeping proteins have long half-lives of hours to days. In addition, eukaryotic cells adapt to environmental stresses such as ischemia, nutrient starvation or serum deprivation by degrading proteins and organelles thereby eliminating damaged or superfluous proteins and providing amino acids for the synthesis of response proteins. These events are essential for the survival of the cell. However, enhanced proteolysis may result in cachexia and cell death (e.g., apoptosis) while suppressed proteolysis can result in abnormal growth (e.g., cancer).

Proteinases have been localized to virtually every cellular compartment of eukaryotic cells. Selective limited proteolysis mediated by endopeptidases is essential for processing pre and pro proteins to their mature forms, modulating signal transduction pathways, regulating cell death, and presenting antigens. These endopeptidases include the family of calpains and caspases. On the other hand, complete proteolysis is mediated by both exopeptidases and endopeptidases resulting in the degradation of proteins to dipeptides and amino acids. This process may be selective or nonselective. Selective degradation is necessary for regulating the cell cycle and for removing damaged or nonessential proteins, while nonselective degradation provides amino acids that are used to maintain protein synthesis. Protein degradation also provides the cell a defense mechanism against environmental stresses and intracellular pathogens. Complete proteolysis is performed by proteinases assembled within the proteasome complex or compartmentalized within the lysosome.

2. Pathways of selective proteolysis

The proteinases within the cytosol regulate many cellular events including signal transduction, differentiation and development, DNA metabolism, cytoskeletal structure, transcription, cell cycle, and cell death. In addition, these proteinases are essential for protein quality control by eliminating damaged and nonessential proteins.

2.1. Calpain-mediated proteolysis

2.1.1. Calpain structure

The active calpain proteinase is a heterodimer composed of a large subunit of approximately 80 kDa and a smaller subunit of 30 kDa. There exist at least 14 genes that encode the large subunit and five genes that encode the small subunit [1,2]. The large subunit contains a cysteine-proteinase domain and may have a calmodulin-like Ca^{2+} binding region. The small subunit contains only the calmodulin-like Ca^{2+} binding region, which assists in activating the large subunit. For example, calpain 4 is the small subunit required for the activation of calpains 1, 2 and 9. Although the expression of this family of proteins in tissues is largely ubiquitous, some of the large subunits are tissue specific. For example, p94 (calpain 3) is found exclusively in skeletal muscle while calpains 11 and 13 are present in the testis and calpain 8 in the stomach.

2.1.2. Substrates and recognition

A variety of protein targets for calpain-mediated degradation have been identified including cytoskeletal proteins, membrane proteins, enzymes, and transcription factors. It is projected that a PEST (rich in prolines, glutamates, serines, and threonines) domain targets proteins for calpain degradation by presenting calcium to the calpain. Indeed, this domain is found in many soluble proteins that are short-lived and the calpain cleavage sites within protein kinase C, amyloid precursor protein, Ca^{+2}-ATPase, and hydroxymethylglutaryl-CoA are either within or near PEST sequences [3]. In addition, the degradation of IκBα by calpains requires the phosphorylation of a PEST domain within IκBα [4]. However, some proteins that are susceptible to calpain hydrolysis lack PEST domains while others that have PEST sequences are either resistant to calpains or susceptible to calpain digestion in a PEST independent manner [3].

2.1.3. Regulation and inhibitors

Calcium levels within the cell regulate the calpains. Calpain 1 is also referred to as μ-calpain because it is activated by micromolar concentrations of calcium. Meanwhile, calpain 2 or m-calpain is activated when calcium concentrations reach millimolar levels. A variety of cellular insults including ischemia and oxygen stress cause an elevation of cellular calcium levels. In addition, the calcium levels tend to oscillate in neurons and muscle cells, which require calcium for function. The primary sources of cytosolic calcium are extracellular, endoplasmic or sarcoplasmic reticulum, and mitochondria.

There exists an endogenous inhibitor of calpains known as calpastatin, which is expressed in virtually all cells. Calpastatin binds to the calcium binding domains of both large and small subunits. Calpain p94 is not inhibited by calpastatin but instead digests this protein. It is presumed that since p94 does not require the 30 kDa subunit for activity calpastatin binds poorly and becomes a substrate.

E64 and leupeptin inhibit calpains *in vitro* and *in vivo*, but these drugs also inhibit the proteasome and lysosomal proteinases. Calpain inhibitors I and II are also cell permeable, but they inhibit the proteasome and some caspases. More specific inhibitors of calpain activities include PD150606, which binds to the Ca^{+2} binding sites and AK275, calpeptin, and E-64d, which bind to the active site of these proteinases [2,4].

2.2. Caspase-mediated proteolysis

2.2.1. Caspase structure
There exist at least 14 caspases that either activate cytokines (caspases 1, 4, 5, 11, 12, 13, and 14) or regulate apoptosis (caspase 2, 3, 6, 7, 8, 9, and 10) in mammalian cells (Table 1). These cysteine proteinases are synthesized in the pro or inactive forms. Upon an extrinsic or intrinsic stimulus these enzymes are proteolytically activated to p20 and p10 subunits that assemble into the active heterotetramer [5]. When activated, caspases are endopeptidases that cleave at aspartic acid residues. Like other cysteine proteinases, caspases are sensitive to transition metal ions such as Zn^{+2}. Unlike the calpains, caspases are unaffected by cellular calcium levels.

2.2.2. Substrates and recognition
The caspases have been shown to degrade a number of cellular proteins at the peptide bond C-terminal to an aspartic acid [5]. The cleavage sites for a number of caspases have been identified (Table 1). Caspase substrates include cytoskeletal proteins (actin, β-catenin, and lamins), transcription factors (STAT1, NF-κB, and IκBα), cytokine precursors (Pro-IL-1β, Pro-IL-16, and Pro-IL-18), proteins that mediate signal transduction (Akt/PkB, Ras, phospholipase A2, MEKK1, and APC), cell cycle proteins (cyclin A, Cdc27, p21, and p27), and proteins that mediate apoptosis (procaspases, Bcl-2, Bid, and IAPs).

The apoptotic caspases act to either initiate apoptosis (caspases 2, 8, 9, and 10) or affect apoptosis (caspases 3, 6, and 7). The initiator caspases proteolytically activate the downstream effector caspases, which in turn proteolytically activate DNA nucleases such as CAD and inactivate nuclear lamins, DNA repair enzymes (e.g., PARP), cytoskeletal proteins (e.g., gelsolin and α-fodrin), and cell cycle regulators (e.g., Cdc27, Wee1, p21Cip1, and p27Kip1) [5,6].

2.2.3. Regulation and inhibitors
There exists a number of endogenous and viral proteins that stimulate and inhibit cellular caspases [7]. For example, cytochrome C released from the mitochondria interacts with Apaf1 forming an apoptosome that stimulates the autoactivation of caspase 9. Caspase 9 then proteolytically activates caspases 3, 6, and 7, which in turn promote cell death. The phosphorylation of caspase 9 by Akt inhibits caspase activity and the activation of downstream caspases. There also exists a family of cellular (IAPs) and viral (CrmA and p35) inhibitors of caspases (Table 1). The IAP (inhibitor of apoptosis) proteins contain one or more BIR (baculoviral IAP repeat) domains, which bind to and inhibit caspases. The mitochondrial protein Smac/DIABLO also interacts with these domains thereby competitively inhibiting IAP binding to the caspases.

Table 1
The Caspase family of proteinases

Name	Aliases	Substrate recognition	Substrates		Inhibitors	
					Cellular	Viral
Caspase 1	ICE	WEHD	IL-1β			CrmA
			IL-18			p35
Caspase 2	ICH-1	DEHD				
	mNedd2					
Caspase 3	CPP32	DEVD	PARP	ICAD	XIAP	CrmA
	apopain		p21Cip1	p27Kip1	HIAP1	p35
	Yama		MEKK1	MST1	HIAP2	
			PAK2	Cdc17	NAIP	
			Gelsolin	c-Raf	Survivin	
			Akt	Wee1	Livin	
			Cdc17			
Caspase 4	TX	(W/L)EHD	pro-Caspase 1			CrmA
	ICH-2					
	ICErel-II					
Caspase 5	TY	(W/L)EHD				CrmA
	ICErel-III					
Caspase 6	Mch2	VEHD	Lamins			p35
			pro-Caspase 3			
			pro-Caspase 7			
Caspase 7	Mch	DEVD	ICAD		XIAP	CrmA
	ICE-LAP3				HIAP1	p35
	CMH-1				HIAP2	
					NAIP	
					Survivin	
					Livin	
Caspase 8	MACH	LETD	Bid			CrmA
	FLICE		pro-Caspase 3			p35
	Mch5		pro-Caspase 7			
Caspase 9	ICE-LAP6	LEHD	pro-Caspase 3		XIAP	
	Mch6		pro-Caspase 7		Livin	
					Akt	
Caspase 10	Mch4		pro-Caspase 3			p35
			pro-Caspase 7			
Caspase 11	ICH-3		pro-Caspase 1			CrmA
Caspase 12						CrmA
Caspase 13	ERICE					CrmA
Caspase 14	MICE					CrmA

Pan-caspase inhibitors (z-Val-Ala-Asp-fluoromethylkethone and Boc-Asp-fluoro-methylketone) are cell-permeable irreversible inhibitors of all caspases. Other fluoromethylketone derivatives have been designed to inhibit specific caspases *in vitro* and *in vivo* (Calbiochem, Inc., La Jolla, CA). For example, z-Asp-Glu-Val-Asp-fluoromethylketone (Z-DEVD-FMK) specifically inhibits caspase 3, Z-LEHD-FMK inhibits caspase 9, Z-IETD-FMK inhibits caspase 8, Z-VEID-FMK inhibits caspase 6, and Z-VDVAD-FMK inhibits caspase 2.

2.3. Ubiquitin-mediated proteasome pathway

Ubiquitin-mediated protein degradation is essential for cell growth and survival. The pathway is responsible for the degradation of cyclins and their kinases, which is required for mitosis, of short-lived regulatory proteins and of proteins that have been damaged during times of cellular stress [8,9]. In addition, viral pathogens exploit the host's protein ubiquitination machinery for their own replication and survival [10].

2.3.1. Protein ubiquitination

The ubiquitin-mediated pathway is a well-characterized pathway of protein degradation (Fig. 1). Proteins are targeted for degradation by three families of

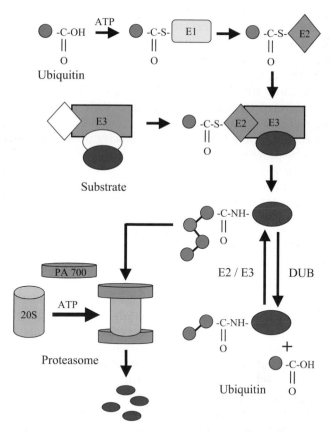

Fig. 1. Ubiquitin-mediated protein degradation. The polyubiquitination of substrate proteins requires the E1 ubiquitin activating enzyme, E2 ubiquitin conjugating enzyme, and E3 ubiquitin ligase. The proteasome recognizes and binds to the polyubiquitin chain and degrades the substrate protein to small peptides. Deubiquitinating enzymes (DUB) remove ubiquitin thereby controlling the length of the polyubiquitin chain, which regulates the affinity of proteasome binding.

enzymes, ubiquitin-activating enzyme (E1), ubiquitin-conjugating enzyme (E2), and ubiquitin-protein ligase (E3) that conjugate the α-carboxyl group of the C-terminal glycine of ubiquitin to the ε-amino group of a lysine moiety in the protein substrate. Ubiquitin is an 8 kDa protein ubiquitously expressed in all cell types. Once a substrate protein is ubiquitinated additional ubiquitins are added to lysine-48 of the protein-bound ubiquitin resulting in a multiubiquitin chain of 2–8 subunits. Multiubiquitinated proteins are recognized by the proteasome and subsequently degraded to small peptides while the ubiquitin moieties escape degradation to be reutilized.

There are two isoforms of E1 that activate ubiquitin for conjugation to proteins and a number of E1-like enzymes that activate ubiquitin-like proteins such as SUMO, Rub1, hAPG12 and MAP-LC3 (hAPG8) for protein conjugation [11,12]. E1 catalyzes the ATP-dependent activation of ubiquitin by forming a thio-ester linkage between a thio group of the active site cysteine and the α-carboxyl group of the C-terminal glycine of ubiquitin. The active site is a highly conserved domain that identifies this family of E1 enzymes. The ubiquitin is then transferred to the cysteine of an E2 ubiquitin-conjugating enzyme. There exists a large family of E2 ubiquitin-conjugating enzymes. These enzymes function independently or with the E3 ligase to conjugate ubiquitin to lysine moieties of substrate proteins and ubiquitin. In addition to the "UBC" domain, the E2's contain N-terminal or C-terminal extensions that mediate their specificities for substrates and E3 ligases.

The E3 ligase is responsible for the high specificity of this degradative pathway [8,9]. This ligase is a protein or a complex of proteins that brings together an E2 ubiquitin-conjugating enzyme with the protein substrate. The E3 ligases include: E3α, the HECT E3s, the ring-finger E3s, the multisubunit cullin-containing E3s, and a chaperon-dependent E3 (Table 2). E3α recognizes the "destabilizing" N-terminus of a substrate protein. The HECT E3s have been implicated in the degradation of p53 and kidney Na^+ channels. The N-terminus contains a binding site for the substrate while the C-terminal HECT domain transfers the ubiquitin from the E2 to the substrate. In yeast, the E2 is Ubc7 while the mammalian homologues are two isomers of UbcH5. The ring-finger E3 c-Cbl is involved in the degradation of membrane receptors for epidermal growth factor and platelet-derived growth factor. This monomeric E3 contains a ring-finger domain that interacts with the E2 ubiquitin conjugating enzyme and an N-terminal region that contains a phosphotyrosine-binding SH2 region, which binds to tyrosine kinases. A second ring-finger E3, Mdm2, regulates the degradation of the p53 tumor suppressor [13]. The SCF (Skp1, Cdc53/Cullin, F box receptor) complex is responsible for the degradation of transcription factors, cyclins and cell cycle kinases, and cytoskeletal proteins [14]. The anaphase-promoting complex (APC) is essential for degrading proteins such as B-type cyclins [15]. A VCB (von Hippel-Lindau, Cullin, Elongin B) complex controls the degradation of two hypoxia-regulated transcription factors, Hif1α and Hif2α. These E3 ligase complexes contain a cullin protein that organizes the complex, an F-box protein or adaptor that recognizes the substrate and a ring-finger protein that interacts with the E2 (Table 2). A newly discovered

Table 2

E3 ubiquitin ligases

E3 ubiquitin ligase	Protein components						Substrates
		Cullin	Adaptor	Skp1/Elongin	Ring finger	Ubc	
E3α	E3α					UbcH1/E2$_{14K}$	Proteins w/ destabilizing N-terminus
U-Box E3	CHIP UFD2a UFD2b CYC4 PRP19		Hsp70 Hsp90 Hsp40			UbcH5A UbcH5B UbcH5C	Immature or aberrant proteins
HECT-domain	E6-AP NEDD4					UbcH5A UbcH5B/UBC4 UbcH5C UbcH7 UbcH8	p53 Blk
Ring-finger E3	c-Cbl BRCA1 Mdm2 Siah1 A07					UbcH5C UbcH7 UbcH8	EGF-R PDGF-R DCC p53
SCF complex		Cullin 1	Skp2 β-TrCP CDC4	Skp1	Roc1 Roc2	CDC34 UbcH5A	E2F-1 Cyclin D p21 Cyclin E p27Kip1 IκB β-catenin CD4 Sic1
VCB complex		Cullin 2	VHL	Elongin B Elongin C	Roc1		Hif1α Hif2α
APC complex	Apc1 Apc3 Apc4 Apc5 Apc6 Apc7 Apc8	Apc2	Cdc20 Cdh1/Hct1 Plk/Plx		Apc11	UBC4/UbcH5B UbcH10	Securins Cyclin A Cyclin B Cdc5/Plk1 Nek2 Cdc20

chaperone-mediated E3 called CHIP mediates the polyubiquitination of unfolded and aberrant proteins that have been captured by molecular chaperones such as Hsp90, Hsc70, and Hsp40 [16,17]. CHIP interacts with a chaperone that acts like an adaptor by binding to the substrate protein. The U box domain of CHIP acts like a ring-finger binding site for E2. There are a number of U box E3 ligases that polyubiquitinate protein substrates. Like the HECT and RING finger E3's, the U box E3's catalyze polyubiquitination at lysine 48.

2.3.2. Proteasome

The proteasome is a large multicatalytic complex, which exists in two forms: the 20S catalytic proteasome and the 26S regulated proteasome [18]. The 20S proteasome is well conserved from yeast to humans and consists of a hollow barrel shaped structure composed of four stacks of heptameric rings of proteinases. These 14–17 different proteinases have been classified into five hydrolytic activities: chymo-trypsin-like activity (cleavage after hydrophobic amino acids), trypsin-like activity (cleavage after basic residues), post-glutamyl peptide hydrolyzing activity, branched-chain amino acid-preferring activity, and neutral amino acid-preferring activity [19]. However, because the proteasome degrades proteins into peptides, further degradation by a tripeptidyl peptidase II and various aminopeptidases are necessary to complete degradation [20]. The 26S proteasome consists of the 20S proteolytic complex with a 19S cap referred to as PA700. This 19S complex contains 17 different proteins including six ATPases and possibly some chaperones. The cap consists of a lid of proteins that recognizes polyubiquitinated protein substrates and a base that caps the end of the proteasome whereby it unfolds the protein substrates and threads them into the core of the proteasome. The 20S proteasome also interacts with an 11S proteasome activator called PA28, which is essential for antigen presentation [18].

The proteasome can also degrade proteins in an ubiquitin-independent manner. For example, the 20S proteasome can recognize and degrade oxidized proteins in the absence of ATP and ubiquitin [21]. In addition, some misfolded proteins are degraded independent of protein ubiquitination. The recognition of these proteins by the proteasome is likely mediated by ATPases that are components of the PA700 cap or by proteins that belong to a family of antizymes [22,23].

2.3.3. Substrates and recognition

The recognition of substrate proteins occurs by its interaction of the protein with the E3 ligase or an adaptor protein of the SCF or APC complex. One such recognition signal is the N-terminal amino acid [24]. The E3α binds to basic and bulky hydrophobic residues of the N-terminus of substrate proteins thereby promoting their ubiquitination. The protooncoprotein c-Mos and a number of viral proteins are degraded by this mechanism. This pathway has a major role in the ubiquitin-mediated degradation in muscle cells, but only a minor contribution in HeLa cells [25]. Interestingly, virtually all proteins contain either an N-α-acetylated methionine or a methionine at their N-termini, which prevents degradation. Nevertheless, modification of the N-terminus can occur by removing the methionine (methionine aminopeptidase) or adding an arginine (arginyl-tRNA-protein transferase) [24]. In addition, some proteins are proteolytically activated thereby exposing a destabilizing N-terminus. For example, the Sindbis virus RNA polymerase is synthesized by a proteolytic cleavage of nsP1234, which exposes a tyrosine moiety thereby targeting it to ubiquitin-mediated degradation. In addition, this pathway may have a role in degrading the protein fragments produced by calpain and caspase proteolysis during cell death. Protein ubiquitination by the SCF$^{\beta TrCP}$ complex requires phosphoryla-tion of IκBα and β-catenin. The WD40 domain of the adaptor β-TrCP constitutes a phosphoserine-binding site for the consensus motif $DS^PG\psi XS^P$ (S^P represents the phospho-serine and ψ a hydrophobic residue) present in these proteins [9]. Phosphorylated Sic1 is recognized and ubiquitinated by the SCF^{CDC4} complex. The Cdc20 adaptor of the APC complex recognizes a destruction box motif (RxxLxxxxN/D/E) present in a number of cyclins [15]. Meanwhile, a second adaptor (Cdh1) of APC recognizes a KEN box (KENxxxD/N) found in Cdc20 and Nek2 [26]. A chaperone-dependent E3 ligase catalyzes the polyubiquitination of immature or aberrant proteins [16,17]. CHIP and Hsc70 target aberrant forms of cystic fibrosis transmembrane conductance regulator (CFTR) for ubiquitination and proteasome degradation [27]. Chaperones can bind to unfolded proteins whereby Hip and Hop regulate protein refolding. If protein refolding fails to proceed, CHIP then interacts with this chaperone complex bringing with it an associated E2 thereby promoting ubiquitination and proteasome degradation.

The quality control system of endoplasmic reticulum (ER) monitors the folded state of newly synthesized proteins. Those proteins that fail to correctly fold or assemble due to protein mutations or environmental stresses exit the ER presumably through the Sec61p translocon to be degraded by the proteasome [28]. As much as 75% of the normal CFTR is degraded before leaving the ER. However, CFTR lacking phenylalanine 508 is misfolded and completely degraded from the ER. The ER-associated degradative pathway requires ER chaperones, BiP and calnexin. Meanwhile, ubiquitination of the unfolded ER protein enhances its degradation, but is not required.

Viral infection promotes ubiquitin-mediated degradation of cellular proteins that may interfere with virus survival [10]. For example, the papillomavirus expresses a protein called E6 that enhances the degradation of the pro-apoptotic p53 by the E3 ligase E6-AP. The Vpu protein from human immunodeficiency virus type 1 stimulates the degradation of CD4 receptor by the $SCF^{\beta TrCP}$ complex. The human cytomegalovirus avoids immune surveillance by stimulating the proteasome-mediated degradation of MHC class1 heavy chain molecules that are normally present in the ER [28]. The Epstein–Barr virus synthesizes the nuclear antigen-1 (EBNA1), which suppresses the immune response by inhibiting antigen presentation [29]. This is accomplished by an internal glycine–alanine repeat that inhibits the proteasome degradation of ubiquitinated proteins (see Chapter 19).

2.3.4. Regulation and inhibitors
The cell cycle requires ubiquitin-mediated degradation that is regulated at the level of protein ubiquitination. The APC E3 ligase activity and specificity is regulated by multiple kinase and phosphatase events that modulate the phosphorylation states of the many subunits of APC [15]. Some subunits of the SCF E3 ligase are phosphorylated and cullin is modified by attachment of the ubiquitin-like protein Rub1 [14]. However, the effects of these modifications on ligase activity have not been clearly defined.

The degradation of p53 is regulated by a number of post-translational modifications of p53 and the E3 ligase Mdm2 [13]. The phosphorylation of p53 can stabilize this protein by preventing its binding to Mdm2. Meanwhile, the phosphorylation of Mdm2 promotes self-ubiquitination and degradation. The conjugation of SUMO, an ubiquitin-like protein, to Mdm2 minimizes self-ubiquitination and promotes the ubiquitination of p53. Truncated forms of Mdm2 produced by alternate splicing of mRNA or by caspase activity can also effectively inhibit Mdm2 thereby causing an increase in p53.

The levels of ubiquitinated proteins and rates of proteasome degradation dramatically increase in stressed cells [30,31]. Transcription of mRNA encoding ubiquitin and several proteasome subunits is enhanced in cells that have been exposed to physiologic stress such as fasting, heat, and γ-irradiation or pathologic stress such as cancer cachexia, diabetes, and sepsis [9,32]. Denervation atrophy in muscles is accompanied by an increase in the synthesis of ubiquitin and proteasome subunits [33]. Lastly, the levels of $E2_{14K}$ (UbcH1), the cognate E2 for E3α increase when cells are starved or exposed to mild oxidative stress [34,35].

An alternative mode of regulating ubiquitin-mediated protein degradation is by deubiquitinating enzymes [36,37]. There exist two classes of deubiquitinating proteinases that cleave at the C-terminal glycine of ubiquitin. The first class serves to provide ubiquitin while the second degrades the polyubiquitin chains on a substrate protein. The ubiquitin C-terminal hydrolases are papain-like thiol proteinases that act on ubiquitin-fusion proteins and on ubiquitin extended by small peptides. This proteinase is essential for the processing of the pro forms of polyubiquitin and two ribosomal proteins (L40 and S27a) that are synthesized as fusion proteins with ubiquitin. This proteinase cleaves at the C-terminal glycine of ubiquitin thereby releasing the mature ubiquitin. The ubiquitin-specific processing proteinases are thiol proteinases that remove ubiquitin from protein substrates and disassemble polyubiquitin chains. This class of proteinases provides the cell the ability to edit protein ubiquitination thereby regulating protein degradation and to recycle the ubiquitin upon degradation of the substrate protein.

There exist a number of synthetic and natural proteasome inhibitors [19]. The lysosomal inhibitor leupeptin and peptide aldehydes, Cbz-leu-leu-leucinal (MG132), Cbz-leu-leu-norvalinal (MG115), (O-t-Bu)-ala-leucinal (PSI), and acetyl-leu-leu-norleucinal (LLnL) inhibit the trypsin-like activity of the proteasome while calpain inhibitors I and II (ALLM) suppress the chymotrypsin-like activity. A second set of inhibitors includes the dipeptide boronic acid PS-341. The boronic acid inhibitors bind irreversibly and have a higher potency than the reversible peptide aldehydes. The natural proteasome inhibitors from bacterial and fungal sources include lactacystin, epoxomicin, TMC-86A, TMC-95 and gliotoxin that act by suppressing the chymotrypsin-like activity of the proteasome. 4-hydroxy-2-nonenal, a product of lipid peroxidation, inhibits both trypsin-like and post-glutamyl peptide hydrolyzing activities. Many of these proteasome inhibitors will cause cell death presumably by suppressing the cell cycle. PS-341 has been shown to reduce tumor growth in both murine and human xenograft models, while lactacystin enhances the antitumor effects of etoposide on colon cancer tumor xenografts. There exists a number of ongoing clinical trails to test the ability of proteasome inhibitors to suppress cancerous growth [19].

2.4. Chaperone-mediated autophagy

2.4.1. Molecular events

Chaperone-mediated autophagy is a pathway for the selective degradation of soluble cytosolic and nuclear proteins [38]. Substrate proteins containing an exposed KFERQ recognition signal bind to Hsc70, which initiates an ATP-dependent unfolding by a chaperone complex that includes Hsp40, Hsp90, Hop, Hip, and Bag-1 [39]. The interaction of the substrate protein with the C-terminus of LAMP2 at the lysosomal surface initiates its translocation into the lysosome (Fig. 2). The chaperone complex presumably stabilizes the Hsc70-substrate at the lysosome membrane thereby assisting in maintaining the protein substrate in the unfolded state. The unfolded protein substrate then traverses the lysosome membrane through a channel that appears to be formed by LAMP2. Protein translocation also requires Hsc70 within the lysosomal matrix.

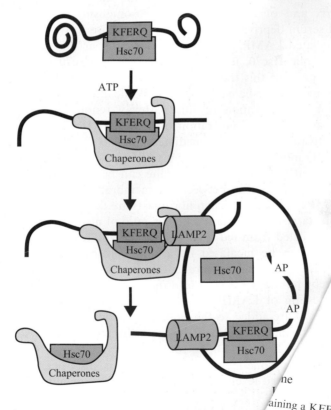

Fig. 2. Chaperone-mediated autophagy. Hsc70 binds to a substrate prop, and Bag-1) proceed to ATP-dependent chaperones and their associate proteins (Hsp40, Hsp90mal surface. The unfolded unfold the substrate protein exposing a binding site for LAMP2 at some where intralysosom substrate protein is then threaded through a LAMP2 channel int' Hsc70 maintains its unfolded state for hydrolysis by acid protein

2.4.2. Substrates and recognition

Approximately 30% of the cytosolic proteins contain one or more KFERQ motifs. RNase A, glyceraldehyde-3-phosphate dehydrogenase, aldolase, dihydrofolate reductase, annexins, c-fos, Pax2, and IκB have been experimentally identified as substrates for this pathway [38,40–43]. Although some proteins have more than one KFERQ motif, the evidence suggests that only one is necessary to promote degradation. RNase A and annexin VI contain only one KFERQ motif and although aldolase B contains three putative motifs, only glutamine 111 was essential for degradation [44]. This motif resides in an extended arm of the aldolase monomer, but appears to be concealed within the functional tetramer. In addition to the KFERQ motif, the folded state of the protein is also critical. For example, the transport of dihydrofolate reductase into isolated lysosomes is reduced when its substrate analogue is present [42].

2.4.3. Regulation

Regulation of this pathway is governed by the amounts of Hsc70 within the lysosome matrix and of LAMP2 at the lysosome surface [45,46]. Amino acid starvation and serum deprivation results in an increase in lysosomal Hsc70 and in membrane associated LAMP2. Although it does appear that the lysosomal Hsc70 arises from cytosolic Hsc70, it remains unclear how Hsc70 is transported to the lysosomes. It is possible this transport may occur by nonselective autophagy (see below) or by chaperone-mediated autophagy. While autophagy is rapidly stimulated upon starvation, the activation of this pathway requires prolonged starvation. Therefore, the autophagic sequestration of cytosolic Hsc70 into lysosomes would subsequently activate the chaperone-mediated pathway. To the contrary, Hsc70 does have two KFERQ motifs that will allow it to be transposed into the lysosome. The levels of LAMP2 at the lysosome membrane can be modulated by changes in its turnover and in its distribution between the membrane and matrix forms [46]. The half-life of LAMP2 is increased when fibroblasts are deprived of serum. Meanwhile, lysosomes from starved rats contain a higher ratio of membrane vs. matrix LAMP2 than lysosomes isolated from fed rats. The proteolytic release of LAMP2 from the membrane appears to be mediated by a serine-protease and a metallo-protease associate with the lysosome membrane. This cleavage occurs near or within the membrane thereby releasing LAMP2 into the matrix where it is then degraded. The proteolytic processing of LAMP2 is suppressed when substrate proteins such as GAPDH and Hsc70 are added to the isolated lysosomes.

Epidermal growth factor treatment dramatically reduces the lysosomal amounts of LAMP2 in RK-52E cells [43]. In turn, the degradation of GAPDH and Pax2 was suppressed. Chaperone-mediated autophagy is also decreased in senescent fibroblasts due to reduced levels of LAMP2 in lysosome membranes [47].

3. Pathways of nonselective proteolysis

The bulk of nonselective acid hydroproteolysis occurs within the lysosomes, which contain a number of acid hydro... that can efficiently degrade proteins to individual amino acids. In addition to chaperone-mediated autophagy, protein substrates can enter the lysosome by microautophagy and macroautophagy pathways. Much of what we know about the molecular events of autophagy comes from the yeast models. There exist more than 25 genes that are essential for autophagy in yeast (Table 3). Many of these proteins have mammalian homologues with similar functions.

3.1. Microautophagy

3.1.1. Molecular events

In microautophagy, the lysosome membrane invaginates and extends arm-like projections that sequester cytoplasm and small organelles (Fig. 3). The opposing arms of the lysosome are brought together and membrane fusion occurs resulting in the formation of a vesicle within the lysosome. The intralysosomal vesicles and their contents are then degraded by lysosomal hydrolases. This process has been best studied in yeast models where it has been shown to be both nonselective and selective. A number of GSA and PAZ genes have been identified and shown to be essential for selective microautophagy of peroxisomes in *P. pastoris*. Many of these proteins are structural homologues of either CVT proteins required for constitutive selective autophagy of aminopeptidase or APG/AUT proteins required for starvation-induced nonselective macroautophagy in *S. cerevisiae* (Table 3). The Apg7/Gsa7-dependent ubiquitin-like conjugation pathways diagrammed in Fig. 5 are required for both glucose-induced selective microautophagy in *P. pastoris* and starvation-induced nonselective microautophagy in *S. cerevisiae* [48,49]. Further studies in *S. cerevisiae*, have revealed that the homotypic fusion of the adjoining vacuole membranes does not require Vam3 or Vam7, which are essential for fusion of the autophagosome with the vacuole during macroautophagy [48].

3.1.2. Regulation

Microautophagy is enhanced in nutrient-starved yeast [48]. In mammalian cells, microautophagy appears to be a constitutive pathway of protein degradation. However, there are conflicting reports as to whether microautophagy is enhanced or suppressed during amino acid starvation [50,51]. Little is known regarding the molecular events that regulate microautophagy in mammalian cells. However ongoing studies in yeast have shown that microautophagy requires ATP [48] and protein kinases Apg1/Gsa10 and Vps15 (unpublished observations, Strømhaug and Dunn). In addition, the *in vitro* uptake of luciferase by isolated yeast vacuoles via microautophagy is inhibited by GTPγS and the V-ATPase inhibitor concanamycin A [48].

3.2. Macroautophagy

3.2.1. Molecular events

Macroautophagy is the primary cellular process for maintaining amino acid levels during times of nutrient deprivation [52]. This degradative pathway also assists in protecting cells from invasive viral and bacterial pathogens [53,54]. In addition,

Table 3

Genes required for autophagy

Yeast name	Aliases	Comments	Mammalian homologue
Apg1	Cvt10, Aut3, Gsa10, Paz1	Serine/threonine protein kinase that complexes with Apg13, Apg17, Cvt9, and Vac8	ULK1
Apg2	Gsa11, Paz7	Soluble protein associated with a "pre-autophagosomal" organelle	AB007864
Apg5		Protein that is conjugated to Apg12	hAPG5
Apg7	Cvt2, Gsa7, Paz 12	E1-like enzyme responsible for the conjugation of Apg12 to Apg5 and Apg8 to phosphatidylethanolamine	HsGSA7
Apg8	Aut7, Cvt5, Paz2	Soluble protein conjugated to phosphatidylethanolamine at the surface of the autophagosome	MAP-LC3
			GATE-16
Apg9	Cvt7, Gsa14, Paz7	Transmembrane protein associated with a "pre-autophagoso-mal" compartment	AK025822
Apg10		E2-like enzyme responsible for the conjugation of Apg12 to Apg5	
Apg12		Ubiquitin-like protein that becomes conjugated to Apg5	hAPG12
Apg13		Regulatory subunit of Apg1 that is dephosphorylated upon the induction of autophagy	
Apg14	Cvt12	Subunit of Vps34 complex that contains Vps15 and Apg6	
Apg16	Paz3	Binds to the Apg12-Apg5 conjugate	
Apg17		Interacts with Apg1 kinase	
Aut1	Apg3, Gsa20	E2-like enzyme responsible for the conjugation of Apg8 to phosphatidylethanolamine	T46276
Aut2	Apg4, Paz8	Cysteine peptidase required to expose the C-terminal glycine moiety of Apg8 to allow its conjugation to phosphatidyletha-nolamine	
Aut10	Cvt18, Gsa12	WD40 protein associated with the vacuolar surface	AL080155
			AF151808
Cvt9	Gsa9, Paz6	Coiled–coil protein found at the vacuolar surface that is required for selective autophagy	
Vps15	Gsa19, Paz13	Serine/threonine protein kinase anchored to a membrane via myristoylation and complexes with Vps34, Apg6 and Apg14	p150
Vps30	Apg6	Soluble protein associated with Apg14, Vps15 and Vps34	Beclin
Vps34		PI3-kinase that interacts with complexes with Vps15, Apg6 and Apg14	Type III PI3-kinase
Vac8	Gsa21	Armadillo-repeat protein required for selective autophagy; anchored to the vacuole surface via myristoylation and palmitoylation	BAB71463.1
			NP_115526
Fkb1	Fpr1, Rbp1	Rapamycin binding protein	FKBP
Tor 2		Protein kinases inhibited by the Fkb1-Rapamycin complex which induces autophagy	mTor
Gcn2		eIF2α kinase	GCN2, PKR, PERK, HRI
Vam3		A member of the t-SNARE family found at the vacuole membrane	Syntaxin 7
Vam7		A member of the SNAP-25 family that interacts with Vam3	
Ypt7	Vam4	A rab GTPase required for fusion events at the vacuole surface	RAB-7

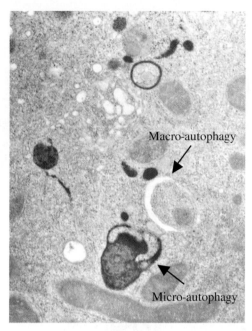

Fig. 3. Microautophagy and macroautophagy in mammalian cells. Amino acid starved mammalian cells were fixed and processed for acid phosphatase cytochemistry. The electron dense reaction product identifies the lysosomal compartments. Profiles of ongoing microautophagy by an acid phosphatase positive lysosome and of ongoing macroautophagy with an autophagosome in close proximity to a lysosome can be observed.

macroautophagy is essential for the removal of damaged organelles, which may occur by environmental stresses or by intracellular pathogens. For example, the loss of mitochondrial permeability stimulates the autophagic removal of these mitochondria [55]. The elimination of leaky mitochondria would protect the cell from stress-induced apoptosis. However, the continued degradation of cellular organelles would ultimately result in cell death.

Macroautophagy proceeds by a series of ordered events: (1) induction and formation of the early autophagosome; (2) maturation into a late autophagosome; (3) docking and fusion of the late autophagosome with a lysosome thereby forming an autolysosome; and (4) degradation and recycling of amino acids (Fig. 4). The induction and formation of the autophagosome is enhanced when cells are deprived of amino acids or serum (see below). The autophagosome arises from ribosome-free regions of the endoplasmic reticulum. This event is suppressed when Ca^{+2}-ATPases present in the endoplasmic reticulum are inhibited by thapsigargin thereby causing the release of calcium [52]. In addition, autophagosome formation requires the assembly of actin and intermediate filaments and the ubiquitin-like conjugation of hAPG12 with hAPG5 [12,50]. HsGsa7 and an Apg10-like E2 protein catalyze the amide linkage between a carboxyl group at the C-terminus of hAPG12 and an ε-amino group of a lysine moiety of hAGP5 (Fig. 5). The hAPG12–hAPG5 complex

528

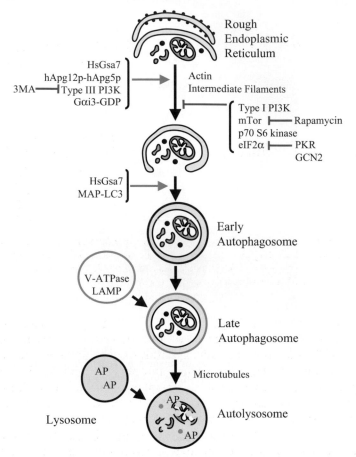

Fig. 4. Model of macroautophagy. Macroautophagy proceeds by a series of ordered events. The first step is the induction and formation of the early autophagosome. This event is highly regulated being stimulated by type III PI3K and HsGsa7 conjugation pathways (see Fig. 5), and inhibited by type I PI3K, mTor, p70 S6 kinase, and eIF2α. 3-Methyladenine (3MA) inhibits autophagy by suppressing type III PI3K while Rapamycin stimulates autophagy by inhibiting mTor. The early autophagosome acquires the membrane proteins LAMP and V-ATPase thereby becoming an acidic late autophagosome. The late autophagosome becomes an autolysosome by fusing with a lysosome and acquiring acid proteinases (AP), which digest the contents of the autophagosome to amino acids and monomeric sugars that are recycled for protein synthesis.

then transiently associates with the forming autophagosome and assists in the targeting of MAP-LC3 (hAPG8) to the autophagosome [56]. MAP-LC3 is required to increase the size of the autophagosome that will not only enable large organelles such as mitochondria to be sequestered but also increase the amount of cellular material sequestered by the autophagosome for degradation. The assembly of MAP-LC3 to the autophagosome is done by an amide linkage between the carboxyl group of the C-terminus of MAP-LC3 (hAPG8) and the α-amino group of phosphatidylethanolamine that is catalyzed by HsGsa7 and an Apg3-like E2 protein

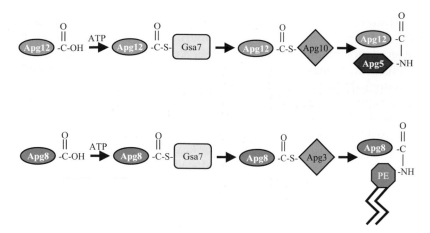

Fig. 5. HsGsa7 conjugation pathways. The ATP-dependent ubiquitin-like conjugation of hApg12 to hApg5 requires the E1-like HsGsa7 and the E2-like enzyme similar to *S. cerevisiae* Apg10. The ATP-dependent ubiquitin-like conjugation of Apg8/MAP-LC3 to phosphatidylethanolamine (PE) requires the E1-like HsGsa7 and the E2-like enzyme similar to *S. cerevisiae* Apg3.

(Fig. 5). The activation of the HsGsa7 conjugation pathways is a major determinant in the formation of the autophagosome. This activation is accompanied by a redistribution of HsGsa7 from the Golgi apparatus to unknown organelles presumed to be newly forming autophagosomes when autophagy is stimulated by amino acid starvation [53]. After the formation of the autophagosome, this vacuole matures into an acidic late autophagosome by acquiring LAMP1, V-ATPase, mannose 6-phosphate receptor, and SNARE-type components required for its docking and fusion with the lysosome. Many of the proteins present in late autophagosomes are also present in autolysosomes. However, the mannose 6-phosphate receptor is not found in autolysosomes, but recycled from the late autophagosome [57]. This recycling can be inhibited by primaquine suggesting it requires vacuole acidification [57]. The maturation of the late autophagosome to an autolysosome can be inhibited by vinblastine or nocodazole thereby requiring microtubules [50,52]. In addition, LAMP2 on the lysosomal surface appears to be required for the fusion of autophagosomes with lysosomes [58]. Finally, the degradation of the sequestered contents requires the many hydrolytic enzymes present within the autolysosome and a functional ubiquitin-activating enzyme, E1 [50]. Leupeptin and E64 have proven to be potent inhibitors of intralysosomal proteinases while weak bases such as NH_4Cl and primaquine neutralize the lysosomal compartment thereby inactivating the hydrolytic enzymes.

3.2.2. Regulation

Autophagy is a highly regulated pathway for protein degradation [51]. Eukaryotic cells adapt to environmental stresses by sequestering proteins and organelles for lysosomal degradation via autophagy. These processes are essential for the survival of the cell especially in times of nutrient adaptation or stress or of bacterial and viral invasion. However, enhanced autophagy has been implicated in cell death (e.g., apoptosis) and suppressed autophagy in abnormal growth (e.g., cancer). Reduced protein degradation and autophagy was observed in ascites hepatoma and primary hepatocellular carcinomas [59]. In addition, fluid from ascites hepatoma effectively inhibited autophagy in normal hepatocytes suggesting that tumor cells are secreting factors that suppress protein degradation [59]. Enhanced protein degradation and autophagy is also believed to contribute to cell death [60]. Autophagic degeneration is thought to play a role in the loss of dopaminergic neurons in patients with Parkinson's disease [61]. Drugs (e.g., cycloheximide and 3-methyladenine) that inhibit autophagy suppress cell death induced by anti-estrogens in mammary carcinoma cells [62]. In addition, TNF-induced death of human T-lymphoblastic leukaemic cells requires autophagy [63]. The evidence suggests that aberrations in the regulation of autophagy may contribute to the cancerous state.

The regulation of protein synthesis and degradation are closely coordinated. Cell growth is accompanied by an increase in protein synthesis and a decrease in autophagy while a decrease in protein synthesis and an increase in autophagy will cause cell death. Autophagy is enhanced by amino acid starvation, which suppresses protein synthesis and is suppressed by insulin and growth factors, which stimulate protein synthesis. At the molecular level, the phosphorylation of ribosomal protein S6 by the p70 S6 kinase stimulates protein synthesis while suppressing autophagy [64]. Those regulatory amino acids (Leu, Tyr, Phe, Gln, Pro, Met, His, Trp) that inhibit autophagy also stimulate the phosphorylation of S6 [64]. In addition, the p70 S6 kinase is activated by type I phosphoinositide 3-kinase and mTOR. The activation of the phosphoinositide 3-kinase by growth factors suppresses autophagy (see below). When mTor is inhibited by rapamycin, autophagy is stimulated [64]. Rapamycin acts as an immunosuppressant and an anti-cancer drug due to its ability to cause cell death of T lymphocytes and various neoplastic cells [65]. However, it remains unclear if rapamycin is killing these cells because of its negative effects on protein synthesis or its ability to stimulate autophagy. Finally, the phosphorylation of eIF2α inhibits protein synthesis and stimulates autophagy [54]. The autophagic response to viral infections is mediated by PKR phosphorylation of eIF2α. However, during amino acid starvation, eIF2α is likely phosphorylated by GCN2 as observed in yeast.

Phosphoinositide 3-kinases (PI3K) and heterotrimeric $G_{\alpha i3}$ proteins regulate autophagy. The products of type I PI3K, phosphatidylinositol 3,4-bisphosphate and phosphatidylinositol 3,4,5-triphosphate, suppress autophagy, while the products of type III PI3K, phosphatidylinositol 3-phosphate, enhance autophagy [66]. Type I PI3K interacts with cell surface receptors to modulate the cellular response to growth factors. Phosphatidylinositol 3,4,5-triphosphate recruits Akt/PKB to the membrane where it is activated by protein kinases PDK1 and PDK2. The Akt/PKB in turn activates the p70 S6 kinase either directly or indirectly through mTor. The phosphorylation of ribosomal S6 stimulates protein synthesis and inhibits autophagy (see above). The type III PI3K is believed to act at the ER, the site of autophagosome formation. In yeast, type III PI3K (Vps34) interacts with Vps15 and Apg6. 3-Methyladenine inhibits autophagy by suppressing type III PI3 kinases.

Wortmannin also inhibits autophagy, but its mechanism is somewhat vague since this drug inhibits types I and III PI3K's. When the activity of the $G_{\alpha i3}$ protein is reduced by pertussis toxin or by transfection with antisense oligos or by overexpression of a GTPase-deficient mutant protein, autophagy is suppressed. The block in macroautophagy due to overexpression of a GTPase-deficient protein could be overcome by increasing the cellular levels of phosphatidylinositol 3-phosphate [66]. Autophagy requires the $G_{\alpha i3}$ protein be membrane bound in the GDP-bound form. The $G_{\alpha i3}$ protein and its GTPase-activating protein (GAIP) are present on the membranes of the Golgi apparatus and the endoplasmic reticulum [67].

Protein phosphorylation appears to be a regulatory event for the formation of the autophagosome [52,58]. Protein phosphatase inhibitors, okadaic acid, microcystin-LR, and calyculin A, suppress autophagy in mammalian cells. In the yeast, the serine/threonine protein kinases Apg1 (ULK1) and Vps15 (p150) have been shown to be necessary for both micro and macro autophagy. In addition, the formation of a complex of proteins that include Apg13, Apg1 kinase, Vac8, Cvt9, and Apg17 requires the dephosphorylation of Apg13 that is regulated by Tor kinases. Interestingly, the kinase that phosphorylates Apg13 thereby inhibiting autophagy has not been identified. In addition, the mammalian homologues of Apg13, Cvt9 and Apg17 have yet to be identified.

4. Summary

Protein degradation in mammalian cells is essential for regulating a variety of cellular processes including the cell cycle and for protecting the cell during times of stress or invasion by pathogens. The selective degradation of proteins may be a constitutive or regulated process resulting in the activation or inactivation of enzymes while nonselective degradation is a highly regulated process that can provide needed amino acids for protein synthesis or can remove damaged organelles. Protein degradation in some neoplastic cells is suppressed resulting in rapid growth while protein degradation in some disease states is enhanced resulting in cell death. In summary, protein degradation is a carefully regulated process that is vital for cell survival.

Acknowledgements

This work was supported by a grant from the National Science Foundation (MCB-9817002).

Abbreviations

APC	anaphase-promoting complex
Apg	protein required for autophagy in *S. cerevisiae*
Aut	protein required for autophagy in *S. cerevisiae*
BIR	baculoviral IAP repeat domain

Cvt   protein required for cytoplasm-to-vacuole transport in *S. cerevisiae*
DUB   deubiquitinating enzymes
E1   ubiquitin-activating enzyme
E2   ubiquitin-conjugating enzyme (Ubc)
E3   ubiquitin-protein ligase
ER   endoplasmic reticulum
Gsa   protein required for glucose-induced selective autophagy in *P. pastoris*
Hsp   heat shock protein
Hsc   heat shock cognate protein
IAP   inhibitor of apoptosis
KFERQ   motif of lysine, phenylalanine, glutamate, arginine, glutamine
PEST   amino acid domain rich in prolines, glutamates, serines, and threonines
Pex   protein required for micropexophagy in *P. pastoris*
PI3K   phosphoinositide 3-kinase
SCF   Skp1, Cdc53/Cullin, F box receptor
VCB   von Hippel-Lindau, Cullin, Elongin B complex
Vps   proteins for vacuolar protein sorting in *S. cerevisiae*.

References

1. Sorimachi, H. and Suzuki, K. (2001) J. Biochem. (Tokyo) 129, 653–664.
2. Sorimachi, H., Ishiura, S., and Suzuki, K. (1997) Biochem. J. 328, 721–732.
3. Carillo, S., Pariat, M., Steff, A., Jariel-Encontre, I., Poulat, F., Berta, P., and Piechaczyk, M. (1996) Biochem. J. 313, 245–251.
4. Shen, J., Channavajhala, P., Seldin, D.C., and Sonenshein, G.E. (2001) J. Immunol. 167, 4919–4925.
5. Chang, H.Y. and Yang, X. (2000) Microbiol. Mol. Biol. Rev. 64, 821–846.
6. Leist, M. and Jaattela, M. (2001) Nat. Rev. Mol. Cell. Biol. 2, 589–598.
7. Holcik, M. and Korneluk, R.G. (2001) Nat. Rev. Mol. Cell. Biol. 2, 550–556.
8. Pickart, C.M. (2001) Annu. Rev. Biochem. 70, 503–533.
9. Ciechanover, A., Orian, A., and Schwartz, A.L. (2000) Bioessays 22, 442–451.
10. Kornitzer, D. and Ciechanover, A. (2000) J. Cell. Physiol. 182, 1–11.
11. Hochstrasser, M. (2000) Nat. Cell. Biol. 2, E153–E157.
12. Ohsumi, Y. (2001) Nat. Rev. Mol. Cell. Biol. 2, 211–216.
13. Alarcon-Vargas, D. and Ronai, Z. (2002) Carcinogenesis 23, 541–547.
14. Deshaies, R.J. (1999) Annu. Rev. Cell. Dev. Biol. 15, 435–467.
15. Page, A.M. and Hieter, P. (1999) Annu. Rev. Biochem. 68, 583–609.
16. Murata, S., Minami, Y., Minami, M., Chiba, T., and Tanaka, K. (2001) EMBO Rep. 2, 1133–1138.
17. Jiang, J., Ballinger, C.A., Wu, Y., Dai, Q., Cyr, D.M., Hohfeld, J., and Patterson, C. (2001) J. Biol. Chem. 276, 42,938–42,944.
18. Bochtler, M., Ditzel, L., Groll, M., Hartmann, C., and Huber, R. (1999) Annu. Rev. Biophys. Biomol. Struct. 28, 295–317.
19. Myung, J., Kim, K.B., and Crews, C.M. (2001) Med. Res. Rev. 21(4), 245–273.
20. Hasselgren, P.O., Wray, C., and Mammen, J. (2002) Biochem. Biophys. Res. Commun. 290, 1–10.
21. Davies, K.J. (2001) Biochimie 83, 301–310.
22. Coffino, P. (2001) Biochimie 83, 319–323.
23. Benaroudj, N., Tarcsa, E., Cascio, P., and Goldberg, A.L. (2001) Biochimie 83, 311–318.
24. Varshavsky, A. (1996) Proc. Natl. Acad. Sci. USA 93, 12,142–12,149.
25. Solomon, V., Lecker, S.H., and Goldberg, A.L. (1998) J. Biol. Chem. 273, 25,216–25,222.
26. Pfleger, C.M. and Kirschner, M.W. (2000) Genes Dev. 14, 655–665.

27. Meacham, G.C., Patterson, C., Zhang, W., Younger, J.M., and Cyr, D.M. (2001) Nat. Cell. Biol. 3, 100–105.

28. Lord, J.M., Davey, J., Frigerio, L., and Roberts, L.M. (2000) Semin. Cell. Dev. Biol. 11, 159–164.

29. Sharipo, A., Imreh, M., Leonchiks, A., Imreh, S., and Masucci, M.G. (1998) Nat. Med. 4, 939–944.

30. Mitch, W.E. and Goldberg, A.L. (1996) N. Engl. J. Med. 335, 1897–1905.

31. Ciechanover, A. (1994) Cell 79, 13–21.

32. Lecker, S.H., Solomon, V., Mitch, W.E., and Goldberg, A.L. (1999) J. Nutr. 129, 227S–237S.

33. Wing, S.S., Haas, A.L., and Goldberg, A.L. (1995) Biochem. J. 307, 639–645.

34. Gomes-Marcondes, M.C. and Tisdale, M.J. (2002) Cancer Lett. 180, 69–74.

35. Wing, S.S. and Banville, D. (1994) Am. J. Physiol. 267, E39–E48.

36. Rajapurohitam, V., Bedard, N., and Wing, S.S. (2002) Am. J. Physiol. Endocrinol. Metab. 282, E739–E745.

37. Wilkinson, K.D. (2000) Semin. Cell. Dev. Biol. 11, 141–148.

38. Cuervo, A.M. and Dice, J.F. (1998) J. Mol. Med. 76, 6–12.

39. Agarraberes, F.A. and Dice, J.F. (2001) J. Cell. Sci. 114, 2491–2499.

40. Cuervo, A.M., Hu, W., Lim, B., and Dice, J.F. (1998) Mol. Biol. Cell. 9, 1995–2010.

41. Cuervo, A.M., Gomes, A.V., Barnes, J.A., and Dice, J.F. (2000) J. Biol. Chem. 275, 33,329–33,335.

42. Salvador, N., Aguado, C., Horst, M., and Knecht, E. (2000) J. Biol. Chem. 275, 27,447–27,456.

43. Franch, H.A., Sooparb, S., Du, J., and Brown, N.S. (2001) J. Biol. Chem. 276, 19,126–19,131.

44. Susan, P.P. and Dunn, W.A., Jr. (2001) J. Cell. Physiol. 187, 48–58.

45. Cuervo, A.M., Dice, J.F., and Knecht, E. (1997) J. Biol. Chem. 272, 5606–5615.

46. Cuervo, A.M. and Dice, J.F. (2000) Traffic 1, 570–583.

47. Cuervo, A.M. and Dice, J.F. (2000) Exp. Gerontol. 35, 119–131.

48. Sattler, T. and Mayer, A. (2000) J. Cell. Biol. 151, 529–538.

49. Yuan, W., Strømhaug, P.E., and Dunn, W.A., Jr. (1999) Mol. Biol. Cell. 10, 1353–1366.

50. Dunn, W.A., Jr. (1994) Trends Cell. Biol. 4, 139–143.

51. Mortimore, G.E., Poso, A.R., and Lardeux, B.R. (1989) Diabetes Metab. Rev. 5, 49–70.

52. Kim, J. and Klionsky, D.J. (2000) Annu. Rev. Biochem. 69, 303–342.

53. Dorn, B.R., Dunn, W.A., and Progulske-Fox, A. (2002) Cell. Microbiol. 4, 1–10.

54. Talloczy, Z., Jiang, W., Virgin, H.I., Leib, D.A., Scheuner, D., Kaufman, R.J., Eskelinen, E.L., and Levine, B. (2002) Proc. Natl. Acad. Sci. USA 99, 190–195.

55. Elmore, S.P., Qian, T., Grissom, S.F., and Lemasters, J.J. (2001) FASEB J. 15, 2286–2287.

56. Mizushima, N., Yamamoto, A., Hatano, M., Kobayashi, Y., Kabeya, Y., Suzuki, K., Tokuhisa, T., Ohsumi, Y., and Yoshimori, T. (2001) J. Cell. Biol. 152, 657–668.

57. Dunn, W.A. (1993) Adv. Cell. Mol. Biol. Memb. 1, 117–138.

58. Abeliovich, H. and Klionsky, D.J. (2001) Microbiol. Mol. Biol. Rev. 65, 463–479.

59. Kisen, G.O., Tessitore, L., Costelli, P., Gordon, P.B., Schwarze, P.E., Baccino, F.M., and Seglen, P.O. (1993) Carcinogenesis 14, 2501–2505.

60. Bursch, W., Ellinger, A., Gerner, C., Frohwein, U., and Schulte-Hermann, R. (2000) Ann. N.Y. Acad. Sci. 926, 1–12.

61. Anglade, P., Vyas, S., Javoy-Agid, F., Herrero, M.T., Michel, P.P., Marquez, J., Mouatt-Prigent, A., Ruberg, M., Hirsch, E.C., Agid, Y. (1997) Histol. Histopathol. 12, 25–31.

62. Bursch, W., Ellinger, A., Kienzl, H., Torok, L., Pandey, S., Sikorska, M., Walker, R., and Hermann, R.S. (1996) Carcinogenesis 17, 1595–1607.

63. Jia, L., Dourmashkin, R.R., Allen, P.D., Gray, A.B., Newland, A.C., and Kelsey, S.M. (1997) Br. J. Haematol. 98, 673–685.

64. Blommaart, E.F., Luiken, J.J., Blommaart, P.J., van Woerkom, G.M., and Meijer, A.J. (1995) J. Biol. Chem. 270, 2320–2326.

65. Gingras, A.C., Raught, B., and Sonenberg, N. (2001) Genes Dev. 15, 807–826.

66. Petiot, A., Ogier-Denis, E., Blommaart, E.F., Meijer, A.J., and Codogno, P. (2000) J. Biol. Chem. 275, 992–998.

67. Petiot, A., Ogier-Denis, E., Bauvy, C., Cluzeaud, F., Vandewalle, A., and Codogno, P. (1999) Biochem. J. 337, 289–295.

S.C. Makrides (Ed.) *Gene Tra'nsfer and Expression in Mammalian Cells*

CHAPTER 19

Stabilization of proteasomal substrates by viral repeats

Nico P. Dantuma and Maria G. Masucci

*Microbiology and Tumor Biology Center, Karolinska Institutet, Box 280, S-171 77 Stockholm, Sweden,
Tel.: +46 8 7286755. fax: +46 8 331399. E-mail: nico.dantuma@mtc.ki.se*

1. Introduction

The steady state levels of intracellular proteins are determined by their rates of synthesis and destruction. It has become widely recognized that both processes are tightly regulated energy-dependent events that are critical for cell viability. Indeed a large fraction of cellular proteins appears to be involved in the identification, labelling and destruction of carefully selected proteins [1] (see Chapter 18). Although our insights in the molecular processes that dictate protein tagging and destruction has grown enormously, the tools that can be used to selectively block protein degradation are limited. Protection of proteins from intracellular degradation may be desired, especially in the context of gene transfer, in order to prolong protein half-life or avoid cellular immune response, which relies on antigenic peptide fragments generated by proteolysis. The tool that is discussed in this review was not developed by molecular biologists but by viruses striving to prosper in the hostile environment of the infected host. Although our knowledge of the mode of action of these viral signals is limited, we have shown that this viral strategy can be easily applied to many different proteins. The purpose of this review is to summarize our present knowledge of this viral signal and outline the features that should be considered for its use in gene transfer protocols.

2. The ubiquitin–proteasome system

While many surface proteins find their end in the lysosomal compartment, soluble cellular proteins are primarily, if not exclusively, turned over by a sophisticated machinery known as the ubiquitin–proteasome system. Although this proteolytic pathway was originally described as being responsible for the regulated degradation of cytosolic and nuclear proteins, it is presently well documented that also transmembrane proteins, integral membrane proteins and proteins residing in the lumen of the endoplasmic reticulum can be subject to ubiquitin/proteasome-dependent proteolysis [2,3]. Many cellular events are orchestrated by this type of regulated proteolysis such as cell cycle progression, induction of apoptosis, intracellular signalling and other processes that are relying on temporal degradation of their key regulators [1]. The peptide fragments that are the end product of proteasomal degradation are the primary source of the peptides that are presented at

536

the cell surface in association with major histocompatibility complex (MHC) class I molecules [4].

Substrates for the ubiquitin–proteasome system are recognized by the presence of degradation signals [5]. These signals can be either constitutive or depend on additional modification such as phosphorylation. Alternatively the presence of misfolded domains within proteins can direct them into this proteolytic pathway. The presence of these motifs guides the conjugation of the C-terminus of a 76 amino acid protein, ubiquitin, to the ε-amino group of an internal lysine residue within the target protein (Fig. 1, left-hand panel) [1]. Because lysine residues within ubiquitin are also targets for ubiquitin conjugation, successive ubiquitination steps result in the formation of a polyubiquitin tree. Ubiquitination is accomplished by an enzyme relay in which the ubiquitin activase (E1) activates ubiquitin through the formation of a thiol–ester bond between the C-terminus of the ubiquitin and a cysteine residue in the E1. The activated ubiquitin is then passed on to a cysteine residue within the ubiquitin-conjugase (E2), which, in concert with a substrate-specific ubiquitin ligase (E3), accomplishes ubiquitination of the substrate.

Polyubiquitinated substrates are recruited to a large barrel-shaped multi-subunit proteolytic complex, the proteasome, which harbours inside its cavity two sets of three proteolytic subunits displaying chymotrypsin-like, trypsin-like and caspase-like activity (Fig. 1, right-hand panel) [6]. Recruited substrates are unfolded by specialized subunits flanking its entry upon which the proteins are tethered into the cavity where they are processed into small peptide fragments [7].

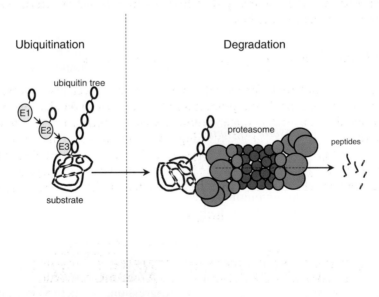

Fig. 1. The ubiquitin–proteasome system. Substrates of the ubiquitin–protein system are ubiquitinated through an enzyme relay of E1, E2 and E3 enzymes, which results in the covalent linkage of a polyubiquitin tree (right-hand panel). The ubiquitinated proteins are recognized by the proteasome that unfolds and degrades the substrate resulting in the generation of small peptide fragments (left-hand panel).

3. Viral repeats that block proteasomal degradation

3.1. Epstein–Barr virus nuclear antigen (EBNA)-1

The Epstein–Barr virus (EBV) is a lymphotropic γ-herpesvirus that causes persistent infection in man [8]. With more than 90% of the population being carriers, EBV can be considered as one of the most widely spread human pathogens. Like many viruses that cause life-long infection, EBV establishes in the host a latent cell reservoir that is characterized by the absence of viral replication and restricted expression of only a small subset of viral proteins. The EBV nuclear antigen (EBNA)-1 is one of those latently expressed viral proteins and, because of its crucial role in replication and episomal maintenance of the viral genome, this protein is a prerequisite for EBV-induced B-cell transformation [9]. In addition, EBNA1 is also a transcriptional regulator that acts on two latent promoters present within the viral genome.

The blood of healthy EBV carriers contains relatively high numbers of cytotoxic T cell (CTL) precursors that recognize peptide fragments of various latent EBV proteins in association with MHC class-I [10]. This pool of EBV-specific CTLs is of great importance for keeping the virus under control as illustrated by the occurrence of EBV-associated lymphoproliferative diseases in immunosuppressed individuals. Several laboratories involved in the characterization of EBV-specific CTL responses reported a surprising lack of EBNA1 specific responses even though this is the only protein present in all latent infected cells [10]. From a viral perspective, lack of recognition of its most crucial protein is clearly beneficial and it was for this reason that we probed into the possibility that EBNA1 may somehow resist the antigen presentation machinery in the host cell.

3.2. Identification of a repetitive sequence that blocks degradation

EBNA1 contains different functional domains that reflect its pivotal role in maintenance of the episomal viral genome and transcriptional regulation, including a DNA-binding domain and a dimerization domain, which comprise most of its carboxy terminal half (Fig. 2) [9]. Strikingly the amino terminal half of EBNA1 contains a long repetitive sequence of glycine and alanine residues flanked by two short positively charged repeats, involved in DNA binding. Although the length of the glycine–alanine repeat (GAr) varies between different virus isolates, it is generally

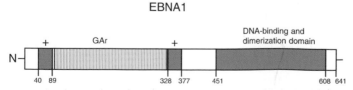

Fig. 2. The primary structure of EBNA1. Schematic representation of EBNA1 from the EBV strain B98.5 strain. The positively charged repeats (+), the GAr and the DNA-binding and dimerization domain are indicated. The length of the GAr varies between different EBV isolates.

long, ranging from approximately 60 to over 300 amino acids. The function of the repeat remained elusive and functional studies showed that this region could be deleted without major consequences for the EBNA1-mediated replication or transcriptional activation in tissue culture [11].

Using an experimental strategy based on the construction of recombinant vaccinia viruses containing EBNA1 variants we have shown that the GAr is directly implicated in the CTL resistance of this viral protein [12]. Cells expressing an EBNA1 containing an identified CTL epitope were not recognized by the matched CTL clone, while removal of the GAr resulted in full restoration of antigen presentation [13]. Moreover, introduction of the GAr in one of the latent EBV proteins that is a primary target for CTLs, EBNA4, completely abrogated the CTL response towards two epitopes present in EBNA4 [13]. Using a similar experimental approach, Mukerjee and co-workers demonstrated that murine CTLs generated against EBNA1 epitopes lysed *in vivo* only target cells expressing EBNA1 lacking the GAr [14]. These key experiments clearly showed that the GAr blocks the generation of epitopes from proteins that harbour the repeat.

The route from full-sized viral protein to triggering of CTL responses by peptide fragments loaded onto MHC class I is long, and there have been reports of viral interference with many of the individual steps, like peptide transport, peptide loading on MHC class I and transport to the cell surface [15]. Studies on the inhibitory activity of the GAr revealed a viral strategy that targets a very early step in the antigen presentation process. Using an *in vitro* assay for protein degradation it was demonstrated that the GAr blocks the degradation of EBNA1 by the proteasome and hence avoids the generation of potential antigenic peptides [16]. Importantly, the repeat did not affect ubiquitination of proteins but rendered the ubiquitinated protein resistant to proteasomal degradation. The viral repeat challenged at that point the paradigm of ubiquitination being the sole decisive step in the ubiquitin–proteasome system and indicated that events downstream of the recognition and tagging of proteasome substrates can strongly influence the outcome of and even abrogate proteolysis.

3.3. Cis-stabilization of cellular proteins by the GAr

Recognition of the substrates by ubiquitin ligases determines the specificity of the ubiquitin–proteasome system while, after this step, the fate of ubiquitinated substrates merges in the interaction with the proteasome and subsequent degradation [1,6]. Thus, the interference of the GAr with an event downstream of the selective step, suggests that this repeat can potentially be used for stabilization of other proteasome substrates. This idea is also supported by the above-mentioned demonstration that introduction of the GAr in EBNA4 blocked degradation of this viral protein [13].

The transferability of the repeat was challenged in a series of studies using various natural and artificial substrates of the proteasome (Table 1). The inhibitor of NF-κB, IκBα, is subject to both constitutive and inducible proteasomal degradation [17]. The inducible ubiquitination is regulated by phosphorylation of two serine residues,

Table 1

Effect of the GAr on different proteasomal substrates

Substrate	Repeat		Inhibitory effect		Reference
	Length	Location	Proteolysis	Antigen presentation	
EBNA1	239	I	+ +	+ +	[13]
EBNA4	39	I	+ +	ND	[16]
	239	I	+ +	+ +	[13]
IκB-α	8	N	+ +	ND	[18]
	24	N, C, I	+ +	ND	
	239	N, C, I	+ +	ND	
P53	25	N, C	+	ND	[21]
	239	C	+ +	ND	
Ub–R–GFP	25	C	±	ND	[20]
	239	C	+	ND	
Ub–P–GFP (modest UFD)	25	C	+	ND	[20]
Ub^{G76V}–GFP (strong UFD)	25	C	±	ND	[20]
	239	C	+	ND	
ErbB2	24	N, I	–	–	[22]

The length and position of the repeats are indicated. N, N-terminal; C, C-terminal; I, internal.

which in turn induces ubiquitination of IκBα by a TrCP-β associated with the Skp1/ Cullin/F-box protein (SCF) ubiquitin ligase complex. Introduction of the GAr in IκBα resulted in the generation of stable functional IκBα chimera that was refractory to constitutive as well as to inducible ubiquitin-dependent degradation [18]. The IκBα proved to be an extremely helpful model system since even very small repeats resulted in a full stabilization of this protein (see Section 3.5.1). A set of green fluorescence protein (GFP)-based proteasome substrates was also generated as a tool to study the protective effect of the GAr [19]. Proteasomal degradation of GFP substrates that contained either a destabilizing N-end rule signal (Ub-R-GFP) or a ubiquitin fusion degradation (UFD) signal (Ub-P-GFP, Ub^{G76V}-GFP) was blocked or at least severely impaired by introduction of the GAr [20]. Even though the primary sequences of these GFP constructs are very similar, the introduced N-end rule and UFD degradation signals are recognized by different ubiquitin ligases and require different components of the ubiquitin–proteasome system. Chimeras of the tumor suppressor p53 with the GAr were also shown to resist ubiquitin/proteasome-dependent proteolysis [21] (see Section 4.2). Also in this case, two unrelated ubiquitin ligases were unable to target the chimeras for degradation. Together these studies suggest that the GAr has a broad inhibitory potential and can block or delay proteasomal degradation of a variety of proteasome substrates that are recognized by different unrelated ubiquitin ligases. However, it should be noted that Piechocki and co-workers recently reported that introduction of a 24 amino acid GAr in the cell surface protein ErbB-2 did not result in abrogation of proteasomal degradation of ErbB-2 [22]. Although the length-dependent effect of the GAr may provide an explanation for the lack of inhibition of this short repeat in the context of ErbB-2 [20], further studies are required to address the question whether ErbB-2 is fully resistant also to the inhibitory effect of longer GArs and, if so, whether this property is related to specific characteristics of ErbB-2, such as its membrane association.

3.4. Related viral repeats

The presence of long repetitive sequence is not restricted to EBNA1 as related sequences have been found in EBNA1 homologues of the Old-world primate γ-herpes viruses, baboon and rhesus monkey lymphocryptoviruses (LCV). Interestingly these EBNA1 homologues are also responsible for the maintenance of viral genomes and the conservation of repetitive sequences within these proteins suggests an important function of the repeats [23]. The baboon and rhesus LCV encode EBNA1 homologues that contain repeats consisting of glycine, alanine and serine residues. These repeats are approximately 50–70 amino acids long and hence relatively short compared to the EBV GAr. These glycine–alanine–serine repeats can inhibit proteasomal degradation, though less efficiently than the GAr [24] (see Section 3.5.2). However, in contrast to EBNA1, the inhibition of ubiquitin/proteasome-dependent proteolysis was not sufficient to abrogate the presentation of epitopes derived from these EBNA1 homologues [23]. Hence, prolongation of protein half-life through delaying proteasomal degradation may reflect the ancestral function of these repeats while the immune evasion function of the EBNA1 GAr may have evolved later through repeat amplification.

Another human γ-herpes virus, the Kaposi sarcoma herpes virus expresses a functional EBNA1 homologue, LANA1, which contains a repetitive sequence of approximately the same length and position as the EBNA1 GAr [25]. Strikingly the LANA repeat encodes predominantly glutamic acid and glutamine residues instead of glycines and alanines found in the EBV EBNA1. Because of its position in a functional related protein it will be interesting to analyse whether the LANA repeat affects proteasomal degradation. Notably, the expanded polyglutamine repeats found in various neurodegenerative disorders have been reported to inhibit proteasomal degradation [26].

Recent findings suggest that presence of the EBNA1 homologues with repetitive sequences is not an absolute prerequisite for latent infection by γ-herpes viruses. An evolutionary more distant relative of EBV, which is endemic in the New-world primate marmosets, contains an EBNA1 homologue lacking any repetitive sequence [27]. No functional data regarding this repeatless EBNA1 homologue have been reported and it is presently unknown whether this protein can resist proteasomal degradation.

3.5. Structural constraints

3.5.1. Length of the repeat
The GAr domains found in natural EBV isolates are long repetitive sequences that are composed of shorter imperfect repeats. Why does the virus require such a long repeat? With the usage of different GFP-based proteasome substrates we have shown

that there is a positive correlation between the length of the repeat and its inhibitory activity. Hence, the half-lives of GAr-containing proteins are determined by the length of the inhibitory repeat and by the efficiency and nature of their degradation signals. The efficiency of a UFD signal is regulated by the ubiquitin-protein fusion linkage i.e. in the modestly destabilized Ub–Pro–GFP fusion the ubiquitin is slowly cleaved from GFP while the strongly destabilized Ub^{G76V}–GFP is not cleaved [19]. The Ub–P–GFP was completely stabilized by the insertion of a recombinant GAr of only 24 amino acids, while the more strongly destabilized Ub^{G76V}–GFP was only partially protected even after insertion of a repeat of 239 amino acids (Table 2) [20]. That this is not just a matter of the efficiency of the degradation signal became clear from studies with IκBα, which turned out to be a substrate particularly sensitive to the inhibitory effect. Even though IκBα has a short half-life of less then 15 min in tumor necrosis factor (TNF)-α-stimulated cells, a minimal repeat of only 8 amino acids protected the protein from degradation [18]. That the same length-dependent effect holds true for IκBα was shown with an 8-amino acid repeat containing one serine residue (see Section 3.5.2). This suboptimal motif affects only marginally the constitutive turnover of IκBα but strong inhibition was achieved by insertion of three copies of the motif (Table 2). Yet, even the longer repeat did not affect substantially the more dramatic TNFα-inducible degradation. Ubiquitin/proteasome-dependent proteolysis can also be positively regulated by the amount of ubiquitin ligase. Co-transfecting p53-GAr chimeras with a plasmid encoding the ubiquitin ligase Mdm2 showed that accelerating the rate of p53 degradation by increasing the amount of ubiquitin ligase counteracts the protective effect of the repeat, which could be overcome by increasing the length of the GAr (Table 2) [21]. Thus the half-life of a substrate-GAr fusion is determined by the length of the repeat on one side and the ubiquitination kinetics, regulated through the efficiency of the degradation signal or the amount of ubiquitin ligase, on the other side. The fact that the presence of GAr sequences does not necessarily fully block but can also retard proteasomal degradation suggests that natural substrates of the proteasome may contain related sequences that regulate protein turnover.

Table 2

Correlation between repeat composition and protective effect

Repeat	Variation	TNFα-inducible	Constitutive
GGAGAGAG	Spacing 1	+ +	+
AGGAGGAG	Spacing 2	+ +	ND
AGAGGGAG	Spacing 3	+	ND
AGAGGGGA	Spacing 4	±	ND
GGGGGGGG	Ala → Gly	−	−
GGVGVGVG	Ala → Val	+	ND
GGSGAGAG	Ala → Ser	−	−
3 × (GGSGAGAG)	3 × octamer	−	+

A selection of the repeats that have been tested for inhibition of constitutive and TNFα-inducible degradation of IκBα [24].

3.5.2. Composition of the repeat insert

The observation that IκBα is very susceptible to the inhibitory effect of the repeat and protected by an 8-amino acid long sequence provided us with an excellent model to study the impact of repeat composition. Early studies with IκBα suggested that the alanine residues within the repeat are especially important as a stretch of only glycine residues lacked any protective effect while a proline-alanine repeat also blocked degradation [18]. In a more detailed study various repeat compositions were tested in the context of IκBα (Table 2) [24]. This revealed that the hydrophobic nature of the alanine residue is an important determinant and substitution with valine did not affect the protection while changing only one out of the three alanines in the 8-amino acid repeat to aspartic acid strongly impaired the protective effect. The natural GAr in EBNA1 lacks a clear repetitive motif but the alanine residues are very regularly distributed with predominant spacing of 1, 2 or 3 glycines while adjacent alanines or alanines spaced by more than 4 glycines are absent. Using different variants repeats composed of 3 alanines and 5 glycines it was indeed shown that the spacing between the alanine residues is a major determinant of the inhibitory activity [24]. Increasing the spacing to more then 3 glycines or positioning the 3 alanines directly adjacent resulted in a dramatic decrease of the protection (Table 2). Longer alanine stretches have inhibitory activity but the protective length appears to be roughly similar to that of the spaced counterpart. Thus a stretch of 8 alanines protected almost as efficient as a GGAGAGAG repeat. It is known that long poly-alanine repeats render proteins prone to aggregation because of the hydrophobic nature of the repeats [28]. Interspersing the intercalation of alanine and glycine residues may well serve to obtain maximal protective effect without rendering the proteins susceptible to aggregation. Indeed, it has been shown with the GFP–GAr fusions that chimeras containing a 239-amino acid GAr did not form aggregates in cells, despite their strong hydrophobicity [20]. Thus the presence of regularly spaced hydrophobic residues (alanine, valine) appears to be the main determinant for obtaining soluble functional proteins that resist proteasomal degradation.

3.5.3. Location of the repeat insertion

While the natural location of the GAr and GAr-like sequences in EBNA1 and its homologues is close to the N-terminus, an artificial GAr chimera revealed that N-terminal, C-terminal and internal repeats have a very similar protective effect (Table 2) [18,21]. Nevertheless, there are reasons to believe that the N-terminal insertion may be of importance for the immune evasion function by the GAr (see Section 4.1). Unfortunately N-terminal insertion of the natural GAr often results in very low expression levels compared to other insertions (S. Heessen, A. Sharipo, N.P. Dantuma, M.G. Masucci, unpublished observations). This may be due to the fact that the natural repeat is extremely GC-rich, which could result in the formation of 5′ hairpin structures in the transcripts obstructing translation. Recently, we have generated a set of recombinant GAr variants of length up to 215 amino acids that are encoded by less GC-rich sequences, which may circumvent this problem (N.P. Dantuma, unpublished observations). The GAr does not form clear secondary structures [29], and does not necessarily affect the function of the host protein

[18,21]. Thus the location of the insertion seems to be of minor concern for the protective effect of the repeat but maybe important in the context of the desired effect, i.e., to avoid recognition of peptides derived from a newly introduced protein.

3.6. Mode of action

An intriguing finding is the observation that the GAr affects a step downstream of ubiquitination [16]. Until recently ubiquitination has been viewed as a "point of no return" in the ubiquitin–proteasome pathway but it is now becoming increasingly clear that events downstream of ubiquitination, such as for example deubiquitination [30], can strongly affect the fate of the ubiquitinated protein. The induction of premature deubiquitination could explain the inhibitory effect of the repeat. Although we cannot formally exclude this possibility, at least one observation strongly argues against this mode of action. A substantial proportion of the p53-GAr chimeras accumulating in Mdm2 over-expressing cells were ubiquitinated, suggesting that the protein is also protected in its ubiquitinated form [21]. It should be emphasized though that this phenomenon requires overexpression of the ligase since, under normal circumstances, the polyubiquitinated species rapidly disappear through the action of the deubiquitination enzymes. Hence, the stable GAr chimeras may be caught in a perpetual cycle of ubiquitination, failed proteasomal degradation and deubiquitination. Importantly, in contrast to wild type IκBα, the ubiquitinated IκBα-GAr chimera is unable to achieve a stable interaction with the proteasome as assessed by coprecipitation [18]. Using a more sensitive assay we have recently shown that the polyubiquitinated p53-GAr chimera could still interact with the S5a subunit of the proteasome, which is known to bind specifically polyubiquitin [31]. Together these findings suggest that the interaction between GAr-containing substrates and the proteasome may have a lower affinity or higher dissociation rate resulting in premature release of the substrate.

The length-dependent protective effect of the GAr can be partly inactivated by accelerating ubiquitination through either increasing the amount of ubiquitin ligase [21] or modifying the degradation signal [20]. Accelerated ubiquitination is likely to increase the number of encounters between the protected substrate and the proteasome since the substrate will be retargeted by the ubiquitin ligases. If the accuracy of the protective effect of the GAr is depending on its length this would predict that increasing the substrate–proteasome interactions will on one hand increase the likelihood of failed protection and degradation during one of those encounters while on the other hand the introduction of longer repeats will improve the accuracy of the protective signal.

An important post-ubiquitination event is the unfolding of the recruited substrate by the regulatory subunits of the proteasome. It has recently been shown that some structures resist the unfoldase activity of the proteasome resulting in protein stabilization [32]. The crystal structure of a stretch of alternating glycine and alanine residues as found in silk revealed a tight antiparallel β-sheet [33]. However, structural analysis of the short protective repeat inserted in IκBα indicated that the GAr has a flexible structure [29]. Moreover it was shown that introduction of a sufficiently strong degradation signal resulted in degradation of proteins carrying a 239 amino acid long GAr [20]. Biochemical studies with designed stable domains have shown that the position of the domain is very important. When inserted at the proper position, these stable domains also functioned as general inhibitors of the proteasome [32]. The position-independent effect of the GAr and its cis-acting protective effect argue against a model based on domain stability. Thus it seems unlikely that the GAr forms an exceptionally stable structure that resists unfolding.

Together these studies suggest that the GAr has a length-dependent inhibitory effect on a late event in the ubiquitin–proteasome pathway. The observation that polyubiquitinated GAr chimeras can interact with the isolated polyubiquitin-binding subunit of the proteasome but fail to interact with the assembled complex suggests that the GAr may destabilize the binding, triggering the release of the substrate from the proteasome. It is noteworthy that recruitment of unfolded non-ubiquitinated proteins to the regulatory subunit of the proteasome may result in their refolding rather than degradation [34].

4. Possible applications in gene transfer settings

For gene transfer applications it may be desirable to protect the newly introduced protein from ubiquitin/proteasome-dependent proteolysis. Protection can be accomplished through inactivation of degradation signals or the introduction of a stabilizing sequence. Endogenous degradation signals can be removed by mutation of the binding motifs for specific ubiquitin ligases [35], or by substitution of the lysine residues that are targeted for ubiquitination [36]. Both approaches have been successfully attempted in tissue culture. However, while specific lysine residues are used for ubiquitination in some substrates, in other substrates the ubiquitin ligases target several lysine residues, which would demand multiple substitutions of lysine residues [17,36]. An important drawback of this strategy is that the mutation of ubiquitin ligase-recognition signals or substitution of lysine residues could cause structural changes that impair the activity of the target protein. Furthermore, peptides presented by MHC class I appear to be mainly derived from proteasomal degradation of newly synthesized defective ribosomal products whose degradation is most likely triggered by misfolding rather then endogenous degradation signals [37,38].

The introduction of structures that resist unfolding could also be used for protection of target proteins from ubiquitin-dependent proteolysis [32,39]. However, the stable structure may "clog" the proteasome resulting in a general inhibition of proteolysis [32]. Due to its involvement in a broad array of biological processes, general impairment of the ubiquitin–proteasome system by small compounds that block the catalytic subunits usually results in induction of apoptosis [40]. The use of the viral stabilizing sequences offers, in this context, several advantages. First, the repeat provides an efficient way for avoiding recognition of antigenic epitopes [13]. Second, introduction of the repeat does not impair the function of the protein

[18,21]. Third, the GAr appears to have a strictly cis-acting effect under physiologic conditions [18,21].

4.1. Avoiding immune recognition

Gene transfer is often accompanied by strong immune reaction against products of the transfer vector or the transferred gene itself. The first problem can be avoided by using naked DNA encoding only the gene-of-interest but reactivity against the transferred product may still be a problem, in particular when the introduced gene is lacking in the host. Indeed, injection of naked DNA encoding dystrophin in dystropin-deficient animals, a model for Duchenne muscular dystrophy, triggers an efficient cellular immune response that limits the duration of expression of the transferred gene [41]. Tolerization and systemic immunosuppression protocols have been successfully used in animals to prolong expression of the newly introduced genes but may be more problematic in clinical settings [42].

The GAr was originally identified by its immune-evasion promoting properties [13] and is therefore an attractive candidate for immunoprotection in gene transfer settings (Fig. 3). This application remains to be tested but the available data on the effect of the GAr in model systems and our knowledge of the mechanisms that regulate immune recognition suggest some important considerations. First, the location of the GAr close to the N-terminus of EBNA1 may be relevant to its immune evasion strategy. Since the bulk of the peptides presented by MHC class I molecules appears to be derived from newly synthesized defective ribosomal products generated by premature translational termination [37,38], positioning of the GAr close to the N-terminus may prevent the generation of antigenic peptides from truncated proteins that lack the protective repeat. Second, the length of the protective repeat appears to be particularly important in the context of immune

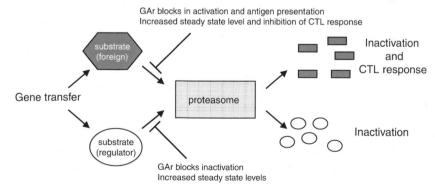

Fig. 3. Applications of the GAr in gene transfer. By introducing the GAr in a gene designed for gene transfer, proteasomal degradation can be blocked or severely impaired. In case a new (foreign) gene is introduced in a deficient patient, the GAr can be used to stabilize the protein and avoid CTL recognition of cells expressing the transferred gene (upper half). Introduction of the GAr into any transferred gene (for example regulators like p53) can be used to generate proteasome-resistant variants with prolonged half-lives.

evasion since partly protected proteins will still trigger CTLs despite their prolonged half-lives [20]. Third, the GAr does not block cross-priming, as internalisation of EBNA1 by specialized antigen-presenting cells results in the generation of antigenic peptides from EBNA1 [43]. Hence, CTLs specific for the transferred gene may be generated but these CTLs will probably fail to recognize cells that express the GAr chimera.

4.2. Stabilizing proteasome substrates

In some gene therapy settings protein variants that resist proteasomal degradation may have additional benefits as they may bypass negative regulatory mechanisms or because the anticipated increased steady state levels of the protein-of-interest may allow more of the desired activity (Fig. 3).

An interesting example for this approach is the tumor suppressor p53, a transcription factor that induces cell cycle arrest and apoptosis in response to DNA damage [44]. The levels of p53 are kept low in normal cells due to rapid ubiquitin-dependent proteolysis but in response to DNA damage degradation is attenuated, which results in accumulation of p53 [45]. The regulation of the p53 tumor suppressor is functionally impaired in malignant cells. While many tumors express a functionally inactive mutant p53, p53 is inactivated in the majority of the remaining tumors through accelerated proteasomal degradation [45]. Accelerated degradation is accomplished either through increased levels of the natural ubiquitin ligase Mdm2 or through retargeting of the cellular ubiquitin ligase E6-AP to p53 through the action of the E6 proteins of the human papilloma virus (HPV)-16 or -18. P53 is an attractive candidate for gene therapy, and animal studies and Phase I clinical trials have shown that treatment with p53-encoding adenovirus vectors can curtail tumor progression [46]. However, while this approach may be especially helpful in tumors that express mutant p53, when p53 is inactivated by accelerated degradation the newly introduced p53 is also subject to rapid inactivation. We have recently reported on the generation of a set of p53-GAr chimeras and tested their stability and activity in the context of Mdm2 or E6/E6-AP-mediated ubiquitination and degradation [21]. Introduction of the GAr resulted in a dramatic prolongation of the half-life of p53 in the presence of either Mdm2 or HPV-E6. The chimeras carrying the short GAr were equally capable of inducing the expression of p53-target genes as wild-type p53. Insertion of the full-length repeat impaired the transcriptional activity of p53 but the chimera carrying this repeat was nevertheless more efficient in inducing apoptosis than wild type p53. Importantly, the p53-GA chimeras had improved growth inhibitory potential in an HPV18 carrying cervix carcinoma cell line and an osteosarcoma cell line expressing elevated levels of Mdm2.

This approach may be applicable to a large number of proteins whose turnover is aberrantly accelerated in various human pathologies such as infection, inflammation and oncogenesis. Although short repeats may not be sufficient to avoid immune recognition this may not be much of an issue when the general aim is to supply the cell with a stable variant of an already present and therefore tolerated endogenous protein. Short repeats are easier to handle, less likely to disturb the biological activity of the protein while still retaining a profound effect on protein turnover.

5. Concluding remarks

Although many questions regarding the mode of action of the GAr remain unresolved, introduction of the repeat appears to be a robust and powerful method for protecting a variety of proteins from proteasomal degradation. To our knowledge, the EBV GAr and related repeats are the only transferable motifs that can be used to selectively stabilize individual substrates in cells, without affecting the overall ubiquitin/proteasome-dependent proteolysis. The next step will be to test the applicability of the repeats to gene transfer settings by analysing antigen presentation, stability, and functionality of the chimeras in animal models.

Acknowledgements

N.P.D is supported by a fellowship from the Swedish Research Council. Our research is supported by grants from the Swedish Cancer Society (N.P.D., M.G.M.), the Swedish Foundation of Strategy Research (M.G.M.) and the Karolinska Institute, Stockholm, Sweden.

Abbreviations

CTL	cytotoxic T cell
EBNA	EBV nuclear antigen
EBV	Epstein–Barr virus
GAr	glycine–alanine repeat
GFP	green fluorescent protein
HPV	human papilloma virus
LCV	lymphocrypto virus
MHC	major histocompatibility class
SCF	Skp1/Cullin/F-box protein
UFD	ubiquitin fusion degradation

References

1. Hershko, A. and Ciechanover, A. (1998) Annu. Rev. Biochem. 67, 425–479.
2. Plemper, R.K. and Wolf, D.H. (1999) Trends Biochem. Sci. 24, 266–270.
3. Strous, G.J. and Govers, R. (1999) J. Cell Sci. 112, 1417–1423.
4. Rock, K.L. and Goldberg, A.L. (1999) Annu. Rev. Immunol. 17, 739–779.
5. Laney, J. and Hochstrasser, M. (1999) Cell 97, 427–430.

6. Baumeister, W., Walz, J., Zuhl, F., and Seemuller, E. (1998) Cell 92, 367–380.

7. Voges, D., Zwickl, P., and Baumeister, W. (1999) Annu. Rev. Biochem. 68, 1015–1068.

8. Rickinson, A.B. and Kieff, E. (1996) In: Fields, B.N., Knipe, D.M. and Howley, P.M. (eds.) Virology, Lippincott-Raven, Philadelphia, pp. 2397–2446.

9. Leight, E.R. and Sugden, B. (2000) Rev. Med. Virol. 10, 83–100.

10. Rickinson, A.B. and Moss, D.J. (1997) Annu. Rev. Immunol. 15, 405–431.

11. Yates, J.L., Warren, N., and Sugden, B. (1985) Nature (London) 313, 812–815.

12. Dantuma, N.P., Sharipo, A., and Masucci, M.G. (2002) Curr. Top. Microbiol. Immunol. 269, 23–36.

13. Levitskaya, J., Coram, M., Levitsky, V., Imreh, S., Stegerwald-Mullen, P.M., Klein, G., Kurilla, M.G., and Masucci, M.G. (1995) Nature (London) 375, 685–688.

14. Mukherjee, S., Trivedi, P., Dorfman, D.M., Klein, G., and Townsend, A. (1998) J. Exp. Med. 187, 445–450.

15. Wiertz, E.J., Mukherjee, S., and Ploegh, H.L. (1997) Mol. Med. Today 3, 116–123.

16. Levitskaya, J., Sharipo, A., Leonchiks, A., Ciechanover, A., and Masucci, M.G. (1997) Proc. Natl. Acad. Sci. USA 94, 12,616–12,621.

17. Karin, M. and Ben-Neriah, Y. (2000) Annu. Rev. Immunol. 18, 621–663.

18. Sharipo, A., Imreh, M., Leonchiks, A., Imreh, S., and Masucci, M.G. (1998) Nat. Med. 4, 939–944.

19. Dantuma, N.P., Lindsten, K., Glas, R., Jellne, M., and Masucci, M.G. (2000) Nat. Biotech. 18, 538–543.

20. Dantuma, N.P., Heessen, S., Lindsten, K., Jellne, M., and Masucci, M.G. (2000) Proc. Natl. Acad. Sci. USA 97, 8381–8385.

21. Heessen, S., Leonchiks, A., Issaeva, N., Sharipo, A., Selivanova, G., Masucci, M.G., and Dantuma, N.P. (2002) Proc. Natl. Acad. Sci. USA 99, 1532–1537.

22. Piechocki, M.P., Pilon, S.A., Kelly, C., and Wei, W.Z. (2001) Cell. Immunol. 212, 138–149.

23. Blake, N.W., Moghaddam, A., Rao, P., Kaur, A., Glickman, R., Cho, Y.G., Marchini, A., Haigh, T., Johnson, R.P., Rickinson, A.B., Wang, F. (1999) J. Virol. 73, 7381–7389.

24. Sharipo, A., Imreh, M., Leonchiks, A., Branden, C., and Masucci, M.G. (2001) FEBS Lett. 499, 137–142.

25. Schwam, D.R., Luciano, R.L., Mahajan, S.S., Wong, L., and Wilson, A.C. (2000) J. Virol. 74, 8532–8540.

26. Cummings, C.J., Reinstein, E., Sun, Y., Antalffy, B., Jiang, Y., Ciechanover, A., Orr, H.T., Beaudet, A.L., and Zoghbi, H.Y. (1999) Neuron 24, 879–892.

27. Cho, Y.G., Ramer, J., Rivailler, P., Quink, C., Garber, R.L., Beier, D.R., and Wang, F. (2001) Proc. Natl. Acad. Sci. USA 98, 1224–1229.

28. Bao, Y.P., Cook, L.J., O'Donovan, D., Uyama, E., and Rubinsztein, D.C. (2002) J. Biol. Chem. 277, 12,263–12,269.

29. Leonchiks, A., Liepinsh, E., Barishev, M., Sharipo, A., Masucci, M.G., and Otting, G. (1998) FEBS Lett. 440, 365–369.

30. Li, M., Chen, D., Shiloh, A., Luo, J., Nikolaev, A.Y., Qin, J., and Gu, W. (2002) Nature (London) 416, 648–653.

31. Young, P., Deveraux, Q., Beal, R.E., Pickart, C.M., and Rechsteiner, M. (1998) J. Biol. Chem. 273, 5461–5467.

32. Navon, A. and Goldberg, A.L. (2001) Mol. Cell 8, 1339–1349.

33. Lotz, B. and Keith, H.D. (1971) J. Mol. Biol. 61, 201–215.

34. Braun, B.C., Glickman, M., Kraft, R., Dahlmann, B., Kloetzel, P.M., Finley, D., and Schmidt, M. (1999) Nat. Cell Biol. 1, 221–226.

35. Lin, J., Jin, X., Page, C., Sondak, V.K., Jiang, G., and Reynolds, R.K. (2000) Cancer Res. 60, 5895–5901.

36. Nakamura, S., Roth, J.A., and Mukhopadhyay, T. (2000) Mol. Cell. Biol. 20, 9391–9398.

37. Reits, E.A., Vos, J.C., Gromme, M., and Neefjes, J. (2000) Nature (London) 404, 774–778.

38. Schubert, U., Anton, L.C., Gibbs, J., Norbury, C.C., Yewdell, J.W., and Bennink, J.R. (2000) Nature (London) 404, 770–774.

39. Lee, C., Schwartz, M.P., Prakash, S., Iwakura, M., and Matouschek, A. (2001) Mol. Cell 7, 627–637.

40. Lee, D.H. and Goldberg, A.L. (1998) Trends Cell Biol. 8, 397–403.
41. Ferrer, A., Wells, K.E., and Wells, D.J. (2000) Gene Ther. 7, 1439–1446.
42. Ilan, Y., Prakash, R., Davidson, A., Jona, V., Droguett, G., Horwitz, M.S., Chowdhury, N.R. and Chowdhury, J.R. (1997) J. Clin. Invest. 99, 1098–1106.
43. Blake, N., Lee, S., Redchenko, I., Thomas, W., Steven, N., Leese, A., Steigerwald-Mullen, P., Kurilla, M.G., Frappier, L., Rickinson, A. (1997) Immunity 7, 791–802.
44. Levine, A.J. (1997) Cell 88, 323–331.
45. Ashcroft, M. and Vousden, K.H. (1999) Oncogene 18, 7637–7643.
46. Horowitz, J. (1999) Curr. Opin. Mol. Ther. 1, 500–509.

S.C. Makrides (Ed.) *Gene Transfer and Expression in Mammalian Cells*

CHAPTER 20

Architecture and utilization of highly expressed genomic sites

Jürgen Bode[1], Sandra Götze[1], Ellen Ernst[1], Yves Hüsemann[1],
Alexandra Baer[1], Jost Seibler[2], and Christian Mielke[3]

[1]*Gesellschaft für Biotechnologische Forschung/German Research Centre for Biotechnology,
Epigenetic Regulation, D-38124 Braunschweig, Mascheroder Weg 1, Germany
Tel.: +49 531 6181 251, Fax: +49 531 6181262. E-mail: jbo@gbf.de*
[2]*Artemis Pharmaceuticals GmbH, Neurather Ring 1, D51063 Köln, Germany*
[3]*Medizinische Poliklinik Dept. Clinical Chemistry University of Wuerzburg, Klinikstraße 6-8,
D-97070 Würzburg, Germany. Tel.: (+49) 931 201 7008 (lab: 7039) Fax: (+49) 931 201 7120*

1. Introduction

A major reoccuring problem in molecular biology, bioengineering and gene therapy is the variable expression of transgenes. Presently, one solution to this problem is sought in the use of nonviral episomal (extrachromosomal) vectors that replicate and segregate in synchrony with the host cell. Although progress has recently been reported [1], the development of this promising system is still in its infancy. If a stable long-term expression is desired, integrating systems are therefore still standard. Unfortunately, according to conventional transfection protocols, integration occurs into random chromosomal loci and at variable copy numbers, which together causes unpredictable gene expression characteristics due to position-dependent inactivation and repeat-induced silencing processes [2]). To circumvent problems of this kind, methodologies have been developed whereby the gene of interest can be inserted into a pre-characterized chromosomal site at the location of a unique genomic tag. Retroviral vectors ([2–4] and references therein) are considered particularly useful vehicles for a gene of interest (GOI) since the retroviral integration machinery is well suited to introduce a single intact copy into a subclass of apparently favorable genomic sites. In the following it is our intention to delineate the architecture of *loci* enabling stable and high-level expression, the methods for their characterization and-most importantly-principles which can be applied to guide a transgene into a characterized site with known properties.

2. Chromatin domains and transgene integration targets

In higher eukaryotes the regulation of transcription is determined, to a large extent, by an intricate interplay of nuclear position and chromatin structure [5]. This becomes especially important in cases where the expression of foreign genetic material is studied or where transgenes are introduced for the efficient production of pharmaceutically relevant proteins. Consequently, the chromatin field has matured from mere basic science to a level where it can contribute major improvements to any eukaryotic expression system.

The eukaryotic genome is organized in a hierarchical fashion within the nucleus. DNA first interacts with the core histones to form the 10 nm fiber. This basic repetitive structure is then folded by the addition of chromatin proteins including histone H1, into a range of more highly condensed superstructures. The resulting 30 nm fiber in turn is organized into looped domains, which are attached to the nuclear scaffold or matrix by the interaction of specific DNA sequence elements (scaffold/matrix attachment regions, S/MARs) with protein components of the subnuclear scaffold or matrix (reviewed in Ref. [6]).

S/MARs of over 1 kb in size have been considered as chromatin domain borders and as such to play a critical role in nuclear architecture and nuclear function [7]. They are the common elements that organize the 25 million nucleosomes into loops among which the small ones (4–20 kb) embed the active genes and the large ones (up to 200 kb) contain the inactive regions [8]. A number of assays have been developed to search for S/MARs all of which identify the same class of anchorage sequences ([9] and references therein). At present, progress is stimulated by the emergence of three independently developed bioinformatical tools, which consider and predict this particular kind of DNA–protein interaction ([10] and references therein). Due to their strategic position and unique chromatin structure, S/MARs are not only necessary for efficient transcription but also required for correct chromosomal replication, condensation, and recombination (reviewed in [1,11]).

2.1. Transcriptional functions of S/MARs

S/MARs increase transcriptional initiation rates in a way that is distinguishable from (but supplementary to) enhancers (Section 2.3 and [7]). It may even represent a superimposition of several contributions. One effect for which there is increasing evidence is a genuine cis-(domain-opening) action, mediated, for instance, by a hyperacetylation of core histones and a suppressive effect on DNA–methylation [12,13]. Another action which has been implied to explain the sometimes dramatic overall influence on transcription initiation rates is that of a targeting element by which S/MAR constructs are guided to a subset of favorable, chromosomal integration sites which is not or only partially overlapping with those hit by the S/MAR-free controls (Fig. 1A). While these would appear to be the most obvious contributions, other possibilities like the 'independent-domain model' and 'prevention of antisense formation by readthrough' cannot be excluded.

Fig. 1B shows a prototype experiment in which a luciferase (Luc) reporter has been transfected with or without flanking S/MARs. From each of the three groups (Luc, S/MAR-Luc and 'minidomain' situation) 9–10 clones with authentic single-copy

A

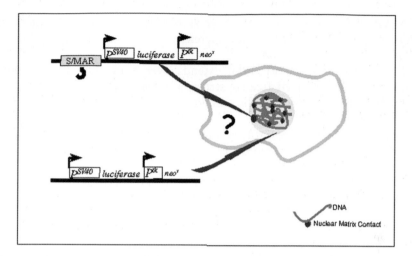

B

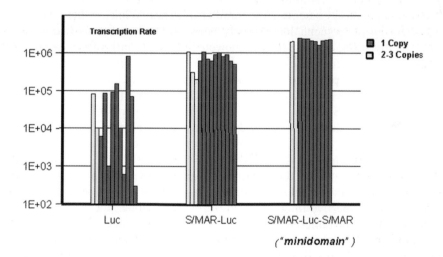

("minidomain")

Fig. 1. Nature of transcriptional augmentation processes. (A)Two potential contributors to transcriptional augmentation. Traditionally, augmentation is derived from the parallel transfection of constructs containing or not containing an active element (S/MAR). The apparent effect is either due to a classical *cis*-action (resembling an enhancer) or because the S/MAR directs the transgene into a distinct population of favorable genomic sites. (B)Comparison of transcriptional levels in luciferase constructs (Luc) that have been supplemented with one (S/MAR-Luc) or two (S/MAR-Luc-S/MAR) bordering scaffold/matrix attachment elements. Note the effect of S/MARs regarding the average expression level and the clone-to-clone variation (uniformity of expression). Obviously, multiple copy integrations are not superior to single copy integration events.

integration events have been isolated and characterized for their transcriptional activities, leading to the following conclusions (Fig. 1B):

- a S/MAR element can increase average stable expression levels by one or two orders of magnitude due to an increase of transcription-initiation rates ('*transcriptional augmentation*');
- while the clone-to clone variation extends over more than three orders of magnitude for the S/MAR-free construct, a much more balanced distribution is obtained, particularly for the 'minidomain' situation ('*insulator*', and/or '*targeting*' action);
- multiple copies do not necessarily grant higher expression levels (a phenomenon termed 'cosuppression' or 'repeat-induced silencing' [2]), cf. the dark and the light bars in Fig. 1B);
- finally, the minidomain is well suited to maintain long-term activity (i.e. to *resist silencing*; shown in reference [7].

2.2. *Silencing phenomena are counteracted by nuclear matrix association*

The nuclear matrix exhibits a pronounced affinity for particular DNA structures as it attracts single-strands, accommodates (and enzymatically converts) DNA supercoils and associates with S/MAR-type DNA [6,7]. The latter binding type is clearly limiting and it has been estimated that the sites, which are available in a cellular matrix equivalent, closely resemble the number of genes. While, in the past, repeat-induced silencing phenomena have mostly been ascribed to heterochromatization and DNA methylation, we have recently detected another contribution, which might actually be the trigger for these processes: lack of S/MAR-binding sites.

For Fig. 2 we have prepared a number of mouse L-cell clones carrying a 39 kb piece of human DNA which includes the human interferon-β (huIFN-β) gene and its 2.2 kb long upstream S/MAR element. FISH analyses (performed on metaphase spreads) established that each of these clones had inserted a concatemer of transgenes at a single genomic position whereby each copy becomes effectively organized as a 'minidomain' (5'S/MAR-IFNβ-5'S/MAR-IFNβ...).

Figure 2A demonstrates the distribution of expression level as a function of copy number. While we cannot totally exclude that transcription rates are modulated by the nature of the respective integration site in a given clone, according to Fig. 1B such an effect should be buffered by effective S/MAR-matrix interactions, at least for clones which have to accomodate a small number of transgenes only. For a deeper understanding of this phenomenon we have subjected clones that differ in the number of transfected IFN-β copies to a novel halo-FISH technique [14] and visualized aspects of their interaction with the nuclear matrix (Fig. 2B, C). For this approach interphase nuclei are fixed on a microscopic slide, gently extracted in situ and stained with DAPI to reveal a circular internal portion (remnants of the nuclear structure harboring the nuclear matrix) and a brightly-fluorescing halo (loops emanating between functional S/MAR-matrix contacts). Finally, DNA probes specific for the insert (red fluorescence) and for the endogenous murine IFNβ (green fluorescence) are used to visualize the respective location of the genes.

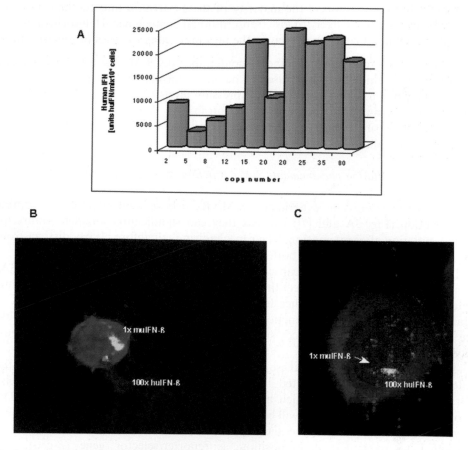

Fig. 2. Transcriptional levels do not necessarily reflect the number of integrated transgenes. (A) Murine L-cell lines with distinct numbers of a huIFN-β S/MAR-construct were established either by electroporation or by transfection. Their expression was monitored and their distribution over nuclear matrix and halo were studied by FISH (below). (B) For a multi-copy cell line of this type only a minority of members (red signals) can be accommodated by the nuclear matrix. This becomes particularly evident if compared to the organization of the endogenous muIFN-β gene locus (metaphase spreads proved that the green signal corresponds to a single locus). This tight association with the matrix is also typical for cells containing 1–10 copies of the transgene (H.H.Q Heng *et al.*, submitted). (C) As B but using a small probe, which covers only the huIFN-β coding region (in contrast to the BAC used for B).

Statistical analyses on numerous nuclei have been performed to evaluate the distribution of transgenes between nuclear matrix and halo and to account for the position of the integration site relative to the nuclear remnant. This procedure has led to the following conclusions (H.H.Q. Heng *et al.*, submitted):

- although the association of S/MARs with the matrix is clearly a dynamic process, at low copy numbers (1–10) genes can be accommodated at the matrix most of the time;

- at higher copy numbers the majority of these partition into the halo portion whereby genes are likely to lose transcriptional competence. Heterochromatization and methylation may then be the inevitable consequences of this situation.

These analyses resolve a longstanding contention why S/MARs meet only one out of two criteria which have been used to define an 'insulator'-function: they clearly reduce the clone-to clone variation of expression levels (Fig. 1B) while they do not mediate a copy-number dependent expression in case they have co-integrated at a single genomic site (which is a common consequence of standard transfection protocols and valid for all clones in Fig. 2).

2.3. The augmentation phenomenon: S/MAR actions are context-dependent

In a transient expression experiment, S/MARs behave neutral or even suppress transcription (Fig. 3A and [7]) whereas they can significantly augment transcriptional initiation after integration. Their mechanism is therefore clearly distinct from enhancers, which are active in both the transient and stable expression phases.

If the sometimes dramatic transcriptional effects of S/MARs (Fig. 1B, Fig. 3A) were cis-actions, we would expect that transgenes profit from these at most genomic integration sites. If on the other hand the predominant function of S/MARs in transgenesis were a targeting function (Fig. 1A), removal of a S/MAR in situ should be without further consequences since integration has occurred into an a priori favorable genomic position.

The tools for investigating these alternatives have become available during the late 80s when geneticists started to use the site-specific recombinases to excise (or invert) DNA cassettes which are flanked by recombination target sites. The molecular basis of these events will be discussed in a broader context below (Section 4 Fig. 6). Fig. 3B includes the general design of a minidomain construct which could be subjected to sequential decomposition: a reporter/selector gene (β-geo), was surrounded by two identical S/MARs and each of these flanked by either FRT or loxP recombinase target sites. Since the relevant monomolecular excision reactions are both entropically and kinetically favored [4] they occur with efficiencies up to 100% in cells which express the respective enzyme [15]. When we investigated a clonal mixture using this technology, we monitored a slight overall decrease of expression levels after excising the S/MAR between either the FRT or loxP sites [7] which would be in line with its operation as a cis-active element. Later studies (exemplified by the data in Fig. 3) demonstrated that the situation is clearly more complex since the respective effects depended, to an extreme extent, on the nature of the integration site, the orientation of the transgene within the site, interactions of the promoter with its genomic surroundings and on the cell line revealing a wide scope of sometimes weak and sometimes dramatic S/MAR-related activities:

- in some clones there is no (immediate) effect of excision (K1);
- in a number of clones there is a more or less pronounced positive effect [revealed by a drop of expression subsequent to the excision of one (K3, K8, K22) or both (K8) S/MARs];

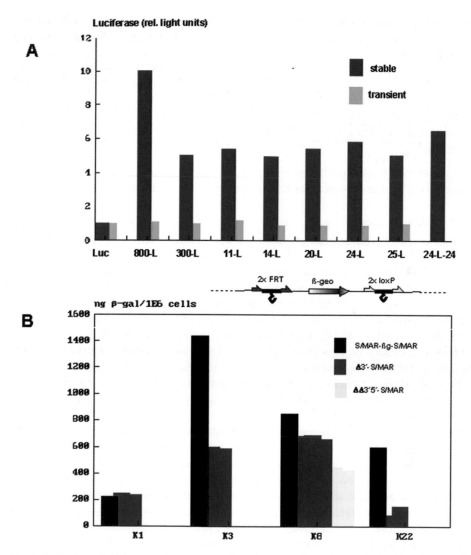

Fig. 3. S/MAR-mediated augmentation: overall and context-dependent actions. (A) S/MARs have been recovered from retroviral integration sites and compared with authentic standard elements (800 and 300 bp for transfectants 800L and 300L, respectively) regarding their transcriptional potential (cf. Fig. 1). Designations 11L, 14L etc. refer to 300-bp fragments constituting the integration sequences for clones 11 and 14, resp. 24-L-24 is a 'minidomain' construct in which luciferase is flanked on both sides by the clone 24 sequence. Note that S/MARs are generally characterized by the fact that they exert no effect on transient expression (which has been set to 1). (B) Effect of excising S/MARs from a single copy of the β-geo transgene at a given chromosomal localization. Excision analyses were performed according to Fig. 6A for several subclones resulting from such a treatment. While a transgene at sites K1 and K3 responds strongly to the loss of a single domain boundary, the effect is moderate for site K8 and absent at site K1. A subsequent excision of the 5′ S/MAR at site K8 adds to the effect of the 3′excision.

- in certain situations (clones, cell lines) the removal of a S/MAR initiated silencing in a process that continued over many generations (Seibler and Baer, unpublished);
- remarkably, in these experiments, the first excision step occurs at a high efficiency while that of the second one is usually reduced (Table 1). This observation is in line with a function of S/MARs in the opening of chromatin domains (due to an increased histone acetylation and an interference with DNA methylation) or a bordering function (preventing the spread of heterochromatization).

As a whole, these results strengthened the impression that 'transcriptional augmentation' is a composite, context-dependent and highly complex phenomenon.

3. Comparison between gene transfer by transfection and retroviral infection

Decades ago, McClintock noted that broken chromosome ends can rejoin efficiently because of an apparent stickiness of the newly created ends. Related observations showed that transfected DNA molecules also become covalently joined into multimers prior to or following their integration into a chromosome [16,17]. More recent *in vivo* studies of DNA end joining in mammalian cells have revealed an unexpected diversity of possible mechanisms (reviewed in Ref. [11]). It is thought that foreign DNA is predominantly integrated at pre-existing DNA breaks and fixed by the nuclear repair machinery. Such a mechanism would explain why integration events subsequent to transfection are so rare and integration sites are distributed apparently at random.

Since single-stranded DNA is known to be much more sensitive to damage than double-stranded DNA, S/MARs have been postulated to be somehow linked to DNA breaking events owing to their strand-separation potential [11]. On the other hand, the energy stored in a single-stranded region is the prerequisite for the formation of alternative secondary structures such as nearby cruciforms or slippage structures. These alternate structures and single-stranded regions themselves may be recognizable features for DNAses, topoisomerases or other structure-converting enzymes (reviewed in Ref. [11]). In fact, S/MARs have been shown to have significant potential for alternate secondary structure formation and they have proven to be recombinogenic [7]. These properties may explain the observation

Table 1

Efficiency of the first and second S/MAR-excision events, respectively (% of recombinase-expressing clones with the authentic event).

Clone	Excision rate 3′ (%)	excision rate 5′ after 3′ (%)
C18	80	17
C37	80	57
C40	67	33

(reproduced in Fig. 4 and Table 2, below) that endogenous cellular S/MARs constitute the dominant (possibly exclusive) integration sites for proviruses and perhaps also for other transgene constructs.

3.1. Genomic sites targeted by transfection, electroporation and retroviral infection are distinct

In order to obtain a predictable expression, the regulatory properties of the genomic integration site should be reasonably well understood. This statement becomes especially relevant considering the tools that have recently become available for a targeted integration of transgenes, which is feasible at almost any pre-characterized genomic site.

We have mentioned above that the integration of transgenes usually requires free DNA ends, both on the side of the host cell and the transfected DNA. An obvious exception is cultivated cell lines (BHK, CHO), which have been selected for decades for their capacity to integrate circular plasmid-based vectors [18]. Here the

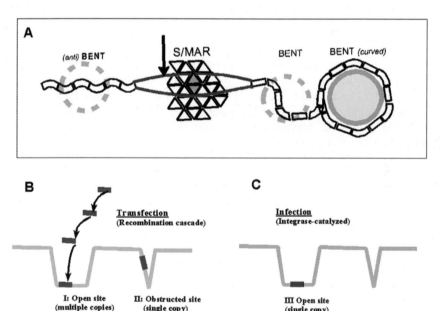

Fig. 4. Retroviral integrase is thought to insert a single copy of the transgene into a highly expressed ('open') genomic site. (A) Retroviral integration sites (arrow) conform to a variety of structural criteria: they occur into a nucleosome-free region of bent DNA with the properties of a 'fragile site' (Table 1). Here, non-B type structures are supported by the association of proteins (triangles) recognizing separated DNA strands or strands with an unwinding potential. (B, C) Single copy clones generated by classical transfection techniques or by retroviral infection are not comparable. Note that in our model single-copy transfectants arise by deposition into a rather occluded site (II) while open sites (I) invite the entry of multiple copies, supposedly [17] by a recombination process following circularization of incoming DNA. In contrast, infection-mediated integration involves a preselection of open (and nuclear-matrix-associated) sites (III).

Table 2

Molecular Characteristics of Retroviral Integration Sites

General:

– there is no consensus sequence

Integrase-mediated insertion in vitro:

– preferential integration into intrinsic or protein-induced bent DNAs (major groove of nucleosomally organized DNA)

Integration in vivo:

– there is a DNAse hypersensitive site within 0.1–0.7 kb (expected average: 6.5 kb);
– this hypersensitive site may coincide with a topo II cutting site (indicative for nuclear-matrix association, see Fig. 4)
– actively transcribed regions (CpG islands) are commonly close to the integration site.
– nuclear matrix association has been demonstrated for 8/8 productively infected cells;
   although of a special type, these insertion regions share a number of S/MAR-typical features
   (bends, repeats, biological activities; see Fig. 4 and [20]);
– adjacent chromosomal regions are nucleosome-free due to bent (or anti-bent) stretches of DNA.
   Note that nucleosomal organization is only compatible with a certain curvature (or bendability)
   but not with pre-existing bends (Fig. 4A and [20]).

---

integration mechanisms are particularly complex and frequently accompanied by duplications and genomic rearrangements which preclude their precise investigation.

In contrast, retroviral integration has permitted the detailed analysis of sequences flanking proviral inserts due to the precise enzymatically catalyzed insertion mechanism [19]. These studies demonstrate that integration sites have markers in common with highly expressed endogenous genes (Fig. 4A). Among these is the apposition of a DNAse I-hypersensitive site, the presence of CpG islands, and adjacent stretches of bent DNA, which are structured in a way that discourages nucleosomal assembly (Table 2; [20] and references therein). While we have described this integration preference for common cultivated cell lines (among these a BHK cell line, which can be infected by common murine proliferative sarcoma virus-derived retroviruses [21]), others have noted that integration of small tumor viruses uses the same strategy as a prelude to cancer development [22]. Interestingly, these sequences not only share important S/MAR-related structural features [7] but they also mediate the transcriptional augmentation phenomenon (Fig. 3A), which has been considered to be one of the factors contributing to an efficient expression of selection markers in artificial constructs or of oncogenes in a natural setting.

One rationale for the preferred use of retroviral vectors as gene transfer vehicles is an early report [23] according to which a gene introduced by viral infection is expressed at 10–50-fold higher levels than a copy of the same DNA which has entered the cell by Calcium phosphate transfection. Although we observed the same general tendency, the extent of this difference is at variance with studies from this laboratory in which infection had been compared with electroporation [20,21].

Unlike other transfection techniques, electroporation can easily be optimized such that one or only few copies of a gene are introduced, overcoming the now established phenomenon of repeat-induced silencing [2]. We have therefore compared the stable expression of a provirus with an electroporated single-copy template, which had been trimmed to the same size. Under these conditions, our experiments yield only a slight (1.3-fold) higher average transcription rate for infected genes [21].

Different integration forms arising from electroporation and Ca-phosphate transfection [21] lead to the question whether or not the same class of genomic sites is hit by both gene transfer techniques. Electroporation involves exposure of cells to a pulsed electric field, which creates pores in the plasma membrane whereby this process becomes less dependent on cell type than other transfection procedures. DNA transfer with Ca-phosphate as a facilitator on the other hand has very specific features. Ca-phosphate is thought to cause a local solubilization of phagocytic vacuoles allowing the passage of DNA into the cytoplasm and finally the nucleus. There, the integration of head-to-tail concatemers and (where applicable also the co-integration of carrier DNA) is probably a consequence of the high local concentration of DNA molecules:

Folger et al. [17] have demonstrated by microinjection that single linear molecules become inserted into the host genome through their ends. When more than 5 linear molecules are injected per cell, the transformants contain multiple copies in the typical head-to-tail array. By injecting two vectors, which are similar but not identical, it was found that head-to-tail concatemers arise from homologous recombination (HR) rather than replication/amplification mechanisms. This would imply that subsequent DNA molecules have to become re-circularized before entry into the concatemer. Therefore, it appeared that in the nucleus there is competition for the ends of donated DNA, which becomes ligated to host DNA (integration), to themselves (circularization) or to other exogenous DNA sequences like carrier DNA (end-to-end ligation).

In higher eukaryotes, HR can occur only before the final chromatin structure is established, i.e., prior to or immediately following DNA transfer. This process will therefore be favored if the first integrating template finds itself in a transcriptionally permissive, open chromatin context (Fig. 4B). Conversely, the few single-copy clones obtained from Ca-phosphate-mediated gene transfer may be the consequence of a rather occluded genomic site and selection for these rare events could mean selection for mediocre sites. Given that this is true, the model suggested in Fig. 4B resolves the apparent discrepancy raised above: while previous workers have compared the expression of a single transfected copy in an obstructed site (II) with a single infected copy in an open site (III), we have more likely established a dominance of single electroporated copies in situation I and related these to situation III.

Although the immediate transcriptional advantage of the infection-mediated clones is rather moderate, we could nevertheless establish that in the cases where a long term inactivation occured to a notable extent, this was mostly true for the electroporated series of clones (Fig. 5A and B). Here the onset of inactivation occurred earlier and to a larger extent than for the infectants. Using the halo-FISH-techniques introduced above (Fig. 2) we could obtain additional evidence for the supposition that S/MAR-targets are more important for the integration of infected rather than transfected DNAs: Fig. 5 contains two extreme, but typical snapshots for

562

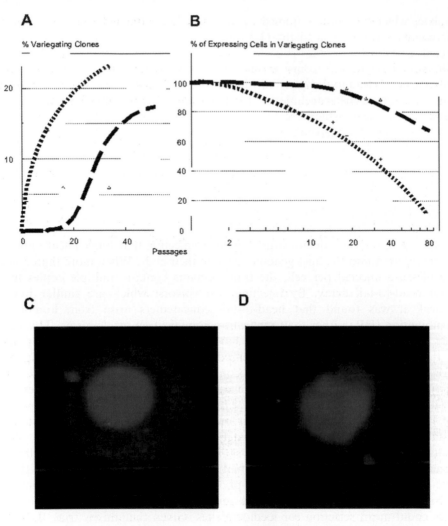

Fig. 5. A discriminating feature of infectants relative to transfectants concerns inactivation characteristics. (A, B) The onset of inactivation occurs later (left) and to a lesser extent (right) in case of electroporation (dotted line, — +—) rather than infection (broken line, – – – –). At certain intervals, unstable clones were identified by a blue-white screening whereby clones containing < 10% of cells not staining with X-gal were classified as 'unstable'. The percentage of variegating (blue/white) clones is shown in A and the percentage of expressing (blue) cells in these clones in B.(C, D) a prototype infectant (C) shows its transgene in association with the nuclear matrix (this is true for 70–90% of all clones at any moment) whereas 50–70% of transfectants are found in the halo-fraction (see example in D).

this situation showing a copy inserted by integration which became associated with the matrix (part A) and a transfected copy which has partitioned into the loop structure. It was interesting to note that the infected group of clones (55–70% matrix association) formed a set, which was perfectly separated from the electroporants with only 0–35% of association at any time [41]. Within each group, the unstable clones were those which were least attached even before the time inactivation started.

Since S/MARs associate with the matrix in a dynamic manner (H.H.Q Heng et al., submitted) this could mean that features associated with particularly strong S/MARs are sensed by the integrase whereas strand breaks can also occur in regions with a weaker propensity to be matrix-associated. Central conclusions from this chapter are:

- expression is superior when a single gene copy can be placed into a site with high transcriptional potential; this is best achieved by retroviral infection but can also be approached by an optimized electroporation protocol using a linearized vector and no carrier DNA;
- if multiple-copy integration cannot be avoided, a system would be desirable which reduces the multicopy-concatenate to the single-copy level. Such a system will be introduced below and discussed in the context of Fig. 6A.

4. Re-use of established highly expressed genomic sites

Introduction into a defined genomic site can-in principle-be achieved by HR and this is the procedure that is commonly applied for embryonic stem cells. However, due to the dominance of illegitimate recombination (IR), HR is quite inefficient in transformed cell lines where the ratio between HR and IR may be as low as 1:1000 (reviewed in Ref. [11]). These facts explain why considerable efforts have to be invested to develop methods for the re-use of integration sites in cell clones with the desired properties. In an increasing number of cases the site-specific recombination apparatus from prokaryotes and lower eukaryotes has been adopted as the solution of choice. The associated site specific recombinases have a number of interesting features and, by utilizing these, a variety of highly specific genomic modifications has become routinely possible (summarized in Fig. 6).

The phage recombinase Cre has become the most common tool to excise or invert DNA cassettes-a process that occurs readily if these cassettes are flanked by loxP sites. Fig. 6 emphasizes that the outcome depends on the relative orientation of the wild type recombination target (here: loxP-) sites. Subsequently, the technology was used to engineer inducible translocations and somatic alterations of the genome and also to reduce multicopy-concatenates to the extent where a single copy is left [24].

The mechanistically similar recombinase from yeast, Flp, was initially expected to be more effective for an operation in the eukaryotic nucleus and therefore some laboratories started to use the Flp/FRT system instead [25]. Today this aspect does not cause any further concern since nuclear transfer is also observed for the prokaryotic 38 kDa Cre, not only because of its small size, but also due to the unexpected presence of nuclear targeting determinants. Since the introduction of more thermostable Flp variants by Buchholz [26] we now have available two efficient targeting systems (and several more to come, cf. [27]), which perform equally well in the fundamental excision and inversion ('flipping') reactions.

A: F-F, L-L

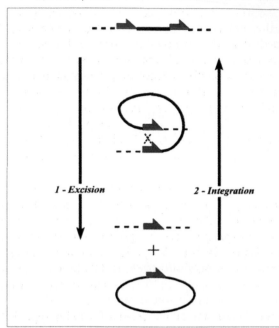

B: F-F_n

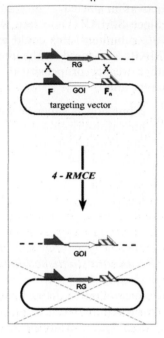

C: F-F, L-J

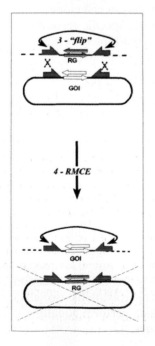

D: FF_n-FF_n

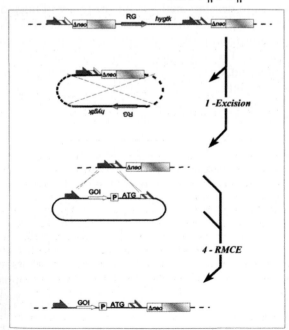

Fig. 6 suggests that it should be possible to reverse the excision pathway in order to achieve the desired *integration* of a transgene into a tagged genomic site (pathway 2). Although a favorable situation of this type has been described [25], we and others have demonstrated that this process is very ineffective in the absence of a highly efficient trap system since the reaction is both entropically and kinetically disfavored [15,28]. To be successful, an incoming selection marker has to integrate behind a pre-existing promoter ("promoter trap") and a translation initiation codon ("initiation trap") in order to become specifically activated. Even more important, in the absence of an appropriate timing of recombinase activity any integration will be immediately reversed. Such a timing has been achieved by the use of ligand-inducible promoters or fusion proteins (reviewed in Ref. [4]).

While the first site-specific integrations in mammalian cells were in fact based on integration at a single recombinase target site, these pioneering methods had one more disadvantage in addition to their low efficiency: since in such a procedure the entire plasmid becomes integrated, an expressed selector gene is left behind in the chromosome (Fig. 6A). That vector sequences of prokaryotic origin or co-expressed genes can severely perturb expression of the gene of interest (GOI) by methylation-dependent inactivation or cosuppression has been amply documented leading to the search for effective alternatives.

Our laboratory has exploited the functional architecture of a wild type FRT sequence and we developed an independent and highly efficient solution to this problem.

4.1. Recombinase-mediated cassette exchange (RMCE)

For the development of the recombinase-mediated exchange principle depicted in Fig. 6B the recombinase target sites had to be modified in a particular way. A complete Flp recombinase target (FRT) is 48 bp long and consists of two inverted 13-bp repeats around an 8-bp spacer and a third 13-bp direct repeat (in addition to one isolated base pair). While the minimum size of an FRT sequence is 34bp

---

Fig. 6. Site-specific recombination procedures for targeting (and modifying) tagged genomic loci. A-The 'in-and-out' principle according to which excision removes the sequence intervening two equally directed recombinase-target sites. Formally, reversion of this reaction results in integration-a process which is highly inefficient under the conditions of a persisting recombinase activity. (B) Recombinase-mediated cassette exchange (RMCE) between sets $(F-F_n)$ of recombinase target sites. These sites differ in the composition of their 8-bp spacer such that there is no cross-interaction between the wild type and a mutant (F, F_n) or between two mutants (F_m, F_n). If these requirements are met, a double-reciprocal crossover event can be triggered with high efficiency. (C) Oppositely oriented target sites mediate a 'flipping' (inversion) reaction of the intervening sequence. Although this may represent a complication in certain experimental setups, this situation also permits a cassette-exchange reaction since intervening sequences are maintained (in either of two orientations) but not lost. (D) Twin sites (F/ F_n) permit that (in contrast to A) an excision reaction can be succeeded by an integration reaction, essentially according to the RMCE principle. In the example shown, the target is an integrated provirus for which the retroviral replication cycle has duplicated the twin sites which, in the original vector, have been present only in the 3'-LTR.

(containing only two of the 13-bp recombinase binding sites), this sequence is a non-substrate for intermolecular recombination while excision according to Fig. 7A can still be detected [3].

In 1994 we first described the engineering of the full 48-bp FRT_{wt} (F) sequence in its 8-bp spacer region in order to obtain FRT_{mut} (F_m, F_n,...) sites which recognize each other with maximum efficiency but show no cross-interaction whatsoever [29]. This is a stringent criterion since any interaction would lead to the loss of the sequence intervening two (equally oriented) FRT sites according to the excision principle shown in Fig. 6A. Since this fortuitous loss would lead to ganciclovir-resistance, the respective clones would survive because of the excision rather than the exchange of the selector cassette (see below). Therefore, a perfect non-compatibility of two sites was required which was achieved with FRTs differing in at least four (better five) positions within the 8-bp spacer (reviewed in Ref. [3]).

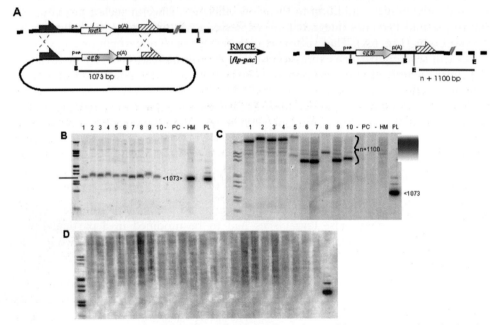

Fig. 7. The efficient production of RMCE-mediated site-directed recombinants may require selection for cells that have received the recombinase expression construct. (A) Outline of an RMCE reaction between the chromosomal target (F_3-*hygtk*-F) and the exchange vector (F_3-*egfp*-F). (B) Two parental clones have been co-electroporated with the circular *egfp*-exchange vector and a *flp-pac* expression construct. Preselection was performed 5 days after an initial (53 h) exposure to puromycin and medium change, before negative selection (for the loss of the *tk*-function) was initiated. It is seen that *all* derivative clones contain the critical parts of the exchange construct (evidenced by the internal 1073-bp BamH1 fragment). (C) Investigation of bordering fragments (i.e. Southern analyses on an n + 1100 bp EcoRI fragment which initiates in the exchange vector and extends to genomic DNA) reflects integration into the predefined target for 8 out of 10 derivative clones. The two outliers (5, 8) originate from random-integration of the exchange vector. (D) Same as in C except that the puromycin-selection step has been omitted. Absence of any specific Southern signal indicates that ganciclovir-resistance has arisen by rearrangements or modifications which have led to the loss of the tk-function.

It is seen in Fig. 6B that the RMCE principle allows the precise exchange of a pre-existing cassette for a corresponding cassette, presented on a circular construct. RMCE is driven by the excess of this "targeting vector" whereby the reaction becomes effectively irreversible according to mass action principles. To enrich for the desired exchange event selection is naturally required. Meanwhile the selection principles could be continuously refined to an extent that close to 100% recoveries are within reach.

First we applied a combination of positive and negative selection steps to this end. As part of the resident cassette we have used a *hygtk* fusion gene or even an *sgtn* (SV40-controlled *egfp-tk-neo*) triple fusion, which includes a fluorescent reporter gene (RG) by which the properties of potential genomic targets can be monitored before introducing the cassette, which comprises the GOI (Baer, unpublished). Selection for hygromycin (or neomycin) resistance is necessary to recover clones that harbored the respective target and it is continued to the very timepoint of RMCE as this procedure enforces maintenance of a functional fusion protein (encoding the positive marker and the HSV- thymidine kinase activity, which is now required for negative selection, see [30]).

Subsequent to cotransfer of the exchange vector and the Flp-expression construct, successful RMCE events can be enriched in the presence of ganciclovir. In the continued presence of the *tk* function, this drug would be converted into a toxic nucleoside and abolish all parental clones that have not undergone exchange (or loss/inactivation of *tk*). This procedure permitted 54–100% exchange rates in cases where negative selection was coupled to the selection for a marker on the incoming cassette [30] which, however, does not comply to our primary goal: expression of the GOI in the absence of a co-expressed selector gene. Sole negative selection was possible for some cell lines but it generally suffered from the propensity of certain cell types to develop spontaneous ganciclovir resistance. The molecular basis of this has been discussed before [30] and further precautions had to be taken to overcome this phenomenon.

We [15] and others [31] have found that one limiting factor in any recombinase-based protocol is the efficiency with which target cell can be transfected by recombinase expression plasmids (and exchange vectors where this is applicable). We therefore constructed a vector, which carries both Flp, and eGFP functions [15] and recovered the successfully transfected cells by fluorescence-activated cell sorting (FACS). This approach, however, has some potential limitations:

(1) Standard eGFP is the most efficient *in vivo* reporter. If this is used for sorting out transfected cells as suggested above, a GFP variant (for instance yellow fluorescent protein, YFP) has to be used in the original target to serve as a probe for the expression characteristics of genomic sites (see above). This variant has to have fluorescent characteristics sufficiently different from eGFP in order to permit separate sorting procedures.
(2) Although an application for ES cells has been reported [32], FACS may not be tolerated by all cell types due to shear forces.

Recently, puromycin-acetyltransferase (*pac*) has been proposed as a transient selection marker, which permits the recovery of (transiently) Flp-expressing clones.

The puromycin system has several convenient features distinguishing it from any other common selection marker: in the absence of an expressed *pac* gene the toxic effects begin to appear just several hours after adding puromycin to the culture medium and selection can be performed at low concentrations. Therefore, transfer of a vector which co-expresses the recombinase and *pac* emerged as an alternative and even more universal tool for the recovery of cells, which receive and express the recombinase [33]. The efficiency of this principle is highlighted by data from this laboratory (Fig. 7) that demonstrate a case for which negative selection due to ganciclovir is clearly insufficient. If, on the other hand puromycin (for transfer of Flp-expression construct) and ganciclovir selection (for loss of the parental cassette) are combined in a refined protocol, 80% of derivative clones contain the authentic RMCE event (Fig. 7B and C).

Until recently there was no combination of loxP sites available which had the same favorable properties. Cre-RMCE as originally introduced by Bouhassira *et al.* [34] used heterospecific loxP sites with a residual crossinteraction capacity, probably because these differed at just a single position in the spacer region. When a chromosomal cassette was flanked by a set of these sites, exchange for another cassette could be traced but the stringency of the system was not high enough to permit negative selection for the efficient recovery of RMCE-clones. This may change after a recent report, which indicates that loxP sites with a double spacer mutation do not recombine with the wild type site [35] or according to the results of a systematic approach for the selection of non-crossinteracting sites in bacteria [36].

While these efforts are in progress, the problem of recombinational deletion has been overcome by using two oppositely oriented Cre sites in a system called "L1-1L" [31] which causes cassette inversion in addition to cassette exchange (Fig. 6C, reactions 3 and 4). This approach produces both orientations of the GOI, which has the disadvantage that additional screening may be required. On balance, it enables studies on the dramatic orientation-dependence of expression parameters at a given genomic site [37].

4.1.1. Twin sites: excision does not exclude subsequent integration

We started the chapter mentioning that we are in the process of developing small circular, non-viral episomal vectors with a maximum cloning capacity and an optimum combination of regulatory elements [1]. To this end we apply the concept of FRT-twin sites. The vector, in which the prokaryotic sequence parts are flanked by twin sites F/F_n [38], each consisting of a wild type (F) and a mutant (F_n) component, is prepared as a plasmid in *E. coli*. After transfer into the recipient eukaryotic cell, a pulse of Flp activity suffices to excise the intervening sequence whereby an episomal self-propagating minicircle is created. After these preparatory steps, other regulatory elements (enhancer, S/MAR) can be introduced by the RMCE principle, between the F and F_n parts of the F/F_n-twin to improve the expression characteristics. This example illustrates that the restrictions of the in-and-out principle of Fig. 6A can be elegantly circumvented.

There are many other experimental designs that profit from the utilization of twin sites in related or different ways. Fig. 6D depicts a published [39] protocol where a twin-site is cloned into a retroviral 3' LTR. During the retrovirus-provirus conversion, this site is copied in a way that it recurs also in the 5'position [40]. If used for tagging, the genomic sites in which the construct becomes integrated are likely to mark a highly expressed locus (Chapter 3). After its thorough characterization via a reporter gene, excision between the twin sites can be used both to remove auxiliary sequences and to create a unique F/F_n -tag. A subsequent RMCE-type reaction can then be used to introduce the GOI in a predefined orientation and at the single-copy level at this very location (reviewed in Ref. [4]).

4.1.2. RMCE multiplexing

Figure 6B and D illustrate the fact that each set of $F_m - F_n$ sites (or each F_m/F_n twin) represents an independent targeting address, suited to guide a targeting vector to the corresponding genomic tag. Therefore, by placing different sets of sites at different genomic locations, a single pulse of Flp activity can catalyze independent DNA recombination events at multiple *loci* in the same cell without encountering the danger of interlocus recombinations ([3] and references therein). Following this multiplexing principle, we have devised a protocol which enables us to elaborate, in several subsequent steps, a complete chromatin domain at a predetermined chromosomal location [3].

5. Lessons

In this brief review we have tried to establish some important molecular details of highly expressed genomic sites (Table 2 and Fig. 4) and have suggested the systematic use of the responsible elements for the design of new classes of vectors. Methods for the characterization, re-use and elaboration of sites which have been pre-tested for their expression characteristics and genomic stability have concluded the chapter during which we have put emphasis on the following principles:

- highly expressed sites can be provided with single copies of a transgene by retroviral infection or single copy electroporation (rather than transfection) techniques; retroviral constructs integrate at an important class of elements, the scaffold/matrix-associated regions (S/MARs)-a principle which can be exploited with advantage;
- for alternative transgenic approaches available S/MAR-elements can be used as essential vector components in order to increase expression levels and/or to maintain a long-term high-level expression (Fig. 1);
- site specific recombinases perform excision reactions with high efficiency (Table 1) for a DNA segment flanked by either loxP or FRT-sites ('floxing' or 'flirting' principles, respectively). They represent powerful tools to reduce any head-to-tail concatemer to the single-copy level in case the transgene has been provided with a recombinase target site;
- RMCE of cassettes which are flanked by a set of FRT-spacer mutants (F_m-F_n) is a favorable and effectively irreversible reaction which does not require the precise timing of recombinase activity (Fig. 6B–D);

- twin target sites (F_m/F_n) can be used to both excise prokaryotic vector parts and to expand a transgene by regulatory elements to improve the characteristics of a site (Fig. 6D);
- twin-sites as parts of a retroviral vector permit an RMCE procedure by which most viral sequence parts can be removed from the final target (Fig. 6D). The only sequence left behind is a single LTR. If the self-inactivating (SIN) principle [40] is part of the protocol, this LTR will be inert, i.e., devoid of enhancer and promoter sequences minimizing the danger of aberrant regulation phenomena;
- each set of $F_m - F_n$ mutants constitutes a unique genomic tag permitting, by multiplex RMCE, the independent introduction of transgenes at multiple predetermined addresses.

We anticipate increased use of these tools and principles not only for the design of more efficient biotechnological expression systems but also for the rational creation of transgenic animals where multiple copy integration at random sites and the subsequent silencing phenomena are still a pertinent problem.

Acknowledgements

We thank all the colleagues who, over the years, have contributed to the development of the RMCE principle, most notably Thomas Schlake, Michaela Iber and Dirk Schübeler. Special thanks go to Henry Heng (Center of Mol. Medicine and Genetics, Wayne State University) for his introduction to the halo-FISH-technology. Work in the author's lab was supported by a grant from the DHGP II-program (BMBF-No. 01 KW 0003) and DFG-grants Bo-419/6-1 and /6-2.

Abbreviations

β-geo	fusion between the functions of β-galactosidase and the neomycin resistance-marker
FISH	fluorescence *in situ* hybridization
FRT	Flp-recombinase target site
GOI	gene of interest
HR	homologous recombination
huIFN-β	human interferon β (encoded by the IFNB1 gene)
hygtk	hygromycin-resistance/HSV-thymidine kinase fusion gene
IR	illegitimate recombination
kb	kilobasepairs
loxP	Cre recombinase target site
ORI	origin of replication
pac	puromycin-acetyl-transferase gene
RG	reporter gene

RMCE   recombinase-mediated cassette exchange
S/MARs   scaffold/matrix attachment regions (also called SARs or MARs)

References

1. Bode, J., Fetzer, C.P., Nehlsen, K., Scinteie, M., Hinrich, B.-H., Baiker, A., Piechazcek, C., Benham, C. and Lipps, H.J.(2001) Int. J. Gene Ther. Mol. Biol. 6, 33–46 (http://www.gtmb.org).
2. Garrick, D., Fiering, S., Martin, D.I.K., and Whitelaw, E. (1998) Nat. Genet. 18, 56–59.
3. Bode, J., Schlake, T., Iber, M., Schübeler, D., Seibler, J., Snezhkov, E., and Nikolaev, L. (2000) Biol. Chem. 381, 801–813.
4. Baer, A. and Bode, J. (2001) Curr. Opin. Biotechnol. 12, 473–480.
5. Cremer, T. and Cremer, C. (2001) Nat. Rev. Genet. 4, 292–301.
6. Bode, J., Schlake, T., Ríos-Ramírez, M., Mielke, C., Stengert, M., Kay, V., and Klehr-Wirth, D. (1995) In: Berezney, R. and Jeon, K. W. (eds.) International Review of Cytology,162A, Academic Press, pp. 389–453.
7. Bode, J., Bartsch, J., Boulikas, T., Iber, M., Mielke, C., Schübeler, D., Seibler, J. and Benham, C. (1998). Gene Ther. Mol. Biol. 1, 551–880 (http://www.gtmb.org/volume1/29_bode.htm).
8. Gasser, S.M. and Laemmli, U.K. (1987) Trends Genet. 3, 16–22.
9. Kay, V. and Bode, J. (1995) In: Papavassiliou, A.G. and King, S. L. (eds.) Methods in Molecular and Cellular Biology, Inc. Wiley-Liss, pp. 389–453.
10. Dietz-Pfeilstetter, A., Arndt, N., Kay, V. and Bode, J. (2003) Transgenic Res. 12, 83–99.
11. Bode, J., Benham, C., Ernst, E., Knopp, A., Marschalek, R., Strick, R., and Strissel, P. (2000) J. Cell. Biochem. 35, 3–22.
12. Dang, Q., Auten, J., and Plavec, I. (2000) J. Virol. 74, 2671–2678.
13. Fernandez, L.A., Winkler, M., and Gosschedl, R. (2001) Mol Cell Biol. 21, 196–201.
14. Heng, H.H.Q and Tsui, L.C. (1998) J. Chromatog. A 806, 219–229.
15. Iber, M., Schübeler, D., Seibler, J., Höxter, M. and Bode, J. (1999) Trends Genet.; TTO, 1999, 1:70:T01668 (http://tto.trends.com).
16. Perucho, M., Hanahan, D., and Wigler, M. (1980) Cell 22, 309–317.
17. Folger, K., Thomas, K., and Capecchi, M.R. (1984) CSH Sym. Quant. Biol. 49, 123–138.
18. Mielke, C., Kohwi, Y., Kohwi-Shigematsu, T., and Bode, J. (1990) Biochemistry 29, 7475–7485.
19. Milot, E., Belmaaza, A., Rassart, E., and Chartrand, P. (1994) Virology 201, 408–412.
20. Mielke, C., Maass, K., Tuemmler, M., and Bode, J. (1996) Biochemistry 35, 2239–2252.
21. Baer, A., Schübeler, D., and Bode, J. (2000) Biochemistry 39, 7041–7049.
22. Shera, K.A., Shera, C.A., and McDougall, J.K. (2001) J. Virol. 75, 12339–12346.
23. Hwang, L.-H.S. and Gilboa, E. (1984) J. Virol. 50, 417–424.
24. Lakso, M., Pichel, J.G., Gorman, J.R., Sauer, B., Okamoto, Y., Lee, E., Alt., F.W., and Westphal, H. (1996) Proc. Natl. Acad. Sci. USA 93, 5860–5865.
25. O'Gorman, S., Fox, D.T., and Wahl, G.M. (1991) Science 251, 1351–1355.
26. Buchholz, F., Angrand, P.O., and Stewart, A.F. (1998) Nat. Biotechnol. 16, 657–662.
27. Ringrose, L., Angrand, P.O., and Stewart, A.F. (1997) Eur. J. Biochem. 248, 903–912.
28. Schübeler, D., Maass, K., and Bode, J. (1998) Biochemistry 37, 11907–11914.
29. Schlake, T. and Bode, J. (1994) Biochemistry 33, 12746–12751.
30. Seibler, J., Schübeler, D., Fiering, S., Groudine, M., and Bode, J. (1998) Biochemistry 37, 6229–6234.
31. Feng, Y.Q., Seibler, J., Alami, R, Eisen, A., Westerman, K.A., Leboulch, P., Fiering, S., and Bouhassira, E.E. (1999) J. Mol. Biol. 292, 779–785.
32. Gagneten, S., Le, Y., Miller, J., and Sauer, B. (1997) Nucl. Acids Res. 25, 3326–3331.
33. Taniguchi, M., Sanbo, M., Watanabe, S., Naruse, I., Mishina, M., and Yagi, T. (1998) Nucl. Acids Res. 26, 679–680.
34. Bouhassira, E.E., Westermann, K., and Leboulch, P. (1997) Blood 90, 3332–3344.
35. Kolb, A.F. (2001) Anal. Biochem. 290, 260–271.

36. Siegel, R.W., Jain, R., and Bradbury, A. (2001) FEBS Lett. 499, 147–153.
37. Eszterhas, S.K., Bouhassira, E.E., Martin, D.I.K., and Fiering, S. (2002) Mol. Cell. Biol. 22, 469–479.
38. Karreman, S., Hauser, H., and Karreman, C. (1996) Nucl. Acids Res. 24, 1616–1624.
39. Verhoeyen, E., Hauser, H., and Wirth, D. (2001) Hum. Gene Ther. 12, 933–944.
40. Yu, S.-F., von Rüden, T., Kantoff, P.W., Garber, C., Seiberg, M., Rüther, U., French Anderson, W., Wagner, E.F., and Gilboa, E. (1986) Proc. Natl. Acad. Sci. USA 83, 3194–3198.
41. Göetze, S., Hüesemann, Y., Baer, A., and Bode, J. (2003) Functional Characterization of Transgene Integration Patterns by Halo-FISH: Electroporation versus Retroviral Infection. Biochemistry, in press.

CHAPTER 21

Intracellular targeting of antibodies in mammalian cells

Quan Zhu and Wayne A. Marasco

Department of Cancer Immunology and AIDS, Dana-Farber Cancer Institute;
Department of Medicine, Harvard Medical School, 44 Binney St., Boston, MA 02115, USA,
Tel.: +1-617-632-2153; Fax: +1-617-632-3889. E-mail: Wayne_Marasco@dfci.harvard.edu

1. Introduction

Antibodies produced by the humoral immune system are so diverse that they are essentially able to bind target molecules of any nature whether they are proteins, nucleic acids, carbohydrates, or lipids. While diverse, each individual antibody produced maintains high specificity and affinity to its antigen. Based on their unique properties, antibodies have been widely used as powerful tools in diagnostic and therapeutic applications as well as in basic biomedical research. It is now possible to manipulate genes encoding antibodies into different forms utilizing recombinant DNA technology. Antibodies can be expressed not only as full-length intact forms but can also be expressed as antigen-binding domains such as Fab fragments, consisting of the entire light chain and partial heavy chain, as single-chain variable region fragments (sFv), consisting only of the heavy and light variable regions linked by a short interchain linker of usually 15 amino acids [1–3] or even as domain antibodies consisting of only the heavy chain variable region [4] (Fig. 1). When fused to well-characterized intracellular protein localization/trafficking signal peptides, antibodies can be expressed in different subcellular compartments depending on the trafficking signals that are used. Intracellular antibodies, termed intrabodies, represent a new family of molecules that can be expressed within the context of a cell to define or mediate function(s) of a particular gene product [1,5]. Since its first applications a decade ago, intrabody technology has been utilized in a variety of research areas such as signal transduction, cancer, neurodegenerative disease as well as AIDS and may hold great potential in functional genomics as well as in gene therapy applications [1,5–7]. In this chapter, we will provide an overview the concept and principles of intrabody technology. Further discussion will be focused on the critical parameters that have been identified that result in an effective intrabody molecule as well as strategies in the selection of functional intrabodies. The chapter ends with examples of recent intrabody applications.

2. Concept of intrabodies and their mechanism of action

By definition, intrabodies are antibodies that are directed against target molecules that are inside a cell and expressed within a particular cellular compartment as

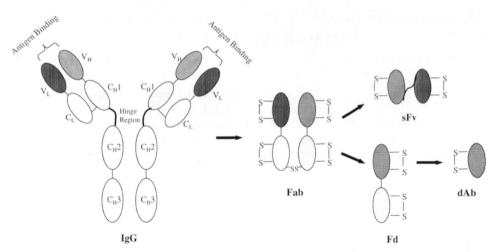

Fig. 1. Schematic diagram of an intact antibody (IgG) and its antigen binding fragments. V_H, heavy-chain variable region; C_H, Heavy-chain constant region; V_L, light-chain variable region; C_L, light-chain constant region; sFv, single-chain variable region fragment; dAb, domain antibody; Fd, partial heavy chain ($V_H + C_H1$).

directed by the intracellular localization signals genetically fused to the N- or C-terminus of the antibody [5,8]. It has been demonstrated that, as a covalently-linked minimal antigen binding form, a single-chain Fv (sFv) antibody fragment generally can maintain the binding specificity and affinity of its parental full length IgG molecule [2,3]. This finding has made sFv antibodies the most common and favorite form of intrabodies. The reason is obvious—a sFv antibody can be expressed from a single cDNA, thus, it can be genetically manipulated with ease, especially when addition or deletion of an intracellular localization signal peptide is required. Expression of an intrabody from a single gene also simplifies the regulation of its expression as well as ease of using established gene transfer procedures for their transient or stable expression *in vitro* or *in vivo*. The classical intracellular trafficking signals that have been identified by molecular cell biologists are an indispensable part of the intrabody technology. When genetically fused to an intrabody, these short polypeptide "signals" can direct the intrabody to specific subcellular compartments where its target antigen is localized. These trafficking signals can also help an intrabody to re-direct or interfere with the subcellular localization of its target antigen upon their interaction. To date, intrabodies have been targeted to the secretory pathway where they can be specifically retained in either the endoplasmic reticulum (ER) or *trans*-Golgi network [9,10]. Intrabodies have also been targeted to the nucleus, mitochondria, cytoplasmic leaflet of the plasma membrane, peroxisomes, and, of course, the cytosol itself ([11–15], reviewed in [5]).

Although intrabodies can be theoretically generated against any intracellular molecule, it is those intrabodies that modulate function of target proteins that have generated the most interest and are the main focus of this chapter. In principle, upon interaction with its target, an intrabody can modulate target protein function or achieve functional knockout by one of the following mechanisms (reviewed in [1,5,7]):

- Modulate protein enzymatic activities by inhibiting an activity directly, by sequestering substrate, or by maintaining the catalytic site in an active or inactive conformation.
- Disrupt the signaling pathway of a target protein through interference with its protein–protein, protein–DNA, or protein–RNA interactions.
- Divert a target protein from its normal site of action such as sequestering nuclear localized proteins in the cytosol, targeting of cytosolic proteins to the nucleus, and retention of secreted or cell surface-expressed proteins in the ER.
- Accelerate degradation of target protein.

In addition to functioning as neutralizing antibodies to inactivate the target function, it has been recently reported that intrabodies can also supply a gain of function for their target. De Fromentel *et al.* [16] showed that anti-p53 sFv intrabodies were able to partially restore the transactivating activity of a p53 mutant. Furthermore, intrabodies have been shown to act as transcription activators and induce gene expression in a sFv-mediated manner when fused to a DNA-binding domain [17]. These examples reinforce that intrabody technology has great potential as well as diversity and are finding an increasing number of new applications within the field, especially in the coming age of functional genomics.

3. Critical parameters for a successful sFv intrabody

What are the most important criteria for choosing a sFv for use as an intrabody? In the past, the same primary considerations were assumed to apply to those sFvs to be used as intrabodies as they would be for any other applications, namely, high specificity and antigen-binding affinity of the sFv binding site. However, when expressed inside a cell, the behavior of sFv antibodies is often unpredictable and does not always correlate with their *in vitro* binding affinities [15–19]. For example, in the cases of anti-HIV-1 Rev (regulator of virion protein) [18] and anti-p53 sFv [16] intrabodies, the authors speculated that the reason the lower affinity sFv intrabodies exerted more potent activity might be due to differences in the binding site (epitope). In other words, the intrabody with lower affinity blocked the activation domain of Rev more directly [18] or activated p53 by inducing a more favorable conformational change [16]. With our study of anti-caspase-7 intrabodies, we found a strong correlation between intrabody efficiency and the turnover rate as well as the related steady-state protein accumulation of a sFv intrabody [15]. The results suggest that intracellular stability of an intrabody, not affinity, is highly critical for its efficacy. This observation is further supported by a later study where a panel of anti-caspase-3 sFv intrabodies was compared for their affinity, intracellular stability and efficacy [19]. Taken together, a useful description of an effective intrabody should include specificity, affinity, and intracellular stability where the stability is the determining factor for intrabody efficacy. When utilized for inactivation of a protein's enzymatic activity or disruption of intracellular protein–protein interactions, the specific target site to which the intrabody binds should be carefully considered if structural/functional information is available. However, in the absence of any knowledge of these parameters, an alternative approach is to screen a panel of intrabodies directly for the desirable biological effect.

4. The source and selection of sFv antibody for intrabody construction

4.1. Source of sFv antibody fragment for intrabody construction

The starting material for construction of intrabodies has traditionally come from the genes encoding the V_H and V_L fragments of immunoglobulins within well characterized hybridoma cell lines. The V_H and V_L fragments are genetically joined together as sFv antibodies through use of recombinant PCR techniques or step-wise subcloning. In most cases, these antibodies are of rodent origin and have known specificity (epitope information) and affinity. However, their availability to any given target protein is limited, and murine monoclonal antibodies could elicit human anti-mouse antibody response (HAMA response) if used in clinical therapeutic applications.

Recent advances in recombinant antibody engineering have made sFv genes available from combinatorial libraries of human heavy and light chain genes. Through reverse transcription and recombinant PCR, sFv genes are engineered by amplifying rearranged human antibody V-genes from B-cells of either immune or naïve donors (reviewed in [7,20–22]). Once established, sFv genes can be cloned into expression vectors to construct sFv cDNA libraries. The choice of vectors is dependent upon the formats of which sFv is displayed and selected (discussed in the next section). The large unimmunized (naïve) human sFv libraries offer a rich source for isolation of antibodies of human origin to virtually any antigen, including self-antigens. The immune libraries created from individuals with autoimmune diseases, tumor transformations or viral infections allow isolation of high-affinity disease-specific antibodies on a relatively small scale. The availability of sFv cDNA display libraries has greatly expanded the capability of immune-based technologies, intrabody technology for target validation and functional genomics applications in particular, with its unlimited sFv resources and rapid sFv isolation.

4.2. The selection of sFv antibody for intrabody construction

Since the intracellular stability is a critical determinant for intrabody efficacy, one has to take it into consideration when designing the selection strategy for sFv intrabodies. The main factors contributing to sFv stability are disulfide bond formation and correct protein folding. The primary amino acid sequence of a sFv fragment also contributes directly to its stability in a yet unpredictable way [23]. In its natural environment, an antibody is directed to the secretory pathway where it is synthesized and processed to allow correct folding and disulfide formation. Once released outside a cell, an antibody maintains its correct conformation and thus stable in the oxidized extracellular environment. However, in many applications, a intrabody is expressed in the cytosol and targeted to subcellular compartments other than the secretory pathway. In the reducing environment of the cytosol, formation of intrachain disulfide bonds (Fig. 1) does not occur, and maintenance of correct folding is inefficient [24], which, in turn, could result in rapid protein turnover or protein aggregation. In either case, an insufficient amount of functional intrabody accumulates to exert the desired biological activity.

Because of the high expectations and broad demands for the intrabody technology, efforts have been made recently to develop stable Fv frameworks that can fold efficiently in the reducing environments of a cell. Two strategies involving a rational approach and a selection approach have been reported. With the rational approach, special sFv-phage display libraries are genetically engineered based on known sequence information and current knowledge of sFv antibodies, with the hope that such libraries may consistently yield effective sFv intrabodies. Two recent reviews [23,25] present detailed discussions on the rational approach, which relies mainly on recombinant antibody engineering. The selection approach is based on the current availability of different sFv display libraries [23,25]. The strategy is to bias the selection of sFv towards increased stability and/or *in vivo* performance. The available sFv display technologies and the sFv intrabody selection strategies will be briefly reviewed below.

4.2.1. Selection by phage display

Phage display technology has proven to be a powerful means for isolation of desired sFv [26,27]. Upon cloning into a phage vector, sFv antibodies are fused to a phage coat protein, allowing the display of sFv antibodies on the surface of phage particles. Selection of sFv antibodies with high affinity and specificity can be achieved through multiple rounds of panning of antibody-expressing phage particles against antigens that are presented on the cell surface, immobilized on resins or blotted on nitrocellulose membranes, or as biotinylated proteins/peptides in solution ([28–30], reviewed in [20–23]). An *in vivo* selection procedure in the mouse has recently been reported for selection of sFv against a tissue-specific antigen from a phage-sFv library [31]. Functional sFv (or Fab) antibodies have been selected from the phage display library and used in the intrabody technology [15,19,32–34]. However, the suitability of each selected sFv as an intrabody cannot be predicted directly from phage selection since the selection does not take place within the intracellular environment. Therefore the selection strategy and optimal conditions must be carefully considered and experimentally determined for each individual antigen. Two recent studies revealed that the intracellular stability and performance of sFv antibodies correlate with their *in vitro* stability, measured by denaturant-induced protein unfolding [19,35]. Thus, in order to obtain sFv antibodies that are stable for intracellular expression, reducing conditions could be applied during panning as well as during *in vitro* characterization of protein stability. Recently isolation of intracellularly stable sFv fragments was accomplished by introducing an additional intracellular screening step following the classical ge panning [23,36]. This and other intracellular selection methods will be discussed further in Section 4.2.3.

4.2.2. Selection by in vitro display

Recently, other display technologies have been developed as alternatives to the phage display methods. One of these technologies is *in vitro display*. Two major *in vitro* display technologies are ribosome display and its variation—mRNA display, reviewed by Amstutz et al. [37]. Since their development, these technologies have been extended from displaying of peptides to proteins. The starting material for both technologies is usually dsDNA fragments that are amplified by PCR from a cDNA library and both involve *in vitro* transcription followed by RNA isolation and *in vitro* translation. The differences are in the methods used to couple the newly translated peptide/protein to its genetic information. In the case of ribosome display, peptides are non-covalently associated with the ribosomes and encoding mRNAs. This is accomplished by the absence of a stop codon as well as by keeping the complexes at low temperature and high magnesium concentration [38]. In the case of mRNA display, the nascent polypeptide is coupled to its mRNA through a covalent linkage introduced by a short DNA linker with a puromycin moiety at the 3′ end of mRNA [39]. The peptide–mRNA–ribosome tertiary complexes in ribosome display or the mRNA–protein fusions in mRNA display can be used to select displayed protein of interest such as sFv by direct binding to its immobilized target. Following selection, the genetic information linked to the bound complexes is PCR-rescued and can be used as the starting material for the next round of selection.

Using ribosome display, Hanes and Pluckthun [38] were able to select a sFv fragment with a 10^8-fold enrichment through five cycles of transcription, translation, affinity-selection, and PCR. The main advantages of *in vitro* display technology include: (a) fast selection cycles; (b) very large libraries, up to 10^{14} members can be handled; and (c) *in vitro* directed protein evolution can be accomplished through multiple rounds of randomization by error-prone PCR. Depending on the selection criteria, one can achieve maturation of protein affinity [40] and general protein stability by selecting under reducing conditions [41]. The technology may be limited by the stability of mRNA. In addition, stable intrabodies cannot be specifically selected due to the lack of intracellular environment.

4.2.3. Selection by intracellular expression

Selecting sFv antibody fragments directly in an intracellular environment would be the best way to isolate both stable and functional intrabodies. To this end, several different strategies have been utilized.

In a model system, Gargano and Cattaneo [36] used a phage display library to isolate an antibody, D7αRT, which neutralizes the reverse transcriptase (RT) of HIV-1 and other retroviruses. Upon cloning into a mammalian expression vector as a partial heavy chain Fd fragment (Fig. 1), D7αRT was then mixed with plasmids encoding a small pool of other anti-RT sFv fragments, and transfected into Balb/c-3T3 cells. The transfected cells were challenged with retroviral infections. Cells that harbor anti-RT neutralizing antibody would have a survival advantage and be enriched. PCR analysis proved that D7αRT Fd fragment could be rescued from the surviving cells, thus providing a selection procedure by which an intrabody can be isolated through intracellular functional selection. Due to the involvement of the transfection step, only a small pool of sFv fragments can be expressed and analyzed in mammalian cells. A pre-selection step becomes necessary.

Several groups have adapted the yeast two-hybrid system for screening of sFv fragments that are intracellularly soluble, stable, and with efficient antigen-binding capacity [42–44]. The sFv fragments are fused to transcription activation domain such as VP16 or Gal4 activation domain. The antigens are fused to a DNA-binding domain, such as LexA or Gal4 DNA-binding domain. Activation of a reporter gene expression inside a cell is mediated through the interaction between the antigen and antibody. Using a known sFv antibody against the artichoke mottled crinkle virus (AMCV), Visintin et al. [42] were able to demonstrate selection efficiency of 1 in half a million. It was further shown that the two-hybrid system could be used in mammalian cells by co-transfection [42]. A variation of this method, termed intra-cellular antibody capture technology (IACT), is to isolate functional sFv intrabodies by using phage display as a preselection step followed by yeast two-hybrid screening [44]. Again, a pre-selection step would be necessary to accommodate the trans-fection/transformation efficiency as well as to reduce the number of cells handled. A sFv library established using viral vectors could potentially be used directly for mammalian intracellular screening of sFv intrabody without a size limitation.

Mossner et al. [45] adapted the protein fragment complementation assay (PCA) [46] for selection of antigen–antibody pairs in the cytoplasm of Escherichia coli. In this system, sFv antibody and target antigens are genetically linked to two dissected halves of the bacterial dihydrofolate reductase (DHFR). Interaction between the antibody and antigen will restore DHFR activity and allow cell growth on minimal medium. As compared to the phage display technique, the advantage of this system is that no antigen isolation is required and the stability of a sFv may be assessed directly because a sFv has to be reasonably stable and soluble within the reducing environment of the bacterial cytoplasm in order for the assay to work. Direct selection of sFv based on a specific function such as activating β-galactosidase activity in E. coli has also been reported [47]. However, such a selection is application-specific. In general, eukaryotic cells differ significantly from E. coli. Hence, the yeast system and ultimately a mammalian system would be the preferred choice for intrabody selection.

5. Recent applications

The ability of antibodies to function inside cells as intrabodies has greatly expanded the utility of antibodies in basic research and potentially in clinical applications. Intrabody technology has been applied in a wide range of biomedical research settings and has shown great potential and exciting results in studies of infectious disease, cancer, neurodegenerative research, autoimmunity, and in plant biology. In the next section, we will provide some samples of intrabody application in each field.

5.1. Intrabodies in AIDS research

The main strategy of utilizing intrabodies in anti-HIV research is to achieve inhibition of the virus replication and/or viral infection. This can be accomplished by either interference with function of the viral proteins or by modulation of cellular proteins that are involved in viral pathogenesis.

Many HIV-1 proteins have been the targets of intrabodies (reviewed in [5,7,48,49]). In fact, study of the HIV-1 envelope protein gp120 was one of the early "proof of principle" examples for the utility of intrabodies [9]. It has been demonstrated that anti-gp120 (or gp41) intrabodies could block normal processing of the viral envelope protein gp160 and its incorporation into virions. As a result, virus-mediated syncytium formation was diminished and the infectivity of progeny HIV particles was reduced. Intrabodies against other viral structural proteins such as matrix (p17) and nucleocapsid proteins also show ability to inhibit HIV-1 infection in vitro. Sequestration of HIV-1 regulatory proteins Tat (transactivator) and Rev within the cytosol by intrabodies has lead to the inhibition of transcription activation of the HIV-1 LTR by Tat or the inhibition of Rev-mediated nuclear export of viral RNA. Anti-HIV-1 reverse transcriptase (RT) intrabodies were shown to exhibit their anti-viral effects by either inhibiting DNA polymerase activity of HIV-1 RT or by providing a steric hindrance on the movement of RT along the RNA template. The presence of anti-HIV-1 integrase (IN) intrabodies provided resistance to HIV-1 infection in human T-lymphocytes, possibly through neutralization of IN activity prior to integration. HIV-1 virion infectivity factor (Vif) is a recently reported intrabody target [50]. It was shown that nonpermissive cells expressing anti-Vif sFv intrabody produced viral particles with a defect in infectivity at the level of reverse transcription.

The other arm of anti-HIV-1 strategy using intrabodies is to target cellular proteins involved in viral pathogenesis. Compared to the HIV genome, cellular targets are less prone to somatic mutations and hence are less likely to escape from gene therapeutic reagents. The cellular targets utilized in anti-HIV applications include CCR5 and CXCR4, two major HIV-1 co-receptors involved in HIV-1 entry [7,51,52]. Expression of intrabodies lead to decreased cell surface expression of either CCR5 or CXCR4, which, in turn, resulted in significant protection of the intrabody-expressing cells from challenge with the corresponding co-receptor-dependent R5 and X4 virus strains, respectively. Inhibition of Tat-mediated transactivation and HIV replication has also been accomplished by anti-human Cyclin (hCyclin) T1 sFv intrabodies [34]. Tat binds cooperatively with hCyclin T1, a regulatory partner of cyclin-dependent kinase 9 (cdk9) in the positive transcription elongation factor (P-TEFb) complex, to the transcription response element (TAR) and promotes HIV transcription elongation. Expression of hCyclin T1-specific sFv intrabodies selectively inhibits Tat-mediated transactivation and HIV replication. More importantly, the normal interaction between hCyclin T1 and cdk9 is maintained. These intrabodies may provide new and useful reagents for anti-AIDS gene therapy.

Like combination chemotherapy, it has been hypothesized that HIV-1 inhibition could be established more efficiently if a combination of antibodies against different targets is used to affect different stages of the viral life cycle [48]. Indeed, it has been recently reported that, when used in combination, sFv intrabodies against CXCR4, RT, and IN provide improved protection to SupT1 human T lymphoma cells from HIV-1 challenge than when used individually [53].

5.2. Intrabodies in cancer research

Oncogene products are generally components of signal transduction pathways - cell surface receptors, guanine nucleotide binding proteins, transcription factors, to name a few, that are directly involved in deregulation of cell growth through their over expression or deregulated activity as a result of mutations in these genes. Intrabodies, due to their ability to specifically modulate the function of endogenous proteins, are effective tools to investigate whether knockout of the oncogene product may lead to reversal of the transformed cell phenotype (reviewed in [54–56]). To date, ER-targeted sFv intrabodies have been utilized to inhibit surface expression of the α subunits of the human interleukin 2 receptors (IL-2Rα) that are overexpressed in adult T-cell leukemia (ATL) cells from human T-lymphotropic virus type I (HTLV-1) infected patients. In these studies, complete phenotypic knockout of IL-2Rα on HTLV-1 transformed cells did not cause any change in the growth properties of these cells thereby demonstrating that IL-2Rα overexpression was not centrally involved in the transformation process. Overexpression of erbB-2 transmembrane protein, a member of the epidermal growth factor receptor family, has been documented in approximately 30% of ovarian, breast, and prostate cancers. ER-directed erbB-2 intrabodies have clearly been shown to reverse erbB-2 transformation of cells [57] or cause targeted tumor eradication both *in vitro* and *in vivo* [58]. Based on this observation, the first clinical application of intrabody technology in the context of human disease was conducted in a Phase I human trial utilizing recombinant adenovirus encoding anti-erbB-2 sFv intrabodies [59]. Although no clinical benefit was demonstrated in this Phase I study, which was perhaps limited by the adenovirus gene transfer vehicle used, intrabody expression was clearly demonstrated and there were no serious adverse events reported.

Another category of oncogenes encodes cytosolic proteins involved in signal transduction (p21^{ras}), transcription regulation (c-Myb) and tumor suppression (p53), or regulation of DNA replication (cyclin E). The Ras protein is a guanine-binding protein (p21^{ras}) located in the cytoplasmic leaflet of the plasma membrane, which is critical for its function in the signal transduction pathway. The use of intrabodies to inhibit p21^{ras} was first demonstrated in 1993 [12] when the insulin-induced, p21^{ras}-dependent meiotic maturation was inhibited in *Xenopus laevis* oocytes following intracellular expression of anti-p21^{ras} intrabody by microinjection of intrabody mRNA. Inhibition of p21^{ras} activity by the anti-p21^{ras} intrabodies was proposed to occur either directly [12,13] or by diverting p21^{ras} protein from the plasma membrane to the cytosol [60]. Importantly, it was later shown that anti-p21^{ras} intrabodies could induce apoptosis in human cancer lines and elicit sustained tumor regression in nude mice [61], thus demonstrating that intrabodies against Ras could represent a potentially attractive clinical approach for cancer therapy. The p53 protein is a tumor suppressor that induces cell cycle arrest and/or programmed cell death. Intrabodies against p53 have been reported to be able to either inactivate its transactivation function [62] or restore transcriptional activity of the mutant p53 [16]. However, the latter approach has more clinical relevance because inactivation of normal p53 activity by either mutation or through its interaction with other oncogenic proteins has been associated with many human cancers. Restoration of normal p53 activity would be required for an anti-p53 intrabody-based cancer therapy. Other important oncogenic proteins targeted by intrabodies include c-Myb and cyclin-E [63,64]. In both cases, expression of nuclear-targeted intrabodies has been reported to inhibit growth of cancer cell lines. This could be a significant first step in utilizing intrabody technology for controlling hematopoietic malignancies where c-Myb overexpression has been reported, or introducing an alternative therapy for breast cancer where both c-Myb and cyclin-E overexpression have been observed.

A new application of intrabody technology in cancer therapy is to achieve functional knockout or down regulation of the multidrug resistance (MDR) gene product since MDR proteins appear to be the main cause for resistance of cancer cells to chemotherapy. In an attempt to mediate MDR gene function, Heike *et al.* [65] showed that cytoplasmic expression of a sFv intrabody against P-glycoprotein, a MDR gene product, increased sensitivity of the transfected cells to adriamycin in a 10-day colony formation assay but not in a 3-day standard colorimetric assay using 3-(4,5-dimethylthiazol-2-yl)-2,5-diphenyltetrazolium bromide (MTT). The authors hypothesized that the difference in results may be due to growth retardation of the transfected cells caused by a cross-reactivity of the intrabody employed with p815$^{c\text{-erbB2}}$. Nevertheless, this initial study indicates that use of intrabodies to target MDR could be a valid approach to increase the effectiveness of cancer chemotherapy.

5.3. Intrabodies in neurodegenerative disease research

Neurodegenerative diseases represent another exciting field for intrabody application. Abnormal protein aggregation in affected neurons is a hallmark of several central nervous system disorders including Huntington's (HD), Alzheimer's and Parkinson's diseases. For example, intracellular protein aggregates are found in human HD brains at autopsy. It is believed that protein aggregation and HD progression are associated with the length of a polyglutamine (polyQ) track encoded by the huntingtin gene. Normal huntingtin protein (Htt) has a polyQ track of 35 amino acids or less. When the polyQ track is extended over 40 amino acids in length, Htt starts to form aggregates [66]. When the abnormal Htt-GFP fusion protein was coexpressed with a human sFv intrabody specific to the N-terminal 17 residues of Htt protein, Lecerf *et al.* [33] were able to show a dramatic reduction in the number of mutant Htt protein aggregates in a cellular model of HD disease. It was hypothesized that the binding of intrabody may help to maintain Htt protein in a soluble state. The antibody–antigen complex is then subjected to the normal protein degradation. Khoshnan *et al.* [67] also reported inhibition of Htt aggregation as well as cell death by a sFv intrabody. Interestingly, such inhibition can only be observed with the sFv intrabody interacting with the polyproline (polyP) domain of Htt. Intrabodies recognizing the polyQ domain actually stimulate Htt aggregation and promote cell death. Whether the anti-Htt intrabodies act by excluding binding of other protein partners of Htt, by accelerating protein degradation, or by stabilizing Htt in a soluble state, thus inhibiting aggregate formation, remains unclear at the present time. Use of intrabodies in these studies demonstrated that intrabodies not only could be valuable tools in dissecting the function of Htt protein but also could be potential therapeutic agents in treating neurological disorders. However, testing the effect of intrabodies in HD *in vivo* will require the development of an efficient central nervous system gene delivery vehicle.

5.4. Intrabodies in transplantation

For the immune system to recognize a foreign antigen, the antigen has to be processed and presented on the surface of cells, a process called antigen presentation. Antigen presentation is also a main cause for rejection in organ transplantation when allogeneic material/organs are used or in autoimmune diseases where "self" antigens are recognized as "foreign" and induce immune responses. There clearly exist clinical needs for regulation of antigen presentation. Intrabodies can be utilized to modulate directly either the expression of cell surface molecules that are involved in antigen presentation or components of the antigen processing system that are responsible for the cell surface display of the antigenic moieties.

In an attempt to reduce/eliminate hyperacute rejection of porcine tissues transplanted into primates, two groups [68,69] have investigated intrabody-mediated phenotypic knock out of the α1,3Galactosyltransferase, the enzyme responsible for generating the major carbohydrate target epitope Galα1-3Gal of human xenoreactive antibodies to pig cells. With the use of intracellular sFv antibodies specific for α1,3Galactosyltransferase in two different cell lines, both groups reported the reduced expression of the Galα1-3Gal epitope on the cell surface and increased resistance to cell lysis by human serum. These findings indicate the potential of intrabodies to protect transplanted cells from antibody responses.

The major histocompatibility complex class I (MHC I) antigens are targets for skin allograft rejection and it has been proposed that the allograft rejection can be controlled through reduction of MHC I molecules on the surface of unrelated donor cells. To this end, Mhashilkar *et al.* [70] were able to show a marked down-regulation of MHC I expression in a variety of human cell lines, including primary human keratinocytes, through intracellular expression of an ER-targeted anti-human MHC I sFv antibody. It remains to be seen if the intrabody-mediated reduction of MHC I expression could lead to prolonged survival of keratinocyte allografts *in vivo*. It should be noted that the anti-human MHC I intrabodies or intrabodies against other components involved in antigen presentation should also prove to be beneficial in regulating human autoimmune diseases.

6. Future perspectives

Space limitations do not permit the listing of every intrabody application to date. It is apparent from the above examples, however, that intrabody technology has been utilized in a wide range of research studies. With the near completion of the human genome sequencing, genes encoding more than 30,000 open reading frames have been revealed [71,72] and it is likely that alternative splicing of these genes will greatly increase the number of proteins expressed in a cell or organism. There is an increasing need to characterize the function of these gene products and to understand how these proteins interact/communicate with each other. With such an understanding, one may eventually gain knowledge into the mechanism(s) by which any given protein contributes to a normal cell life cycle and/or to a disease process, which, in turn, could lead to the validation of such protein as a potential target in therapeutic applications [15]. Intrabody technology presents itself as a very attractive tool with great potential in functional genomics at the protein level (i.e. proteomics) by modulating/blocking the activities of these newly identified proteins and dissecting the interactions among them. Compared to other gene manipulation methods, such as gene knock out or RNA-based technologies (ribozyme, antisense RNA or RNA interference), intrabodies have the unique advantage of targeting proteins in different cellular compartments as well as specific structural or functional motifs of a protein. These advantages are particularly useful in evaluating the post-translational modifications of a protein as well as in studying proteins that are multifunctional, have different protein binding partners [34] and are involved in different signal transduction pathways.

As discussed previously in Sections 3 and 4, selecting a sFv intrabody that is soluble, stable and functional is a key to a successful application. In addition to establishing stable sFv cDNA libraries through rational design in antibody engineering and/or applying appropriate selection strategies, another possible strategy that can increase correct folding and the overall stability of sFv intrabodies, is to co-express molecular chaperons [73] in the cytosol of mammalian cells. In E. coli, it has been shown that co-expression of a molecular chaperon-like molecule such as a derivative of the disulfide-bond isomerase DsbC [74] or expression of a sFv-maltose-binding protein fusion [75] promotes correct sFv folding and/or increases their solubility and stability.

Gene therapy application of intrabodies in animal studies or eventually as therapeutic agents in human diseases needs to overcome other hurdles that are common for gene therapy, such as development of efficient gene transfer vehicles and high level gene expression that is not only long-term, but also regulatable before intrabodies find widespread use. For example, a tetracycline-inducible system was used to regulate anti-c-Myb intrabody expression, and the results showed that the intrabody achieved regulatable inhibition of protein function [63]. Gene regulation will certainly allow intrabodies to have wider use in the future, especially when the intrabody target is an essential cellular protein that is required for normal cell growth. These advances combined with recent improvements of viral and non-viral vectors that are safer, less immunogenic and capable of achieving tissue-specific gene transfer through cell surface targeting should enable intrabody technology and its broad applications to flourish.

Acknowledgements

This work was supported by National Institute of Health Grants (5 R01 AI28785 and 5 R01 AI41954) to W. Marasco. Q. Zhu is an NRSA fellow (NIH T32 AI 07386). This work was also supported by the Dana-Farber Cancer Institute-Beth-Israel Deaconess Medical Center and Children's Hospital Center for AIDS Research (CFAR) Grant. The authors would like to express their gratitude to Dr. Aimee Tallarico for her help in designing Fig. 1 and careful reading of the manuscript, and to Ms. Christina Karunaratne for her administrative assistance.

Abbreviations

AIDS	acquired immunodeficiency syndrome
AMCV	the artichoke mottled crinkle virus
ER	endoplasmic reticulum
HD	Huntington's disease
HIV	human immunodeficiency virus
HTLV-1	human T-lymphotropic virus type I
IgG	an immunoglobulin type G molecule—an antibody
MDR	the multidrug resistance gene
MHC-I	the major histocompatibility complex class I molecules
MTT	3-(4,5-dimethylthiazol-2-yl)-2,5-diphenyltetrazolium bromide
sFv	single-chain variable region fragment of IgG
V_H	heavy-chain variable region of IgG
V_L	light-chain variable region of IgG

References

1. Chames, P. and Baty, D. (2000) FEMS Microbiol. Lett. 189, 1–8.
2. Bird, R.E., Hardman, K.D., Jacobson, J.W., Johnson, S., Kaufman, B.M., Lee, S.M., Lee, T., Pope, S.H., Riordan, G.S., Whitlow, M. (1988) Science 242, 423–426.
3. Huston, J.S., Levinson, D., Mudgett-Hunter, M., Tai, M.S., Novotny, J., Margolies, M.N., Ridge, R.J., Bruccoleri, R.E., Haber, E., Crea, R., Oppermann, H. (1988) Proc. Natl. Acad. Sci. USA 85, 5879–5883.
4. Ward, E.S., Gussow, D., Griffiths, A.D., Jones, P.T., and Winter, G. (1989) Nature 341, 544–546.
5. Marasco, W.A. (1998) In: Marasco, W.A. Intrabodies—basic research and clinical gene therapy applications, Springer, New York, pp. 1–22.
6. Jones, S.D. and Marasco, W.A. (1998) Adv. Drug Delivery Rev. 31, 153–170.
7. Marasco, W.A. (2001) Curr. Top. Microbiol. Immunol. 260, 247–270.
8. Biocca, S. and Cattaneo, A. (1995) Trends Cell Biol. 5, 248–252.
9. Marasco, W.A., Haseltine, W.A., and Chen, S.Y. (1993) Proc. Natl Acad. Sci. USA 90, 7889–7893.

586

10. Zhou, P., Goldstein, S., Devadas, K., Tewari, D., and Notkins, A.L. (1998) J. Immunol. 160, 1489–1496.
11. Mhashilkar, A.M., Bagley, J., Chen, S.Y., Szilvay, A.M., Helland, D.G., and Marasco, W.A. (1995) EMBO J. 14, 1542–1551.
12. Biocca, S., Pierandrei-Amaldi, P., and Cattaneo, A. (1993) Biochem. Biophys. Res. Commun. 197, 422–427.
13. Biocca, S., Pierandrei-Amaldi, P., Campioni, N., and Cattaneo, A. (1994) Biotechnology 12, 396–399.
14. Duan, L., Bagasra, O., Laughlin, M.A., Oakes, J.W., and Pomerantz, R.J. (1994) Proc. Natl. Acad. Sci. USA 91, 5075–5079.
15. Zhu, Q., Zeng, C., Huhalov, A., Yao, J., Turi, T.G., Danley, D., Hynes, T., Cong, Y., DiMattia, D., Kennedy, S., Daumy, G., Schaeffer, E., Marasco, W.A., Huston, J.S. (1999) J. Immunol. Methods 231, 207–222.
16. de Fromentel, C.C., Gruel, N., Venot, C., Debussche, L., Conseiller, E., Dureuil, C., Teillaud, J.-L., Tocque, B., and Bracco, L. (1999) Oncogene 18, 551–557.
17. Mary, M.-N., Venot, C., de Fromentel, C.C., Debussche, L., Conseiller, E., Cochet, O., Gruel, N., Teillaud, J.-L., Schweighoffer, F., Tocque, B., Bracco, L. (1999) Oncogene 18, 559–564.
18. Wu, Y., Duan, L., Zhu, M., Hu, B., Kubota, S., Bagasra, O., and Pomerantz, R.J. (1996) J. Virol. 70, 3290–3297.
19. Rajpal, A. and Turi, T.G. (2001) J. Biol. Chem. 276, 33,139–33,146.
20. Poul, M.A. and Marks, J.D. (1998) In: Marasco, W.A. (ed.) Intrabodies—basic research and clinical gene therapy applications, Springer, New York, pp. 23–46.
21. Griffiths, A.D. and Duncan, A.R. (1998) Curr. Opin. Biotechnol. 9, 102–108.
22. Hoogenboom, H.R. (2002) Methods Mol. Biol. 178, 1–37.
23. Cattaneo, A. and Biocca, S. (1999) Trends Biotechnol. 17, 115–121.
24. Biocca, S., Ruberti, F., Tafani, M., Pierandrei-Amaldi, P., and Cattaneo, A. (1995) Biotechnology 13, 1110–1115.
25. Worn, A. and Pluckthun, A. (2001) J. Mol. Biol. 305, 989–1010.
26. McCafferty, J., Griffiths, A.D., Winter, G., and Chiswell, D.J. (1990) Nature 348, 552–554.
27. Lowman, H.B., Bass, S.H., Simpson, N., and Wells, J.A. (1991) Biochemistry 30, 832–838.
28. Osbourn, J.K., Derbyshire, E.J., Vaughan, T.J., Field, A.W., and Johnson, K.S. (1998) Immunotechnology 3, 293–302.
29. Liu, B. and Marks, J.D. (2000) Anal. Biochem. 286, 119–128.
30. Chames, P., Hoogenboom, H.R., and Henderikx, P. (2002) Methods Mol. Biol. 178, 147–157.
31. Johns, M., George, A.J.T., and Ritter, M.A. (2000) J. Immunol. Methods 239, 137–151.
32. Gargano, N., Biocca, S., Bradbury, A., and Catteaneo, A. (1996) J. Virol. 70, 7706–7712.
33. Lecerf, J.M., Shirley, T.L., Zhu, Q., Kazantsev, A., Amersdorfer, P., Housman, D.E., Messer, A., and Huston, J.S. (2001) Proc. Natl. Acad. Sci. USA 98, 4764–4769.
34. Bai, J., Sui, J., Zgu, R.Y., Tallarico, A.S., Gennari, F., Zhang, D., and Marasco, W.A. (2003) J. Biol. Chem. 278, 1433–1442.
35. Worn, A., Auf der Maur, A., Echer, D., Honegger, A., Barberis, A., and Pluckthun, A. (2000) J. Biol. Chem. 275, 2795–2803.
36. Gargano, N. and Cattaneo, A. (1997) FEBS Lett. 414, 537–540.
37. Amstutz, P., Forrer, P., Zahnd, C., and Pluckthun, A. (2001) Curr. Opin. Biotech. 12, 400–405.
38. Hanes, J. and Pluckthun, A. (1997) Proc. Natl. Acad. Sci. USA 94, 4937–4942.
39. Roberts, R.W. and Szostak, J.W. (1997) Proc. Natl. Acad. Sci. USA 94, 12,297–12,302.
40. Hanes, J., Schaffitzel, C., Knappik, A., and Pluckthun, A. (2000) Nature Biotechnol. 8, 1287–1292.
41. Jermustus, L., Honegger, A., Schwesinger, F., Hanes, J., and Pluckthun, A. (2001) Proc. Natl. Acad. Sci. USA 98, 75–80.
42. Visntin, M., Tse, E., Axelson, H., Rabbitts, T.H., and Cattaneo, A. (1999) Proc. Natl. Acad. Sci. USA 96, 11,723–11,728.
43. Auf der Maur, A., Zahnd, C., Fischer, F., Spinelli, S., Honegger, A., Cambillau, C., Escher, D., Pluckthun, A. and Barberis, A. (2002) J. Biol. Chem. 277, 45075–45085.

44. Visintin, M., Settanni, G., Maritan, A., Graziosi, S., Marks, J.D., and Cattaneo, A. (2002) J. Mol. Biol. 317, 73–83.

45. Mossner, E., Koch, H., and Pluckthun, A. (2001) J. Mol. Biol. 308, 115–122.

46. Pelletier, J.N., Campbell-Valois, F.-X., and Michnick, S.W. (1998) Proc. Natl. Acad. Sci. USA 95, 12,141–12,146.

47. Martineau, P., Jones, P., and Winter, G. (1998) J. Mol. Biol. 280, 117–127.

48. Rondon, I.J. and Marsco, W.A. (1997) Annu. Rev. Microbiol. 51, 257–283.

49. Marasco, W.A., LaVecchio, J., and Winkler, A. (1999) J. Immunol. Methods 231, 223–238.

50. Gocalves, J., Silva, F., Freitas-Vieira, A., Santa-Marta, M., Malho, R., Yang, X., Gabuzda, D., and Barbas, C. (2002) J. Biol. Chem. 277, 32,036–32,045.

51. Steinberger, P., Andris-Widhopf, J., Buhler, B., Torbett, B.E., and Barbas, C.F. (2000) Proc. Natl. Acad. Sci. USA 97, 805–810.

52. BouHamdan, M., Strayer, D.S., Wei, D., Mukhtar, M., Duan, L.-X., Hoxie, J., and Pomerantz, R.J. (2001) Gene Ther. 8, 408–418.

53. Strayer, D.S., Branco, F., Landre, J., BouHamdan, M., Shaheen, F., and Pomerantz, R.J. (2001) Mol. Ther. 5, 33–41.

54. Richardson, H.R. and Marasco, W.A. (1998) In: Marasco, W.A. (ed.) Intrabodies—basic research and clinical gene therapy applications, Springer, New York, pp. 47–60.

55. Beerli, R.R., Graus-Porta, D., and Hynes, N.E. (1998) In: Marasco, W.A. (ed.) Intrabodies-basic research and clinical gene therapy applications, Springer, New York, pp. 61–74.

56. Curiel, D.T. (1998) In: Marasco, W.A. (ed.) Intrabodies—basic research and clinical gene therapy applications, Springer, New York, pp. 97–128.

57. Beerli, R.R., Wels, W., and Hynes, N.E. (1994) J. Biol. Chem. 269, 23,931–23,936.

58. Deshane, J., Siegal, G.P., Alvarez, R.D., Wang, M.H., Feng, M., Cabrera, G., Liu, T., Kay, M., and Curiel, D.T. (1995) J. Clin. Invest. 96, 2980–2989.

59. Alvarez, R.D., Barnes, M.N., Gomez-Navarro, J., Wang, M., Strong, T.V., Arafat, W., Arani, R.B., Johnson, M.R., Roberts, B.L., Siegal, G.P., Curiel, D.T. (2000) Clin. Cancer Res. 6, 3081–3087.

60. Lener, M., Horn, I.R., Cardinale, A., Messina, S., Nielsen, U.B., Rybak, S.M., Hoogenboom, H.R., Cattaneo, A., and Biocca, S. (2000) Eur. J. Biochem. 267, 1196–1205.

61. Cochet, O., Kenigsberg, M., Delumeau, I., Virone-Oddos, A., Multon, M.-C., Fridman, W.H., Schweighoffer, F., Teillaud, J.-L., and Tocque, B. (1998) Cancer Res. 58, 1170–1176.

62. Cohen, P.A., Mani, J.-C., and Lane, D. (1998) Oncogene 17, 2445–2456.

63. Kasono, K., Heike, Y., Xiang, J., Piche, A., Kim, H.-G., Kim, M., Hagiwara, M., Nawrath, M., Moelling, K., and Curiel, D.T. (2000) Cancer Gene Ther. 7, 151–159.

64. Strube, R.W. and Chen, S.-Y. (2002) J. Immunol. Methods 263, 149–167.

65. Heike, Y., Kasono, K., Kunisaki, C., Hama, S., Saijo, N., Tsuruo, T., Kuntz, D.A., Rose, D.R., and Curiel, D.T. (2001) Int. J. Cancer 92, 115–122.

66. Wanker, E.E. (2000) Biol. Chem. 381, 937–942.

67. Khoshnan, A., Ko, J., and Patterson, P.H. (2002) Proc. Natl. Acad. Sci. USA 99, 1002–1007.

68. Vanhove, B., Beatrice, C., Cassard, A., Pourcel, C., and Soulillou, J.-P. (1998) Transplantation 66, 1477–1485.

69. Sepp, A., Farrar, C.A., Dorling, T., Carins, T., George, A.J.T., and Lechler, R.I. (1999) J. Immunol. Methods 231, 191–205.

70. Mhashilkar, A.M., Doebis, C., Seifert, M., Busch, A., Zani, C., Soo Hoo, J., Nagy, M., Ritter, T., Volk, H.-D., Marasco, W.A. (2002) Gene Ther. 9, 307–319.

71. International Human Genome Sequencing Consortium (2001) Nature 409, 860–921.

72. Venter, J.C., Adams, M.D., Myers, E.W., Li, P.W., Mural, R.J., Sutton, G.G., Smith, H.O. et al. (2001) Science 291, 1304–1351.

73. Hartl, F.U. and Hayer-Hartl, M. (2002) Science 295, 1852–1856.

74. Jurado, P., Ritz, D., Beckwith, J., de Lorenzo, V., and Fernandez, L.A. (2002) J. Mol. Biol. 320, 1–10.

75. Bach, H., Mazor, Y., Shaky, S., Shoham-Lev, A., Berdichevsky, Y., Gutnick, D.L., and Benhar, I. (2001) J. Mol. Biol. 312, 79–93.

S.C. Makrides (Ed.) *Gene Transfer and Expression in Mammalian Cells*

CHAPTER 22

Inducible gene expression in mammalian cells

Wilfried Weber and Martin Fussenegger

*Institute of Biotechnology, Swiss Federal Institute of Technology, ETH Hoenggerberg,
CH-8093 Zurich, Switzerland, Tel.: +41 1 633 3448; fax: +41 1 633 1051.
E-mail: fussenegger@biotech.biol.ethz.ch*

1. Introduction

Artificial control systems for adjusting transgene expression in mammalian cells, animals and eventually humans have generated tremendous impact on different areas of modern biomedical engineering ranging from basic gene-function analysis, drug discovery, drug testing in animals, the design of animal-based human disease models and biopharmaceutical manufacturing to gene therapy and tissue engineering strategies. Heterologous gene regulation systems have become an integral part of current spearhead therapeutic technologies focusing on production and delivery of therapeutic proteins where they are needed in the human body. Current gene therapy settings combining *in vivo* production and delivery of desired protein therapeutics require precise control of desired transgenes delivered by transfection, transduction or transplantation of engineered cells to: (i) titrate them into the therapeutic window; (ii) adjust them to the daily changing dosing regimes of the patient; and (iii) enable timely control of the overall therapy. Several heterologous gene regulation systems have been developed in the past (reviewed in [1]). Only a few of these systems have the assets for human therapeutic use including absence of any interference with endogenous regulatory networks, and graded as well as rapid response characteristics showing low basal and high maximum expression levels following administration of a clinically licensed drug (e.g., antibiotics, immunophilins and steroid hormones). These human-compatible systems which are currently competing in preclinical studies for optimal performance in adjusting expression of a single therapeutic (model) gene (e.g., erythropoietin, insulin or human growth hormone) stand at the center of this review. Gene regulation systems which are unlikely to meet with clinical standards such as those responsive to environmental stimuli (heat, oxygen, metal ions, IPTG, dioxin) have been extensively reviewed elsewhere [1]. Recent initiatives to combine several compatible heterologous gene regulation systems have exemplified that complex artificial gene control configurations such as regulatory cascades and networks will develop in the next years from concept studies to a therapeutic reality. Such multiregulated multigene metabolic engineering will enable optimal integration of next-generation gene interventions in endogenous proliferation-, differentiation- and apoptosis-regulatory networks to achieve cell phenotypes designed to improve the understanding and therapy of currently untreatable human diseases.

In this chapter we review different human-compatible heterologous gene regulation systems, outline their adaptation to specific expression configurations and highlight their potential to be integrated into higher order control systems to achieve next-generation gene therapy and tissue engineering strategies.

2. Artificial transcription control systems

A variety of strategies have been designed in the past decade to adjust heterologous protein production to desired levels [1]. Artificial transcription control systems seem to prevail, particularly when designed in a binary configuration consisting of a repressor/transactivator, which modulates transgene expression by binding to a chimeric operator–promoter configuration in a drug-dependent manner. Binary gene regulation modalities have been constructed using well-evolved immunophilin-based dimerization phenomena, bacterial antibiotic response regulators or hormone-dependent DNA-binding characteristics.

2.1. Dimerizing technology

2.1.1. Inducible dimerization systems
Inducible heterodimerization of human immunophilins FKBP and FRAP by the immunosuppressive, clinically licensed drug rapamycin was used for the design of several mammalian gene regulation concepts [2]. The rapamycin-inducible dimerizing technology is based on the observation that the DNA-binding and activation domains of transcription factors function independently and retain their activity in a fusion protein. In its pilot configuration, the rapamycin-inducible gene control design was based on a heterodimeric transactivator consisting of FKBP fused to an artificial DNA binding domain ($ZFHD1_{BD}$; $ZFHD1_{BD}$–FKBP) and a FRAP domain (designated FRB) fused to a human NF-κB-derived transactivation domain (p65; FRB–p65) (Fig. 1A). Reconstitution of this transactivator was achieved by rapamycin-dependent heterodimerization ($ZFHD1_{BD}$–FKBP–rapamycin–FRB–p65) which enabled binding and transgene expression from chimeric promoters ($ZFHD1_n$–P_{min}) assembled by cloning tandem modules of $ZFHD1_{BD}$-target sequences ($ZFHD1_n$) to minimal eukaryotic promoters [2] (P_{min}, Fig. 1). The rapamycin-induced dimerizing technology enables stringent transgene control modalities in mammalian cells and has been validated in a variety of animal models including non-human primates [3].

2.1.2. Repressible dimerization technology
A variation of the rapamycin-inducible gene expression system is the ligand-reversible dimerization strategy, which is based on self-sufficient homodimerization of mutated FKBP (called F_M) in the absence of the immunosuppressive agent FK506 [4]. F_M fused to $ZFHD1_{BD}$ (F_M–$ZFHD1_{BD}$) and p65 (FM–p65) assembles to form a transactivator ($ZFHD1_{BD}$–F_M–F_M–p65) which initiates transcription from $ZFHD1_n$–P_{min} promoters (Fig. 1). Addition of FK506 disrupts the dimerized transactivator ($ZFHD1_{BD}$–F_M–FK506; FK506–F_M–p65) and results in dose-dependent repression of $ZFHD1_n$–P_{min}–driven transgene expression [4] (Fig. 1B).

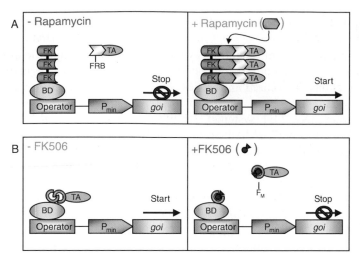

Fig. 1. Dimerizering-based gene regulation technology. (A) Rapamycin-inducible dimerization-controlled gene expression system. In the absence of rapamycin three FKBP domains (FK) fused to a DNA-binding domain (BD) are tethered to an operator module placed 5′ of a minimal promoter (P_{min}). Only upon addition of rapamycin (+ Rapamycin), the FKBP domains heterodimerize with FRB fused to a transactivation domain (TA) which initiates expression of the gene of interest (*goi*). (B) FK506-repressible dimerization-controlled gene expression system. In the absence of the immunosuppressive agent FK506 (-FK506) a F_M–TA fusion protein homodimerizes with another F_M-domain fused to a BD and binds this chimeric transactivator to an operator module thereby inducing a 3′-located P_{min}. Addition of FK506 disrupts the F_M–F_M interactions by competitive binding resulting in *goi* repression.

2.1.3. Controlled release technology

Almost all systems designed for adjusting protein production levels are based on modulating promoter activity in response to a heterologous small molecule drug. Recently, controlled release technology has been developed, which adjusts secretion of insulin by reversible FK506-dependent aggregation of this protein therapeutic in the rough endoplasmic reticulum (rER) [5]. FK506-adjustable secretion of desired proteins requires their fusion to several copies of the F_M-homodimerization domain, which aggregates this fusion protein in the rER in a secretion-incompetent complex. However, administration of an FK506-derived ligand dissolves the aggregates by binding to the individual F_M domains thus enabling secretion of the fusion protein. Concomitant with secretion of the F_M-insulin fusion protein the F_M domain is cleaved off the target protein by an endoplasmic furin protease. The key advantage of controlled release technology is its rapid secretion kinetics reaching maximum values within 2 h following administration of the de-aggregating drug [5].

All aforementioned control technologies based on modulation of immunophilin–drug interactions have been humanized to contain besides some artificial genetic elements only determinants of human origin, which is expected to preclude induction of any immune responses in humans. However, immunophilin-binding drugs are highly immunosuppressive and engineered dimerizer drugs devoid of immuno-modulatory characteristics have yet to prevail in clinical trials [6].

2.2. Antibiotic-responsive systems

Current incompatibilities of regulating agents with human therapeutic use as seen for the dimerizing system could be alleviated by designing binary regulation systems responsive to clinically licensed antibiotics. Three different antibiotic-adjustable gene regulation systems have been developed so far which are responsive to antibiotics from the tetracycline (TET systems), streptogramin (PIP systems) and macrolide (E.REX systems) class [7–9]. All antibiotic-adjustable expression control systems have been designed in an inducible and repressible configuration and were shown to be fully compatible with each other supporting independent regulation of up to three transgenes [7,9].

2.2.1. Tetracycline-responsive gene expression

Tetracycline-repressible gene regulation. The first and most widely used hetero-logous mammalian gene regulation system is responsive to tetracycline antibiotics and consists of components derived from the *E. coli* Tn*10*-encoded tetracycline resistance operon. The pioneering basic tetracycline-repressible (TET$_{OFF}$) gene regulation system has been assembled by fusing the *E. coli* tetracycline resistance gene repressor, TetR, to the *Herpes simplex*-derived VP16 transactivation domain. This tetracycline-dependent transactivator (tTA; TetR$-$VP16) binds and activates a tetracycline-responsive promoter (P$_{hCMV*-1}$, *tetO$_7$*$-$P$_{hCMVmin}$), constructed by cloning a heptameric tetracycline operator module (*tetO$_7$*) adjacent to a minimal version of the human cytomegalovirus immediate early promoter (P$_{hCMVmin}$) [8], in the absence of tetracycline antibiotics. However, in the presence of tetracycline antibiotics, tTA dissociates from P$_{hCMV*-1}$ and transgene expression is shut down (TET$_{OFF}$) (Fig. 2A). Several improved versions of the basic TET$_{OFF}$ system have been designed in the past decade which have been successfully validated in diverse *in vitro* and *in vivo* studies including transgenic animals [1,10–15].

Tetracycline-inducible gene regulation. In certain configurations, particularly therapeutic ones, inducible rather than repressible expression systems would be highly desirable. In order to reverse tetracycline-dependent regulation characteristics to enable transgene expression exclusively in the presence of tetracycline antibiotics, two different tetracycline-inducible expression arrangements have been constructed. (i) Deuschle and coworkers [16] pioneered tetracycline-inducible expression control by replacing the VP16 transactivation domain by a KRAB-type transsilencing domain. In the absence of tetracycline, this tetracycline-dependent transsilencer (tTS) binds to a heptameric *cis*-acting *tetO$_7$* operator and silences a 3′-placed CMV promoter (P$_{hCMV}$). Upon addition of tetracycline, tTS is released from its cognate operator and P$_{hCMV}$ is derepressed [16] (Fig. 2B). (ii) An alternative approach to generate tetracycline-inducible expression control was based on reversing the allosteric characteristics of tTA in a way that interaction between the reverse tetracycline-dependent transactivator (rtTA) and P$_{hCMV*-1}$ and concomitant transgene induction occurs only in the presence of the tetracycline derivative doxycycline (dox), which shows higher affinity to rtTA compared to tetracycline.

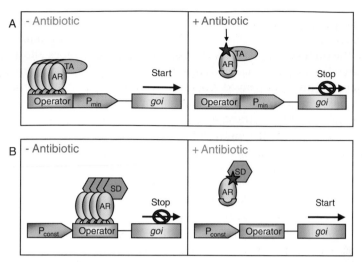

Fig. 2. Generic antibiotic-responsive gene regulation technology (E.REX, PIP and TET systems). (A) Antibiotic-responsive gene control systems of the "OFF" type (E_{OFF}, PIP_{OFF}, TET_{OFF}). In the absence of regulating antibiotics, an antibiotic-dependent transactivator, a fusion protein between a prokaryotic antibiotic response regulator (AR) and a transactivation domain (TA), which binds to a cognate operator thereby inducing a *cis*-located minimal promoter (P_{min}) as well as the gene of interest (*goi*) (AR = E, PIP or TetR for the E.REX, PIP and TET systems). This configuration results in transcription of the gene of interest (*goi*). Following addition of regulating antibiotics (erythromycin, E.REX; pristinamycin, PIP system; tetracycline, TET system), these drugs bind AR and induce an allosteric change which abolishes AR-operator interaction as well as *goi* transcription. (B) Antibiotic-inducible gene regulation technology (E_{ON} and PIP_{ON} systems). In the absence of regulating antibiotics AR (optionally fused to a silencing domain) binds an operator module and blocks (silences) a 5′-located constitutive promoter (P_{const}). Derepression is induced by disruption of the AR-operator interacting in the presence of regulation antibiotics.

As rtTA dissociates from $P_{hCMV*-1}$ in the absence of dox attendant transgenes remain silent [17]. The rtTA-based tetracycline-inducible expression concept (TET_{ON}) has been successfully used to adjust different transgenes in a variety of cell contexts as well as in transgenic animals [1,11,18,19]. Since TET_{ON}-based gene induction requires dox concentrations of over 100 ng/ml, which is difficult to attain in certain organs such as the brain, the use of this system for *in vivo* studies may be limited [10,19]. However, rtTA2^S-M2 an rtTA derivative emerging from random mutagenesis showed improved induction profiles and dox-response characteristics which enabled brain-specific expression control in mice following administration of dox in their drinking water [10,19].

2.2.2. Streptogramin-responsive gene expression (PIP systems)
The streptogramin-adjustable gene control systems (PIP systems) are derived from the *Streptomyces coelicolor* and *Streptomyces pristinaespiralis* pristinamycin resistance operons, which confer resistance to antibiotics of the streptogramin class (pristinamycin (PI), virginiamycin, Synercid® (quinupristin-dalfopristin); [7]). These *Streptomyces* operons consist of the pristinamycin-induced protein PIP, which

binds to an operator (*pir*) thereby repressing the overlapping promoter of the pristinamycin resistance gene (P_{PTR}) in the absence of streptogramin antibiotics. However, in the presence of this class of antibiotics, PIP undergoes allosteric changes and dissociates from P_{PTR}, thus enabling transcription of streptogramin antibiotic resistance genes. This well-evolved prokaryotic streptogramin response regulation arrangement has been adapted to a mammalian expression context by constructing the pristinamycin-dependent transactivator PIT, a fusion of PIP and the *herpes simplex* VP16 transactivation domain, which binds, in the absence of streptogramins, to a chimeric promoter (P_{PIR}) assembled by cloning the PIP operator module *pir* adjacent to a minimal eukaryotic promoter (P_{PIR}) [7]. PIT retains its antibiotic-responsive characteristics and binds to P_{PIR} only in the absence of streptogramins. Graded response profiles of P_{PIR}-driven transgenes are achieved by addition of different concentrations of the regulating compound (PIP_{OFF} system) (Fig. 2A).

The streptogramin-adjustable gene expression technology has also been designed in an "ON"-type configuration, which mediates adjustable gene expression following addition of streptogramins. In this so-called "PIP_{ON}" system, the PIP protein binds to tandem *pir* operator arrangements (pir_n) placed downstream of a strong constitutive promoter such as P_{SV40}. Following binding of PIP, promoter activity is therefore abolished, whereas addition of pristinamycin (PI) results in dissociation of PIP and thus fully reconstitutes P_{SV40}-mediated expression levels [7] (Fig. 2B).

Since the first report of the Pip systems, their components have been repeatedly optimized for enhanced regulation performance in a wide variety of mammalian and human cells (see below). Also, streptogramin-responsive expression technology has been integrated in sophisticated expression configurations including multicistronic, autoregulated and lentiviral arrangements as well as in higher order control settings such as artificial regulatory cascades and dual-regulated expression concepts [9,20,21].

2.2.3. Erythromycin-responsive gene expression
The family of antibiotic-adjustable gene control systems has recently been expanded by the development of the erythromycin-responsive gene regulation system (E.REX), which can be fine-tuned by macrolide antibiotics (e.g., erythromycin, clarithromycin or roxithromycin; [9]). Similar to the generic construction principle of the antibiotic-responsive expression systems, E.REX is derived from a macrolide resistance operon cloned from clinical *E. coli* isolates. This bacterial survival determinant consists of the repressor protein E, which binds to its cognate ETR operator-containing promoter (P_{mphA}) only in the absence of macrolide antibiotics. The heterologous human-compatible macrolide-responsive system (E_{OFF}) consists of a macrolide-dependent transactivator (ET, a fusion of the E protein and different human and viral transactivation domains) and its cognate promoters (P_{ETR}), which have been designed by linking the ETR binding sequence to different eukaryotic minimal promoters [9] (Fig. 2A). The E_{OFF} configuration enables adjustable gene expression by fine-tuning the ET–P_{ETR} interaction and thereby modulating P_{ETR}-driven transgene induction profiles [9]. Analogous to the PIP_{ON} system, the E.REX technology was configured such that it enabled inducible gene expression following

addition of macrolide antibiotics (E_{ON}). Erythromycin-inducible promoters ($P_{ETR}ON$) consisting of tandem ETR modules placed 3′ of strong constitutive promoters were efficiently repressed by E optionally fused to a silencing domain of the KRAB type, in the absence of macrolides. Derepression initiated by addition of macrolides induced $P_{ETR}ON$-driven transgenes in a dose-dependent manner [9]. Both the E_{OFF} and E_{ON} systems have been modified at the promoter and transactivator/transrepressor level to enable improved response characteristics and compatibility with other members of the antibiotic-responsive expression system family in different mammalian cell lines [22]. The erythromycin-inducible as well as - repressible gene regulation systems have been validated in mice. Human cells transgenic for E_{ON}- and E_{OFF}-controlled erythropoietin (EPO) expression were encapsulated in coherent alginate-poly-L-lysine-alginate capsules and implanted intraperitoneally into immunocompetent mice. Administration of different erythromycin doses resulted in adjustable serum EPO levels in mice [9].

2.3. Receptor-based systems

2.3.1. Hormone receptor-modulated systems

A large family of mammalian gene regulation systems has been designed based on endogenous mammalian signaling pathways and their corresponding intracellular receptors. The generic set-up of these intracellular receptor-modulated systems is based on the finding that uninduced receptors are sequestrated in the cytoplasm by interaction with specific proteins (e.g. HSP90 for steroid), whereas addition of the inducing ligand (e.g. steroid hormones like estrogen) disrupts this interaction resulting in dimerization and translocation of the receptor dimers into the nucleus where transcriptional modulation of specific promoters occurs. Artificial hormone-responsive promoters consist of receptor-specific operator modules fused to minimal eukaryotic promoters. In the presence of appropriate hormones, steroid-dependent transcription factors translocate to the nucleus and transactivate artificial promoters whereas these transactivators remain bound in a cytosolic HSP90-containing complex unable to induce heterologous transgene expression in the absence of the regulating molecules (reviewed in [1,14]) (Fig. 3). Despite excellent regulation performance of a variety of hormone-responsive systems they elicit undesired side effects due to intrinsic co-regulation of endogenous hormone-dependent regulatory network [1,15,23].

In order to prevent any interference between endogenous and artificial hormone-regulatory systems some gene control arrangements have been designed based on receptor–DNA interactions of heterologous origin. One prominent example includes the conditional gene expression concept responsive to the insect molting hormone ecdysone [24]. Potent heterodimeric ecdysone-responsive transcription factors were constructed by fusing the ecdysone-binding domain of the *Drosophila* ecdysone receptor to an artificial DNA-binding domain as well as to the *Herpes simplex* VP16 transactivation domain (VgEcR). Ecdysone-dependent heterodimerization of VgEcR with RXR fused to a complementary DNA binding domain generates a chimeric transactivator, which translocates to the nucleus and induces cognate

596

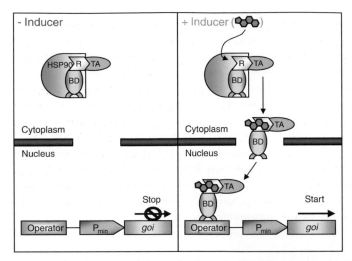

Fig. 3. Hormone receptor-based gene regulation technology. Artificial hormone-responsive transcription factors consist of fusion proteins assembled from a DNA-binding domain (BD), a receptor-derived hormone-binding domain (R) and a transactivation domain (TA). In the absence of regulating hormones, the artificial transcription factor is complexed by the heat shock protein 90 (HSP90) complex and remains in the cytoplasm. Binding of regulating hormones (e.g., estrogen, tamoxifen, progesterone, mifepristone) to the receptor portion of the transactivator release it from the HSP90 complex, induce dimerization as well as translocation into the nucleus where the transactivator binds to cognate operator modules and induces gene of interest (goi) expression driven by a minimal promoter (P_{min}).

promoters [24]. These promoters consist of a chimeric operator specific for the DNA-binding domains of VgEcR and RXR which is fused to a minimal promoter. Although the ecdysone-responsive expression systems show tight regulation characteristics even in transgenic mice, concerns about side effects of insect molting hormones will likely prevent their successful application in human therapy. Also, clinically licensed ecdysone derivatives have not yet appeared in basic research pipelines.

2.3.2. Synthetic steroid-responsive expression

In order to minimize interference of hormone-response regulators with endogenous regulatory networks while maintaining a high degree of human compatibility, a variety of synthetic steroid-responsive expression systems have been designed. Wang et al. [25] pioneered such systems by modifying the human progesterone receptor in a way that it was responsive to the progesterone antagonist mifepristone (RU486) but not to natural progesterone (hPRB891). Fusion of hPRB891 to the yeast Gal4 DNA-binding module and the *Herpes simplex* VP16 transactivation domain resulted in an artificial transcription factor, which activates gal4-containing chimeric promoters in an RU486-inducible manner. Regulation-effective concentrations of RU486 have been reported to be up to 50-fold below the abortive concentration of this progesterone antagonist [15] (Fig. 3). The RU486 control system has been validated in different transplantation and transgenics configurations in which it showed excellent regulation of the human growth hormone (hGH) [26].

Similar to the RU486-responsive system, the murine estrogen receptor has been modified to be responsive to 4-hydroxytamoxifen (4-OHT) instead of its natural ligand estrogen [23]. This mutant estrogen receptor has been fused to the VP64 transactivation domain as well as to artificial DNA-binding zinc finger domains generating 4-OHT-responsive transcription factors [27]. Depending on different modules (transactivator, receptor, DNA-binding domain) and their relative order target gene expression could be induced by up to three orders of magnitude in the presence of 4-OHT. Beerli *et al.* [27] suggest that by combining their technology with standard zinc finger domain engineering, artificial transcription factors could be tailored to virtually any DNA target sequence. Further work showed that the 4-OHT-responsive receptor fused to the E2F1 transcription factor induced apoptosis in several human cancer cell lines and restrained tumor growth in mice following administration of 4-OHT [28].

3. Generic strategies to improve key characteristics of basic gene regulation systems

All aforementioned technologies have a proven track record in adjusting desired transgene expression in a drug-dependent and dose-responsive manner. However, specific regulation characteristics including basal expression and maximum expression levels depend on the cell type and the overall configuration and modules used for construction of a particular gene regulation system. Critical parameters which define the overall performance of a heterologous gene regulation system are: (i) the type and the number of transactivation/transrepression domains fused to small molecule response regulators; (ii) the choice of the minimal promoter; (iii) the spacing between the operator module and the minimal promoter; and (iv) the number of binding sequences in the operator module.

Besides their transactivation/transrepressing potential, two major issues are associated with transactivation/transrepression domains: their capacity to titrate essential endogenous transcription-modulating factors away (a phenomenon also known as squelching), and the potential to elicit an immune response. Recent studies have compared several transactivation/transrepression domains (VP16, p65, Oct-2, Sp1, AP-2, E2F4/KRAB-silencing domains from *kox1* and other zinc finger proteins) of different origin with a focus on optimizing the expression levels, minimizing squelching and preventing induction of immune responses [11,12,29–32]. In essence, the KRAB domain demonstrated near optimal transsilencing capacity in several cell lines whereas the VP16 domain showed the best regulation performance in a variety of cell lines, although maximum expression levels and tightness could be outperformed by other transactivation domains in specific cell lines [31]. Baron *et al.* [11] have assessed in great detail the microstructure of the VP16 domain in order to eliminate superfluous sequences potentially associated with squelching. Their work has revealed a minimal VP16 domain which displayed significantly reduced squelching characteristics and outperformed the full-length VP16 when fused in tandem repeats to antibiotic response regulators of the Tet, Pip and E type [11,31].

Tandem repetition of transactivation domains was also shown to be successful in the dimerizing systems in which up to four p65 domains were bundled to a single DNA-binding heterodimerizing partner in the presence of rapamycin and its analogues [2,29,33].

Although a variety of minimal promoters have been used in the current heterologous gene control arena, comparative studies of the impact on the overall regulation performance are rare. The pioneers of antibiotic-responsive gene regulation systems, Bujard and coworkers, directly compared minimal promoters of the thymidine kinase gene (P_{TKmin}) and the human cytomegalovirus ($P_{hCMVmin}$) showing that $P_{hCMVmin}$ outperforms P_{TKmin} in maximum expression levels and tightness [8]. Comparative studies on the *Drosophila* heat-shock protein 70 promoter ($P_{hsp70min}$) and $P_{hCMVmin}$ in the PIP and E.REX configurations validated the outstanding expression performance of $P_{hCMVmin}$ with respect to maximum expression levels [31].

A parameter which has so far attracted little attention when it comes to improving basic gene regulation configurations is the spacing between the operator and the minimal promoter. Rigorous assessment of operator-$P_{hCMVmin}$ distance in the PIP and E.REX systems revealed dramatic differences in the maximum as well as basal expression levels when isogenic promoters containing 2, 4, 6, 8 and 10 bp linker between operator and minimal promoter were used in a variety of different cell lines [31]. Although the overall regulation performance of these promoters varied tremendously between cell lines, the overall data suggested distances of 0 or 2 bp for E.REX and 8 bp for the PIP system to combine higher maximum expression levels with lower leaky expression [31].

Intuitively, the number of transactivator binding sites in a particular promoter module should increase maximum expression levels as more transactivator may bind and initiate transcription at cognate promoters. However, there is a direct correlation between increased maximum expression levels and enhanced leaky expression and the art of balancing these two parameters determines the performance of a particular gene regulation setting in the desired cell line [1,7].

Although scientists typically speak about using *THE* TET, PIP or E.REX technology, there is no generic system available but rather a selection of transactivators (containing different transcription-modulating modules) and promoters (harboring different tandem operator repeats and minimal promoters with variable relative spacing); thus, only a well-balanced combination of these will guarantee optimal performance in a particular cell line and regulation set-up [34].

4. Complex regulation systems

4.1. Combining regulatory elements for superior performance

Although recent developments in the heterologous gene regulation field have resulted in improved gene control modalities for the majority of the aforementioned systems, certain specific gene adjustment scenarios require sophisticated combination of single gene regulation systems for improved performance.

4.1.1. Enhanced regulation characteristics

The basic TET_{OFF} system is known for the rather increased basal expression levels, which may limit its use in experimental settings that require absolute tightness of desired transgenes. In order to minimize the leakiness associated with the TET_{OFF} system and extend its regulatory window, rtTA and tTS derivatives were combined to act on $P_{hCMV*-1}$ and modulate its expression levels in a mutually exclusive manner. In the absence of tetracycline, tTS binds and actively silences $P_{hCMV*-1}$ thus preventing or further reducing basal expression. However, following addition of tetracycline, tTS is released from $P_{hCMV*-1}$ and replaced by rtTA, which subsequently induces chimeric promoters to desired levels correlating with the concentration of the antibiotic. A major challenge of this set-up was to prevent formation of non-functional tTS/rtTA heterodimers by engineering the dimerization domain of the transmodulators for exclusive homodimerization [18,35]. This combined binary repressor/transactivator concept has been successfully tested in transgenic mice engineered to adjust interleukin 13 expression into the therapeutic window [36]. Also, tight expression control of the combined tTS/rtTA systems was validated by regulated expression of the cyclin-dependent kinase inhibitor (CKI) p16, which resulted in conditional growth arrest of mammalian cell lines. However, controlled proliferation technology has previously been implemented by p27^{Kip1}-based engineering of CHO cells using the basic TET_{OFF} system, and has recently been extended for dual-regulated p27^{Kip1} sense and antisense expression by combining the PIP_{OFF} and TET_{OFF} systems for positive and negative growth control [20,37].

4.1.2. Rheostat vs. ON/OFF switch

The heterologous gene regulation systems developed so far enable graded gene expression characteristics in response to the regulating agent. However, many natural regulation phenomena follow an all-or-nothing kinetic profile typical for an ON (fully induced)/OFF (fully repressed) switch [38]. Rossi et al. [38] converted the basic TET_{OFF} rheostat into an ON/OFF switch with digital expression profiles. To this end, heterodimerization-incompetent tTS and rtTA were combined to modulate expression of $P_{hCMV*-1}$ [18]. The competition between tTS and rtTA for the same operator sites at intermediate doxycycline concentrations enables their binding in a mutually exclusive manner resulting in a true ON/OFF switch rather than a graded response profile [38].

4.2. Interconnected regulation systems

Most endogenous genes are adjusted for optimal integration into the cell's global regulatory networks to ensure orchestration of vital processes in space and time as well as coordination of molecular responses to metabolic, hormonal and environmental signals. While a particular gene regulation system is sufficient for analysis of a single gene or correcting one gene-based deficiencies, complex multi-level regulated molecular interventions involving higher order control modalities will require combination of several heterologous gene regulation systems for rational reprogramming of proliferation, apoptosis and differentiation regulatory pathways [7,20,39,40].

4.2.1. Multiple regulation technology
Mutually exclusive regulation of two genes. The availability of gene control systems specific for distinct DNA-binding sites and responsive to different regulating stimuli enables control of multiple genes within a single cell [7,20,27,39]. Baron *et al.* [39] engineered heterodimerization-deficient tetracycline-dependent (tTA2-1) and reverse tetracycline-dependent (rtTA2-1) transactivators with distinct DNA-binding motifs specific for different promoters assembled by fusing the respective cognate sequences to $P_{hCMVmin}$ (P_{tet4} and P_{tet6}). In this configuration, only P_{tet4} is activated by tTA2-1 in the absence of doxycycline, whereas P_{tet6} remains repressed. At a dox concentration of 30 ng/ml, tTA2-1 is released from P_{tet4}. As this dox concentration is insufficient to mediate binding of rtTA2-1 to P_{tet6} both promoters remain shut down. However, at dox concentrations above 3000 ng/ml, rtTA2-1 binds and activates P_{tet6}, whereas P_{tet4} remains silent. The different dox response characteristics of tTA2-1 and rtTA2-1 enable mutually exclusive adjustment of two independent genes. Yet, this configuration has only three expression settings (ON/OFF, OFF/OFF, OFF/ON) while ON/ON expression profiles cannot be achieved. Also, it remains to be seen whether the system could be tuned *in vivo* to a dox concentration of 30 ng/ml required for silencing both genes simultaneously [1,39].

Independent regulation of several sets of genes. In order to achieve all four expression settings (ON/ON, ON/OFF, OFF/ON, OFF/OFF) Beerli *et al.* [27] combined the tamoxifen- and mifepristone-responsive technology for completely independent regulation of two transgenes within a single cell. Another more human-compatible multiregulated multigene set-up was constructed by coalescing the PIP_{OFF} and TET_{OFF} systems in a dual-regulated expression format configured in a two- or one-vector setting and encoded by a classical, onco- or lentiviral expression vector [40–42]. Dual-regulated expression technology has been successfully used to pioneer complete proliferation management by targeted interventions in cell-cycle regulatory networks. Placing the cyclin-dependent kinase inhibitor p27[Kip1] in sense orientation under control of the TET_{OFF} system and in antisense orientation under PIP_{OFF} control enabled induction of a G1-specific growth arrest in the $+PI/-Tet$ configuration when p27[Kip1]-sense is fully induced and increased proliferation in the $-PI/+Tet$ setting mediating an exclusive p27[Kip1] antisense expression state [20]. While dual-regulated expression technology will have to prove its impact in the clinic by providing expansion and reimplantation modalities for modern tissue engineering, pilot expression configurations which enable independent regulation of up to three transgenes are under development by combining the E.REX, PIP and TET systems (Weber *et al.*, unpublished).

4.2.2. Artificial regulatory networks

Besides adjustment of multiple sets of transgenes multiregulated multigene metabolic engineering will also enable complex artificial regulatory networks such as artificial regulatory cascades, oscillating systems and epigenetic gene switches to mediate higher order control configurations, which have previously been inaccessible to monogenic adjustment strategies ([40,43] for reviews).

Artificial regulatory cascades. The majority of mammalian signaling pathways are mediated by regulatory cascades, in which one control module triggers the activity of the consecutive one. Regulatory cascades enable rapid amplification of small differences in signal levels while maintaining the flexibility to intervene and fine-tune responses at any level of the cascade [1,40].

Several artificial regulatory cascades have been described that combine different heterologous gene regulation systems in line to achieve signaling cascade-like regulation profiles [42,44,45]. These configurations enabled extremely precise target gene control due to two or more control points for fine-tuning target gene expression. Recently, an artificial regulatory cascade has been used in a therapeutic context [46]. It has been assembled by combining a hypoxia-responsive gene control system with TET_{OFF} technology in a way that mediated oxygen-dependent expression of a modified tTA followed by tetracycline-adjustable control of the transgene. This artificial regulatory network provided three control levels: (i) the hypoxia-responsive element controls expression of the tetracycline-repressible transactivator (tTA, TET_{OFF} system); (ii) rapid elimination of the modified tTA under normoxic conditions due to an oxygen-degradation domain fused to the original tTA; (iii) tetracycline-adjustable expression of the $P_{hCMV*-1}$-driven erythropoietin gene. The combined oxygen/TET_{OFF} expression system enabled high EPO expression in mice whereas low hormone expression levels could be observed under normoxia or in the presence of tetracycline. This multilevel gene regulation configuration has been suggested for the treatment of anemia and to eliminate cancer as tumor centers are often hypoxic and may be eradicated by expression of cytotoxic genes [46].

Higher order control—artificial epigenetic switches and oscillators. The aforementioned artificial regulation networks have successfully been used in establishing novel opportunities in molecular reprogramming of cells to achieve sophisticated gene therapy and tissue engineering goals. However, complex gene regulation modalities such as epigenetic gene switches and oscillating regulation networks have been designed so far only for prokaryotes ([47] for a review). Artificial multistable epigenetic elements enable all-or-none expression profiles which are locked in one expression configuration even in the absence of the switching agent and are inherited in an epigenetic-like manner. One of the most complex naturally evolved regulation scenarios generate oscillating gene expression profiles and are expected to be difficult to assemble from single artificial gene control modules. Although yet science fiction in the mammalian cell engineering arena, artificial oscillators will beyond doubt generate tremendous impact in treatments which require periodic drug

602

administration (e.g., insulin) and may even be projected as inheritable molecular pacemakers for the treatment of heart insufficiencies.

5. Conclusions

Heterologous gene regulation concepts have gathered momentum in the past years and will develop from a mere gene-function analysis tool of basic researchers to a molecular surgery device to design custom-made molecular interventions for the treatment of prominent human diseases. Recent progress along this line has indicated that there is no generic gene regulation system able to comply with any therapeutic need. Rather gene regulation systems have to be viewed as general concepts that require specific adaptation to cell or other therapeutic contexts in order to achieve optimal control modalities. Progress in reducing immune responses of current gene control technology makes it reasonable to expect a bright future for heterologous gene regulation systems in the clinic.

Acknowledgements

We thank Stefan Schlatter for critical comments on the manuscript and Bettina Keller for designing the figures. Work in the laboratory of M.F. was supported by the Bundesamt für Bildung und Wissenschaft (BBW) within the 5th Framework Program of the European Commission.

Abbreviations

EM	erythromycin
EPO	erythropoietin
E.REX systems	erythromycin-responsive expression systems
ET	erythromycin-responsive transactivator
FKBP	FK506-binding protein
PI	pristinamycin
PIP	pristinamycin-induced protein
PIP-systems	pristinamycin-responsive expression systems
PIT	pristinamycin-responsive transactivator
P_{ETR}	erythromycin-repressible promoter
$P_{ETR}ON$	erythromycin-inducible promoter
$P_{hCMV*-1}$	tetracycline-responsive promoter
P_{PIR}	pristinamycin-repressible promoter
$P_{PIR}ON$	pristinamycin-inducible promoter
rtTA	reverse tetracycline-responsive transactivator
TET	tetracycline
TET-systems	tetracycline-responsive expression systems

TetR	tetracycline-responsive repressor
tTA	tetracycline-responsive transactivator
tTS	tetracycline-responsive transsilencer

References

1. Fussenegger, M. (2001) Biotechnol. Prog. 17, 1–51.
2. Rivera, V.M., Clackson, T., Natesan, S., Pollock, R., Amara, J.F., Keenan, T., Magari, S.R., Phillips, T., Courage, N.L., Cerasoli, F., Jr., Holt, D.A., Gilman, M. (1996) Nat. Med. 2, 1028–1032.
3. Ye, X., Rivera, V.M., Zoltick, P., Cerasoli, F., Jr., Schnell, M.A., Guang-ping, G., Hughes, J.V., Gilman, M., and Wilson, J.M. (1999) Science 283, 88–91.
4. Rollins, C.T., Rivera, V.M., Woolfson, D.N., Keenen, T., Hatada, M., Adams, S.E., Andrade, L.J., Yaeger, D., van Schravendijk, M.R., Holt, D.A., Gilman, M., Clackson, T. (2000) Proc. Natl. Acad. Sci. USA 97, 7096–7101.
5. Rivera, V.M., Wang, X., Wardwell, S., Courage, N.L., Volchuk, A., Keenan, T., Holt, D.A., Gilman, M., Orci, L., Cerasoli, F., Jr., Rothman, J.E., Clackson, T. (2000) Science 287, 826–830.
6. Iuliucci, J.D., Oliver, S.D., Morley, S., Ward, C., Ward, J., Dalgarno, D., Clackson, T, and Berger, H.J. (2001) J. Clin. Pharmacol. 41, 870–879.
7. Fussenegger, M., Morris, R.P., Fux, C., Rimann, M., von Stockar, B., Thompson, C.J., and Bailey, J.E. (2000) Nat. Biotechnol. 18, 1203–1208.
8. Gossen, M. and Bujard, H. (1992) Proc. Natl. Acad. Sci. USA 89, 5547–5551.
9. Weber, W., Fux, C., Daoud-El Baba, M., Keller, B., Weber, C.C., Kramer, B.P., Heinzen, C., Aubel, D., Bailey, J.E., Fussenegger, M. (2002) Nat. Biotechnol. 20, 901–907.
10. Urlinger, S., Baron, U., Thellmann, M., Hasan, M.T., Bujard, H., and Hillen, W. (2000) Proc. Natl. Acad. Sci. USA 97, 7963–7968.
11. Baron, U., Gossen, M., and Bujard, H. (1997) Nucl. Acids Res. 25, 2723–2729.
12. Akagi, K., Kanai, M., Saya, H., Kozu, T., and Berns, A. (2001) Nucl. Acids Res. 29, e23.
13. Rossi, F.M.V., Guicherit, O.M., Spicher, A., Kringstein, A.M., Fatyol, K., Blakely, B.T., and Blau, H.M. (1998) Nat. Genet. 20, 389–393.
14. Saez, E., No, D., West, A., and Evans, R.M. (1997) Curr. Opin. Biotechnol. 8, 608–616.
15. Clackson, T. (1997) Curr. Opin. Chem. Biol. 1, 210–218.
16. Deuschle, U., Meyer, W.K.-H., and Thiesen, H.-J. (1995) Mol. Cell. Biol. 15, 1907–1914.
17. Gossen, M., Freundlieb, S., Bender, G., Muller, G., Hillen, W., and Bujard, H. (1995) Science 268, 1766–1769.
18. Blau, H.M. and Rossi, F.M.V. (1999) Proc. Natl. Acad. Sci. USA 96, 797–799.
19. Mansuy, I.M. and Bujard, H. (2000) Curr. Opin. Neurobiol. 10, 593–596.
20. Fux, C., Moser, S., Schlatter, S., Rimann, M., Bailey, J.E., and Fussenegger, M. (2001) Nucl. Acids Res. 15, e19.
21. Moser, S., Schlatter, S., Fux, C., Rimann, M., Bailey, J.E., and Fussenegger, M. (2000) Biotechnol. Prog. 16, 724–735.
22. Weber, W., Marty, R.R., Keller, B., Rimann, M., Kramer, B.P., and Fussenegger, M. (2002) Biotechnol. Bioeng. 80, 691–705.
23. Littlewood, T.D., Hancock, D.C., Danielian, P.S., Parker, M.G., and Evan, G.I. (1995) Nucl. Acids Res. 23, 1686–1690.
24. No, D., Yao, T.-P., and Evans, R.M. (1996) Proc. Natl. Acad. Sci. USA 93, 3346–3351.
25. Wang, Y., O'Malley, B.W., Jr., Tsai, S.Y., and O'Malley, B.W. (1994) Proc. Natl. Acad. Sci. USA 91, 8180–8184.
26. Wang, Y., DeMayo, F.J., Tsai, S.Y., and O'Malley, B.W. (1997) Nat. Biotechnol. 15, 239–243.
27. Beerli, R.R., Schopfer, U., Dreier, B., and Barbas, C.F., III (2000) J. Biol. Chem. 275, 32,617–32,627.
28. Putzer, B.M., Stewe, T., Crespo, F., and Esche, E. (2000) Gene Ther. 7, 1317–1325.

29. Natesan, S., Molinari, E., Rivera, V.M., Rickles, R.J., and Gilman, M. (1999) Proc. Natl. Acad. Sci. USA 96, 13,898–13,903.
30. Urlinger, S., Helbl, V., Guthmann, J., Pook, E., Grimm, S., and Hillen, W. (2000) Gene 247, 103–110.
31. Weber, W., Kramer, B.P., Fux, C., Keller, B., and Fussenegger, M. (2002) J. Gene Med. 4, 676–686.
32. Freundlieb, S., Schirra-Muller, C., and Bujard, H. (1999) J. Gene Med. 1, 4–12.
33. Pollock, R., Issner, R., Zoller, K., Natesam, S., Rivera, V.M., and Clackson, T. (2000) Proc. Natl. Acad. Sci. USA 97, 13,221–13,226.
34. Senner, V., Sotoodeh, A., and Paulus, W. (2001) Neurochem. Res. 26, 521–524.
35. Foster, K., Helbl, V., Lederer, T., Urlinger, S., Wittenburg, N., and Hillen, W. (1999) Nucl. Acids Res. 27, 708–710.
36. Zhu, Z., Ma, B., Homer, R.J., Zheng, T., and Elias, J.A. (2001) J. Biol. Chem. 176, 25,222–25,229.
37. Fussenegger, M., Schlatter, S., Datwyler, D., Mazur, X., and Bailey, J.E. (1998) Nat. Biotechnol. 16, 468–472.
38. Rossi, F.M., Kringstein, A.M., Spicher, A., Guicherit, O.M., and Blau, H.M. (2000) Mol. Cell 6, 723–728.
39. Baron, U., Schnappinger, D., Helbl, V., Gossen, M., Hillen, W., and Bujard, H. (1999) Proc. Natl. Acad. Sci. USA 96, 1013–1018.
40. Fussenegger, M. (2001) Gene Ther Reg. 1, 233–264.
41. Mitta, B., Rimann, M., Ehrengruber, M.U., Ehrbar, M., Djonov, V., Kelm, J., and Fussenegger, M. (2002) Nucl. Acids Res. 30, e113.
42. Moser, S., Rimann, M., Fux, C., Schlatter, S., Bailey, J.E., and Fussenegger, M. (2001) J. Gene Med. 3, 529–549.
43. Weber, W. and Fussenegger, M. (2002) J. Biotechnol. 98, 161–187.
44. Aubrecht, J., Manivasakam, P., and Schiestl, R.H. (1996) Gene 172, 227–231.
45. Imhof, M.O., Chatellard, P., and Mermod, N. (2000) J. Gene Med. 2, 107–116.
46. Payen, E., Bettan, M., Henri, A., Tomkiewitcz, E., Houque, A., Kuzniak, I., Zuber, J., Scherman, D., and Beuzard, Y. (2001) J. Gene Med. 3, 498–504.
47. Hasty, J., McMillen, D., Isaacs, F., and Collins, J.J. (2001) Nat. Rev. Genet. 2, 268–279.

S.C. Makrides (Ed.) *Gene Transfer and Expression in Mammalian Cells*

CHAPTER 23

Protein production by large-scale mammalian cell culture

Liangzhi Xie[1], Weichang Zhou[1], and David Robinson[2]

[1]*Fermentation and Cell Culture, Merck Research Laboratories,*
Merck & Co., Inc. West Point, PA 19486, USA
[2]*Biocatalysis and Fermentation Development, Bioprocess R & D, Merck Research*
Laboratories, Merck & Co. Inc., Mail Stop R80Y-110, P. O. Box 2000, Rahway,
NJ 07065, USA, Tel.: 732-594-5828; fax: 732-594-4400.
E-mail: david.robinson@merck.com

1. Introduction

In the past two decades, more than a dozen recombinant proteins have been approved and successfully commercialized as therapeutic medicines. Currently, over a hundred recombinant protein molecules, including humanized monoclonal antibodies, cytokines, and growth hormones, are in clinical trials, targeting a variety of diseases such as cancer, psoriasis, asthma, and viral infections. These therapeutic proteins offer patients new treatment options that are more effective, safer, and/or more convenient than traditional treatments.

The typical dose for therapeutic proteins falls in the range of 20–500 mg. In combination with chronic usage and/or a large patient population, this translates into production requirements that can exceed 100 kg/annum in some cases. A good example is Enbrel®, a dimeric fusion protein that treats chronic Rheumatoid Arthritis with a bi-weekly dose of 25 mg. The annual consumption per patient is calculated to be about 2.6 g. It is therefore not surprising that the current demand based on rapid and broad acceptance in the medical community exceeds the manufacturer's existing production capacity, and a large mammalian cell culture facility with several bioreactors at a scale greater than 10,000 liter is being constructed. Development of such a large-scale process, construction of the facility, and factory startup are no small tasks, which may take more than five years and need to be conducted according to regulations by agencies such as the US Food and Drug Administration (FDA). Careful considerations need to be given with respect to selection of the host cell line, culture medium, process scale-up, culture vessel design, and so forth. Clearly it is not possible to cover the in-depth technology know-how of large-scale protein production in this short chapter; instead, the authors will strive to provide a comprehensive overview of the state-of-the-art of large-scale protein production technology with a focus on the process aspects.

2. Cell Lines

2.1. Commonly used cell lines in industry

Mammalian cells are the preferred production vehicle for many therapeutic proteins due to the requisite post-translational processing, in particular appropriate glycosylation and proper folding of the proteins produced. Continuous cell lines that are capable of indefinite growth are often chosen as the hosts for recombinant protein production [1]. These cell lines can either grow directly (such as hybridoma cell lines) or be adapted to grow (such as Chinese hamster ovary (CHO) cell lines) in suspension culture, which offers process scale-up advantages over adherent cell cultures. The most commonly used cell lines for commercial therapeutic protein production are CHO, SP2/0, and NS0 [2,3]. CHO cells have been well studied and characterized through decades of research that generated a large database of process know-how, product quality, and product safety in humans [4]. Therapeutic proteins produced in CHO cells include tissue plasma activator (tPA), erythropoietin (EPO), coagulation factor IX, interferon β-1a, and several monoclonal antibodies (such as Herceptin^{TM}, and Rituxan^{TM}). Both SP2/0 and NS0 are derived from murine myeloma cells that can readily grow in suspension cultures. The extensive studies conducted to generate hybridoma cells for monoclonal antibody production in the last two decades created a wealth of literature data to support their use in recombinant protein production.

2.2. Cell banking systems

Regulatory guidelines require the establishment of a master cell bank (MCB), which ensures sustained production from a commonly derived and fully characterized cell source. The typical size of a MCB is in the range of 100–1000 ampules, which is often not sufficient for long-term manufacturing if the MCB ampules are used directly in each production lot. Owing to this consideration, a two-tier banking system is widely used and recommended by regulatory agencies for industrial applications to ensure a long life of the MCB [5]. In a two-tier cell bank system, Working Cell Banks (WCBs) are derived from the MCB on a regular basis and only the WCB ampules are used for the production batches. The typical size of a WCB is also in the range of hundreds of vials that may last several years, depending on the number of batches produced per year. Although testing of the WCB is less extensive and hence less costly than that of the MCB, one should still try to make a relatively large WCB whenever possible to avoid the need to prepare and release WCB's too frequently.

2.3. Cell substrate characterization

A full battery of tests are required to fully evaluate the safety of the cell substrate and to ensure that the MCB and WCBs are free of adventitious and endogenous agents. These tests should be developed based on a knowledge of the history and potential exposure to contaminating agents of the cell line in question, and can be guided by a thorough understanding of the regulatory requirements and guidelines. These requirements typically include testing for cell line identity, stability, sterility, mycoplasma, mycobacteria, various viruses that are capable of infecting the host cell line and human cells, tissue culture safety, and tumorigenicity. Some of the tests are conducted directly on the MCB and/or WCBs but some tests such as host cell genetic stability and tumorigenicity require passaging the cells to a population doubling level (PDL) beyond the projected routine manufacturing level or end of production [6]. In addition, tests for cell line characterization are required to ensure their quality and consistency. These tests include genetic stability, growth characteristics, and protein expression.

3. Cell culture media and raw materials

3.1. Serum-free and protein-free media

Mammalian cells require a comprehensive mixture of nutrients in order to survive and grow *in vitro*. Since the publication of Eagle's medium in the 1950s [7], significant progress has been made towards understanding the nutritional requirements of various cell types. Many basal media containing glucose, amino acids, vitamins, and various salts are available commercially. However, supplementation of animal sera to these media is often required for cell growth and protein expression. More recently, serum-free media have been developed and successfully employed in large-scale biological manufacturing [8–11]. These media are more complex and typically include additives such as growth factors (e.g., insulin), iron-transporting reagents (e.g., transferrin), reducing reagents (e.g., glutathione), shear-protecting reagents (e.g., Pluronic F-68), trace metals (e.g., selenium and zinc), lipids and/or lipid precursors and carriers (e.g., linoleic acid and lipoic acid), and protein hydrolysates (e.g., peptone). In some cases, the low levels of proteins can be replaced completely to yield a protein-free medium. The application of protein-free media in routine manufacturing remains to be demonstrated but there is an increasing emphasis to remove animal sera and other animal derived components from culture media in large-scale protein production. The inclusion of animal sera and other animal-derived raw materials in the culture media carries a potential risk of adventitious agent contamination. These concerns have been heightened in recent years with the identification of prions as the causative agent of transmissible spongiform encephalopathies (e.g., "mad cow disease"). In addition, animal sera increases the raw material cost in cell culture and cost of downstream protein purification in large-scale manufacturing. Not surprisingly, serum-free media have been widely employed in the protein production industry using mammalian cell culture.

3.2. pH control reagents

Large-scale mammalian cell culture is often carried out under well controlled conditions. Mammalian cells produce by-products including lactic acid and carbon dioxide. Both species need to be removed and/or neutralized to maintain a constant pH. This can be achieved at a low cell concentration simply by stripping off CO_2 from the culture medium with air sparging if a bicarbonate buffer is used. At a high cell concentration, the CO_2 removal rate may not be sufficient to offset CO_2 production from cell metabolism, let alone to maintain the pH in light of the lactate produced. Hence, addition of base is always needed at large-scale or high-cell concentrations. Commonly used pH control reagents are concentrated $NaHCO_3$, Na_2CO_3, and NaOH solutions. The advantage of $NaHCO_3$ and Na_2CO_3 is that the pH of the concentrated base solution is modest as compared to that of NaOH solution, which may reduce the magnitude of a local pH spike from base addition due to poor mixing. On the other hand, the addition of bicarbonate could result in additional CO_2 accumulation in the culture which could lead to inhibition of cell growth and/or deterioration of protein glycosylation [12,13]. The effect of pH control reagents needs to be carefully studied before a suitable pH control reagent is selected for a specific production system. When $NaHCO_3$ or Na_2CO_3 is selected, the effect of CO_2 accumulation should be thoroughly evaluated and measures should be taken to prevent CO_2 accumulation as discussed in more detail below. Should NaOH be chosen, mixing time in large-scale vessels should be considered with an emphasis on the prevention of local pH spikes at the addition spot. Subsurface addition may also be used.

3.3. Release testing of culture media and raw materials

All culture media and raw materials used in cGMP production need to be tested and released prior to use. These include basal culture media, medium additives, gases, pH control reagents, and cleaning reagents. Release tests vary depending on the type of materials but usually include identity, sterility, strength, purity, use test, and other physical quality measures (such as pH, color, etc.) when appropriate. Adventitious agent testing needs to be included for animal-source components. An expiration date needs to be assigned to all released materials with supporting stability data to justify the length and conditions of storage (such as container, closure, storage temperature, light exposure, and moisture). Containers also need to be qualified for integrity, closure system, leachables, and adsorption of active components. For complex culture media, a successful use test is usually the most reliable measure of the quality of the medium.

3.4. Prevention of microbial and viral contamination

Mammalian cells in culture are susceptible to microbial and viral contamination [14–16]. Although some medium components can be reliably heat-sterilized, the labile nature of many of the medium constituents requires sterilization by filtration or irradiation. Microbial contamination can be largely prevented by sterile filtration of all culture media and raw materials using 0.22 micron filters and by following good aseptic practices. Microbial contamination, when it occurs, often manifests itself readily by changes in the culture performance and visual characteristics owing to the relatively faster growth rate of most microbes. Viral contamination is frequently much more difficult to detect and can potentially be a more severe safety hazard. Most viruses are about one order of magnitude smaller in diameter than bacteria, and hence are not effectively removed by filtration through 0.22 micron filters. A low level of viral contamination may not be detected until a significant period of virus propagation has elapsed, which may allow spreading into overlapping batches. Viral contamination incidents in Genentech's large-scale CHO cell culture process led to the development of a viral barrier system for all culture media and raw materials [15–18]. This has been shown to be very effective in preventing viral contamination in large-scale protein production. Establishment of a microbial and viral entry barrier should be included as an important aspect of large-scale bioreactor design and construction.

4. Cell expansion process

Each production lot typically starts by thawing an individual ampule of cryopreserved cells containing 10^7 to 10^8 cells. Multiple passages of cell expansion in progressively larger vessels (also called a cell expansion train) are needed prior to the inoculation of the final production vessel. There are many different choices in terms of process scheme and vessel type for the early stages of cell expansion at relatively small scales. Considerations are usually given to process reproducibility, operation complexity, and cell passaging scheme during the design of the cell expansion process as discussed below.

4.1. Process reproducibility

The consistency of a biological production process must be demonstrated as a means of ensuring consistent product quality. The vessel type and size at each passage are always pre-specified and not subject to ad hoc changes from batch to batch. It is also often a good idea to specify a fixed inoculation cell concentration for each passage. The inoculation volume at each passage can be fixed as well, but sometimes it may be desirable to have some flexibility within the working volume range of the vessel. The flexibility allows one to avoid frequent process deviations due to unexpected slow cell growth and yet minimize the discard of excessive cells. An alternative approach is to design the process according to the slowest cell growth rate so that there is always a sufficient number of cells to inoculate the subsequent passage with cells to spare, but this could lead to an increase in the number of days required for cell expansion.

    Batch culture is the most widely chosen operation mode for cell expansion owing to its simplicity. Fed-batch, perfusion, or continuous culture modes are rarely used in cell expansion. Occasionally one may choose a batch with refeed mode, especially when a process needs to be retrofitted into a pre-existing facility. For example, if the number of cells in a particular passage is not sufficient to inoculate the subsequent passage at the minimum inoculation cell concentration at the full working volume, it is possible to inoculate at a fraction of the maximum working volume as long as it is above the minimum required working volume (dictated for example by the need to provide adequate mixing), followed by an addition of fresh medium to increase the working volume after sufficient cell growth has occurred to maintain the cells above the minimum inoculation cell concentration at all times. This is essentially a mini-split within a single passage.

4.2. Operational complexity

Consideration should be given with respect to operational complexity when choosing a certain type of vessel for cell expansion. For example, at a scale less than 10 liter working volume for suspension culture, one could choose disposable vessels (such as roller bottles, spinner flasks, and shake flasks) or stainless steel bioreactors. In cases where full process control is not critical, disposable vessels, such as Wave Bioreactors are an excellent choice due to the elimination of the need of cleaning, sterilization, and associated validation requirements. However, the use of disposable cultureware (except for Wave Bioreactors) may increase the number of aseptic operations, and thus increase the possibility of contamination of the culture. In cases where full process control is needed, stainless steel bioreactors, spinner flasks, and disposable cultureware that facilitate process monitoring (for example the Wave Bioreactor) can be used. A general rule in choosing the type of culture vessels is to minimize the opportunities for contamination by minimizing the number of aseptic operations and to minimize the potential for operational errors by simplifying the process steps as long as process requirements can be met.

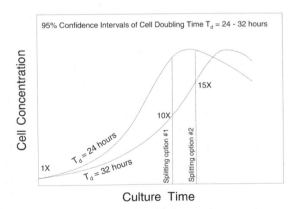

Fig. 1. Effect of cell growth rate variation from batch to batch on the design of cell passaging scheme. Showing on the graph are the cell growth curves at the lower and upper 95% confidence intervals of the cell growth rate range and the two optional target cell expansion schemes ($10 \times$ and $15 \times$ cell expansion) and corresponding splitting options. The splitting times are determined by the low cell growth curve which corresponds to a 32 h doubling time. With a $10 \times$ target, cells are still in the exponential growth phase at the specified splitting time (splitting option #1) even when cells grow at the upper range of the cell growth rate. With a more aggressive target of $15\times$ cell expansion (splitting option #2), cells may have entered the stationary growth phase (which is not desirable), should they grow at the upper range of the cell growth rate (lower doubling time).

4.3. Cell passaging scheme

The scheduling of a batch is often an important factor to consider, especially for routine manufacturing. For easy planning and operation coverage, it is highly desirable that every passage has a fixed splitting schedule. A general rule in cell passaging is to split cells prior to their entering the stationary growth phase, which will minimize the lag phase in the subsequent passage. With a fixed time for each passage, it is very important to gather sufficient cell growth data to be able to accurately predict the batch-to-batch variations in terms of cell growth rate and peak cell concentration. In the absence of sufficient data, it is often safe to assume no greater than a ten-fold cell expansion for each passage for process design purposes. An amplification factor of greater than 10-fold may run into the risk of passaging cells in stationary phase should cells grow at the high end of the growth rate range, as illustrated in Fig. 1. This can be circumvented if cell concentration is monitored frequently, but a flexible splitting schedule is needed that may not always be feasible in routine manufacturing.

5. Cell culture process development and optimization

5.1. Process development

With the exception of a few well established biotechnology companies where platform cell culture technology has been established, the majority of the small biotechnology companies need to go through the entire process development cycle for product candidates. In the early stages of product development, the highest priority is often to bring the product candidate to human testing as quickly as possible and hence manufacturing cost is not as important, considering the relatively small quantities required for early phase trials. It is therefore more important to produce a small amount of high quality product in the shortest period of time, although with a higher unit cost, than to develop a sophisticated and efficient process. A simple process is often preferred that requires minimum process development and can be easily implemented in cGMP production.

Once the product candidate is in the clinical testing, further development of the process and scale-up occur concurrently with the clinical trials. As the product candidate moves to larger late stage testing, the required quantities also increase. In addition, the production process needs to resemble the final manufacturing process as much as possible. Delays in the clinical trials due to process development and clinical manufacturing issues should be avoided to minimize the product development time. This requires careful planning and risk management in process development. Decisions on the timing of investment in process development need to be made without knowing the probability of success of the product candidate in clinical testing. It is especially difficult in small companies where resources and budgets may be limited. Adequate allocation of resources early on in the program to ensure development of a robust and consistent process is critical to minimize potential difficulties in routine manufacturing or in obtaining regulatory approval.

5.2. Process optimization strategies

Protein production in cell culture is a function of the cell concentration, cell specific productivity, and culture duration. If the cell specific productivity is constant over the course of the culture, the final protein concentration is proportional to the time integral of the viable cell concentration (IVC). In general, there are two distinct but non-competing approaches to optimize protein production. One is to increase the cell specific productivity and the other is to increase IVC. The specific productivity can either be cell growth-associated or growth rate-independent, but can often be increased by genetic means, such as promoter optimization, gene amplification, clonal selection, as well as by manipulating the cell culture environment, in particular increasing the osmotic pressure, adding sodium butyrate, optimizing culture temperature, and increasing shear stress [19–22]. The IVC can be increased by maintaining a high viable cell concentration for an extended period of time through process optimization by maximizing cell growth and viability. When choosing a certain strategy for process optimization, one needs to consider the impact of the process changes on product stability, process scale, genetic stability of the host cells, and operational complexity.

6. Large scale cell culture process choices

6.1. Batch culture

Batch operation remains one of the simplest cell culture methods. It is most frequently used in cGMP production in the early phase of clinical trials where optimizing the process yield and efficiency is not as important as producing material for clinical trials as early as possible. Owing to its simplicity, the development and implementation of a batch process is relatively straightforward and rapid. On the other hand, due to the onset of nutrient limitations and accumulation of inhibitory by-products, batch culture typically leads to a maximum cell concentration in the range of 1 to 6×10^9 cells/l, and a protein concentration less than 500 mg/l. Although batch culture was widely used for commercial manufacturing of FDA-approved protein products prior to the 1990s, it has been increasingly replaced with more productive culture methods in the last decade. This is particularly true for monoclonal antibodies, which usually require a high dose and efficient production methods.

6.2. Fed-batch culture

Fed-batch culture is a simple variation of the batch culture. A fed-batch culture is initiated with a basal medium, followed by feeding of a concentrated nutrient solution at a certain stage of the culture to increase the cell concentration and to extend the culture life, while minimizing the dilution effect to the culture. The basal medium can be the same as that used in batch culture, in which case feeding is only needed at the late stage of a typical batch. By addressing the nutrient limitation issues in most batch cultures, this type of fed-batch culture has the advantage of simplicity in operation and yet can extend cell growth to reach a significantly higher product concentration than batch cultures [23–27].

In addition to nutrient limitation, accumulation of by-products such as lactate and ammonia can also inhibit cell growth. This can be addressed effectively by using a modified fed-batch mode. The nutrient level in the basal medium can be reduced significantly as long as the cell growth rate is not significantly compromised [28,29]. For example, most commercial media contain glucose at concentrations greater than 20 mM so that it is not depleted too early in a batch culture. However, for most mammalian cell lines, the maximum cell growth rate can be reached at glucose concentrations less than 2 mM. The benefit of a reduced glucose concentration is that lactate production is dramatically reduced. Because of the minimum amounts of nutrients present in the optimized basal medium, they are not sufficient to sustain cell growth for a significant period of time, and hence feeding of a nutrient mixture is needed from the beginning of the culture. The goal of the feeding is to provide all of the nutrients needed for the synthesis of cell mass and product so that the optimal culture environment can be maintained for as long as possible, despite cell metabolism and growth [28,30]. This is certainly not an easy task because a large number of nutrients are consumed for cell growth. It is critical to ensure that the formulation of the supplemental feed solution is stoichiometrically balanced to match the requirements for cell growth, which can be achieved by solving the stoichiometric equation that governs the cell growth through understanding of the cell metabolic network [31]. The feeding rate of the feed solution can be controlled by projection of cell growth (simulation) and actual measurements, or on-line monitoring of critical process parameters, such as cell concentration, oxygen consumption rate, and nutrient consumption [26,32]. The control of the culture environment minimizes the chances of nutrient depletion or accumulation and delays the onset of cell growth inhibition from lactate and ammonia accumulation, and ultimately leads to significantly improved process performance.

Since the 1990s, fed-batch culture has been extensively studied and has become a platform technology for large-scale manufacturing of therapeutic proteins [2,25,26]. It is simple to operate and can be easily retrofitted into existing facility previously designed for batch operation. No major modification of the bioreactors and other equipment is required. Also, it allows one to obtain a high product concentration and a high medium utilization efficiency for reduction of bulk manufacturing costs. In a 550 h hybridoma cell culture, the application of an optimized stoichiometric feeding strategy achieved a total cell concentration of 5×10^{10} cells/l and a monoclonal antibody concentration of 2.4 g/l [29]. Through the tight control of the nutritional environment, the reduction of lactate and ammonia production rate by 83-fold and 6-fold, respectively, from a typical batch culture played an important role in achieving improved process performance. Fed-batch cultures using different feeding strategies have also been developed for NS0 cell lines and CHO cells in industrial applications, reaching product concentrations around 3 g/l [27,33,34].

6.3. Continuous culture

Unlike batch and fed-batch cultures where the final product is harvested in a batch fashion, cells and product are continuously removed from the culture vessel and fresh culture medium is added to maintain a constant culture volume in a continuous culture. In this mode of operation, the dilution rate, defined as the ratio of fresh medium added per day to the volume of the culture, equals the cell growth rate and must be less than the maximum cell growth rate in order to reach a steady state and avoid a washout of the cells. Continuous culture has been shown to be a powerful tool for studying cell metabolism, physiology, and genetic stability in research. However, its utility in commercial protein production is less attractive, mainly because of low cell and product concentration, large volume for downstream processing, and difficulties in defining a production lot [35]. Regulatory concerns on cell line stability and process validation are also significant challenges to overcome [36–38].

6.4. Perfusion culture

A more attractive continuous operation mode is perfusion culture. Unlike simple continuous culture, cells in the harvest stream are retained or recycled back to the production vessel. This leads to a very high cell concentration and higher product concentration than in a continuous culture [39,40]. Usually a maximum perfusion rate of 1–3 volume exchanges per day is used at the peak cell concentration and cell growth is maintained at a low rate due to cell retention. Sometimes, a portion of the cells have to be bled from the culture in order to achieve a stable cell concentration in the bioreactor. Perfusion culture is most suitable for cell lines with non-growth-associated production rates because cell growth is usually very low [41].

Various devices have been designed for cell retention or recycling as reviewed by Su [42]. These include internal and external spin filters, continuous centrifugation, cell settlement devices by gravity and by acoustic waves, membrane separation, entrapment devices (such as fibers, ceramics, microcarrier), and hollow fiber systems [42,43].

Perfusion operation is significantly more complex than batch, fed-batch, and continuous cultures as special devices for cell retention or recycle are needed. The efficiency of medium utilization is typically lower than batch and fed-batch cultures but higher than a simple continuous culture. Owing to the complexity in operation, perfusion culture has not been attempted at a scale greater than 1000 L, as compared to scales greater than 10,000 liter for batch and fed-batch cultures. However, perfusion culture offers significant advantages in terms of higher product concentration and volumetric bioreactor productivity than a continuous culture [40,44,45]. In cases where the protein product is not very stable in the culture environment, perfusion culture is particularly attractive due to the low residence time of the product in the culture vessel as compared to a fed-batch culture.

7. Process scale-up challenges

Owing to the sensitivity of mammalian cells to the physical and chemical environment, achieving equivalent mass transfer (oxygenation and CO_2 removal) and mixing while preventing cell damage from shear becomes an increasingly challenging task as the reactor scale increases. It is well known that shear from mechanical agitation and gas sparging can damage cell membranes, lyse cells, and alter cell metabolism [43,46–49]. Culture pH, temperature, dissolved oxygen and carbon dioxide levels, and nutrient concentrations have significant impact on cell growth, metabolism, and product synthesis. As the scale of the culture vessel increases, it becomes increasingly more challenging to ensure rapid mixing to reach homogeneity after addition of a concentrated base solution for pH control. The same is true to ensure sufficient oxygenation and carbon dioxide removal, while the surface-to-volume ratio decreases with scale, and to ensure similar exposure of cells to mechanical shear as in a small scale [50–52], discussed below.

7.1. Mixing

Mammalian cell growth and metabolism are significantly slower than those of most microbes and the rates of chemical reactions, and hence the supply of nutrients is generally not limited by the rate of mixing even at large scales. In some rare cases of poor mixing, it can be problematic, especially when a concentrated solution is added to the culture. Addition of a concentrated base solution is often needed to neutralize the acidic substances such as lactate and carbon dioxide generated from cell metabolism. Local pH spikes occur at the point of the base addition [53]. Poor mixing can aggravate this problem and could adversely impact cell growth and lead to product degradation. Poor mixing at a large scale can also be problematic in fed-batch cultures where a concentrated nutrient mixture is added to the culture. Gradients in nutrient concentrations that result from poor mixing in a large-scale vessel can lead to low nutrient utilization efficiencies at the point of nutrient addition and cell starvation at the distant locations.

Mixing time is affected by various operating conditions, such as the agitation speed and gas sparging rate, as well as vessel configuration in terms of the agitator design, baffles, and vessel aspect ratio (the ratio of liquid height to vessel diameter). Mixing can be improved by increasing the agitation rate but often the maximum agitation rate is limited by potential foaming from gas entrainment and the sensitivity of cells to mechanical shear. Installation of baffles can reduce vortex flow, which improves mixing. In addition, subsurface addition has been shown to provide better dispersion of concentrated base and nutrient solutions than above surface addition.

7.2. Oxygenation

Mammalian cells require dissolved oxygen for growth and metabolism, but due to the low solubility of oxygen in culture medium, a continuous supply of oxygen from the gas phase is required. The oxygen uptake rate (OUR), is proportional to cell concentration and the specific oxygen consumption rate that is often relatively constant for a given cell line. To maintain a certain level of the dissolved oxygen, the oxygen transfer rate (OTR) from the gas phase to the liquid phase needs to be equal to or greater than OUR. The OTR is proportional to the mass transfer coefficient ($k_L a$) for oxygen and the driving force that increases with the partial pressure of oxygen in the gas phase. At a large scale, mass transfer through the head-space liquid surface is often not sufficient. It can be increased significantly by moving the top impeller closer to the gas–liquid interface [50], but often gas sparging through the bulk liquid is still required.

The oxygen transfer rate can be enhanced by increasing the mass transfer coefficient and/or the oxygen driving force. The overall mass transfer coefficient, $k_L a$, increases with increasing agitation and gas sparging rate. Under a constant gas sparging rate, the gas–liquid interface area increases as the bubble size is reduced, but smaller gas bubbles have been shown to be more detrimental to fragile mammalian cells, presumably due to the higher shear force generated upon rupture of smaller bubbles [47]. As a result, gas bubbles greater than 2 mm in diameter are often used in sparged cell cultures. The mass transfer driving force can be manipulated in two ways. One is to reduce the set point of the dissolved oxygen level in the bulk liquid as long as cell growth and product synthesis are not affected. The other is to increase the oxygen partial pressure in the gas phase either by increasing the head-space pressure and/or enriching the oxygen percentage in the inlet gas.

7.3. Carbon dioxide accumulation

Carbon dioxide is an end-product from glucose catabolism in mammalian cell culture. Carbon dioxide accumulation results in an increased osmolarity and has been shown to be detrimental to mammalian cells and to alter protein glycosylation [12,13]. Multiple factors affect the balance of CO_2 in cell culture, including its production rate, removal rate through the gas phase, bicarbonate addition for pH control, and the culture pH.

The CO_2 production rate, also called CO_2 evolution rate (CER), is proportional to cell concentration and is typically about 100–120% of the oxygen uptake rate [54,55]. Hence the CO_2 removal rate needs to be about 100–120% of the oxygen transfer rate to remove all of the CO_2 produced from cell metabolism. Carbon dioxide removal depends on the mass transfer coefficient ($k_L a$) for CO_2 and the concentration of dissolved CO_2 in the liquid phase that is affected by the bicarbonate (HCO_3^-) concentration and pH. At a low pH, the dissolved CO_2 concentration is significantly higher at the same bicarbonate concentration, which favors the removal of CO_2 due to a higher driving force. In a culture containing 25 mM bicarbonate, the maximum driving force for CO_2 removal at pH 7.0–7.3 is 1.5–2.8 mM, as compared to a maximum driving force for oxygen of ~ 0.2 mM, assuming a 0.21 atm oxygen partial pressure in the gas phase and a minimum dissolved oxygen level in the culture of 10% air saturation. The mass transfer coefficient for CO_2 is about 90% of that for oxygen. Under typical cell culture conditions where oxygenation is provided by air, the maximum CO_2 removal rate is usually significantly higher than the oxygen transfer rate and CO_2 production rate. As a result, a net reduction in CO_2 from culture medium is sufficient for pH control in batch cultures where modest amounts of lactate are produced.

The discussion above is based on the assumption that air is used for oxygenation and the gas phase CO_2 percentage can be maintained close to zero. At large scale, the residence time of gas bubbles in the bulk liquid increases with the liquid height, which leads to the accumulation of CO_2 in the gas bubbles as the bubble rise through the bulk liquid. This essentially reduces the driving force for CO_2 removal. In addition, oxygen enrichment in the inlet gas may be required to sustain a high cell concentration. When the oxygen percentage in the inlet gas increases to a certain level, it is possible that the driving force for oxygenation exceeds that for CO_2 removal. In other words, the CO_2 removal rate could become less than the CO_2 production rate. A net accumulation of CO_2 from cell metabolism could result. To make things worse, lactate production also increases with cell concentration, and bicarbonate is widely used to neutralize the lactate accumulation in order to control pH. This further increases the likelihood of CO_2 accumulation in the culture medium.

Several measures can be taken to avoid CO_2 accumulation in large-scale high cell concentration cultures. A simple approach is to reduce the culture pH as long as the cell growth and protein synthesis and quality are not adversely affected. A low pH not only increases the driving force for CO_2 removal but also could potentially reduce lactate production significantly, which reduces the need for bicarbonate addition [56–58]. Another measure is to use NaOH for pH control instead of $NaHCO_3$. The addition of concentrated NaOH needs to be carefully designed to avoid local pH spikes due to poor mixing. The combination of the use of a low concentration NaOH solution (0.3–0.5 M) and subsurface addition helps to reduce this problem. In high cell concentration cultures, one should avoid using an extremely high percentage of oxygen in the inlet gas. Instead, one should increase the mass transfer efficiency (such as agitation rate and impeller position) and/or gas sparging rate, which improves oxygenation as well as CO_2 removal. When appropriate measures are taken, CO_2 accumulation in large-scale mammalian cell culture can be effectively prevented.

8. Large scale bioreactor design

8.1. Choices of culture device

Disposable devices such as roller bottles, Nunc Cell Factories, Cell Cubes, and the Wave Bioreactor have been used in mammalian cell culture. The scale of each individual disposable device is often quite small, and hence a large number of units are often needed to produce a large quantity. Not surprisingly, these devices are often used in relatively small-scale production for clinical trials, rather than in routine manufacturing where large-scale stirred-tank bioreactors are often required.

8.2. Large scale stirred-tank bioreactor design

Mammalian cells lack a protective cell wall and are susceptible to damage from intensive shear from agitation and gas sparging. Considerations must be given to minimize cell damage from shear. The agitator type, size, and agitation rate need to be determined to ensure the maximum shear rate is far below the level that can damage cells. Gas bubbles are known to trap cells in the foam and release an enormous amount of energy that can damage cells when the bubbles rupture [59,60]. Although Pluronic F-68 (PF-68) is widely and successfully used in cell culture to reduce cell attachment to and damage from bubbles, it is still necessary to minimize the gas sparging rate.

The design of a large-scale stirred-tank bioreactor is a complex task. One needs to consider operating conditions, safety, mass and heat transfer, mixing, cleaning, sterilization, system integrity (mechanical seals and valves), material input and output, process monitoring and control, and so on [61]. Some engineering and process aspects are discussed briefly below.

8.2.1. Aspect ratio

Stirred-tank bioreactors for mammalian cell culture are generally cylindrical with aspect ratios (height to diameter) somewhat greater than 1. The aspect ratio is one of the most important parameters in vessel design for mammalian cell culture [62]. As the vessel scale increases, both the diameter and height increases if a constant aspect ratio is maintained. However, the mass transfer through the headspace gas-liquid interface decreases with the liquid height, owing to a reduced surface to volume ratio. On the other hand, mass transfer through gas sparging becomes more efficient as the residence time of gas bubbles in the bulk liquid increases with the liquid height. At a fixed sparging rate (volume of gas per volume of liquid per minute, VVM), the mass transfer coefficient, $k_L a$, increases with scale.

At a fixed scale, a low aspect ratio tends to increase the contribution of mass transfer through the headspace but decrease the contribution of the gas sparging. A high aspect ratio tends to decrease mass transfer through headspace but increase that of gas sparging. Obviously, the overall mass transfer coefficient does not always increase or decrease with the aspect ratio. It depends on the scale and the exact operating conditions. However, it is usually true that the headspace contribution to mass transfer in a typical large-scale stirred-tank bioreactor is relatively small as compared to gas sparging, and hence the overall mass transfer coefficient usually increases with an increased aspect ratio. The disadvantage of a slim tall vessel (high aspect ratio) is that it is more difficult to ensure axial mixing and hence the aspect ratio is typically maintained in a narrow range of 1 to 3.

8.2.2. Impellers and baffles

The impeller design is also very important to ensure optimal mixing while minimizing the mechanical shear exposure to cells. Many different types of impellers are designed and applied in mammalian cell culture, but marine propellors, Rushton turbine impellers, and pitch-blade impellers are the most common ones. The ratio of the tank diameter to the impeller diameter is typically between 2 and 3: about 3 for marine and Rushton impellers and about 2 for pitch-blade impellers [61]. Rushton impellers provide excellent mass transfer but poor axial mixing, while marine impellers provide good axial flows but are less efficient for mass transfer. Pitch-blade impellers presumably provide good mixing and mass transfer, while generating less shear, but they are not as well characterized as marine and Rushton impellers. For tall vessels (aspect ratio greater than 1), multiple impellers (2–3) are typically used to ensure good mixing and they are spaced one impeller diameter apart.

It is advantageous to have the impeller propel downward (for marine and pitch-blade impellers), which helps to increase the residence time of gas bubbles and hence improve mass transfer. In addition, less time is needed to reach homogeneity after addition of concentrated base or nutrient feed solutions, especially from above surface addition.

Baffles are often used in stirred-tank bioreactors to minimize fluid swirling and vortexing, which results in air entrainment, cell damage, poor vertical mixing, and variable power consumption. Three to four baffles are typically installed (either permanently welded to vessel or removable) per vessel and the width is typically 1/10 of the vessel diameter.

8.2.3. Sparger design

Several types of spargers, including fritted spargers with micropores, ring spargers with pore sizes on the order of millimeters, and point spargers with a single opening, have been employed in mammalian cell culture vessels. The size of the bubbles is affected by the pore size of the sparger as well as the fluid dynamics. For example, large gas bubbles exiting a point sparger can be broken up into smaller and more stable gas bubbles. It is known that more severe shear is generated from rupture of smaller bubbles and as a general rule, bubble sizes are maintained at greater than 2 mm in mammalian cell cultures [47]. The sparger is usually placed underneath the bottom impeller to improve bubble distribution and breakup of large gas bubbles.

9. Process validation

Validation of the protein production process is required to demonstrate consistency of the process and to obtain market approval from regulatory agencies worldwide [63]. Although regulatory guidelines were originally developed primarily for pharmaceutical products, it is even more critical to have a well characterized and validated process for the more complex and less defined biological molecules.

Validation of the process allows one to define the operation boundary of the process conditions and hence to ensure the consistency of the final product quality [64,65].

Different validation approaches have been applied in different companies, but the general concept remains the same. The goal is to identify all critical process parameters (CPP) that have a potential impact on the product quantity and quality and to define the acceptable operating ranges for all CPPs so that pre-determined specifications on all critical quality attributes (CQA) can be met. The identification of CPPs and CQAs requires extensive documented experimental evidence for justification.

Process development usually occurs concurrently with the clinical trials. In early phases of clinical trials (Phase I and II), not all process conditions have been locked-in and the process scale for clinical manufacturing is usually small. It is therefore not feasible or required to conduct a comprehensive process validation. Typically, a thorough process validation exercise is conducted during the full-scale consistency lot manufacturing. A validation protocol that specifies all CPPs and CQAs needs to be written using data generated from scale-down experiments and pilot lot manufacturing. The protocol needs to be pre-approved prior to the start of what is typically a set of three consecutive consistency lots. Upon completion of the consistency lots, the results are analyzed according to the pre-approved protocol to demonstrate the consistency of the process.

10. Conclusion

Large scale mammalian cell culture technology is a comprehensive piece of art that only a handful of companies have been able to practice in routine manufacturing. With more than one hundred protein molecules in various stages of clinical development, the need for large-scale mammalian cell culture will certainly increase dramatically in the years to come. In this chapter, various process aspects of state-of-the-art large-scale mammalian cell culture technology for protein production have been briefly reviewed.

Abbreviations

CER	Carbon dioxide evolution rate, mmole/l/h
cGMP	current good manufacturing practices
CHO	Chinese hamster ovary cell line
CQA	Critical quality attributes
CPP	Critical process parameters
FDA	Food and drug administration
IVC	time integral of the viable cell concentration
k_La	mass transfer rate, 1/h
MCB	Master cell bank
OTR	Oxygen transfer rate, mmole/l/h

OUR	Oxygen uptake rate, mmole/l/h
PF-68	Pluronic F-68
T_d	cell doubling time, h
VVM	Volume of gas per volume of liquid per minute
WCB	Working cell bank

References

1. Stacey, G. (2000) In: Spier, R.E. (ed.) Encyclopedia of cell technology, Vol. 1. John Wiley & Sons, Inc, New York, pp. 79–82.
2. Chu, L. and Robinson, D.K. (2001) Curr. Opin. Biotechnol. 12, 180–187.
3. Ganguly, S. and Shatzman, A.R. (1999) In: Flickinger, M.C. and Drew, S.W. (eds.) The encyclopedia of bioprocess technology: fermentation, biocatalysis, and bioseparation, Vol. 2. John Wiley & Sons, Inc, New York, pp. 1134–1145.
4. Wurm, F. and Bernard, A. (1999) Curr. Opin. Biotechnol. 10, 156–159.
5. Wiebe, M.E. and May, L.H. (1990) In: Lubiniecki, A.S. Large-scale mammalian cell culture technology, Marcel Dekker Inc, New York, pp. 147–160.
6. Doyle, A. and Stacey, G. (2000) In: Spier, R.E. Encyclopedia of cell technology, Vol. 1. John Wiley & Sons, Inc, New York, pp. 285–293.
7. Eagle, H. (1955) Science 122, 501–504.
8. Casnocha, S.A. and Wolfe, R.A. (1999) In: Flickinger, M.C. and Drew, S.W. (eds.) The encyclopedia of bioprocess technology: fermentation, biocatalysis, and bioseparation, Vol. 3. John Wiley & Sons, Inc, New York, pp. 1660–1676.
9. Iscove, N.N. and Melchers, F. (1978) J. Exp. Med. 147, 923–933.
10. Jayme, D.W. (1991) Cytotechnol. 5, 15–30.
11. Tsao, Y., Gould, S.L., and Robinson, D.K. (2000) In: Spier, R.E. (ed.) Encyclopedia of cell technology, Vol. 1. John Wiley & Sons, Inc, New York, pp. 35–41.
12. Kimura, R. and Miller, W.M. (1996) Biotechnol. Bioeng. 52, 152–160.
13. Zanghi, J.A., Schmelzer, A.E., Mendoza, T.P., Knop, R.H., and Miller, W.M. (1999) Biotechnol. Bioeng. 65, 182–191.
14. Froud, S.J., Birch, J., McLean, C., Shepherd, A.J., and Smith, K.T. (1997) In: Carrondo, M.J.T., Griffiths, B. and Moreira, J.L.P. (eds.) Animal Cell technology, Vol. 1. Kluwer Academic Publishers, London, pp. 681–686.
15. Garnick, R.L. (1998) Dev. Biol. Stand. 93, 21–29.
16. Saluzzo, J.F. (1997) In: Carrondo, M.J.T., Griffiths, B. and Moreira, J.J.P. (eds.) Animal cell technology, Vol. 1. Kluwer Academic Publishers, Boston, pp. 153–157.
17. Liu, S., Carroll, M., Iverson, R., Valera, C., Vennari, J., Turco, K., Piper, R., Kiss, R., and Lutz, H. (2000) Biotechnol. Prog. 16, 425–434.
18. Kiss, R., Widrig, R., Magreta, P., Quan, C. and Cacia, J. (1997) Oral presentation at ACS national meeting, San Francisco, CA, April, 1997.
19. Bebbington, C.R., Renner, G., Thomson, S., King, D., Abrams, D., and Yarranton, G.T. (1992) Biotechnol. 10, 169–175.
20. Oh, S.K.W., Vig, P., Chua, F., Teo, W.K., and Yap, M.G.S. (1993) Biotechnol. Bioeng. 42, 601–610.
21. Oh, S.K.W., Nienow, A.W., Al-Rubeai, M., and Emary, A.N. (1989) J. Biotech. 12, 45–62.
22. Robinson, D. K., DiStefano, D., Gould, S. L., Cuca, G., Seamans, T. C., Benincasa, D., Munshi, S., Chan, C. P., Stafford-Hollis, J., Hollis, G. F., Jain, D., Ramasubramanyan, K., E., M. G. and Silberklang, M. (1995) In: H. Wang and T. Imanaka (eds.), Antibody Engineering, ACS, Worthington.
23. Bibila, T.A. and Robinson, D.K. (1995) Biotechnol. Prog. 11, 1–13.
24. Bibila, T.A., Ranucci, C.S., Glazomitsky, K., Buckland, B.C., and Aunins, J.G. (1994) Biotechnol. Prog. 10, 87–96.

622

25. Sauer, P.W., Burky, J.E., Wesson, M.C., Sternard, H.D., and Qu, L. (2000) Biotechnol. Bioeng. 67, 585–597.
26. Xie, L. and Zhou, W. (2002) In: S. Ozturk and W.S. Hu (eds.), "Cell culture technology for pharmaceutical and cell based therapies", accepted.
27. Zhou, W., Chen, C., Buckland, B., and Aunins, J. (1997) Biotechnol. Bioeng. 55, 783–792.
28. Xie, L. and Wang, D.I.C. (1994) Cytotechnol. 15, 17–29.
29. Xie, L. and Wang, D.I.C. (1996) Biotechnol. Bioeng. 51, 725–729.
30. Xie, L. and Wang, D.I.C. (1994) Biotechnol. Bioeng. 43, 1175–1189.
31. Xie, L. and Wang, D.I.C. (1994) Biotechnol. Bioeng. 43, 1164–1174.
32. Zhou, W., Rehm, J., and Hu, W.-S. (1995) Biotechnol. Bioeng. 46, 579–587.
33. Chang, D.Y.H. (2000) Cell Culture Engineering Conference VII. Santa Fe, NM.
34. Tsao, E. (2001) Oral presentation at 2001 Waterside Monoclonal Conference, Norfolk, VA.
35. Europa, A.F., Gambhir, A., Fu, P.-C., and Hu, W.-S. (2000) Biotechnol. Bioeng. 67, 25–34.
36. Barnes, L.M., Bentley, C.M., and Dickson, A.J. (2001) Biotechnol. Bioeng. 73, 261–270.
37. Schenerman, M.A., Hope, J.N., Kletke, C., Singh, J.K., Kimura, R., Tsao, E.I., and Floena-Wasserman, G. (1999) Biologicals 27, 203–215.
38. Sinacore, M.S., Charlebois, T.S., Drapeau, D., Leonard, M., Harrison, S., and Adamson, S.R. (2000) In: Spier, R.E. (ed.) Encyclopedia of cell technology, Vol. 1. John Wiley & Sons, New York, pp. 458–472.
39. Ryll, T., Dutina, G., Reyes, A., Gunson, J., Krummen, L., and Etcheverry, T. (2000) Biotechnol. Bioeng. 69, 440–449.
40. Yang, J.-D., Angelillo, Y., Chaudhry, M., Goldenberg, C., and Goldenberg, D.M. (2000) Biotechnol. Bioeng. 69, 74–82.
41. Robinson, D.K., Widmer, J., and Memmert, K.W. (1992) J. Biotechnol. 22, 41–50.
42. Su, W.W. (2000) In: Spier, R.E. (ed.) Encyclopedia of cell technology, Vol. 1. John Wiley & Sons, New York, pp. 230–242.
43. Griffiths, J.B. (1999) In: Flickinger, M.C. and Drew, S.W. (eds.) The encyclopedia of bioprocess technology: fermentation, biocatalysis, and bioseparation, Vol. 3. John Wiley & Sons, Inc, New York, pp. 1594–1607.
44. Chuppa, S., Tsai, Y.-S., Yoon, S., Shackleford, S., Rozales, C., Bhat, R., Tsay, G., Matanguihan, C., Konstantinov, K., Naveh, D. (1997) Biotechnol. Bioeng. 55, 328–338.
45. Konstantinov, K.B., Tsai, Y., Moles, D., and Matanguihan, R. (1996) Biotechnol. Prog. 12, 100–109.
46. Chalmers, J. (2000) In: Spier, R.E. (ed.) Encyclopedia of cell technology, Vol. 1. John Wiley & Sons, New York, pp. 41–51.
47. Chisti, Y. (2000) Trends Biotechnol. 18, 420–432.
48. Chisti, Y. (1999) In: Flickinger, M.C. and Drew, S.W. (eds.) The encyclopedia of bioprocess technology: fermentation, biocatalysis, and bioseparation, Vol. 5. John Wiley & Sons, Inc, New York, pp. 2379–2406.
49. Thomas, C.R. (1990) In: Windler, M.A. Chemical Engineering problems in biotechnology, Vol. 5. Elsevier Applied Science, London, pp. 23–93.
50. Aunins, J.G. and Henzler, H.-J. (1993) In: Rehm, H.-J., Reed, G., Puhler, A. and Stadler, P. (eds.) Biotechnology, Vol. 3. VCH, New York, pp. 219–281.
51. Chisti, Y. (1999) In: Flickinger, M.C. and Drew, S.W. (eds.) The encyclopedia of bioprocess technology: fermentation, biocatalysis, and bioseparation, Vol. 3. John Wiley & Sons, Inc, New York, pp. 1607–1640.
52. Hu, W.S. and Aunins, J.G. (1997) Curr. Opin. Biotechnol. 8, 148–153.
53. Langheinrich, C. and Nienow, A.W. (1999) Biotechnol. Bioeng. 66, 171–179.
54. Bonarius, H.P.J., Gooijer, C.D.d., Tramper, J., and Schmid, G. (1995) Biotechnol. Bioeng. 45, 524–535.
55. Xie, L. and Wang, D.I.C. (1996) Biotechnol. Bioeng. 52, 591–601.
56. Miller, W.M., Blanch, H.W., and Wilke, C.R. (1988) Biotechnol. Bioeng. 32, 947–965.
57. Ozturk, S.S. and Palsson, B.O. (1991) Biotechnol. Prog. 7, 481–494.
58. Schmid, G., Balnch, H.W., and Wilke, C.R. (1990) Biotechnol. Lett. 12, 633–638.

59. Chalmers, J.J. (1994) Cytotechnol. 15, 311–320.

60. Trinh, K., Garcia-Briones, M., Hink, F., and Chalmers, J.J. (1994) Biotechnol. Bioeng. 43, 37–45.

61. Charles, M. and Wilson, J. (1994) In: Lydersen, B.K., D'Elia, N.A. and Nelson, K.L. (eds.) Bioprocess engineering: systems, equipment and facilities, Vol. 3. John Wiley & Sons, Inc, New York, pp. 4–67.

62. Palomares, L.A. and Ramirez, O.T. (2000) In: Spier, R.E. Encyclopedia of cell technology, 1. John Wiley & Sons, New York, pp. 183–201.

63. Sofer, G. and Hagel, L. (1997) Handbook of process chromatography: a guide to optimization, scale-up and validation. Academic Press, New York.

64. Moran, E.B., McGowan, S.T., McGuire, J.M., Frankland, J.E., and Oyebade, I.A. (2000) Biotechnol. Bioeng. 69, 242–255.

65. Zabriskie, D.W., Smith, T.M., and Gardner, A.R. (1999) In: Flickinger, M.C. and Drew, S.W. (eds.) The encyclopedia of bioprocess technology: fermentation, biocatalysis, and bioseparation, Vol. 4. John Wiley & Sons, Inc, New York, pp. 2070–2079.

S.C. Makrides (Ed.) *Gene Transfer and Expression in Mammalian Cells*

CHAPTER 24

Protein production in transgenic animals

Yann Echelard and Harry M. Meade

GTC Biotherapeutics, Inc, Five Mountain Road, Framingham, MA 01701, USA
Tel.: (508) 370-2420; fax: (508) 370-5299. E-mail: yann.echelard@gtc-bio.com

1. Introduction

Since insulin became available for the treatment of diabetes in 1922, protein therapeutics have been derived from fluids or tissues of animal origin. The biotechnology and genomic revolutions have led to the identification of an increasing number of human proteins and their development for therapeutic uses. Concurrently, the realization during the 1980s of the consequences of the use of human blood or tissue sources contaminated with adventitious infectious agents has provided added incentives for the development of recombinant production strategies for human polypeptides. A number of methods are employed to manufacture recombinant proteins for pre-clinical and clinical uses. Bioreactors using bacterial, yeast and mammalian cell cultures are routinely used for production of approved protein therapeutics. However, there is a need for alternative methods. For example, cost-effective recombinant production of complex molecules such as antithrombin, α_1-antitrypsin, or serum albumin, all used in high doses and currently extracted from human plasma, has not yet been achieved in microbial or mammalian cell bioreactors. In addition, since capital investments associated with production plants represent a significant portion of the development cost of new recombinant drugs, the inherent risk associated with the regulatory approval process is another stimulus for the development of flexible and inexpensive approaches for the manufacture of therapeutic proteins.

The introduction of foreign DNA, first into the mouse [1] and later into the livestock genome [2], as well as the characterization of tissue-specific regulatory sequences have provided the opportunity to express a wide range of proteins of biopharmaceutical interest in farm animals. An expression vector, comprising a gene encoding the human or humanized target protein fused with tissue-specific regulatory sequences, is inserted into the germline of the selected production species. When integrated, the expression construct becomes a dominant genetic characteristic that is inherited by the progeny of the founder animal. This general strategy makes it possible to harness the ability of animal tissues to produce large quantities of complex proteins. It has been utilized to target expression of therapeutic proteins to the milk, blood, urine and even the seminal fluids of various species. In addition, there is intense research on the targeted expression of foreign proteins to chicken eggs. Among these transgenic animal production systems, the transgenic mammary gland is the most advanced. Although most of this chapter will be dedicated to reviewing this system, the strategies and challenges relevant to the production of recombinant proteins in the milk of transgenic animals are largely applicable to other systems.

2. Production of recombinant proteins in the milk of transgenic animals

2.1. General overview

Following the first descriptions of the targeting of heterologous proteins to the mammary gland of transgenic mice [3–5] there was a flurry of articles reporting the generation of transgenic sheep, goats, cows and pigs carrying milk-specific transgenes (reviewed in Refs. [6–9]). Production of human recombinant protein pharmaceuticals in the milk of transgenic farm animals solves many of the problems associated with either microbial or animal cell expression systems. Bacteria often improperly fold complex proteins, leading to involved and expensive refolding processes, and both bacteria and yeast lack adequate post-translational modification machinery for mammalian-specific N- and O-linked glycosylation, gamma-carboxylation, and proteolytic processing. Cell culture systems require high initial capital expenditures, lack scale-up (or down) flexibility, use large volumes of culture media, and often result in relatively low yields. Transgenic livestock can be maintained and scaled-up in relatively inexpensive facilities, use cheap raw materials (animal feed), and can achieve impressive yields of recombinant proteins. Limitations of the transgenic expression systems are related to time-lines and the potential adverse effects of bio-active heterologous proteins on the health of the production animals. In addition, although transgenic expression systems are able to perform complex N- and O-linked glycosylation and gamma-carboxylation, there are species- and tissue-specific characteristics for these modifications that may affect the appropriateness of a given system for the expression of specific proteins or class of proteins (see Section 2.2.3). This latter situation is also an issue with mammalian cell culture.

To target a recombinant protein to the milk of a transgenic animal (Fig. 1), an expression vector containing a gene encoding the protein of interest fused to milk-specific regulatory elements is usually introduced in the germline of the chosen production species. Pronuclear microinjection of one-cell embryos or, alternatively, transfection into a primary cell population suitable for somatic cell nuclear transfer have both been used to generate transgenic founders. Following integration into the germline, the mammary gland-specific transgenes are predictably inherited by the offspring of the founder animal. Often, transgenic animals will express the protein(s) of interest at concentrations surpassing 2 mg/ml (Table 1) depending on the mammary-specific regulatory sequences, the gene expressed, and the integration site of the transgene. Milk can easily be obtained using established large-scale technologies of the dairy industry, and is an excellent starting material from which recombinant therapeutic proteins can be purified.

2.2. Production species

The choice of the production species is largely driven by the expected quantity of the therapeutic protein needed. As summarized in Table 2, there is usually a trade-off between milk yield and time to natural lactation. Another consideration may be a

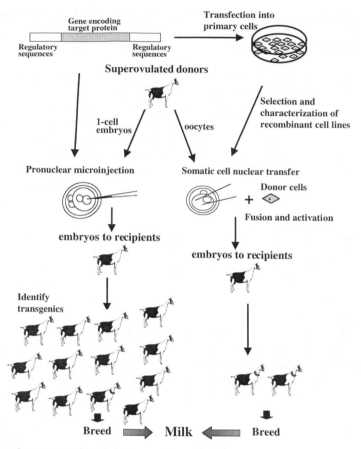

Fig. 1. Schematic representation of the transgenic production process.

species-specific ability to perform specialized post-translational modifications more efficiently.

2.2.1. Mice

Transgenic mice are mainly used for the testing of expression constructs prior to or concomitant with the generation of larger founder transgenic animals. The milk promoter sequences from all mammals tested appear capable of expression in the mouse mammary gland [6,7,9,10]. This allows the relatively inexpensive and rapid evaluation and optimization of transgene constructs. The access to this rapid, reliable test system has proven crucial to the development of milk expression technology, allowing the definition of regulatory sequences that efficiently target expression of heterologous genes to the mammary gland (reviewed in Ref. [10]). Transgenic mice can also provide an early warning of potential adverse effects on the health of animals caused by the production of the recombinant protein in milk.

Table 1

Summary of published data reporting the expression of recombinant proteins in the milk of transgenic large animals. Expression of human growth hormone and human factor IX in the milk of transgenic goats was obtained by introducing the transgene directly into the mammary gland using a "gene therapy" approach

Species	Therapeutic protein	Regulatory sequences	Maximum expression	References
Pig	Human protein C	Mouse WAP	2 mg/ml	[19]
	Human factor VIII	Mouse WAP	2.7 µg/ml	[18]
	Human factor IX	Mouse WAP	0.2 mg/ml	[19]
Sheep	Human α_1-antitrypsin	Sheep β-lactoglobulin	37 mg/ml	[6]
	Human factor IX	Sheep β-lactoglobulin	0.025 ng/ml	[6]
	Human fibrinogen	Sheep β-lactoglobulin	5 mg/ml	[6]
	Human protein C	Sheep β-lactoglobulin	0.3 mg/ml	[6]
	Human factor VIII	Sheep β-lactoglobulin	4-6 ng/ml	[58]
Goat	Human long-acting tPA	Goat β-casein	8 mg/ml	[7]
	Human antithrombin	Goat β-casein	14 mg/ml	[7]
	Human α_1-proteinase inh.	Goat β-casein	20 mg/ml	[7]
	Anti-Lewis Y IgG1	Goat β-casein	14 mg/ml	[8]
	Human G-CSF	Goat β-casein	50 µg/ml	[59]
	Human growth hormone	MoMLV LTR	118 ng/ml	[48]
	Human factor IX	Bovine β-casein	13. 7 ng/ml	[49]
Cattle	Human lactoferrin	Bovine α_{S1}-casein	3.5 mg/ml	[25]
	Human α-lactalbumin	Human α-lactalbumin	2.4 mg/ml	[6]
	Human serum albumin	Goat β-casein	48 mg/ml	[26]

Table 2

Summary of relevant reproductive and lactation characteristics of species used in the transgenic production of recombinant proteins. The data represent average numbers and usually apply to dairy breeds for the livestock species. There are often wide variations according to breed (or strain) within a species

Species	Reproductive maturity (months)	Length of gestation (months)	Average number of offspring	Average% of transgenic birth per total offspring	Average yield per natural lactation (l)	Total protein content in the milk (%)[a]
Mouse	1	0.75	10	10–20	0.0015	9.0
Rabbit	6	1	8	5–15	1.5	13.9
Pig	8	4	9	5–15	80–120	4.8
Sheep	8	5	2	3–5	300–500	5.5
Goat	8	5	2	3–5	600–800	2.9
Cattle	15	9	1	0.5–3	8–10,000	3.4

[a]Data taken from Ref. [60].

Obviously, the very limited milk yield from transgenic mice restricts expression of recombinant proteins to milligram amounts. But in some instances, this is sufficient to obtain meaningful data on the therapeutic protein of interest. As an example, it was possible to purify enough Malaria antigen $MSP1_{42}$ from transgenic mouse milk to test for immune protection in a primate model [11].

2.2.2. Rabbits

Transgenic rabbits have several advantages over other transgenic bioreactors: generation of transgenic animals by microinjection is straightforward and inexpensive relative to ruminants; rabbits have a short gestation interval that allows up to 8 lactations/year; and up to 1.5 l of milk can be obtained per lactation, a 1000-fold what can be obtained from mice. Rabbits have been used successfully for the production of gram quantities of proteins [12–15]. Labor-intensive milking and high husbandry costs may limit the use of transgenic rabbits to the production of therapeutic proteins needed in small quantities. One example of such an indication is the rare and severe lysosomal storage disorder, Pompe disease. In the severe infantile cases of this disease, characterized by a deficiency of acid α-glucosidase needed for the degradation of lysosomal glycogen, patients do not survive beyond the age of one year. In a small trial, recombinant α-glucosidase produced in the milk of transgenic rabbits was used to treat four young patients [16], stabilizing what would normally be a rapidly declining clinical condition. This trial indicated that protein therapy could be beneficial for Pompe disease patients. However, scale-up of the rabbit production system for α-glucosidase may still be problematic, and other systems (or production animals) with higher yields may be more adequate for the larger patient population.

2.2.3. Pigs

The production in the milk of transgenic sows of human Protein C [17], factor VIII [18], and factor IX [19] has been reported. Although pigs are not ordinarily viewed as dairy animals, a surprising amount of milk (100–200 liters) can be obtained from a lactating sow. With these production levels, pigs could be useful for the production of some of the vitamin K-dependent proteins such as factor IX and Protein C, which have estimated needs in the low kilogram range. Interestingly, pig mammary gland cells appear to carry out the complex post-translational modifications (γ-carboxylation, proteolytic processing) on factor IX and Protein C at rates higher than those encountered with mammalian cell and transgenic mouse milk systems [20].

2.2.4. Sheep

The first published report of production of therapeutic proteins in the milk of transgenic dairy farm animals was the targeting of factor IX (0.025 ng/ml) and α_1-antitrypsin (5 μg/ml) to the milk of transgenic ewes [21]. More recently, other proteins such as fibrinogen and factor VIII have also been expressed in the mammary gland of transgenic sheep (Table 1). The most extensively characterized of all sheep milk-expressed recombinant proteins is α_1-antitrypsin. An expression level in excess of 30 g/l was achieved in the milk of a founder transgenic ewe [6]. This was obtained using a chimeric transgene, which consisted of a genomic minigene encoding α_1-antitrypsin fused to the sheep β-lactoglobulin promoter sequences. The transgene integration in this particular line proved unstable, and another transgenic sheep line was identified from which heterozygous females produced α_1-antitrypsin at levels of 13–16 g/l [6]. A herd generated from this line has been propagated, producing a large amount of purified material for use in clinical trials [6].

Aerosolized recombinant α_1-antitrypsin derived from sheep milk is currently in phase II trials for both the cystic fibrosis and the hereditary emphysema indication (PPL Therapeutics Inc., [22])

2.2.5. Goats

A number of recombinant therapeutic proteins has been expressed in transgenic goats (Table 1). Dairy goats, with an average milk output per doe on the order of 600–800 liters per natural lactation, are very well adapted to the production of therapeutic proteins. Concentrations of heterologous therapeutic proteins in excess of 5 g/l of milk have been achieved reproducibly. The time-line from initiation of transgene transfer to natural lactation of resulting transgenic does is 16–18 months for goats. A large number of production females can be easily generated from a transgenic male using standard artificial insemination or embryo transfer techniques. Relatively small herds of a few hundred transgenic does can then easily yield several hundreds of kilograms of purified product per year. This level of production can meet the manufacturing needs of the large number of recombinant antibodies currently in development, several having been expressed in the milk of transgenic goats [8].

Of all recombinant proteins produced in transgenic expression systems, milk-derived antithrombin has gone farthest down the path of regulatory approval [23]. Clinical trial material was obtained using a process that achieves greater than 99.99% purity, with a cumulative yield in excess of 50% [24]. Structurally, the recombinant human antithrombin derived from transgenic goat milk was indistinguishable from antithrombin obtained from human plasma fractionation, with the exception of the carbohydrates. However, the milk-derived antithrombin showed equivalent bioactivity in various pre-clinical and clinical situations [23]. Two phase-III clinical trials, evaluating the effects of recombinant antithrombin in treating heparin resistance in cardiac surgical patients, have been completed, attaining their primary endpoint (Genzyme Transgenics Corp., [24]). Further clinical development of recombinant antithrombin is currently underway.

2.2.6. Cows

In terms of transgenic production, cattle have important advantages and one major drawback. Dairy cows have a yearly milk output in the range of 10,000 liters; consequently, with concentrations routinely achieved with most mammary gland-specific proteins, yields of tens of kilograms of recombinant proteins can be produced by one lactating transgenic cow. In addition, due to the economic importance of bovine agriculture, embryo culture and transfer technologies are well established for cattle breeds. This, combined with the large availability of oocytes derived from abattoirs, allows for the efficient generation of transgenic cows by somatic cell nuclear transfer. However, with cattle it takes almost 3 years from the onset of transgene transfer to obtain milk from a natural lactation. Strategies aiming at reducing this interval will probably need to be implemented for most projects. For example, somatic cell nuclear transfer to generate several identical founder animals combined with lactation induction, can significantly increase the speed at which large amounts of recombinant proteins are available [25,26]. Calves can be hormonally induced to lactate as early as 2 to 3 months after birth, making it possible to obtain recombinant proteins approximately one year after embryo transfer. Since 6–8-month-old calves can produce several liters of milk per day following hormonal induction (Echelard *et al.*, unpublished data), it is theoretically possible to start generating early clinical trial material less than 18 months after embryo transfer. Although these time-lines are not as short as normally achieved with mammalian cell culture, the tremendous scale-up potential offered by transgenic cattle may compensate for this drawback, especially for indications that necessitate large quantities of protein.

2.3. Engineering of mammary gland-specific transgenes

Transgenes that target the expression of heterologous proteins to the mammary gland are usually constructed by fusing the gene encoding the target protein to regulatory sequences of highly expressed milk-specific proteins. Transgenes (reviewed in Refs. [7,10]) have been constructed with the regulatory sequences of: the mouse- and rabbit-WAP genes; sheep β-lactoglobulin; goat, guinea pig, and bovine α-lactalbumin; rat, rabbit, goat, and bovine β-casein; and bovine-alpha$_{s1}$ casein. Depending on the regulatory sequences employed and the architecture of the target gene, varying levels of expression of the protein of interest were observed (see Table 1). In general, for the same gene, genomic sequences were expressed at higher levels than cDNAs [27,25], although the presence of regulatory elements within the introns of the target gene can be counter-productive in some instances [28].

The success of the transgenic production approach is dependent on high-level, well-controlled expression of the heterologous gene. The design of the transgene is crucial. It is particularly important to remove extraneous enhancers that may interfere with expression of the transgene in the mammary gland, or lead to undesirable ectopic expression. Cryptic splice sites have also been shown to have potentially deleterious effects on the RNA level of the target gene [29]. The mechanisms that govern the silencing of an artificially constructed chimeric transgene are not always well understood. Particularly useful is the ability to verify that a specific combination of mammary gland regulatory sequences and heterologous gene will lead to high-level secretion of the target protein in milk in the mouse system, in advance of generating transgenic large animals.

Since the generation of large transgenic animals can be technically challenging and very costly, efforts have been expanded on strategies aiming at decreasing position-effect sensitivity of transgenes to increase the proportion of productive lines. One strategy is to design very large transgenes using yeast [30] and bacterial artificial chromosomes [31], insuring the presence of most mammary gland-specific regulatory elements and shielding the transgene transcription unit from the surrounding chromatin. One drawback is that precise insertion of the foreign gene in such constructs may be challenging and that the co-expression of several proteins, as for the production of recombinant antibodies, may be impractical. Another strategy is to supply compact, well-defined insulator elements (see Chapter 11) to the transgene.

Experiments in transgenic Drosophila [32], as well as in transgenic mice [33,34], have shown that flanking a transgene with the chicken β-globin 5'cHS4 control element can significantly reduce silencing due to chromosomal position effects and promoter interference. We have used the chicken 5'cHS4 β-globin insulator sequence to shield the mammary-expression cassette from the surrounding chromatin and from the neomycin resistance cassette in constructs used to derive cows by somatic cell nuclear transfer [26].

2.4. Generation of transgenic animals

A number of approaches have been investigated for the development of transgenic animals. However, the great majority of transgenic large animal bioreactors have been derived either by pronuclear microinjection or, more recently, by somatic cell nuclear transfer. Other techniques are in development [35], and may eventually facilitate further transgenic modifications. An alternative approach is to introduce transgenes transiently in the mammary glands of lactating animals.

2.4.1. Pronuclear microinjection

To date, pronuclear microinjection is the method employed to produce most of the transgenic animal bioreactors. Direct microinjection of foreign genes into pronuclei was first reported for mice [1] and then adapted to gene transfer into rabbits, pigs and ruminants [2,36,37]. In general, fertilized eggs are collected from superovulated donors and at least one pronucleus is microinjected with a diluted DNA solution, introducing a few hundred copies of linear transgene(s) per embryo. This is followed by either uterine or oviductal transfer to a recipient, depending on the extent of *in vitro* culture after microinjection. After transfer, manipulated embryos are allowed to develop to term, and offspring are then screened for the presence of the transgene(s). Although this method is technically simple, the main challenge is to devise an efficient embryo collection and transfer system for the chosen species.

While successful, the technique of pronuclear microinjection has limitations. The frequency of transgene integration into the genome of large animals is low, with less than 0.1% (cattle) to up to 2% (pigs) of the microinjected embryos producing transgenic offspring [2,6,7,9,12,21,25,35–37]. Transgene integration into the genome is a random process, and the number of integrations, their position in the genome, and the number of transgene copies per cell, are unpredictable. The frequent generation of mosaic founder animals, carrying the transgene only in some of their cells, is another problem associated with pronuclear microinjection (reviewed in [38]).

2.4.2. Somatic cell nuclear transfer

The demonstration in sheep [39], cattle [40], goats [41], pigs [42,43], and rabbits [44] that differentiated somatic cells can function as karyoplast donors in nuclear transfer has expanded the range of possibilities for germline modification in transgenic livestock. Transgenes can be introduced into primary cell lines by transfection methods such as lipofection or electroporation. This allows for the characterization of the transgenic cell line before the generation of transgenic embryos. Recombinant primary cell lines can be screened for transgene copy-number, integrity of the transgene, and even chromosome localization [45]. Another advantage is the absence of transgene mosaicism, as all animals generated by somatic cell nuclear transfer should be fully transgenic, unless the primary cell line from which they are derived is not pure.

One issue with nuclear transfer is that it is more technically complex than microinjection. Appropriate primary cell culture and transfection conditions must be determined and there are infinite variations on karyoplast/cytoplast fusion and activation conditions. However, the main problem associated with somatic cell nuclear transfer is that, in contrast with microinjection, it is associated with the generation of a high proportion of animals suffering from various developmental abnormalities [46]. Despite these drawbacks, it is clear that for some systems, somatic cell nuclear transfer is more efficient. For example, in the generation of transgenic cattle for the production of recombinant human serum albumin, our group obtained 16 healthy founder calves by nuclear transfer [26], as compared to one transgenic calf by microinjection [47].

2.4.3. Transient approaches

Inspired by somatic cell gene therapy approaches, these strategies aim to bypass the germline in obtaining recombinant protein production in animals. In an early attempt, the mammary glands of goats hormonally induced to lactate were infected with replication-defective retroviral vectors carrying the human growth hormone (hGH) cDNA regulated by the Moloney murine leukemia virus (MoMLV) long terminal repeat (LTR) [48]. Initial expression levels of up to 118 ng of hGH/ml of milk stabilized, by day 3, to 3–12 ng/ml. Another group [49] directly introduced liposome-plasmid complexes into the mammary gland of lactating goats. In these constructs, the factor IX gene was regulated by bovine β-casein sequences. Production of human factor IX was 13.7 ng/ml of goat milk at day 3, with the majority of the recombinant factor IX active and correctly γ-carboxylated.

The advantage of the transient methods is that expression of the recombinant protein is obtained quickly, without the delay caused by the generation interval required to produce and breed transgenic animals. Moreover, the ability to quickly generate gram amounts of a recombinant protein in the mammary gland system would be invaluable. This material could be used to predict post-translational modifications of a specific protein in the prospective large-scale bioreactor, and possibly to generate pre-clinical and early-clinical phase data. The simplicity of the system could allow relatively cost-effective production of small amounts of many potential protein drug candidates and facilitate the evaluation of each drug. However, the production levels reported have been uniformly very low. This would preclude manufacturing use of this approach until a significant increase (at least 100-fold) in expression levels is achieved.

2.5. Protein purification from the milk of transgenic animals

The usual challenge of recombinant protein purification from mammalian cell culture is that high volumes of media must be processed due to low expression levels of the protein (often in the mg/l range). Typical contaminants in serum-free media are cell debris, lipids, host DNA, host cell proteins, viruses and pyrogens. In serum-supplemented cultures, other major impurities include bovine serum albumin, transferrin, and bovine IgG.

Recombinant proteins produced in the milk of transgenic goats at the relatively modest level of 2 g/l represent 7% of the total protein. Since recombinant proteins are usually expressed as whey proteins, the initial separation steps aim to achieve casein-free, fat-free and lactose-free protein concentrates in which the recombinant protein is at 30–60% purity, depending on the expression level. Fat may be separated from milk using standard centrifugal procedures or by membrane filtration [6–8,24]. Membrane filtration can also be used to eliminate most of the caseins and small molecules. The resulting clarified whey is an excellent starting material for chromatographic purification.

Several groups [6,24] have described purification processes for the isolation of clinical grade therapeutic proteins from the milk of transgenic large animals. For the purification of recombinant antithrombin from goat milk, the following steps result in a product with purity greater than 99.99%: clarification through a 500-kd tangential flow membrane filtration unit, capture and then elution from a heparin affinity chromatography column, anion-exchange chromatography and hydrophobic interaction chromatography [24]. The cumulative yield, after a number of production runs, was greater than 55% and there were no detectable contaminating proteins.

3. Other transgenic bioreactors

3.1. Production of recombinant proteins in the blood of transgenic animals

Fractionation of human and animal plasma is a well-established technology. In addition, porcine factor VIII has been used extensively for the treatment of hemophiliacs who develop inhibitors to human Factor VIII. Although blood collection is more invasive than milking, it can be applied several times during the lifespan of the production animal. These factors, and the availability of well-characterized blood-specific regulatory sequences, have led to the development of transgenic animals that produce recombinant proteins in their blood.

In 1992, the production of recombinant human hemoglobin A (HbA) was reported in transgenic pigs [50] carrying a transgene composed of the human β-globin locus control region linked to the human α- and β-globin genes. The objective was to produce recombinant human hemoglobin to be used as a cell-free oxygen carrying red-blood cell substitute. Levels of up to 9% human HbA dimers (human α/human β) were formed with the remainder of the hemoglobin being composed of hybrid dimers (12% human α/pig β) and of fully pig dimers (79%). In a follow-up report [51], a higher expression level of human hemoglobin (up to 24% human dimers) was achieved. This corresponds to 32 g of human hemoglobin per liter of transgenic pig blood. The improvement in expression appeared to be related to the use of the porcine β-globin promoter to drive the human β-globin gene. Animals appeared to tolerate well the high levels of human HbA in their erythrocytes. Structural and functional investigations of recombinant human HbA isolated from the erythrocytes of transgenic pigs confirmed correct expression, post-translational processing, and assembly of human HbA [52]. However, even with this high expression level, purification of human hemoglobin from the hybrid and pig hemoglobin may be too costly. With the advent of porcine somatic cell nuclear transfer, a strategy aiming at generating transgenic pigs producing solely human hemoglobin is now attainable, potentially decreasing production costs.

Another potential use of the recombinant hemoglobin expression system takes advantage of the property of terminally extended α-globin to remain functional. Carboxy-terminal fusions of human α-globin with α-endorphin and magainin were expressed in the erythrocytes of transgenic mice [53]. Fusion proteins were expressed at levels corresponding to 25% of the total hemoglobin, seemingly without ill effects on the transgenic mice. This strategy could become an alternative to the chemical approach for the large-scale synthesis of therapeutic peptides, especially if cost-efficient methods to cleave away the peptide of interest from the fusion protein were to be devised.

3.2. Transgenic chickens

The concept of targeting the expression of foreign proteins to the eggs of transgenic chickens has generated considerable enthusiasm worldwide, as judged by the large number of biotechnology companies engaged in research towards that goal. The total US egg production surpassed 85 billion in 2001 (National Agricultural Statistics Service, Agricultural Statistics Board, U.S. Department of Agriculture). It is easy to understand that even with transgenes that would allow expression of only a few hundreds of milligrams of therapeutic proteins per egg, transgenic chickens could be scaled up to commercial levels. Technologies for the manipulation of egg by-products are available, and vaccines, often injected into millions of people, are routinely produced from chicken eggs. Purification of proteins for pharmaceutical applications should then be readily achievable in this system. Another potential advantage, if tightly regulated egg-specific regulatory sequences can be identified, would be the possibility of producing very bio-active molecules such as insulin or erythropoietin without affecting the health of the transgenic hens.

At the time this chapter was completed, only one peer-reviewed article describing the production and yield of heterologous proteins in the eggs of transgenic hens has appeared [54]. One explanation is the relative inaccessibility of chicken embryos. Production of germline-transmitting transgenic chickens has been technically challenging. Techniques such as direct microinjection, stem cell- and primordial germ cell-based strategies, somatic cell nuclear transfer, sperm-mediated transgenesis, and viral-based methods have all been applied to chicken, and some progress has been reported [55,36]. In the only report on transgenic production in eggs [54], transgenic chickens were generated using replication-deficient retroviral vectors (derived from the avian leukosis virus) carrying the β-lactamase marker gene under the control of the CMV promoter. Expression of concentrations of up to 1.34 µg/ml of β-lactamase was recorded in egg whites over several generations. Predictably, due to the non-specific nature of the CMV regulatory sequences, even higher levels of β-lactamase (up to 6.7 µg/ml) were observed in the serum of transgenic hens. Once regulatory sequences that can efficiently target the expression of heterologous genes to the egg are discovered, commercially viable levels of recombinant proteins should be achievable.

3.3. Bladder-specific expression

Although the bladder does not come to mind as an obvious source of recombinant proteins, urine has long been used as a source of biologicals. For example, FSH purified from the urine of post-menopausal women is still used clinically for the treatment of anovulation and in assisted reproduction technologies such as *in vitro* fertilization (IVF) and intra-cytoplasmic sperm injection (ICSI). Additionally, several drug preparations used in estrogen replacement therapy are isolated from pregnant mare urine. In 1998, one group [56] described the use of the mouse uroplakin II (a bladder-specific gene) promoter to target the expression of hGH to the urine of transgenic mice. Low expression levels (up to 0.5 µg/ml) were observed. At these levels, commercial transgenic production of recombinant proteins would be impractical. If it were possible to significantly increase protein expression levels, the bladder-specific system may have the advantages that urine can be harvested soon after birth and throughout the life of the animal, and that urine is a less complex medium than milk or blood, potentially decreasing purification costs. However, it is not clear that it is physiologically possible to increase the protein secretion characteristics of the urothelium cells, nor that it would be compatible with the health of the producing animals.

3.4. Seminal vesicle

The potential of the seminal vesicle as a source for recombinant proteins was examined by a group that produced transgenic mice using a chimeric construct linking the murine P12 promoter to the hGH gene [57]. One line was found to express up to 0.5 mg of hGH/ml of pre-ejaculatory fluid obtained from seminal vesicles of sacrificed transgenic males. Application of this technology to pigs which can be ejaculated 2–3 times per week (200–300 ml per ejaculate), could lead to the production of several grams of recombinant protein per year.

4. Conclusions

Expression of recombinant proteins by transgenic animals represents an opportunity to achieve cost-effective, large-scale production of a wide variety of therapeutics. During the last 15 years we have seen the evolution of mammary-gland specific production from proof-of-concept to late clinical stage. At this juncture, it is clear that the mammary gland expression system can be used to produce large quantities of highly purified, appropriately modified complex proteins. The clinical efficacy of the proteins furthest along in clinical trials, i.e., antithrombin and α_1-antitrypsin, will determine regulatory approval, rather than issues related to the recombinant production system. Other transgenic production systems, such as transgenic chickens and blood-derived proteins, appear very promising but have not yet overcome basic technical issues or produced compounds in human clinical trials.

Acknowledgements

We would like to thank all our colleagues at Genzyme Transgenics Corporation for their contribution to the work reviewed in this chapter, as well as Dr. Yuko Fujiwara for her help with the figure.

References

1. Gordon, J.W., Scangos, G.A., Plotkin, D.J., Barbosa, J.A., and Ruddle, F.H. (1980) Proc. Natl. Acad. Sci. USA 77, 7380–7384.
2. Hammer, R.E., Pusrsel, V.G., Rexroad, C.E., Jr., Wall, R.J., Bolt, D.J., Ebert, K.M., Palmiter, R.D., and Brinster, R.L. (1985) Nature 315, 680–685.
3. Simons, J.P., McClenaghan, M., and Clark, A.J. (1987) Nature 328, 530–532.
4. Gordon, K., Lee, E., Vitale, J.A., Smith, A.E., Westphal, H., and Hennighausen, L. (1987) Bio/Technology 5, 1183–1187.
5. Meade, H., Gates, L., Lacy, E., and Lonberg, N. (1990) Bio/Technology 8, 443–446.
6. Clark, A.J. (1998) J. Mammary Gland. Biol. Neopl. 3, 337–350.
7. Meade, H.M., Echelard, Y., Ziomek, C.A., Young, M.W., Harvey, M., Cole, E.S., Groet, S., Smith, T.E., and Curling, J.M. (1998) In: Fernandez, J.M. and Hoeffler, J.P. (eds.) Gene Expression Systems: Using Nature for the Art of Expression, Academic Press, San Diego, pp. 399–427.
8. Pollock, D.P., Kutzko, J.P., Birck-Wilson, E., Williams, J.L., Echelard, Y., and Meade, H.M. (1999) J. Immunol. Methods 137, 147–157.
9. Houdebine, L.M. (2000) Transgenic Res. 9, 305–320.
10. Maga, E.A. and Murray, J.D. (1995) Bio/Technology 13, 1452–1457.
11. Stowers, A.W., Chen, L.H., Zhang, Y., Kennedy, M.C., Zou, L., Lambert, L., Rice, T.J., Kaslow, D.C., Saul, A., Long, C.A., Meade, H.M., Miller, L.H. (2002) Proc. Natl. Acad. Sci. USA 99, 339–344.
12. Stinnackre, M.G., Massoud, M., Viglietta, C., and Houdebine, L.M. (1997) In: Houdebine, L.M. (ed.) Transgenic Animals, Generation and Use, Harwood Academic Publishers, Amsterdam, pp. 461–463.
13. Korhonen, V.P., Tolvanen, M., Hyttinen, J.M., Uusi-Oukari, M., Sinervirta, R., Alhonen, L., Jauhiainen, M., Janne, O.A., and Janne, J. (1997) Eur. J. Biochem. 245, 482–489.

14. Stromqvist, M., Houdebine, L.M., Andersson, J.O., Edlund, A., Johansson, T., Viglietta, C., Puissant, C., and Hansson, L. (1997) Transgenic Res. 6, 271–278.

15. McKee, C., Gibson, A., Dalrymple, M., Emslie, L., Garner, I., and Cottingham, I. (1998) Nat. Biotechnol. 16, 647–651.

16. Van den Hout, J.M.P., Reuser, A.J., de Klerk, J.B., Arts, W.F., Smeitink, J.A., and Van der Ploeg, A.T. (2001) J. Inherit. Metab. Dis. 24, 266–274.

17. Velander, W.H., Johnson, J.L., Page, R.L., Russell, C.G., Subramanian, A., Wilkins, T.D., Gwazdauskas, F.C., Pittius, C., and Drohan, W.N. (1992) Proc. Natl. Acad. Sci. USA 89, 12,003–12,007.

18. Paleyanda, R.K., Velander, W.H., Lee, T.K., Scandella, D.H., Gwazdauskas, F.C., Knight, J.W., Hoyer, L.W., Drohan, W.N., and Lubon, H. (1997) Nat. Biotechnol. 15, 971–975.

19. Van Cott, K.E., Butler, S.P., Russell, C.G., Subramanian, A., Lubon, H., Gwazdauskas, F.C., Knight, J., Drohan, W.N., and Velander, W.H. (1999) Genet. Anal. Biomol. Eng. 15, 155–160.

20. Lubon, H. and Paleyanda, R.K. (1997) Thromb. Haemost. 78, 532–536.

21. Clark, A.J., Bessos, H., Bishop, J.O., Brown, P., Harris, S., Lathe, R., McClenaghan, M., Prowse, C., Simons, J.P., Whitelaw, C.B.A., Wilmut, I. (1989) Bio/Technology 7, 487–492.

22. Tebbutt, S.J. (2000) Curr. Opin. Mol. Ther. 2, 199–204.

23. Levy, J.H., Weisinger, A., Ziomek, C.A., and Echelard, Y. (2001) Semin. Thromb. Hemost. 27, 405–416.

24. Edmunds, T., Van Patten, S.M., Pollock, J., Hanson, E., Bernasconi, R., Higgins, E., Manavalan, P., Ziomek, C., Meade, H., McPherson, J.M., Cole, E.S. (1998) Blood 91, 4561–4571.

25. Brink, M.F., Bishop, M.D., and Pieper, F.R. (2000) Theriogenology 53, 139–148.

26. Echelard, Y., Destrempes, M.M., Koster, J.A., Blackwell, C., Groen, W., Pollock, D., Williams, J.L., Behboodi, E., Pommer, J., Meade, H.M. (2002) Theriogenology 57, 779.

27. Whitelaw, C.B.A., Archibald, A.L., Harris, S., McClenaghan, M., Simons, J.P., and Clark, A.J. (1991) Transgenic Res. 1, 3–13.

28. Hurwitz, D.R., Nathan, M., Barash, I., Ilan, N., and Shani, M. (1994) Transgenic Res. 3, 365–375.

29. Yull, F., Harold, G., Wallace, R., Cowper, A., Percy, J., Cottingham, I., and Clark, A.J. (1995) Proc. Natl. Acad. Sci. USA 92, 10,899–10,903.

30. Fujiwara, Y., Miwa, M., Takahashi, R., Hirabayashi, M., Suzuki, T., and Ueda, M. (1997) Mol. Reprod. Dev. 47, 157–163.

31. Stinnakre, M.G., Soulier, S., Schibler, L., Lepourry, L., Mercier, J.C., and Vilotte, J.L. (1999) Biochem. J. 339, 33–36.

32. Chung, J.H., Whiteley, M., and Felsenfeld, G. (1993) Cell 74, 505–514.

33. Taboit-Dameron, F., Malassagne, B., Viglietta, C., Puissant, C., Leroux-Coyau, M., Chereau, C., Attal, J., Weill, B., and Houdebine, L.M. (1999) Transgenic Res. 8, 223–235.

34. Potts, W., Tucker, D., Wood, H., and Martin, C. (2000) Biochem. Biophys. Res. Comm. 273, 1015–1018.

35. Wall, R.J. (2002) Theriogenology 57, 189–201.

36. Bondioli, K.R., Biery, K.A., Hill, K.G., Jones, K.B., and De Mayo, F.J. (1991) Biotechnology 16, 265–273.

37. Ebert, K.M., Selgrath, J.P., DiTullio, P., Denman, J., Smith, T.E., Memon, M.A., Schindler, J.E., Monastersky, G.M., Vitale, J.A., Gordon, K. (1991) Bio/Technology 9, 835–838.

38. Echelard, Y. (1997) In: Houdebine, L.M. (ed.) Transgenic Animals, Generation and Use, Harwood Academic Publishers, Amsterdam, pp. 233–235.

39. Wilmut, I., Schnieke, A.E., McWhir, J., Kind, A.J., and Campbell, K.H. (1997) Nature 385, 810–813.

40. Cibelli, J.B., Stice, S.L., Golueke, P.J., Kane, J.J., Jerry, J., Blackwell, C., Ponce de Leon, F.A., and Robl, J.M. (1998) Science 280, 1256–1258.

41. Baguisi, A., Behboodi, E., Melican, D.T., Pollock, J.S., Destrempes, M.M., Cammuso, C., Williams, J.L., Nims, S.D., Porter, C.A., Midura, P., Palacios, M.J., Ayres, S.L., Denniston, R.S., Hayes, M.L., Ziomek, C.A., Meade, H.M., Godke, R.A., Gavin, W.G., Overstrom, E.W., and Echelard, Y. (1999) Nat. Biotechnol. 17, 456–461.

42. Onishi, A., Iwamoto, M., Akita, T., Mikawa, S., Takeda, K., Awata, T., Hanada, H., and Perry, A.C. (2000) Science 289, 1188–1190.

43. Polejaeva, I.A., Chen, S.H., Vaught, T.D., Page, R.L., Mullins, J., Ball, S., Dai, Y., Boone, J., Walker, S., Ayares, D.L., Colman, A., Campbell, K.H. (2000) Nature 407, 86–90.

44. Chesné, P., Adenot, P.G., Viglietta, C., Baratte, M., Boulanger, L., and Renard, J.P. (2002) Nat. Biotechnol. 20, 366–369.

45. McCreath, K.J., Howcroft, J., Campbell, K.H., Colman, A., Schnieke, A.E., and Kind, A.J. (2000) Nature 405, 1066–1069.

46. McEvoy, T.G., Robinson, J.J., and Sinclair, K.D. (2001) Reproduction 122, 507–518.

47. Behboodi, E., Groen, W., Destrempes, M.M., Williams, J.L., Ohlrichs, C., Gavin, W.G., Broek, D.M., Ziomek, C.A., Faber, D.C., Meade, H.M., Echelard, Y. (2001) Mol. Reprod. Dev. 60, 27–37.

48. Archer, J.S., Kennan, W.S., Gould, M.N., and Bremel, R.D. (1994) Proc. Natl. Acad. Sci. USA 91, 6840–6844.

49. Zhang, K., Lu, D., Xue, J., Huang, Y., and Huang, S. (1997) Chin. J. Biotechnol. 13, 271–276.

50. Swanson, M.E., Martin, M.J., O'Donnell, J.K., Hoover, K., Lago, W., Huntress, V., Parsons, C.T., Pinkert, C.A., Pilder, S., Logan, J.S. (1992) Bio/Technology 10, 557–559.

51. Sharma, A., Martin, M.J., Okabe, J.F., Truglio, R.A., Dhanjal, N.K., Logan, J.S., and Kumar, R. (1994) Bio/Technology 12, 55–59.

52. Manjula, B.N., Kumar, R., Sun, D.P., Ho, N.T., Ho, C., Rao, J.M., Malavalli, A., and Acharya, A.S. (1998) Protein Eng. 11, 583–588.

53. Sharma, A., Khoury-Christianson, A.M., White, S.P., Dhanjal, N.K., Huang, W., Paulhiac, C., Friedman, E.J., Manjula, B.N., and Kumar, R. (1994) Proc. Natl. Acad. Sci. USA 91, 9337–9341.

54. Harvey, A.J., Speksnijder, G., Baugh, L.R., Morris, J.A., and Ivarie, R. (2002) Nat. Biotechnol. 20, 396–399.

55. Pain, B., Chenevier, P., and Samarut, J. (1999) Cells Tissues Organs 165, 212–219.

56. Kerr, D.E., Liang, F., Bondioli, K.R., Zhao, H., Kreibich, G., Wall, R.J., and Sun, T.T. (1998) Nat. Biotechnol. 16, 75–79.

57. Dyck, M.K., Gagne, D., Ouellet, M., Senechal, J.F., Belanger, E., Lacroix, D., Sirard, M.A., and Pothier, F. (1999) Nat. Biotechnol. 17, 1087–1090.

58. Niemann, H., Halter, R., Carnwath, J.W., Herrmann, D., Lemme, E., and Paul, D. (1999) Transgenic Res. 8, 237–247.

59. Ko, J.H., Lee, C.S., Kim, K.H., Pang, M.G., Koo, J.S., Fang, N., Koo, D.B., Oh, K.B., Youn, W.S., Zheng, G.D., Park, J.S., Kim, S.J., Han, Y.M., Choi, I.Y., Lim, J., Shin, S.T., Jin, S.W., Lee, K.K., Yoo, O.J. (2000) Transgenic Res. 9, 215–222.

60. Fox, P.F. and McSweeney, P.H.L. (1998) In: Houdebine, L.M. (ed.) Dairy Chemistry and Biochemistry, Blakie A & P, London, pp. 146–237.

S.C. Makrides (Ed.) *Gene Transfer and Expression in Mammalian Cells*

CHAPTER 25

Strategies for the purification of recombinant proteins

Steven L. Giardina

Purification Laboratory, Biopharmaceutical Development Program, SAIC Frederick, Inc., National Cancer Institute at Frederick, Building 320, PO Box B, Frederick, MD 21702-1201, USA
Tel.: 301-846-1821; Fax: 301-846-6886. E-mail: giardina@ncifcrf.gov

"The end must justify the means."

— *Hans Carvel (1700)*

1. Introduction

An understanding of the function of a gene requires that the investigator produce, isolate, and characterize the gene product. This is often a source of apprehension, if not outright anxiety, since proteins, unlike nucleic acids, are generally not amenable to a "one-size-fits-all" approach to purification. Recombinant proteins may be used for a variety of purposes ranging from laboratory reagents to therapeutics and vaccines intended for human or veterinary use. It is the intended use of the product, therefore, that will define what is meant by purity and the strategies that will lead to a successful outcome. No one wants to expend precious time and resources on what appears to be an unnecessary effort. However, focusing on short-term research needs without considering the possibility of future commercialization with its associated regulatory requirements can lead to costly mistakes and delays down the road. The requirements of the regulatory agencies begin at the time of discovery, and to ignore them is folly. While the bulk of the discussion will focus on current trends and methods for the purification of recombinant proteins, I also take the opportunity to point out regulatory concerns and issues, where appropriate. Consistent with the title of this book, the emphasis here is on the purification of recombinant proteins from mammalian cells.

2. Developing a rational approach for protein purification

The Holy Grail of protein purification would be defined as being able to purify large amounts of intact protein after a single-step purification process. The process would be rugged and be able to handle variations in the quality of the source material. However, for a biological drug that will be introduced into interstate commerce,

"Products shall be free of extraneous material except that which is unavoidable in the manufacturing process described in the approved biologics application (21 CFR 610.13)". These processes are typically orthogonal, that is, they consist of several process steps, each of which takes advantage of a different biochemical attribute of the target molecule. The latter definition requires that the product be free of not just product-related contaminants such as aggregates or proteolytic fragments, but other non-proteinaceous contaminants derived from cellular components and the media within which the cell were grown. Further, there are generally accepted limits for non-proteinaceous contaminants such as endotoxin and DNA.

The design of a purification strategy becomes more complicated when the recombinant protein is produced in mammalian cells. The removal and/or inactivation of endogenous viruses present in the cell line or adventitious viruses introduced into the product feedstream during cell culture must be a component of the purification strategy. It is easier and more cost-effective to be able to prove to the FDA that the source materials one starts with are "clean". Clearly, therefore, qualification of raw materials and maintaining accurate records of the source materials used for cell culture and the origin of the cell lines used for expression should be considered when starting to formulate an expression and purification strategy.

3. Assay development

Prior to any efforts aimed at expressing and purifying a recombinant protein, the assays required for assessing the quality and quantity of the product should be in place. It is not possible to develop a purification strategy, however simple, without the methodologies in place to analyze the structural and functional integrity of the product or determine the process and overall step yields. It is beyond the scope of this discussion to elaborate on the many options available and their relative strengths and weaknesses, however, the following outline may be useful in getting started.

3.1. Evaluation of product purity and identity

Sodium dodecyl sulfate-polyacrylamide gel electrophoresis (SDS-PAGE) analysis is an essential assay in determining purity and presumptive identity based on the apparent molecular weight. However, it is a denaturing technique and should be supplemented with other analytical, non-denaturing methods such as high-pressure liquid chromatography (HPLC) using a column that separates the product and potential contaminants based on size, charge, and/or hydrophobicity. A common analytical method in wide use and based on the hydrophobicity of the product is reverse-phase HPLC (RP-HPLC). Proteins may be applied in an aqueous solution and then eluted from the alkylated resin based on their hydrophobic nature using an organic solvent such as acetonitrile. The eluted proteins are monitored as peaks based on their absorbance at $OD_{280\,nm}$. Such methodologies have a great deal of resolving power and are relatively fast and inexpensive to perform.

It is also important to establish that the purified protein is, in fact, the target protein. This can be accomplished biochemically by N-terminal sequencing and immunologically using an adequately characterized antiserum or monoclonal antibody. The assays that are ultimately selected should be relatively rapid and as sensitive as possible in order to avoid using large amounts of material.

There are a variety of approaches for quantitating the amount of protein present in a solution. Colorimetric assays are available in kits and are easy to perform. Protein quantity is established based on a comparison to an irrelevant protein standard, e.g., bovine serum albumin, supplied by the manufacturer. These methods have an inherent error of as much as $\pm 10\%$ and are not reliable when the protein concentration is low (typically less that $100\,\mu g/mL$) or in a buffer system that interferes with color development. The use of an extinction coefficient is advantageous, as it is quick and easy, and employs a parameter based on the inherent character of the protein itself. The extinction coefficient can be estimated based on the theoretical amino acid content or determined empirically by amino acid analysis. However, since it is based on the presence of absorbing amino acids such as tyrosine and tryptophan, the presence of contaminating proteins can lead to overestimating the amount of material actually in the protein solution. This approach is usually reserved for highly purified protein solutions.

Equally important, if not more so than establishing the proper analytical assays, is the development of an appropriate and reliable bioassay. This may be an *in vitro* assay using an established and well-characterized indicator cell line or an *in vivo* model. The latter is far less desirable due to its inherent variability, high cost, and the long lag times required to generate meaningful data. It is possible to expend a great deal of time and effort generating a highly purified recombinant protein that, in the end, is the wrong protein or is biologically inactive. Therefore, establishing meaningful analytical assays and reliable bioassays to evaluate the results of "scouting" for each purification step is critical for the development of a rational, high-yield process.

4. Product recovery and initial purification

The link between the choice of expression system and the requirements for the initial purification or recovery of the protein is an important one that must be carefully considered early on in the process. The selection of a suitable expression system should be based on a reasonable estimate of the amount of material that will be needed both in the short-term and with a view towards potential commercialization. Thus, a significant effort should be invested in the selection of stable clones that have high specific productivities. Irrespective of its sophistication, the "downstream process" is in fact entirely dependent on the abundance of the product in the source material. Even for the initial laboratory-scale screening process, the less abundant the protein of interest, the longer and more expensive is the process of accumulating sufficient material. Further, while some proteins are very stable, many are subject to degradation through proteolysis, aggregation, and deamidation. In short, protein separation and purification is more rapid and cost effective if the feedstream is highly enriched.

There is a wide range of methods available for the recovery and initial purification of recombinant protein in a stable crude form. These early steps depend on the form in which the product is generated by the expression system, for example, soluble and insoluble states. The source material must be processed immediately to a stable form or stored at $-70°C$ to maintain its integrity. Also, care must be taken to avoid the introduction and growth of bioburden over time if the product is stored at 2–8°C. This can be addressed easily by storing the cell paste in self-sealing plastic bags for subsequent processing, and by filtering the cell culture media through 0.2 or 0.22 μm filters to remove microorganisms that could contaminate and degrade the source material.

5. Purification of proteins from culture media

The optimal situation is one where the product is soluble and secreted into the culture medium. The cell mass can be removed rapidly by centrifugation or tangential flow filtration, depending on the amount of material and the equipment available. This approach avoids the viscosity problems or the need to remove particulate matter associated with cell lysates. It should be noted that contaminants that are undetectable at different stages in the purification process may be concentrated in subsequent steps. Further, even a well-considered purification process may not be capable of compensating for errors or oversights made during the generation of the source material.

In addition to avoiding animal-derived raw materials, the use of serum-free or, preferably, protein-free media will reduce the presence of similar molecules in the feedstream, which may be difficult to detect and remove during the purification process. Normally, pH indicators and antibiotics are included in the final culture media formulation. These can complicate the purification process by either interfering with subsequent downstream chromatographic purification or by co-purifying and concentrating in the product. Unless essential for stable growth and expression, they should be avoided.

6. Purification of proteins from cell lysates

In the case of intracellular expression, whether the recombinant protein is soluble or insoluble, it will be necessary to recover the cell mass and release the intracellular contents into a medium that stabilizes the product. There are many methods for cell harvesting and disruption, and these should be tailored to the requirements established by the host cell and the stability of the recombinant protein. A number of excellent textbooks have been published in recent years, detailing the relative advantages and disadvantages of the various approaches [1–5]. The harvesting of the "cell paste" may be accomplished by centrifugation at the bench-scale for mammalian or bacterial systems. Given their more durable nature, bacterial cells can also be recovered by continuous-flow centrifugation or tangential flow filtration. There are many different types of filtration systems available commercially, for example, Pall Corporation (East Hills, NY) and Millipore Corporation (Bedford, MA) offers a wide range of options. The harvested cells should be washed free of media components prior to proceeding with cell disruption or placing the pellets into frozen storage, as discussed below. The buffer system should minimize cell lysis and be compatible with the next phase of the recovery and purification process. Protease inhibitors are added to the wash or lysis buffers, with the caveat that some chemical inhibitors, such as PMSF, can derivatize both the target protein as well as the proteases. It is preferable to work at low temperatures (2–8°C or lower), and to rapidly process the material to a stable intermediate form. It is also best to work with manageable batches than with large amounts of material, a practice that is less likely to effect product degradation during biomass processing.

The process of cell disruption, like all parts of a purification process, should be cost effective and scalable. For mammalian cells, gentle disruption techniques are usually adequate. Methods such as freeze/thaw, detergent treatment, or osmotic shock may be satisfactory. Sonication is also a popular choice for small-scale operations. However, for reproducible results, the energy input and number/duration of cycles should be monitored along with providing adequate chilling of the cell suspension. Hand-held or small blade homogenizers that disrupt cells using shear forces are usually sufficient for handling gram quantities of biomass. Homogenization also shears the DNA that is present in the lysate, thereby reducing lysate viscosity and increasing the ease of handling throughout the remainder of the recovery and purification process. This can also be achieved using DNAse, e.g., Benzonase® (Calbiochem, San Diego, CA), which, though expensive, is effective and acceptable by regulatory agencies for the manufacture of biopharmaceuticals.

For larger-scale operations or one where the host organism is resistant to such relatively gentle approaches, a more vigorous method for cell disruption may be required. In this case, a method should be chosen based on the quantity and type of biomass that needs to be processed. Methods that utilize shear forces, such as the French Press or Mouton-Gaulin homogenizer are available for both bench-scale and larger bioprocess scale operations. Another approach is to mix a slurry of the cell suspension with glass beads and vigorously vibrate the mixture. These bead-mills lyse cells by disrupting their cell walls [3].

7. Volume reduction and partial purification

A hypothetical purification process for a recombinant protein is represented schematically in Fig. 1. A more specific process plan for the purification of a recombinant humanized monoclonal antibody expressed in mammalian cells for human clinical trials is presented in Fig. 2. Most purification strategies for material

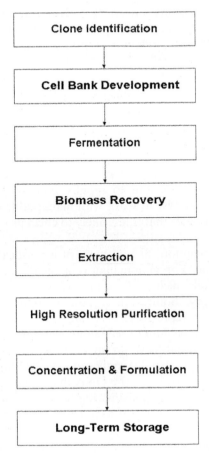

Fig. 1. General scheme for the purification of a soluble recombinant protein from a mammalian cell.

intended for initial characterization in an **R&D** environment will use some or all of the concepts outlined in Fig. 1. At the very least, cells will be cultured and the product recovered in a stable form. If the product advances beyond the initial discovery phase and into product development, then the final biopharmaceutical manufacturing process may look very much like that outlined in Fig. 2. The initial step in any purification process should accomplish at least two things: volume reduction and partial purification.

Reducing the volume of the product feedstream is necessary to make the downstream process more efficient, cost effective, and easier to perform. The cost of chromatography resins can be very substantial, in the thousands of dollars per liter, depending on the manufacturer and type of resin. Therefore, the amount of resin that is needed for the downstream-purification process is lowered substantially by removing as many possible competing impurities feeding into that phase of the purification. Further, the overall cost of handling large amounts of volume can be

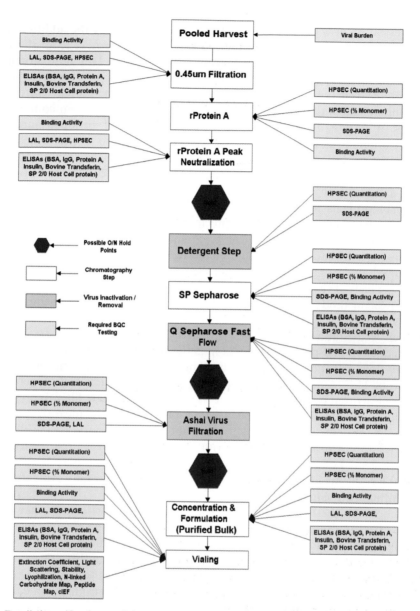

Fig. 2. Detailed purification and in-process testing scheme for a humanized monoclonal antibody expressed in a mammalian cell line.

substantial. During the time it takes to process large volumes of dilute feed material, the product may be prone to degradation or loss of activity. Therefore, it is essential that the product be rapidly recovered in a stable form.

An initial precipitation step to fractionate the source material and reduce the volume of the feedstream is a reliable and simple means to an important end.

The sequestration of water from the surface of the molecule by the addition of salts such as ammonium sulfate leads to the separation of the protein from the solvent phase. This precipitate can be recovered by centrifugation. Different proteins will "salt-out" at different concentrations of salt, affording an opportunity to separate target proteins from contaminants. By carefully evaluating the precise concentration of ammonium sulfate that is necessary to separate the product from contaminating molecules, it is possible to achieve a substantial level of product purification. Typically, attention is paid to the temperature at which this is performed and, if necessary, the control of the pH during the slow addition of the ammonium sulfate to the crude bulk product. An additional advantage of fractionation is that most proteins are stable in the precipitated form. In my laboratory, I have stored recombinant proteins for several years at −70°C with no effect on the ability to purify active protein from the archived precipitates. Other precipitating salts and low molecular weight forms of polyethylene glycol may also be used to partially purify a target protein and reduce the volume of the feedstream. Organic solvents such as acetone or ethanol can also be used when the protein is known to be resistant to the denaturing effects of these solvents. These reagents, however, are essentially reserved for analytical methods and should be evaluated carefully before being used for bioprocess purposes.

Another well-established approach to satisfying the two goals stated above, volume reduction and partial purification, is the use of filtration and, in particular, tangential flow filtration (TFF). Space limitations do not permit an in depth discussion of TFF. An excellent review of the theory and practice of tangential flow filtration was written by Michaels [6]. The use of the appropriate filter or a "train" of appropriately graded filters can be used to clarify and significantly purify the recombinant protein from a cell lysate or fermenter harvest [8–10]. Many laboratories appear to be comfortable with the ubiquitous pressure-driven terminal filtration system. For smaller volumes, spin filtration systems may be useful, but significant loss due to adsorption may occur and should be carefully evaluated before experimenting with valuable material.

In practice, this method is limited to solutions free of particulates that clog the filters. Further, in an effort to accelerate this process, investigators often increase the pressure without considering the potential for pressure-induced protein denaturation. TFF establishes a solute gradient over a stationary filter that concentrates the product as the fluid moves in parallel to the membrane using a simple peristaltic pumping system. Depending on the membrane pore size, the product can be retained or passed through the membrane in the permeate. Placing a clarifying filter ahead of a more selective ultrafiltration membrane can facilitate the processing of relatively small volumes of crude material easily and efficiently. Nanofiltration using a Viresolve® membrane (Millipore Corporation) or a Planova® filter (Asahi Kasei Corporation, Tokyo, Japan) can be used to remove aggregates or viruses from the protein solution. TFF also has the advantage of being scalable and is compatible for the very large volume handling requirements of a commercial process. Other companies that produce a range of filtration systems include Pall Corporation and Sartorius (Goettingen, Germany).

8. Chromatographic purification of recombinant proteins

All chromatographic separations are based on the interaction of the target molecule with the chromatographic medium. As we will examine in the remainder of this chapter, interaction can be based on a variety of characteristics—both of the molecule and the resin itself. Regardless of what approach is taken as the initial step in the post-recovery phase of the process, it is essential that the investigator prepare the material appropriately. The feed solution should be of the appropriate ionic strength and pH. The rate at which the process can be performed is limited by the resin selection, the column, and system used to perform the purification process and the viscosity of the feed solution. An additional limitation is the inherent stability of the protein while it is being processed. Working at lower temperatures can reduce the potential for degradation but can also increase the viscosity of the protein solution. While column chromatography is a precise science, it is also an art that requires practice to master. For these reasons, for small-scale operations where time and a high-throughput capability are of greater importance than the resin cost, the use of pre-packed columns is the best option. For larger-scale operations, several of the leading manufacturers of resins offer training classes on the theory and practice of column chromatography, that can save an investigator time in mastering the nuances of selecting and using the appropriate resin. An excellent series of handbooks and CD-ROM media on the principles and practice of the major chromatographic methodologies is available from Amersham Biosciences (Piscataway, NJ).

8.1. Affinity chromatography

Although proteins are not amenable to a generally applicable purification method, some basic rules do apply. One is to take advantage of the "specific" interaction between the protein of interest and an immobilized ligand to adsorb the product from a heterogeneous protein solution. This can be an interaction that is native to the target molecule or, as we will see below, a conferred affinity that is added on through the attachment of a "tag" to the molecule. Affinity chromatography can be performed using a standard column-chromatography system but is also amenable to simple batch adsorption in a tube or other container. With relatively little effort, under appropriate buffer conditions, the degree of purification from this one step alone can be on the order of several 1000-fold [3]. Further, given sufficient time, affinity chromatography enables the extraction of a relatively scarce protein from a complex protein mixture. The matrix is then washed and the protein is eluted under the appropriate buffer conditions. Thus, affinity chromatography can also be used to purify and concentrate a protein of interest from large volumes of source material.

In addition to the interaction specificity between product and ligand, it is important that the interaction be reversible if the goal is the quantitative recovery of biologically active protein. A classic example of the principle of affinity chromatography is based on the affinity of an antibody for its cognate antigen. The immunoglobulin fraction is chemically immobilized onto a stationary matrix. A number of kits available from commercial sources (e.g., Pierce Endogen, Rockford,

IL) now make this process straightforward. Source material can be an antiserum or a monoclonal antibody. There are a number of potential disadvantages that need to be considered before using this method to purify protein from a complex mixture. For example, the affinity of the immobilized antibody for the target protein may be particularly high, complicating recovery and necessitating the use of harsh conditions that may potentially denature the antigen of interest. Another potential downside to this approach is that despite its chemical coupling to the immobilized resin, the immunoglobulin can leach from the matrix and contaminate the protein mixture. The development of an efficient immunoaffinity purification process can be complicated by other "specific" interactions between the antigen and immobilized antibody, e.g., carbohydrate–carbohydrate interactions.

Detailed experimental protocols for immunoaffinity chromatography can be found in a number of excellent texts [1–5] and brochures (Amersham Biosciences). The purification of antibodies and antibody fragments can be performed using immobilized recombinant Protein A. The binding motif for the interaction of an antibody to recombinant Protein A resides in the heavy chain variable region of molecules bearing the VHIII segment [7]. Protein A has five homologous IgG-binding domains [11]. IgG immunoglobulins from human, rabbit, and guinea pig bind to Protein A with high affinity. Murine IgGs bind with lower affinity, as do most other immunoglobulin classes. Typically, the Ig-containing solution is adjusted to a neutral or slightly basic pH and allowed to interact with recombinant Protein A. It is then washed with an appropriate buffer, and the specifically bound antibody is eluted at a low pH (~ 3.5) and immediately neutralized to avoid hydrolysis. The recovered immunoglobulin is typically more that 95% pure using this single-step method.

A variation on the antibody–antigen theme is the use of immobilized lectins for the affinity purification of carbohydrate-modified target molecules. The advantage of this approach is that the ligand is eluted under milder conditions using the appropriate sugar to compete with and disrupt the interaction between the lectin and the sugar moiety on the recombinant protein surface.

8.1.1. Affinity chromatography of fusion proteins

The use of affinity "tags" is a rapid and powerful method for the screening, selection and single-step purification of clones. Although column-mode purification is a viable option, it is relatively cumbersome for the evaluation of a large number of potential isolates. Recently, the use of microtiter plates for the rapid screening and purification of a large number of isolates has been reported [12]. In addition, Draveling *et al.* [13] have recently described a new affinity technology designed for this purpose.

Immobilized metal-affinity chromatography (IMAC) was originally described by Porath *et al.* [14]. Typically, a hexahistidine tag is fused in-frame to the N- or C-terminus of the recombinant protein. The latter fusion polarity has the advantage that only fully translated proteins can be purified (including fragments, of course). The IMAC stationary phase chelates metal ions that possess selectivity for certain ligands, e.g., histidine. Since histidine is relatively rare in most proteins, the adsorption and retention of the tagged protein onto the immobilized matrix is reasonably specific, and the method also satisfies the recovery requirements for affinity purification since the buffer conditions for elution are mild [15]. This method is now widely used, but it too has its own disadvantages. Thus, although the presence of the polyhistidine tag does not usually interfere with protein folding, such is not always the case. Further, the presence of the tag on a therapeutic protein may be problematic due to the potential for increased immunogenicity [16]. The tag can be removed by site-specific proteolysis. Obviously, the use of a protease is not desirable either when isolating small amounts of recombinant proteins for structure/function analysis or when aiming towards developing a marketable product. The use of so-called "self-cleaving" affinity tags has also been described [17,18].

There is a large number of affinity tags that have been used successfully for the construction of fusion proteins for purification as well as for many other applications. Some recent examples include the use of hexahistidine, chitin, oligonucleotides, maltose, streptavidin-binding peptide, and glutathione-s-transferase. The reader is referred to recent comprehensive reviews that describe fusion tags and their applications in both mammalian and bacterial cells [19–21].

Another affinity-based approach is based on the Protein A-immunoglobulin interaction described above. In this scenario, the immunoglobulin-binding motif is inserted upstream or downstream of the target open-reading frame. The expressed fusion protein is amenable to standardized purification protocols for antibody concentration and purification. Brown et al. [11] recently reported the generation of a modified B-domain from Protein A that permits elution of the Protein A–immunoglobulin complex at a less acidic pH. Although streptococcal Protein G interacts with human immunoglobulins and human serum albumin in an analogous manner [22], it has not been as extensively in a large-scale bioprocess situation as it has in small-scale laboratory settings. However, it should be considered as an option when designing a recombinant fusion protein for affinity purification. In any case, the tag should not adversely affect the proper folding of the gene product, and it should be removed prior to initiating structure/function or animal toxicity studies.

8.2. High-resolution chromatography

While the power of affinity chromatography is evident, by itself it may not lead to the recovery of product of adequate purity. Any fragment or aggregate that displays the tag will be concentrated and purified along with the intact monomer. This may not necessarily be of concern, depending on the intended use of the product. Fig. 2 indicates that for a biopharmaceutical agent, additional purification steps are required to meet the regulatory standard of purity. Laboratory-grade material need not rise to this level of rigor, but additional purification steps based on other physico-chemical attributes of the product may be needed to obtain a well-purified, i.e., >95% pure, molecule. Indeed, in those cases where affinity purification is not an option, the first step in the purification process may include one of these other methods (Fig. 2). While this chapter cannot begin to cover the resin options or variables that can affect the success of the purification effort, a more detailed discussion of the labyrinthine world of protein purification may be found in a number of excellent texts [1–5]. However, a brief discussion of the major chromatographic strategies and their potential uses is essential.

As in affinity chromatography, the physical and chemical interaction of the target protein with a chromatographic medium provides the basis for purification by column chromatography. Even this statement is an oversimplification of the myriad factors that may affect the outcome of a particular chromatographic purification step. Chromatographic purification is based on the size or hydrodynamic volume of the protein, as in size-exclusion chromatography; the net positive or negative charge in a particular solvent, as in ion-exchange chromatography; or the solvation effects on hydrophobic moieties, as in hydrophobic interaction chromatography. The more information one has on the biochemistry and stability of the product and the identity of contaminants or impurities, the greater the likelihood of designing a highly selective process with a successful outcome. The use of currently available databases to identify molecules similar to the one of interest can aid greatly in limiting the purification strategy to those chemistries most likely to work.

Similarly, simply stringing together a series of purification steps because "the resin is available in the lab at that moment", is a process probably doomed to failure. Further, as the number of steps increases, the yield decreases [2]. For example, assuming that the recovery in each step of a ten-step purification procedure is 95%, then the final overall recovery would be approximately 60%. With only four process steps in the purification scheme, the yield would increase to 80%. Assuming the step yield to be 80%, then after ten purification steps, the investigator would be in possession of about 10% of the starting material! Thus, it is imperative to design purification processes that use a minimal number of steps in order to conserve material and reduce costs. The chromatographic approach should be logical, with each step complementing the previous phase of the process. That is, there should not be a string of redundant purification steps based on the same physicochemical attribute of the recombinant protein. The removal of non-proteinaceous contaminants should also be considered a priority if the product is to be used for safety studies in small mammals or non-human primates. What follows is a brief discussion of the dominant bioprocess methodologies and their potential uses.

8.3. Hydrophobic interaction chromatography

It would seem to be counterintuitive that the solvated surface of a protein molecule displays patches of hydrophobic amino acids. Nevertheless, our current knowledge of the surface topology of globular proteins has confirmed that the molecular surface can have extensive hydrophobic patches. Hydrophobic domains will tend to self-associate. This characteristic is a major driving force behind protein refolding of globular proteins. It is also the basis for the binding of a protein to a hydrophobic ligand immobilized to a stationary matrix [23,24]. The extent of the interaction between these non-polar patches and the ligand increases with higher concentrations of neutral salts, and elution is achieved by lowering the salt concentration [25]. Hydrophobic interaction chromatography is applicable to most, if not all, proteins. A wide variety of ligand-matrix chemistries are now available from a number of commercial vendors. These vary in their relative hydrophobicities. This fact allows the investigator to modulate the strength of the protein–ligand association. Adjusting the concentration of lyotropic salt, e.g., ammonium sulfate, can also vary the effect of this association. The salt concentration must be high enough to facilitate the association of the protein with the matrix but not initiate protein precipitation. Although the pH is not generally considered to be a critical factor in column performance, this must be evaluated on a case-by-case basis. Temperature, however, can dramatically affect the reproducibility of this method. The higher the temperature, the more tightly the protein will associate with the matrix. Consequently, the salt concentration needed to elute the protein will decrease as the temperature increases. Gradual reduction of the salt concentration is less likely to denature the adsorbed protein. A gradual salt reduction may be able to separate misfolded isoforms from the native form by taking advantage of the differences in surface topography. Step-elution or "bumping" the protein off the resin by employing an abrupt shift downward in the salt concentration can be an excellent method for concentrating the product.

Given their lipophilic nature, endotoxins tend to bind more tightly to the hydrophobic matrix than most proteins. Therefore, HIC can also be useful for the removal of endotoxin from the product. However, it is also possible that undetectable levels of endotoxin in the feedstream can concentrate with the product on the resin and elute with the product at levels unacceptable for animal or human studies. Until recently, despite its obvious power, HIC has not achieved the same widespread use in the laboratory or in bioprocess manufacturing strategies as have some of the other chromatographic approaches. As our understanding of the forces that drive protein folding has increased, the utility of HIC has found an increasing role in large-scale protein purification schemes [26,27].

8.4. Ion-exchange chromatography

Ion-exchange chromatography has been a dominant, high-resolution protein purification method for over 50 years. Proteins interact with ion-exchangers by electrostatic forces between patches of charged residues on the protein surface and dense clusters of charged groups on the resin [3]. This method of purification has found wide applicability within the biopharmaceutical industry since it offers high-resolution separation under relatively mild binding and elution conditions. Ion-exchange chromatography satisfies the dual requirements of high-capacity and ruggedness. The amphoteric nature of most proteins ensures that virtually any soluble protein is amenable to purification using this method. Further, ion-exchangers can be cleaned and depyrogenated repeatedly with strong alkali without loss of binding capacity or leaching of the ionic moiety into the product.

Two general types of ion-exchangers are currently available: anion- and cation-exchange resins. Within these broad categories, a variety of resins with varying attributes are available that can be tailored to a particular use. The selection of an appropriate combination of resins and buffers for a particular application is based on the pI or zero-charge point of the protein and the investigator's objectives. A theoretical pI value based on the predicted amino acid content is useful but may not necessarily be a useful guide of what to use since it is the clustering of charged groups on the molecular surface that defines the interaction of the protein with the resin. Indeed, even a pI value based on standard gel methodologies may not predict success if the charge densities are too small or weak. Nevertheless, once a pI value is known, it is a relatively straightforward process to begin identifying the binding and elution conditions.

The chemical properties of all ion-exchangers are determined by the charged ligand groups. These are classified as "weak" or "strong", based on the effect of pH and not, as one might assume, on the strength of association [28]. The buffer pH affects the charge on weak ion-exchangers but not the charge on strong exchangers. The most common weak anion exchanger, diethylaminoethyl (DEAE) loses its charge above pH 9, and the weak cation-exchanger using carboxymethyl groups (CM) loses its charge below pH 4. Strong ion-exchangers have a wide pH range where they remain completely ionized, and these are most often found in larger-scale bioprocess operations [29].

A protein in a solution that has a pH below the protein pI will have a net positive charge and can be paired with a cation-exchanger. Conversely, if the solution pH is above the protein pI, then the protein will have a net positive charge and can be purified using an anion-exchange resin. Optimal binding conditions can be quickly determined in microcentrifuge tubes in batch mode where the protein solution is mixed with a candidate resin under a variety of pH and ionic conditions. The resin is pelleted and the supernatant fluid assayed for the presence of protein. Those conditions that result in the disappearance of the product from the supernatant fluid can be applied to the larger-scale purification process [30]. This approach can also be applied to identifying the parameters required for elution of the product from the resin. Implicit in all this is that a sensitive quantitation assay is available to permit mass balance calculations. An ELISA assay or activity assay should also be used in parallel to ensure that the eluted protein is the target product and not a contaminating protein with similar molecular mass.

The adsorbed protein is eluted from the resin by increasing the salt concentration or by decreasing the pH of the mobile phase. This situation is the opposite of the HIC method described above where the mobile phase transitions from a high-salt to a low-salt buffer. If a step-elution is possible, a shift in the salt concentration or pH will work well. However, it is difficult to translate a pH-based method to a larger scale, and it is more difficult to repeat a pH gradient reliably from batch to batch. In contrast, a salt-gradient or step-elution is easier to reproduce and perform at larger scales of operation.

Additional considerations in resin selection, other than cost, are the size and uniformity of the resin bead. The smaller and more uniform the bead size, the greater the resolving power. For many purposes, larger bead sizes are the most appropriate choice if the process is a simple on-off step process. Like affinity chromatography, a batch-mode operation is quick and inexpensive when high resolution is not required. If the goal is to remove protein dimers, aggregates, or misfolded isoforms, then a shallow salt gradient utilizing the more expensive small, monodisperse-bead size is required. Additional resolution can be achieved by increasing the bed height and/or reducing the amount of the load solution, and by increasing the viscosity of the mobile phase through addition of non-ionic excipients, such as glycerol, which reduce diffusion as the protein peak migrates down the column. Unfortunately, none of this can be determined a priori, and a series of carefully designed bench-scale experiments is necessary.

An additional benefit of ion-exchange chromatography is its ability to remove lipopolysaccharide (LPS) and residual DNA from a protein solution. DNA binds very tightly to anion-exchange resins requiring 0.5–1.0 M NaCl for elution. LPS binds less tightly but can be separated from proteins by using a shallow salt gradient. For this reason, even when the recombinant protein does not bind to an anion-exchange resin, a small anion-exchange "guard column" can be positioned upstream of the cation-exchange resin to which the protein will actually bind. These columns can be linked in tandem making the overall process no more complicated than using a single column alone. The larger, less expensive large bead resins are very useful for this purpose.

In our program, we have also used ion-exchange chromatography to concentrate dilute solutions of protein or to increase the concentration of the product prior to size-exclusion chromatography. It is a simple matter to dilute the product solution to an ionic strength low enough to permit binding. To avoid dilution, the high-salt solution is applied to the column in reverse flow (bottom to top). The product will elute rapidly and in a concentrated form as soon as the high-salt solution reaches the bound protein at the top of the resin.

A feed solution free of cells and cellular debris is typically required for ion-exchange chromatography. However, centrifugation or filtration of the feedstream prior to standard column-mode chromatography may add time and expense to an already complicated process. Recently, use of fluidized or expanded bed adsorption chromatography has been gaining a larger role in both laboratory- and bioprocess-scale chromatography [31,32]. In this scenario, the bed is not tightly packed or immobilized within a glass cylinder. Rather, the medium is kept suspended while a crude feed solution is pumped through it from the bottom inlet of a column specially designed for this purpose. Streamline™ resins (Amersham Biosciences), specifically designed for this application, contain quartz or metal alloys that have been incorporated into the bead structure. As a result, the beads are sufficiently buoyant to allow the bed to expand upward as the feedstream moves through the column without being transported out of the column itself. The ability to omit the traditional clarification steps prior to initiating chromatographic purification removes at least one and, possibly, two steps from the purification process.

Small-bead, anion-exchange chromatography can also be used to purify plasmid DNA and separate intact supercoiled plasmid from a nicked contaminant. Finally, anion-exchange chromatography can be a critical element in virus removal when developing a validation strategy for the manufacture of biopharmaceuticals.

8.5. Size-exclusion chromatography

Size-exclusion chromatography (SEC) has been in existence for over 40 years and is also known as gel permeation chromatography, gel filtration chromatography, or molecular sieving. SEC separates proteins based on their hydrodynamic volume. It is a ubiquitous method at both analytical- and process-scales, and is useful for determining the size heterogeneity of a protein solution under non-denaturing conditions. In the past, it was often used for estimating molecular weights and calculating equilibrium constants. There are a wide variety of resin types and commercial sources from which to choose [33].

SEC is also a relatively gentle purification step, and there are few instances where protein inactivation is a significant problem. Resolution between different molecules in the load solution is a function of the bed height, flow rate, the mobile phase, and sample size. SEC is a relatively low-resolution chromatographic step and is usually reserved for the later steps in a purification process where the product has already been subjected to some of the higher resolution methods described above. SEC is also very useful for buffer exchange, a process also known as desalting. When used for desalting, a short, wide column is used that can be loaded with 15–20% of the bed volume at a time. If, however, the goal is to separate dimers or aggregates from monomers or other contaminating proteins, then a sufficiently tall column loaded with less than 5% of the total column volume is preferred. The most critical parameter, however, is the selection of the appropriate resin [34].

Each of the commercially available resins functions best within a specified fractionation or separation range. This range and, therefore, the resolving power of the particular resin are a function of the pore size of the chromatographic media. Essentially, larger molecules elute earlier than smaller ones, depending on the fraction of the pores that are accessible to the protein migrating through the resin. Thus, larger molecules are excluded more often than smaller ones and, therefore, can traverse the column with fewer "side-trips". As a result, the largest molecules elute first, and the smallest molecules elute last. Intermediate-sized molecules will elute somewhere between the two based on their relative effective sizes. Again, this is an oversimplification of the actual process, and each of the many commercially available resins exploits it in different ways.

The "plate count" is another term that is often discussed when considering the resolving power and ability of a column to discriminate between different solutes in a particular solution. The plate count is expressed as the number of theoretical plates per meter [35]. For a particular column packed with a specific resin, a plate count is an insight into the resolving ability of that column. A higher number means greater resolution and, therefore, a greater probability of separating the target molecules. Column packing itself is as much an art as a science, requiring mastery of the many factors that can influence the acceptability of a hand-packed column for a given purpose. Therefore, unless one is operating at a bioprocess-scale, purchasing prepacked columns is a safer and more reliable option. For desalting small samples, many of the commercial vendors sell small, prepacked columns that serve this purpose very well.

SEC is also a powerful analytical method for assessing the homogeneity of a protein in an aqueous solution. Whereas low flow-rate chromatography in plastic or pharmaceutical-glass columns will have a plate count in the thousands, HPLC columns can have plate counts in the tens of thousands. To operate at faster flow rates, these are purchased already packed into a steel column in order to withstand the higher operating pressures. For the processing of milligram quantities of protein, larger diameter HPLC columns are available and are a viable option for the purification of relatively small quantities of product.

Like many of the other resins available for bioprocessing, most SEC resins can withstand harsh cleaning regimes and are resistant to prolonged exposure to sodium hydroxide. Indeed, in our program, we have found that parts of the column may be more susceptible to the effects of base than the resin itself. SEC is typically reserved as the process step immediately prior to vialing and long-term storage of the product. Fractionation by size affords an opportunity to exchange the product into the final buffer formulation and provides one last opportunity to ensure that the protein is free of aggregates or process-related contaminants that would affect product stability or safety.

9. Product stability and long-term storage

A review of protein purification, however general and brief, cannot conclude without a discussion of the importance of preserving what one has worked so hard to produce and purify. Yet, the concept of identifying a buffer formulation and storage conditions for preserving the physical and functional integrity of the protein is often not a primary focus of concern in many laboratories. While many proteins, e.g., immunoglobulins, can tolerate a wide spectrum of storage conditions with respect to buffer, temperature, or concentration at which they are stored, this cannot be assumed for all proteins. Suitable assays should be used to assess the storage stability of the product as soon as sufficient material is available for this purpose.

Identification of a buffer formulation that preserves functionality and suppresses product degradation is crucial before starting large-scale or even small-scale purification operations. The product should be stored in single-use aliquots to avoid repeated entry into the container and the risk of contamination. The containers should be resistant to breakage and non-reactive with the product or the buffer. The containers should be pyrogen-free and sterile if the material is to be used in animal studies. In order to avoid modulations in pH during storage, the "headspace" over the product solution should be kept to a minimum and the container itself should be resistant to gas exchange.

An analysis of the physical stability of the product following repeated freezing and thawing or prolonged exposure to elevated temperatures is also prudent. In our laboratory, we have noticed that for a number of recombinant products, simply being located at the front or top of the freezer can lead to significant loss of structural integrity and function over time. And while $-20°C$ storage may be what is available and convenient, product is not easily shipped to a collaborator or testing laboratory without experiencing a temperature fluctuation in transit. Most proteins are more stable at $-70°C$ than at higher temperatures, and shipping on dry ice ensures product stability. Finally, periodic testing of archived material provides valuable historical data in case the protein makes the transition from an interesting gene product to a biopharmaceutical destined for clinical trials.

Acknowledgements

This project has been funded in whole or in part with Federal funds from the National Cancer Institute, National Institutes of Health, under Contract NO1-CO-12400.

Disclaimer

The content of this publication does not necessarily reflect the views or policies of the Department of Health and Human Services, nor does mention of trade names, commercial products, or organization imply endorsement by the US Government.

Abbreviations

HIC	hydrophobic interaction chromatography
HPLC	high pressure (performance) liquid chromatography
IMAC	immobilized metal-affinity chromatography
LPS	lipopolysaccharide
SDS-PAGE	sodium dodecyl sulfate-polyacrylamide gel electrophoresis
SEC	size-exclusion chromatography
TFF	tangential flow filtration.

References

1. Olson, W.P. (ed.) (1995) Separations Technology—Pharmaceutical and Biotechnology Applications, Interpharm Press, Inc, Buffalo Grove, IL.
2. Sofer, G. and Hagel, L. (1997) Handbook of Process Chromatography—Guide to Optimization, Scale-up and Validation, Academic Press, San Diego, CA.
3. Scopes, R.K. (1994) In: Cantor, C.R. (ed.) Protein Purification—Principles and Practice, 3rd ed., Springer-Verlag, New York, NY.
4. Janson, J.-C. and Rydén, L. (eds.) (1998) Protein Purification—Principles, High-Resolution Methods, and Applications, 2nd ed., John Wiley & Sons, Inc, New York, NY.
5. Wheelwright, S.M. (1991) Protein Purification—Design and Scale up of Downstream Processing, John Wiley & Sons, New York, NY.
6. Michaels, S.L. (1995) In: Janson, J.-C. and Rydén, L. (eds.) Separations Technology: Pharmaceutical and Biotechnology Applications; Tangential Flow Filtration, 2nd ed., Interpharm Press, Inc, Buffalo Grove, IL, pp. 57–75.
7. Sasso, E.H., Silverman, G.J., and Mannik, M. (1991) J. Immunol. 147, 1877–1883.

8. Bailey, S.M. and Meagher, M.M. (1997) J. Membrane Sci. 131, 29–38.
9. Bailey, S.M. and Meagher, M.M. (2000) J. Membrane Sci. 166, 137–146.
10. Bailey, S.M. and Meagher, M.M. (1997) Biotechnol. Bioeng. 56, 304–310.
11. Brown, N.L., Bottomley, S.P., Scawen, M.D., and Gore, M.G. (1998) Mol. Biotechnol. 10, 9–16.
12. Schuster, M., Wasserbauer, E., Einhauer, A., Ortner, C., Jungbauer, A., Hammerschmid, F., and Werner, G. (2000) J. Biomol. Screen 5, 89–97.
13. Draveling, C., Ren, L., Haney, P., Zeisse, D., and Qoronfleh, M.W. (2001) Protein Expres. Purif. 22, 359–366.
14. Porath, J., Maisano, F., and Belew, M. (1975) Nature 258, 598–599.
15. Yip, T.-T. and Hutchens, T.W. (1996) In: Doonan, S. (ed.) Protein Purification Protocols, 2nd ed., Vol. 59. Humana Press, Inc, Totowa, NJ, pp. 57–75. Methods in Molecular Biology.
16. Olson, W.P. (ed.) (1995) Separations Technology; Affinity Chromatography, 2nd ed., Vol. 59. Interpharm Press, Buffalo Grove, IL, pp. 303–351.
17. Chong, S., Mercha, F.B., Comb, D.G., Scott, M.E., Landry, D., Vence, L.M., Perler, F.B., Benner, J., Kucera, R.B., Hirvonen, D.A., Pelltier, J.J., Paulus, H., Xu, M.-Q. (1997) Gene 192, 271–281.
18. Leong, L.E.C. (1999) Mol. Biotechnol. 12, 269–274.
19. Hearn, M.T.W. and Acosta, D. (2001) J. Mol. Recognit. 14, 323–369.
20. Makrides, S.C. (1999) Protein Expr. Purif. 17, 183–202.
21. Makrides, S.C. (1996) Microbiol. Rev. 60, 512–538.
22. Linhult, M., Binz, H.K., Uhlén, M., and Hober, S. (2002) Protein Sci. 11, 206–213.
23. Eriksson, K.-O. (1998) In: Janson, J.-C. and Ryden, L. (eds.) Protein Purification, Principles, High-resolution Methods and Applications—Hydrophobic Interaction Chromatography, 2nd ed., Vol. 59. Wiley-VCH, Inc, Weinheim, Germany, pp. 283–309.
24. Wheelwright, S.M. (1991) Protein Purification: Design and Scale-up of Downstream Processing, J. Wiley & Sons, Inc, New York, NY, pp. 172–181.
25. Queiroz, J.A., Tomaz, C.T., and Cabral, J.M.S. (2001) J. Biotechnol. 87, 143–159.
26. Gagnon, P., Grund, E., and Lindback, T. (1995) Biopharm 8, 21–27.
27. Karlsson, G. and Winge, S. (2001) Protein Expres. Purif. 21, 149–155.
28. Desai, M.A., Rayner, A., Burns, M., and Berminham, D. (2000) In: Desai, M.A. (ed.) Downstream Processing of Proteins: Methods and Protocols; Application of Chromatography in the Downstream Processing of Biomolecules, 2nd ed., Vol. 59. Humana Press, Inc, Totowa, NJ, pp. 73–95.
29. Karlsson, E., Ryden, L., and Brewer, J. (1998) In: Janson, J.-C. and Ryden, L. (eds.) Protein Purification, High-resolution Methods and Applications, 2nd ed., Vol. 59. Wiley-VCH, Inc, Weinheim, Germany, pp. 145–205.
30. Sheehan, D. and FitzGerald, R. (1996) In: Doonan, S. (ed.) Methods in Molecular Biology. Ion-Exchange Chromatography, 2nd ed., Vol. 59. Humana Press, Inc, Totowa, NJ, pp. 145–150 .
31. Roy, I., Sharma, S., and Gupta, M.N. (2002) Protein Expres. Purif. 24, 181–187.
32. Graslund, T., Hedhammer, M., Uhlen, M., Nygren, P.-A., and Hober, S. (2002) J. Biotechnol. 96, 93–102.
33. Wu, C.-S. (ed.); Column Handbook for Size-exclusion Chromatography; Academic Press; New York, NY.
34. Williams, A. and Hagel, L. (1999) In: Wu, C.-S. (ed.) Column Handbook for Size-exclusion Chromatography; Size Exclusion for Analysis and Purification of Aqueous Macromolecules, Academic Press, New York, NY, pp. 145–150.
35. Hagel, L. (1998) In: Janson, J.-C. and Ryden, L. (eds.) Protein Purification: Principles, High-resolution Methods and Applications; 2nd edition, Gel Filtration, Wiley-VCH, Inc., Weinheim, Germany.

Subject Index

672

680